The Bauers
Joseph, Franz & Ferdinand

Masters of Botanical Illustration

An Illustrated Biography

HANS WALTER LACK

PRESTEL
Munich • London • New York

To Eva, the companion of my life

This book has been published in conjunction with the exhibition on the Bauers commencing at the Natural History Museum, London in November 2015.

This edition is published by Prestel Verlag by arrangement with the Natural History Museum

 Address of author: h.w.lack@bgbm.org

Prestel Verlag, Munich
A member of Verlagsgruppe Random House GmbH

Prestel Verlag
Neumarkter Strasse 28
81673 Munich
Tel. +49 (0)89 4136-0
Fax +49 (0)89 4136-2335

www.prestel.de

Prestel Publishing Ltd.
14-17 Wells Street
London W1T 3PD
Tel. +44 (0)20 7323-5004
Fax +44 (0)20 7323-0271

Prestel Publishing
900 Broadway, Suite 603
New York, NY 10003
Tel. +1 (212) 995-2720
Fax +1 (212) 995-2733

www.prestel.com

Library of Congress Control Number is available; British Library Cataloguing-in-Publication Data: a catalogue record for this book is available from the British Library; Deutsche Nationalbibliothek holds a record of this publication in the Deutsche Nationalbibliografie; detailed bibliographical data can be found under: http://www.dnb.de

ISBN 978-3-7913-5489-7 (trade edition)

Designed by Mercer Design, London
Reproduction by Saxon Digital Services, UK
Printed by C&C Offset, China

FSC www.fsc.org MIX Paper from responsible sources FSC® C008047

CONTENTS

PREFACE

There are moments in life one does not forget. To me such a moment occurred in the afternoon of a summer day in 1974 whilst a visitor at the Natural History Museum in London. Together with Dr. William T. Stearn, principal senior scientific officer in the Department of Botany, great connoisseur of botanical illustration and one of the key figures in the history of science at that time, we had discussed the preliminary version of a joint paper on the typification of several Linnaean names in composites. Realizing that there was still some time left before the gates of the museum closed, my host opened a cupboard and showed me a very large format, bound volume of watercolours. Never before had I seen such exquisite plant illustrations, perfectly true to nature, recording in utmost precision many minute flower details and painted in strikingly vivid colours. Stearn told me that this material originated from Franz and Ferdinand Bauer, two brothers who had worked for the cream of the naturalists of their time, but that little was known about their lives. He added that they would merit an in-depth study and that I would be suited to write such a biography, being both a botanist and a native of Vienna where he believed Franz and Ferdinand had spent their formative years.

This was the beginning of my research into the lives and work of three outstanding personalities, since it turned out that there existed a third brother, Joseph, who also made a career as an artist, although with a quite different profile. It soon became clear that they had left behind a vast oeuvre consisting not only of botanical and zoological illustrations but also of landscapes and of the pictorial documentation of dissections and microscopic preparations. In addition to being an accomplished botanical illustrator, there is evidence that Franz was also an early experimental plant pathologist, who had successfully attempted to cultivate snow algae, unsuccessfully tried to observe sperm cells in deer and studied blood coagulation in man. He was also the first to draw nuclei in plant cells, although, admittedly, he was unaware of the importance of his discovery. Rather unexpectedly Ferdinand, a botanical illustrator of similar accomplishment, turned out to be also a collector of many plant specimens gathered mainly in Australia and, in his late years, in Austria, almost all of them held today in the Naturhistorisches Museum in Vienna. Joseph, so far almost totally unknown, became director of the Gallery of the Reigning Prince of Liechtenstein and was found to have been responsible for the first public presentation of the rich collection of fine paintings and sculptures exhibited in the family's summer residence in the Rossau quarter of Vienna.

In parallel with the increased understanding of the manifold activities of the Bauer brothers it gradually became evident that their various pencil drawings, finished watercolours, oil paintings, manuscripts and their few extant letters were widely scattered. For me this resulted in repeated travels all over Europe. Again there were special moments not to be forgotten: discovering by chance, in the archive of the Real Jardín Botánico in Madrid among the miscellaneous papers of Thaddaeus Haenke, the early colour chart of the Bauer brothers explaining the colour code they had used; spending days without end in the Sherardian Library of Oxford University marvelling at Ferdinand's watercolours documenting plant and animal life in the Levant and correlating them with the specimens kept in the Sibthorpian Herbarium; rediscovering fourteen massive volumes of plant illustrations in a pristine state of preservation in the Fürstliche Sammlung in Vaduz thought lost during the Second World War and in part painted by Joseph, Franz and Ferdinand in their early years; finding in the Staatsbibliothek in Berlin an exceedingly rare Russian journal containing the publication of some of the letters on the earliest heliographs in existence sent to Franz and kept in the archive of the Russian Academy of Sciences in St. Petersburg; locating in the Biblioteka Jagellońska in Cracow, Ferdinand's swan-song, his last botanical illustrations datable with the help of watermarks and documenting passionflowers with utmost attention to anatomical detail; studying in the Niedersächsische Staats- und Universitätsbibliothek in Göttingen those parts of Franz's estate which had not ended up in the British Museum, where they had been deposited by order of Queen Victoria, but in the hands of her uncle Ernest Augustus, King of Hanover.

Over the decades it became evident that the Bauer brothers had travelled widely. Joseph had worked for seven years in Rome, Franz had taken part in a conventional *grand tour* accompanying Joseph Franz Edler von Jacquin, the son of his then employer, criss-crossing Central Europe. Later he settled down as a resident botanical illustrator near the Royal Garden at Kew, living on an annuity from Sir Joseph Banks, President of the Royal Society, and leading for the following fifty years a sedentary life in what was then a village on the River Thames way out of London. Ferdinand joined John Sibthorp on his first voyage of exploration in the Ottoman Empire and later became, together with the naturalist Robert

Brown and the gardener Peter Good, one of the 'botanical gentlemen' taking part in the first circumnavigation of Australia headed by Captain Matthew Flinders. Those were adventurous undertakings–with moments of triumph, like climbing Mount Parnassus as one of the earliest tourists in the region, while others came close to catastrophe, like the daredevil voyage on the leaky *Investigator* from Timor along the southern coast of Australia to Port Jackson, Sydney with a contagious disease on board. In Kew, Franz went on journeys of a different kind. With the help of microscopes, primitive by modern standards, he ventured into worlds then largely unknown, examining urinary sediments, legs of insects and the germination of pollen cells.

While Kew, London, Rome and Vienna were places known to me for decades, Feldsberg (now Valtice in the Czech Republic), the home town of the Bauer brothers and place of their formative years, became the destination of several excursions only much later. No doubt it turned out to be helpful that I was already familiar with many classical sites in Greece and Turkey, notably Athens and Istanbul, documented by Ferdinand in several views, some with outstanding precision. By contrast I had no chance to see with my own eyes most of Ferdinand's numerous collecting sites on the southern, eastern and northern coast of Australia and, in particular, on Norfolk Island in the Pacific, where he had spent no less than seven months. At the same time it became clear that Franz underwent a metamorphosis late in life from illustrator to scientist in his own right, publishing a few papers as the sole author and independently from his employers. Elected Fellow of the highly prestigious Royal Society, he became known as the Nestor of Kew, continuing his extended researches into old age.

So far several books on the Bauer brothers have been published, albeit all focused either on only one of them or on only a single aspect of their lives and work. My book *The Flora Graeca Story*, Oxford, 1999 deals exclusively with Ferdinand's travels in the Levant together with Sibthorp and the subsequent publication of the botanical results. David Mabberley's *Ferdinand Bauer. The nature of discovery*, London, 1999 is the first short biography of Ferdinand, just like my *Franz Bauer. The painted record of nature*, Vienna, 2008 (also available in a German version) is a first short biography of Franz. By contrast, Marlene Norst's *Ferdinand Bauer. The Australian Natural History Drawings*, London, 1989 deals only with Ferdinand's work on the *Investigator* voyage. My book *A Garden for Eternity. The Codex Liechtenstein*, Berne, 2003 (also available in a German version) analyses the early work of the Bauer brothers, whereas Rudi Palla's *Der Kapitän und der Künstler*, Cologne, 2013 is focused on Flinders, Ferdinand and the ill-fated voyage to Australia. As clearly evident from the title Joyce Stewart & William T. Stearn's book *The orchid paintings of Franz Bauer*, London, 1993 deals only with this part of Franz's oeuvre. Sandro Fusina's *Bauer, Bauer & Co.*, Lodi, 2002 (in Italian, privately published and with extremely few copies in public libraries) concentrates on the watercolours by Ferdinand kept in the Natural History Museum, London and is heavily biased towards his work on the flora and fauna of Australia. Now the time seems to have come for a first synthesis.

Needless to say, this is not the definite biography of the three brothers, but only a summary of what is currently known. It has been impossible to follow up all ramifications and connections between the lives and work of Franz and Ferdinand with the circles of naturalists, physicians and gardeners who acted as their employers, commissioners or colleagues and sometimes also as the suppliers of the botanical and zoological materials they were documenting in a most sophisticated way. By contrast Joseph moved largely in a different environment – among artists, art dealers and connoisseurs of art – in Rome and in Vienna.

Looking back on the last four decades of my life I continue to feel the pleasure of having been able to see almost the complete oeuvre of the Bauer brothers with my own eyes, which even in this digital age continues to offer a special meaning to me. Among the highlights were the superb wreath of roses, almost completely covered by butterflies, painted by Franz and discovered by me in the Österreichische Nationalbibliothek in Vienna and his portrait of the tree peony, possibly the first specimen grown outside its native China, kept in the Natural History Museum, London; the illustration of Persian cyclamen from the coast of Cyprus by Ferdinand in the Sherardian Library in Oxford and his portrait of two koalas captured south of Botany Bay in Australia, conserved in the Natural History Museum, London. Unfortunately, I was unable to study the most precious object of Franz's estate: the first heliograph exposed to light in a camera, the icon of the early history of photography, now in a xenon-filled security showcase of the Gernsheim collection of the University of Austin, Texas.

For good reason Ferdinand has been praised as the Leonardo of natural history illustration. However, nobody found words as appreciative as Johann Wolfgang von Goethe did, when he wrote in his essay *Über Blumen-Mahlerei* with enthusiasm on the coloured copper engravings published in *The genus Pinus* based on watercolours by Ferdinand and Franz "Accomplished in his [their] art and applying it ingeniously to his [their] purposes, Ferdinand [and Franz] Bauer present[s] the different species of the genus

Pinus and the manifold transformations of their branches, twigs, needles, leaves, buds, blossoms, fruit, involucre, and seeds to our greatest satisfaction . . . For this reason one is enchanted at the sight of these leaves, nature is evident, art is hidden, the precision is great, the execution mild, the presence determined and satisfying, and we must consider ourselves lucky to be able to repeatedly lay this exemplary work from the treasures of the Grand Ducal library [in Weimar] before us or our friends."

Over the last decades and thanks to the digital revolution the exceptionally fine works of Ferdinand have become very popular for a simple reason, in particular in Australia: the Natural History Museum, London made Ferdinand's finished watercolours on the flora and fauna of Australia available on the internet, and the Sherardian Library in Oxford followed with Ferdinand's finished watercolours on the flora, fauna and topography of the Levant. For several of these watercolours this effectively meant first publication, in particular for the Levantine animals and landscapes, but also for some of the Australian works. On the other hand, the greater part of Franz's rich and equally superb botanical representations and the preliminary works by Ferdinand, continue to be accessible only to the small circle of the initiated with a few published exceptions. Meanwhile, very many publications of their time which include printed illustrations based on works by Franz and Ferdinand have also been made available via the internet by the Biodiversity Heritage Library.

In contrast to Richard Dawkins's *Unweaving the rainbow*, a dictum popular among analytically inclined biologists, my text aims at the opposite - at weaving the rainbow, attempting in a sense a 'triple biography', a great synthesis on the lives and achievements of three outstanding men who dedicated themselves exclusively to the sciences and arts.

ACKNOWLEDGEMENTS

These Bauerian studies brought me into contact with many colleagues, created lasting friendships and resulted in several co-authored publications. David Mabberley (Sydney) shared his immense knowledge of the *Investigator* voyage and the subsequent publication of the botanical results, as well as the life and work of Brown with me, and helped to open many doors at Oxford. Victoria Ibáñez (Madrid) became my co-author when making the early colour chart of the Bauer brothers and their ingenious working methods widely known. Anne-Marie Catteral (Oxford) was invariably helpful when I was working for several months in the Sherardian Library, just like the staff of the Botany Library at the Natural History Museum, London notably Malcolm Beasley, Andrea Hart, Judith Magee and Armando Mendez, and the staff of the Department of Manuscripts at Niedersächsische Staats- und Universitätsbibliothek. Gina Douglas (London) traced Ferdinand's unpublished letters from Australia in the archive of the Linnean Society, Christa Riedl-Dorn (Vienna) made the extensive collection of colour-coded pencil drawings by Franz and Ferdinand, kept in the archive of the Naturhistorisches Museum, accessible to me, and S. Lindemann (Potsdam) kindly communicated his unpublished transcripts of the Itzenplitz correspondence. Thanks to the platform 'Virtual Herbaria' and Ernst Vitek (Vienna) it was easy to trace herbarium specimens collected by Ferdinand deposited in several herbaria, while Milada Rigasová (Mikulov) helped to organize photocopies of the earliest works of Joseph and Franz then still kept in the defunct convent of the brothers of St. John of God in Feldsberg. Hans Kräftner (Vienna) arranged for the study of archival material kept in the Fürstliche Sammlungen and pointed out the importance of Joseph Bauer, effectively one of his predecessors as director of the Gallery of the Liechtenstein family, Cornelia Reiter (Vienna) made the treasures of the Kupferstichkabinett, Akademie der Bildenden Künste, Wien available for the study and provided helpful guidance to literature on Austrian artists resident in Rome. Martin Rodeweld (Berlin) prepared all the maps.

A final word of thanks goes to my home institution, the Botanischer Garten und Botanisches Museum Berlin-Dahlem, Freie Universität Berlin where, in parallel with my official obligations, I could undertake, my extensive, though repeatedly interrupted, studies on the Bauer brothers and their circle over the years.

My wife Eva Lack (Berlin) and James Compton (Salisbury) have kindly read a draft version of the text and commented on it. Having donated so much of their valuable time to my text I feel particularly obliged to them and grateful for their help. Several chapters were also read by Nick Turland, Botanischer Garten and Botanisches Museum Berlin-Dahlem.

PRELIMINARY NOTES

It is highly recommended that the following notes are read. They help explain the structure of the book and the reasons for the unusual high number of notes.

QUOTATIONS AND REFERENCES

All quotations from sources in English are given in their original spelling, irrespective of whether they are taken from unpublished or published sources, i.e. diaries, notes and letters, or from the literature. The orthography, though sometimes odd, is not corrected; nor is the inconsistent use of capital letters. Consequently the traditional addition of '[sic]' is only very rarely used. Illegible words are indicated by square brackets thus [...]; any additions to the quotations are given in square brackets.

All quotations from sources in other languages, i.e. Czech, French, German, Italian, Polish and Russian, have been translated by the author, except the letters by Ferdinand sent to his brothers where the translation given in *Ferdinand Bauer. The Australian Natural History Drawings* (Norst, 1989) is followed. The sources of the quotations are given in the endnotes of the respective chapter.

References are given as endnotes; however, this is done only once in every chapter. As a consequence quotations in the final portions of a chapter often relate to references in the earlier sections of that chapter. In several cases references to the literature are given in a summary at the beginning of the endnotes, such as 'If not otherwise indicated, this chapter is based on Lack (2000a, 2008b)'. In the endnotes 'OXF' is used as an abbreviation for the Sherardian Library, one of the Bodleian Libraries of Oxford University.

DATES AND LOCALITIES

In any book of travels, it is important to know precisely the localities visited and the dates of arrival and departure. These dates are always given. Localities are indicated as spelled in the original documents, not in quotes; their current names are added, always in square brackets on their first appearance in the text; later they are not repeated. In contrast, only contemporary names are used for major modern cities and islands frequently referred to. For Ferdinand's travels in the Ottoman Empire the toponyms used in *The Flora Graeca Story* (Lack, 1999c) are followed; for Ferdinand's circumnavigation of the globe in the *Investigator*, the toponyms used in *Nature's Investigator: The diary of Robert Brown in Australia 1801–1805* (Vallance, Moore & Groves, 2001) are adhered to. Special attention is given to the political geography of the time. As a consequence modern denominations like Germany or Italy are avoided. In the same vein, the toponym Feldsberg is used consistently for this small town, which until 1920 belonged to the Archduchy of Austria below the River Enns, and only later became part of Czechoslovakia and now the Czech Republic.

NAMES OF INSTITUTIONS

The names of institutions are neither abbreviated nor translated. Natural History Museum always refers to the institution in London, Naturhistorisches Museum always to the institution in Vienna. For the repository of the various illustrations and collections see Appendix 1.

SCIENTIFIC NAMES

Every attempt has been made to provide modern scientific names for the plants, animals and minerals referred to in the legends and the text, sometimes added in square brackets inserted in a quotation. In order to determine correct scientific names a plethora of reference works, monographs, revisions and Floras has been consulted, which are not always quoted. For Ferdinand's travels with Sibthorp *Flora Graeca. The annotated re-issue* (Strid & Strid 2009, 2010, 2012, 2013) turned out to be extremely helpful; for Ferdinand's circumnavigation of the globe the *Catalogue of the holdings of the Natural History Museum (London) of the Australian botanical drawings of Ferdinand Bauer* (Mabberley & Moore, 1999) was largely followed. No attempt is made to list the taxonomic novelties based on the collections by Ferdinand nor the illustrations based on the works of Franz or Ferdinand, which formed components of the protologues of names new to science.

BIOGRAPHICAL INFORMATION

Whereas biographical information on the core figures of the story is always given in full, this is not the case for the more peripheral persons. For these, standard biographical references have been consulted, among them *Australian Dictionary of Biography, Dizionario Biographico degli Italiani, Oxford Dictionary of National*

Biography, *Neue Deutsche Biographie*, *Österreichisches Biographisches Lexikon*, and many others. For the considerable number of Fellows of the Royal Society dealt with in this book, biographical information found in the archives of the Royal Society and available at www.royalsociety.org/library/collections/biographical-records/ turned out to be extremely helpful; it is never cited. As a rule errors relating to the Bauer brothers encountered in the literature are not discussed in any detail.

BIBLIOGRAPHICAL INFORMATION

Bibliographical information, in particular publication dates, is exclusively based on the standard bibliography for botany, the second edition of *Taxonomic literature* with its supplements. For the *Philosophical Transactions of the Royal Society* and the *Transactions of the Horticultural Society* the dates of publication are taken from the title pages of the respective volumes, while for the *Transactions of the Linnean Society* those given in a standard work (Raphael, 1970) are used.

DRAMATIS PERSONAE

This list gives the dates of the more important persons mentioned in this book, who were in contact with the three principals – Joseph, Franz and Ferdinand Bauer. There is a focus on their patrons and employers, their commissioners and their travel companions. For all others reference should be made to standard biographical sources. As a rule, information given in this list is not repeated in the text.

THE BAUERS

Baker–Bauer, Elizabeth *fl.* c. 1800–1841
Bauer, Maria Anna Dorothea 1750–1768
Bauer, Ferdinand 1760–1826
Bauer, Franz 1758–1840
Bauer, Johann 1754–1833
Bauer, Joseph 1756–1831
Bauer, Lucas 1708–1762
Bauer, Therese 1730–1790

THEIR PATRONS AND EMPLOYERS

Banks, Sir Joseph, 1st Baronet 1743–1820
Boccius, Norbert, OH 1729–1806
Dietrichstein-Leslie-Proskau, Maria Christina Princess 1735–1788
Dundas, Henry, 1st Viscount Melville, First Lord of the Admiralty 1742–1811
Grenville, Thomas, First Lord of the Admiralty 1755–1846
Grey, Charles, 2nd Earl Grey, First Lord of the Admiralty 1764–1845
Jacquin, Nicolaus Joseph Baron 1727–1817
Jervis, John, 1st Earl St Vincent, First Lord of the Admiralty 1735–1823
Liechtenstein, Aloys I Joseph, 9th Reigning Prince 1759–1805
Liechtenstein, Franz Joseph I, 8th Reigning Prince 1726–1781
Liechtenstein, Johann I Joseph, 10th Reigning Prince 1760–1836
Liechtenstein, Joseph Wenzel, 7th Reigning Prince 1696–1772
Middleton, Charles, 1st Baron Barham, First Lord of the Admiralty 1726–1813
Phipps, Henry, 1st Earl of Mulrave, First Lord of the Admiralty 1755–1831
Sibthorp, John 1758–1796
Yorke, Charles Philip, First Lord of the Admiralty 1764–1834

THEIR COMMISSIONERS

Brown, Robert 1773–1858
Granville, August Bozzi 1783–1872
Hawkins, John 1761–1841
Home, Sir Everard, 1st Baronet 1756–1832
Hooker, Sir William Jackson 1785–1865
Lambert, Aylmer Bourke 1761–1842
Smith, Sir James Edward 1759–1828
Thomson, James 1777–1850

THEIR TRAVEL COMPANIONS

Allen, John 1755–
Asaros, Asfaroos *fl.* 1786–1787
Brown, Robert 1773–1858
Flinders, Matthew 1774–1814
Ganzarowitz, Joseph *d.* 1787
Good, Peter *d.* 1802
Gruber, Leonhard 1740–1810/1811
Hawkins, John 1761–1841
Imrie, Ninian *d.* 1820
Jacquin, Joseph Franz Baron 1766–1839
Westall, William 1781–1850

INTRODUCTION

The *Oxford English Dictionary* defines a biography as 'an account of someone's life by someone else'. A scientific biography may be regarded as a biography based on methods and principles of science, i.e. an account containing a maximum of hard evidence and a minimum of speculation and hypotheses. In this context hard evidence clearly consists of two different elements: primary sources provided by the subject of the biography itself, like diaries, letters and other written documents, and secondary sources provided by others on the subject of the biography, again as a rule, in written form. There is a straightforward correlation: the more sources there are available, the more detailed and colourful a scientific biography can be. The oversupply of sources on an acting political figure like, for example, Napoleon makes it very easy to write his biography, while it is quite impossible to produce a biography of an analphabetic peasant who has left behind no written record and may at best have been mentioned in passing by his literate contemporaries.

Clearly, the Bauer brothers did not belong to the analphabetic sector of society in the second half of the eighteenth century, but to the literate artists or, to be more precise, to the literate draughtsmen. However, their social rank should not be overestimated. In their youth Joseph, Franz and Ferdinand were regarded as belonging to the artisans rather than to the artists, with Franz's and Ferdinand's commissioner in Vienna, Nikolaus Joseph Edler von Jacquin, henceforth Jacquin, recording in retrospect the arrangements for their successor in the job – 'whenever he works, he has lunch and one guilder for the day' (see Chapter 3). Back from Rome Joseph entered the service of the Reigning Prince of Liechtenstein as a badly paid servant, but only years later he was advanced in the hierarchy to court painter (see Chapter 12). Indeed, the Bauer brothers grew up in a clearly stratified society and their place was within the layer of those serving the upper strata of the system. After all, their commissioners like Jacquin, Sibthorp, Banks and Home were established figures of society and rich compared to Franz and Ferdinand, not to mention the almost indescribable wealth of the Liechtenstein family in Vienna who were to employ Joseph for life.

As botanical, or rather natural history illustrators, Franz and Ferdinand served their superiors by documenting what was requested by them. Only later in their lives did they become somewhat more independent and were able to select their study objects more freely, with Franz enjoying a particularly generous arrangement with Banks. Clearly, all three did not belong to that class of gentlemen, but to the upper tiers of the working classes. At the same time their lives tell the story of three young men, who made their way from their native little town of Feldsberg to capital cities like Vienna or London, both centres of art and science. Joseph and Ferdinand managed to establish themselves in the system: late in life Joseph possessed stocks, Ferdinand owned a house with garden in Hietzing, then already an elegant suburb of Vienna near Schönbrunn, the preferred summer residence of the Habsburgs. Franz on the contrary was financially less successful; and it is not clear whether the auction of his estate brought sufficient money to pay off his debts. Considering that the Bauer brothers are not known to have attended school or received a formal education, the total balance of their achievements is remarkable indeed.

When contemplating a biography of Joseph, Franz and Ferdinand Bauer every potential author is faced with a striking discrepancy: the primary information for such a work consists of a negligible number of written documents by the Bauer brothers themselves, particularly when considering that all three reached old age. This paucity of written evidence is in contrast with an extraordinary rich oeuvre of pencil drawings, watercolours and a few paintings in oil. Regrettably, many of these do not carry an annotation, signature or date in the hand of the Bauer brothers, a complete nightmare for every researcher studying this material. In addition, the secondary information on the Bauer brothers is widely scattered among the writings of their contemporaries, such as diaries or buried in unpublished letters and miscellaneous notes. Furthermore only the travels by Franz and Ferdinand are recorded in any detail. In the case of Franz a series of letters exists by Joseph Franz Edler von Jacquin, henceforth Jacquin the son, to members of his family, which record their joint movements in the years 1788-1789. In the case of Ferdinand the incomplete diary and miscellaneous letters by his superior Sibthorp survived their joint tour through the Levant in 1786-1787. By contrast more ample sources exist for Ferdinand's share in the voyage on the *Investigator*: the official report by Flinders, which appeared in print in London in 1814, the recently published diaries by Brown and Good, which, however, mention Ferdinand only in a few cases and in passing, plus several letters, including a few by Ferdinand to his brothers in Kew and Vienna. These well-documented months are in contrast with many years of an extremely meagre record. A good example is the first surviving letter written by Ferdinand dating from 1793, when he was thirty-three years old and had completed already more than 1,300 plant and animal illustrations for Sibthorp alone. This totally unbalanced situation is in itself hardly encouraging to a biographer and inevitably results in a more detailed account of the years *en route* compared with the much longer periods in the studio.

There are further difficulties: Franz and Ferdinand have produced an unusually narrowly specialised oeuvre. While they documented anything living from conifer cones to red snow, from beetles to kangaroos, and from ergot to ferns on several thousand sheets of paper, they omitted completely many fields to which other draughtsmen and painters dedicate themselves with devotion, namely the presentation of humans and of anthropogenic objects. Where a human being appears in Ferdinand's work, it serves best as a comparison of measurement and is never meticulously elaborated; as a result his landscapes are almost always deserted. This narrow focus on living nature void of man and artefacts has brought Franz and Ferdinand the highest esteem from botanists and zoologists, but it has largely prevented their appreciation by the general public. There is a simple reason: what matters to the general public are as a rule images of humans and artefacts, even landscapes need a human touch, like a deliberately cut tree or a human figure at the distance. In this respect the work of Joseph is much more conventional, depicting for example in Rome exactly what every artist would record, human figures, antiquities and other artefacts.

Every student of a drawing or watercolour representing a plant or an animal is confronted with a whole series of basic questions, six of them being of particular relevance.

The first of these concerns what may be called the authenticity of the work: has the given botanical or zoological illustration indeed been based on a specimen or is it simply a copy of a pre-existing illustration? In contrast with established opinion, copying pre-existing botanical and zoological illustrations was a widespread phenomenon for a variety of reasons. In an age before photocopies and scanners, copying was a standard technique and also part of the curriculum in the training of artists. In their youth the Bauer brothers produced many copies of copper engravings and even as a resident artist at Kew, Franz continued to produce at least a few copies based on printed illustrations.

A second question refers to the author of a drawing or watercolour. Who prepared it? In the case of Franz and Ferdinand an answer can only by found after a detailed comparison between those extremely few productions they have signed with their own hand, or which are known from circumstantial evidence to have been prepared by them, and the unsigned representation in question. Since the two brothers had not only many genes in common but also shared the same education in Feldsberg and in Vienna, it is extremely difficult to distinguish their unsigned works. Additionally it should be noted that they again collaborated later in life when Franz was permanently based in Kew. Having said this it may be helpful to keep in mind that Franz showed a more pronounced inclination to microscopic work and, late in life, provided an extensive record of animal and human anatomy. In this he markedly differed from Ferdinand.

Quality is another major concern. But how on earth can one define quality in a botanical or zoological illustration? In our context it is not what might be called aesthetic considerations, i.e. the sum total of what pleases the eye, but truth-to-nature, in particular the documentation of what the specialist requires to see in a representation of a natural object. For this a capacious memory of the illustrator, an ability to analyse and synthesize impressions, as well as a capacity for constantly comparing observations in order to find a statistical standard, for example of a plant species, are key qualifications. In short, we have to imagine the illustrator as the specialist on the lookout for a respective standard specimen defined by statistics, not for an abstract ideal based on theoretical considerations.

Franz and Ferdinand have been rightly praised for precisely this matter-of-fact approach in their illustrations, but regrettably we have no record of how they actually worked and how they communicated with their commissioners. Did the latter adhere to the principle of four-eyed sight, i.e. did they consistently and persistently control Franz's and Ferdinand's work, or did they simply rely on their superior expertise in recording objects of nature? Apart from their early work the latter seems the case, which was definitely the exception in natural history illustration at that time. Being so focused on truth-to-nature, is it just a coincidence that Franz owned the first photographs ever, the famous heliographs produced by Nicéphore Niépce, and that his last acquisition in life was a daguerreotype showing the Pont Neuf in Paris dated 1840, the year of his death?

A fourth question when addressing a botanical or zoological illustration is the identity of the organism represented, indicated almost never by Franz or Ferdinand in writing. Considering the immense diversity of plant, fungal, algal, and animal life, specialists are needed to examine a given illustration in order to determine the identity of the respective organism and to attribute a scientific name to it. In accordance with established museum tradition many botanical and zoological illustrations prepared by Franz and Ferdinand were later annotated with scientific names by specialists, unfortunately largely without disclosing their identity and without giving a date. As a consequence expertise in graphology is necessary to interpret these annotations, with some remaining un-attributable.

Finding out the provenance of the specimen depicted is fifth problem: again specialist knowledge is mandatory. Only in the case of locally occurring plants and animals is it possible to find out the geographic origin of an illustration that has not been annotated with a locality. However, when the Bauer brothers worked, many animals from all four corners of the world were already kept in captivity in zoos, and in a similar way very many plants from all five continents were already cultivated in specialist gardens. A sixth concern is the precise dating of a given drawing or watercolour: datable watermarks, when available, offer at best a *terminus post quem*, the precise dating of a published print based on a particular drawing or watercolour may offer a *terminus ante quem*. Circumstantial evidence, like information taken from diaries or travelogues written by others, may be helpful in dating individual sheets.

For a variety of reasons, the terms 'botanical artist' and 'painter of plants' are avoided in this text. They imply, at least indirectly, a degree of artistic freedom and independence which simply did not exist, whereas in the real world many aspects of the work were highly standardised. Instead the term 'botanical illustrator' is consistently used, stressing the technical and analytical aspects of the job and the interrelationship with a commissioner. As a matter of fact it is the comprehensive documentation of botanical and zoological characters which count in science and nothing else.

The interpretation of the largely undated, unsigned, and un-annotated watercolours is further complicated by the fact that the more or less finished watercolours are often based on preliminary pencil drawings, which, however, are often kept in another place. Some pencil drawings of the early work of the Bauer brothers are conserved in the Naturhistorisches Museum in Vienna, while the finished works are in the Fürstliche Sammlungen in Vaduz. In the same way very many pencil drawings by Ferdinand ended up also in the Naturhistorisches Museum with the finished watercolours being conserved in the Natural History Museum, London. Only in the case of Ferdinand's work on the plants and animals of the Levant did both pencil drawings and watercolours remain together at Oxford. Most notably Ferdinand's preliminary pencil drawings often carry at least some cryptic annotations, like 'Cybrus' [Cyprus] or 'XII' [Head of Spencer Gulf, South Australia], not found in the finished watercolours. Unfortunately, not a single one of these preliminary pencil drawings kept in Oxford and Vienna has ever been made available on the internet, making detailed comparisons and the allocation of collecting sites difficult.

This brief introduction exemplifies the complexities of bringing the extensive, almost completely undated oeuvre into some sort of a coherent, let alone chronological sequence.

Every biographer tries to describe and analyse the personality of his subject. In the case of the Bauer brothers we possess almost zero personal evidence. For Joseph we have a single and brief note in a leaving certificate issued by the Vereinigte K. K. Akademie der Künste [United Imperial Royal Academy of Arts] in Vienna, describing the young man as 'always. . . very well behaved, quiet, peaceable, virtuous' (see Chapter 3). Sibthorp reports on Ferdinand from Naples 'his good Temper & honest Countenance endear Him me much' (see Chapter 5) and later Ferdinand is described by Banks as 'indefatigable' in his work. Franz's and Ferdinand's productions were highly appreciated by sophisticated minds like Brown who was full of praise for both of them (see Chapter 20) and the same is true for Home, who had collaborated over fifteen years with Franz (see Chapter 15).

Enveloped in their sublime world of recording the multifaceted manifestations of life, Franz and Ferdinand became totally obsessed by their work. Otherwise they remain largely enigmatic personalities. We have not the slightest knowledge of their sexual lives, the only thing we know for certain is the fact that Franz and Ferdinand, like their less eminent elder brother Joseph, remained single. The epilogue of Mabberley's biography of Ferdinand quite appropriately ends by noting 'Combined with this talent we have other traits of a plain-speaking man, who was perhaps angular and gauche in his dealings with the rich or more intellectual people who were his employers. And there is an earthy robustness with a canny financial flair too: [Ferdinand] Bauer may have been gifted, brilliant and obsessed with his work, but he was not a saint'. On a similar line is the statement in the epilogue in my biography of Franz 'Bauer had seen much, penetrated deeply into the unknown and produced a permanent record of extraordinary precision, comparable only to that of his circumnavigating brother. In a sense he had been single-minded, an anonymous obituary noting 'Few men, perhaps, devoted their existence so entirely to an uninterrupted observation of natural objects, and enriched science with so many discoveries; and still fewer have been so indifferent to all ordinary considerations; for [Franz] Bauer regarded his talents merely as a stepping-stone to that tranquillity and happiness, which ought to be the ultimate object of all science and all acquirements.' '

This triple biography tries to unravel the lives of three largely unknown personalities, to analyse their vast oeuvre, of both superb quality as well as striking diversity, and to set it into the context of their times.

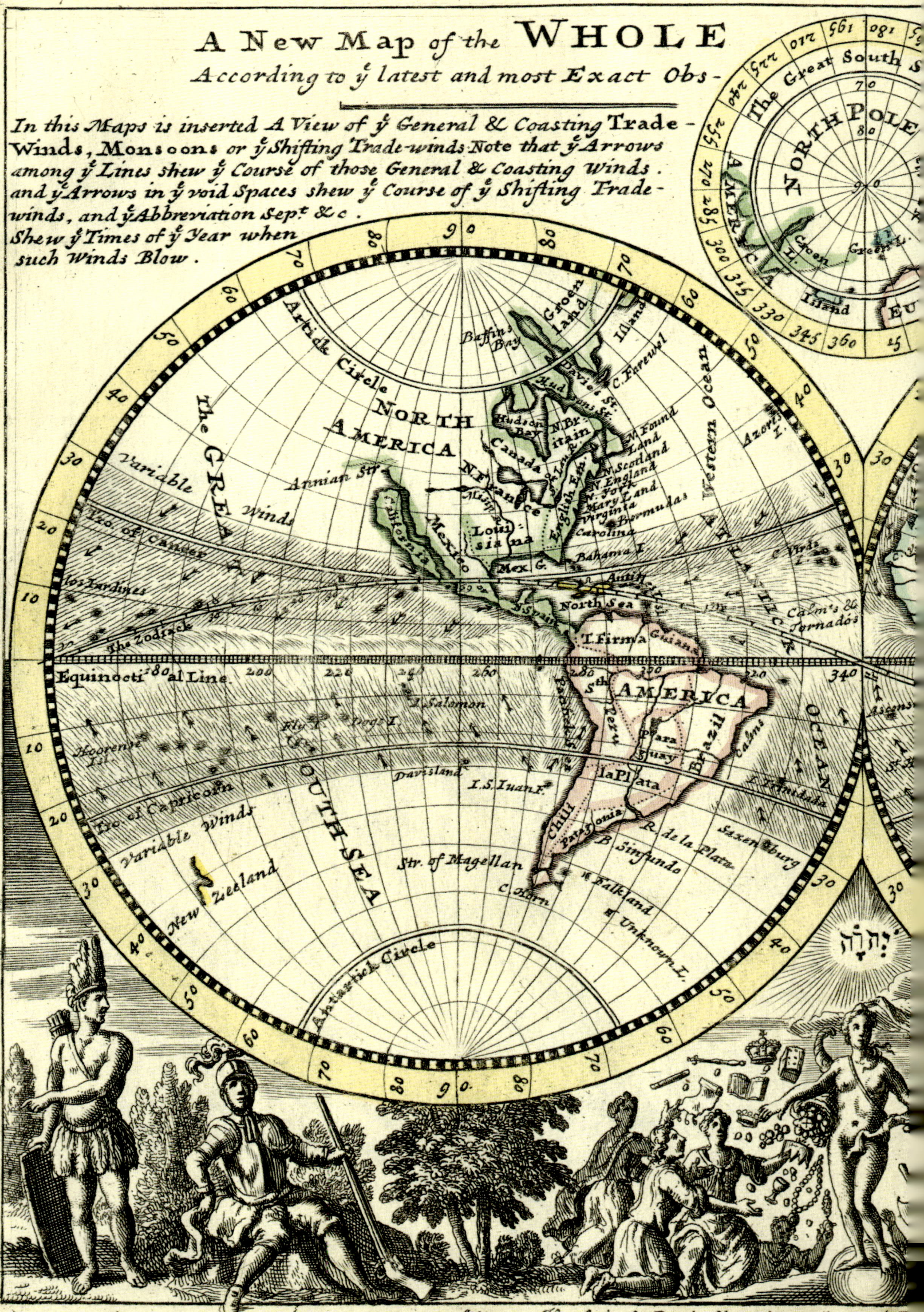

Printed for Thos. Bowles Print and Map seller next ye Chapter House in St. Pauls Churchyard; and J

CHAPTER 1

The incompletely known world

Time is a strange thing. Every historian would agree, and even more so every historian of science and technology. The reason for this is very simple: a period of time without information freely available in the internet, communication without e-mail, orientation without global positioning system is extremely difficult to imagine in the early twenty-first century. But all these technical commodities of modern everyday life were not only totally unknown but completely undreamt of facilities in 1750, when our story begins. Therefore it seems reasonable to briefly reflect the general circumstances in which the lives of the Bauer brothers unfolded.

As already pointed out, the eighteenth century was a period characterized by a clearly stratified society with the upper classes refraining largely from contact with the lower ranks. The strata differed markedly not only in wealth, manner, dress, language and habits but also in access to information, the latter available only to the literate elite and largely stored away in libraries and archives kept and maintained in the centres of power or of learning. Access to these facilities was strictly controlled by the system and could easily be limited even further when found suitable or necessary. In short, the modern concept of a basically free flow of information accessible to everyone simply did not exist.

Conventional communication was either by word of mouth or in the form of letters written by hand, entrusted to a messenger travelling to the recipient of the information. This kind of transfer was not only slow, in particular when involving people on the other side of the globe, but also largely unreliable, resulting in what was subsequently stored in archives and later became available to the historian, only being those letters which had happened to reach their destination or had not been discarded. As a consequence the individual letters were often numbered in order by their writers to enable the recipients to quickly spot a letter lost in transit. Fragmentary transfer of information was therefore the rule, not the exception and as a consequence gaps in our story are easy to understand.

Compasses were of course long known and so was simple orientation on the basis of the movements of the sun. Otherwise only rather primitive maps existed, printed, available to the literate elite only and to be interpreted by the initiated. Navigation at high seas was still in its infancy and specialist knowledge was necessary to make the calculations based on the movements of sun, moon and the stars with no reliable clock yet available. In short any traveller was at that time utterly ignorant about his precise geographic position.

All this went hand in hand with a very incomplete knowledge of the world's topography.

When re-examining the world maps available in 1750 it quickly becomes evident that, although the contours of four continents were roughly known, only the most fragmentary and incorrect information existed on the western and northern coasts of Australia (Fig. 1.2.), very little on New Zealand and Japan, nothing on the

FIG. 1.1. Western hemisphere. A new map of the whole world with trade winds, H. Moll, London, 1736. – Munich, Bayerische Staatsbibliothek, map 74 f1.

islands in the Pacific – neither Hawaii, nor Tahiti, nor New Caledonia had yet been discovered by Europeans. Australia's eastern and southern coastlines remained unknown and so were the high southern latitudes, with absolutely no concept existing on what is now known as Antarctica. The regions in the interior of Africa, Asia and the Americas (Fig. 1.1.) still remained largely unknown. Even maps documenting the topography of Europe were far from reliable, not only, for Russia or the Balkan Peninsula but also for the much better known islands in the Mediterranean Sea. Mount Blanc, the highest elevation in the Alps, had not yet been climbed, let alone Mount Elburs, Europe's highest mountain.

The incomplete knowledge of the world's topography went in parallel with the even more fragmentary knowledge of the diversity of the world's fauna and flora, the latter understood here as comprising plants, fungi and algae. What was known around 1750 by the cognoscenti, and summarized in a most admirable fashion by Linnaeus in his great works, was just the tip of the iceberg. By then not a single animal and only two plant species were known from the continent of Australia, and only the vaguest notion of plant and animal diversity in Africa, Asia, and the Americas existed without the slightest idea of the actual numbers of plant and animal families and genera. In addition, forms of life observable only with the help of a microscope were largely ignored and so was much of animal, plant, fungal and algal life in freshwater and in the high seas. The notion of life in the deep sea did not even exist as a hypothesis. More was known about Europe's fauna and flora, but again huge gaps of knowledge existed, both spatially and temporally. Not only did substantial white areas exist on the botanical and zoological maps of Europe, there were also seasons of the year when only few studies of animal and plant diversity had taken place, all this resulting again in a meagre and totally unbalanced state of knowledge.

Any analysis and description of animal and plant diversity is linked to specimens, which had to be permanently preserved in order to be available for re-examination at a later day. In the eighteenth century this was easier said than done. Herbarium specimens, loose or glued on paper, were difficult to defend against insects without the help of potent insecticides or deep freeze, both unavailable then, and the same problems applied to insect specimens. Skeletons of vertebrates were safer and therefore more popular, but the skins and feathers pertaining to them were again prone to attack by insects. Specimens preserved in liquid posed other problems, among them availability of sufficient quantities of concentrated alcohol, evaporation, fragility of the container, incomplete mode of preservation, the latter difficulties also applying to all forms of anatomical preparations. In addition, in 1750 relatively few public depositories existed for animal and plant specimens in Europe. Most of them ended up in private collections, with places like the British Museum or the Kunstkammer of the Imperial Academy of Sciences in Saint Petersburg being among the rare exceptions.

When looking back on the fragmentary zoological and botanical knowledge in the eighteenth century one must not forget that fundamental principles of biological thought simply did not exist – neither the all-embracing concept of evolution nor the no less all-embracing concept of genetics, neither the idea of plant and animal geography nor the fundamentals of comparative anatomy, embryology and cytology. Basic phenomena like sex, reproduction, photosynthesis, respiration, muscle contraction, blood coagulation were not yet or only very incompletely understood. Disasters like mass infections of wheat by rusts or contagious disease in humans were still largely seen as a punishment by God. Only the few initiated, mainly naturalists and physicians, would regard them as natural phenomena open to study and analysis - while others, for example in the field of pharmacology, continued to rely on dogma and tradition, rejecting the concept of experiment which is now regarded as the cornerstone of the natural sciences as well as medicine. This is one of the reasons why the interpretation of the medicinal plants mentioned in the writing of Dioscorides, a Greek physician, pharmacologist and botanist who had lived in the first century, continued to be regarded as relevant by many. Put simply, in the first half of the eighteenth century the enlightenment had not yet penetrated deeply into society.

In addition to what has just been said about the general availability of information in the twenty-first century, the efficiency of present-day communication and the ease of geographical orientation, one more aspect necessitates our special attention – the modern omnipresence of digital images. They have become so dominant in our lives, and are today so easily and quickly made, stored and transmitted, even by children, that many have stopped reflecting on what an image really is. For the younger generation the time before the existence of the digital image is remote past, although the digital camera, an apparatus to encode digital images and store them for later reproduction, became commercially available only in the early 1980s. Before that period conventional photography existed, which basically meant that light was induced in a simple camera with a lens to produce a latent image on a carrier, which in a second step had to be transformed into a permanent image. This had been a revolutionary invention, going back to the pioneer works of Nicéphore Niépce and Louis Daguerre, who had changed the world in a similar way to how the invention of digital images did about 140 years later. Most importantly, both a digital camera and a conventional camera are recording devices everyone could learn to use quickly and easily.

Even more remote for many is the time when photography in whatever form had not yet been invented, i.e. the period when an image could only be produced by the hand of a specialist. The specialist does not rely on an apparatus but solely on his eyes, brain and hands as well as the drawing and painting equipment. Basically, with the help of the specialist's eyes, the subject to be documented creates a latent image in the brain, which, by emitting electrical impulses, makes the muscles contract to move the hand, which draws a line on the carrier of the image with the help of a pencil or applies liquid paint to it with a brush. While a digital camera or a conventional camera is basically nothing more than an optical as well as an electronic/chemical device to record an image, a drawing or painting made by hand is a direct expression of a person's personality, i.e. of his or her brain, without the intermediary of an apparatus. Personalities differ, and so do images produced by them, either with the help of an apparatus or without. However, under all circumstances and in all cases these images are forms of self-expression, like writing, singing or dancing.

It seems sensible to reflect briefly about the more general circumstances of preparing a drawing, watercolour or painting in oil in the mid-eighteenth century, which differ markedly from the situation in the early twenty-first century. Every artist would confirm that light is the precondition of his or her work. Yet, only daylight or candlelight was available when our story starts, which severely restricted the numbers of hours available for work and resulted in a totally different time regime. In addition, candlelight possessed a different colour temperature that had to be taken into account when preparing a watercolour. Secondly, the number of drawing aids was still rather limited – the camera obscura, mirrors, magnifying glasses and also spectacles had been known for centuries, and simple microscopes existed, but taking precise measurements with their help was still impossible. Thirdly, a sheet of plain glass in a light box continued to be an important item when it came to take tracings, which enabled the more efficient production of copies, or duplicates, of a pre-existing image, no other devices being then available. Fourthly, no rubber erasers existed and bread was used instead to delete pencil lines. Fifthly, paper had to be used with economy. This applies to thin drawing paper as well as to solid watercolour paper, the latter too heavy and expensive to be taken along on expeditions and only to be used in the studio, with plain-air painting becoming fashionable only much later. Sixthly, the spectrum of pigments available for watercolours or paintings in oil was markedly smaller than the enormous range of hues available in the twenty-first century. In addition, the time-consuming preparation of pigments was largely the job of the illustrator or artist, another extra burden on his or her time budget, often not properly considered by art historians.

Specialisation in natural history illustration aiming at truth-to-nature on the one hand implies a mixture of training, experience, dedication and focus on the part of the person preparing the drawing or painting, while on the other hand the end product begs its price, which the specialist is entitled to ask for as compensation for his or her labours in the form of money, food or other commodities. However, the market for natural history illustrations was small and remained so because only a tiny fraction of society is focused on the accurate documentation of plant and animal diversity. Although this may seem banal and straightforward, it needs to be stressed that any drawing or painting is a unique item, a unicum, and by itself has a special value.

Replication with the aid of graphical methods, basically the art of copper engraving, in the eighteenth century and in the natural sciences, since lithography had not yet been discovered, is a totally different matter, involving a graphical artist, a printer and, in case of coloured engravings, a colourist, i.e. three different persons who as a rule have never seen the subject originally documented. Needless to say, there is always some loss of quality and precision when a drawing or watercolour true to nature undergoes a graphical process and is replicated, irrespective of the method applied. Also needless to say, the print-run or edition, i.e. the end result of a graphical process, is never completely homogeneous, in particular when copies coloured by different colourists are compared. More import is the fact that such a reproduced, i.e. printed, image was no longer a unicum and consequently less elitist than the drawing or water-colour.

In the incompletely known world of the mid-eighteenth century images aiming at truth-to-nature and precision were particularly relevant and sought after by naturalists and physicians. This is due to multiple reasons: their objects were often ephemeral – the flower of a poppy will last only shortly, a freshly collected mushroom will soon transform itself into a slimy discoloured mass, within hours a fish will lose its shiny colours, the exanthemata caused by measles or shingles may change within less than a day. Furthermore, often objects were so rare that no one could presume to be able to observe them again soon – it may take a plant in a conservatory many years until it flowers for the first time with no gardener being able to tell when it would do so next time, a rare bird may be observed for the first time and is in need of recording, in a post mortem an unconventional, very rarely occurring malformation of an internal organ may have been found that needed documentation. In many cases the respective objects were relevant to science – recording a potential medicinal plant in a precise image may be a prerequisite for finding it again, just like the documentation of an intestinal worm or of a human foetus developed in the ovary instead of in the uterus may be very useful. In addition, the objects of naturalists and physicians often showed sequential phases of development best recorded in precise serial images – the dynamics of germination, anthesis and fructification, of metamorphosis in amphibians and insects being good examples to the point, not to mention the profound changes some parasitic fungi undergo when passing from one host to another. Finally, with scientific terminology still in its infancy and standards of scientific communication largely missing, an accurate image would be much better suitable for an exchange of opinion than a descriptive text, and this applied to plants, fungi and algae in the same way as it applied to animals. In short, for a wide spectrum of reasons a wide spectrum of objects offered itself to documentation, in the form of drawings or watercolours, aiming at the strictest truth-to-nature possible.

There is a notable paradox in this approach: in the ideal world the documentation of such objects aiming at truth-to-nature leads to images which are neither datable nor attributable to the hand of an illustrator, they become in a sense de-personalized, neutral and very similar to electronic documentation made by a digital camera or photographic documentation made with the help of a conventional camera.

To the historian of science several collections of such de-personalized images aiming at strictest truth-to-nature are known, owned by naturalists or medical doctors and often brought together over an extended period of time. For good reason some of these have been denominated paper museums, like the Museo assembled by Cassiano dal Pozzo in Rome, which documented his collections including not only antiquities but also natural history objects in precise images, with both the paper museum and the collections now widely dispersed. Along the same lines, is the painted documentation of the plants cultivated over centuries in the Jardin du Roi in Paris and the animals kept in the menagerie in Versailles, kept today in the Bibliothèque Centrale of the Muséum National d'Histoire Naturelle in Paris, with several repeats conserved in the Österreichische Nationalbibliothek in Vienna. Is it surprising that many of these de-personalised, timeless images were left unsigned and undated by the illustrators? Being unaffected by time is a special feature of many of these natural history images focussed on truth-to-nature. For some, this old-fashioned opticality offers a special aura and in a sense mystifies the illustrator, for others it is a constant frustration – how to distinguish the famous iris painted, possibly in Venice, by Albrecht Dürer from the iris documented in the Levant and later painted in Oxford by Ferdinand? How to distinguish the watercolour of a male imago of the stag beetle, *Lucanus cervus*, recorded by Dürer in the J. Paul Getty Museum in Malibu from the watercolour by Franz in the Niedersächsische Staats- und Universitätsbibliothek Göttingen showing the same species?

In a sense Dürer *c.* 1500 stands at the beginning of the era of natural history illustration aiming at truth-to-nature and the Bauer brothers in the 1820 to 1840s at its end, when photography started to become an alternative. Neither Franz nor Ferdinand were ever superseded in their truth-to-nature or replaced by others. It was the approach to preparing images that was to change, in fact in the year before Franz died – from images made by the brain and executed by hand to images made by an apparatus.

FIG. 1.2. Eastern hemisphere. A new map of the whole world with trade winds, H. Moll, London, 1736. – Munich, Bayerische Staatsbibliothek, map 74 f1.

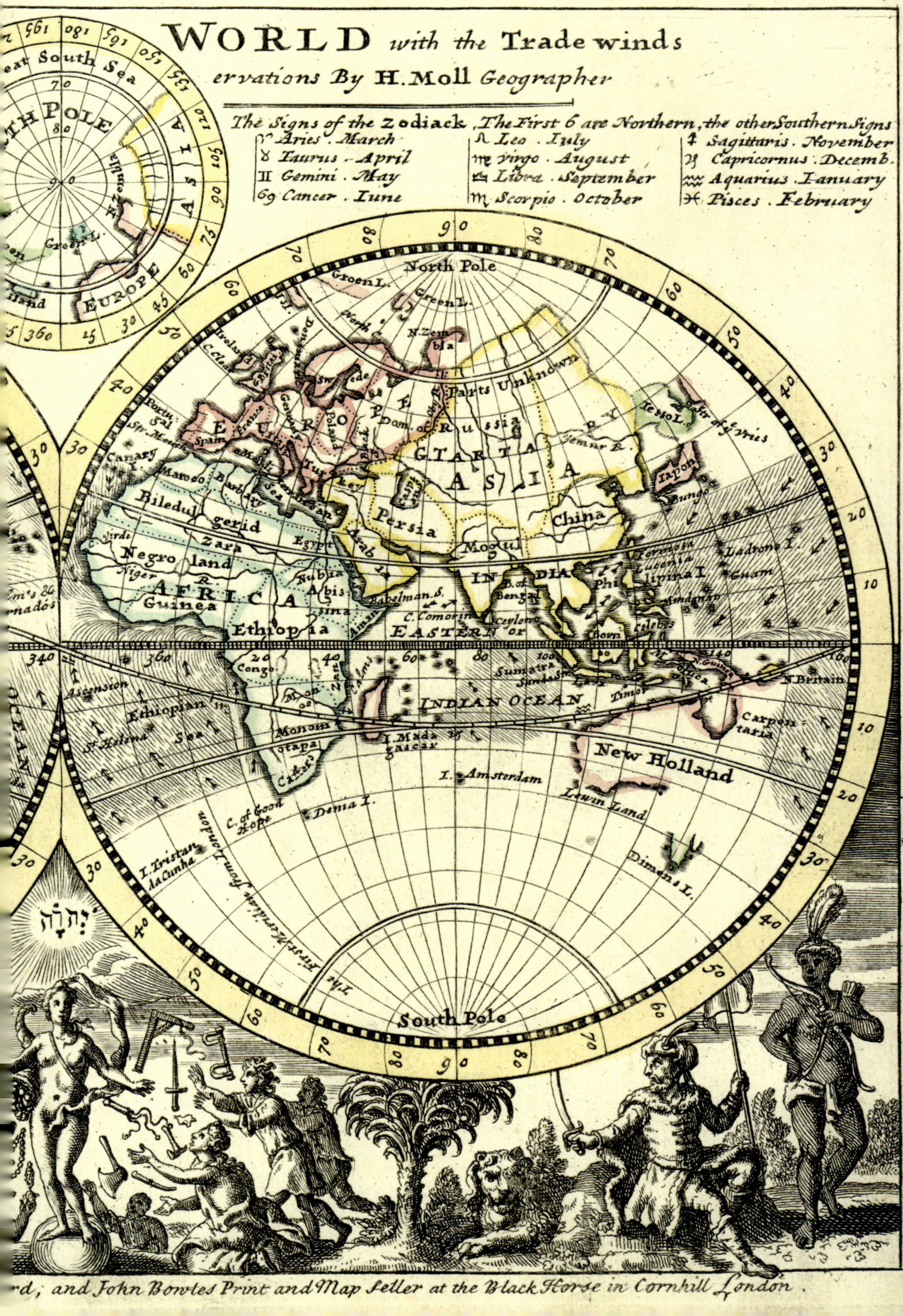
WORLD with the Trade winds
ervations By H. Moll Geographer
The Signs of the Zodiack, The First 6 are Northern, the other Southern Signs
Aries . March
Taurus . April
Gemini . May
Cancer . Iune
Leo . Iuly
Virgo . August
Libra . September
Scorpio . October
Sagittaris . November
Capricornus . Decemb.
Aquarius . Ianuary
Pisces . February
North Pole
South Pole
EUROPE
ASIA
AFRICA
Persia
China
INDIA
Tartaria
Parts Unknown
Dom. of Russia
Iapon
Arabia
Mogul
Persia
Biledulgerid
Zara
Barbary
Egypt
Negroland
Nubia
Abissina
Guinea
Ethiopia
Congo
Monomotapa
Caffers
C. of Good Hope
Ethiopian Sea
St. Helena
Ascension
INDIAN OCEAN
I. Madagascar
I. Amsterdam
Dema I.
I. Tristan da Cunha
New Holland
Carpentaria
Lewin Land
Dimens L.
N. Britain
Timor
Sumatra
Ceylon
C. Comorin
Ladrone I.
Guam
Philipina I.
Borneo
Celebes
Jesso I.
Nova Zembla
Greenl.
The First Meridian from London
Great South Sea
South Pole
Europe
Asia
rd; and John Bowles Print and Map seller at the Black Horse in Cornhill London.

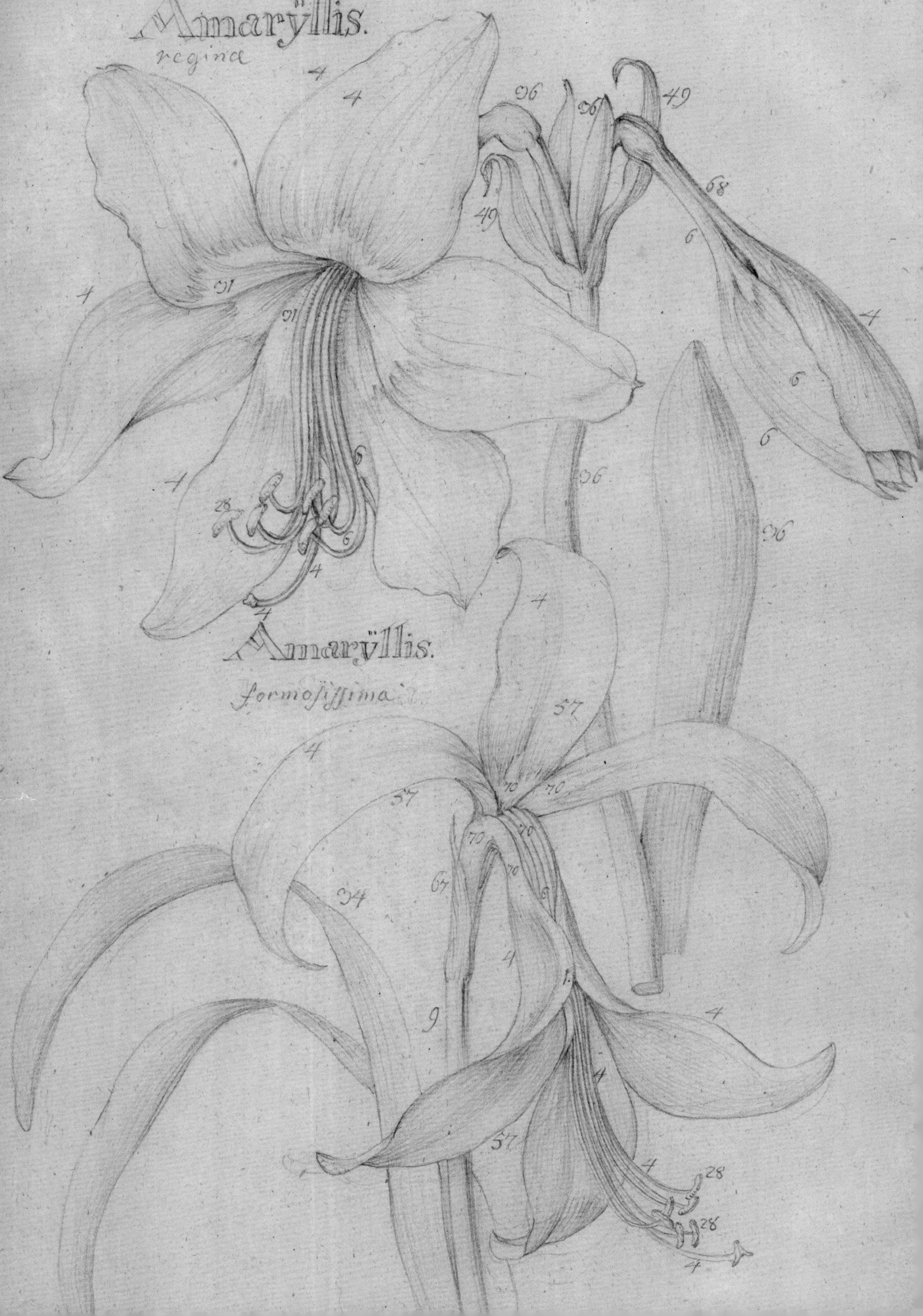
Amaryllis.
regina
Amaryllis.
formosissima

CHAPTER 2

The Bauer brothers: formative years in Feldsberg[1]

RURAL COUNTRYSIDE

SOFTLY UNDULATING HILLS alternating with wide plains extend from Vienna northward as far as Brno, the capital of Moravia. It is peaceful, rural countryside, with fertile fields where a variety of crops, mainly wheat, barley and maize, are grown. A land characterized by large vineyards, rich meadows and wetlands along the rivers, and deciduous broad-leaved forests dominated by sessile oak, field maple, European hornbeam and small-leaved lime. To the south, this region is delimited by the Danube, to the east by the River Morava, and to the west by a long chain of hills called Manhartsberg; only northwards is there no natural border. A benign climate favours this country, characterized by long, warm and dry summers alternating with relatively short, cold winters.

Since medieval times the southern part of this area had belonged to the Archduchy of Austria below the River Enns, forming the so-called Quarter below the Manhartsberg, now informally and for good reason called Weinviertel (Wine Quarter). The northern part had for many centuries belonged to the Margraviate of Moravia, forming part of the lands of the Bohemian Crown, while the country to the east of the Morava had belonged to the lands of the Hungarian Crown. However, following the momentous battle of Mohács in 1526 and the resulting political changes, the Archduchy of Austria, the Margraviate of Moravia and the Kingdom of Hungary were ruled by the same family, the Habsburgs. As a consequence the area was divided by administrative lines, one of them meandering west to east and dividing Moravia, with Brno as its capital, from the Archduchy of Austria, with Vienna as its capital. The small, equally meandering River Thaya (Dyje) traverses this region in the same direction, with an unconventional consequence: here and there it flows into the Archduchy of Austria, then into Moravia, then into the Archduchy of Austria again, forming the border between the two territories for a few kilometres before zigzagging into a much larger river, the Morava. Today the northern part of this area belongs to the Czech Republic and the southern part to Austria, while the region east of the Morava is part of Slovakia.

In medieval times several small towns had been founded along the Thaya, from Raabs, Drosendorf, Hardegg and Frain (Vranov nad Dyjí) in the west, to Znaim (Znojmo), Laa and Lundenburg (Břeclav) further east. Others lie at some distance from the river, which is known for its repeated inundations in late spring and early summer. Among them, about twelve kilometres apart, are Nikolsburg (Mikulov) and Feldsberg, both now in the Czech Republic, the latter about sixty-five kilometres to the north of Vienna. Feldsberg is the birthplace of the three principal figures of our story and where the most famous of them, Ferdinand, stayed longer than anywhere else in his life. The Feldsberg years were formative to all three brothers and they are dealt

FIG. 2.1. Barbados lily, *Hippeastrum puniceum*, and Jacobean lily, *Sprekelia formosissima*. Drawing by the Bauer brothers, before 1778. Pencil. – Vienna, Naturhistorisches Museum, Archiv.

with here in detail, since the professional orientation of Joseph, Franz and Ferdinand took place there.

The aristocracy and the landed gentry, the owners of often large estates, shaped and determined life in this rural area for centuries. Over generations several of these families had managed to amass considerable wealth, which enabled them to live in grand style, to build large country houses and to collect works of art. On the lower Thaya no other families equalled the Dietrichsteins, who were based in Nikolsburg and then in Moravia, and the Liechtensteins, who were based in Feldsberg and then in the Archduchy of Austria, the latter with a fine summer residence at a distance of seven kilometres in Eisgrub (Lednice) in adjacent Moravia. Both families possessed monumental city residences in Vienna, among them those adjacent to each other on Minoritenplatz, and both belonged to the high echelons of the aristocracy: Princes of the Holy Roman Empire, having fiefs that had no suzerain except the emperor. As a consequence they enjoyed imperial immediacy and special privileges, among them the right to collect taxes and tolls, and to mint coins. The Prince of Dietrichstein and the Prince of Liechtenstein were entitled to sit on the secular bench in the Council of Princes at the Imperial Diet, which met as a rule in Ratisbon. For good reason the Liechtensteins regarded the title of Prince of the Holy Roman Empire to be higher than their older title of Duke of Troppau (Opava) and Jägerndorf (Krnov), two towns in Silesia, which were titles referring to the Kingdom of Bohemia. For generations members of both families have been introduced to the Order of the Golden Fleece, awarded exclusively to Catholic royals and to the inner circle of Catholic high aristocracy. They played a leading role during the counter-reformation in the lands of the Bohemian Crown, which helped them to augment their wealth and extend their large properties even further. The two families owned sumptuous burial chapels – the Dietrichsteins in Nikolsburg, the Liechtensteins in Wranau (Vranov) near Brno.

The rise of the Dietrichsteins had started with Franz Cardinal Dietrichstein (František z Ditrichštejna), Bishop of Olmütz (Olomouc), Baron of the Holy Roman Empire, who crowned Archduke Matthias and later Archduke Ferdinand Kings of Bohemia in Prague. Franz was also the first of his family to receive the hereditary title of Prince of the Holy Roman Empire in 1624 under Emperor Ferdinand II. Upon the death of the great cardinal the title passed to his nephew and remained for generations in the Dietrichstein family.

The rise of the Liechtensteins started with Karl Count Liechtenstein-Nikolsburg zu Feldsberg, created prince in 1608 and thereby becoming Karl 1st[2] Reigning Prince of Liechtenstein (Karel I z Lichtenštejna), who was later made Viceroy of Bohemia under Emperor Ferdinand II. Two generations later, Anton Florian, the 4th Reigning Prince of Liechtenstein, obtained the title of Prince of the Holy Roman Empire in 1719 during the reign of Emperor Karl VI (Charles VI). The earlier acquisition of the small, distant and economically irrelevant territories on the upper Rhine, which received the name of the family and today form the Principality of Liechtenstein, had been the prerequisite for this promotion in rank. However, for the following two centuries the members of the Liechtenstein family remained focused on their vast estates in the Archduchy of Austria and in the lands of the Bohemian Crown, that is in Moravia, Silesia and Bohemia, with the princely administration based in Vienna.

The Dietrichsteins, who also possessed city palaces in Brno and Graz, and the Liechtensteins, who owned an additional city palace in Prague, differed in wealth and family success, the latter outrivalling the former. The Dietrichsteins were to die out in the male line in 1864, whereas the Liechtensteins are still continuing to flourish and thanks to Napoleon (see Chapter 12) were later to become sovereigns in their possessions on the upper Rhine. In contrast to the Dietrichsteins they were also able to maintain their city and summer palaces in Vienna (see Chapter 3) and continue to own large properties in the Weinviertel and elsewhere. It is difficult to get a realistic grasp of the richness of the Liechtenstein family at the time when our story begins, but in 1805 – when Aloys I Joseph, 9th Reigning Prince of Liechtenstein, died – an obituary reported 'These estates together have 756 places, including 24 towns and 35 market towns, with a population of 302.000 souls. The princely properties include important hammer works, and glass factories, and 164 dairy farms . . . The [annual] proceeds of the aforementioned properties amount to some 1,200.000 guilders'. In brief, the Liechtenstein estates, like those of the Princes of Schwarzenberg (Schwarzenbergové) in Bohemia, effectively formed an entity of remarkable size within the Hereditary Lands of the Habsburgs. This was almost a state within the state, with the Reigning Princes of Liechtenstein sometimes jokingly referred to as the bankers to the imperial family, to whom they, however, remained loyal for centuries.

In the same way as the Dietrichsteins, in particular Franz Cardinal Dietrichstein, had modelled Nikolsburg with its princely residence, parks and gardens according to their personal taste, Feldsberg and Eisgrub were modelled by succeeding generations of the Liechtensteins. They left a lasting mark on both places, which formed part of their Feldsberg estate, one of the many they owned, comprising some three hundred and fifty square kilometres, situated just a few kilometres south of the dense riparian wetlands and ponds of the Thaya valley, with Feldsberg set amongst extensive vineyards. In 1770 Feldsberg consisted of just 243 houses and was still surrounded by a defensive wall. A contemporary account of the local geography reads 'the town has . . . five markets each year . . . the church is dedicated to the Assumption of the Virgin Mary . . . in addition to the church there are two monasteries in Feldsberg, one Franciscan and one belonging to the Brothers of Mercy . . . On a hilltop . . . is the attractive castle of the prince [of Liechtenstein]. This residence . . . contains a marvellous chapel, lavishly decorated chambers, a magnificent garden, and an impressive riding school built in two storeys, with marbled stalls. The associated domain is extensive . . . The wine of Feldsberg is amongst the best in Austria. The hunting is also excellent, especially for wild boar'.

A few kilometres north of Feldsberg and connected by a straight avenue lies the village of Eisgrub, described by the same source as 'a market with church, a newly built, attractive, splendidly furnished summer-castle, and nearby an equally impressive stable, close to a very large, unusually fine, artistically and tastefully designed garden . . . on the right bank of the Taya [Thaya], from which the water is lifted artificially for the grottoes and fountains of the garden . . . 554 Christian, 13 Jewish houses'. What is here called an 'impressive stable' has become known as the 'Palace for the horses',[3] a baroque building designed by the first architects of their time, including Johann Bernhard Fischer von Erlach and Domenico Martinelli (see Chapter 3), with construction works continuing for more than two centuries. The stables are documented in copper engravings by Johann Adam Delsenbach, who also represented the garden of the summer residence in Eisgrub, an extensive parterre in baroque style. Whereas this country house, also built in baroque style, was later demolished and substituted by an extensive neo-Gothic building, the significantly older Liechtenstein castle in Feldsberg[4] has survived intact, although it has lost a large part of its furniture (see Chapter 19). Nevertheless, it continues to be a particularly fine residence, mirroring the elevated status of the princely family with an enfilade of richly decorated rooms, among them an exquisite cabinet where Archduke Karl, later Emperor Charles VI, the father of Maria Theresia (Maria Theresa), is reported to have stayed. The cabinet is adorned with a painting of the goddess Flora, her followers, and lengthy flower garlands on the ceiling. An eyewitness who visited Feldsberg in 1782 marvelled at the exquisite rooms, their splendid furniture and the extensive collection of paintings, counting no less then 568 pieces in the galleries[5]. At the same time he estimated the value of the paintings in one of the festive halls alone to approach 30,000 guilders, a considerable sum of money. On the ground floor a fabulous chapel can still be admired, a perfect amalgam of architecture, painting and sculpture, rightly praised as a refined *Gesamtkunstwerk* of the late baroque period and among the finest in the Archduchy of Austria and adjacent Moravia. Today the altar piece is a copy of the famous painting 'The adoration of the shepherds' by Guido Reni, of which two original versions are known – *in situ* in the Certosa di S. Martino in Naples and in the National Gallery in London[6], the latter from Feldsberg. Nothing illustrates the luxurious decoration of this chapel better than the fact that even the seating area of the pews is richly worked in intarsia.

In the fine state rooms the over-doors (decorative paintings set over the doors) are more relevant in our context, since they comprise exquisite flower and fruit pieces, painted among others by Franz Werner von Tamm. These show, for example, a composition of different grape cultivars and melons; there are also individual still lives by Tamm and by Philipp Peter Roos (better known by his alias Rosa de Tivoli) presenting mainly animal life-like deer, fowl and horses. Considering the special status of the Liechtenstein family, it is no surprise that Maria Theresa, like her father before, is reported to have visited this fine residence several times,[7] followed later by her sons, the Emperors Joseph II[8] and Leopold II.

For a tiny town such as Feldsberg the parish church, dedicated to the Assumption of the Virgin Mary, was and continues to be hopelessly over-sized, with the two towers reaching the height of thirty-four metres. It took several decades to complete the building, which had been planned on the basis of severe, early baroque models in Rome by Giovanni Jacomo Tencalla. Like at St Peter's in Rome there is an extremely large horizontal inscription band over the main facade reading 'Carolus Eusebius princeps de Liechtenstein de Nicolsburg Opp Carnoviae dux' (Karl Eusebius Prince of Liechtenstein of Nikolsburg, Duke of Troppau and Jägerndorf), referring to Karl Eusebius, the 2nd Reigning Prince and son of Karl 1st. 'Money has only one function – to leave behind monuments for eternal and immortal memory' was one of Karl Eusebius's dicta, which characterized him and several of his successors in an appropriate way.

It was Karl 1st who had invited the brothers of the hospital order of St John of God, often referred to as the Brothers of Mercy, to found a convent in Feldsberg, the first north of the Alps and the first in the Holy Roman Empire. Karl Eusebius and his wife Johanna Beatrix, née Princess Dietrichstein zu Nikolsburg, continued to support the confraternity of this mendicant order with lands and funds, a tradition followed by their successors.

Apart from the usual vows of poverty, chastity and obedience, members of this order also take the vow of hospitality – that is, they commit themselves for the rest of their lives to serve the sick in a hospital attached to the convent. In this context a hospital of the Brothers of Mercy has to be understood as a hospital for men only, an emergency hospital where patients are cured, and not just nursed and cared for, as in many hospitals of other orders.[9] Most significantly treatment was for free and available to all, including non-Catholics. The confraternity was headed by a prior, comprised male nurses and surgeons, all laymen, and included mostly only a single ordained priest. In accordance with the rule of this order, the convent in Feldsberg, which had been founded as early as 1605, consisted of a church, a hospital, a pharmacy and a medicinal garden, the latter two of particular relevance for our story. The complex was built beyond the city walls on the road to Eisgrub, with later additions, and survives largely unchanged.

In the beginning the convent may have been very modest and remained so for decades, but it was from Feldsberg that the foundation of the convents of Vienna, Graz, Prague, Görz (today a divided town consisting of Gorizia in Italy and Goricia in Slovenia), Pozsony (Bratislava) and others followed in quick succession. From 1718 onwards a school for male nurses formed part of the convent in Feldsberg, which at the time was one of the very few places outside universities where anatomy, simple surgery, wound-dressing, nursing and botany – effectively pharmaceutical botany – were taught. Feldsberg was the mother convent for the province 'Zum Heiligen Erzengel Michael' (To the Holy Archangel Michael), in 1781 divided in two provinces, one covering the Hereditary Lands of the Habsburgs renamed 'Zum Hl. Karl Borromaeus' (To Holy Charles Borromaeus). Each of them was an administrative entity of the order headed by a Father Provincial, who had as his superior the Prior General of the order in Rome. As in other convents, for example in Vienna, there was also a well-stocked library in Feldsberg, including herbaria (collections of pressed and dried plant specimens), which in 1763 had received new rooms within the building.

Our story starts with a baptism on 29 December 1750 in the church of the Assumption of the Virgin Mary in Feldsberg. At that time Benedict XIV (Prospero Lorenzo Lamberti), aged seventy-five, was Pope in the sixth year of his pontificate. Franz I Stephan (Francis I Stephen) of Lorraine, aged forty-two, was Emperor of the Holy Roman Empire in the sixth year of his reign, residing in Vienna. His wife Maria Theresa, aged thirty-three, was Queen of Hungary and Queen of Bohemia in the tenth year of her reign. She also held many other titles, such as Archduchess of Austria and Margrave of Moravia. In London George II, King of Great Britain and Ireland and Prince Elector of the Holy Roman Empire, aged sixty-seven, was in his twenty-third year on the throne. In Nikolsburg the key figure was Karl Maximilian, the 6th Reigning Prince of Dietrichstein-Proskau zu Nikolsburg; in Feldsberg his equivalent was Joseph Wenzel, the 7th Reigning Prince of Liechtenstein.[10]

No doubt Joseph Wenzel was the most brilliant figure in the long history of this family and the quintessence of a prince in the late baroque period. In a letter to one of her sons, the Archduke Ferdinand, Maria Theresa was to call Joseph Wenzel, who had died just three days before 'my good old friend', continuing 'he is one of those figures which you do not find anymore'. And in her letter of condolences to his successor, Franz Joseph I, the 8th Reigning Prince of Liechtenstein, she mentions that Joseph Wenzel had 'fulfilled with great honour various military and political honours'. This indeed is no exaggeration: in 1735 he had been sent as imperial envoy to Berlin, where he befriended the crown prince Friedrich, later Friedrich IV Margrave of Brandenburg and Friedrich II (Fredrick II) King in Prussia (later of Prussia). From 1738 to 1741 Joseph Wenzel was imperial ambassador in Paris, where he held an *entrée publique* in grand style, which had to be repeated in Versailles on the explicit wish of Louis XV, King of France.[11] However, when Emperor Charles VI died, Joseph Wenzel was unable to maintain peace between the Bourbons and the Habsburgs and was obliged to return to Vienna. His next position was that of governor general of Lombardy, then commander-in-chief of the imperial army in Italy, winning the battle of Piacenza, followed by commander-in-chief for Hungary. More relevant for the imperial couple, however, was his engagement for the modernization of the imperial artillery, since Maria Theresa was waging a series of wars against Frederick II. In theory this was a feud between the Queen of Bohemia and the Margrave of Brandenburg, but since the two competitors also possessed other crowns and titles and had formed a network of coalitions with other powers, this effectively turned into a global conflict. In 1763 it ended in the peace of Hubertusburg, which for Maria Theresa meant the loss of the greater part of Silesia, but at the same time the consolidation of her political power in Central Europe and the beginning of a period of peace. This was praised by Mathias Claudius in his famous poem, beginning with the verses 'She made peace! That is my song. She was her people's happiness and her people's blessing'.

Joseph Wenzel's earlier advance financing of artillery production had resulted in a significantly improved situation for the Habsburg forces on the battlefield, which brought him a most unconventional honour: a monument including a bust in bronze by Balthasar Ferdinand Moll, placed in the Imperial Arsenal in Vienna in 1758, when Joseph Wenzel was sixty-two years old. Today this fine fire-gilt statue is exhibited in the Österreichisches Barockmuseum in Vienna.[12] In his later years Joseph Wenzel was chosen to accompany Isabella Princess of Parma, the bride of Archduke Joseph, from Parma to Vienna, and to act as imperial ambassador at Frankfurt, when this very Archduke was to be elected and crowned Emperor of the Holy Roman Empire, subsequently called Joseph II. At the same time Joseph Wenzel was an important Maecenas, connoisseur, bibliophile and collector of art, who was the first to have a catalogue of all his paintings and sculptures printed (see Chapter 3). He seems to have possessed an early *gusto antico*, since he acquired from his friend Eugene Prince of Savoy a most spectacular statue – the 'Praying Boy', a bronze from Greek antiquity originally found on the island of Rhodes. Later Joseph Wenzel sold this trophy to Frederick II, who had it placed on the terrace of Sanssouci, his newly built summer residence in Potsdam. Later copied *ad nauseam*, this iconic object was transferred first by Napoleon to Paris and later came back to Berlin, where it is a key piece in the famous Antikensammlung[13] in the Altes Museum today.

THE BAUER FAMILY

The baby baptized on 29 December 1750 in Feldsberg and almost certainly born in Joseph Wenzel's residential town, was Maria Anna Dorothea Bauer, the first child and only daughter of Lucas Bauer, aged forty-two, and Therese Bauer, née Hirsch, aged twenty. We know very little about this couple: Lucas Bauer (spelt Baur in the birth register) had been born on 17 January 1708 in Ehrenbreitenstein on the Rhine, then the permanent residence of the Archbishop of Trier and Prince Elector of the Holy Roman Empire. Lucas Bauer had become associated with Joseph Wenzel as his court painter in 1744 at the latest. Therese had been born in Feldsberg on 12 October 1730. Nothing is known of when and where the couple got married, but Joseph Wenzel must have agreed to it, as that was the standard requirement in those feudal times. Subsequently Lucas rose to 'inspector' (curator) of the princely gallery in Vienna, but not even a signature or a portrait of him or his wife has been traced. Although it has been stated in writing, it is most unlikely that Lucas was later promoted to director of the Princely Gallery, since the position was then held by Gaetano Fanti,[14] previously court painter of Eugene Prince of Savoy, and after Gaetano's death in 1759 by his son Vincenzo (see Chapter 4). For certain, the job of an inspector of that gallery was not purely administrative, but involved artistic work on the interior decoration of the many properties of the Liechtenstein family, including maintenance. It seems safe to assume that Lucas knew the palaces of the family in Vienna, Feldsberg, and elsewhere, but nothing seems to be on record about his work.

We are somewhat better informed about Lucas's altar pieces, probably all in oil – painted for example in 1753 for the chapel in the castle in Eisenberg (Ruda), in 1755 for the church in Hilbethen (Hilvany), in 1756 for the church in Lundenburg,[15] and around 1757 in Hof[16] (Dvorce u Bruntálu), all situated on the estates of Joseph Wenzel. A hand-written inventory of the paintings then kept in the castle of Feldsberg entitled 'Lucas Bauerische Hoffmahlerey' (Court painting of Lucas Bauer) lists several of his works, among them for example 'no. 10 flower piece, no. 11 wild goose . . . no. 60 deer, shot at Lundenburg by his serene highness, no. 149 small flower pot'; of these nothing seems to have survived. It is likely that Lucas led the peripatetic life of a court painter, required of him by the variety of jobs at the properties of his employer, with Feldsberg as his permanent home. In our context three points are relevant. Although employed by Joseph Wenzel, Lucas Bauer did not belong to the inner circle of the princely household, but to its periphery – the upper stratum of the servants, ranging below people such as the chaplain, the chamberlain or the private secretary of the reigning prince. Among other duties Lucas was commissioned to document objects in a life-like way, much as a photographer would do today. It seems very likely that he knew the art works kept in the various residences of his employer, in Vienna, Feldsberg, and elsewhere. In the neatly stratified society of the late baroque period, Lucas Bauer was a simple employee who could be dismissed at any time by the reigning prince, his employer.

Maria Anna Dorothea did not remain an only child. At a time before any kind of birth control was available, six brothers were born in quick succession, almost certainly in Feldsberg. All were baptized in the sumptuous parish church of the small town – Antonius Lucas on 16 April 1753, Joannes Lucas on 27 June 1754, Joseph Antonius on 5 March 1756, Franciscus Andreas on 14 March 1758, Ferdinandus Lucas on 20 January 1760 and Carolus on 10 March 1762. The birth register does not list the domicile of the Bauer family, but local history records give Hauptplatz (main square) 7. In conversation with Henriette Charlotte Countess von Itzenplitz,[17] Franz mentioned many years later that he had been born in Feldsberg castle, which is a possibility, since his father may have had the privilege to live in rooms there. Among the godparents we find the names of employees of the Liechtenstein family. The church of the Assumption of the Virgin Mary, where the Bauer children were baptized, was at the time adorned with a huge painting of the Assumption by Peter Paul Rubens, now one of the more famous pictures of the Fürstliche Sammlungen in Vienna. Thus a work of art by a famous painter hung above the font that witnessed the baptism of three boys who were to continue on their father's line, with two of them destined for fame.

Infant mortality was high – Anton Lucas died on his seventh day and Carolus on his eighth – but these were the harsh realities of life at that time and were not atypical events. By contrast, the death of their father Lucas in Feldsberg on 28 June 1762,[18] aged fifty-four, just three months after the birth and death of his youngest son, must have come as a shock for the family, in particular for Therese. At that moment her surviving children were still very young – eleven, eight, six, four and two years old – so their future looked grim. Therese, who was probably illiterate because at that time no school is known to have existed in the small town, did not remarry, but concentrated on the education of her five children and was lucky to lose only one of them later in her life. Most importantly for our story, she engaged her children in copying the works of their father, thus sowing seed on an extremely fertile ground in the case of three of her sons. Years later Joseph was to follow his father as court painter in the service of the Liechtensteins, eventually becoming director of the Princely Gallery in Vienna, while the younger two became natural history illustrators of outstanding reputation. For good reason the Bauer brothers have therefore long been seen as an impressive example of the inheritance of remarkable abilities, just like in the family of composer Johann Sebastian Bach.[19]

One has to assume that the Bauer children grew up speaking German, and the few letters sent many decades later by Ferdinand to his brothers Franz and Joseph from his voyage with Flinders and Brown to Australia were written in this language, just like the early letters sent by Franz to Vienna. Considering the fact that Feldsberg lies on the fringe of the German-speaking area, it is almost certain that they heard some Czech; when Joseph died in 1831, his belongings included a Czech dictionary.[20] Since the nearby village of Oberthemenau (Charvátská Nová Ves) had a considerable Croatian population, the Bauer children may also have come into contact with that language. In addition, they must have got to know members of the substantial Jewish communities living in the nearby town of Nikolsburg and the villages of Lundenburg and Eisgrub.

Like everyone else in Feldsberg, the Bauer family was Catholic and probably pious. Maria Anna Dorothea joined the convent of the Sisters of St Elizabeth (Ordo Sorores Hospitalariae Sanctae Elisabethae or OSE) in Vienna, where she died aged eighteen. This may be interpreted as an act of piety, since this order is devoted to the care of the female sick and needy and focused, like the Brothers of Mercy, on running hospitals. Ferdinand's testament contains expressions of deep religious feelings and a letter describing the last days of Franz in Kew near London includes the following statement 'He [Franz] complains rather of his prolonged existence in such a miserable state, but said it was the Will of God – he had lived in His fear all his life, and His name be Blessed and praised' (see Chapter 16).

Franz kept very positive memories of his mother. In an anonymous obituary published in *The Athenaeum*, the clearly very well informed author (see Chapter 20), writes of Therese as an 'excellent mother', who had initiated Franz and his brothers in the 'ready use of the pencil'. This is corroborated by Franz, who recorded in the conversation with countess Itzenplitz mentioned in her letter cited above his only toys, 'old brushes and alpine plants'. An obituary for Ferdinand (see Chapter 20), which integrated information passed on to its author by Franz,[21] is even more detailed: 'in his earliest youth . . . [he] copied plants and birds from the designs of his late parent', the superlative being most notable. It is easy to imagine what lasting effect the mother's stimulus to copy the drawings and paintings prepared by their recently deceased father aroused in the boys. What kind of works were these? Fortunately the inventory mentioned above gives us some clues, but more relevant is the small art collection in the possession of Therese. When she died in Feldsberg of a stroke on 19 October 1790, aged sixty, her belongings were auctioned – among them were no less than 53 pictures ranging from flower pieces, landscapes to portraits, comprising even two fruit pieces and 'a bird piece, unframed'. This seems to have been the type of material copied and studied by the Bauer brothers in their early youth. In addition, it is likely that they had seen the brilliant over-doors and still lifes in Feldsberg castle, which may well be called an exquisite art gallery. An early source reports, 'An innate talent for painting, apparent in all four brothers [Johann, Joseph, Franz, Ferdinand] from a very early age and in a splendid way, led them all in the same direction, and in the case of Ferdinand and Franz developed into genius'. It is noteworthy that both younger sons devoted their whole life to the same kind of plant and animal subjects that their father had prepared for Joseph Wenzel. Only Johann is known not to have followed the artistic path.

The five children must have grown up in a rural world shaped by the alternation of the seasons – in particular the varying works in the fields, meadows and the extensive vineyards surrounding the small town, with the vintage being the peak season of the year, exactly as it is today in Feldsberg. At an early date they must have become familiar with the logging in the forests, the sowing of wheat, and fishing dry of the carp ponds. They will have watched the changing of the princely guards wearing uniforms in the blue-red colours of the family, identical to those of the Principality of Liechtenstein's flag today. They will have attended religious manifestations like the procession on Corpus Christi Day and the training of the horses, which formed part of the traditional representation of the Liechtenstein family in Feldsberg until 1793. The Bauer children must have also come in contact with the realities of the feudal system, which meant on the one hand an elevated and privileged status for the princely family, including their courtiers. On the other hand there was socage (unpaid labour) for the common people, wherever and whenever this was felt necessary by their superiors – hard work in the forests, in the fields, in the vineyards.

The members of the reigning family, including Aloys Joseph, the heir presumptive, and his younger brothers Johann Joseph and Philipp,[22] of which the Bauer children may have heard, led a totally different life: there was the arrival in late spring in Feldsberg; the leisure time spent in their exquisite stately homes and luxurious gardens during summer; the hunting season, when the inner circle of the high aristocracy from the Hereditary Lands, including the Kingdom of Hungary, spent several weeks in and around Feldsberg as guests of Joseph Wenzel and later of Franz Joseph I; and the return to their city residence in Vienna, a lifestyle that the Liechtensteins shared with the Dietrichsteins. Both families were regularly accompanied by members of the respective princely courts, including footmen, gentleman's valets, lady's maids, and the princes' tutor. From 1773 to 1783/1784 this well-paid position was held in the case of the Liechtensteins by Angelo Soliman, a multilingual African born in what is now Nigeria, sold as a slave in Messina, who later became a respected figure at the imperial court in Vienna.[23] The Bauer children may have seen him in Feldsberg, but they hardly entered into personal contact. Though similar in several respects, the Liechtenstein and Dietrichstein families differed in one aspect: the former was to employ the elder two Bauer boys for life, while the latter engaged Franz for six years only.

More relevant in our context, the five children must have become familiar with the extremely diverse plant and animal life of the region, ranging from the wetlands along the Thaya to the calcareous cliffs of the Pollau Mountains (Pavlovske vrchý) near Nikolsburg, with their dry meadows. At the same time they must have realized the complete dependence of man on the injurious effect of the weather and of animal and plant diseases with their wide-reaching consequences. Decades later Franz was to depict the germination, growth, flowering and ripening of wheat and to document diseases of agricultural crops affected by parasitic fungi, nematodes and insects. Maybe the Bauer children were also permitted to visit the gardens and conservatories of the Liechtensteins – among them the orangery in Eisgrub, with a length of one hundred and eighty metres (then one of the largest on the continent), containing a considerable quantity of ornamental plants, specifically mentioned by the eyewitness of 1782.

We know nothing about the schooling of the Bauer children, and no records of a school in Feldsberg could be traced in any of the pertinent archives. An early report reads 'Ferdinand Bauer . . . and his three brothers . . . educated themselves'; this may have been the reason for their lifelong reserve towards expressing themselves in writing, resulting in the paucity of letters (and then with phonetic spelling) and the lack of any kind of diaries. At the same time, this is the very cause of why the personalities of all three brothers remain elusive and why their lives have never been studied in a way similar to those of their commissioners.

NORBERT BOCCIUS, OH ARRIVES

In May 1763, when the half-orphaned Bauer children were twelve, nine, seven, five and three years old, a change took place in the convent of the Brothers of Mercy in Feldsberg. A young surgeon joined the fraternity, Johann Adam Boccius, a native of Temesvár (Timişoara in Romania) and son of Johann Adam Georg Boccius (Boccio, Boccy, Poccius, Poccy), Head Physician of the Banat region and Military Physician at Orsowa (Orşova in Romania) on the River Danube.[24] After the early death of his father the young Johann Adam had lived in Vienna and attended the Jesuit gymnasium in Znaim before moving to Prague. There he had taken the vows on 27 July 1749 and become a member of the Brothers of Mercy (Ordo Sancti Ioannis de Deo or Ordo Hospitalorum, or OH) adopting the name Norbert.[25] For the rest of his life he was to wear the black habit of the fraternity with scapular and hood, held together by a thin leather belt. And Norbert was to obey the rules of the order and to attend to the sick and needy in the hospital forming part of the respective convent.

Very little is known about his subsequent medical training, but it is recorded that Norbert studied the works of the famous Herman Boerhaave, professor at Leiden University, and of his equally famous pupil Gerard van Swieten. In 1745 the latter had been appointed by Maria Theresa, then pregnant for the seventh time, as her Protomedicus, or Principal Head Physician, overseeing all the physicians-in-ordinary and pharmacists at the imperial court in Vienna. At the same time Swieten had also been made prefect of the Imperial Library under very favourable conditions and was soon to become the great reformer of the faculty of medicine at Vienna University. Swieten, a courtier in a key position and at the same time a highly influential personal advisor of Maria Theresa, who in 1758 created him Freiherr (Baron), acted as Norbert's medical examiner in Vienna, but we do not know the date of the exam. In the archival documents kept in the Konvent der Barmherzigen Brüder in Vienna, Boccius is listed as examined surgeon as well as 'D. tor Medic.' (Doctor of medicine). However, this title is not substantiated by the record conserved in the archives of Vienna University, and he may have actually received his degree from Prague University.

After six years as Prior of the convent in Görz, Boccius was transferred to the mother convent of the province in 1763. This was the year when a disastrous fire had destroyed no less than a third of the houses in the small town of Feldsberg. Aged thirty-two, he started to teach human anatomy and natural history in the nursing school and soon dominated education in Feldsberg more or less directly and indirectly all through the second part of the 18th century. Norbert has been described as a very pious brother, regularly praying the rosary, but at the same time an indefatigable man prepared to help the sick at any time of the day or night and a remarkably successful surgeon. Among his patients were the poor of the Feldsberg region, mainly peasants and artisans, but also Joseph Wenzel, who died in 1772 in Vienna, Franz Joseph I and other members of the princely family, probably including the latter's elder, sickly son Aloys Joseph (also Alois or Lois), who succeeded his father in 1781 aged twenty-one; later Boccius was to receive an annuity from Franz Joseph I for his medical services.

It was almost unavoidable that Boccius rose in the hierarchy of his order: in 1766, three years after his arrival in Feldsberg, he was elected the 29th Prior of the convent. After three years in this position and three more years in another function, Norbert was elected 31st Prior in 1772, which a second time meant the responsibility for about forty hospital beds, the pharmacy, the medicinal gardens and all the economic affairs of the convent. Subsequently he was several times confirmed by election in this position.

More importantly, in 1784 Boccius was elected Father Provincial of the province 'Zum Hl. Karl Borromaeus'. The most northern convent of Norbert's province was located in Kukus (Kuks), the most eastern in Lviv, now Ukraine, the most western in Görz, and the most southern in Trieste. As head of a considerable number of hospitals and pharmacies, including medicinal gardens scattered all over Central Europe, for thirteen years (until 1797) he was an important figure in the medical establishment of the period. Six years before the end of his time in office no less than 369 Brothers of Mercy had been counted in the province.

Yet, in our context other aspects of Boccius' life are more relevant. Even before his arrival in Feldsberg he had acquired considerable botanical knowledge and started to collect plants. He had them pressed, dried and glued into three massive large-sized volumes inscribed 'Herbarium vivum', which are kept today in the library of the Konvent Milosrdných bratří in Brno. The flat specimens, which had lost their three-dimensional structure, were fixed with curved paper strips reminiscent of scrolls in baroque engravings on paper and had the scientific name of the plant, determined by Boccius, indicated.

Preparing such herbaria had an extremely long tradition among physicians and pharmacists, going back to the Renaissance period. Their relevance for the pharmacist is aptly described by Jacquin, the second professor of chemistry and botany at Vienna University and director of the botanical garden in that city (see Chapter 3). Jacquin noted in 1792 'The physician prescribes the plant, the pharmacist gives it, [but] the latter usually buys it from the herb-dealer and rhizotomist . . . The pharmacist will not transfer his rights to the herb-dealer, but will wish to know the plants on his own comprehension in order to be able to judge with certainty the authenticity of the plants brought to him by the herb-dealer, or favourable circumstances permitting he [the pharmacist] will grow them on his own [in his garden] or go and collect them'.[26] Though devoid of collecting localities, Boccius's herbarium acted as a permanent record of the plants he had encountered before 1766, first in Görz and later in Feldsberg. It was a snapshot of the natural vegetation enriched by several not yet fully established crop plants, such as sunflower, *Helianthus annuus*, pineapple, *Ananas comosus*, castor-oil plant, *Ricinus communis*, and ornamentals including African marigold, *Tagetes erecta*, and Cupid's dart, *Catananche caerulea*. Even a few pressed animals, such as butterflies and dragonflies, are included in the three tomes.

However, in several cases the plants are not herbarium specimens in the modern sense, often having been manipulated in certain ways. Thus Boccius removed pressed and dried leaves and glued these back in different positions on the stem, and shortened internodes and petioles. Most noteworthy is the treatment of the sunflower: the very large flower head was found unsuitable for inclusion in the book-herbarium and not only were the stem leaves removed, their petioles shortened and attached to an unsuitably thin stem, but the flower head was replaced by a brown, round cardboard disc with the radial florets and the apical parts of the involucral bracts stuck around the margin. Clearly Boccius wished to give a convincing representation of this species on a single page of his book-herbarium, which may indicate that he intended to use this 'specimen' in his botanical teaching in Feldsberg. In a similar way he constructed a 'specimen' of pineapple by slicing and drying pieces of the infructescence and gluing them together with the apical rosette leaves on a page of his 'Herbarium vivum'.

The portrait of Boccius kept in the Konvent der Barmherzigen Brüder in Vienna, a canvas in oil, offers further evidence for his botanical interests: he is shown holding a book labelled on the spine 'Genera plantar[um] Lin[n]aei', one of the more important publications of the Swedish naturalist Linnaeus, with a flowering specimen of a *Cyclamen* painted above his hand. There are indications that Boccius was also able to record plants as botanical illustrations. His portrait in the Fürstliche Sammlungen has in its lower part an agglomeration of different objects probably intended to illustrate his personality and activities, among them a painter's palette with a number of brushes sticking out of the hole, and a large piece of paper. On the same line is a fern illustration, admittedly of rather inferior quality, forming part of the *Codex Liechtenstein* (see below) and signed 'APN', almost certainly standing for Admodus Poccius Norbert (the honourable Norbert Boccius).

In addition to his commitment to the convent's pharmacy and medicinal garden, about which we know almost nothing apart from its substantial extension, Boccius was also a field botanist, who managed to find the tartar bread plant, a rare crucifer then new to science, in the vineyards near the modern village of Hustan (Kurdějov) near Auspitz (Hustopeče) in Moravia. He sent specimens to Vienna, where the plant got the name *Crambe tataria*, published in a thesis in 1779 submitted to Vienna University by Alexander Sebeók de Szent-Miklós, which resulted in Boccius being called much later an 'indefatigable discoverer of Moravian plants'. The text must have passed through the hands of Jacquin, who had obtained his combined chair at Vienna University thanks to Swieten, whose son had been his school mate at the gymnasium in Antwerp (see Chapter 3). There is also evidence for Norbert's interests in horticulture (for example a bill for '11 workers to dig the big garden' or the statement that he 'grew medicinal herbs for use in the hospital and for teaching') and in botany (for example a bill for binding the works of Linnaeus). In addition Boccius is reported to have grown the 'noblest exotic fruit' in his orchard in Feldsberg, and to have dug the ground 'in the company of orphaned children'.

In the small world of this tiny town the half-orphaned Bauer children, maybe encouraged by their mother, came into contact with Boccius, but we have no knowledge of how this happened. It seems to have been at an early moment, as a chronicle of the Brothers of Mercy states that Boccius educated the three brothers when they were still children. In any case the meeting was decisive for Joseph, Franz and Ferdinand, since Boccius was able to focus their early inclination to drawing on the documentation of plant and animal life and may have taught them also the art of writing. Clearly he became their tutor and mentor, the first in the line of several more medical men who were later to collaborate with Ferdinand and Franz – although only one of them, Sir Everard Home, a Vice President of the Royal Society, actually practised (see Chapter 15). We may hypothesize that the Bauer boys were influenced by the young surgeon also in other respects: at a very early date they came into contact with the realities of suffering, pain and death, easily experienced in every hospital, and they seem to have remained pious (see above). Many years later Franz would be asked by Home to record in very precise drawings pus, seminal fluid, blood coagulation and a bone tumour of the skull (see Chapter 15). In any case the Bauer boys will have experienced early what it means to work round the clock, in an atmosphere of Christian belief, humility and in caring for the sick. Later Ferdinand was to be described as indefatigable by Banks and Flinders, a character trait also observable in the long-serving prior of the convent in Feldsberg.

Boccius seems to have been a knowledgeable, well-read person, although we neither possess his reading list nor the catalogue of the library of his convent in Feldsberg. Judging from the manuscript catalogue of the Konvent der Barmherzigen Brüder in Vienna for that time, which lists no less than 700 medical books – among them works by Boerhaave, Anton de Haen, Marcello Malpighi and Swieten – Boccius must have had access to the standard medical and pharmaceutical literature of the time. The anonymous eulogy delivered in 1799 states that Norbert knew Jacquin (see Chapter 3), as well as Joseph Gottfried Mikan, the first director of the botanic garden and first professor of chemistry and botany at Prague University (see Chapter 7).

Obviously Boccius intended to create a painted complement to his herbarium – a florilegium consisting of plant illustrations very true to nature and painted in watercolour, similar to two illuminated manuscripts he may have seen in Vienna. The first of these was the *Receuil des plantes* (*Selection of plants*), formerly in the library of Eugene Prince of Savoy, and therefore also called the Florilegium of Prince Eugene of Savoy, who had been a friend of Joseph Wenzel's. Later this collection, bound in ten volumes, was acquired by Emperor Charles VI for the Imperial Library, under the care of Swieten since 1745. Through this chain of contacts Boccius may have seen this spectacular collection of most exquisite plant illustrations prepared by the inner circle of botanical painters at the Jardin du Roi in Paris, documenting the plants then cultivated in this famous botanical garden. In a sense they formed a shoot of this most famous florilegium ever prepared – in Vienna. These illustrations were painted in gouache on a particular fine parchment, the individual plant illustrations framed in gold and annotated in gold and red lead, and the volumes provided with indexes. The second florilegium that Boccius probably knew was the *Phytanthologia eikonike* (*Illustrated anthology*), commissioned by Johann Jakob Well, the owner of a pharmacy in the centre of Vienna and professor of natural history at Vienna University. Painted in watercolour on paper, the individual plant illustrations unframed and annotated in black ink, and the volumes provided with indexes, this florilegium was started in 1768, whereas the luxurious florilegium of Prince Eugene had been prepared in the late-seventeenth century. Judging from the information on the painted title pages, the production of the *Phytanthologia eikonike* continued until 1780 and comprised a total of eight volumes, probably kept in Well's pharmacy. They first passed to the monastery of the Order of the Visitation of Holy Mary (Ordo Visitatio Mariae, or OVM) in Vienna and in 1938 to the Nationalbibliothek (now Österreichische Nationalbibliothek), also in that city.

Several aspects of the *Phytanthologia eikonike* are of special interest in our context since they were emulated in the florilegium prepared by Boccius: the plant illustrations are in part based on specimens, in part on printed illustrations, in the latter case often simplified and taken from the works of Well's colleague Jacquin. The eight title pages combine text and image, often a view or a landscape. As will become evident later, Boccius largely followed the earlier example set by the *Phytantologia eikonike*, but he also took up one element from the *Receuil des plantes* – the frame surrounding each and every one of the plant illustrations of his florilegium. As a member of a mendicant order with a black habit he probably preferred a double frame, both frames in black ink, over the double frame, with one frame covered in gold leaf, appropriate to the kings of France. Another, more important, difference between the two works is that the number of copies of printed plant illustrations is significantly higher in the *Phytanthologia eikonike*, which makes the florilegium commissioned by Boccius so much more relevant to science. In addition, the quality of illustrations in Boccius' florilegium is distinctly superior and approaches that of the *Receuil des plantes*.

It seems helpful to pause for a moment to reflect about the reasons for Well and Boccius to have a florilegium prepared and to consider the advantages and disadvantages of a florilegium compared with those of a herbarium. A florilegium consists of plant images (that is, artefacts very true to nature), prepared by a botanical illustrator instructed by a botanist, whereas a herbarium consists of specimens (that is, plants or parts of plants, collected solely by the botanist). Specimens are always part of nature and are consistently and persistently preferred by scientists. However, specimens – in particular dried botanical ones – are much more vulnerable to the attacks of insects than illustrations on paper, and hardly any protective measures against herbarium beetles were available at the time. This is best evidenced by the deplorable state of preservation of Boccius' Herbarium vivum, partly destroyed by insects, in Brno. Botanical specimens are easily ruined, and being brittle they need much more care when it comes to using them in demonstrations, for example when giving botanical lessons, which is exactly what Well and Boccius did. Furthermore, herbarium specimens quickly lose their colours and become totally unattractive for the non-specialist, whereas botanical illustrations keep their colours, at least as long as they are kept dry. Most important, however, is another factor: often a herbarium specimen can be only a tiny fragment of a plant – for example a leaf, a single flower of an inflorescence or a piece of a fruit – whereas the botanical artist is always able to prepare an image of the complete plant, as well as any part of it if necessary. However, the preparation of a herbarium specimen is much cheaper, easier and quicker than the preparation of a plant illustration true to nature prepared by a botanical illustrator.

Such florilegia are always the result of a close collaboration between a botanist and a botanical illustrator, with the botanist choosing the specimen to be recorded and the illustrator documenting it in such a way that it is very true to nature. It is up to the botanist to direct the work, to give instructions on which characters to focus upon, to control the end product and have it corrected. In short, the botanical artist is totally subservient to the authority of the naturalist[27] and enjoys zero freedom in his work. The botanical draughtsman acts, so to speak, as the botanist's hand, hired to produce the images that the botanist desires. In a sense the naturalist serves as the draughtsman's eye, selects the object to be depicted, and takes full responsibility for the latter's productions, before they pass into his possession. For the botanist the desired images are almost scientific objects, not artistic creations, and in his view the botanical illustrator must strictly avoid any painterly style, which is the botanist's constant concern. As a rule naturalists even considered themselves to be the true authors of these images, with the illustrators as necessary, but subordinate, amanuenses. Consequently the name of the botanical illustrator is largely irrelevant for the botanist and sometimes ignored or suppressed, as this was done by Jacquin (see Chapter 3). In agreement with this general line the botanical illustrator almost never added the scientific name of the plant portrayed, which belonged to the realm of the botanist. He also rarely noted the date, which makes the establishment of precise chronology in botanical illustration so very difficult: indeed, out of 2748 sheets of the florilegium prepared for Boccius only a single one is dated.

For every botanical illustrator utmost precision combined with acute capacities in observing and comparing as well as concentration are key qualifications (see Chapter 1), in particular when it comes to recording easily confusable medicinal plants. In addition, a docile attitude towards the commissioning and controlling botanist, not particularly widespread among artists, is useful.

For a single image the production process took on average several days and involved multiple expert observations and decisions, sometimes including discussions and clarifications with the botanist who had commissioned the work. As a consequence, the painted record of a single specimen, let alone of all the plants in a garden or conservatory, was exceedingly expensive and could be considered only by the fortunate few.

Boccius may have wished to possess a collection of botanical illustrations similar to those in the hands of Swieten and Well, prepared with the help of the Bauer brothers, whom he was to employ, train and supervise. Nothing is on record about his motivation: Was his florilegium to be used in his pharmaceutical teaching at the school for male nurses? Was it intended as a substitute for his herbarium, to which nothing seems to have been added after 1766? Was it meant as a standard of reference for the incoming herbs acquired by the convent's pharmacy? Or was it simply a labour of love for Boccius? We simply do not know. Maybe this project of long duration served all these ends.

It has been rightly said that the botanist's 'way of seeing involved a constant triangulation among image, text and specimen, using books to interpret what they saw in the field' and that they 'used images as working tools, to record or work out what exactly they were recording'. Indeed this may also have applied to Boccius, since botany in its traditional sense 'was fundamentally a visual discipline, based on the observation and representation of specimens that sometimes were "out there" in the field, other times "in here" in collections [in a herbarium or a botanical garden], and yet other times in the hybrid domesticated space of illustrations, in which . . . the field and the collection were collapsed into a single paper nature that was always and perfectly available'.

At any rate, Boccius's Herbarium vivum got a painted successor, entitled by him *Liber regni vegetabilis retinens plantas ad vivum pictas* (*Book of the plant kingdom containing plants painted from life*), effectively a collection of painted plant illustrations bound in fourteen large volumes. It formed at the same time a visual archive and a painted encyclopaedia, with a few landscapes, views and painted title pages as well as indexes attached. Since the first twelve volumes were given in 1799 by Boccius to Aloys I Joseph, on the latter's explicit wish in exchange for a donation to the convent, the thirteenth as a gift to the same recipient and the fourteenth as a gift to his successor Johann I Joseph, this work has become better known as the *Codex Liechtenstein*. This donation – 'out of Our love for the sick and poor' as worded in the foundation charter – consisted of five percent interest per annum on a sum of four thousand florins, which remained with the Liechtenstein family in order to finance two beds in the hospital of the convent in Feldsberg permanently. First kept in the famous library built for Aloys I Joseph in his city residence in Vienna's elegant Herrengasse, a spectacular neoclassical hall fifty-six metres long divided into three aisles by two rows of Ionic columns, it was moved a century later to the summer palace of the family in the Rossau quarter of Vienna (see Chapter 3). All fourteen volumes are now permanently stored in the Fürstliche Sammlungen in Vaduz (see Chapter 19).

The close relationship between the *Phythanthologia eikonike* and the *Codex Liechtenstein* may best be exemplified by the illustration of the tartar bread plant, which is included in both works: in the first an illustration based on a copper engraving published in Jacquin's *Icones plantarum rariorum* (*Illustrations of rare plants*; see Chapter 3); in the second an illustration based on a totally different living specimen. This indicates one of the ties between Well, Boccius and Jacquin, but there are also others: Well, Jacquin and Stoerck (see below) were co-authors of the *Pharmacopoea Austriaca* (*Austrian pharmacopoiea*). At the time this was the reference work for pharmaceutical drug specifications prepared by the authority of the government in Vienna, required reading for every pharmacist in the Hereditary Lands, a book which Boccius must have seen and consulted in the library of the Feldsberg convent.

Clearly Boccius was the initiator and mastermind of the *Liber regni vegetabilis* project, which – judging from the information on the painted title pages – was started in 1776 at the latest, but probably earlier, sometime after 1770. It ended in 1805, a year before the death of its founder. In a sense it was a mammoth undertaking, resulting in 2748 sheets of plant illustrations in a format of 35 × 30 cm, with a considerable proportion showing more than one plant species per sheet. Boccius, the botanist, must have planned the complete production: he would have been the one to select the plants to be illustrated, determine the identity of the plants and lay down their names, decide the sequence of the watercolours painted on single-folded sheets when it came to binding the volumes, and also make the necessary arrangements for the indexes. Right from the beginning this was an open-ended project aiming at a pictorial documentation of plant diversity, or to be more precise, of what was known of plant diversity in the late-eighteenth century in Central Europe. We know nothing about the finances of the great project; not a single receipt seems to have survived, from the Bauer brothers nor their successors; similarly, no bills exist from the scribers who were to produce the registers, nor from any book binders. Boccius may have used his annuity from Franz Joseph I to finance this great work, but he also may have used funds from the convent for his project. The half-orphaned Bauer brothers may have been paid by Boccius or may just have received a daily warm meal from the convent's kitchen.

Unfortunately, we do not possess detailed information on the precise share of the Bauer brothers in the production of the *Codex Liechtenstein* either; on stylistic grounds the first six volumes can be ascribed to them in their integrity, just like the bulk of the plant illustrations in volumes seven to nine; by contrast the volumes ten to thirteen comprise very few watercolours by their hand. Taking into consideration their very early training to become botanical artists (see above), they may have started working for Boccius at the age of about fourteen: Joseph may have begun working for the florilegium in 1771 and must have finished in early 1781; Franz may have joined his brother in 1773 and continued until early in 1788; whereas Ferdinand may have started to illustrate plants for Boccius in 1775 and broken off early in 1786. Of course all this is highly conjectural. In particular, the role of the eldest brother (see below) is unclear, and so is the extent of the later commissions from Jacquin (see Chapter 3) and those from Marie Christine Princess Dietrichstein-Proskau (see below). However, this rough estimate demonstrates the enormous time budget, effectively about thirty-five years of work, which went into the project.

On the basis of this hypothetical calculation we may presume that each brother had to produce about fifty bodycolours per year, a heavy burden of work requiring concentration, reliability and continuity. In addition, all the work described in the following paragraphs seems to have been achieved without any assistants.

AN EARLY MASTERPIECE

Nothing is known about the sequence of events in Feldsberg leading to the florilegium, but in an anonymous obituary of Franz (see Chapter 20) the clearly well-informed author reports that an early botanical illustration by Franz was published in 1771 as

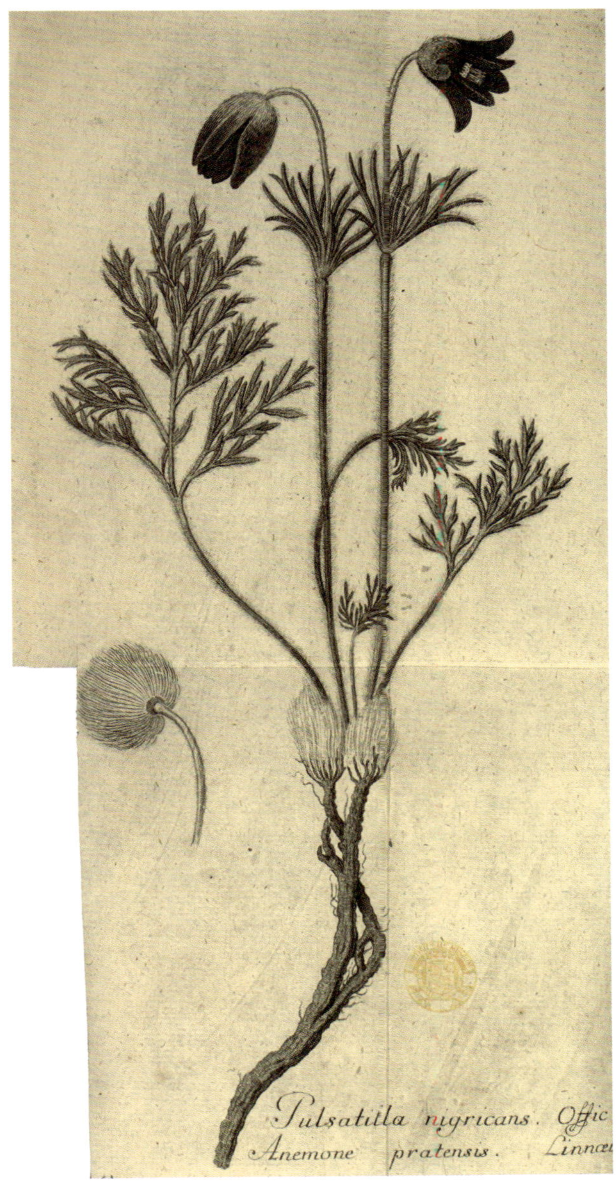

FIG. 2.2. ***Pulsatilla pratensis*** **subsp.** ***nigricans*****. Copper engraving based on Franz Bauer. – A. Stoerck, Libellus de uso medico Pulsatillae nigricantis, Vindobonae, 1771. – London, British Library.**

a copper engraving (Fig. 2.2.) in a pharmaceutical text. Its author was Anton Freiherr von Stoerck, professor at the medical faculty of Vienna University, one of the physicians-in-ordinary at the court in Vienna, later Swieten's successor as Protomedicus of the imperial family and third co-author of the *Pharmacopoea Austriaca*. Stoerck, today regarded as one of the fathers of experimental pharmacology, was well connected: he had accompanied Archduke Joseph to the election and coronation in Frankfurt staged by Joseph Wenzel and had been elected Rektor (Vice Chancellor) of Vienna University. The engraving showed a flowering specimen of a plant Franz must have known well: *Pulsatilla pratensis* subsp. *nigricans*, a relative of the pasque flower, common in early spring on the calcareous slopes of the Pollau Mountains near Nikolsburg. The specimen is shown with its characteristic flower nodding already at anthesis and the deep-reaching rootstock, which means that it must have been dug up and put into water so that the fine roots would become visible.

This is a masterpiece of critical observation, representing a plant true to nature: complete, matter-of-fact, without any embellishment or 'improvement' by the hand of the botanical illustrator, stretched out like a specimen before being pressed between sheets of paper to become a herbarium specimen. At the same time this representation of a decontextualized and isolated specimen placed on the white background of the page is typical for botanical illustration, devoid of any explanatory information on its provenance and time of collection. Unfortunately, the drawing prepared by Franz that formed the basis of this copper engraving could not be traced and seems to have been lost, which is often the case when a botanical illustration is passed to a graphical artist.

Having a production published at the age of 13 is remarkable, but not a record. In early 1764 Mozart, being eight years old, had his *Sonates pour clavencin* (K. 6, 7) printed in Paris. However, it was clearly the ambitious father who had made the necessary arrangements for his son, whereas Franz had already lost his father nine years before publication. We may hypothesize that Franz had accompanied Boccius on his botanical collecting tours in the surroundings of Feldsberg, and that Boccius later passed the drawing prepared by his protégé directly or indirectly on to Stoerck.

The second proof of work done by the Bauer brothers for Boccius has nothing to do with botany. It is a large-format drawing showing the rehabilitation house of St Theresia of the Brothers of Mercy in Vienna, annotated 'Ioseph Bauer fecit 1773': a pen-and-ink drawing on card with monochrome wash, with a graphite pencil outline visible in many places. It was found by the present author in 1993 in the deserted building of the convent in Feldsberg in a store room among broken glass and frames. Prepared in bird's-eye view when Joseph was seventeen years old, it is likely that this drawing is a copy of an older image. Among the Brothers of Mercy, as in other spiritual orders, there is a tradition of having such views of the brother convents or monasteries exhibited in their corridors. Two similar bird's-eye views, both then conserved in the convent in Feldsberg and showing (1) the convent of the Brothers of Mercy in Letowitz (Letovice) in Moravia and (2) the Imperial Royal Military House for the Disabled in Pest (Budapest) in Hungary, are signed 'Franc Bauer 1775' and 'Franc Bauer fecit A. 1775', respectively, the latter being of particular interest. The border of the inscription is crowned by a double-headed eagle, the old imperial symbol, and military motifs, among them cannons, drums, flags and bayonets. The latter are almost certainly taken in great detail from an anonymous copper engraving,[28] a copy of which is kept in the Fürstliche Sammlungen. It documents the monument erected in 1758 on orders from Maria Theresa in recognition of the outstanding services of Joseph Wenzel for the imperial armed forces. Clearly Franz (then aged seventeen), Boccius, or both, were well informed and must have had access to the copper engraving, the monument itself or the medal produced on the occasion of Joseph Wenzel's death showing this very scene. Boccius, Franz, or both of them may well have been to Pest in order to visit this military hospital, or alternatively Franz may have copied the view from a pre-existing drawing.

Twenty more large-format bird's-eye views of convents and rehabilitation houses, without a single exception all from the order's provinces 'Zum Hl. Karl Borromaeus' and 'Zum Hl. Erzengel Michael', all undated and unsigned, and found in the Feldsberg convent by the present author in 1993, are so very similar in style that they have been attributed to the Bauer brothers. Typical for them, there is outstanding attention to detail – the buildings with their towers and clocks as well as their gardens with beds, hedges, fences and fountains are depicted, as are the associated convent cemeteries. The Bauer brothers may well have seen some of these convents with their own eyes, maybe when accompanying their mentor on his inspecting tours within the order's province, though hardly all of them – this would have meant extensive travels within the Hereditary Lands and beyond. It seems plausible to assume that all these views, very recently transferred to the Konvent milosrných bratří in Brno, were prepared in the convent in Feldsberg. This is also the most likely place for the numerous plant illustrations that were produced by the Bauer brothers over the following years, ending in April 1788 at the latest. In the library of the convent they were near to Boccius, the medicinal garden and the botanical books, and probably found sufficient space as well as ideal peace and quiet for the concentrated work that was about to begin.

THE *CODEX LIECHTENSTEIN* TAKES SHAPE

In agreement with the standard practice of the time[29] it would have been the logical next step for Boccius to induce the Bauer brothers to copy botanical illustrations, and there is ample proof that they did so. The *Codex Liechtenstein* contains a certain number of very precise copies based on several sources, among them the lavishly illustrated botanical works of Jacquin fresh from the press.

These copies are identical in size to the originals and so very accurate that straightforward copying by sight would have been impossible. Rather, a piece of frosted glass was used with a light source behind. By placing a copper engraving on the glass, then a sheet of paper on top of the engraving, it is easy to take a tracing with the help of a pencil. In a second step, paint (usually gouache) was applied to the copy. Clearly this copying process in two steps was a standard technique among graphical artists, and was extremely practical and efficient. When studying such copies in the *Codex Liechtenstein*, the graphite lines incompletely covered by paint are occasionally visible.

When copying the coloured copper engravings included in the works of Jacquin, such as his *Hortus Botanicus Vindobonensis* (*Vienna Botanic Garden*) or his *Florae Austriacae icones* (*Illustrations of the Austrian flora*), sometimes even the volume and plate number were noted by the Bauer brothers in pencil on the margin of the sheet and later only incompletely erased. One of these very precise copies shows nettle-leaved speedwell, *Veronica urticifolia* (Fig. 2.3.), a species first described by Jacquin in his *Hortus Botanicus Vindobonensis*. We do not know how Boccius got access to these printed works, but we know for certain that the *Florae Austriacae icones* were kept in the library of the Liechtenstein family in Vienna, although as a consequence of the Second World War these very volumes are now in the Library of the Botanical Institute of the Russian Academy of Sciences in St Petersburg. Copies of two copper engravings showing scarlet Turk's cap lily, *Lilium chalcedonicum*, and the American *Lilium superbum* (Fig. 2.4.) have also been included in the *Codex Liechtenstein*. The first is based on a plate in Johann Wilhelm Weinmann's

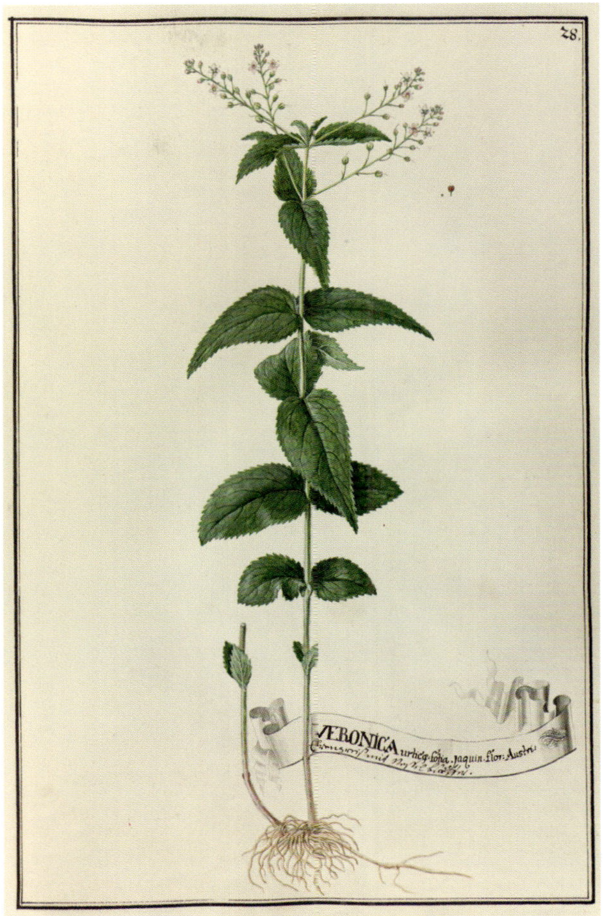

FIG. 2.3. Nettle-leaved speedwell, *Veronica urticifolia*. Watercolour by the Bauer brothers after Franz Scheidl, c.1775. Watercolour, ink. Codex Liechtenstein volume 1: f. 28. – Vienna, Fürstliche Sammlungen.

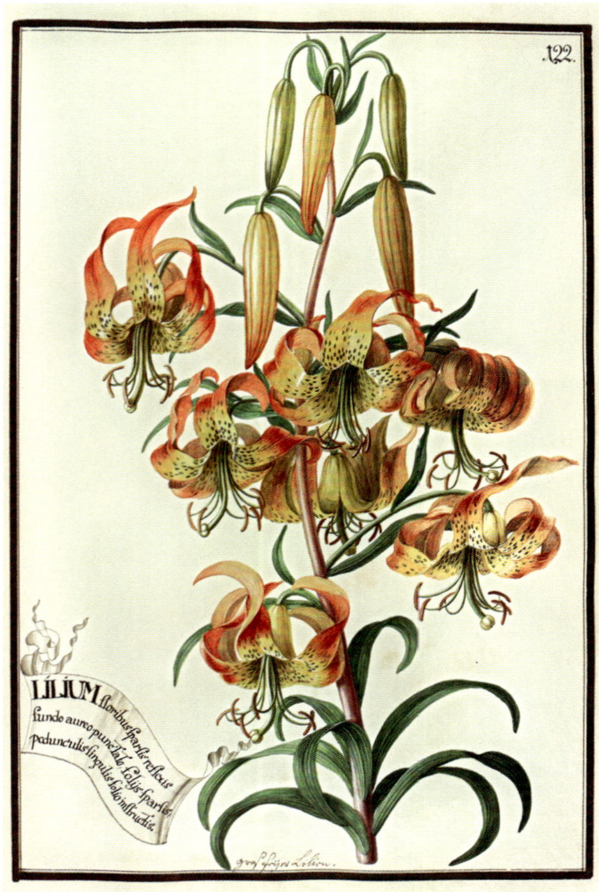

FIG. 2.4. *Lilium superbum.* **Watercolour by the Bauer brothers after Georg Dionysius Ehret, c.1778. Watercolour, ink.** ***Codex Liechtenstein*** **volume 4: f.122. – Vienna, Fürstliche Sammlungen.**

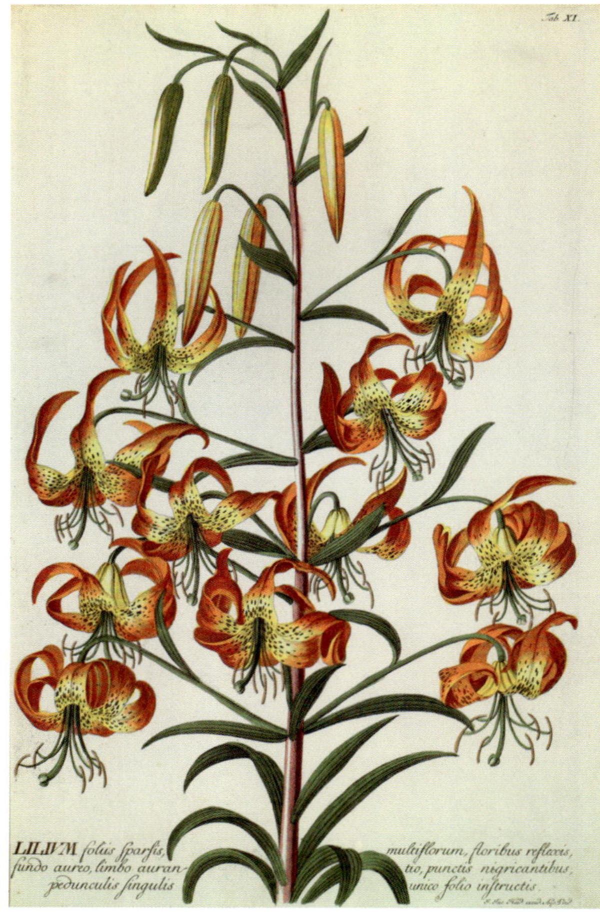

FIG. 2.5. *Lilium superbum.* **Coloured copper engraving based on Georg Dionysius Ehret. – C. J. Trew, Plantae selectae, f. 11, Nuremberg, 1751. – London, Natural History Museum, Library and Archives.**

Phytanthoza iconographia (*Iconography of Flowering Plants*), published in Ratisbon from 1737 to 1745, the second on a plate in Christoph Jakob Trew's *Plantae selectae* (*Selected plants*) (Fig. 2.5.), published in Nuremberg in 1751. The latter plate was based on a watercolour by the famous plant illustrator Georg Dionysius Ehret. When the copper engraving showing *Lilium superbum* was copied, the Bauer brothers could not know that Ferdinand, like Ehret, would later work for several years in Oxford – under the patronage of John Sibthorp, the third Sherardian Professor of Botany, a son of Ehret's patron Humphrey Sibthorp, the second Sherardian Professor (see Chapter 6).

In contrast to copying a pre-existing image, the preparation of an illustration from nature, more precisely from a living plant, was a much more complex process, and more than two-thirds of the first nine volumes of the *Codex Liechtenstein* are based on these. We have no reports from eyewitnesses or contemporaries on how work proceeded, apart from general statements such as, that the brothers 'chose nature as a model which they strove to imitate' or Ferdinand 'took to painting from nature, and followed her as his chief guide throughout life'. Fortunately it is possible to reconstruct the sequence of events, in particular since a few preliminary pencil illustrations have survived, all kept in the Naturhistorisches Museum. They refer to several illustrations in the *Codex Liechtenstein*, and we may hypothesize that a similar procedure was regularly followed. We cannot tell what steps in the production process were the special responsibility of Joseph, Franz and Ferdinand. Probably they worked as a team, since their names are indicated on the title page of the first volume of the *Codex Liechtenstein* (see below), and we may assume that they were familiar with the various steps involved. However, it has not been possible to distinguish the three hands in the first nine volumes of the *Codex Liechtenstein*.

As the first step the plant specimen to be depicted had to be chosen by Boccius. Various factors had to be taken into consideration by him. The plant had to be in the right condition and complete, exhibiting its essential features, particularly flowers, seeds and fruits. Most importantly, the specimen also had to be as 'typical' as possible – that is, it had to correspond to the statistical average. We may hypothesize that the Bauer brothers gradually learned how to select suitable (representative) specimens, but no doubt Boccius continued to maintain a tight control over this process, in particular on what species still had to be recorded.

On rather thin, cheap paper the Bauer brothers prepared as a first step a pencil drawing, to which they added light and shade. In 1786 Johann Christian Jacob Friedrich, botanical illustrator in Dresden, wrote in his *Anweisung zum Zeichnen und Blumenmalen* (*Instruction for drawing and flower painting*) about hatching: 'the pencil strokes one uses to give any object its appropriate light and shadow . . . one can make the strokes either heavy or light . . . the heavier and closer together they are, the darker the shadow . . . and the shadow can be made even darker if one uses cross-hatching. . . . The finer and the further apart [the strokes] are, the lighter the effect, until they eventually lose themselves in the light, because eventually the stroke has to be so fine that it becomes barely visible. . . . When drawing flowers one must take particular care with broad leaves, not to let the hatching slant across, or go in a curve. The hatching must rather follow the outer contour [outline] or edge of the leaf and extend as far as its base. So for example when one is drawing a flower with many petals, the hatching must always follow the middle of the flower down to the base of each leaf where it emerges from the base [axis]. And this goes for every leaf, whatever its orientation, position, shape or elevation: the hatching follows its shape down to its base. The hatching must eventually fade out gradually, and not only where there are folds or twists: a stroke must always be allowed to come to an end, so that the shadow of a wide leaf tapers into the slenderest pedicel'. We have to assume that the Bauer brothers added light and shade in a similar way to their outline sketches, which were part of the raw material for further work: for example, a pencil drawing showing Barbados lily, *Hippeastrum puniceum*, and Jacobean lily, *Sprekelia formosissima*, side by side (Fig. 2.1., p. 18).

Nothing is on record about the use of drawing aids such as mirrors or the camera obscura by the Bauer brothers at this stage of their career, but taking the precision and accuracy of the pencils in the Naturhistorisches Museum into account, this would not be surprising (see Chapter 3).

RECORDING AND CODING COLOURS

For every botanical illustrator time is an important factor: the preparation of a pencil drawing even when light and shade are added is considerably less time-consuming than that of a watercolour. In addition, there is another fact: a watercolour can be safely prepared only in the studio, not in a conservatory with its humid atmosphere or in the field under the conditions of an excursion, where it is faced with a broad spectrum of risks. After all, a finished water colour is an extremely vulnerable and fragile structure, irreversibly ruined within seconds. As a consequence, the Bauer brothers set out to solve a problem every botanical illustrator working in the field is faced with – how to record colour. Indeed, this poses a very big problem, since human memory is limited and recalling a particular hue of the rainbow as exactly observed in a given plant is often impossible. In addition, colours are not only very difficult to describe in words, with no two persons agreeing on the name of a given hue, but it is also often difficult for a botanical artist to recreate a colour by preparing and mixing the different pigments a second time. For a botanical artist working in a seasonal climate another problem exists: during the flowering season a great number of plants have to be recorded, but far fewer in the non-flowering season. Whoever has seen the Pollau Mountains in southern Moravia in full flower in May will immediately understand the Bauer brothers' predicament, even more so when considering their peripatetic tours to make permanent records of plants cultivated in the various gardens and conservatories. Thus a system for recording colours accurately that would also serve as a guide for recording the pigments to be used became necessary.

The Bauer brothers developed a colour code by devising a colour chart (Fig. 2.6.), which correlates the various hues of colour to numbers. These numbers

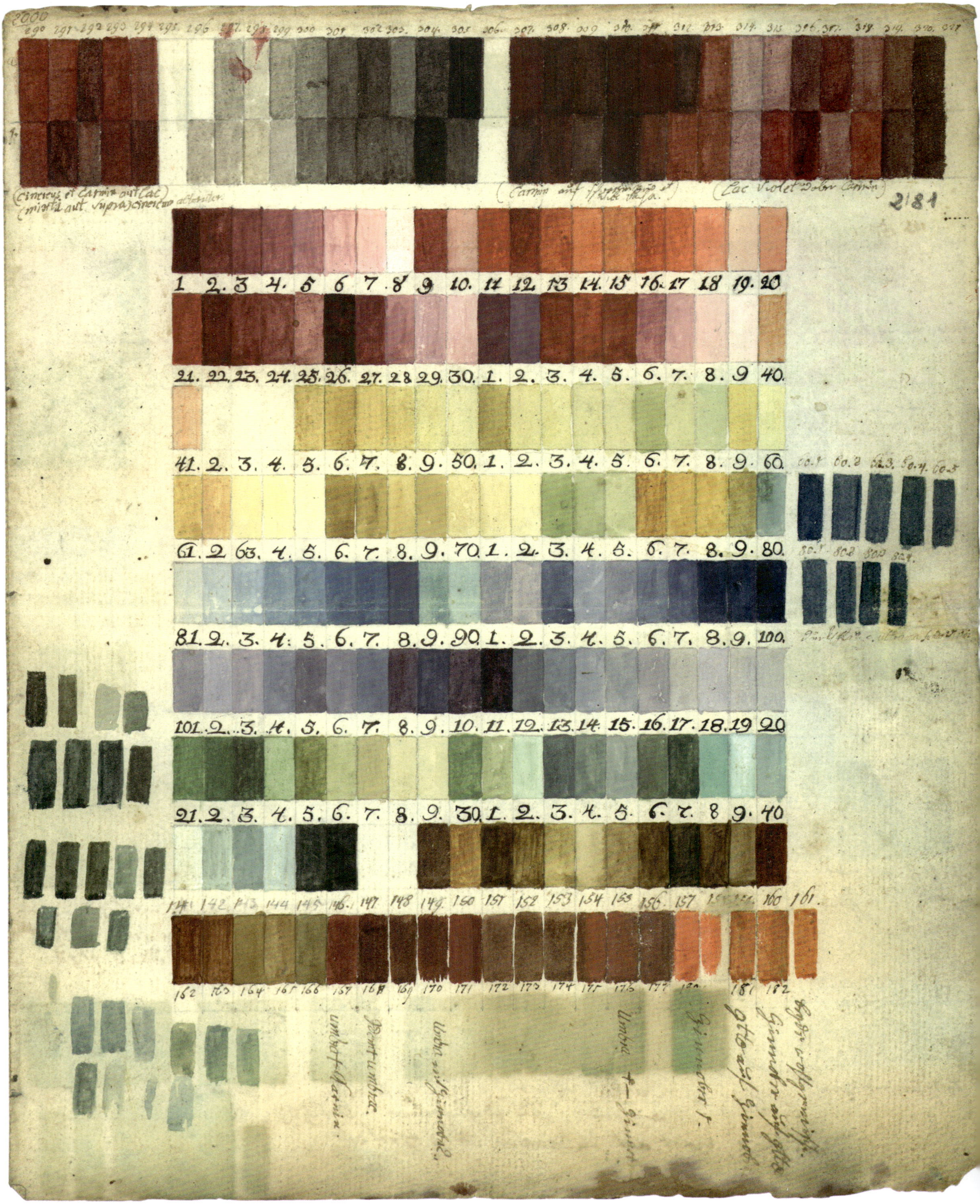
2181
1 2. 3 4. 5 6 7. 8 9 10. 11 12 13 14. 15 16. 17 18 19. 20
21. 22. 23. 24. 25. 26. 27. 28 29. 30. 1. 2. 3. 4. 5. 6. 7. 8. 9 40.
41. 2. 3. 4. 5. 6. 7. 8. 9. 50. 1. 2. 3. 4. 5. 6. 7. 8. 9. 60.
61. 2 63. 4. 5. 6. 7. 8. 9. 70. 1. 2. 3. 4. 5. 6. 7. 8. 9. 80.
81. 2. 3. 4. 5. 6. 7. 8. 9. 90. 1. 2. 3. 4. 5. 6. 7. 8. 9. 100.
101. 2. 3. 4. 5. 6. 7. 8. 9. 10. 11. 12. 13. 14. 15. 16. 17. 18. 19. 20
21. 2. 3. 4. 5. 6. 7. 8. 9. 30. 1. 2. 3. 4. 5. 6. 7. 8 9. 40
141 142 143 144 145 146. 147 148 149. 150 151 152 153 154 155 156. 157 160 161.
162 163 164 165 166 167 168 169 170 171 172 173 174 175 176 177 181 182

were noted down by them on the pencil drawings, for example a number in pencil for the scarlet red of the tepals in the Barbados lily (Fig. 2.1) referring to the finished watercolour (Fig. 2.7.). Whereas it was rather straightforward to correlate this pencil drawing kept in the Naturhistorisches Museum with the finished body colour in the *Codex Liechtenstein* conserved in the Fürstliche Sammlungen in Vaduz, locating the colour chart was another matter.

By lucky chance a colour chart was found by the present author in the archive of the Real Jardín Botánico in Madrid,[30] bound into a manuscript (now available as a facsimile with introductory commentary[31]) belonging to the estate of Thaddaeus Haenke. He had been an assistant of Mikan in Prague and was later supported by Jacquin in Vienna (see Chapter 3), where he had arrived in September 1786. He left Vienna in June 1789 for Madrid with the intention of meeting up with the Malaspina expedition, destined to visit the Spanish possessions on the coast of the Pacific Ocean. Hard to believe but true, this piece of paper to which Haenke had made substantial additions on the margins had survived the shipwreck of the *Nuestra Señora del Buen Viaje* in the mouth of the Río de la Plata in 1789, had been to California, Alaska, Kamchatka and Australia, and after a stop of about two decades in what is now Bolivia ended up in the Spanish capital in 1820. On a piece of badly water-stained paper, almost certainly because of the wreck, we find 140 rectangular squares painted in different colours arranged in seven rows each comprising 20 strips, the whole forming a square, which the present author has called Bauers' square. A number from 1–140 has been added in black ink to each coloured rectangle, with the reds forming 1–40, the yellows 41–80, the blues 81–120 and the greens and browns 121–140. The handwriting of the numerals is identical in the colour chart in Madrid (Fig. 2.6) and the pencil drawings in Vienna (Fig. 2.1), although the correlation of the colours is not absolutely perfect. It is impossible to determine if the handwriting on this colour chart is that of Joseph, Franz or Ferdinand. However, Ferdinand is the most likely candidate, since only he continued to use this method later in life and developed colour codes of increasing complexity (see Chapters 5 and 10).

Making use of this or a similar colour chart, the Bauer brothers annotated their pencil drawings with the numbers 1–140, again in pencil, resulting in a cloud of numbers surrounding their works. By preparing these colour-coded sketches they achieved a perfect, reliable and permanent record of something quite ephemeral – plant colours. The colour chart is undated; either this chart or a variant of it must have been in use from the early 1770s onwards. The use of a colour code on the pencil drawings of plants seems to have been a speciality of the Bauer brothers, the present author being unaware of any other botanical illustrator making use of this approach. It seems that the Bauer brothers applied this technique consistently during their work for Boccius. All pencil drawings kept in the Naturhistorisches Museum referring to their early work are colour coded.

FIG. 2.6. The Bauer brothers' colour code. Watercolour by Ferdinand (?) Bauer with later additions by Thaddaeus Haenke on the margins. Watercolour, ink. F. 281 r of note book 'Systema Colorum' (Div. VI, 3, 2, f. 278-285). – Madrid, Real Jardín Botánico, Archivio.

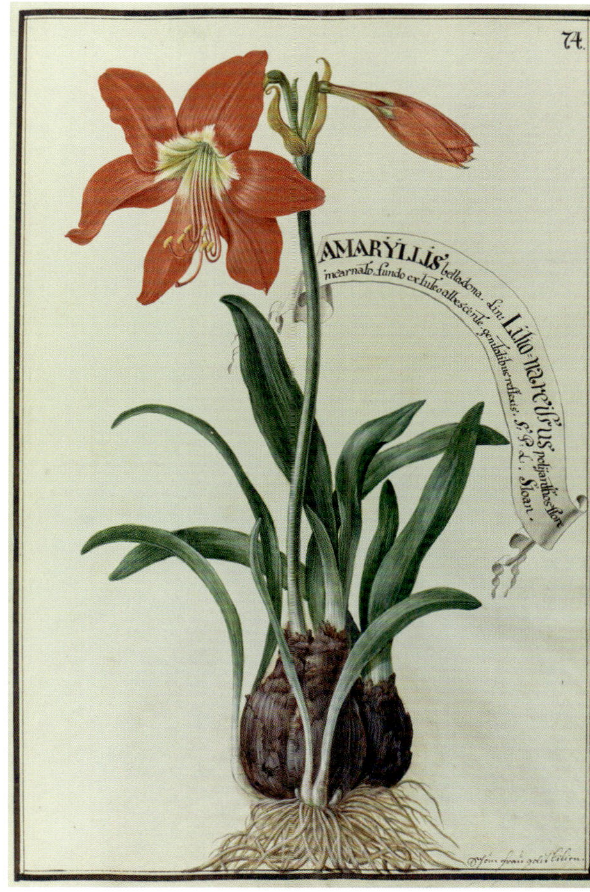

FIG. 2.7. Barbados lily, *Hippeastrum puniceum*. Watercolour by the Bauer brothers, c.1778. Watercolour, ink. *Codex Liechtenstein* volume 4: f. 74. – Vienna, Fürstliche Sammlungen.

It is not known if the Bauer brothers invented the colour code or simply followed a pre-existing example, such as the concept laid down in Jacob Christian Schaeffer's rare *Entwurf einer* [sic] *allgemeinen Farbenverein*[32] (*Plan of a universal relationship of colours*) or Ignaz Schiffermüller's even rarer *Versuch eines Farbensystems* (*Attempt at a system of colours*). Both correlate a system of letters and numbers with the different hues and attempt a standardization of 'the wavering designations of colours' – with plate 1 of the latter work even using the same sequence of colours as the Bauer brothers did in their colour chart. One of the copper plates illustrating Schiffermüller's treatise shows rectangles painted with different colours annotated by lower case and capital letters. In short, the Bauer brothers may have simply substituted running numbers for those letters. The great surprise is the date and place of publication of the two slim volumes: Schaeffer's in Ratisbon in 1769, Schiffermüller's in Vienna in 1771. The latter was in contact with Jacquin, and the idea could quickly have been passed via Boccius to the Bauer brothers, who, at the time of the publication of Schiffermüller's text, were fifteen, thirteen and eleven years old.

This method of colour-recording has several consequences. Firstly, even after decades, when the memory of a particular colour of a particular plant had long faded, a watercolour very true to nature (for example of *Hippeastrum puniceum*, Fig. 2.7.) could easily be prepared on the basis of the colour-coded pencil drawing. For this the use of a light-box or a piece of glass was again very likely: by placing the pencil drawing prepared on thin paper onto a piece of glass and above a sheet of the thicker watercolour paper, either a tracing could be made or paint could be applied directly, taking the underlying colour code into account. Secondly, provided the colour chart and the colour-coded pencil drawing are kept, a second copy, effectively a repeat, could be made at any time without the necessity of going back to the first production. As a matter of fact Ferdinand and Franz kept some of these coloured-coded pencil drawings from their very early years. When they died, all of their surviving preliminary pencil drawings for the *Codex Liechtenstein* ended up in the Naturhistorisches Museum, with the exception of a single one, which passed via Franz to the Natural History Museum and shows an illustration of the stinking granadilla, *Passiflora foetida*, combined with *P. holosericea*. Appropriately these coloured-coded pencil drawings were described in a note among Franz's manuscripts in the Niedersächsische Staats- und Universitätsbibliothek in Göttingen (see Chapter 17) as 'a kind of short hand Drawing consisting of very correct Pencil sketches, with the natural colours of the several parts marked on the margin so that at any time finished drawings might be made from them'. One is inclined to add, 'achieving perfect fidelity in colour'.

As a rule these preliminary pencil drawings were annotated by the Bauer brothers in pencil with scientific names provided by Boccius, often in calligraphy. It seems that an alphabetic arrangement following the generic names was intended on the large sheets of thin paper folded over, not a Linnaean sequence. This resulted in a kind of alphabetically arranged Thesaurus of pencil drawings. No indication exists that the pencil drawings, which may well be called intermediates, were ever coloured – the paper would have been too thin and rather unsuitable to accept the paint.

The preparation of a finished coloured plant illustration began by purchasing the watercolour paper and then, unlike today, was continued by preparing the paper with an emulsion to receive paint without letting it seep through. In the earlier volumes of the *Codex Liechtenstein* the only type of paper used was watermarked 'C & I Honig', an expensive high-quality product imported from the Low Countries. This again points to professional contacts with Jacquin, a native of Leiden, for he too used this solid, rather thick paper for the coloured copper engravings of his own works. The pencil drawings were then copied on this watercolour paper – either, we have to assume, with the help of a light box or by sight – as a sketch, which is virtually always invisible as the drawing would then be covered with a heavy layer of opaque paint.

The next step was to prepare the pigments, a laborious and time-consuming process, which moreover varied from colour to colour. Friedrich (see above) describes the preparation of opaque white as follows: 'Take some Kremnitzer [i.e. from Kremnica, a mining place in present-day Slovakia] white, powder it finely and mix it with water. Then add some gum tragacanth [a gum made of resin from *Astragalus* species] and a very small gum Arabic [a gum made of hardened sap from *Acacia senegal*] and dissolve them in water so that they are still workable even when they are dry. You should take care never to apply white straight to the paper because it easily turns black. You should apply another colour underneath it first, and this will hold the white when you apply it wet, and it will not run. When you are building up a body of colour – that is, when you mix another colour with white – you must apply another colour which has not been mixed with white first . . .'. According to Friedrich the procedure for yellow is different 'You should always use the gummigutti [gamboge, a resin from *Garcinia* trees] straight from the block, or else put onto the palette only as much as you are going to use immediately, because the more often you dissolve it the browner it gets', and for orange 'This must always be powdered very finely and mixed with water, like cinnabar, otherwise it takes on an ugly, burnt appearance. When it is ready you should leave it to stand on the piece of stone on which the pigment is prepared for a while until it has lost any excess moisture; then you should add enough gum Arabic so that it does not run. I ought also to add that if you want to be fully confident when using these two colours, as soon as you dissolve them in water but before you add any gum you should put them into aqua fortis [nitric acid], and also allow to burn them off, and then refresh them with water again until no aqua fortis remains; only then should you add the gum'. Shades of red are no less difficult to produce. Friedrich says that 'Lacquer red' is hard to dissolve and recommends that it is prepared on a piece of glass.

In short, in addition to being botanical illustrators the Bauer brothers had to engage themselves in the work of microchemists in order to prepare their pigments. They used rather opaque colours, which made corrections easier. A quote from Friedrich illustrates the situation of a botanical illustrator when painting and shading a yellow plant part, like the surface of a pumpkin (see below) 'Dark yellow . . . first shade it with aurora yellow, and then with ox gallstone. If you do not have any you can always use dark yellow ochre. High yellow . . . first shade it with dark yellow, and then with dark yellow ochre. Pale yellow . . . first shade it with high yellow, and then with blackish, and then go over the shades with greenish. Aurora yellow, or yellow arsenic . . . first mix them with cinnabar or carmine, and then shade them with ox gallstone . . . Light aurora yellow . . . shade it with the same colours, only paler . . . Pale aurora yellow . . . first simply shade it with ox gallstone, and then crosshatch it with greenish-yellow'. The Bauer brothers probably shaded their paintings in a similar way. They frequently used opaque white when adding lights, as on the surface of a ripe aubergine, *Solanum melongena* (Fig. 2.8.), and sometimes they applied a thin film of egg tempera or gum arabic to the upper layer of the paint in order to give extra lustre to a particular feature.

Where did the preparation of the finished watercolours take place? For a variety of reasons almost certainly in Feldsberg.

As far as we know the plant illustrations were not done on single sheets of paper but on larger sheets folded over, so that each sheet carries two illustrations, on pages one and three. It was Boccius' aim to arrange the *Codex Liechtenstein* according to the Linnaean system, so it was always necessary for him to determine the exact position that each watercolour was to occupy. New illustrations would have to be inserted as close as possible to their correct place in the ever-growing collection so as to maintain the correct (that is, Linnaean) sequence. A collection of half-finished singly folded sheets would thus lie awaiting completion. We may assume that Boccius was directing the additions to his great work and that they were under his supervision. Since he is not known to have left Feldsberg, even when he was Father Provincial, this stock of incomplete sheets must have been kept in the convent, which is also the most likely place where the completed sheets were conserved.

When the watercolour was finished and the sheet left to dry, the annotation began. At this point the sheet was probably shown to Boccius, who controlled the accuracy of the finished work and had to decide which scientific names were to be used. This was an immense task: in the late-eighteenth century there were few reference works to guide those assigning scientific plant names, and both pre-Linnaean names and Linnaean binomials were applied by Boccius. He would certainly have made use of Linnaeus's *Genera plantarum*, which he is shown to hold in his hand in one of his portraits (see above). When one takes into account the state of botany before 1800, it is clear that he did an excellent job, not dealt with here in any detail. This is particularly so for the native vegetation of the Archduchy of Austria and of Moravia, and for plants that had long been known for their practical usefulness or decorative qualities. After all, the work Boccius had initiated was to become a veritable 'Flora universalis' of its time.

The actual addition of the scientific name – always in Indian ink, with the generic name frequently stencilled – was probably done sometimes by the Bauer brothers, sometimes by other members of the convent of the Brothers of Mercy who could write, since several different

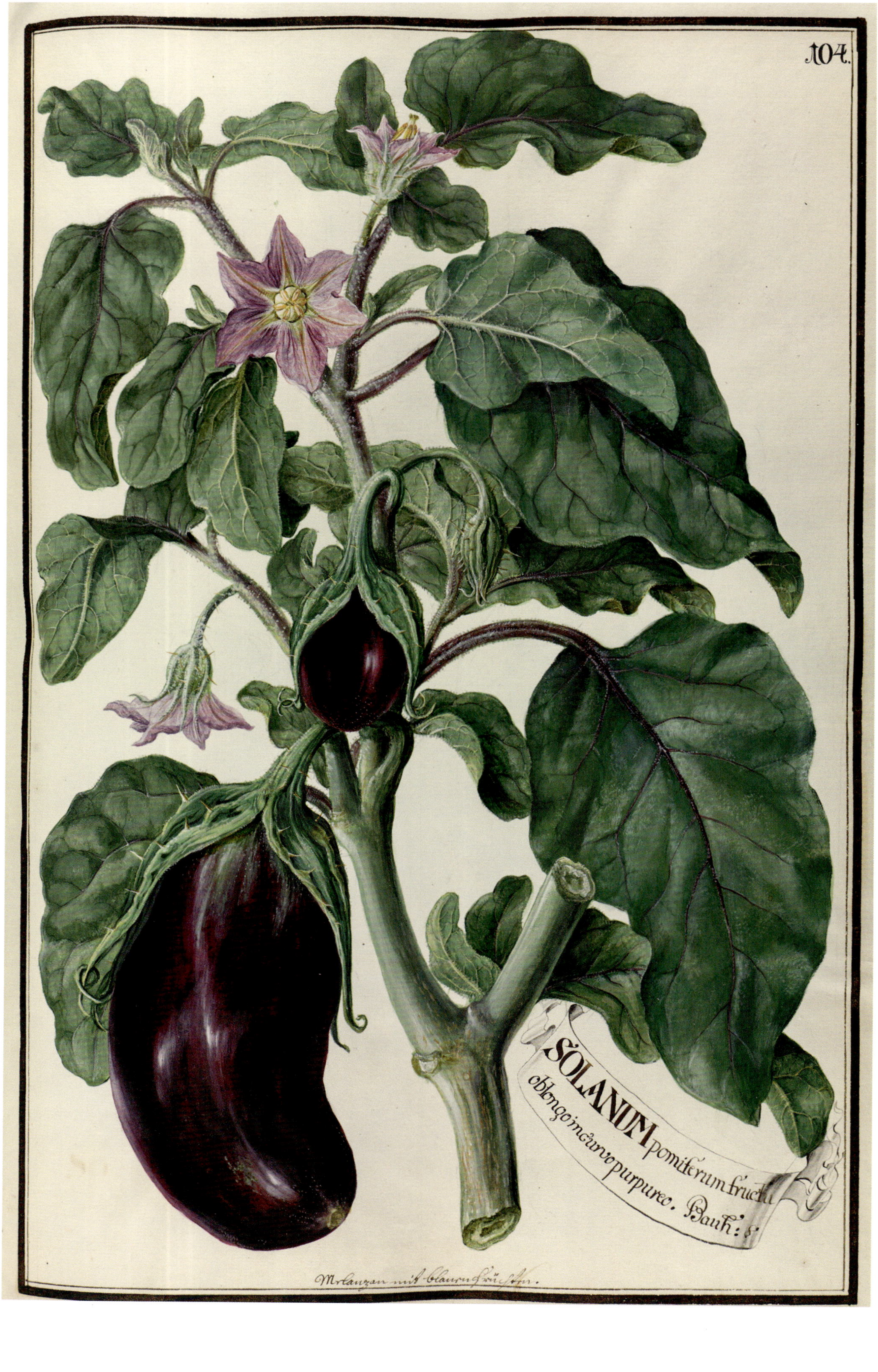
104.
SOLANUM pomiferum fructu
oblongo incurvo purpureo. Bauh:
Melanzan mit blauen früchten.

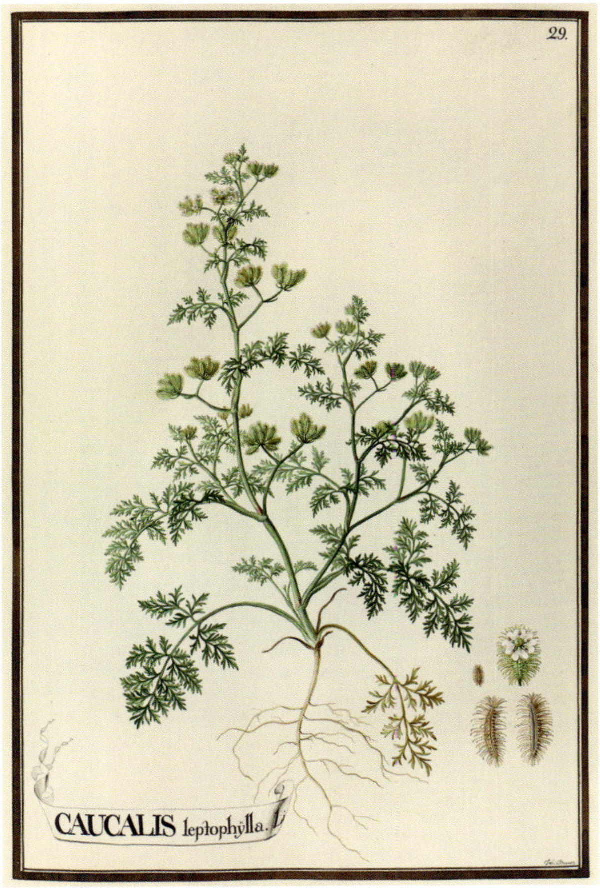

FIG. 2.9. *Caucalis leptophylla*. Watercolour by Ferdinand Bauer, c.1783. Watercolour, ink. *Codex Liechtenstein* volume 6: f. 29. – Vienna, Fürstliche Sammlungen.

hands can be distinguished in the *Codex Liechtenstein*. Consistently these names were put into the baroque scrolls added to the plant illustrations, echoing the paper strips in the Herbarium vivum used for fixing specimens. There is a considerable number of orthographic errors, many of them distorting the intended meaning, and one suspects that some of the scribes, and definitely the Bauer brothers, did not know Latin.

Next, in a separate operation, the striking double frame would be inked in, and in a very few cases the illustration, for example of *Caucalis leptophylla* (Fig. 2.9.), would be signed. Possibly as the final step Boccius would add the vernacular name in freehand in Indian ink, since most of the German names seem to be in his handwriting, vernacular names in other languages such as Czech are missing. After the compilation of the index, again prepared by scribes, and the addition of the title page and a garden view or landscape (see below), the book binder was asked to unite the single-folded sheets into a volume according to the instructions given by Boccius, who attempted to follow the Linnaean system in the sequence of the illustrations. In a final step the book block was painted in bright red by the book binder. Clearly the key innovation of the Bauer brothers in this long sequence was the preparation of the colour-coded pencil drawing in the field, that is a colour-coded intermediate, which acted as the template for the preparation of the finished watercolour in the studio.

FIG. 2.8. Aubergine, *Solanum melongena*. Watercolour by the Bauer brothers, c.1778. Watercolour, ink. *Codex Liechtenstein* volume 5: f. 104. – Vienna, Fürstliche Sammlungen.

The arrangement of the illustration on the sheet of paper is another matter, which has nothing to do with the precise documentation of a specimen, but rather with aesthetics – a branch of philosophy dealing with the nature of art, beauty and taste. Pertinent considerations are totally neglected in this book, which is focused on science and its history, rather than on philosophical questions. In addition, zero information is available on the aesthetical considerations to which the Bauer brothers and their contemporaries subscribed. There is only a single fact: they were able to gather stimuli from many sources – from the works they copied, from the botanical illustrations kept in the library of the convent, and from the works of art in the country houses of the Liechtensteins in Feldsberg and Eisgrub with which they were familiar.

THE PLANT AND ANIMAL ILLUSTRATIONS OF THE *CODEX LIECHTENSTEIN*

The plants illustrated come from different sources, forming a kind of painted encyclopaedia of what was known to Boccius, although surprisingly some plant groups are totally lacking, such as gymnosperms, non-vascular plants (with the exception of a single moss), and fungi. Others, such as the grasses, are hopelessly underrepresented. Furthermore the number of trees and shrubs depicted is rather small. It also should be remembered that less than a third of the illustrations included in the *Codex Liechtenstein* are straightforward copies based on coloured copper engravings (see above); these are not dealt with here.

Within these limitations, three clearly distinguishable groups of plants may be recognized, although with such a rich array of illustrations it is only possible to give examples.

A significant proportion of plants depicted came from the natural vegetation of Feldsberg and its surroundings: they include common and widespread species such as great burdock, *Arctium lappa*, musk thisle, *Carduus nutans* (Fig. 2.10.), and stemless carline thistle, *Carlina acaulis*, to name but three examples of composites. However, species rare in the area are also included, such as pink convolvulus, *Convolvulus cantabricus*, a sub-Mediterranean element, woolly clary, *Salvia aethiopis*, and Boccius's trophy, the tartar bread plant, *Crambe tataria*, the latter characterized by a Pannonian and Pontic distribution pattern. We find an excellent cross-section of the aquatic plants native in the wetlands and ponds of the Thaya valley, among them white water-lily, *Nymphaea alba*, yellow water-lily, *Nuphar lutea*, and water caltrop, *Trapa natans* (Fig. 2.11.). The latter is a floating annual with ornate shaped fruits, and the Bauer brothers paid attention to its striking leaf dimorphism, showing the submerged, feather-like and the floating, undivided, somewhat succulent leaves with saw-toothed edges. They also recorded the similarly conspicuous leaf dimorphism in common water crowfoot, *Ranunculus aquatilis*, which is also found in aquatic habitats. We find an image of the ubiquitous broad-leaved common pondweed, *Potamogeton natans*, rooting in stagnant waters, of bog-bean, *Menyanthes trifoliata*, and of the much rarer free-floating water soldier, *Stratiotes aloides*. In the equally free-floating greater bladderwort, *Utricularia vulgaris*, another plant from the Thaya wetlands, the characteristic bladders are depicted with great care. From areas more distant from Feldsberg we find representations of montane plants such as globe flower, *Trollius europaeus*, or bilberry, *Vaccinium myrtillus*, and from even higher altitudes alpine snowbell, *Soldanella alpina*, and the obligatory edelweiss, *Leontopodium*

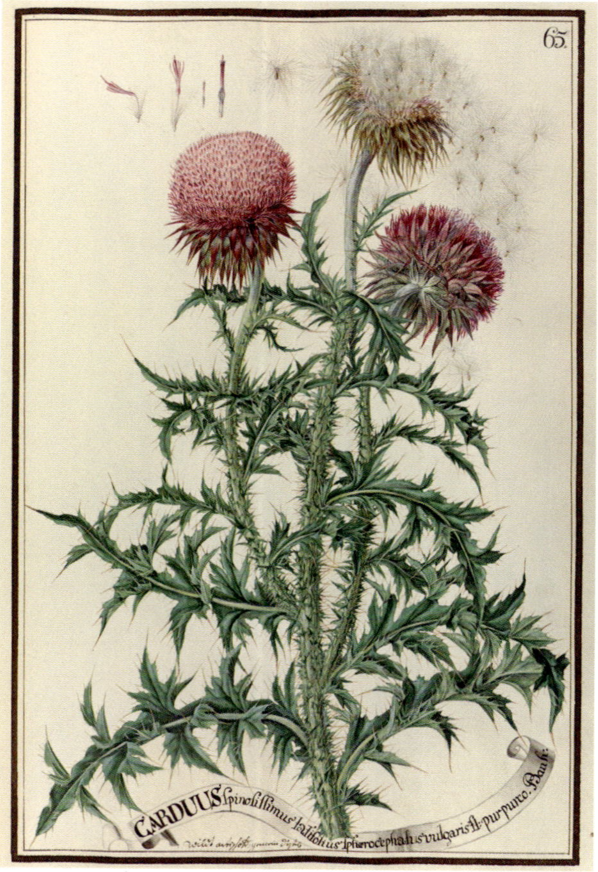

FIG. 2.10. Musk thistle, *Carduus nutans*.
Watercolour by the Bauer brothers, c.1777.
Watercolour, ink. *Codex Liechtenstein* volume 3: f. 65. – Vienna, Fürstliche Sammlungen.

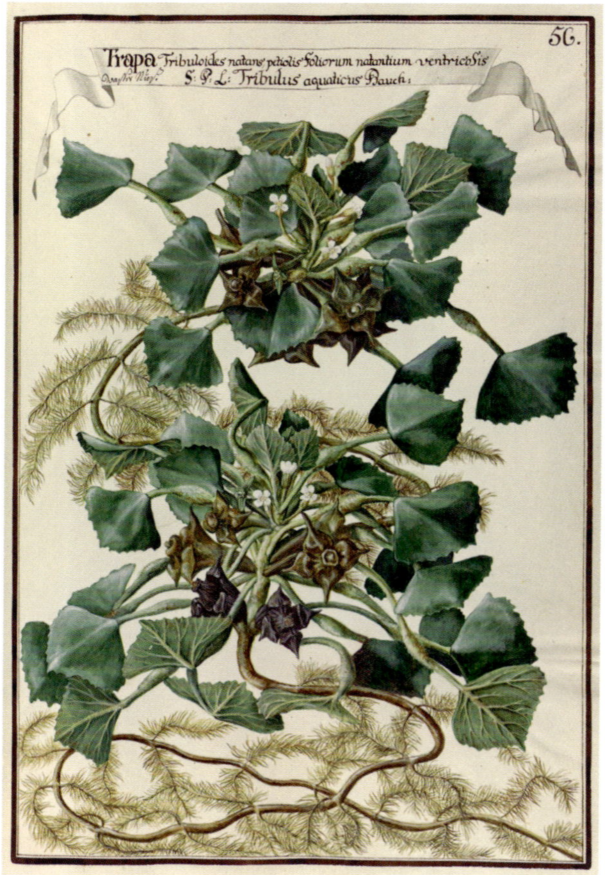

FIG. 2.11. Water caltrop, *Trapa natans*.
Watercolour by the Bauer brothers, before 1788.
Watercolour, b ink. *Codex Liechtenstein*, volume 8: f. 56. – Vienna, Fürstliche Sammlungen.

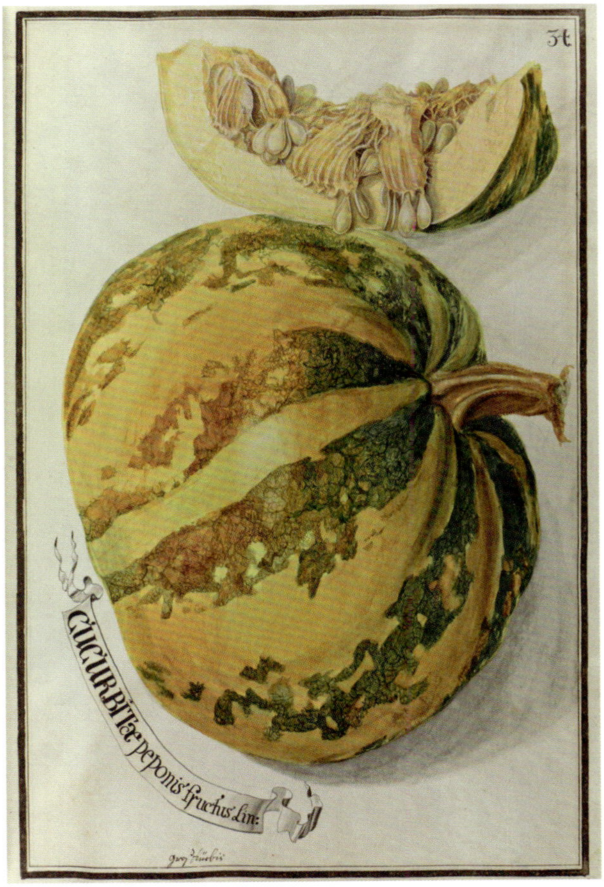

FIG. 2.12. Pumpkin, *Cucurbita pepo*. Watercolour by the Bauer brothers, c.1778. Watercolour, ink. *Codex Liechtenstein*, volume 4: f. 34. – Vienna, Fürstliche Sammlungen.

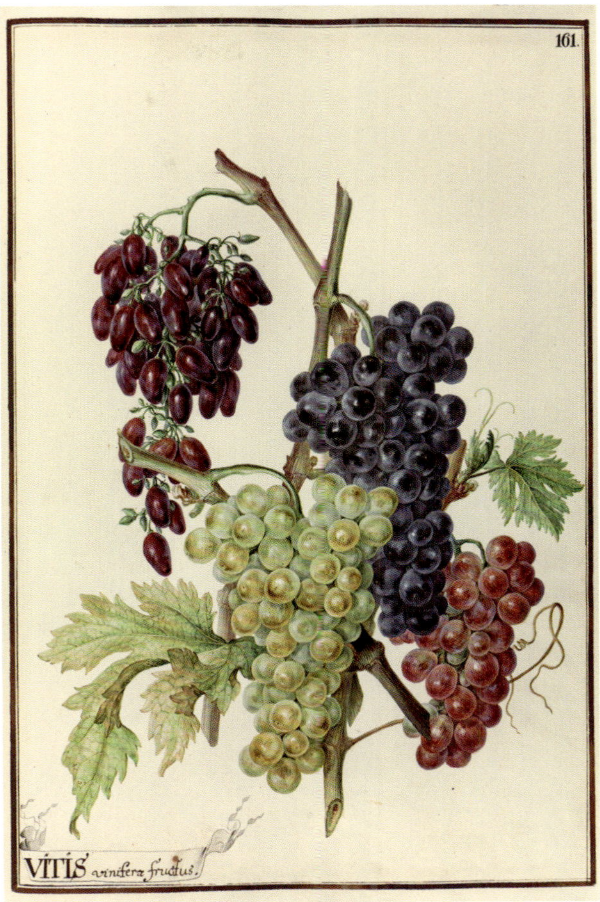

FIG. 2.13. Grape, *Vitis vinifera*. Watercolour by the Bauer brothers, c.1783. Watercolour, ink. *Codex Liechtenstein*, volume 6: f. 161. – Vienna, Fürstliche Sammlungen.

alpinum. This is precisely what one would expect: from Feldsberg the Carpathians can be seen with the naked eye, and even the Eastern Alps with mountains like Schneeberg, often snow-capped from October until May, are visible on very clear days.

The second group are the useful and decorative plants that had long been in cultivation in the region. These include crop plants such as pumpkin, *Cucurbita pepo* (Fig. 2.12.), sunflower, *Helianthus annuus*, and aubergine, and ornamentals such as madonna lily, *Lilium candidum*, blue passion flower, *Passiflora caerulea*, and calla lily, *Zantedeschia aethiopica*. In several cases the Bauer brothers depicted different cultivars of such plants – for example oleander, *Nerium oleander*, with differently coloured petals, tulips, *Tulipa gesneriana*, with variegated tepals, and marvel of Peru, *Mirabilis jalapa*, with a labile gene. Many species are shown with double flowers or inflorescences, as in the common daisy, *Bellis perennis*, in albino forms (white-flowered cultivars, as in borage, *Borago officinalis*), in variants with variegated, white-edged, or anthocyan-rich leaves, and there is even a spineless mutant of thorn apple, *Datura stramonium*. Boccius was particularly careful to include cultivars of food plants – thus there are four different cultivars of grape, *Vitis vinifera* (Fig. 2.13.), reminiscent of Tamm's over-doors in the state rooms of Feldsberg castle, a whitish-fruited cultivar of redcurrant, *Ribes rubrum*, a yellow-fruited cultivar of aubergine, and at least seven kinds of cabbage, *Brassica oleracea*, including cauliflower, broccoli, red cabbage, kohlrabi and ornamental cabbage. He also selected species where the underground parts differed in different cultivars, for example in parsnip, *Pastinaca sativa*, which possesses roots with either yellow or violet flesh, and there are several cultivars of the tomato, *Solanum lycopersicum*, differing in shape and colour of their fruits. We also find rarities like the Chilean strawberry, *Fragaria chiloensis*, which had contributed

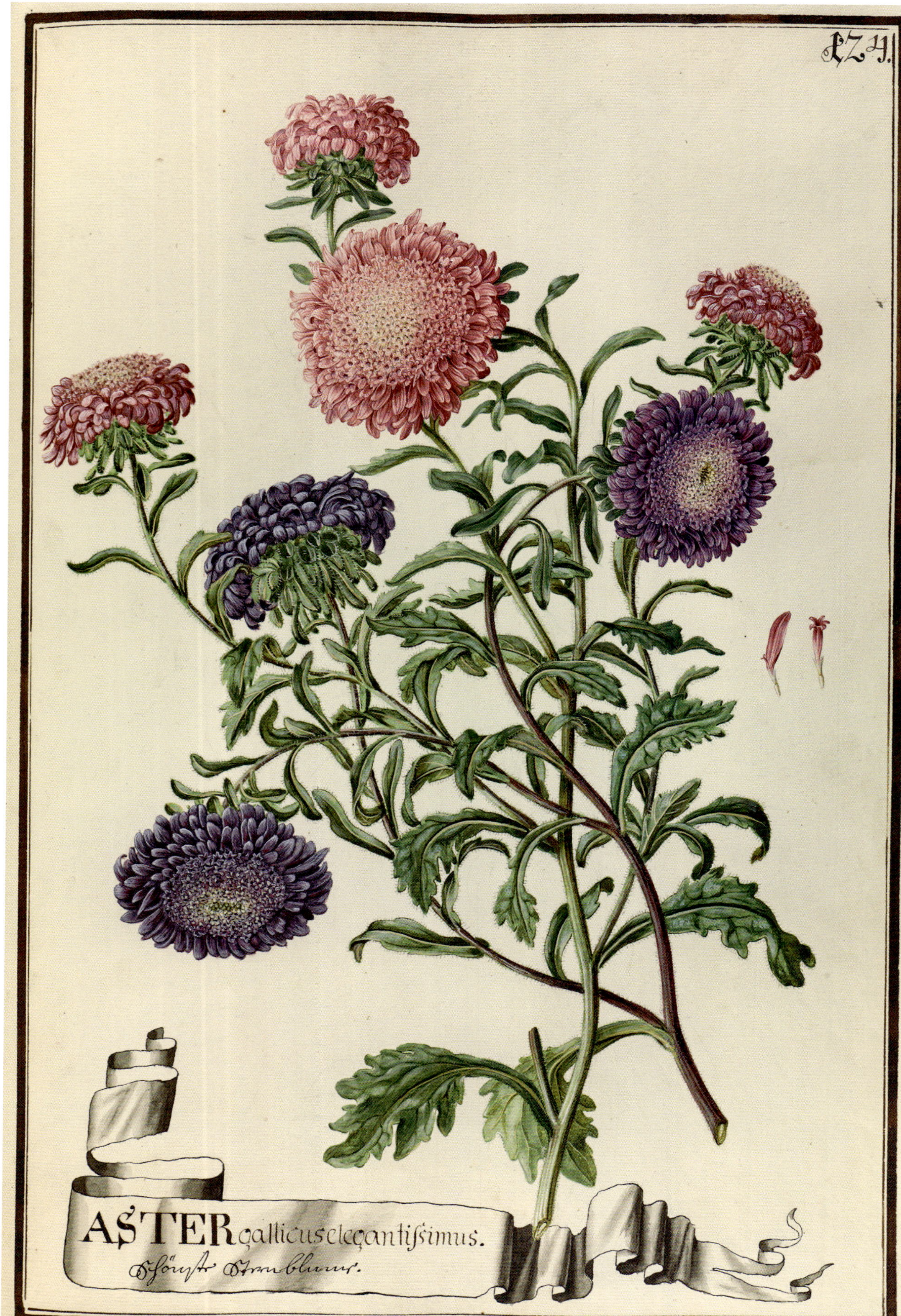
ASTER gallicus elegantissimus.

its genome to the now globally cultivated garden strawberry, *Fragaria × ananassa*. These illustrations make the *Codex Liechtenstein* a unique record of food and garden plants cultivated in the last decades of the eighteenth century in central Europe.

More interesting for historians of botany is a third group: plants that only rarely had been illustrated in printed works before the Bauer brothers portrayed them, for example the Barbados lily from tropical America, and the China aster, *Callistephus chinensis* (Fig. 2.14), which flowers late in the year and had been cultivated for centuries in China. It seems, however, that the first six volumes of the *Codex Liechtenstein* do not contain any image of a plant that was then new to science and had not yet received a proper name.

Naturally, a physician such as Boccius also turned his attention to medicinal plants, of which the opium poppy, *Papaver somniferum*, and deadly nightshade, *Atropa belladonna*, had been known since antiquity for their pharmaceutical properties, whereas foxglove, *Digitalis purpurea*, would only become famous in 1785 for its potential in the treatment of the early phases of congestive heart failure. All three species are recorded in the *Codex Liechtenstein*, but more surprising is a pink-blossoming plant restricted in its natural distribution to Madagascar – the Madagascar periwinkle, *Catharanthus roseus*. It synthesizes vincoblastine, today an important component of chemotherapy regimes; neither Boccius nor any physician or pharmacists of his time could have dreamed about the potential of this plant to combat cancer, in particular leukaemia in infants. Less surprising are the representations of several species of the nightshade family containing psychoactive components, for example bell henbane, *Scopolia carniolica*, described by Jacquin as a species new to science only two decades before, and species of the genus *Brugmansia* from the New World tropics. Boccius must have been aware of the effects of these substances, since he called the first species 'Zigeiner krauth, Raasewurtz' (gipsy herb, raging root) and the second 'Indianisches Tollkrauth' (Indian furious herb), and must have known them as living plants.

FIG. 2.14. China aster, *Callistephus chinensis*. Watercolour by the Bauer brothers, before 1788. Watercolour, ink. *Codex Liechtenstein*, volume 8: f. 124. – Vienna, Fürstliche Sammlungen.

Unfortunately, Boccius never gave a provenance for the plants illustrated, and therefore only intelligent guesses are possible. It seems safe to assume that for the non-native plants the conservatories in the Eisgrub gardens belonging to the Liechtenstein family were one of the more important sources for the tender plants. The anonymous traveller from Vienna to Brno in 1782 had noted the existence of three stoves in this village containing three hundred specimens of pineapple and many exotic species, plus an orangery (see above) for the cultivation of seven hundred large and three hundred medium-sized citrus trees as well as twenty myrtle trees. He added that two hundred fathoms of wood are used for heating these conservatories during winter and lists the garden staff responsible for this part of the Feldsberg domain. The conservatories of the local aristocracy, like the Dietrichsteins, may have provided further plants, but the bulk of the cultivated material seems to have originated from gardens and fields in the archduchy of Austria and Moravia, among them the medicinal garden of the convent in Feldsberg. This type of work may have necessitated travels for which the working procedure in two steps (that is, the preparation of a colour-coded pencil drawing followed by the preparation of a watercolour based on the former in the studio) was ideally suited. Indeed, Ferdinand, the traveller, maintained this approach for the rest of his life, whereas the sedentary Franz in his studio only very rarely seems to have produced colour-coded pencil drawings.

Although rightly praised for their accuracy and precision, a few critical remarks on these plant illustrations seem appropriate. There are indications that Boccius insisted on representations of any given plant on a single page in order to give a general impression, but at the same time avoiding any cuts of the specimen. In several cases this resulted in images of 'jolted' specimens reminiscent of the manipulated specimens in Boccius's herbarium. Foxglove in the flowering stage is column-like and has been painted as a bent, jolted plant with internodes far too short, and the same applied to the image of canna, *Canna indica*, white veratrum, *Veratrum album*, and black henbane, *Hyoscyamus niger* (Fig. 2.15.). Individual flower and fruit parts are only occasionally represented, which seems to concur with this approach of emphasizing the general impression and may also be due to Boccius. On the other hand we find a negligible number of mistakes, for example flower parts associated with the wrong representation of the complete plant, as in the genus *Stapelia*.

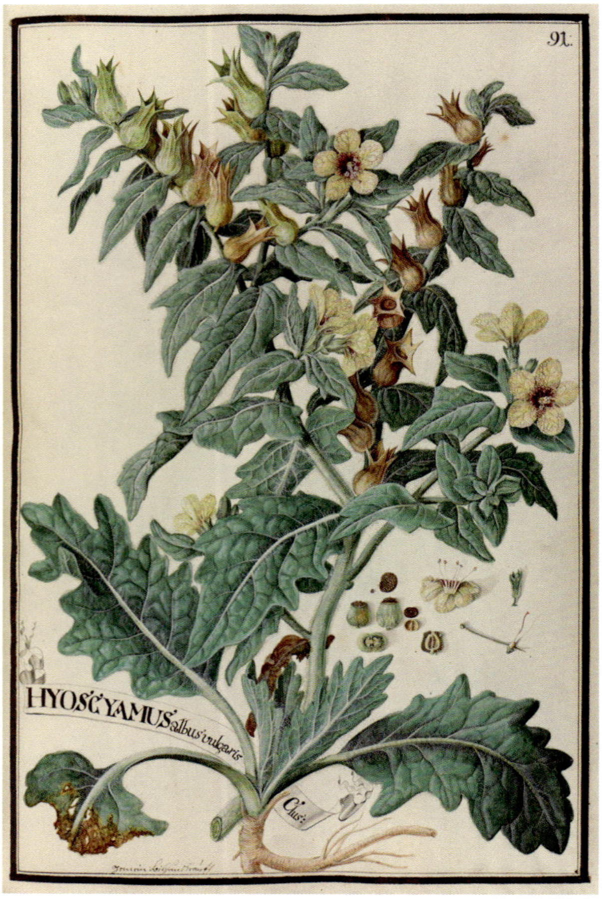

FIG. 2.15. Black henbane, *Hyoscyamus niger*. Watercolour by the Bauer brothers, c.1779. Watercolour, ink. *Codex Liechtenstein*, volume 5: f. 91. – Vienna, Fürstliche Sammlungen.

There are several exceptions to the general lack of such detail, notably in the composites, where we find individual florets depicted with and without a corolla as in great burdock, the fruit with its pappus in milk thistle, *Silybum marianum*, or a longitudinal section of the flower of the cactus *Cereus* cf. *alacriportanus*. On the other hand, Ferdinand's work on Australian plants with its numerous dissections is foreshadowed in the representation of the ripe capsule of black henbane (Fig. 2.15.) shown in transverse and longitudinal section, plus a ripe seed with particular attention given to its surface, and several other dissected flower parts. As immediately realized by Ferdinand's future commissioner John Sibthorp (see Chapter 5), the images of species of the umbel family are particularly outstanding in quality, since they show the complex architecture of their inflorescences by subtle differences in the various shades of green; examples include celery, *Apium graveolens*, fennel, *Foeniculum vulgare*, lovage, *Levisticum officinale*, parsley, *Petroselinum crispum*, and greater water parsnip, *Sium latifolium*. These illustrations anticipate the fine hues of green and grey that make the portraits of the various species of the genus *Pinus* prepared by Ferdinand and Franz for Aylmer Bourke Lambert such outstanding specimens of scientific illustration, praised by a connoisseur such as Goethe (see preface and Chapter 9). At the same time Sibthorp's praise of the *Codex Liechtenstein* examined by him in Feldsberg (see Chapter 5) is the only contemporary report on this outstanding document recording plant diversity.

Equally notable is the frequent stress on the underground parts of the plants depicted, which must have been dug up and put into water just like in the famous *Pulsatilla pratensis* subsp. *nigricans* (see above). This is indicative of the traditional concept of the rhizotomists, maybe still held by Boccius, that the pharmaceutical principle of a plant is always located in the underground parts. This attitude is the reason for the many excellent depictions of: rhizomes, such as calla lily, *Zantedeschia aethiopica*; of root tubers, such as Jerusalem artichoke, *Helianthus tuberosus*, and beetroot, *Beta vulgaris* var. *vulgaris*; and of roots, such as Spanish salsify, *Scorzonera hispanica*, shown by the Bauer brothers in cross-section with its brown corky outer layer and milky sap extruding. Several bulbs are depicted in an exemplary manner, like those of the Barbados lily (Fig. 2.7.) and madonna lily. The representations of a healthy specimen of cypress spurge, *Euphorbia cyparissias*, and a specimen of this species infected with the rust fungus *Uromyces pisi* resulting in distorted growth are excellent pieces of observation; even the aecia on the yellowish leaves of the latter are depicted. Less than two decades later Franz would devote his attention to fungal parasites of wheat and prepare fine drawings of his microscopic preparations for his long-term patron Sir Joseph Banks (see Chapter 8).

The *Codex Liechtenstein* also contains a few illustrations of animals true to nature, with a single exception all subordinate to the plant illustrations. Apart from a house sparrow, *Passer domesticus* (see below) and a tree frog, these are mostly insects, usually just the imagines – flies, hymenopterans, grasshoppers, beetles, dragonflies, butterflies and bugs. One particularly fine illustration is of a mole cricket, *Gryllus gryllotalpa*, which the Bauer brothers must have often encountered in the vineyards around Feldsberg. Even the insect's shadow is depicted, which is only occasionally undertaken in the botanical illustrations, where it is used to increase the contrast when flower parts are shown dissected, as in the case of black henbane (Fig. 2.15.). The tanner beetle (possibly *Prionus coriarius*) on the spathe of the calla lily is another representation of the local fauna known for its mainly crepuscular and nocturnal activity. Probably the Bauer brothers worked also in this case with pencil intermediates kept for the production of later repeats. The many, very precise illustrations of dragonflies (one with only five legs instead of six), however, are no surprise: whoever has visited the wetlands along the Thaya in summer will have encountered them in the thousands. In representing butterflies on individual plates, the Bauer brothers followed the pattern set in the *Receuil des plantes*, which include occasional images of butterflies. Some of these capture the exact posture of the insect, as in the scarce swallowtail, *Iphiclides podalirius*, in which even the antennae are correctly shown and lead over to the magnificent wreath of roses visited by many butterflies, a masterpiece of Franz prepared for his second employer (see Chapter 3). Other animal subjects include an illustration of a copulating pair of insects, a fly stuck in the viscid secretion of sticky catchfly, *Lychnis viscaria* subsp. *viscaria*, and an occasional spider or snail shell. No animals from outside Europe are depicted, indicating that the *Codex Liechtenstein* was prepared without reference to zoological collections.

The representation of the house sparrow is noteworthy: it is dated 1775, signed by Franz when he was seventeen years old, and also shows a lifelike representation of the maidenhair fern, *Adiantum capillus-veneris*. This combination of a bird illustration with a depiction of unobtrusive or non-flowering plants such as grasses, ferns, algae or mosses is also – with only a few exceptions – a feature of the illuminated manuscript entitled *Receuil d'Oyseaux de la Ménagerie Royale du Parc de Versailles*, which was held in the Imperial Library in Vienna. It forms the ornithological equivalent to the *Receuil des plantes*, and Swieten may have shown this painted record of the birds in the menagerie in Versailles to Boccius. Alternatively, this could have been also done by his son Gottfried Freiherr van Swieten, who had been first imperial envoy in Berlin and subsequently succeeded his father as prefect. Equally, the Bauer brothers could have seen this illuminated manuscript when they visited Vienna (see Chapter 3). A few years later Ferdinand was to follow this approach in his animal illustrations from the Levant produced for Sibthorp in Oxford (see Chapter 6) and later in his animal illustrations from Australia prepared for the Admiralty in London (see Chapter 11).

THE TITLE PAGES, GARDEN VIEWS AND LANDSCAPES OF THE *CODEX LIECHTENSTEIN*

The volumes of Well's *Phytanthologia eikonike* (see above) comprise painted title pages, which include landscapes and views. Among them we find a view of the ruin of Rauhenstein near Baden bei Wien. All title pages are signed or have been attributed to Franz Scheidl, who had also worked as botanical illustrator for Jacquin in Vienna (see Chapter 3). Boccius also followed Well's model in this point, although in a slightly different way: the title pages include garlands and flower pieces of utmost refinement and are separate from the garden views and landscapes, with the title pages following the latter group of views.

Four groups of title pages have been distinguished in the fourteen volumes of the *Codex Liechtenstein*. Only title pages of the first group, comprising volumes 1–6, carry signatures of Franz or Ferdinand or have been attributed to them. These are works of art of unsurpassed elegance, continued in the painted title pages of the luxury edition of Jacquin's *Selectarum stirpium americanarum historia* (*Selected American plants*) (see Chapter 3). The Bauer brothers painted mixed flower garlands, either pendent or wrapped on oval frames, and flower pieces (Fig. 2.16.), probably the finest ever prepared, combined with text containing a chronogram. In this way the individual title pages can be dated: that of the first volume of the *Codex Liechtenstein* and also the first of those by the Bauer brothers is from 1776, interpreted here as the *terminus ante quem* for all the illustrations bound into this volume. At that moment Joseph, Franz and Ferdinand were twenty, eighteen, and sixteen years old, respectively. All three names are explicitly given on this first title page, being the only indication for a contribution of Joseph to the great project.

The title pages of the other volumes are dated 1777, 1778, 1779, 1780 and 1784, also to be interpreted as *terminus ante quem* for the illustrations of the respective volume, which give us an additional (albeit rough) time frame for the engagement of the Bauer brothers. However, they must have started work significantly before 1776 and certainly continued after 1784; in addition, volumes 10–13 contain a few plant illustrations that must be by the superior hand of the Bauer brothers. Their signatures on the title pages are microscopic, hidden in the ornate letters, which comprise several

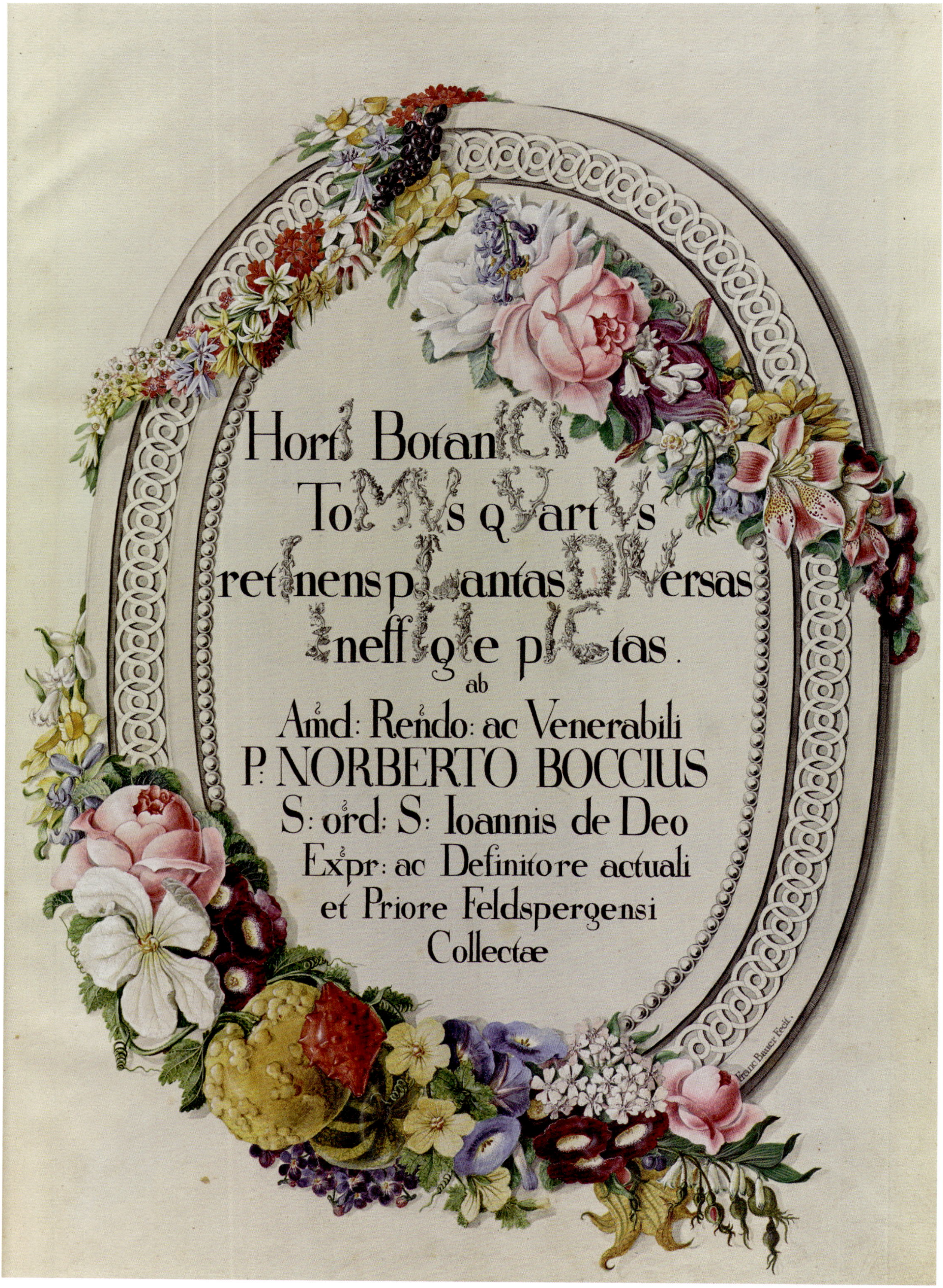
Horti Botanici
Tomus quartus
retinens plantas diversas
ineffigie pictas.
ab
Amd: Rendo: ac Venerabili
P: NORBERTO BOCCIUS
S: ord: S: Ioannis de Deo
Expr: ac Definitore actuali
et Priore Feldspergensi
Collectæ

Franc. Bauer Fecit.

equally microscopic drawings in black ink referring to the convent in Feldsberg and the brethren living there.

The six title pages merit the special attention of art historians. They show a profound change of style: the pendent flower garlands dated 1776 and 1778, respectively, supported by delicate cloth ribbons can be attributed to the rococo period, while those dated 1784 and 1788 contain elements that are safely to be interpreted as indicative of neoclassicism. The latter lead over to the title pages with flower garlands painted by Ferdinand for John Sibthorp a few years later in Oxford (see Chapter 6) and, in a sense, to Joseph's neoclassical painting in Rome (see Chapter 4). The title page of volume 8 is a hybrid: the painting is by the Bauer brothers, but the text including the date 1793 is almost certainly by their successor Jacob Walter, who seems to have started to work for Boccius after the engagement of the Bauer brothers had come to an end.

At the front of the first volume is a remarkable garden vista (Fig. 2.17.), a painting best analyzed with a lens. It is signed by Franz and shows a portal with Corinthian columns and the inscription 'Hortus Botanicus' (Botanical Garden), which, so to speak, serves as the entrance to the *Liber regni vegetabilis*. To the left a faun is painted with one of Jacquin's works in its hand, to the right a monkey with a palette in one hand, a plant in the other, and a paint brush in its mouth. This allegory seems to refer to the initial phase of the project – copying (or aping) printed illustrations, for example those published by Jacquin. This is confirmed by the plant images on the left of the faun and the right of the monkey – they show *Iris pumila*, a native of the Pollau Mountains, and the yellow foxglove, *Digitalis lutea*, both copied from Jacquin's publications. The view goes towards a second garden entrance in baroque style decorated with flower garlands, with the goddess Flora in a medallion, and cut hedges and a pavilion in the background. On the left and right there are two more medallions showing a head with the barely readable letters 'BAU' added, almost certainly an abbreviation of the name Bauer, standing for Franz and Ferdinand. The two medallions may stand for the future fame the two brothers may have dreamed of (see Chapter 16). The two portals may be interpreted as double triumphal arches and give a precise impression of the typical formal garden of the late baroque era, as known to the Bauer brothers from Feldsberg, Eisgrub and elsewhere.

With a single exception all other garden views are unsigned, undated, portray scenes of the Feldsberg estate and are of special relevance for garden historians, since the period around 1780 is poorly recorded for this area and no other garden views are known to exist. The alley leading from Feldsberg to Eisgrub with the convent of the Brothers of Mercy is documented in the *Codex Liechtenstein*, and so is another garden view. The latter is dated c.1776 and depicts a baroque parterre including fountains, monumental vases and two *corbeilles en fleures*, flat flower baskets fashionable at that time, with a plain Chinese pavilion at the end of an alley in the background and a garden pavilion to the right. The existence of these garden elements is confirmed by the anonymous traveller who had visited the castle of Feldsberg in 1782 and mentioned a parterre, four vases, two flower baskets, a fountain, a green, latticed garden pavilion, and in the forest a Chinese pleasure house.

However, all these garden views belong to an art form called capriccio (see also Chapter 4 for the bodycolour that includes a vista of the ruined temple of Castor and Pollux in Rome); they are definitely not realistic representations and, like the garden view with the double portal, belong to the earliest phase in the work of the Bauer brothers. The capriccio with the Chinese pavilion is also best interpreted as such an amalgam – the flat ground seems to refer to Eisgrub, but all other garden structures to Feldsberg, including the pavilion, which has recently been identified as the Chinese gloriette erected in the latter garden under Joseph Wenzel and long demolished. The same applies to the capriccio, including the view of the convent, where the foreground with the balustrade topped by monumental vases does not agree with the building in the background. This view also includes a *corbeille en fleures* with plants no gardener would ever consider for this purpose, such as Jacobean lily, exactly what is expected for this type of art. Within these limitations the few capricci of the *Codex Liechtenstein* offer an impression of the Eisgrub and Feldsberg gardens before Aloys I Joseph commissioned his various improvements in the second part of his reign. These were certainly done on the basis of scarcely paid labour, which is reported to have led in 1786, three years before the outbreak of the French Revolution, to civil unrest in Feldsberg. It was not much later that all these

FIG. 2.16. Title page. Watercolour by Franz Bauer, 1779. Watercolour, ink. *Codex Liechtenstein*, volume 4: f. iv. – Vienna, Fürstliche Sammlungen.

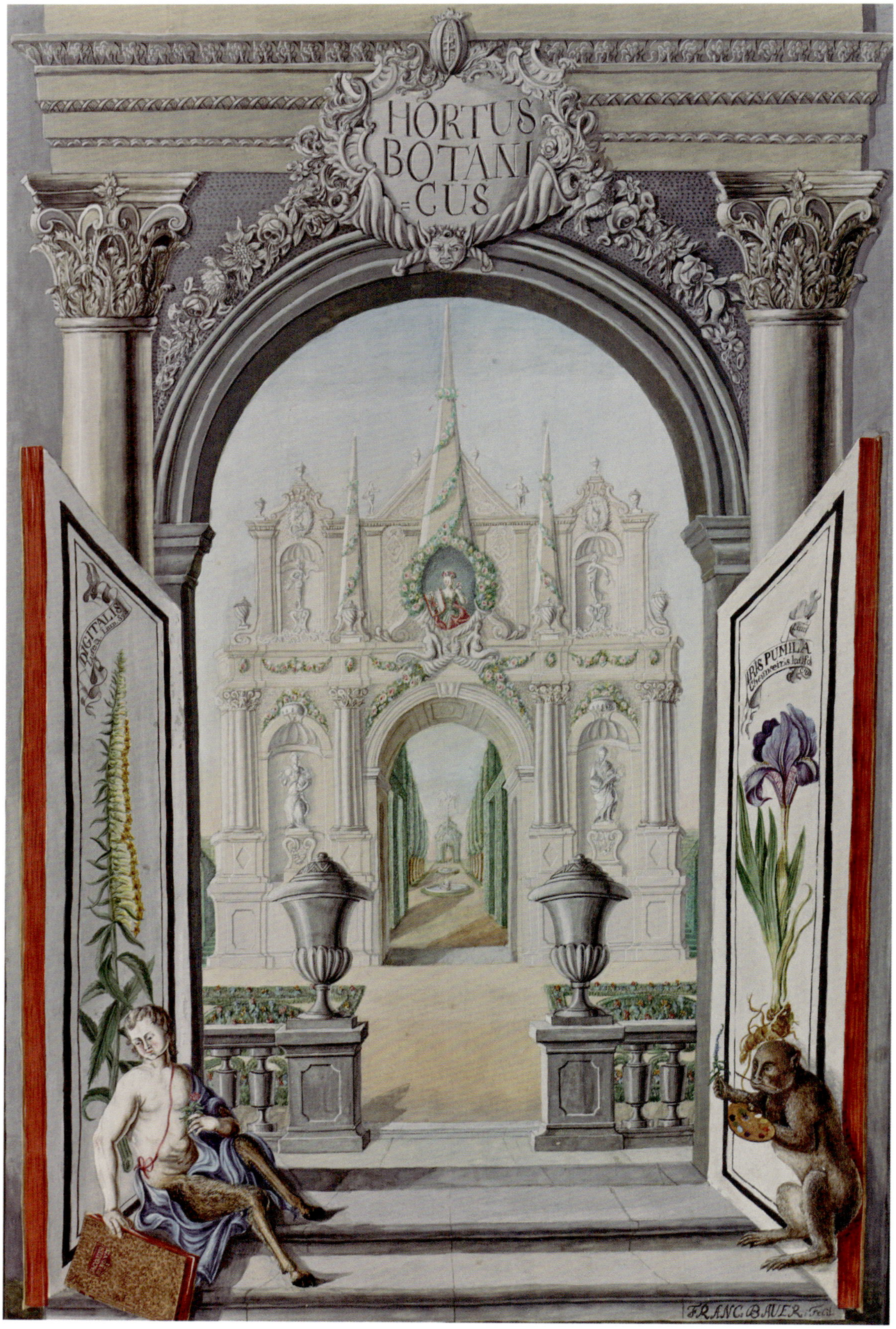

FIG. 2.17. Garden vista with faun and monkey. Watercolour by Franz Bauer, c.1775. Watercolour, ink. *Codex Liechtenstein*, volume 1: f. iii. – Vienna, Fürstliche Sammlungen.

FIG. 2.18. (*opposite*) The town of Feldsberg (Valtice). Watercolour by Ferdinand Bauer, c.1783. Watercolour, ink. *Codex Liechtenstein*, volume 6: f. iii. – Vienna, Fürstliche Sammlungen.

garden improvements were fundamentally altered or deleted by his younger and politically extremely active brother Johann I Joseph, the 10th Reigning Prince of Liechtenstein, who preferred a different garden style in his possessions.

Two landscapes, both signed by Ferdinand, are included in the *Codex Liechtenstein*. One of them shows his home town (Fig. 2.18.), the second the view from Skalky to the Pollau Mountains with Nikolsburg and the Holy Mountain (Svatý Kopeček). They stand in sharp contrast to the capricci – very precise and full of detail, the first with a Roman ruin in the foreground, adorned with two nosegays. In the first landscape (Fig. 2.18.) from left to right the castle, the parish church, the convent of the Brothers of Mercy and the alley leading from Feldsberg to Lundenburg, another Liechtenstein

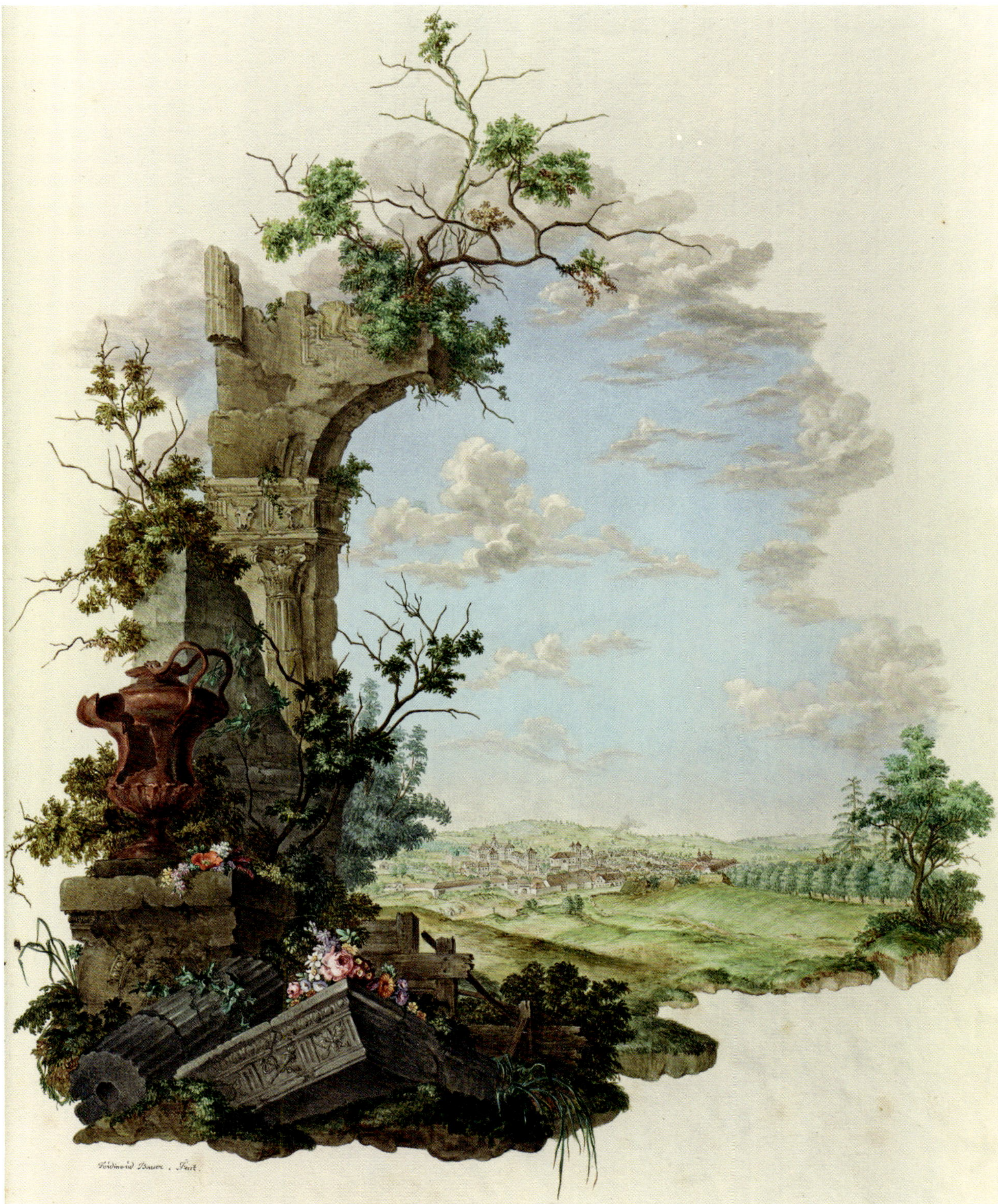

estate, then planted with linden trees,[33] are clearly visible. Both landscapes are masterpieces, documenting the region where the formative years of the Bauer brothers had taken place.

The view from Skalky belonging to volume 5 is notable for another reason – the lower border contains a short garland consisting of single fruits of aubergine and Chinese lantern, *Physalis alkekengi*, and several fruits of black nightshade, *Solanum nigrum*, thereby referring to the nightshade family, which is represented in this very volume of the *Codex Liechtenstein*. The idea of combining a landscape with naturalistic representations of several plants illustrated in the particular volume was a new one and a model for the well-known frontispieces Ferdinand would prepare several years later in Oxford for John Sibthorp (see Chapter 6). When these two landscapes were painted Ferdinand could not have imagined that he would soon abandon these familiar hills on the border of the Archduchy of Austria for more than a quarter of a century.

THE BAUER BROTHERS BREAK OFF

Work for the *Liber regni vegetabilis* was clearly a never-ending story, again and again Boccius must have encountered new plant material that had not yet been documented. So the project went on, effectively until 1805. It may be supposed that the Bauer brothers wished for a change and looked out for opportunities that offered themselves; however, we have no documents to support this hypothesis. Clearly all three intended to continue along the same line and may have searched for other commissions or for a chance for additional qualifications, such as those offered at the United Imperial Royal Academy of Arts in Vienna. Commissions came from Jacquin (see Chapter 3) and as mentioned by Franz in a conversation with Countess Itzenplitz (see above) from Maria Christina Princess Dietrichstein-Proskau zu Nikolsburg, née Countess Thun und Hohenstein. She was the wife of Karl Johann Baptist, the 7th Prince Dietrichstein-Proskau zu Nikolsburg, formerly the imperial ambassador to the King of Denmark, and the mother of two of his successors, the younger of them Moritz I Joseph von Dietrichstein-Proskau-Leslie, the 10th and last Prince of Dietrichstein. According to Franz he had been the princess's court painter during the last six years of her life. Since Marie Christina, an enlighted princess and supporter of the Nikolsburg born political reformer and writer Joseph Baron Sonnenfels,[34] died on 4 March 1788, Franz's engagement must have started in 1782; this is plausible, especially as he left Vienna in early May 1788 (see Chapter 7). However, not a single work commissioned by her from Franz, among them possibly flower-pieces, can be traced, although they may still exist in private hands. In addition nothing is known of where Franz worked for Marie Christina – in Nikolsburg as indicated in the chronicle of Feldsberg referred to above, in Vienna or in one of the several country houses owned by the Dietrichsteins, among them Seelowitz (Židlochovice) in Moravia, just south of Brno.

The first of the Bauer brothers to leave Feldsberg was Joseph, probably in 1779 or the following year, when he was about twenty-three and heading for Vienna (see Chapter 3), then Rome in 1781 (see Chapter 4). Consequently his share in the *Codex Liechtenstein* must have been significantly smaller than that of his two younger brothers. Much less is known about the moment when Ferdinand and Franz followed him to the imperial city, where they were commissioned by Jacquin to contribute to his projects (see Chapter 3). However, it is quite possible that they continued to work at least temporarily for Boccius in Feldsberg. This period came to a definite end when Ferdinand and Franz took up other commissions in March 1786 and April 1788, respectively, leaving Vienna altogether (see Chapters 5 and 7).

What remained in Feldsberg were loose plant illustrations by the Bauer brothers on single-folded sheets to which their successor Jacob Walter added. This material was subsequently bound together by Boccius into the later volumes of the *Codex Liechtenstein*, a magnificent document of strictly focused dedication, thereby creating an amalgam of images that are difficult to disentangle. And Joseph, Franz and Ferdinand left behind in Feldsberg their elder brother Johann, who may have already been recruited to the administration of the princely estates and was to outlive Ferdinand, Joseph and their mother Therese. She had been in the lucky position of following closely the training and artistic development of three of her sons. Therese possibly even had the chance to see with her own eyes the results of the immense labours that they had undertaken for Boccius.

When Ferdinand and Franz finally left Feldsberg, in 1786 and 1788 at the latest, they were twenty-six and thirty years old. Thanks to Boccius they had qualified themselves as highly competent plant illustrators familiar with all technical aspects of the job, excluding engraving, and had become familiar with a rich selection of vascular plant families from all continents, Australia excepted. They were later to observe and record plants in the wild that they had already depicted for Boccius, among them e.g. white henbane, *Hyoscyamus albus*, which Ferdinand saw only one year later on the island of Rhodes, as well as several other species. In short, Franz and Ferdinand were perfectly prepared for further botanical projects, probably knew how to prepare herbarium specimens and had learned what dedicated, concentrated well-organized work can achieve – about 1600 plant illustrations of outstanding quality and precision, an absolutely staggering achievement for three young men. This early imprint on their minds led to their willingness to devote themselves totally to the recording of biological subjects with utmost precision. The immense later output for which Franz and Ferdinand are rightly famous, is similar to that of another young man they probably met in the home of their next employer – Mozart.

No doubt Franz and Ferdinand possessed brilliant talents in their field like Mozart in music, but there was also this special blend of diligence, consistency and perseverance on the one side and the mixture of maternal encouragement, support by Boccius and carefully selected models of top quality to follow, such as the engravings based on Ehret, on the other. These aspects were the prerequisites to continue to produce an extremely rich, diverse oeuvre of almost photographic quality. Feldsberg also left an imprint on Franz and Ferdinand in another respect: they had become focused on gardens of which they had experienced fine examples in Eisgrub, Feldsberg, Nikolsburg and elsewhere. In 1790 Franz, almost thirty-two, took his permanent residence in the immediate vicinity of the Royal Garden at Kew near London, and Ferdinand after his return to Vienna many years later settled down in a house very near to the Dutch Botanic Garden in Schönbrunn (see Chapter 12). After two strenuous voyages he was to live again, as in his youth, near conservatories, gardeners and exotic plants.

The Vienna period of Joseph, Franz and Ferdinand that was now to begin promised new challenges and, more importantly, new contacts. However, it turned out to be only a short interval in the long and eventful lives of the three brothers.

ENDNOTES

1 Unless otherwise stated this chapter is based on Lack (1999c, 2000a, 2008b).
2 Through this text the standardized numbering of the Reigning princes of Liechtenstein is followed.
3 Lorenz (1990)
4 Rogasch (2001)
5 S. E. v. R. (1782)
6 Inv. No. NG 6270
7 Kippes (2000)
8 Körner (2005)
9 Watzka (2008)
10 Baumstark (1990), Press (1990)
11 Haupt (1990)
12 Inv. No. 4245
13 Inv. No. Sk 2
14 Kurdiovsky (2003)
15 Wolný (1858)
16 Becker (1887)
17 Potsdam, Landeshauptarchiv, Rep 37, (Alt) Friedland, No. 382. – H. C. Itzenplitz to H. C. von Friedland, Kew, 4 iv 1793, unpublished letter.
18 Brno, Moravský zemský archiv, Matriky Valtice, Sign. 16/3530, f. 110.
19 Wittmack (1924)
20 Wien, Stadt- und Landesarchiv, Mag. Zivilgericht, Abh., Fasz. 2 7425/41, Verlassenschaft Joseph Bauer
21 London, Linnean Society, Archive. – Franz Bauer to John Lhotsky, 17 iii 1839, unpublished letter
22 Wolf (1875)
23 Firla (2004)
24 Anon., Deutsche und deutschsprechende Persönlichkeiten aus dem Banater Bergland. – www.dfbb.ro/content/pdftext/BB_Persoenlichkeiten.pdf (accessed 28 xii 2014)
25 Anon. (1799)
26 Jacquin (1792)
27 Bleichmar (2011)
28 Inv. No. GR 2622
29 Daston & Galison (2007)
30 Lack & Ibáñez (1997)
31 Mabberley & San Pío Aldarén (2011)
32 Nickelsen (2006)
33 Zatloukal, Krejčiřík & Zatloukal (2012)
34 Brichtová (2002)

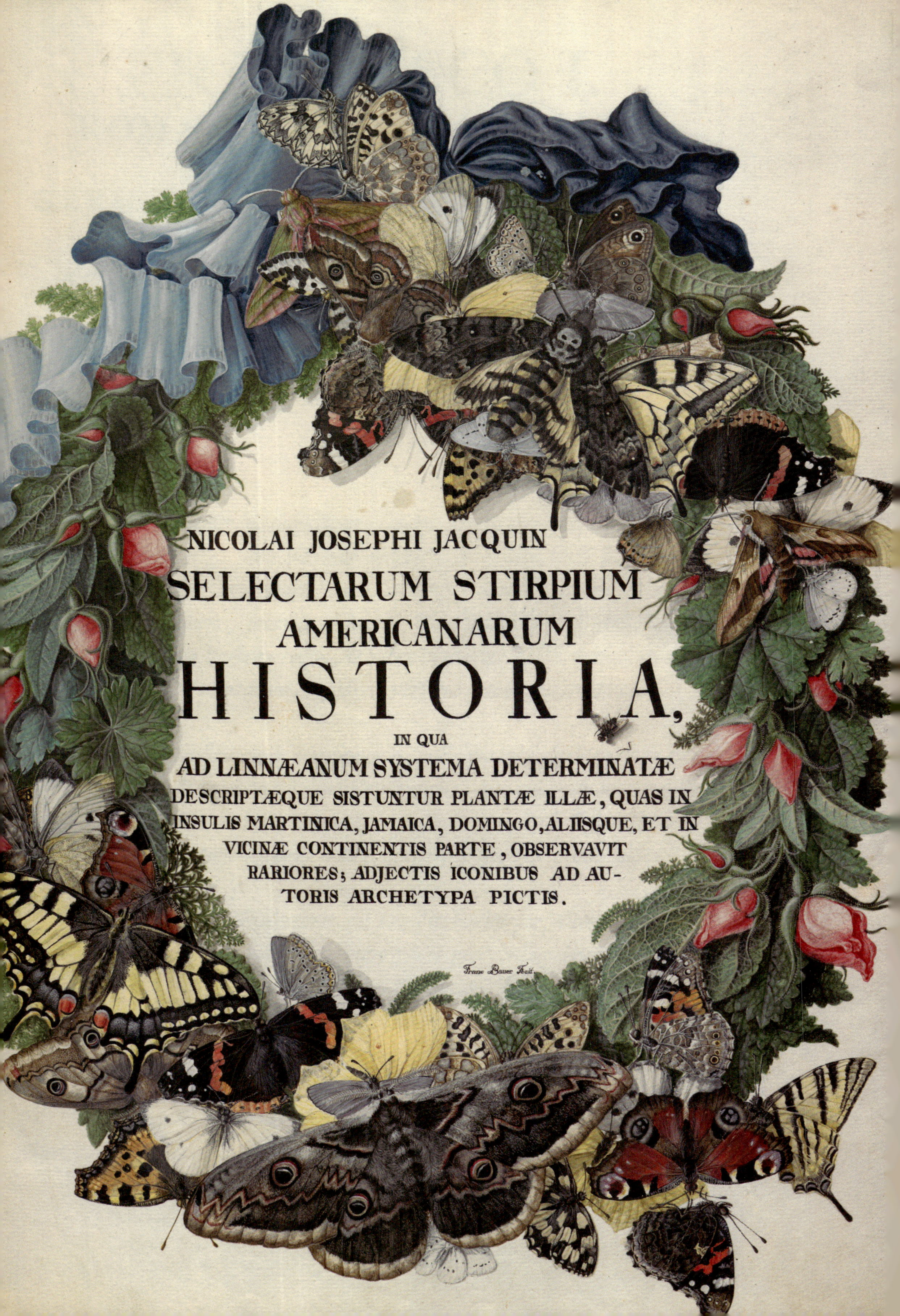

NICOLAI JOSEPHI JACQUIN

SELECTARUM STIRPIUM AMERICANARUM HISTORIA,

IN QUA

AD LINNÆANUM SYSTEMA DETERMINATÆ DESCRIPTÆQUE SISTUNTUR PLANTÆ ILLÆ, QUAS IN INSULIS MARTINICA, JAMAICA, DOMINGO, ALIISQUE, ET IN VICINÆ CONTINENTIS PARTE, OBSERVAVIT RARIORES; ADJECTIS ICONIBUS AD AUTORIS ARCHETYPA PICTIS.

Franc Bauer fecit.

CHAPTER 3

The interval in Vienna[1]

THE IMPERIAL CITY

If it were possible today for the Bauer brothers to stroll through the very centre of Vienna, now the innermost part of the city's first district, they would not need a map. Several buildings and the network of roads they had first seen when visiting Vienna, in around 1780 at the latest, have remained unchanged or only slightly altered.

The Bauer brothers must have been to St Stephen's Cathedral, a major structure of the late Gothic age, then like today the main landmark of the city, dominating its skyline. St Stephen's was the seat of the Archbishop of Vienna, then Christoph Anton Cardinal Migazzi Count Waal and Sonnenthurm, a native of Trient (Trento), then in the Tyrol, formerly imperial ambassador in Madrid and bishop of Waitzen (Vác) in Hungary. In addition, the cathedral served also as the burial site of a long list of key figures of the state, among them the great Eugene Prince of Savoy (see Chapter 2). The Bauer brothers may have visited his grave and epitaph, which had strong associations because of the family owning the Feldsberg estate: it had been erected by Maria Theresia, Duchess of Savoy – Carignan, née Princess of Liechtenstein – for her uncle. She had died childless, with her large properties passing to Franz Joseph I, then the Reigning Prince of Liechtenstein. The cathedral was also the burial site for the intestines of the Habsburgs, a tradition the family maintained until 1878. When Franz died some sixty years later in Kew, his estate was auctioned in November 1841 (see Chapter 17) and included a contemporary printed account of this famous church. Furthermore he had owned no less than 31 engravings showing views of Vienna, which seems to have made a lasting impression on him. This may well also apply to Joseph and Ferdinand: Joseph was to live the greater part of his life in Vienna, and Ferdinand lived there for his last 11 years (see Chapter 12).

It seems plausible to assume that having been Catholics the Bauer brothers also visited some of the other churches, such as St Augustine, the church where several daughters of Maria Theresa and Emperor Francis I Stephen had been married off by proxy like figures on the political chessboard. Archduchess Maria Amalia was sent to Parma, Archduchess Maria Carolina to Naples and Archduchess Maria Antonia (Marie Antoinette) to Paris, where her life ended under the guillotine. However, St Augustine was also the church where the hearts of a long list of Habsburgs are buried and Swieten (see Chapter 2) had found his final resting place – on the explicit orders of Maria Theresa, his most prominent patient. Adjacent to St Augustine and linked to it by corridors was the Hofburg, the residence of the imperial family and nerve centre of Habsburg politics. Over centuries this building, which effectively made Vienna the capital of the Hereditary Lands, had been enlarged to become one of the most extensive city residences in Europe.

FIG. 3.1. Titlepage by Franz Bauer. Jacquin, N. J., Selectarum stirpium americanarum historia, ed. 2, Vienna, c. 1780. – Vienna, Österreichische Nationalbibliothek.

A new wing containing the Imperial Library had been added to the Hofburg according to a plan by an architect the Bauer brothers must have known from the 'Palace of the Horses' in Eisgrub – Johann Bernhard Fischer von Erlach – and completed by his son Joseph Emanuel (see Chapter 2). For good reason this building is universally regarded as the apotheosis of a baroque festive hall, with a pompous statue of Emperor Charles VI as Roman Emperor at its centre, then as now used for solemn acts of state. At the time of the arrival of the Bauer brothers in Vienna, Swieten the son was prefect of the Imperial Library, which contained the two *Receuils* from Paris that Boccius, and probably also Franz and Ferdinand (see Chapter 2), seem to have known.

In the 1760s an estimated 52,000 inhabitants[2] lived within the fortress walls of Vienna, and the population certainly had increased by c.1780. Accommodation pressures had long forced city builders to think vertically, which resulted in an average of c.40 people per dwelling, residing above shops, taverns and stables. Many others lived in the suburbs, separated by the city walls and the so-called glacis, a large open space intersected by roads leading in and out of the city centre. The anthill structure of Vienna and its cosmopolitan nature is admirably described by an eyewitness in 1786, the year Ferdinand left the city: 'Here you can often meet the Hungarian, striding stiffly, with his fur-lined dolman, his close-fitting trouser reaching almost to his ankles, and his long pigtail; or the round-headed Pole with his monkish haircut and flowing sleeves . . . Armenians, Wallachians and Moldavians, with their half-Oriental costumes are not uncommon. The Serbians with their twisted moustaches . . . Greeks smoking their long-stemmed pipes in the coffee-houses . . . bearded Muslims in yellow mules, with their broad, murderous knives in their belts. The Polish Jews, all swathed in black, their faces bearded and their hair all twisted in knots . . . Bohemian peasants with their long boots; Hungarian and Transylvanian waggoners with sheepskin greatcoats; Croats with black tubs balanced on their heads – all provide entertaining accents in the general throng'.[3] Only a few of these figures were foreigners, whereas most of them were living within the extensive Hereditary Lands of the Habsburgs, the Bohemians together with the Hungarians, Poles, Croats and Serbs, the Muslim population in the Bukowina, with the Wallachians and Moldavians their near neighbours.

No evidence exists of where exactly the Bauer brothers stayed upon their arrival in Vienna, but it is most unlikely that as nascent artisans they could afford a room in the city centre, where the upper classes had their homes. They were more likely to have had a bed in somewhere like the convent of the Brothers of Mercy, which comprised, as in Feldsberg, a church, a hospital (Vienna's oldest), a pharmacy and a medicinal garden. Such a place might be a plausible guess for their first accommodation, possibly organized by Boccius in a letter to the prior in Vienna. Although outside the fortifications, the convent was situated within walking distance from the city centre and may have provided the Bauer brothers with an atmosphere familiar to them. Housing conditions were difficult in those days, with Haydn, later Mozart's mentor, reporting two decades earlier that he was staying in 'a poor garret . . . with no heating, with . . . hardly any protection from the rain'.[4]

Towards the end of the feudal period Viennese society was extremely stratified, with enormous differences in wealth and influence. Clearly the imperial family, in 1780 comprising the widowed Maria Theresa and her eldest widowed son Emperor Joseph II, co-regent in the Hereditary Lands, stood at the top of the social pyramid. They were followed by the aristocracy, coming from all corners of the Hereditary Lands to Vienna, among them the Liechtenstein, Dietrichstein and Schwarzenberg families from Moravia and Bohemia, as well as the Pálffy and Esterházy families among those from the Kingdom of Hungary. Noble families from other parts of the Holy Roman Empire also resided in Vienna, like the Fürstenberg or Fries families, based in Donaueschingen and Mülhausen (Mulhouse), respectively. Others, such as the Collalto family – in whose city residence Mozart gave his first concert in Vienna aged six – came from the Republic of Venice. They all tried to outrival each other, spending huge sums of money on an extravagant lifestyle and on their fine city residences and exquisite gardens, with those belonging to the Reigning Prince of Liechtenstein being particularly luxurious (see Chapter 2).

We might hypothesize that the Bauer brothers gazed at the magnificent palace of the Liechtensteins in Herrengasse, now no longer in existence, as well as at the no-less magnificent palace of this family in Schenkenstraße, now Bankgasse. The latter had been built on the basis of a plan by Martinelli, with two pompous entrances crowned by over-sized escutcheons of the family with representations of the Golden Fleece, both completely covered in gold. Their father Lucas Bauer must have already known these city residences and the fine art collections that they then contained (see Chapter 2). We might also suspect that his sons were aware of the new catalogue describing the gallery, in 1780 hot from the press, the anonymous *Description des tableaux et des pieces de sculpture que renferme la gallerie de son altesse François Joseph chef et prince de Liechtenstein* comprising no less than 279 pages. It superseded Vincenzo Fanti's *Descrizzione completa di tutto ciò che ritrovasi nella galleria di pittura e sculttura di sua altezza Giuseppe Wenceslao*, printed in 1767 in Vienna. Both catalogues would later become important reference works for Joseph (see Chapter 12).

On a square then called Spitalsplatz the Bauer brothers might also have gazed at the splendid palace of the Lobkowicz family, previously owned by the Dietrichsteins. It had been constructed according to plans by two architects they also knew from Feldsberg and Eisgrub, Tencalla and Johann Bernhard Fischer von Erlach, containing some paintings by Gaetano Fanti. Among the more recently completed buildings the Bauer brothers could have seen the New Aula, the seat of Vienna University, with its astronomical observatory, erected in 1753–1755 on the basis of a proposal by Swieten. Bernardo Bellotto, named Canaletto, had documented this monumental structure, built on orders from Maria Theresa and opened by her, in a fine painting now in the Kunsthistorisches Museum in Vienna. In the same way he had been commissioned by Joseph Wenzel to paint the summer residence of the Liechtenstein family in the Rossau quarter in two views, both now in the Fürstliche Sammlungen in Vienna. Two wings had been added to the Imperial Library erected on the basis of plans by Nicolaus Pacassi, forming a fine square that Joseph, Franz and Ferdinand must also have seen, later adorned with the equestrian statue of Emperor Joseph II (see Chapter 4 and 12). Opposite the Imperial Library construction works were under way when Joseph left for Rome – on the site of a former monastery the banker Johann Count Fries had an impressive city residence built for his family. It was the first neoclassical palace in Vienna, now called Pallavicini Palace after the family who owns it today, with Johann Ferdinand Hetzendorf von Hohenberg (see Chapter 4) responsible for the planning. During the winter of 1785–1786 Ferdinand's later employer John Sibthorp spent a few months there as guest of the Fries family (see Chapter 5).

If the Bauer brothers strolled today through the districts surrounding the centre of Vienna, they would, however, need a map. With this guide they would be able to find a few of the summer residences of the aristocracy that they must have seen at their time, for example that of the Liechtenstein family, which was many years later converted into the family's gallery by Joseph (see Chapter 12). Other monumental summer residences still extant include that belonging to the Schwarzenberg family, and the Upper and Lower Belvedere built for Eugene Prince of Savoy. At the time of the arrival of the Bauer brothers, the latter had already been purchased by Maria Theresa, who had given a great farewell party in its extensive gardens for her daughter Maria Antoinette, when she was about to leave for France. This is regarded by many as the last baroque feast of the *ancien régime* in Vienna. The panoramic view on the city from the Upper Belvedere has also been documented by Canaletto in a painting commissioned by Maria Theresa, now in the Kunsthistorisches Museum. In the year that Joseph left for Rome, the former summer residence of Eugene Prince of Savoy would become the receptacle of the collection of paintings belonging to the Habsburgs.

Other summer residences, like that of the Harrach family, are difficult to locate today, having been largely demolished. Several more have totally disappeared, like that which had belonged to Anton Wenzel Prince Kaunitz-Rietberg, Prince of the Holy Roman Empire (see Chapter 4), documented again in utmost detail in a painting by Canaletto, now in the Szépművészeti Múzeum in Budapest. By contrast the Bauer brothers would be surprised to find a largely unchanged garden adjacent to the Belvedere, hidden between tall walls and today, as then, strictly private. This is the garden belonging to the monastery of the Sisters of the Visitation of Mary (see Chapter 2) on a road now called Rennweg, once the main connection to Hungary. Another garden very near to the Belvedere and later substantially enlarged was soon to become the working place for Franz and Ferdinand, the Botanical Garden of Vienna University. It had also been founded during the reign of Maria Theresa on a piece of land in the 'Hungarian Quarter', i.e. part of the present-day third district, purchased by her on the basis of a proposal from Swieten.

Our story continues with the arrival of Joseph in Vienna in c.1780, when he was around 24 years old. At that time the life of Maria Theresa was approaching its end. She had successfully opposed a coalition of several European powers, had lost only a portion of a single province to her arch enemy, Frederick II, King in Prussia, whom she called 'monstre' (monster). However, she had added the extensive Kingdom of Galicia to the Hereditary Lands and managed to stabilize Habsburg power in Central Europe. Under her reign the agglomeration of territories she had inherited from her father started to become a state, where many languages were spoken and where the inhabitants adhered to several creeds. She conversed in French, Italian and German and was for good reason called 'Landesmutter' (mother of the country), having given birth to sixteen children and buried six of them. Although already in bad health in early 1780 she continued to exert herself, responding to correspondence and working on state papers with her usual efficiency – despite being short of breath, especially when walking, even on a flat surface, and suffering greatly from oedema. Maria Theresa died on 29 November 1780 of congestive heart failure, the responsibility for her terminal care resting on the shoulders of the successor to Swieten, her Protomedicus Störck, who had published the drawing of *Pulsatilla pratensis* subsp. *nigricans* prepared by Franz nine years earlier (see Chapter 2).

Stormy years followed. After the death of his more conservative mother, who had adhered to the concepts of baroque Catholicism, Emperor Joseph II decided to start very many, often long overdue, reforms in the Hereditary Lands. His impatient personality and his early death rendered a good deal of these measures unsuccessful. Nevertheless, and following the principles of enlightened rule, he replaced the traditional forced serf labour system by a system of rent-paying tenants, reduced the power of state censorship, removed restrictions against the practice of Protestant, Orthodox and Jewish religion in the Edict of Tolerance and demolished many legal barriers against Jews. At the same time he initiated a series of drastic steps against the power of the Catholic Church, closing numerous monasteries, mainly of contemplative orders, disrupting established religious traditions, reshaping the configuration of dioceses as well as parishes and regarding the clergy more or less as his civil servants. However, the convents of the Brothers of Mercy were exempted from these draconian measures, and Boccius was confirmed by Joseph II as prior and as Father Provincial.

Domestic resistance from several quarters forced Joseph II to rescind many of these reforms towards the end of his short reign, a policy that resulted in disappointment in parts of society. However, Joseph II laid the foundation for many promising institutions that were to become longstanding, among them the famous 'Allgemeines Krankenhaus' (General Hospital) in Vienna opened in August 1784. In addition he ordered the first complete mapping of his territories, which ranged from present-day Belgium to the western parts of present-day Ukraine. Most importantly, he and his enlightened reforms may have been the main reason why the ideals of the French Revolution never found substantial support in the Hereditary Lands.

Lucas Bauer (see Chapter 2) is reported to have frequented the K. K. Hofakademie der Maler, Bildhauer und Baukunst (Imperial Royal Court Academy of Painters, Sculptures and Architecture) in Vienna from 1728 until 1733.[5] In a sense it was only logical that his artistic sons Joseph, Franz and Ferdinand would follow this tradition. Therese, Boccius or Jacquin may have made the suggestion for them to follow in their father's footsteps. Indeed there is evidence, albeit extremely fragmentary, that they received tuition at this institution, in those years based in the New Aula of Vienna University. Kaunitz was the curator of the United Imperial Royal Academy of Arts, while Sonnenfels, later the rector (Vice Chancellor) of Vienna University, acted as permanent secretary, and Schiffermüller was among the honorary members.

For Joseph only a leaving certificate dated 13 June 1781 has been preserved in the archive of the Academy reading 'the present Joseph Bauer . . . has himself visited the school there [i.e. in Vienna] and especially our own, with great attentiveness '. This note was written when Joseph was about to leave for Rome (see Chapter 4). The names of Franz and Ferdinand are also found among the student lists of the Academy, with entry dates 1781 for Franz, when he was 23 years old, and 1784 for Ferdinand, when he was 24 years old. Very few details are available: Franz attended the copper engraving school; Ferdinand won the Gundel-Prize in 1785, which had been first awarded in 1781 and came with a medal – not a bad recognition for a man of 25 years. However, not a single work of the three brothers can be associated with their stay at the Academy.

Knowing the art of copper engraving had little relevance in Franz's later life, although it probably enabled him to better control the transfer of his drawings onto copper plates. We can hypothesize that Franz might have been in contact with Johann Christian Brand, then professor of landscape drawing at the Academy, who is known to have also painted flower pieces; for Ferdinand this is stated as fact in his obituary (see Chapter 20). The two brothers had already produced a few such flower pieces for Boccius serving as title pages (see Chapter 2) and more were to follow in Vienna. Several years later Franz was to paint a few flower pieces (see Chapter 8), and likewise Ferdinand painted patterns for seven title pages to be used for Sibthorp's *Flora Graeca* (see Chapter 6) and later, in London, a few flower pieces in oil (see Chapter 11). Sketches of many landscapes were to be prepared by Ferdinand during the journey in the Ottoman Empire (see Chapter 5). Some of these were later integrated into the patterns of the title pages of the *Flora Graeca*, and many more were worked up into finished *vedute* for John Hawkins (see Chapter 9). Although so very little is on record, the tuition that Franz and Ferdinand received at the Academy seems to have encouraged them to continue on their line of strict specialization, then and now *de rigeur* at academies of art.

Nothing is known about any personal introductions the Bauer brothers may have brought with them from Feldsberg. In theory Boccius could have recommended them to Störck, the principal head-physician at the imperial court and thus a key figure in Vienna society, but this is not likely. The apothecary Well (see Chapter 2) would have been another possibility. It seems more plausible to assume that Franz and Ferdinand were introduced by Boccius to Jacquin (see Chapter 2), a bewigged courtier like Störck, although not that high up in the hierarchy. In 1774 he had been ennobled Edler von Jacquin and it was Maria Theresa in person who arranged for the marriage of his younger sister Agathe to Jan Ingen Housz, another physician-in-ordinary at the imperial court. The latter had successfully inoculated three of Maria Theresa's children against smallpox, after she had lost three others by this highly contagious disease. Herself disfigured by the contagion Maria Theresa also saw her daughter Maria Elisabeth severely affected by the 'speckled moster'; Maria Christina Princess Dietrichstein-Proskau (see Chapter 2) had been one of the latter's ladies-in-waiting.[6] Jacquin, no doubt a brilliant figure at court and in academia, was to exert a decisive influence on Ferdinand and Franz as their second patron, but more importantly by bringing them – directly in the case of Ferdinand, indirectly in the case of Franz – in contact with their future employers. At the same time Jacquin was also the first genuine scientist the Bauer brothers ever met – long before their first contact he had started publishing the results of his many findings in botany and chemistry, something their first mentor Boccius had refrained from doing.

NIKOLAUS JOSEPH EDLER VON JACQUIN, THE AUSTRIAN LINNAEUS

Swieten, his son Gottfried and Gottfried's schoolmate Nikolaus Joseph Jacquin, a descendant of a French family, had all been born in Leiden. The boys had attended the gymnasium in Antwerp, then belonging to the Austrian Netherlands and ruled from Brussels. All three were Catholics; later they gravitated towards Vienna because the religious situation in the Republic of the Seven United Netherlands was unfavourable for Swieten's career. Swieten and his family came to Vienna by invitation from Maria Theresa in 1745, Jacquin by invitation from Swieten in 1752. Earlier Jacquin had started to study philosophy in Leuwen, another town in the Austrian Netherlands, then continued with medicine in Leiden and Paris, where he was strongly attracted by the Jardin du Roi (see Chapter 2). The two Swietens and Jacquin were joined in 1768 by Ingen Housz, another Dutchman, born in Breda, then also belonging to the Austrian Netherlands, who arrived in Vienna by invitation from Maria Theresa and who made his career as one of Swieten's protégés.

In 1752–1753 Emperor Francis I Stephen had started to make arrangements for the layout of a menagerie and a botanical garden including conservatories near the summer residence of the Habsburgs in Schönbrunn, then a village far outside Vienna. All that were needed were tropical plants and animals. While at Schönbrunn on one of his numerous visits identifying plants, Jacquin is reported to have met the Emperor, who, most likely upon the recommendation of Swieten, asked him to conduct an expedition: to bring exotic plants and animals for the conservatories and the menagerie as well as natural history objects for the imperial cabinet. This was a challenge no one could resist, and definitely not Jacquin. The Emperor allocated no less than 30,000 guilders for the expedition, a huge sum of money, about half the price which he had paid for his personal city palace in Vienna, the Toskanische Kanzlei (Tuscan Chancellery), plus a salary of 20 guilders per month for Jacquin. Tropical America was chosen as the most promising region, since it was connected with harbours in Europe via relatively quick and frequent shipping routes. It is hard to believe but substantiated by the archival record kept in the Archivio di Stato in Florence that the detailed instructions for the project were written by Emperor Francis I Stephen in his own hand. He stipulated that 'he [Jacquin] will personally choose these species of flowers that are rare or deserve to be in my garden and will take bulbs or roots or seeds . . . He should select only those that are beautiful or that have a pleasant scent'. Perhaps the Emperor was aware of the sort of unattractive plants that botanists find of interest.

What followed was a voyage to the islands of the Caribbean Sea with a spell of about seven months on the American continent near Cartagena in present-day Columbia. This made Jacquin one of the few botanists of his time to see the flora of the neotropics with his own eyes and not only as miserable herbarium specimens or as poorly developed plants trying to survive in conservatories. Before his departure Jacquin had taken drawing lessons, just as the Bauer brothers did later, at the United Imperial Royal Academy of Arts when well-known painter Paul Troger was rector. This turned out to be extremely helpful when termites had destroyed Jacquin's herbarium specimens on the island of Martinique. Since it was then very difficult to prevent dried specimens from being attacked by insects and mould in a tropical climate, Jacquin and his accompanying gardener and compatriot Richard van der Schot, a native of Delft, later largely refrained from collecting specimens and instead relied on their own drawings. Seven shipments reached Vienna by various routes and brought a considerable number of living plants and animals to the conservatories and the menagerie in Schönbrunn, which made Jacquin famous and welcome in court circles. Minerals, ethnographical objects and some permanently preserved specimens of plants arrived in Vienna in the same shipments.

Upon his return to the imperial city in July 1759, Jacquin embarked on several publications, making his botanical discoveries known to the scientific community. Of these the fully illustrated *Selectarum stirpium americanarum historia* (*History of selected American plants*) published in Vienna in 1763 and dedicated to Emperor Francis I Stephen is remarkable because of the 184 copper engravings based on Jacquin's and van der Schot's field drawings. In the precursor of this work, the *Enumeratio systematica plantarum*, Jacquin had given the mahogany tree its scientific name, *Swietenia mahagoni*, a tribute to his protector, Swieten, and still valid today. The *Selectarum stirpium americanarum historia* and his earlier publications show Jacquin to be a devout follower of the Linnaean system and one of the first to follow the binomial nomenclature introduced by Linnaeus in 1753.

Jacquin's career was not to be stopped. Subsequently he was appointed Professor of Chemistry and Metallurgy at the newly founded Royal Academy of Mining in Schemnitz / Selmecbánya (Banská Štiavnica in Slovakia) and member of the Royal Mining Council for Hungary, with the bonus of receiving two salaries. When the combined Chair of Chemistry and Botany became vacant at Vienna University, his protector Swieten secured him the position, which went with the directorship of the botanic garden and again with two salaries. Jacquin's success rests on a remarkable combination of qualities. On the one hand he was a precise and critical observer, a hard worker, capable of coping with all sorts of challenges, and a man who could fix his discoveries both in word and accurate drawing and publish them. On the other hand, he had what clearly were excellent contacts at the imperial court and with prosperous and influential patrons. The result of all this was that his comprehensive scientific work was able to appear quickly and in editions printed at the highest quality. No other botanist in the eighteenth century came anywhere near his output of large-format volumes, with their total of more than 2,000 coloured copperplate engravings. As put by Frans Antoine Stafleu,[7] 'Jacquin's great achievement, describing and illustrating thousands of species in magnificent folio volumes, was exactly what botany at that time demanded. He provided information on an enormous number of new plants at a time when no survey of flowering plants in all their multiplicity had as yet been undertaken'.

In Jacquin's publications following the *Selectarum stirpium americanarum historia* the copper engravings were no longer based on botanical illustrations by his own hand, but by hired botanical illustrators. The first of these was the prolific Franz Scheidl who had accompanied Jacquin on botanical explorations in the Alps, but had also worked for Well (see Chapter 2) and others producing masses of illustrations true to nature, allegedly of about seven thousand plants, seven thousand shells and numerous animals. These include a collection of shell illustrations recently acquired by the Fürstliche Sammlungen. Jacquin reported that Scheidl had provided him with all illustrations for his *Hortus botanicus vindobonensis* and his *Florae Austriacae icones*, a total of eight hundred botanical illustrations, of which the Bauer brothers had copied a considerable number for the *Codex Liechtenstein* (see Chapter 2). However, after the completion of these two works in 1778, Scheidl is reported by Jacquin to have left for Denmark, and Joseph Hofbauer became his successor. Nothing is known about him apart from Jacquin's note that Hofbauer prepared the first 240 illustrations, which were printed as copper engravings, forming part of the first twelve fascicles of his *Illustrationes plantarum rariorum* (*Illustrations of rare plants*). This was another large-format work comprising no less than six hundred and forty-nine coloured copper engravings published by the productive professor of botany. The scope of this work was the description and illustration of rare plants cultivated in the gardens of Vienna coming from all over the world, many then new to science. For unknown reasons Hofbauer later stopped his involvement with this work.

This was the chance Franz and Ferdinand may have been waiting for. In any case Jacquin hired them, but again we have no details about the time of their engagement, nor about the financial arrangements. It seems likely that they prepared plant illustrations only; we cannot exclude the possibility that they also engraved the copper plates and coloured the prints, but there is no evidence for such an assumption.

Evidence for the activities that were done for Jacquin by the Bauer brothers is meagre. Jacquin himself states that Ferdinand produced several and Franz a few of the illustrations for the *Illustrationes plantarum rariorum*; for Franz this came with the caveat 'other commitments permitting', which may refer to those for Maria Christina Princess Dietrichstein-Proskau (see Chapter 2). It is plausible to assume that the Bauer brothers contributed to the later fascicles of *Illustrationes plantarum rariorum*, although they did not prepare the illustrations for the last one hundred and twenty plates, which are reported by Jacquin to have been based on works by Johannes Scharf. The maximum involvement of the Bauer brothers is therefore estimated to have been around three hundred botanical illustrations; considering Jacquin's statement (see above), however, their number was almost certainly significantly lower than this.

Since the original drawings pertaining to the *Illustrationes plantarum rariorum* are untraceable and the copper engravings do not contain information on the botanical illustrators involved, it is impossible to attribute individual copper engravings to the Bauer brothers with any degree of certainty. However, there are exceptions. For example, because of the representation of accurate bulb details, it is hypothesized that the copper engraving showing *Albuca caudata* (Fig. 3.2.) of the hyacinth family published in this work may have been based on a drawing by the Bauer brothers. The solitary, partially hypogeal bulbs covered by short, papery, brown to grey scales that reach different heights recall the attention given by the Bauer brothers to the bulbs of the Barbados lily, *Hippeastrum puniceum*, in the *Codex Liechtenstein* (see Chapter 2). This very copper engraving exemplifies at the same time the relevance of the Bauer brothers' work for modern plant taxonomy. Jacquin had regarded this bulbous plant from a clearly circumscribed area in South Africa west of Grahamstown as a species new to science and given it a properly published scientific name. However, the material basis of this name is meagre: no pertinent herbarium specimen is extant, and the copper engraving is the only surviving element Jacquin associated with the name *Albuca caudata*. As a consequence this print has been selected as the lectotype of this scientific name,[8] i.e. its permanent standard of reference.

Nothing is known about any drawing aids the Bauer brothers might have used for this kind of work in Vienna. Considering the fact that the camera obscura, mirrors and lenses were standard equipment for artists at the time and that the camera obscura was explicitly mentioned in the new catalogue of the Liechtenstein gallery as having being used by Canaletto in the preparation of his paintings (see above), we may hypothesize that the brothers at least were aware of these aids. However, so far their application is impossible to prove.

Early chronicles indicate a friendly welcome for Ferdinand in Vienna and merit being cited: 'Ferdinand . . . soon made the acquaintance of the famous . . . Jacquin, who cultivated his talent and became like his second father. He now lives in Jacquin's house and under his direction completed a large number of original drawings for his immortal works'; 'In Vienna . . . Jacquin helped him [Ferdinand] fulfil his artistic potential and achieve success as a botanical artist'. This is clearly second-hand information, but plausible. It seems likely that Franz was welcomed in a similar way, although we again have no proof. In any case, the elder of the two brothers was found suitable in April 1788 to accompany Jacquin's elder son on an official journey across central Europe (see Chapter 7).

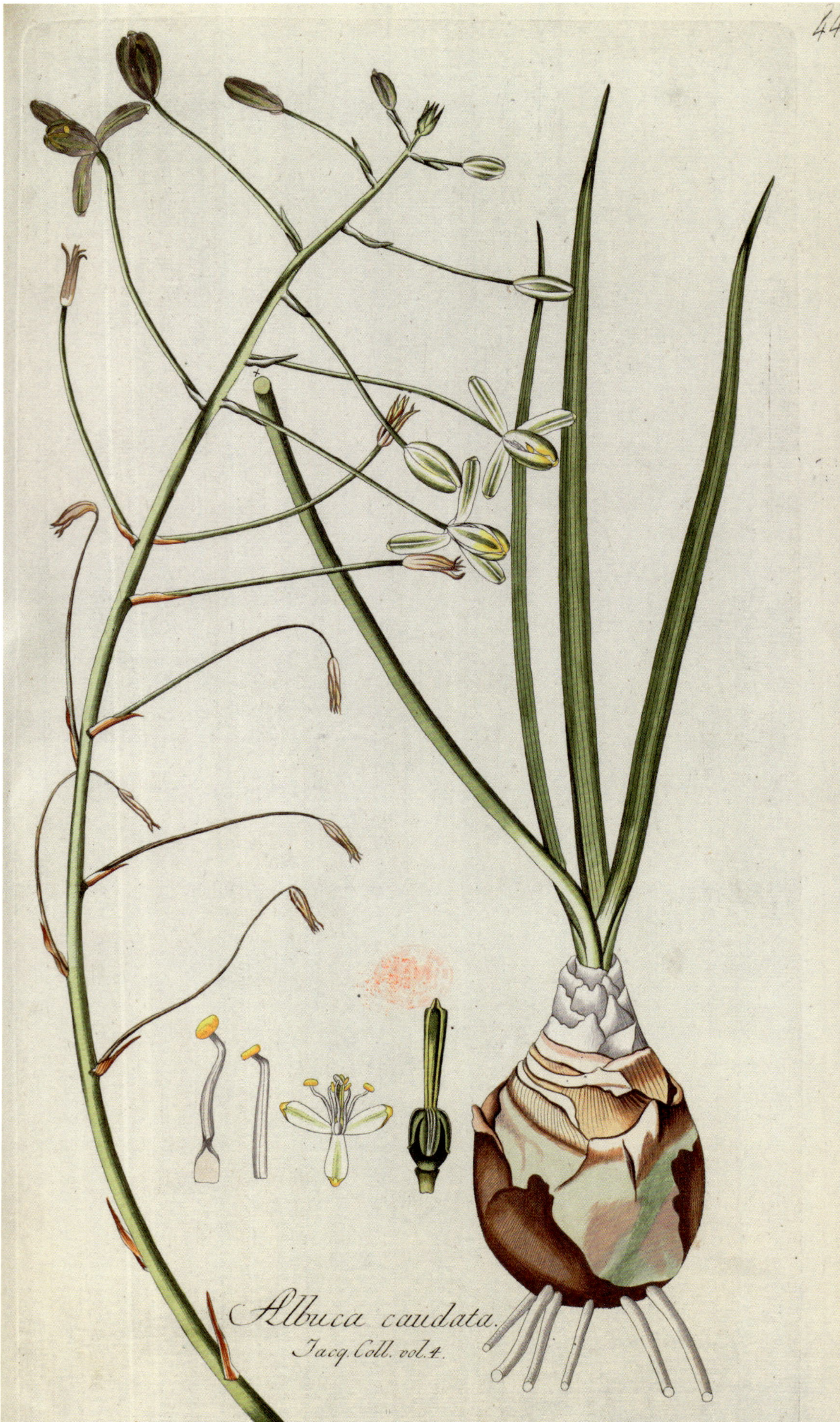

Albuca caudata.
Jacq. Coll. vol. 4.

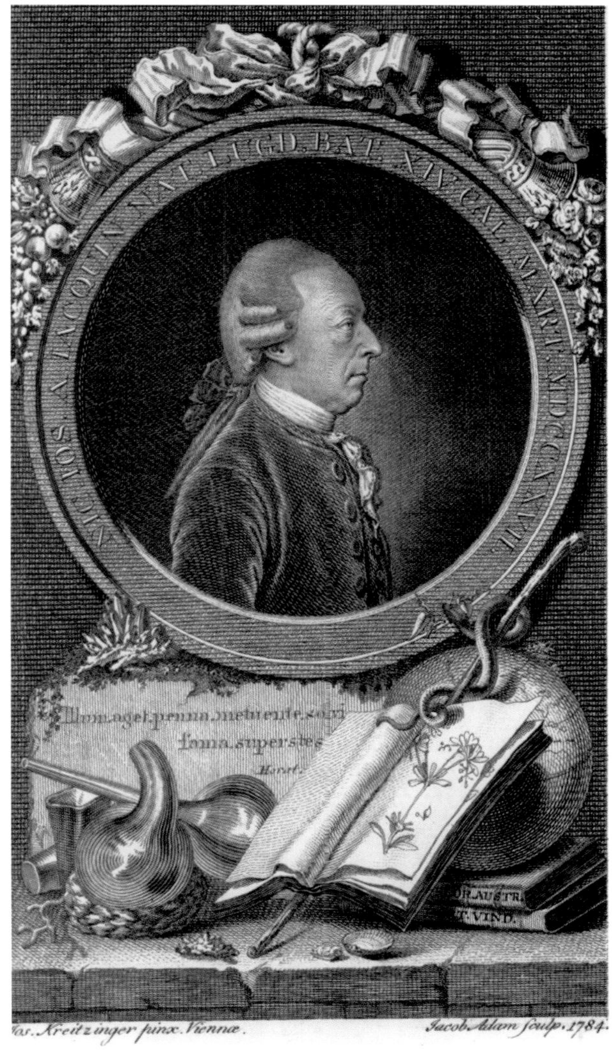

FIG. 3.3. **Nikolaus Joseph Edler von Jacquin. Portrait, 1784. Copper engraving by Jacob Adam based on J. Kreizinger. – Frankfurt, Universitätsbibliothek Frankfurt.**

We may ask how Jacquin could have cultivated the talent of the Bauer brothers. For this there is no convincing answer. It seems that the supervision was similar to that by Boccius, probably with more emphasis on work with a magnifying glass and on dissected flower parts, which are regularly shown in the *Icones plantarum rariorum* and are an advance over the *Codex Liechtenstein*. In any case Jacquin wrote in retrospect about his botanical artists 'Very often I have spent hours at a time – sometimes whole days – sitting with the painters, instructing them or comparing their illustrations with all my descriptions of the plants in question. I have been happy to do this'. This kind of highly specialized and concentrated work under constant supervision by a botanist and without any artistic freedom was not new to the Bauer brothers. Possibly they also were invited to join Jacquin (Fig. 3.3.), who was to celebrate his sixtieth birthday in 1787, on his botanical excursions into the surroundings of Vienna in the same way that this had been done previously by Scheidl, but again we do not have evidence of this. The Bauer brothers may also have assisted Jacquin in collecting plants and preparing herbarium specimens, as they probably had helped Boccius before in doing so.

A travelling naturalist visiting Jacquin in 1784 – thus at a time when Franz and Ferdinand were probably already working for him – wrote 'Professor Jacquin . . . teaches botany and directs the botanic garden . . . the Emperor [Joseph II] appreciates dignified learned people and has bestowed many advantages on Jacquin . . . whose circumstances are considerable, being one of the . . . academics who is not only a decent man but who is able to live in luxury . . . he now lives in the botanic garden, where he has also set up his cabinet. This is as rich as it is dazzling. It consists mainly of minerals, and one can see here opalized marble, wonderful lead crystals, tourmalines, and others such as Tyrolean and Hungarian minerals in large quantity and exquisite perfection'. Jacquin's house, which the brothers frequented, was the official residence of the director of the botanic garden belonging to Vienna University, situated on what is now Rennweg 14. In retrospect his elder son described it as 'a single-storey spacious building in a pleasant style . . . containing thirteen rooms, some large, some small, including a large . . . hall . . . in the wing one large and two small rooms, used as storeroom for seeds and as a library'. In addition the Jacquin family possessed a flat in Bäckerstraße opposite to the New Aula of Vienna University in the centre of the city; it is possible that they made use of it predominantly during the winter months.

Evidence for the other commitments Franz had to fulfil are sparse, but the Niedersächsische Staats- und Universitätsbibliothek in Göttingen keeps a collection of fourteen unsigned botanical drawings relevant in this context.[9] Several of them are to be interpreted as materials used by Jacquin in his lectures in order to demonstrate morphological terms, such as the

FIG. 3.2. ***Albuca caudata* Jacq. Coloured copper engraving, probably based on a watercolour by the Bauer brothers. Jacquin, N. J., Icones plantarum rariorum 2: t. 442, Vienna 1795. – London, Natural History Museum, Library and Archives.**

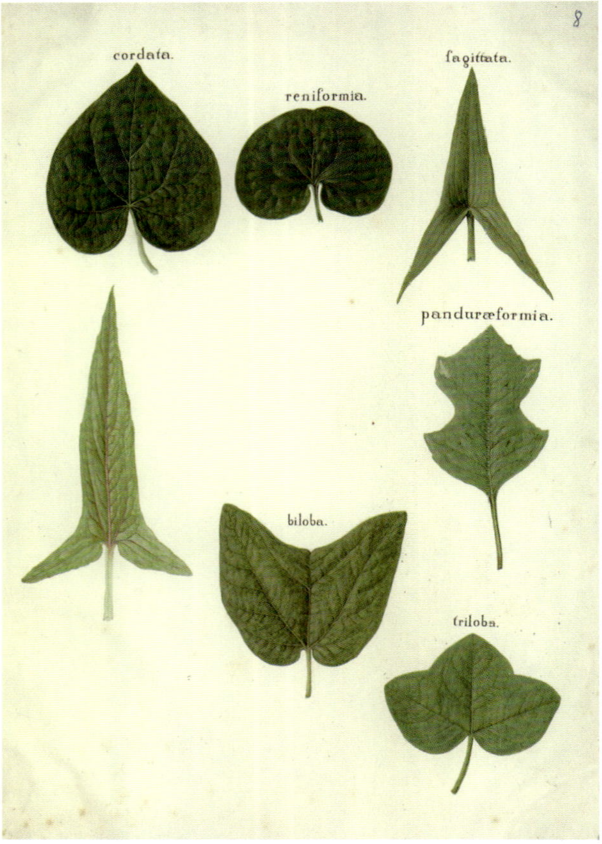

FIG. 3.4. **Leaf forms. Watercolour by the Bauer brothers, c. 1784. gr 2° MS Histor. Nat. 94, VIII: f. 8. – Göttingen, Niedersächsische Staats- und Universitätsbibliothek, Handschriftenabteilung.**

shape of leaves (Fig. 3.4.), echoing a copper engraving published by Linnaeus,[10] one of Jacquin's correspondents. Other sheets in this Göttingen folder document the underground parts of common toothwort, *Lathraea squamaria*, growing as a parasite on the roots of hazel and alder, and of the Turk's cap lily, *Lilium martagon* (Fig. 3.5.), the latter with a longitudinal section of the bulb. In the lower left-hand corner this illustration is annotated '[W]ienn 1784', probably in the hand of one of the Bauer brothers, giving the only precise date for the Vienna interval. These watercolours intended for demonstrations in Jacquin's lectures were clearly an advance over the Herbarium vivum[11] with its manipulated plant specimens (see Chapter 2) probably used by Boccius in his teaching. Five more unsigned illustrations conserved in Göttingen depict various fruits, among them a peach with its velvet-like skin in four different views, one of them with the stone and the adherent juicy flesh visible; as usual they are undated, but may be attributable to the association of the Bauer brothers with Jacquin. The same applies to four illustrations of beetles, all in the Niedersächsische Staats- und Universitätsbibliothek, among them a male stag beetle, *Lucanus cervus*, with its enlarged mandibles, annotated with its scientific name in the calligraphy typical for the Feldsberg pencils. Additional materials of this type, for example an illustration of a horn of plenty, *Craterellus cornucopioides*, an edible mushroom also called the trumpet of the dead, exist in the archive of the Naturhistorisches Museum.[12] These few drawings are indicative of the relevance of botanical illustrations in academic teaching: because of the seasonal climate and their ephemeral nature very often the respective botanical objects are not to be found when they should be available for demonstration during lessons.

Franz also got involved in another project of Jacquin's, the second edition of his *Selectarum stirpium americanarum historia.*[13] Produced in a very limited number for the affluent elite in Vienna and elsewhere, probably by subscription, this luxury work became a collectors' piece for aficionados of very rare books. About thirty copies are known today,[14] all consisting of two hundred and forty pages of printed text, two hundred and sixty-four tables of plant illustrations copied by hand, and a painted title page. Franz was selected by Jacquin as one of the artists to produce individual title pages (unique items) very much along the same line as the title pages that the Bauer brothers had produced previously for Boccius in Feldsberg. Three of these title pages commissioned by Jacquin are signed 'Franc Bauer', two more 'F. Bauer', which of course may refer to either Franz or Ferdinand. Other painted title pages of decidedly inferior quality were prepared by Hofbauer and are not dealt with here.

The title page of the copy previously owned by Karl Joseph Count Firmian, Governor General of Lombardy, then another Habsburg territory, echoes the title page of the first volume of the *Codex Liechtenstein* with its pending garlands. It is signed 'Franc Bauer' and kept today in the Biblioteca Nazionale Braidense in Milan. Another copy has repeatedly changed hands. It bears a note in Jacquin's hand 'Madame la Comtesse du Nord', an incognito for Maria Feodorovna, Grand Duchess of Russia, née Sophia Dorothea Augusta Princess of

FIG. 3.5. **Turk's cap lily (*Lilium martagon*). Watercolour by the Bauer brothers, 1784. gr 2° MS Histor. Nat. 94, VIII: f. 1. – Göttingen, Niedersächsische Staats- und Universitätsbibliothek, Handschriftenabteilung.**

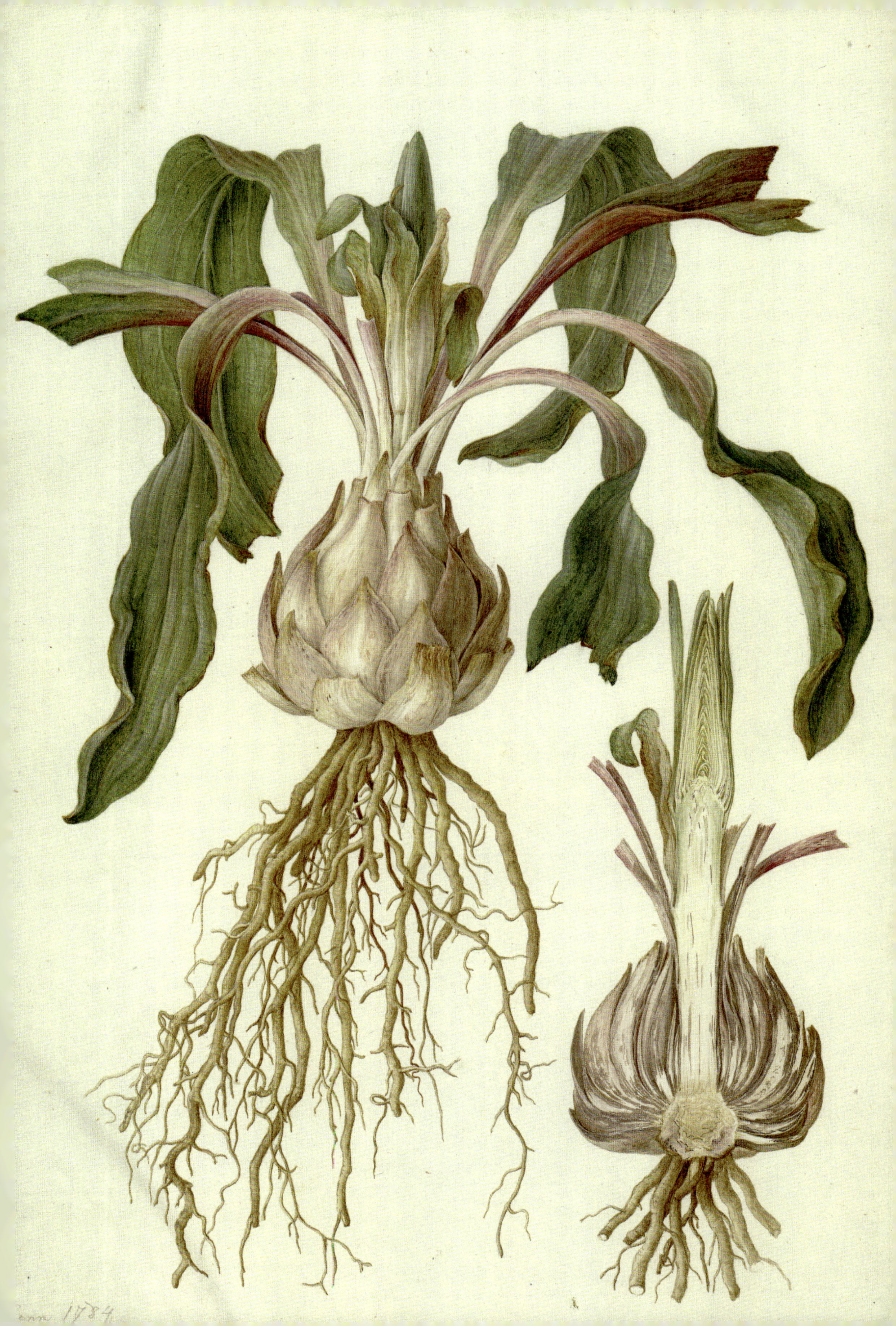
1784

Württemberg. She was the wife of Tsarevitch Pavel, a son of Catherine the Great, later as Pavel I Emperor and Autocrat of all Russias. This annotation indicates that the copy might have been purchased or subscribed when the couple visited Emperor Joseph II in Vienna in late 1781. On that occasion the guests from Saint Petersburg attended a concert given on Christmas Day 1781 in the Hofburg, where Haydn's famous Russian Quartets (Hob. III: 37–42), which represented the final development of this type of chamber music, were premiered. Pavel had visited the laboratory of Jacquin's brother-in-law Ingen Housz on 7 December 1781 and attended two experiments. Maria Feodorovna followed on New Year's Day, with the Grand Duchess presenting Ingen Housz with a diamond ring. The date of these two visits coincides roughly with the publication of the review of the second edition of Jacquin's work in the *Göttingische Gelehrte Anzeigen*, which was based on the copy kept in the library of Göttingen University, now Niedersächsische Staats- und Universitätsbibliothek. Since this copy has a title page attributable to the Bauer brothers, though not signed by them, there is evidence for at least occasional work by them for Jacquin at that early time. Subsequently the copy with the annotation 'Madame la Comtesse du Nord' with a title page signed 'F. Bauer' passed to the Linnean Society of London and then to the international art market. Its title page is a variant of the title page of volume 6 of the *Codex Liechtenstein*.

By far the most spectacular title pages painted and signed by Franz are those belonging to the copies previously owned by Carl Eugen, Duke of Württemberg, an uncle of Sophia Dorothea Augusta, and by his commissioner Jacquin. The former copy is kept in the Württembergische Landesbibliothek in Stuttgart which is is precisely where one would expect this rare book: its title page is a variant of the title page of volume 4 of the *Codex Liechtenstein* adorned with an additional dark velvet ribbon. The latter copy (Fig. 3.1., p. 54) is in the Österreichische Nationalbibliothek in Vienna, forming part of the famous Fideikommissbibliothek founded by Franz I (Francis I), Emperor of Austria, previously as Franz II (Francis II) the last Emperor of the Holy Roman Empire. The background for this merits attention: Jacquin or his elder son had rather unwisely shown this copy to Francis I, Emperor of Austria, who proposed an exchange of his markedly inferior copy to that owned by the Jacquins with the title page painted by Franz. It was felt that this wish expressed by the emperor should not be declined.

The copy in the Fideikommissbibliothek shows a flower-and-butterfly piece of breath-taking opulence and quality – an oval wreath decorated with a blue velvet ribbon and with rose flowers in bud. The wreath also contains leaves of various other plants and is covered by dozens of butterflies, all drawn in utmost perfection and in natural size, making the title page an exceptionally accomplished piece of art. All of the butterflies are common in the surroundings of Vienna and include the swallowtail, *Papilio machaon*, the peacock butterfly, *Vanessa io*, and the great peacock moth, *Saturnia pyri*. Two insects merit special attention – the first of them is the red admiral, *Vanessa atalanta*, with clasped wings, found also precisely in this position on the title pages painted by Hofbauer, which might be a symbol of the Jacquin workshop. The second is a dead house fly, *Musca domestica*, between the letters I and A in the word Austria, clearly an indication of the transience of life. With a gilt book block decorated with a selection of painted plants this precious volume is an example of the refinement of Vienna at its best.[15]

The title page of the copy kept in the Library of Congress in Washington, DC has also been attributed to the Bauer brothers, although it, like the Göttingen copy, is unsigned. It includes a representation of a male timberman beetle, *Acanthocinus aedilis*. The antennae have been very correctly depicted, measuring four times the length of the body.

The refined calligraphy of all title pages painted by the Bauer brothers, including those of the respective volume of the *Codex Liechtenstein* (see Chapter 2), is another proof of their very superior eyes. Every single letter is carefully composed, with attention to size, style and arrangement. Needless to say, an extravagant price was paid for the two copies of the *Selectarum stirpium americanarum historia* that came on the London art market in the late 1990s.

IN A BOTANICAL GARDEN

In Feldsberg Franz and Ferdinand had become familiar with a medicinal garden, and in Eisgrub with the gardens and conservatories of the Liechtenstein family. The interval in Vienna provided the brothers with their first contacts with a botanical garden, or rather two botanical gardens – the one headed by Jacquin near the Belvedere palace and belonging to Vienna University, and the one on the grounds of the summer residence of the imperial family in Schönbrunn, where Jacquin acted as an informal scientific advisor at the time. The main differences between the gardens in Feldsberg and Eisgrub *vs.* those in Vienna must have been the much higher number of unknown or incompletely known plants, many of them received by Jacquin through a network of naturalists, gardeners, physicians and apothecaries from all over the world. Furthermore the botanical garden of Vienna University was characterized by a consistent labelling of pots or individual plants, a system not on record for Feldsberg and Eisgrub.

Jacquin's elder son Joseph Franz noted in retrospect about the Botanical Garden of Vienna University: 'When the garden was first laid out, the outdoor plants were numbered, using oil paint on three-lobed (clover-leaf shaped) convex tin labels, in black on a white background, and, in the taste of the period [c.1755], nailed to three foot high stakes, delicately turned and painted in green and white. These numbers referred to a manuscript list known only to the director of the garden [then Robert-François Laugier]. This was then the situation my father [Nikolaus Joseph Jacquin] found when he took over. Starting in the following year, 1770, he had an alphabetical index printed, with the numbers 1–7815, of all the plant species contained in the second edition of Linnaeus's *Species Plantarum* of 1762 [in fact it was the twelfth edition of Linnaeus's *Systema*], and labelled the plants in the garden with these numbers, firstly using galvanized tin labels . . . with the numbers stamped. The labels were nailed to 5- to 6-foot-high stakes, used in the vineyards and placed in the beds behind the plants. In the case of container plants these numbers were nailed to pieces of wood pushed into the soil . . . New plants, accessioned by the garden and not included in the printed numerical index, were given a running number, without any order, and recorded in a manuscript appendix'. Thus the plants cultivated in the Botanical Garden of Vienna University were first simply numbered, but these numbers were meaningless without reference to the printed index. This was the situation when the Bauer brothers worked for Jacquin. Only later did a change take place, which they most probably did not experience. Jacquin's elder son reported: 'In the section . . . of the garden created by my father for medicinal and economic plants, at first just the Latin names [those used in pharmacies] were marked on metal labels, in the same way as the numbers; later the systematic name [the scientific name] was added to the officinal, and eventually the German names as well, on large metal labels . . . For trees and shrubs, the labels with numbers were initially attached to the branches with wire rings or nailed to the trunk, but later they were attached to nearby vineyard sticks'. Little is known about the conservatory in the Botanical Garden of Vienna University, Jacquin's elder son reporting: 'The large greenhouse consisted of three sections, separated by walls, with the largest, central part being the hot house, and the two outer sections the orangery and the Tepidarium [cold house]. The central part was heated by a subterranean canal . . . the other two parts each by a tilted stove'.

In principle botanical gardens were open to the public, but the word 'public' had a more restricted meaning in the second half of the eighteenth century than it does today. In contrast to the first director of the Botanical Garden of Vienna University, Jacquin opened the garden to the general public, with the exception of the propagation area. Nothing is known about opening hours, which were probably very limited, nor about entrance fees, and closure during the winter months must have been the rule.

Particularly important for the Botanical Garden of Vienna University were the good relations with the Imperial Gardens, for which – a few years after the departure of the Bauer brothers – Jacquin became formally the scientific advisor. This was one of the traditions begun by Laugier and later further developed. We may safely assume that at least occasionally the Bauer brothers visited the Imperial Garden in Schönbrunn, in particular their conservatories: the monumental orangery on the Meidling end, east of the palace, and the conservatories on the Hietzing end, west of the palace. The latter, including the hothouse erected on the orders of Emperor Francis I Stephen, were on the fringe of the Schönbrunn botanical garden, which was more widely known as the Dutch (Botanical) Garden. It had its name from the succession of Dutch gardeners having been in charge of its care, starting with Adriaan van Steckhoven.

This hothouse had received the plants sent by Jacquin from his voyage to the islands in the Caribbean and subsequently described by him, and he continued to keep an eye on them. Much more plant material must have arrived there from other sources after his return and the Bauer brothers might have contributed to the study of these plants by preparing the necessary pictorial documentation for Jacquin. In fact, it is often impossible to decide where a plant described and illustrated in the

Icones plantarum rariorum had been cultivated – in the Botanical Garden of Vienna University, in the Dutch Garden in Schönbrunn, or elsewhere. These were often of the greatest rarity and we have to assume that the Bauer brothers were commissioned to make images of them sometimes *in situ*, inside the conservatories. This might have been already occasionally the case when they were still based in Feldsberg and worked for the *Codex Liechtenstein*.

If the Bauer brothers strolled through the park and gardens of Schönbrunn today in search of the conservatories they would be surprised to see the extensive orangery with a length of around one hundred and ninety metres, very similar in size to the one they knew from Eisgrub, still standing. They would also find still in existence the *Cedratenhaus* – the house designated for the cultivation of citrons, *Citrus medica* – although no longer used as a conservatory. They would wonder about the totally altered heating system in the orangery, but otherwise the use of the building has remained unchanged. Only part of the orangery remains a place for the cultivation of plants during the cold season; nowadays the other part is a location for all sorts of events, notably concerts held throughout the year. At least on a few occasions more than two centuries ago this was already the case: on 7 February 1786, just a month before Ferdinand and John Sibthorp left for Rome, a musical competition sponsored by Emperor Joseph II took place here – pitting a singspiel in German against an opera in Italian. Mozart wrote the short singspiel *Der Schauspieldirektor* (*The impresario*) (K.486), Antonio Salieri the opera buffa (comic opera) *Prima la musica, poi le parole* (*First the music, then the words*). The members of the Jacquin family, all addicted to music, might have attended this performance.

Seven years later an impressed guest from England described the conservatories of Schönbrunn during winter: 'Covered with snow I experienced a marvellous contrast on entering this splendid greenhouse from the freezing cold and biting wind, where, although the plant world "outside" seemed absolutely dead, my eyes enjoyed the sight and my nose the perfume . . . of rare plants'. However, rather than the orangery, this account may refer to the hot house, then named Hortus Adonidis (Garden of Adonis). It was in this conservatory in early 1782 that Ingen Housz had undertaken some of his groundbreaking experiments on photosynthesis later published in the *Philosophical Transactions* in London, while a few years later a less inquisitive visitor wrote about this conservatory as 'a botanical treasure chamber, truly an Imperial living botanical cabinet'.

The Bauer brothers would miss this hothouse today, the building having been replaced by a much larger and higher glass-and-steel construction with art nouveau elements. Called the Palm House, it was erected under the long reign of Franz Joseph I, Emperor of Austria, a grandson of Francis I, Emperor of Austria. Equally they would miss the long row of conservatories on the western fringe of the Dutch garden, of which some were dedicated to the cultivation of pineapples. This plant, documented by the Bauer brothers in a colour-coded pencil drawing kept in the Naturhistorisches Museum,[16] had been explicitly mentioned by Emperor Francis I Stephen in his instructions given to Jacquin to be brought back by him from the Caribbean. However, the Bauer brothers would be able to find Moll's monument, erected by Maria Theresa for her husband, the great supporter of natural sciences, although not precisely in the original position on the grounds of the former Dutch Garden as documented in a drawing by Brand dated 1775.

The house of the gardeners still stands on the margin of the Dutch Garden, and it is possible that before 1783 Franz Boos had lived there. Several years earlier he had worked in Eisgrub for Franz Joseph I. What the Bauer brothers probably did not know during their time in Vienna was the immense maintenance costs of the Dutch Garden, including its conservatories (6000 guilders per annum – whereas the total cost for the very extensive Schönbrunn gardens amounted to 26,000 guilders per annum; thus one-fortieth of the area required a quarter of the budget).

Unfortunately, a painted record from c.1780 has not been found for the contents of the orangery and the *Cedratenhaus* or of the hothouse and adjacent conservatories. Canaletto's two famous *vedute* of Schönbrunn, dated 1759/60, both in the Kunsthistorisches Museum, focus on the main building. One view is taken from the cour d'honneur, the other from a slightly raised point overlooking the extensive parterre. For the conservatories only plans exist, just like for the Dutch Garden, the earliest plan being annotated 'Hortus Imperialis Floriger et Fructuarius' (Imperial Flower and Fruit garden).

However, the Bauer brothers experienced Schönbrunn at a time when it was clearly not at its best. The palace was empty, as the widowed Emperor Joseph II preferred to stay in his Augarten residence, and many tropical plants had been lost in late 1780 when his mother's life approached its end. One cold night the gardener in charge (allegedly not sober) had neglected the control of the heating system. When he realized his mistake the next morning he ignited such a large mass of wood that he lost control over the temperature in the conservatory, resulting in the death of many precious tropical specimens. To outbalance the losses Emperor Joseph II had sent out two more expeditions to the Americas and South Africa, respectively, but the replacements had hardly started arriving in Schönbrunn when the Bauer brothers left Vienna. Nevertheless, the splendour and grandeur of the Schönbrunn gardens must have left a lasting impression on the Bauer brothers. Subsequently neither Ferdinand nor Franz could be impressed by any garden or any conservatory, but only by new plants or new animals that no one had seen and documented before. Of these, however, very many were still waiting for them.

It goes without saying that Franz and Ferdinand may have visited several other gardens in and around Vienna, but for this we have no proof. The garden belonging to the summer residence of the Reigning Prince of Liechtenstein in the Rossau quarter of Vienna, with its orangery, would be the most likely candidate. However, in the area many more fine gardens, maintained by an affluent aristocracy and well documented thanks to detailed plans,[17] would have merited a visit. The same is true of the conservatories adjacent to these gardens and their contents, which contributed substantially to the prestige of their owners.

THE POLISHED BOARDS OF VIENNESE SOCIETY

In her posthumously published recollections, writer Caroline Pichler notes: 'I should mention another household with which my parents' household had very good relations when I was a child. This was the family of the famous Baron [Nikolaus Joseph] von Jacquin. Already at that time (60 or 70 years ago, [c.1785]) it was a brilliant focal point for the scientific world both within and beyond Vienna and visited by many because of its pleasant and friendly atmosphere. The learned and those aspiring to that condition would seek out the famous father and the son who sought to emulate him (Joseph [Franz] Edler von Jacquin, who died a few years ago [1839]), while the younger set gathered around the younger son, Gottfried, a lively and cultured individual with an extraordinary musical gift and a pleasant voice, who formed the centre of a merry circle, and his sister Franziska, now Frau von Lagusius [Hasenöhrl]. She was an excellent pianist, one of the best pupils of Mozart, who wrote the clarinet trio for her, and she also sang very prettily. As far back as I remember Wednesday evenings were open house at the Jacquins', even in winter . . . In the father's rooms learned conversation went on, while we young people chattered, joked and diverted ourselves splendidly'.

The old lady had a good memory: Jacquin's elder son, whom Franz was soon to accompany across Europe as far as London (see Chapter 7), had indeed died in 1839, four years before Caroline Pichler. When the recollections were written Franziska was still alive, and as a girl she had indeed been taught by Mozart. The social gatherings described must have taken place either before April 1788 or in late 1791. Gottfried Edler von Jacquin died on 24 January 1792, and between April 1788 and late summer 1791 his older brother Joseph Franz was on his official tour. If the earlier date is right, Franz Bauer might also have attended these soirées, and Ferdinand as well if they had taken place before March 1786. However, this is by no means certain: Franz and Ferdinand belonged to a different social class, they did not converse in French, which was the language of the Jacquin household, and despite their outstanding talents would probably have been considered to be artisans. In addition they would have had little preparation for the polished boards of Viennese society and possibly did not even possess the dress coat and white tie that were reported to have been the standard at these meetings.[18] Taking these social differences into account, Franz could hardly have envisaged during his time in Vienna that he would be elected Fellow of the Royal Society in 1821, thereby joining the exclusive circle to which Swieten, Ingen Housz and Jacquin had belonged several decades before him.

In our context the various contacts between Mozart and the Jacquins are relevant because of a brief note among Franz's mixed papers kept at the Niedersächsische Staats- und Universitätsbibliothek in Göttingen; it reads 'Mozarts [sic, indicating a German native speaker] celebrated opera Zauberflöte arranged as Quartetto' in an unidentified hand without a date. Clearly this should not be over-interpreted but may be indicative of Franz's interest in Mozart's music, not surprising for someone who had lived in Vienna at that time.

The clarinet trio mentioned by Pichler is clearly K.498, the 'Kegelstatt-Trio', dated 5 August 1786 on the autograph by Mozart, so called because he is said to have composed it while playing skittles, with the piano part specifically written for Franziska. It is reported to have been first performed in the home of the Jacquins. On 14 January 1787 Mozart wrote from Prague about Franziska: 'I must say that I have never had such an eager and hardworking pupil'. For a lesson Mozart is reported to have received one ducat from Jacquin, more than four times the salary Scharf was to receive for a whole day's work. Further evidence is provided by Mozart's letters to Gottfried, also a composer although of a different calibre, telling him about the first performance of *Don Giovanni* in Prague. And there is the manuscript of Mozart's song *Als Luise die Briefe ihres treulosen Liebhabers verbrannte* (K.520) which bears the composer's note '26 May 1787 Landstrasse W. A. Mozart in Mr. Gottfried von Jacquin's room'. It was in that year that Mozart lived within walking distance of the Jacquins in what is now Landstrasser Hauptstraße 75–77. In the garden belonging to this house his timeless *Eine kleine Nachtmusik* (K.525) was written, the serenade in G major, dated 10 August 1787. This piece of music is the younger and more famous sister of the serenade in C minor (K.388), almost certainly ordered by Aloys I Joseph in 1781, soon after becoming the Reigning Prince of Liechtenstein. Another connection exists between the Jacquin family and Mozart: after the return from his official tour Jacquin's elder son was to marry one of Mozart's piano pupils, Maria Barbara (Barbette) Natorp, who had been together with her sister Maria Anna Clara (Nanette) in 1787 the dedicatee of Mozart's piano sonata in C major (K.521). However, at that time Mozart was still on the way to the stars: the iconic *Zauberflöte* with its many associations to the Jacquin family – after all, Jacquin the father played the flute – was to be premiered on 30 September 1791 in Vienna. By then Ferdinand and Franz had long left the imperial city and were working in Oxford and Kew, respectively.

Among the friends and regular guests of the Jacquin family were several interesting figures of Viennese society. One of them was physician Leopold Edler von Auenbrugger, who found out that by applying an ear to the patient's chest and tapping lightly on it one could assess the texture of the underlying tissues and organs. This method had its origins in testing the level of wine in the casks kept in the cellar of his father's inn in Graz. By this percussive diagnosis Auenbrugger was able to plot the outline of the heart and could determine an important aspect of chest disease relatively accurately. Although published in 1763, the new approach was not taken up by the medical fraternity in Vienna or elsewhere until a French translation of Auenbrugger's text appeared in 1808, just one year before his death. Auenbrugger was a highly praised city physician, member of the medical faculty of Vienna University and also an author of clinical texts on mania and suicide. He was created Edler in 1784, as Jacquin had been a few years before, and like Jacquin he had formed a friendship with a well-known composer – Haydn, who dedicated no less than six piano sonatas to Auenbrugger's daughters (Hob. XVI: 35–39, 20). Later Auenbrugger was to write the libretto for the opera *Der Rauchfangkehrer* (*The chimney sweep*) by Antonio Salieri, Mozart's chief competitor in the musical circles of the imperial city. Indeed Vienna remained the place to be for composers over the next decades. Two peripheral figures of our story, which the Bauer brothers may have heard about, had later music pieces by Beethoven dedicated to them: Gottfried van Swieten the symphony No. 1 in C major (op. 21) and Sonnenfels the piano sonata No. 15 in D major (op. 28).

A few years before the arrival of the Bauer brothers in Vienna discussions on a special topic must have taken place in the house of the Jacquins. According to a report of 11 December 1775 the chancellery proposed to Maria Theresa that 'the following individuals be entrusted with initial plans for an Academy [of Sciences]: the physicist Fr Karl Schärfer . . . the astronomer Fr Maximilian Hell . . . Fr Paul Mako von Kerek Gede, also of the Society of Jesus; . . . and the famous botanist and chemist Nikolaus Joseph von Jacquin'. Maria Theresa's decision is written in her own hand: 'I could not possibly resolve to set up an academy of sciences with 3 ex-Jesuits and a professor of chemistry even though he is a good man the whole world would laugh at us . . . I do not think abbé Hell is up to produce something worse [sic] than other academies already in existence . . . is not worth the effort and money'. In fact, the Imperial Academy of Sciences in Vienna, the Austrian equivalent to the Royal Society in London, was founded as late as 1847 by Ferdinand I, Emperor of Austria, a great-grandson of Maria Theresa. Ten years later its seat was transferred to the New Aula in the centre of Vienna.

Another interesting character belonging to the circle of the Jacquins was the precocious, multitalented Thaddaeus Haenke, a student from Prague University who arrived in Vienna in late 1786,[19] several months after the departure of Ferdinand but before the departure of Franz. Thaddaeus played not only the piano and the organ, but also several wind instruments, contributing to the musical aspects of the soirées held in the house of the Jacquins. At the same time Thaddaeus, one year younger than Ferdinand, was a dedicated naturalist and plant collector. He undertook excursions into the mountainous parts of the Hereditary Lands, in his native Bohemia as well as in Styria, in what is now Slovakia as well as in the Tyrol, in part accompanying Jacquin's elder son. At the same time he was a proper scientist with his first papers published by Jacquin in the latter's *Collectanea ad botanicam, chemiam et historiam naturalem*. Like other naturalists, Haenke was also able to prepare plant illustrations true to nature, and as a consequence he seems to have been interested in the works of the Bauer brothers. In Vienna he must have acquired the colour chart that Ferdinand or Franz had left behind when departing from the city and that was of no more use to them (see Chapter 2). When Haenke left Vienna on 26 June 1789 for Madrid, he took it with him and subsequently added further numbered colour strips to the chart in the margins (see Fig. 2.6.).

For the Bauer brothers another feature in the Jacquin household may have been much more important. After all, Jacquin had seen faraway lands and may have told them about his dramatic adventures during his voyage in the tropics, subsequently put on paper by his son in an unpublished manuscript now in the Österreichische Nationalbibliothek and characterized by the latter as 'Salonerzählungen' (Soirée talks). And Jacquin could have shown the Bauer brothers some of his trophies from the neotropics, which he may have kept in his natural history cabinet. All this must have created in the two country lads from Feldsberg the desire to travel abroad, in particular as their elder brother Joseph had successfully ventured as far as Rome (see Chapter 4) in 1781. Franz may have been the less adventurous personality of the two; in any case he remained in Vienna or Feldsberg until April 1788, continuing his work for Jacquin, Maria Christina Princess of Dietrichstein-Proskau and probably also Boccius.

By contrast, when Sibthorp offered Ferdinand eighty pounds per annum in early 1786 on condition that he joined him as his botanical artist in a journey to the Ottoman Empire (see Chapter 5), he was prepared to accept. Among their first destinations was a place that may have been an additional motive for Ferdinand to agree: Rome.

NOTE

Since a vast amount of information on late-eighteenth-century Vienna is available from a multiplicity of sources, the bibliography had to be restricted to the very essential. Useful background information is summarized in many reference works, notably F. Czeike's *Historisches Lexikon Wien* in six volumes and the *Dehio-Handbuch Wien*. For the history of the Schönbrunn gardens the volume *Schönbrunner Gärten* issued by the Bundesdenkmalamt in Vienna is indispensable.

ENDNOTES

1 Unless otherwise stated this chapter is based on Lack (2000a, 2008b).
2 Beale (2011)
3 Pezzl cited in Landon (1991)
4 Griesinger cited in Landon (1981)
5 Römer (1993)
6 Brichtová (2002)
7 Stafleu (1981)
8 Martínez-Azorin et al. (2011)
9 Lack (2003)
10 Daston & Galison (2007)
11 Koukal (1995)
12 Riedl – Dorn (2000, 2014)
13 Lack (1998a)
14 Madriñán (2013)
15 Konig & Kiehn (1998) Lack (2008a)
16 Riedl – Dorn (1989)
17 e.g. Wimmer (2014)
18 Petz-Grabenbauer (2006)
19 Kühnel (1960)

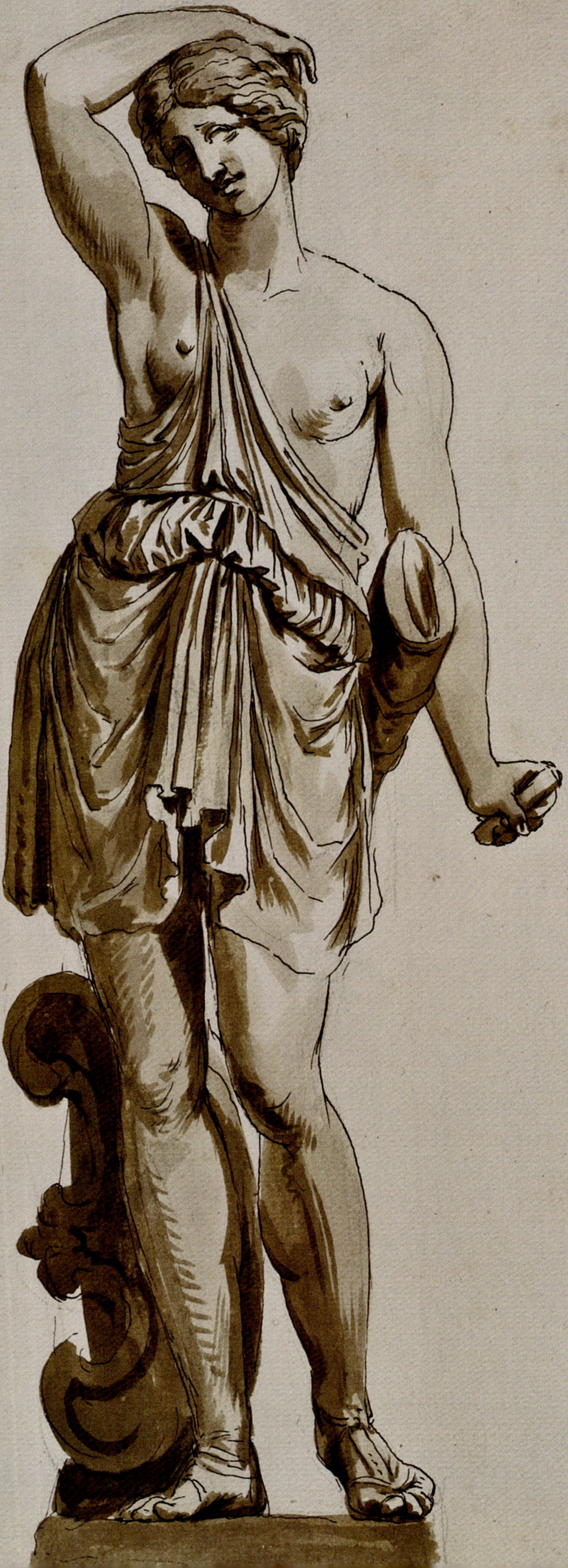

CHAPTER 4

Joseph Bauer in papal Rome

SPECIAL RELATIONSHIPS

For centuries a very special relationship existed between Vienna and Rome. The title 'kaiserliche Reichshaupt- und Residenzstadt' (imperial capital of the empire and residential city) was the name by which Vienna was known, referring to the fact that it acted as the permanent residence for a long chain of Emperors of the Holy Roman Empire, though not an uninterrupted one. Effectively, even though not formally and again with the exception of an interval of a few decades, Vienna was the capital of the Holy Roman Empire, from the mid-fifteenth century until its end was proclaimed in 1806.

Rome had another title – 'urbs urbium orbis terrarum' (city of all cities of the whole world), first recorded under the reign of Emperor Augustus back in classical times. It referred to the fact that it was the largest and most important city of the Roman Empire, residence of the Emperor and centre of administration. Later abbreviations such as 'l'urbe' (the city) and titles such as 'città eterna' (eternal city), or, more famously, 'caput mundi' (capital of the world), stress this key role of Rome. After the collapse of the Roman Empire this central position passed in a sense to the popes, the spiritual leaders of Christianity, and later to the Catholic Church. To this day the Pope is called 'pontifex maximus' (supreme priest), exactly the same title the Roman emperors had possessed previously. This is also the reason the standard opening of ancient Roman proclamations 'urbi et orbi' (to the city and the world) has been used for centuries as the name for the Apostolic Blessing given by the Pope on certain occasions. In addition to being Bishop of Rome and successor of Saint Peter, the Apostle, the Pope also possessed temporal power over the so-called Papal States, a mosaic of territories with different laws, governments and outlooks in the Apennine Peninsula and in France with Rome as its capital, from the eighth century until their end in 1870.

Both the Emperor of the Holy Roman Empire and the Pope were elected political figures, and their election procedures were interwoven in a subtle way. Both took place following long-established traditions in churches, as a rule in St Bartholomew Cathedral in Frankfurt and in a church in the Vatican, respectively, and both always possessed wide-ranging, as well as long-ranging, political consequences. Both elections were followed by coronations, which again took place in churches, in St Bartholomew Cathedral in Frankfurt and in St Peter's Basilica in Rome.

The famous Golden Bull declared law by Emperor Karl IV (Charles IV) in 1356 during the Diet of Nuremberg vested the right of voting an Emperor in electoral princes, the number of whom was fixed at seven. As electors the edict appointed three ecclesiastical princes (that is, the archbishops of Mainz, Trier and Cologne) and four secular princes, the King of Bohemia, the Count Palatinate of the Rhine, the Duke of Saxony and the Margrave of Brandenburg, positions which effectively were hereditary.

FIG. 4.1. Amazone Mattei by Joseph Bauer, 1781–1788. Pen and wash on paper. - Vienna, Akademie der bildenden Künste, Kupferstichkabinett HZ 16645.

Although later modifications of the rules increased the number of secular princes from four to six, in political reality it was evident that a candidate had the best chances for being elected when supported by a coalition including the three ecclesiastical princes. This became even more evident during and after the reformation, since the King of Bohemia, albeit with a single exception, always remained Catholic. As a rule his vote together with those of the ecclesiastical princes could therefore bring about a majority for a Catholic candidate. This very special situation is stressed by the fact that the coronation ritual, including the anointment of the Emperor, was led by the Archbishop of Mainz, the arch-chancellor of the Holy Roman Empire, with formal questions being asked in Latin, such as 'Will he defend the Holy Faith? Will he defend the Holy Church? Will he show due submission to the Pope?' Considering the political implications it was clear that the votes from the ecclesiastical princes were not cast without explicit instructions from the Pope. In short, the Pope had a decisive word to say when it came to electing a new emperor.

The conclave is the meeting of the College of Cardinals convened to elect a new bishop of Rome with regulations concerning the majority necessary for the successful candidate varying over the centuries. From about 1600 certain Catholic monarchs, among them the Emperor of the Holy Roman Empire, claimed a 'ius exclusivae' (right of exclusion), that is, a veto over papal elections, exercised through a crown-cardinal or cardinal-protector. The conclave of 1799–1800, which because of Napoleon took place in Venice and in the end elected Pius VII (Barnaba Gregorio Chiaramonti), may act as an example. Two cardinals were elected Pope one after the other, but one after the other was vetoed by Cardinal Franz de Paula Hrzan von Hrzan-Harras (František de Paula Hrzán z Harasova) and only the third elected candidate was accepted by him. On this occasion he combined in his person the functions of an imperial ambassador to the Holy See and of crown-cardinal of Francis II, the last Emperor of the Holy Roman Empire.

This interdependence of the Emperor and the Pope led to the special relationship referred to above, and it resulted as early as 1529, the year of the first unsuccessful siege of Vienna by the Turks, in the first apostolic nunciature ever, a formal liaison office between the Holy See and the Emperor – needless to say in Vienna. Later the position of a delegate of the Emperor in Rome, among them Karl Count Liechtenstein – Nikolsburg,[1] later the 1st Reigning Prince (see Chapter 2), had been created as a symbol of the close ties between Imperium and Sacerdotium. Significantly, until the end of the Holy Roman Empire the Good Friday Prayers, the 'orationes solemnes', contained not only a section of prayers for the church and for the Pope but also for the most Christian emperor 'that the Lord God may reduce to his obedience all barbarous nations for our perpetual peace'. And when after an interval of several centuries a pope, then Pius VI (Gianangelo Braschi), left the Papal States for the first time, his destination was clear – Vienna, sometimes jokingly even today called 'urbs' by its inhabitants, where he reciprocated an earlier visit of Emperor Joseph II to Rome.

In the long list of representatives of the Emperor of the Holy Roman Empire in Rome Anton Florian, Prince of Liechtenstein,[2] later the 4th Reigning Prince (see Chapter 2), stands out as the first ambassador ever. He arrived in the 'città eterna' on 20 September 1689, just six years after the second unsuccessful siege of Vienna by the Turks, which ended in the liberation of the city by a coalition army led by Jan Sobieski III, King of Poland. And he left Rome on 2 September 1694, representing Emperor Leopold I during two pontificates: that of Alexander VIII (Pietro Ottoboni), who canonized John of God, the founder of the Brothers of Mercy (see Chapter 2), and of Innocent XII (Antonio Pignatelli), who had previously been nuncio in Vienna. Anton Florian's primary goal in Rome was to ensure the continuous financial support of the Pope for the wars Leopold I led against the Ottoman Empire in what is now Hungary, Serbia and Croatia. In this Anton Florian was successful, just like the Holy League of 1684 initiated by Pope Innocent XI (Benedetto Odescalchi), a coalition of the Holy Roman Empire, the Kingdom of Hungary, the Polish-Lithuanian Commonwealth, the Republic of Venice and the Tsardom of Russia, was on the battlefield. Towards the end of Innocent XII's pontificate, this conflict was ended by the conclusion of the treaty of Karlowitz (Sremski Karlovici in modern Serbia), a turning point in European history, in which Eugene Prince of Savoy had played an important role. It marked the beginning of the end of the Ottoman Empire and the rise of the Habsburgs as key political figures in the eastern part of central Europe.

Representative obligations of Anton Florian in Rome included the formal presentation of New Year's congratulations to the Pope, which took the form of an *entrée solennelle* on 27 December 1691. In the ancient tradition of the 'imperator triumphans' the imperial ambassador led a festive procession, consisting of pompous carriages full of allusions to Leopold I's great victories over the Ottoman Empire. Special attention was given to every detail, including the horses sent for this very occasion from the Eisgrub stables to Rome, with the carriages described in a text printed on the occasion in Rome , the *Breve descrizzione e disegni delle Carozze dell' Ecc. Sig. Antonio Floriano.* When Anton Florian received the order to return to Vienna, another important job was waiting for him: he was appointed tutor to Archduke Karl (see Chapter 2), then aged nine, whom he accompanied later to Spain. There he served as his right hand for a throne still to be conquered. Years later Anton Florian's outstanding position at the imperial court was to enable him to raise the status of his family even further (see Chapter 2).

Anton Florian had been neither the first nor the last member of the Liechtenstein family in Rome. His eldest surviving son and successor Joseph Johann Adam, later the 5th Reigning Prince of Liechtenstein, visited the papal city in 1708. Several decades later Karl Borromaeus, Prince of Liechtenstein, an imperial field marshal and brother of Franz Joseph I (see Chapter 2), stayed in Rome from October 1783 to February 1784.[3] He was accompanied by his sons Karl and Joseph Wenzel, the latter a student of theology who stayed until 1787 in the monastery of Santo Stefano del Cacco in Rome and a member of the Pontificia Accademia dell'Arcadia. This was a papal literary academy based in Via Arcione near the Trevi Fountain. And it was Joseph Wenzel, aged nineteen, who in January 1787 introduced Goethe, aged thirty-seven, to this academy, the very man who was to describe in his recollections *Dichtung und Wahrheit* in vivid detail the entry of the former's homologous and far more famous great-uncle Joseph Wenzel in Frankfurt (see Chapter 2). Goethe knew about the young Joseph Wenzel from the latter's sister Josepha Countess Harrach, née Princess Liechtenstein, whom he had met during his first stay in Karlsbad [Karlovy Vary] in 1785 (see Chapter 7).

The list of members of the Habsburg family visiting Rome is equally impressive. Emperor Joseph II stayed for the first time in the papal city from 15 to 29 March 1769 accompanied by his younger brother and successor on the throne Archduke Leopold. Hard to believe but true, Joseph II even took part in the conclave recommending the election of Lorenzo Cardinal Ganganelli, who indeed became Pope Clement XIV. The stay of the two brothers in Rome is documented by a well-known painting in oil by Pompeo Battoni in the Kunsthistorisches Museum, showing St Peters's Basilica and the Castel Sant'Angelo, symbolizing Christianity and the Antique, in the background. The two brothers were later followed by their sister Archduchess Maria Christina accompanied by her husband Albert Prince of Saxony, Duke of Teschen. However, the main objective of their visit to the Apennine peninsula was clearly not politics, but, among others, acquiring works of art for their private collection, which later became the foundation stone of the Albertina in Vienna. A painting in oil by Friedrich Heinrich Füger dated 1776 and kept in the Österreichische Galerie Belvedere in Vienna documents the return of the young couple to Vienna. There it is shown to demonstrate their trophies to their mother Maria Theresa and other members of the imperial family. It is on record that this work brought Füger to the attention of Kaunitz-Rietberg, the long-serving minister of state to a succession of four emperors from Francis I Stephen to Francis II. Later this contact resulted in Füger's scholarship of the United Imperial Royal Academy of Arts, henceforth the United Academy, for a stay in Rome.[4]

ARTS, ARTISTS AND ART-DEALERS

Rome was not only a world-famous city of arts, artists and art-dealers but also of academies of art, the most important in our context being the Accademia di San Luca, an association of artists with the explicit purpose of elevating the work of artists above that of pure craftsmen. With its emphasis on teaching and public exhibition, this academy became a model for many similar institutions to be founded elsewhere. For obvious reasons the papal authorities are reported to have exerted some control over this academy with Pope Urban VIII (Maffeo Barberini) giving the Accademia di San Luca at an early date the right to decide who was to be considered an artist in Rome and to tax all artists and art-dealers in the eternal city, although these measures were poorly enforced. Pius VI was no exception to this tradition; he reaffirmed members' privileges, extended the Academy's right to censor all publications concerning art and artists and specified that no restoration work – cleaning, washing or retouching any paintings of quality – was to be undertaken without the Academy's explicit certification.[5] Since its foundation the Accademia di San Luca had covered the arts of drawing, painting and sculpture, and was headed by directors called 'principi' (princes). Among these in the second half of the eighteenth century were two key figures of the so-called neoclassical movement – Anton Raphael Mengs and Anton von Maron, both from the Hereditary Lands, born in Aussig (Ústí nad Labem) in Bohemia and Vienna, respectively, and both represented by self-portraits in the Gemäldegalerie in Berlin.[6] The two were linked in a special way – in Rome Maron had married Mengs's sister Theresia, who copied paintings in Rome exclusively for Catherine the Great, and Maron was effectively to become Mengs's deputy and successor. Honorary members of the Accademia di San Luca included Maria Theresa, Joseph II, his youngest brother archduke Maximilian Franz, later the Archbishop of Cologne (see Chapter 7), and Kaunitz, the protector of the United Academy and important promoter of the sciences, forming an additional link with the imperial city.

Mengs had become a friend of Johann Joachim Winckelmann, rightly called the prophet and founding hero of modern archaeology, who first articulated the differences between Greek, Greco-Roman, and Roman art, stressing the primacy of Greek art. Living in Rome from 1755 and librarian to a succession of cardinals, he exerted a decisive influence on the rise of neoclassical movement. This was effectively a revival of Greek and Roman art, and at the same time a counter movement to baroque and rococo art. It was Winckelmann who in his writings coined the ideal of 'edle Einfalt, und stille Größe' (noble simplicity, and quiet grandeur), which became the motto for the new style advocated by him, Mengs, Maron and others, in Rome and elsewhere. Another of his key statements reads 'the one way for us to become great, yes, inimitable, if it is possible, is by imitating the ancients'. However, with imitation Winckelmann did not mean slavish copying – 'what is imitated, if handled with reason, may assume another nature, as it were, and become one's own'. Appointed prefect of antiquities by Clement XIII (Carlo della Torre Rezzonico) in 1763, he was tremendously influential by producing a comprehensive chronology of all antique art. In addition he became for years the central figure in the circles of artists, art connoisseurs, philosophers and intellectuals in Rome.

The success of the Accademia di San Luca had resulted in repeats elsewhere, the oldest in the Holy Roman Empire being founded in Vienna in 1688 as a private institution. Shortly afterwards this academy received financial support from the government and, following the wish of Emperor Joseph I, was named the Imperial Academy in 1705. However, it was discontinued after the death of its founder Peter Strudel and reopened much later. The programme of the new United Academy, later attended by the Bauer brothers (see Chapter 3), was shaped by Kaunitz, largely on the advice of Maron, with the latter quite appropriately regarded as the court artist of the Habsburgs residing in Rome. The new profile contained among others the concept of a limited number of imperial scholarships for promising artists in Rome. They were to stay in contact with the imperial embassy[7] but did not form a separate academy like the Académie de France à Rome. Füger was an early recipient of the Rome scholarship; the sculptor Franz Anton Zauner another. During his stay in the papal city, the latter produced a statue of the muse Clio for Kaunitz, now in the Fürstliche Sammlungen, and back in Vienna the famous statue of Emperor Joseph II on horseback emulating the even more famous equestrian statue of emperor Marcus Aurelius in Rome, then on Piazza del Campidoglio, now in the Musei Capitolini. Every visitor of Vienna knows Zauner's masterpiece – it has been placed in front of the Imperial Library opposite the city residences of the Fries and Pálffy families, on a square now named Josefsplatz in the very centre of the city (see Chapter 12). It was Vienna's first equestrian statue and at the same time a typical example of the neoclassical movement. Füger, Zauner and other stipendiaries returning from the papal to the imperial city were instrumental in promoting the change from rococo to neoclassicism, the latter style strongly focused on Rome.

Taking all this into account, it seemed totally sensible to send the young, promising Joseph Bauer not to Paris or London but to Rome. Compared to other places fashionable for nascent artists like Venice, a stay in Rome was then regarded as a special privilege and brought higher prestige. For Franz Joseph I (see Chapter 2) there may have been additional motives, especially as he may have then already contemplated engaging the second son of his predecessor's court painter at a later date. In the Liechtenstein family over decades there was a clear preference for buildings of grave *Romanitas* and for architects from Rome: the monumental, severe

Liechtenstein palace in Bankgasse[8] in the centre of the city had Enrico Zuccalli from Roveredo as author, although the works were later continued by Domenico Martinelli from Rome (see Chapter 3). The no less monumental and severe summer palace of the family in the Rossau quarter[9] of Vienna, in Fürstengasse, was started by Egidio Rossi, later likewise to be continued by Martinelli. In addition, the Bankgasse palace was clearly a significantly enlarged version of Bernini's Palazzo Chigi-Odescalchi in Rome and became the prototype for several later family palaces in Vienna.

Franz Joseph I may have also known about the considerable building activities that had started in Rome under Pope Pius VI, who was to become the last great papal patron of the arts in the renaissance and baroque tradition. In fact, he was the last to extend St Peter's, the last to erect a sumptuous family palace and the first pontiff to transform the Vatican into something approaching a modern museum. His Museo Pio-Clementino, built as a receptacle of the most important Greek and Roman masterpieces in the papal collections, was in fact to become one of his lasting monuments in Rome. Franz Joseph I could also have heard about the Accademia di San Luca, its broad spectrum of activities and the stimulating atmosphere of Rome, between 1760 and 1830 no doubt the centre of the world for arts and artists as well as the ultimate destination of all those undertaking their Grand Tour. From a purely practical point of view, a stipend for Joseph could be easily paid out via the traditional ecclesiastical channels linking Vienna and Rome with the Liechtenstein administration transferring money to the nuncio, to be paid out to Joseph in Rome.

It should be noted that Joseph was not a recipient of the highly prestigious Rome scholarship of the United Academy, which significantly improved the chances of later obtaining one of its chairs.[10] Contrary to Füger, for example, Joseph would not have received written instructions from the United Academy for his stay in Rome to be handed over upon arrival to Hrzan and to Maron, which was the tradition among the recipients of the Rome scholarship. There is explicit proof of payments from the Liechtenstein administration to Joseph, but only for the period from 1786 to 1788 via Cardinal Giovanni Battista Caprara, the nuncio in Vienna, although it is likely that similar amounts of money were transferred during the earlier years. This arrangement echoes that between Franc(o) Caucig (Kavčič) from Görz, another former student of the United Academy, and Philipp Count Cobenzl. The latter was a diplomat and for a short period the successor to Kaunitz as minister of state as well as, for a longer period, protector of the United Academy. Like Joseph, Caucig spent precisely the same years (1781–1788) in Rome; no hard evidence exists that the two artists met there, although this seems very possible.

THE NASCENT ARTIST FROM FELDSBERG

Hard facts on Joseph's seven years in Rome – he was twenty-five when he arrived, and thirty-two when he returned to Vienna – are meagre. Not a single letter written by him or sent to him seems to have survived. No preliminary sketches and no diaries are known or have been traced. A most meticulously researched standard work on German-speaking artists in Rome does not list Joseph, who also seems not to appear in the records of the Accademia di San Luca. The only reference to him is found in Aloys Hirt's manuscript list of the most well-known artists living in Rome, probably prepared for Goethe in 1787 and published only very recently.[11] The relevant entry is very brief, stating correctly that Joseph was from Vienna, about thirty years of age, and a pensioner of the Liechtenstein family.

Again we are faced with circumstantial evidence: the payments recorded in the Liechtenstein archives to 'Joseph Bauer hoch fürstl. Mahler in Rom' (Joseph Bauer high princely painter in Rome) and a handwritten note by an anonymous art collector stating '158 Blätter Original Handzeichnungen von: H: Bauer Historien=Maler und Fürstlich Liechtensteinischer Gallerie=Direktor. welche selber auf seiner Reise in Rom nach dem Anticken [sic] gezeichnet hat' (158 sheets of original drawings by Mr Bauer, historical painter and director of the gallery of the Prince of Liechtenstein, which he has prepared during his journey in [sic] Rome on the basis of the antique). The 158 drawings, all unsigned, without annotations, in large format, and on paper, plus a few over-sized drawings, and the manuscript note cited above are kept in the Kupferstichsammlung of the Akademie der Künste in Vienna.[12] They offer the most significant information on Joseph's activities in the papal city. In addition, three unsigned drawings that have been attributed to Joseph and his stay in Rome are conserved in the Fürstliche Sammlungen.[13] From the archival record it becomes clear that they were sent to Aloys I Joseph as proof of Joseph's continued activity as artist in residence. The statement in a standard dictionary of artists[14] that Joseph produced portraits of 'mainly cardinals' must refer to his seven years in Rome, but has not been substantiated. This is echoed by the assertion in a chronicle of the town of Feldsberg that 'copies of paintings from Rome, in particular of cardinals, are in the galleries of Liechtenstein country houses'.[15] Unfortunately, none of these could be traced.

Furthermore, we possess no indication whatsoever as to how Joseph travelled from Vienna to Rome, though he may well have taken the conventional route via Venice, Bologna and Florence, which was taken five years later by his youngest brother Ferdinand (see Chapter 5). When passing through Venice, Joseph may have contemplated his colleague Canaletto born in this city, who had worked together with his son and assistant Lorenzo from 1759 to 1760 in Vienna.[16] There Canaletto had received commissions for *vedute* from Kaunitz, Joseph Wenzel and, notably, Maria Theresa, who ordered no less than thirteen views, all later kept by her daughter Maria Christina and her son-in-law in the castle at Pozsony (Bratislava), all now in the Kunsthistorisches Museum. The two paintings by Canaletto showing the summer palace of the Liechtenstein family in the Rossau quarter from the garden, including the princely family though not showing Joseph Wenzel,[17] may have been already familiar to Joseph. Near Ferrara Joseph is likely to have entered the Papal States. He may have contemplated in Bologna Gaetano Fanti,[18] born in this town, who had been invited by Eugene Prince of Savoy to come to Vienna in order to work as architectural painter for his summer palace, the Upper Belvedere, and who subsequently decorated together with Johann Michael Rottmayr the grandiose St Charles's Church in Vienna. In later years Fanti had become director of the Liechtenstein Gallery, to be succeeded in this job by his son Vincenzo. In succession they had been the superiors of Joseph's father Lucas (see Chapter 2), and both were predecessors of Joseph, but this he could not yet know. In the Uffizi in Florence Joseph could have had a look at the self-portrait of Vincenzo with his gallery catalogue in the right hand. A copy of this painting was bought in 2007 by Hans-Adam II, the 15th Reigning Prince of Liechtenstein, and is now conserved in the Fürstliche Sammlungen in Vienna.[19]

Like every other visitor coming for the first time to Rome, the country lad from Feldsberg must have been overwhelmed by what he saw, 'tired in the evening and exhausted from looking and being amazed' as put in retrospect by Goethe in his famous *Italienische Reise* for 5 November 1786. At an early moment Joseph must have visited the cathedral of St Peter's, which had made Goethe realize 'that Art, like Nature, can abolish all standards of measurement'. However, Joseph may have been slightly less impressed than Goethe, who had never seen any large city before – after all, the number of inhabitants of Rome equalled that of Vienna in the late-eighteenth century, a considerable number of high baroque churches and palaces adorned the imperial city in a somewhat similar manner, and Vienna like Rome was surrounded by spectacular summer residences of the high nobility.

It is probable that it took Joseph several weeks to get adjusted to his new situation. Like so many others before and after him, it would have been logical for Joseph to rent a room, maybe together with fellow artists, and look out for his first contacts. In the hierarchical society of the late eighteenth century, no doubt Joseph must have found himself in an entirely different situation to that of his benefactors and employers, had they ever visited Rome – Franz Joseph I, Aloys I Joseph and Johann I Joseph were on a par with the Roman nobility. By contrast, the life of a young artist in Rome with no name, probably few or no recommendations, and little money has to be imagined as lacking special appeal, apart from the hope in the long run for better chances of securing an attractive and permanent position. People like Joseph simply did not count to a cardinal, a member of the aristocracy or a diplomat. At least access to churches, monasteries and other ecclesiastical buildings was free, but tips were necessary in the great houses to be permitted to glance at the treasures or to study the private collections.

It is difficult to imagine that a Catholic like Joseph was not attracted to the splendour of the religious ceremonies and manifestations as well as the liturgical drama revived and encouraged during the pontificate of Pope Pius VI, although he was hardly able to grasp the more sublime aspects involved in the theorization of Catholic ritual then at its very peak. A report by Spanish traveller Leandro Fernández de Moratín offers his impression of the solemn benediction 'urbi et orbi' as he experienced it in 1794: 'All is in silence and reverence. Rising to his feet and raising his face and arms to the heavens, the pope [Pius VI] blesses the entire Catholic world from his majestic throne . . . military instruments play, bells sound, voices of mirth rise while the cannons of Hadrian's Tomb [Castel Sant'Angelo] echo in the distance. Perhaps in Asia there is something similar, but in the rest of the world there is no sovereign who presents himself to his people with such grandeur, nor who, uniting power and priesthood, appears to them as father, prince, interpreter of God's will and dispenser on earth of his pardon and his blessings. In the end . . . it is impossible to attend this function without feeling an irresistible surge of wonder and enthusiasm.' Louis-Jean Desperez's canvas 'Easter benediction of Pius VI', dated 1783, kept in a private collection in Rome and showing the magnificent procession on St Peter's square, and his 'Pius VI saying Christmas Mass at St. Peter's in 1783', dated 1787 in the Nationalmuseet in Stockholm, are witnesses of the elaborate choreography of Catholic ritual at that time.

The drawings prepared by Joseph and kept in the Kupferstichsammlung turn out to be an amalgam consisting of several elements, evidently not based on the antique only, though this had been stated by the anonymous collector. Like so many other artists in Rome Joseph seems to have attended classes at an academy, possibly the Scuola libera del Nudo, associated with the Accademia di San Luca, evidenced by a drawing[20] showing a naked male model in two positions (Fig. 4.2.). In accordance with the conventions of academies in Rome at that time, including the Académie de France à Rome, and quite in contrast to the famous erotic, or rather pornographic, drawings prepared just a few years earlier by Johann Heinrich Füssli (Henry Fuseli) in that

FIG. 4.2. Male model by Joseph Bauer, 1781–1788. Pen and wash on paper. – Vienna, Akademie der bildenden Künste, Kupferstichkabinett, HZ 16502.

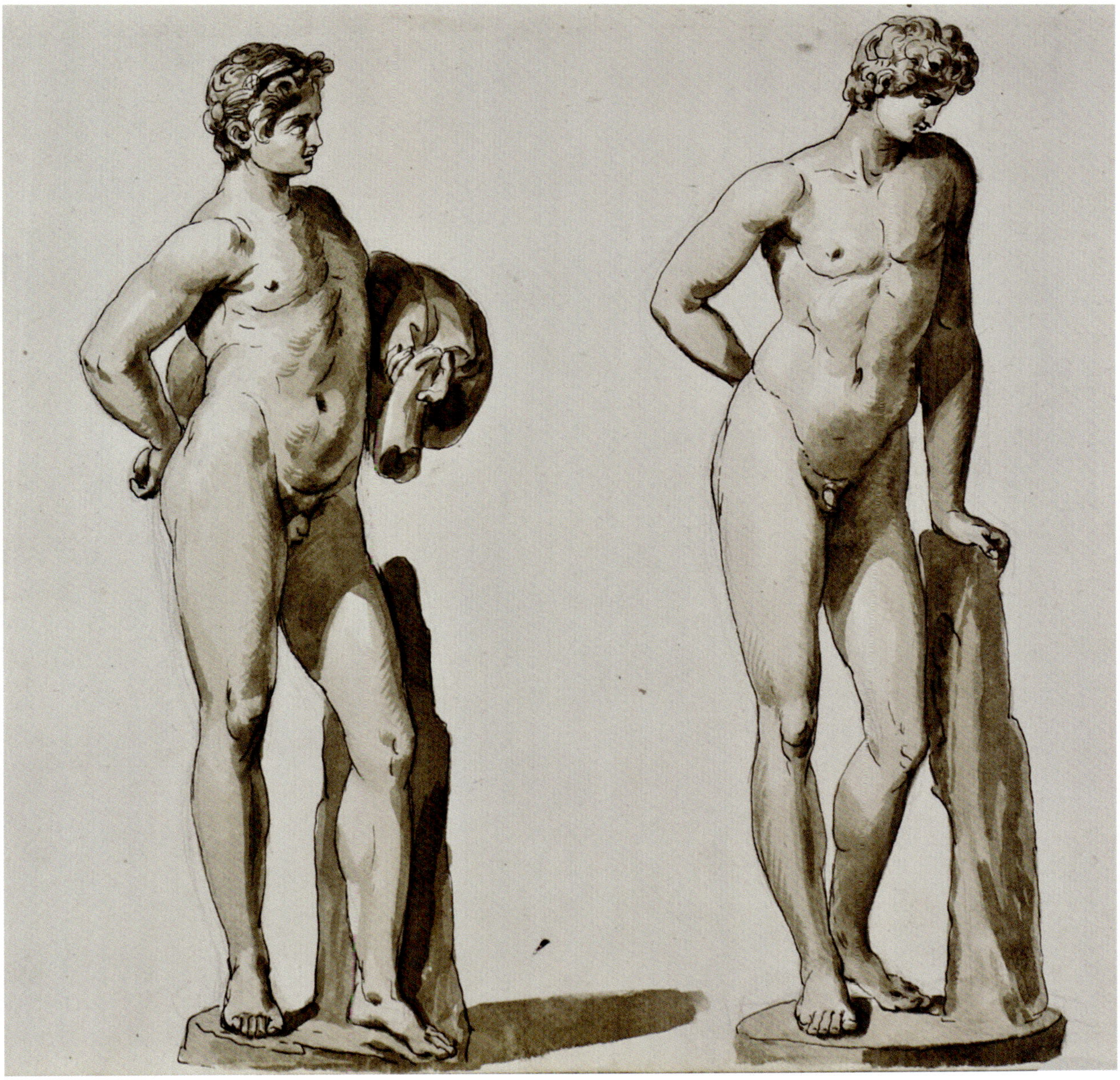

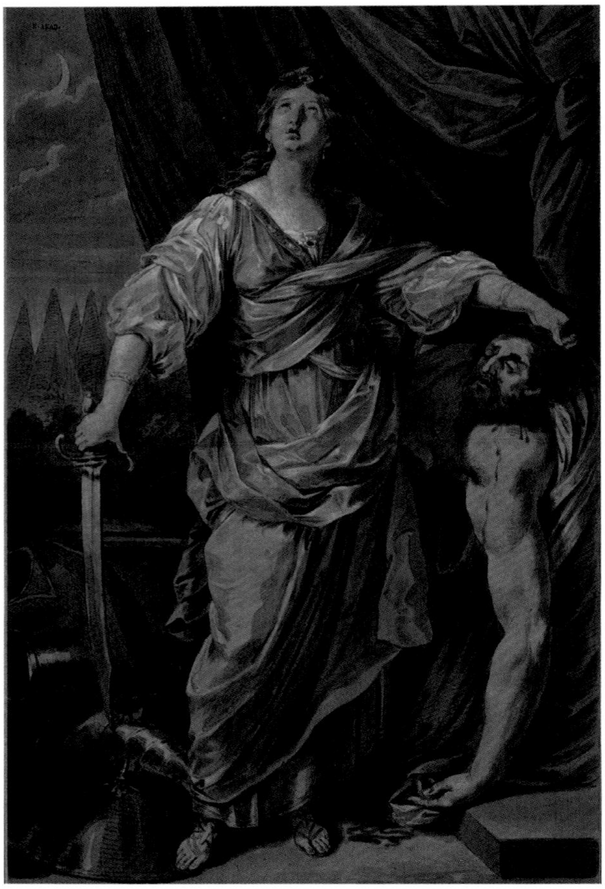

FIG. 4.3. **Judith after the murder of Holofernes. Copy by Joseph Bauer based on Guido Reni, 1781–1788. Pen and wash on paper. – Vienna, Akademie der bildenden Künste, Kupferstichkabinett, HZ 3522.**

city, Joseph's drawings do not show any naked women. However, this may indicate his rather conservative views. After all, Caucig had the courage to prepare at least a single drawing of a nude woman while in Rome, which is kept in the Museo Provinciale in Görz.[21]

Taking copies of well-known paintings was also a traditional activity of young artists, and Joseph was no exception in this. He seems to have had a preference for the works of Guido Reni whose 'Adoration of the shepherds' he must have remembered from the Chapel in the castle of Feldsberg (see Chapter 2). Joseph copied, for example, the latter's 'Archangel Michael pushing the devil to hell', a large-sized painting in oil on canvas placed in the church of Santa Maria della Concezione in Rome, dated c.1636, and originally commissioned by Francesco Cardinal Barberini called Sant'Onofrio, a nephew of Pope Urban VIII. Measuring 643 × 436 mm and worked in pen and brush, this is a particularly convincing copy.[22] Joseph also copied Reni's 'Suicide of Lucretia', a small painting in oil now in the Pinacoteca Capitolina in Rome, and Reni's 'Judith after murdering Holofernes', dated c.1625, now in the Sedelmayer Collection in Geneva, of which a draft in black chalk on blue paper is kept in the Gabinetto Disegni e Stampe degli Uffizi in Florence.[23] These two large copies are rather conventional in approach,[24] one of them is the copy of Reni's 'Judith' (Fig. 4.3.).

OBJECTS FROM CLASSICAL ANTIQUITY

Drawing objects from classical antiquity was one of the principal reasons for sending a young artist to Rome. However, in the eighteenth century in addition to originals from that period many plaster copies were already available to work on (for example, in the Académie de France à Rome, then in Palazzo Mancini), and therefore a particularly critical approach is necessary. Clearly Joseph based his drawings on various sources, copies and originals, on objects in private collections, such as those kept in the Palazzo Barberini in Rome, and in collections made accessible to the general public, such as in the Museo Pio-Clementino in the Vatican, founded by Pope Clement XIV and substantially enlarged by Pope Pius VI.

As examples, three of Joseph's drawings based on objects of classical antiquity are commented on here. A statue of a bearded man, traditionally interpreted as Asclepios, holding in his right hand a long stick clasped by a snail and in his left hand a patera (a broad, shallow dish used for drinking) is reported to have been acquired by a member of the Ruspoli family in 1776. It had subsequently been placed in their palace on via del Corso, one of the most monumental stately homes in Rome. Caucig prepared a drawing[25] of the Asklepios in c.1781 marked by him 'Palazzo Ruspoli' and so did Joseph.[26] In 1826 the original statue was sold and is now part of the Antikensammlung in Berlin, exhibited in the Rotunda of the Altes Museum in that city. A plaster copy of an even more famous statue, the Amazone Mattei, the original until 1770 in the villa of the Mattei family in Rome, was exhibited in the Académie de France à Rome. It was also painted by Caucig[27] in c.1781, annotated by him 'Accademia di Francia', and Joseph[28] (Fig. 4.1. p. 72). In a similar way a plaster copy of a marble statue of a young woman known as Juno Cesi, the original now in the Museo Capitolino in Rome, was on show in the Académie de France à Rome, and again was recorded by Caucig[29] in c.1781 and Joseph.[30]

Joseph's and Caucig's drawings cannot be regarded as precise documentations, but rather as free copies in the sense of Winckelmann. Whereas Caucig annotated his drawings as a rule, for example by adding 'Accademia di Francia' on the back of the Juno Cesi drawing, Joseph consistently refrained from any written comment. This makes the search for the objects documented by him so much more difficult and applies also for two bases from classical antiquity[31] (Fig. 4.4.).

By contrast it has been possible to identify three objects from ancient Egypt[32] (Fig. 4.5.). All were conserved in the late-eighteenth century in the Palazzo Barberini[33], another pompous stately home in the centre of Rome. The falcon-headed statue in the centre is the so-called Osiris Barberini, now in the Staatliches Museum für Ägyptische Kunst in Munich.[34] This well-documented and large object in black granodorite represents, however, the god Horus and had been found in the early 17th century near Santa Maria sopra Minerva in Rome at a place of an ancient sanctuary, the so-called Iseum Campestre. The left-hand object is a granite naophore, more specifically the statue of Ahmosi with a naos containing Osiris, with later additions, now in the Ägyptisches Museum in Berlin, exhibited in the Neues Museum of that city.[35] The right-hand object, a basalt statue of an unknown Ptolemaic king, also ended up in this famous collection,[36] with a similar statue in the Museo Gregoriano Egizio[37] in the Vatican. The naophore, showing a priest named Amasis (jah-mes) from the 27th dynasty, and the Ptolemaic king, which both had subsequently passed from the Barberini to the Colonna family, were later exhibited in the Palazzo Sciarra. Documenting Egyptian artefacts was mainstream at that time in Rome: Caucig for example prepared a drawing of a superficially very similar statue of a Ptolemaic King[38] in the Vatican,[39] although the hypothesis has been put forward very recently that it formed part of the collection then in the Villa Albani.[40]

It seems plausible to assume that Joseph could also have paid a visit to this villa, long since engulfed by the agglomeration of Rome. In any case, he must have missed its creator, Alessandro Cardinal Albani, for many years the imperial ambassador in Rome and Hrzán's predecessor, since he had already died in December 1779. A nephew of Pope Clement XI (Giovanni Francesco Albani), cardinal at the age of 29, formal protector of the Accademia di San Luca, librarian of

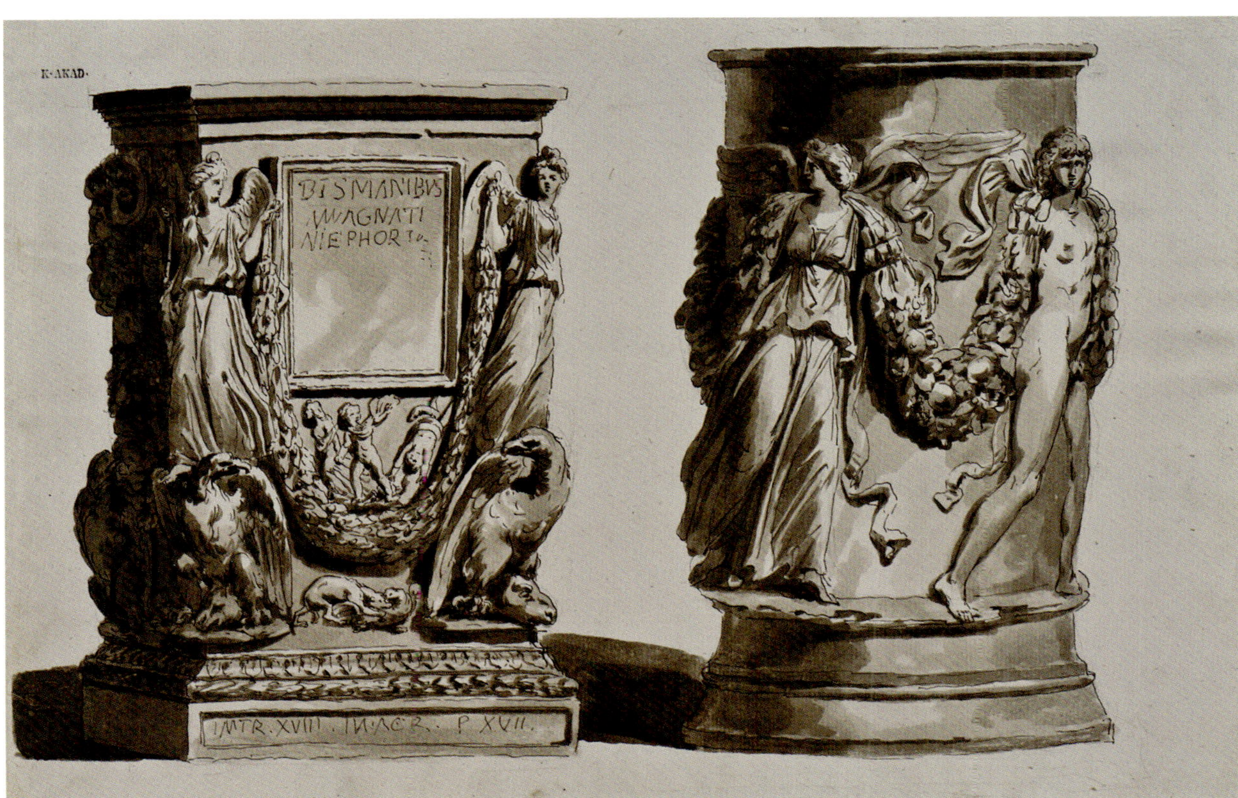

FIG. 4.4. Two antique pedestals by Joseph Bauer, 1781–1788. Pen and wash on paper. – Vienna, Akademie der bildenden Künste, Kupferstichkabinett, HZ 16513.

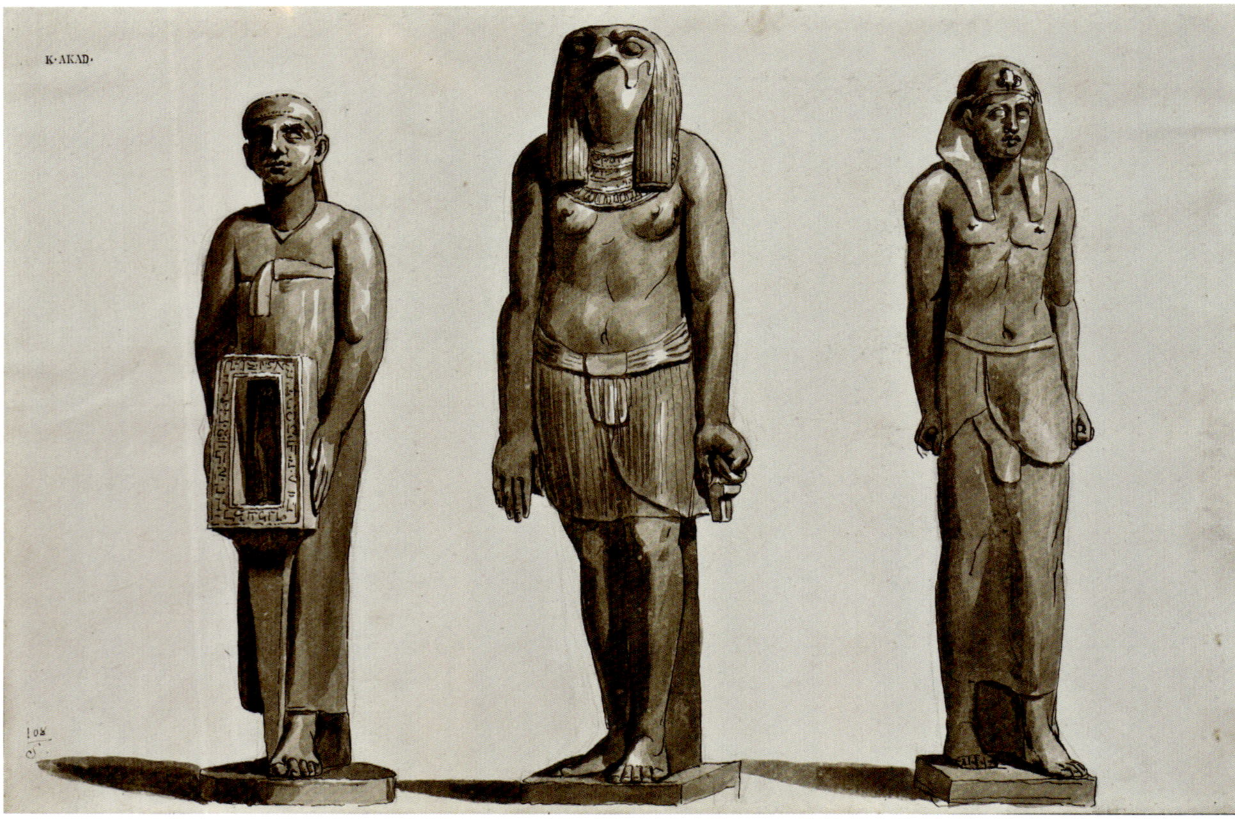

FIG. 4.5. **Three statues from ancient Egypt by Joseph Bauer, 1781–1788. Left to right: Naphor, Osiris Barberini, Ptolemaic king. Pen and wash on paper. – Vienna, Akademie der bildenden Künste, Kupferstichkabinett, HZ 16635.**

the Holy Roman Church and astute antiquarian, he was at the same time one of the great Maecenas in Rome. Albani had employed Winckelmann to catalogue his evolving, constantly replaced and renewed collection of antiquities. On the ceiling of the main hall in Villa Albani Joseph may have admired one of Mengs's masterworks, his fresco 'Apollo, Mnemosyne and the nine muses', called also 'The Parnassus'. It is a piece of art programmatic for early neoclassicism, strongly influenced and subsequently praised by Winckelmann. Decades later Joseph was to copy another of Mengs's paintings kept in the imperial collections in Vienna (see Chapter 12). However, the two artists could not have met in the papal city: Mengs, documented in 1779 as a plaster bust by Christopher Hewetson in the Skulpturensammlung in Dresden,[41] had died in Rome a few months before his protector Albani, and thus also before the arrival of the young man from Feldsberg.

Rome was also the place where any artist could easily find classical objects outside conventional collections, for example when studying the various triumphal arches, columns and other remains of Rome's glorious past. One of Joseph's drawings[42] has been reported to show a scene of the war fought by the Romans against the Dacians illustrated on the Colonna Traiana (Trajan's Column). This is a freestanding column about thirty metres in height, thirty-five metres if its large pedestal is included. Completed in AD 113, it consists of twenty colossal marble drums, each weighing about thirty-two tonnes, with a continuous one hundred and ninety-metre long frieze winding around the shaft twenty-three times. Originally the column was crowned by a statue of Emperor Trajan, though substituted by a bronze statue of St Peter, the Apostle, put there under Pope Sixtus V (Felice Peretti, 1585–1590). The narrative band portrays Trajan's two victorious military campaigns against the Dacians in AD 101–102 and 105–106, with numerous scenes depicting the Roman army in various military activities. Goethe was amazed, writing in retrospect about his visit on 23 July 1787, 'In the evening I climbed the column of Trajan. Seen from the height and at sunset, the Colosseum, with the Capitol close by, the Palatine behind and the city all around, it was a superb sight'. For Joseph Trajan's Column must have brought back memories of Vienna: In 1713, one year after the last great plague epidemic, Emperor Charles VI had made a solemn vow to build a church for his namesake patron St Charles Borromeo in Vienna. This became St Charles's Church, one of the most outstanding baroque church structures in existence, with a huge dome and on the right and left two monumental columns with spiral bas-reliefs modelled on Trajan's column. In addition, the two tower pavilions show the influence of Roman baroque with references to Bernini and Borromini. From his time in Vienna Joseph must have known this very prominent church, with its columns recalling the Pillars of Hercules, another imperial symbol (though at the same time the symbol of the King of Spain, a throne Charles VI had been unable to acquire). It is possible that a few more of Joseph's drawings from Rome, for example one representing a group of Roman soldiers with three military standards (Fig. 4.6.), show other scenes from Trajan's column, which inspired so many later victory columns. However, they may also represent elements of a triumphal arch, something that can be clarified only after considerably more research is done on these drawings.

In 1777, just four years before Joseph left for Rome, architect Hohenberg had the so-called obelisk fountain built on the eastern fringe of the Schönbrunn garden. Among other elements it comprised a column about thirty metres high, consisting of twenty-seven blocks of stone from the quarry of Szentmargitbánya (St Margareten in Austria), crowned by a gilt sphere on which an eagle had been placed. The decoration of this obelisk with ornaments reminiscent of hieroglyphs then not yet deciphered was completely fictional, but the whole iconography referred back to classical Rome, where the eagle had been the traditional emblem of the Emperor. In a similar way the obelisks, first imported as monoliths from Egypt under Emperor Augustus as military spoils, were regarded as symbols of imperial power. Pope Sixtus V was the first to revive the imperial tradition and had four large monoliths erected – the Vaticano, the Esquilino, the Lateranense, all three from Egypt, and the Flaminio from Roman times – in order to create a modern capital worthy of the post-Tridentine Church. For him, re-erecting the fallen monoliths symbolized Catholicism's triumph over paganism – and Protestantism – just as their original arrival in Italy had announced Rome's conquest of Egypt. Needless to say, he and later popes had the obelisks crowned with the cross adding eloquent inscriptions to narrate the story of their re-erection and, of course, the coat of arms of their families.

In 1781, the year of Joseph's arrival in Rome, a large granite obelisk, fire-scarred and broken into three parts, was rediscovered by chance between San Carlo al Corso and the River Tiber. Since no comparable find had been made for centuries, the event generated considerable excitement. Too large for the Museo Pio-Clementino, Pope Pius VI took the decision to have the restored monolith re-erected at Montecavallo, in the piazza next to the Pope's summer palace on the Quirinal. Combining the obelisk, subsequently called Quirinale, with the two horse tamers, colossal groups that survived there since classical antiquity, took considerable time. However, in 1786 the horse tamers had been moved, the monolith erected and crowned with a bronze cross containing relics of saints Peter, Paul, Andrew, Gregory and Pius V, the Pope's personal protector. Joseph may have followed the events leading to the erection of this famous monument, although we have no evidence for this. Many years later he may have witnessed the erection of the

FIG. 4.6. Group of Roman soldiers carrying military symbols by Joseph Bauer, 1781–1788. Pen and wash on paper. – Vienna, Akademie der bildenden Künste, Kupferstichkabinett, HZ 16498.

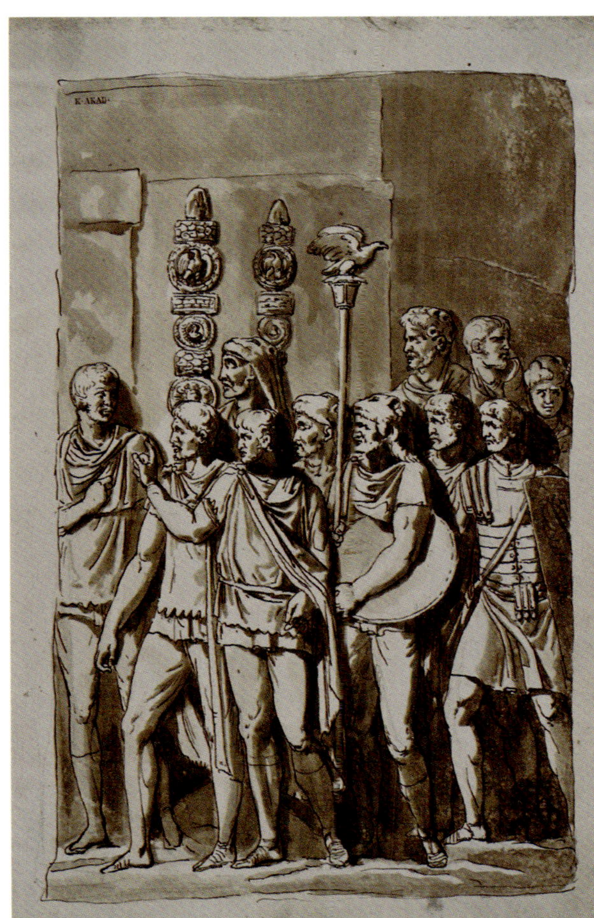

obelisk commemorating the treaty of Campo Formio on the grounds of the Feldsberg estate under the reign of Aloys I Joseph. Small world: this peace agreement with Napoleon had been signed by Philipp Count Cobenzl, then the plenipotentiary of Emperor Francis II, who had been Caucig's benefactor in Rome.

Apart from Trajan's column and the obelisks Joseph may have paid special attention to the many other ruins from classical antiquity right in the centre of Rome, in particular the remains of the temple of Castor and Pollux on the Forum Romanum, and to the various aqueducts. He may have remembered the capriccio painted by his brother Franz for the *Codex Liechtenstein* (see Chapter 2). The latter had combined: (1) a landscape with various representatives of the sunflower family with (2) a monumental oak tree; (3) an alley, possibly the one from Feldsberg to Eisgrub; (4) the three remaining Corinthian columns of the temple of Castor and Pollux copied from Giovanni Battista Piranesi's copper engraving inscribed 'Avanzo del tempio di Castore e Polluce'; and (5) a two-storied, ruinous aqueduct, possibly also taken from Piranesi. About two decades later Joseph could witness the construction of the artificial ruin that his brother had envisaged. Joseph Hardtmuth, later director of architectural works for Johann I Joseph, planned a Roman aqueduct, albeit one-storied, next to the large pond in Eisgrub, with the building finalized in 1805.

The ruins in Rome were genuine, whereas the so-called Roman ruin next to the obelisk fountain in Schönbrunn, which Joseph may have seen before his departure for Rome, was a fake, just like Hardtmuth's later creations. Based on a copper engraving by Piranesi showing two temples on the western margin of the Forum Romanum, this complex structure had also been created by Hohenberg in 1778. During the final years of Maria Theresa, this ensemble was called 'The ruin of Carthage', probably a reference to the victory of imperial Rome over Carthage, another link between Vienna and Rome: the Emperor of the Holy Roman Empire, then Joseph II, possessed a monument within the grounds of his summer residence documenting a decisive victory of the Roman Empire against one of its opponents.

During the extremely long pontificate of Pius VI considerable building activities took place in the Papal States, in particular in Rome, and Joseph could have watched several construction sites, for example the building of the new museum in the Vatican, documented in a canvas entitled 'Pius VI supervising the construction of the Museo Pio-Clementino' by Michael Wutky, kept now in the Philadelphia Museum of Art. He could even have met in Rome his compatriot Wutky from Krems in the Archduchy of Austria, who spent the years 1781–1787 in the papal city and had like him previously attended the United Academy. It seems quite possible that Joseph also visited this museum, which housed among numerous other sculptures from classical antiquity the original Amazone Mattei, acquired from the Mattei family. The museum was a showpiece for Pope Pius VI, who no doubt regarded himself as a temporal monarch, a spiritual leader but also, and – for him, it seems, even more important – a great patron of the arts. The latter role is exemplified by the visit of Gustav III, King of Sweden, to the Pope, who seems to have shown his guest personally round the new museum, a scene recorded in a painting by Bénigne Gagneraux in the Narodní Galeria in Prague. The addition of a new, mannerist-baroque sacristy to St Peter's was another major construction work initiated by Pius VI, and Joseph may well have visited this building site as well.

However, Rome had more to offer to an artist than just churches, papal residences and remnants of the classical past. The black nobility, the Barberinis, Borgheses, Ludovisis, Orsinis, Pallavicinis and several other families, possessed monumental city palaces and luxurious summer residences. In a sense Joseph must have felt like he did in Vienna: the city palace of the Borghese family in Piazza Borghese, for example, referred to the Villa Borghese in Piazzale Scipione Borghese in exactly the same way as the city palaces of the Liechtenstein family in Bankgasse and Herrengasse referred to their villa suburbana in the Rossau district. And when many years later Joseph was commissioned by Johann I Joseph to transfer the family's collection of works of art from the city palaces to the summer palace and arrange them as a private museum (see Chapter 12), recollections from his stay in Rome must have come to his mind.

Joseph may also have paid a visit to the famous Villa Chigi, better known as Villa Farnesina, since one of his drawings (Fig. 4.7.)[43] shows two winged putti with a fruit garland between them, which recalls the decoration on the garden façade of this building, though not in a precise manner. In the loggia of this villa Joseph might have wondered at other garlands: those frescoed in 1515–

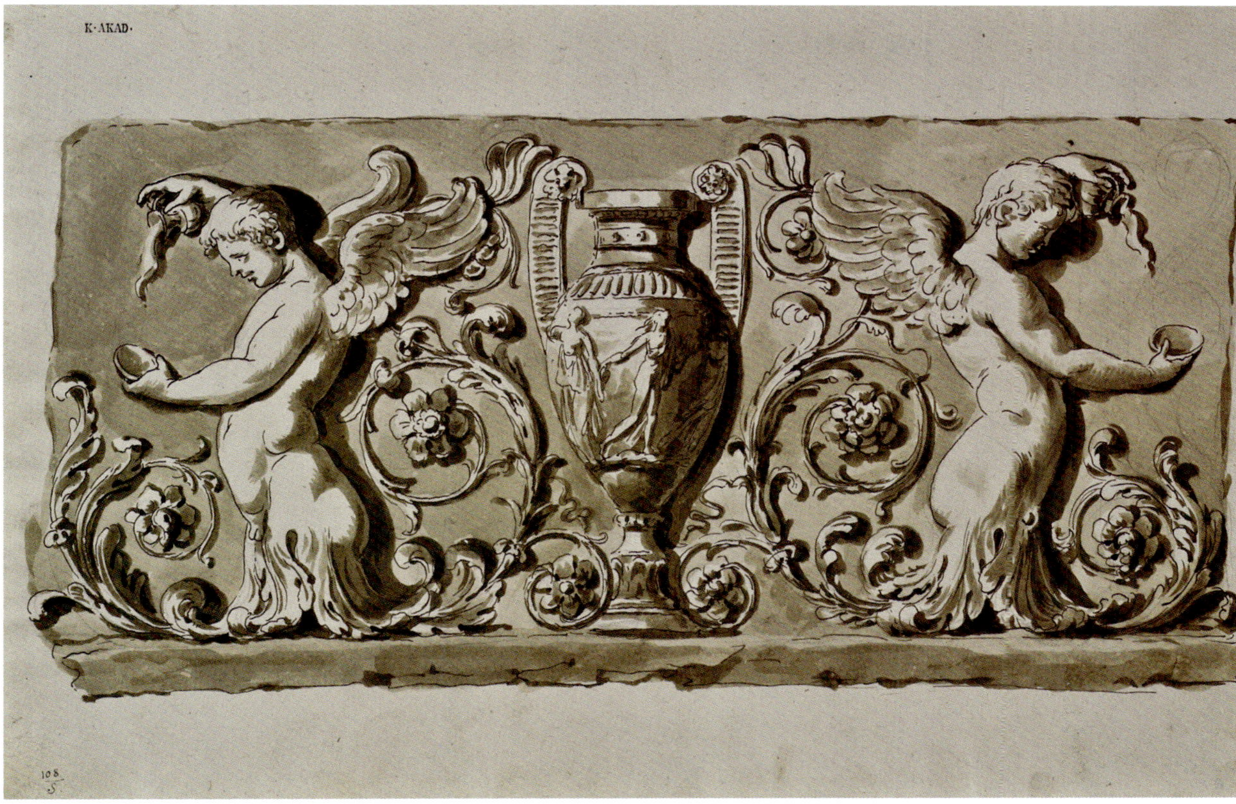

FIG. 4.7. Two winged putti with a fruit-garland by Joseph Bauer. Pen and wash on paper, 1781–1788. – Vienna, Akademie der bildenden Künste, Kupferstichkabinett, HZ 16521.

1517 by Giovanni da Udine,[44] the very first document in existence of the arrival of maize (*Zea mays* L.), a native of the Americas, in the Old World. Joseph must have known this originally Mexican plant from his days in Feldsberg. Small world again: its cultivation in Moravia was promoted at that time by Maria Christina Princess Dietrichstein-Proskau.[45]

NEOCLASSICAL IDEALISM AT ITS BEST

No attempt is made here to list the very many sculptors and painters who stayed in the papal city while Joseph was there and could have met or influenced him. Antonio Canova, another 'principe' at the Accademia di San Luca, is a particularly prominent example: during these years he produced the funeral monuments for Pope Clement XIV in the church Santi Apostoli in Rome and subsequently also for Pope Clement XIII in St Peter's, which made him famous and stood at the beginning of an outstanding career and an impressive list of later master works. These include the reclining statue of Princess Paolina Borghese Bonaparte, a sister of Napoleon, as *Venus Victrix*, now exhibited in the Villa Borghese in Rome, and the funeral monument for Archduchess Maria Christina, in St Augustine's Church in Vienna. Canova, probably the most innovative sculptor of neoclassicism, also prepared three marble busts, which ended up in the Fürstliche Sammlungen: one showing Leopoldine Princess of Esterhazy, née Liechtenstein, the youngest sister of Joseph's long-term superiors Aloys I Joseph and Johann I Joseph, as *Venus Italica*; the second Napoleon, who made Johann I Joseph a sovereign in his possessions on the upper Rhine (see Chapter 2); the third Napoleon Franz, the Roi de Rome, the infant son of Napoleon and grandson of Emperor Francis II. The sculptor Zauner (see above) was in Rome until October 1781, which means that the overlap with Joseph was minimal.[46] Füger (see above) remained in Rome until summer 1783 and had more chance of meeting Joseph. Back in Vienna, the former became Deputy Director of the United Academy and later Director of the imperial picture gallery, then already in the Upper Belvedere.

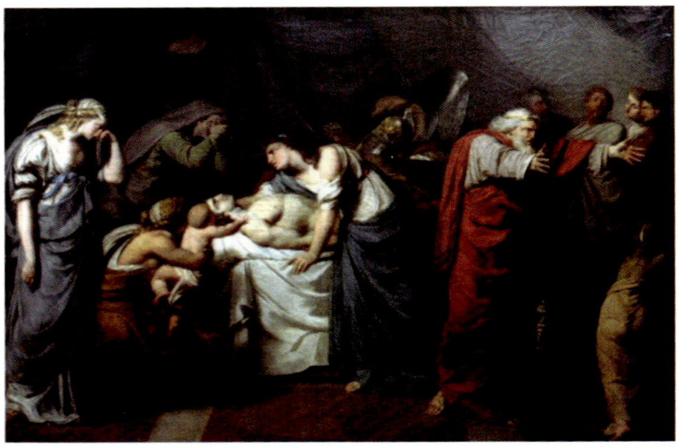

FIG. 4.8. Deploration of Hector by Andromache by Joseph Bauer. Oil on canvas. – Clermont-Ferrand, Musée d'art Roger Quilliot, Inv. No. 73.25.1.

Painter Angelika Kauffmann, another key figure of neoclassicism, born in Chur in Switzerland with strong ties to the Hereditary Lands, settled down in November 1782 in Rome's via Sistina 72, the house formerly inhabited by Mengs. She made her home a meeting place for artists and visitors, among them Emperor Joseph II, Anna Amalia, Duchess of Saxony-Weimar-Eisenach, née Duchess of Brunswick-Wolfenbüttel, Johann Gottfried Herder and, importantly, Goethe. The latter three had come from Weimar and are reported to have had Hirt as their cicerone in Rome, later the first professor for archaeology at the newly founded Friedrich Wilhelms University in Berlin and involved in planning a museum that eventually became the Altes Museum in that city. Goethe is reported to have called Angelika a woman of outstanding talent, although he was not convinced by the portrait she painted of him. In September and October 1782 in Naples, Kauffmann had produced a sketch of Ferdinand IV, King of Naples (as Ferdinand III also King of Sicily) with his wife Maria Carolina, a daughter of Maria Theresa, and her six children. This formed the basis for a modello, in the Fürstliche Sammlungen, whereas the finished large-sized painting dated 1783 is in the Museo di Capodimonte in Naples. In the same year Jacob Philipp Hackert produced his painting 'Ferdinand IV hunting for coot on Lake Fusaro' also in the Museo di Capodimonte. When Goethe visited Rome in 1786, Hackert became his drawing tutor. In the Fürstliche Sammlungen Hackert is represented by one of his many paintings in oil from the Apennine Peninsula showing an 'Arcadian' landscape near Roveto, then a favourite topic for artists in Rome. Another 'Arcadia' was painted by Caucig, now in the Kereszrény Múzeum in Esztergom in Hungary. Goethe had arranged a scholarship for the painter Johann Heinrich Wilhelm Tischbein, with whom he stayed in 1786 in Rome. In the same year the former started to work on a canvas in oil denominated 'Goethe in the Campagna' now in the Staedel in Frankfurt, showing the poet stretched out in his white coat on a fallen obelisk. Later this painting became another icon of both neoclassicism and 'Italiensehnsucht' (the longing for Italy), so widespread in central Europe at the time.

Probably more important for Joseph than the works by Canova, Zauner, Füger, Kauffmann, Hackert and Tischbein was the large painting 'Oath of the Horatii' by Jacques-Louis David, now in the Louvre, painted in 1784 in Rome. It occupies an extremely important place in David's work and in the history of neoclassicism in general and has been praised by many. Since the painting is reported to have attracted the special attention of the general public[47] because of its 'antique simplicity', its existence must have been known in Rome's artistic circles to which Joseph belonged. It has rightly been said that David' s painting shows a morally uplifting story, promoting civil duty over personal considerations, reflecting the values of the age of enlightenment and neoclassical idealism.

Joseph is known to have produced a single canvas in oil of similar quality, clearly his masterpiece, entitled 'Deploration of Hector by Andromache' (Fig. 4.8.) and kept today in the Musée d'art Roger-Quilliot in Clermont-Ferrand.[48] It is undated but signed Bauer and correlates so perfectly with a complex, large-sized drawing[49] in the Kupferstichkabinett listed as 'Briseis and the corpse of Patroklos' attributed to Joseph (Fig. 4.9.), that there is no room left for doubt. The proposal is put forward that this drawing, on light brownish paper with pen and brush in brown, was also prepared in Rome. As in the case of David's famous painting, the subject is taken from classical antiquity – in the case of Joseph from the Ilias, scroll 24. The canvas shows the mourning over Hector killed in battle against the Greeks, with his wife Andromache bending over his corpse, and his son Astyanax trying to approach him. Hector's sister Helena, his mother Hecuba and his father Priam join the mourning widow. A total of eleven people are shown in addition to Hector stretched out on his deathbed. Again we have a morally uplifting story, an example of virtue and courage, of a family mourning in solidarity, reflecting in a sense the values of the enlightenment combined with neoclassical ideals. As a matter of fact this very scene was popular among Joseph's colleagues, as exemplified by a canvas by David kept in the Louvre in Paris, which David based on his drawing, today in the Musée du Petit Palais in Paris.[50] However, in contrast to Joseph's painting only the core scene with Hector, Andromache and Astyanax is shown by David.

The Clermont-Ferrand painting illustrates a general approach to themes from classical antiquities very similar to the oil paintings by another colleague of Joseph in Rome – his compatriot Caucig. In fact some of Caucig' s works, for example 'Esther before king Ahasuerus' in the University of Virginia Art Museum, Charlottesville, 'The judgement of Solomon' in the Narodna Galerija in Ljubljana, or his 'Focion with his wife and a wealthy Ionian woman' in the Akademie der Künste in Vienna, could also have been from Joseph's hand.

One more drawing[51] kept in the Kupferstichkabinett and attributed to Joseph is comparable in detail and size to the drawing of the 'Deploration of Hector': it shows a young satyr with grapes in his right hand and a nymph, both almost in a lying position, the latter nurturing a baby (Fig. 4.10.). Joseph's other subjects such as 'Zeus on his throne surrounded by other gods',[52] 'Three Graces',[53] 'Europa abducted by the bull'[54] and 'Death of Hector',[55] the latter in the Fürstliche Sammlungen have a similar background typical of the neoclassical movement. When the estate of his younger brother Franz came up for auction in London in November 1841, a 'Pythagoras addressing his disciples' and a drawing in chalk 'Theseus recognizing his father',[56] both by Joseph, were listed; their current location are unknown. However, there are also drawings in the Kupferstichkabinett showing scenes from the Bible, such as 'The last victims of the Great Flood',[57] or a rather conventional 'Pietà'.[58] Nothing indicates, however, that Joseph recorded plant and animal life while in Rome, thereby discontinuing his earlier work in Feldsberg, however small it may have been in comparison to that of his brothers Franz and Ferdinand (see Chapter 2). In short, he was fixed on representations of human figures and artefacts, excluding those of natural objects.

There are notable differences between Joseph's work in Rome and that of Caucig: the latter is the author of numerous views from Rome and elsewhere, which

FIG. 4.9. Deploration of Hector by Andromache by Joseph Bauer, 1781–1788. Pen and wash on paper. – Vienna, Akademie der bildenden Künste, Kupferstichkabniett, HZ 3524.

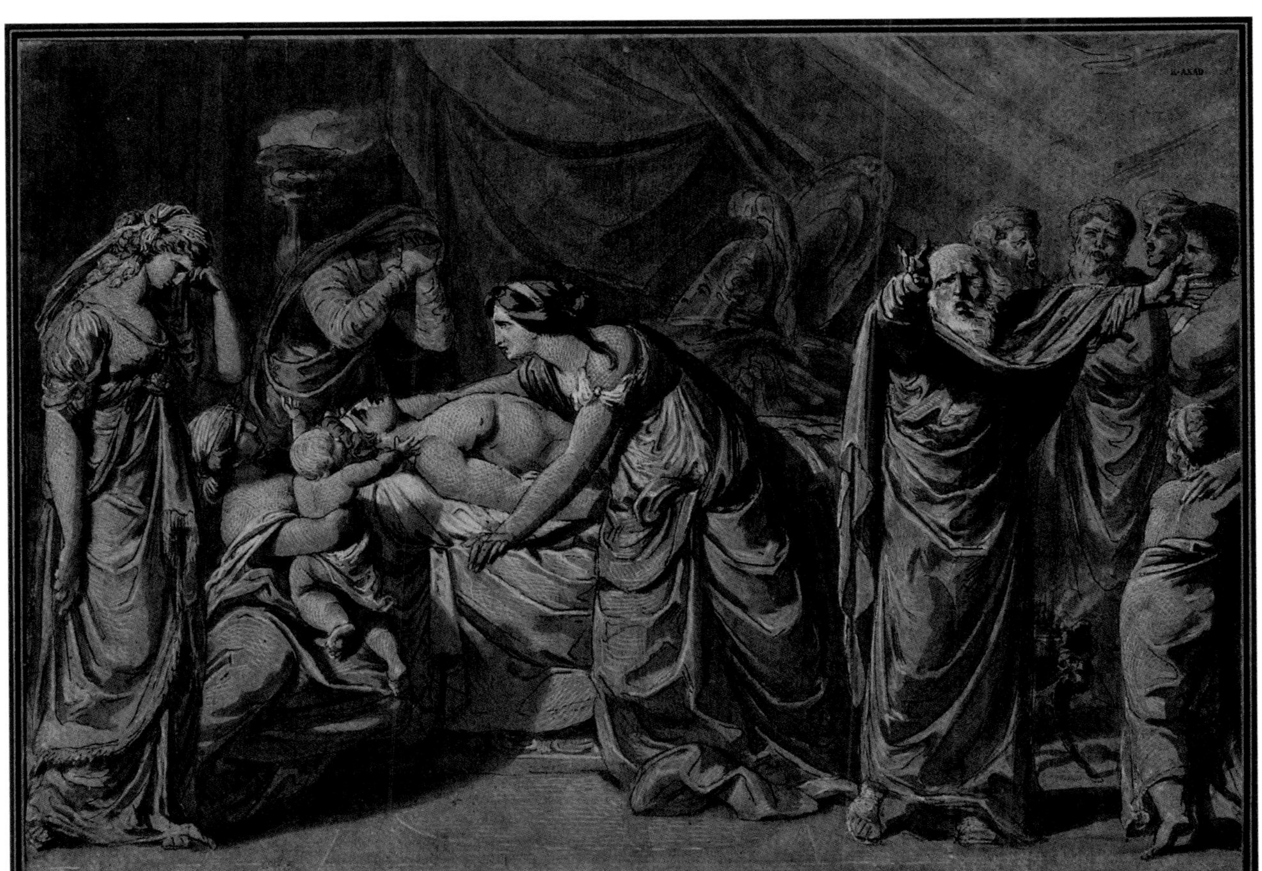

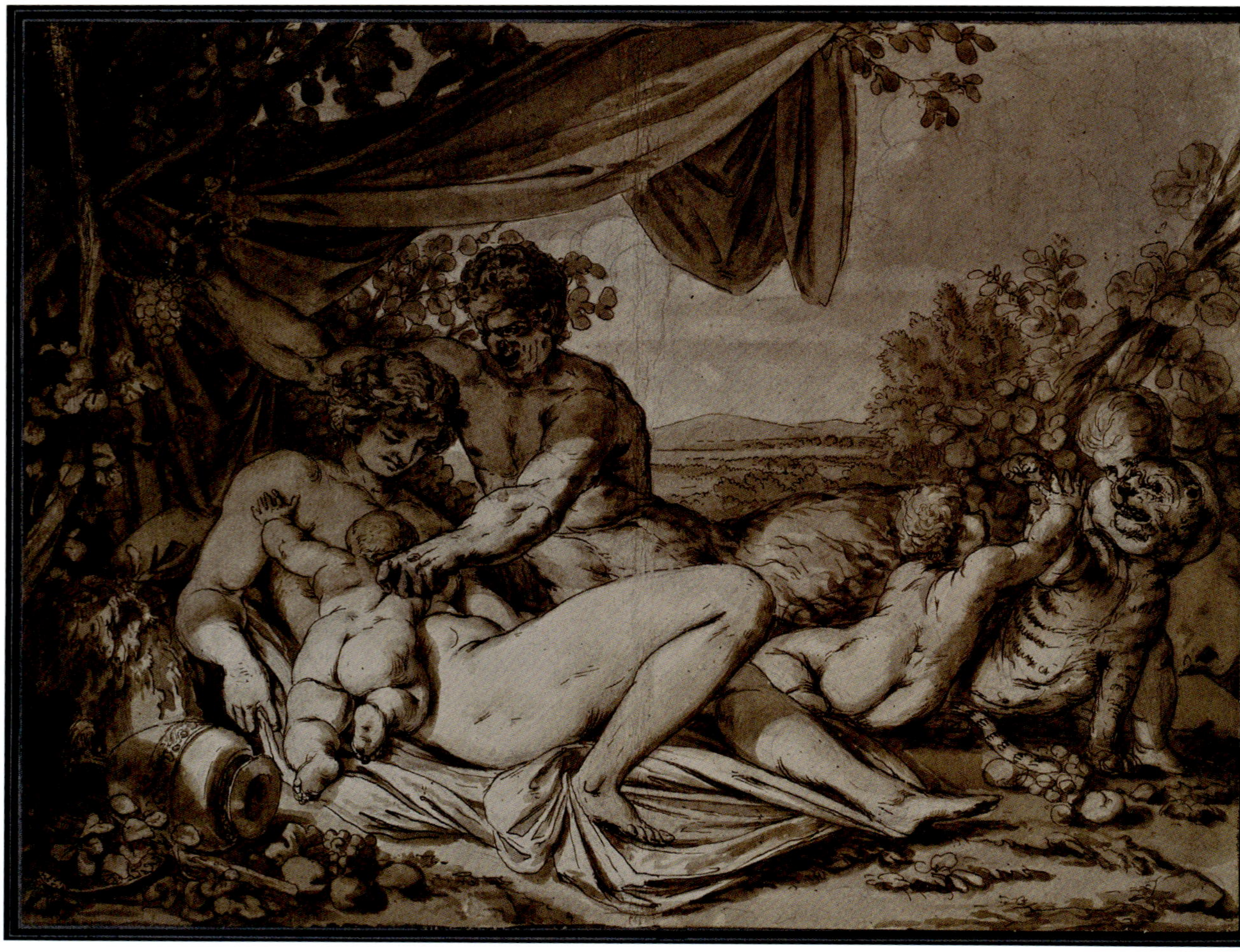

FIG. 4.10. Satyr and nymph by Joseph Bauer, Pen and wash on paper, 1781–1788. – Vienna, Akademie der bildenden Künste, Kupferstichkabinett, HZ 3521.

surprisingly seem to be almost totally missing from Joseph's works, apart from a 'Temple of Venus' in the Fürstliche Sammlungen.[59] In addition, Joseph's oeuvre was very much smaller, numbering maybe one-tenth of Caucig's works (no less than 1948 drawings by Caucig are indicated for the Kupferstichkabinett[60] alone), or has survived only as a fragment. Because of the blatant similarity of their drawings and paintings Joseph may even be regarded as the little brother of Caucig.

Everything therefore seems to indicate that Joseph cannot be regarded as belonging to the inner circle of painters in Rome, but rather to the periphery. 'More diligence and practice than accuracy and sensitivity' was Hirt's comment on Joseph, listing him in a range from 1 to 33 as number 14 among the German-speaking historical painters living in 1787 in the papal city, with Maron, understandably, as number 1.

Joseph and Caucig share a poor youth in provincial towns, an early move to the imperial city, a long stay in Rome and then a return to Vienna, which was final for the former though not for the latter. More importantly in our context, Joseph seems to have largely discontinued working as an artist after his return to Vienna in 1788. By contrast, there was no pause in Caucig's productivity, neither after his return to Vienna, nor during his stays in Mantua and Venice. Later in his life Caucig again lived in Vienna and became a central figure at the United Academy, eventually becoming Director from 1820 until his death.

Looking back on his seven years in the papal city Joseph may have been prepared to subscribe to Goethe's comment on late *settecento* Rome, as put forward in his *Italienische Reise*: 'Rome is a marvellous place. One finds here not only objects of every kind, but also people of every kind who take their work seriously and know exactly what they are doing, and one makes rapid and easy progress in their company. I am beginning, thank God, to be able to learn from others and profit from their experience'.

For the continuation of Joseph's life see Chapter 12.

NOTE

Since a vast amount of information on late-eighteenth century Rome is available from a multiplicity of sources, the bibliography had to be restricted to the very essential. Useful background information is summarized in many reference works. For the arts during the pontificate of Pius VI, Collins (2004) was used as main guidance. Data on the Fürstliche Sammlungen were extracted from www.liechtensteincollections.at, accessed in January 2013.

Thanks are due to E. Czerny, Vienna for information on the three Egyptian statues; to C. Reiter, Vienna for helpful comments on Joseph's drawings kept in the Kupferstichsammlung, in particular on his copies based on Reni; and to Dr K. Rozmann, Ljubljana for information on F. Caucig and the lack of any data on his contacts with Joseph.

ENDNOTES

1 Lack (2000a); www.barmherzige-brueder.at/content/site/oesterreich/orden/unsere_ordensprovinz/unser_mutterhaus/index.html
2 Haupt (1990)
3 Noack (1927), Press (1990)
4 Grabner (2002)
5 Collins (2004)
6 Mengs - Inv. No. 1989, Maron - Inv. No. 1796
7 Frodl (2002)
8 Lorenz (1999 a)
9 Lorenz (1999 b)
10 Gams (1972 a)
11 Tausch & Johannsen (2004)
12 HZ 16443 - 16526, 3521 - 3527
13 GR 119 - 121
14 Wiench & Römer (1993)
15 Becker (1887)
16 Schütz (2005)
17 Załęski (2005)
18 Knall-Brskovsky (1998)
19 Kräftner, pers. comm. (2012)
20 HZ 16502
21 Rozmann (1978)
22 HZ 3525
23 Inv. No. 17573 F; cf. Birke (1981)
24 HZ 3523, 3522
25 HZ 682 ; Müller-Kaspar (2004)
26 HZ 16642
27 HZ 650
28 HZ 16645
29 HZ 672
30 HZ 16472
31 HZ 16513
32 HZ 16635
33 E. Czerny, pers. comm. (2013)
34 Inv. No. WAF 22
35 Inv. No. 14765
36 Inv. No. 14764
37 Inv. No. 103
38 HZ 201
39 E. Czerny, pers. comm. (2013)
40 Kamin - Kajfež (2010)
41 Inv. No. Mgs 27 (ASN 9728)
42 HZ 16456 ; Gams (1972 b)
43 HZ 16521
44 Caneva (1992)
45 Brichtová (2002)
46 Wagner (1972, 1973)
47 Rauch (2006)
48 Inv. No. 73.25.1
49 HZ 3524
50 Resemann (2006)
51 HZ 3521
52 HZ 12855
53 HZ 16617
54 HZ 16467
55 GR 119
56 Lack (2003)
57 HZ 12812
58 HZ 16621
59 GR 121
60 Kofler (2004)

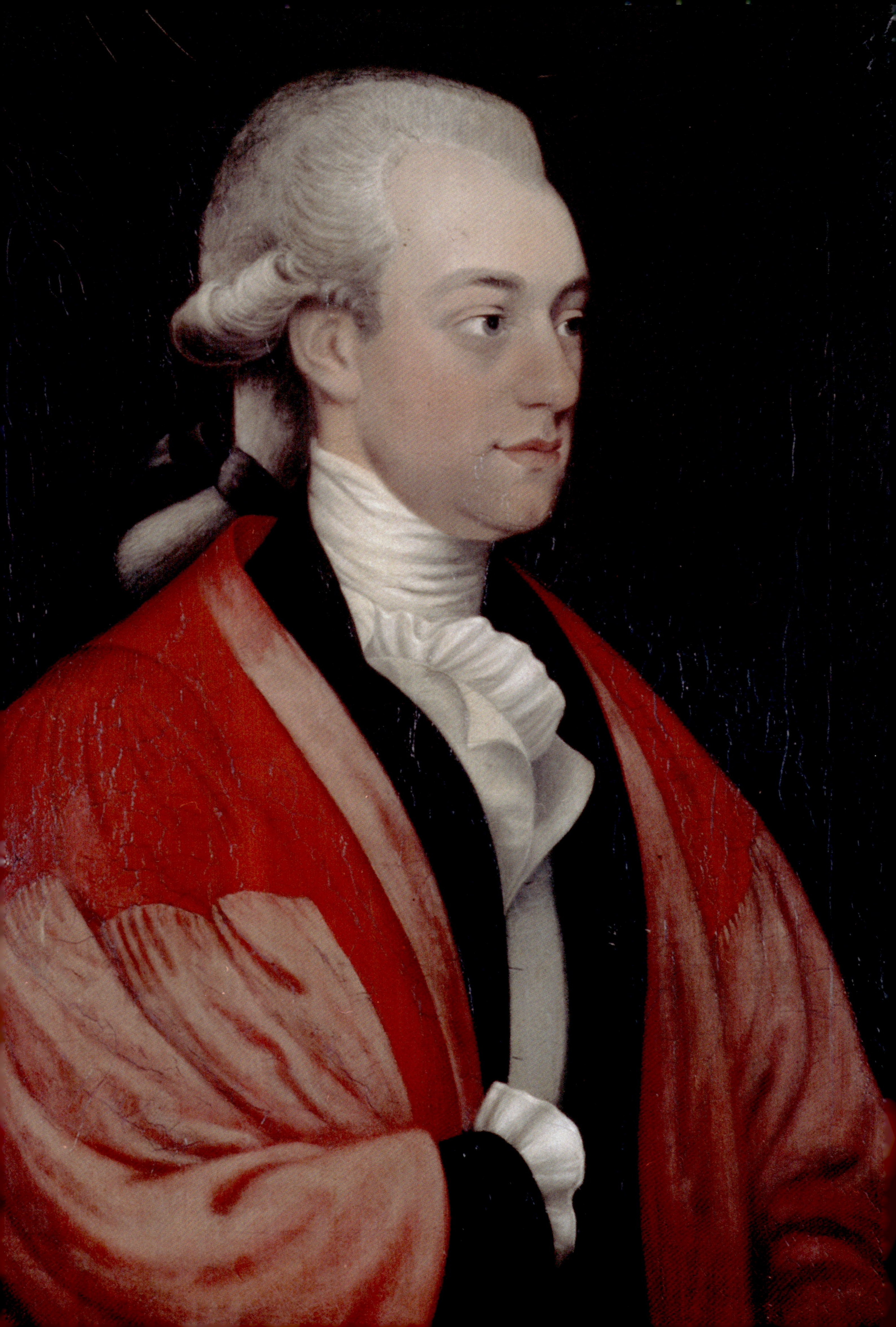

CHAPTER 5

Ferdinand Bauer and John Sibthorp explore the Levant[1]

FERDINAND MEETS JOHN SIBTHORP

IN LATE 1785 JOHN SIBTHORP, the third Sherardian Professor of Botany at Oxford University, descended on Vienna and soon after hired Ferdinand as his drawing amanuensis. As this employment lasted until early 1794, Ferdinand took the decision to leave Vienna and relocate to Oxford. This move resulted in his subsequent career as a botanical illustrator engaged by the prominenti in England. It therefore seems appropriate to start with a brief biographical sketch of his new superior.

Born on 28 October 1758 in Oxford, John Sibthorp was the only son of Humphrey Sibthorp, M.D., second Sherardian Professor of Botany at Oxford University, and his second wife Elisabeth Sibthorp, née Gibbs. The father belonged to an old Lincolnshire family of wealthy landowners with Canwick Hall as one of their homes, while the mother was the daughter of a rich London merchant and his eventual heiress. As a member of the landed gentry, Humphrey had married wealth and stood in the social hierarchy of his time well above Jacquin, later to become his equivalent in Vienna, but far below members of the aristocracy, such as the Duke of Marlborough or the Reigning Prince of Liechtenstein, for example. As we shall see, Humphrey's position in the system later also applied to his youngest son John.

FIG. 5.1. **John Sibthorp. Portrait, 1850. Oil on canvas. – Oxford, University of Oxford, Department of Plant Sciences.**

Humphrey Sibthorp had been elected to the chair in 1747 at the age of 35 without having ever published a single word. There is universal agreement that the new Sherardian Professor was, and more importantly, remained for his subsequent thirty-six years in office, a nonentity in science and the most ineffectual holder of the chair. Persuaded by flattering letters from Humphrey, in 1750 botanical illustrator George Dionysius Ehret, a native of Heidelberg, had accepted the position of gardener at the Physic (Botanical) Garden of Oxford University, which was supervised by the Sherardian Professor. Their relationships seem to have been difficult right from the beginning and 'a clash between the two was inevitable'. Less than a year after his appointment, Ehret complained to correspondents about the state of affairs in the Physic Garden, adding that Sibthorp was 'a Tyrannizing man and Claims a[n] Arbitrary Power in the Garden'. Soon after this Ehret resigned from his Oxford engagement and devoted himself again to his work as a botanical illustrator in London and teacher of the drawing and painting of plants. In our context a single point is of relevance – Humphrey Sibthorp knew the most qualified botanical illustrator of his time in person and was probably aware of Ehret's highly appreciated botanical illustrations, years before his own son John was born.

John Sibthorp must have grown up in the official residence of the Sherardian Professor, 'a spacious first-floor apartment above the main long conservatory on the street front of the Physic Garden' very near to Eastbridge across the river Cherwell. From his earliest days John was thus close to botany and the academic world, with Magdalen College just across the road – manuscripts, books, herbarium specimens and plant illustrations must have accompanied him from childhood. Clearly he grew up in a privileged environment as the son of a well-connected and well-off professor belonging to a famous university. In 1775 Humphrey Sibthorp bought 'a finely-sighted farm-property . . . just across the two branches of the Cherwell', had the farmhouse pulled down and an elegant three-story villa constructed, which became known as Cowley House. No details are available on the date of its completion, nor do we have any information as to when the Sibthorp family, including John's half-brothers and half-sisters, moved in. But this villa was to become the scene of many of the events of the next chapter and was to be accurately recorded by Ferdinand in one of his Oxford vedute entitled 'View from the walk of Christ Church Meadow to Magdalen College at Oxford' – a drawing in pen and black ink and grey washes, washed over with pale yellow, drawn with ruled borders washed in black and grey (Fig. 9.2.). Cowley House still exists, forming the centre of the north block of St Hilda's College, but buildings having been added to the villa, so its original layout is not easy to make out. The garden running down to the river Cherwell survives, although it has considerably changed in appearance.

When he was 15 years old John matriculated at Lincoln College, Oxford. He obtained his Bachelor of Arts in 1777 and subsequently entered the medical profession, following in his father's footsteps. Taking the deplorable state of medicine at Oxford in the eighteenth century into account it is hardly surprising that Sibthorp went to Edinburgh in January 1779,[2] where he lived – thanks to an income of £100 p.a. plus a legacy from an aunt – in great style, noting in typical upper class arrogance 'the Characters of a Gentleman & Student are . . . incompatible'. A specific interest in botany is first mentioned in early 1780, shortly before John obtained the degree of Master of Arts at Oxford University. In the same year John lost his mother, Elizabeth Sibthorp. Her death made him the owner of a beautiful estate at Instow near Bideford in North Devon, later documented by Ferdinand. Thus the young Master of Arts, who was at all events resolved to appear as a gentleman, became a landowner in his own right at the age of 22. The rents from Instow constituted an important source of his income, which was increased very substantially by the Radcliffe Fellowship he obtained in June 1781.

Under the very generous conditions of the Radcliffe Foundation John received the then-colossal sum of £300 p.a. for no less than ten years, with the only stipulation of travelling 'in part beyond sea' (that is, outside Great Britain) for five years for his 'improvement'. The candidates for the Radcliffe Fellowship had to be Masters of Arts and to have entered the 'physical line' (studied medicine), and, typical for Oxford, the jury included the Archbishop of Canterbury, the Bishop of London, the Bishop of Winchester, the Lord Chancellor of Great Britain, the Master of the Rolls and the Chancellor of Oxford University. There is evidence that Humphrey Sibthorp, being accustomed to the realities of academic life, had actively lobbied for his son, travelled to the electors in London and Winchester and managed to have the decision on the scholarship postponed until his son would be eligible. We may also assume that he had spoken to the Chancellor of his university and to his son-in-law Sir Thomas Sewell, who happened to be Master of the Rolls.

The winning of the Radcliffe Fellowship was of decisive importance for John Sibthorp's career – on the one hand he was now obliged to travel to the continent, but on the other hand he was able to do so in the style of a gentleman of leisure. His first tour began in October 1781 and probably ended in autumn 1783, with Paris being his most important destination. Unsurprisingly he was attracted by the Jardin du Roi (see Chapter 2) noting 'Here is the Botanick Garden within view of my Window & I might have sketched some Plan from the Leçons of a Jussieu for a Course which did the College of Physicians flatter my ambition I might at some Day have giving [given] to the University [of Oxford]. I had taken a Catalogue of the Plants, & had entered upon my Collection'. Having encountered Antoine-Laurent de Jussieu, 'sous-démonstrateur de l'extérieur des plantes', and other members of staff, Sibthorp may have been introduced to the botanical materials collected by Joseph Pitton de Tournefort, Andreas von Gundelsheimer and Claude Aubriet in the Levant. These included the specimens collected by Tournefort and Gundelsheimer in Tournefort's herbarium in the Jardin du Roi, with duplicates conserved in various other collections, notably those of the Jussieu family, as well as living material that could still have been in cultivation in the garden. In addition, John definitely studied Aubriet's botanical illustrations, including those based on drawings made during the voyage of the Levant. Although we cannot exclude the possibility that John had seen herbarium specimens collected by Tournefort in the collections under the care of his father at Oxford, his experience in the Jardin du Roi must have been a turning point in his life. Tournefort's expedition became the model for his travels to the east and his visit may have stimulated his ambitions to explore the eastern Mediterranean. Paris also offered various attractions in other ways, such as taking part as 'a fine Gentleman in Domino at the grand masqued Ball at Versailles', watching the arrival of the Queen (Marie Antoinette) or simply 'taking a jolie fille under my arm'. In 1782 John continued to Montpellier, famous for the oldest botanical garden in France, and received his first impressions of the flora and fauna of the Mediterranean area, which were to determine his subsequent activities.

By October 1783 John Sibthorp was back at Oxford obtaining the degree of Bachelor of Medicine on 8 December 1783. In the same month his father vacated the chair of botany and on 20 January 1784 John received the degree of Doctor of Medicine at his university, thus echoing the academic career of his father in every detail. For a man of his class and calibre it was only natural to have his portrait painted in oil soon afterwards, and indeed this was the case. Today a copy of it, inscribed 'J. Smith & Sons 1850', hangs in a prominent position in the Department of Plant Sciences of Oxford University. It shows John Sibthorp nearly in profile facing to the right in his scarlet Doctor of Medicine gown over a black coat and white waistcoat under a powdered wig tied behind by a black ribbon (Fig. 5.1.). Two and a half months later, on 5 April 1784, the Royal College of Surgeons elected John Sibthorp third Sherardian Professor of Botany. He was 25 years old and, like his father, had not published a single work. It seems that John was the driving force behind this carefully staged academic manoeuvre. However, he not only fulfilled all requirements – being 'a MA, not in holy orders and at least registered on the physic line' – he also possessed the perfect social background and, in addition, came from a family of Oxford men. One of his grandfathers, two of his uncles, his father, two of his half-brothers and the husband of one of his half-sisters had studied at that university. An earlier comment on this method of appointment is very appropriate: 'by the standards of today the whole affair reeks of jobbery and nepotism', but 'as it turned out, a very good and supremely keen botanist was appointed'. In fact the resignation of the 72-year-old Humphrey Sibthorp, who had been inactive for many years, and the election of his much more energetic son greatly improved the standard of botany at Oxford.

Like his father, John Sibthorp showed no inclination to devote himself to the routine duties of a professor, nor to take any immediate action concerning the fine Botanical Garden, the oldest in England, which he was to supervise. In order to comply with the regulations of the Radcliffe Scholarship, Sibthorp had to spend three more years 'beyond sea'. The idea of a tour to the Ottoman Empire is not referred to in his correspondence and it may well have developed only during his stay on the Continent. Ample funds were available to him – the generous scholarship, his salary as Sherardian Professor, his income from the Instow estate, an allowance from his father, and possibly still the legacy from his aunt.

While crossing the English Channel in the late summer of 1784 Sibthorp met for the first time a young man, aged 23, who more than two years later would become his travel companion for several months – John Hawkins, the fourth son of Thomas Hawkins and his wife Anne, née Heywood.[3] John's family belonged to the landed gentry of Cornwall with extensive properties, including many mines. For obvious reasons John Hawkins was interested in geology and had been admitted a pensioner at Trinity College, Cambridge, where he had obtained the degree of Bachelor of Arts in 1782.

We first find Sibthorp in the Republic of the Seven United Netherlands, from where he entered the Holy Roman Empire. His first destination seems to have been Göttingen, then ruled by the Prince-Elector of Brunswick-Luneburg, who happened also to be King of England as George III. Two letters were written in this small town, 'a German university of much learned labour and sober science, with a library the richest in modern books of any in Europe'. In Göttingen Sibthorp was elected corresponding member of the Königliche Gesellschaft der Wissenschaften (Royal Society of Science),[4] the equivalent of the Royal Society in London. In addition he seems to have started to plan his tour to the Ottoman Empire, possibly influenced by the classical studies for which the small town was well known.

In Berlin, the residence of the Elector of Brandenburg and King of Prussia, then the aging Frederick II, 'a city too military to please me much', Sibthorp seems to have met Hawkins again, but he continued alone via Potsdam to Leipzig, an important commercial city belonging to the Electorate of Saxony. There he reported having 'dined several times with [Johann] Hedwig & . . . made two or three herborisations together, he showed me most of the Discoveries he had made with the Microscope . . . He has discovered the Organs of Generation in every part of the Vegetable Kingdom & . . . has illustrated his Discoveries with Examples from the Filices [ferns] Algae Musci [mosses] & Fungi' (see Chapter 7). Dresden was his next stop. Sibthorp was not only attracted, like every tourist today, by the Grünes Gewölbe, the world-famous collection of jewellery collected by the Electors of Saxony and Kings of Poland, but by a herbarium of Tournefort, now lost. This seems to indicate his continued interest in this important expedition to the Levant. It is very likely that Sibthorp continued from Dresden via Prague to Vienna, where he must have arrived in early November 1785. On 10 November Sibthorp reports to his father that he had been presented to Emperor Joseph II on the day of the celebration of the Order of St Stephen, which had been created by Maria Theresa. He added that he had dined with Kaunitz (see Chapter 4), who after the death of the Queen of Hungary and Bohemia had lost some of his power but remained Chancellor and Minister of Foreign Affairs. He had also met with the Russian and French ambassadors, and visited the Imperial Garden at Schönbrunn.[5] In Vienna Sibthorp was also introduced to Ignaz Edler von Born and told his father about recent discoveries on the germination of seeds made by Ingen Housz (see Chapter 3).

The stay of about three months in the imperial city – he left Vienna on 6 March 1786 – was of decisive importance for Sibthorp because of five events: the meeting with Jacquin (see Chapter 3), his equivalent at Vienna University; the engagement of Ferdinand; the acquisition of engravings showing plant illustrations from two early Byzantine manuscripts; the development of a more precise itinerary for the journey to the East; and the purchase of the necessary travel equipment.

In Vienna, Sibthorp's contact was the Fries family, wealthy and influential bankers (see Chapter 3). Two letters written by Sibthorp to his father and Hawkins are the main sources of his stay in Vienna, which he seems not to have enjoyed – regarding the inhabitants as 'barbarians' and disliking the 'ennuie with the Gêne of a Court'. Describing a supper with 'Prince Pany' (possibly Lipót Count Pállfy de Erdöd, Fries's neighbour), including 'the profusion of Dishes the Service of massive Gold twenty servants in waiting, Tokay & Grape Wines', he told his father that he would have preferred to have eaten 'hatched Cockles with you at Instow – & drank the Grape Wine made from our own Vineyards'.

In late 1785 Sibthorp must have met Ferdinand, since the former told his father on 3 January 1786 'I have engaged an Artist to accompany me who defigures all Sorts of Subjects in Natural History particularly Plants superior to any Artist I have yet seen. – He is the principal Draughtsman of Jacquin to whose friendship I am . . . indebted'. Typically for Sibthorp he did not even mention the artist's name, and the same is true of his letter of 3 March 1786 to Hawkins writing 'my Painter in each part of Natural History is Princeps pictorum [the prince of painters] he joins to the Taste of the Painter, the Knowledge of a Naturalist & Animal, Plant & Fossil touched by his Hand shew the Master'. For a young man like Ferdinand this is high praise indeed. The reasons for engaging Ferdinand were straightforward, with Sibthorp telling his father 'The Superiority of my Draughtsman will fence me what I shall publish . . . & will entitle me to a Place in the Petersburg or to our Academy [Royal Society] with just Pretensions'. In short, Ferdinand should act as the vehicle for his master's fame.

Sibthorp offered a salary of £80 p.a. for accompanying him as an illustrator and for the subsequent work. This was double the salary of the *praefectus horti* at the Physic Garden at Oxford in that year and probably meant a considerable increase of income for Ferdinand bearing in mind his previous earnings in Vienna. In addition, it meant an engagement for the duration of a project, not any longer the occasional commissions from Jacquin.

Sibthorp seems to have made a visit to the little town of Feldsberg, since he described on the first page of his travel diary the plant illustrations kept by Boccius 'He [Ferdinand] was educated at Feldsberg under the care of Norbert Boccius . . . His Collection of painted plants amounting to upward of two thousands done by le[s] Frères Bauer . . . is the most splendid work of this kind that perhaps exists. the Plants are painted in their Natural Size in large Folio . . . The Plants painted by Aubriet in the Cabinet du Roi at Paris with all their Lustre of Colouring are greatly inferior to the Plants of Norbert. The Umbelliferous Plants will probably remain unrivalled in the accuracy & Elegance of Design'. This brief statement proves that Sibthorp, while in Paris, had indeed seen the famous 'collection des vélins', now in the Bibliothèque Centrale of the Muséum National d'Histoire Naturelle. Secondly it shows that he possessed an excellent visual memory, being able to compare the quality of the watercolours in Paris with those he had seen in Feldsberg.

It should be noted that we have no source on Ferdinand's engagement other than that of Sibthorp. Were the financial aspects of the arrangement with Sibthorp the sole cause? Was he motivated, as so many young people, by wanderlust? Did the chance to meet his elder brother in Rome matter? Did Sibthorp disclose his travel plans in any detail to Ferdinand? We simply do not know.

In line with the traditions of his time at Oxford, Sibthorp was a classicist and must have been thrilled to hear about the existence of two illuminated Byzantine manuscripts kept at the Imperial Library in Vienna, known today as the *Codex Vindobonensis* and the *Codex Napolitanus*, the latter having being transferred from a monastery in Naples to the imperial city. Both manuscripts contained the writings of Dioscorides and were of venerable age, the first produced in a scriptorium in Constantinople in the early-sixth century, the latter in a scriptorium on the Apennine Peninsula in the late-seventh century. It seems very likely that Sibthorp heard about the two manuscripts from Jacquin, who had collaborated with Swieten and Adam František Kollár from Terhely (Terchová in modern Slovakia), scriptor at the Imperial Library, in an abortive project to publish the plant illustrations contained in the two herbals. Tracings of the Byzantine plant illustrations had been made, copper plates had been engraved and five sets of proofs *avant la lettre* prepared, but then the project was discontinued. The unexpected death of Emperor Francis I Stephen, Maria Theresa's husband and supporter of the natural sciences, was among the more important factors. However, one set of proofs annotated with the plant names in classical Greek in Kollár's hand had ended up in Jacquin's library.

We cannot exclude the possibility that Sibthop inspected the two early Byzantine manuscripts in the Imperial Library, just opposite the city residence of the Fries family, in person. However, there is a simple fact: Jacquin's set of proofs turned up in Sibthorp's estate and from much later correspondence it is clear that Sibthorp must have received this material from Jacquin, either as a loan or as a gift. As will be shown in this chapter, these annotated proofs were of considerable relevance during fieldwork in the Levant. They provided Sibthorp with an edge over his model, Tournefort, who is not known to have taken with him an illustrated edition of Dioscorides's writings, let alone one annotated with early Byzantine plant names.

Vienna was also the place where Sibthorp started to plan his tour to the Levant, writing to his father 'It is with a Sort of Enthousiastic Rapture that I dream of tracing the Steps of Tournefort in the Isles of the Archipelago. Mount Olympus [either Olimbos in modern Greece or Uludağ in modern Turkey] towers before me & the Rocks of Crete covered with the Coton Down of their silver Plants. I propose first going to Constantinople [Istanbul] either thro' Hungary or Dalmatia. If my friend Compte Basegli [Tomo Basiljević, Tomo Bassegli] should be at Ragusa [Dubrovnik] in the Spring I shall take the latter Route; if not I propose passing through Hungary . . . From Constantinople I can easily make Excursions to the Shore of the Bosphorus The Black Sea [Karadeniz] & in a few Days reach the Mount Olympus [Uludağ]. – From Constantinople I shall pass thro the Archipelago the shining Cyclades to Crete – Hence to Malta & Sicily – to Naples. My route will be confined to the European Turkey'. In a similar way, Sibthorp's ideas were set out in a letter to Hawkins, which only in the last paragraph is concrete 'I am soon to tread Classick Ground I think I already discover the distant Flames of Etna & in my Dreams which I hope will soon be realized, wander thro' the Thessalian Vale [valley near Tembi] – the Plains of Marathon [Marathonas] or trace the Ruins of Troy [Truva] on the Banks of the Scamander [Küçük Menderes] . . . Dioscorides & Tournefort make an interesting part of my travelling Baggages & in tracing the steps of the One, I hope to discover the obscure March of the Other It will be late in the Summer before I arrive at Constantinople [Istanbul] My Route at present by Trieste Venice, Bologna, Florence Rome & Naples – thence to Sicily, Malte & the Isle Candie the ancient Crete shining with its silvered Plants thence to Athens. – by the Grecian Islands Salonica [Thessaloniki] Mount Athos [Agion Oros] famous in ancient Story. – to Constantinople the Thracian Bosphorus [Çanakkale Bogazi] the Black Sea. & the Heights of Olympus [Uludağ]'. A single aspect of these quotations is of relevance in our context – the emphasis on following Tournefort's model and travel route – in fact Sibthorp had hired Ferdinand on the same understanding on which Tournefort had engaged Claude Aubriet. Like the latter Ferdinand was to document plants, animals, landscapes and even dresses, though the latter material – listed in the auction catalogue of the estate of Franz[6] as 'Costumes in Turkey' – could not be traced and must be regarded as lost. In addition, Sibthorp explicitly refers to the obscure travels of Dioscorides. This is revelatory – the real aim of the Sherardian Professor was to find the plants of Dioscorides *in situ*, that is in the region where he had lived, which in the late-eighteenth century belonged to the Ottoman Empire.

Little is known about the logistical preparation of the proposed tour. Ferdinand applied for permission to leave the country, addressing his application to Aloys I Joseph, but was informed that the decision had to be taken by the government of the Archduchy of Austria to which Feldsberg belonged. It seems plausible to assume that Sibthorp was in contact with Sir Robert Murray Keith the Younger in Vienna, then Envoy-extraordinary of the Court of St James to the Emperor, in order to obtain a permission to enter the Ottoman Empire and a firman (edict) waiting for him in the town of Chania (Hania, then Hanya). At that time Keith's equivalent in Istanbul was Sir Robert Ainslie, great favourite and boon companion of Abdul Hamid I, the 27th Sultan of the Ottoman Empire. During the next winter Ainslie was to host Sibthorp in Istanbul and possibly also Ferdinand. There would have been ample possibilities in Vienna to contact the representatives of the other countries that Sibthorp intended to visit, notably the Republic of Venice, the Papal States, the Kingdoms of Naples and Sicily, but nothing is on record. The political circumstances for the planned tour to the Ottoman Empire were favourable at that time, but nobody could know that only one year later another war would begin – that between the Russian Empire and its ally, the Holy Roman Empire, on the one side and the Ottoman Empire on the other. This war ended inconclusively and led in 1791 to the Treaty of Sistov (Svishtov in modern Bulgaria) between the Holy Roman and the Ottoman empires and in 1792 to the Treaty of Jassy (Iași in modern Romania) between the Russian and the Ottoman empires. For a short interval the window of opportunity had opened, and Sibthorp made use of this circumstance.

Nothing transpires from the two letters on the practical preparations for the tour, which kind of equipment was acquired and which books Sibthorp took with him besides the copper engravings based on the two Dioscorides manuscripts and Tournefort's *Relation d'un voyage du Levant*. Evidently Sibthorp regarded himself the successor of the latter, whose collections he had studied in Paris and Dresden, as well as those that were under his care at Oxford. Only a single indispensable item must have been specially prepared for the tour, and it is very likely that this was done before the departure from Vienna – a new colour chart, unfortunately no longer extant, with a significantly larger number of hues numbered 1–c.305 as compared to the earlier colour chart (see Chapter 2). Needless to say, the preparation of this chart, including the careful selection of pigments and the binding media to be used, were among Ferdinand's first jobs for his new master.

VIENNA TO ROME

The journey now beginning should not be regarded as a proper expedition, since no instructions exist, a concept that would have been at any rate completely unpalatable for a gentleman like Sibthorp. Several groups of sources are available on the route taken, the most important being information in Sibthorp's manuscript diary, a nightmare to decipher considering the poor-quality ink used and his difficult handwriting. It is a particularly valuable text since it was not intended for publication, although more than twenty years after Sibthorp's death some portions appeared in print in *Memoirs relating to European and Asiatic Turkey* and *Travels in various countries of the east*, both edited by the Reverend Horace Walpole. The terminology and orthography used in Sibthorp's diary, including place names, is maintained in this chapter with the modern equivalents added in brackets. In his choice Sibthorp identifies himself as an early partisan of what later would be called philhellenism, since he does not use the Ottoman toponyms of the day but names taken from classical Greek. Practical reasons may also have played a role – after all, Sibthorp was almost certainly unable to read Arabic script, which in the Ottoman Empire was the standard at this time. In short, the idea of Greece existed only in the minds of Western connoisseurs, while in the real world these were some of the western eyalets (provinces) of the Ottoman Empire. In our context the most important of these were the eyalet Girit (Crete), the eyalet Mora (Peloponnese) and the eyalet Jezāyir-i Bahr-i Sefid (Islands in the Aegean Sea), the latter with Gallipoli (Gelibolu in modern Turkey) as capital.

It should also be noted that the part of the diary referring to the journey from Naples to Istanbul is missing and so are accounts of the stays in Istanbul and Athens. Fortunately Hawkins produced a much better readable transcript of Sibthorp's diary, but this refers only to their joint travels from Istanbul to Athens in 1787 and contains Hawkins's own notes, while at the same time parts of Sibthorp's observations were crossed out. The second group of sources comprises the letters from Sibthorp and Hawkins, which are particularly relevant for those portions of the journey for which the diary records are no longer extant. Miscellaneous notes written by Sibthorp, Hawkins and Ninian Imrie, a geologist from Scotland who joined the trio in Istanbul for several months and whose records were published many decades later, form the third group.

Most important in our context is the pictorial record prepared by Ferdinand, effectively the 'cameraman' of the journey. It is a particularly valuable source not only for flora and fauna but also for reconstructing the course of the tour with the help of the views sketched by him. Again it should always be kept in mind that Ferdinand's record is also incomplete: the finished and partly finished watercolours of plants, the finished watercolours of animals, all without information on localities, and the finished landscapes have been preserved, but only a fraction of the pertinent pencil drawings. The latter comprise colour-coded pencils in small format of many but not all plants and animals, often provided with at least some locality data. By contrast, the large-format pencils of plants and animals, which must have also existed, are mostly missing. Similarly the pencil drawings forming the basis for the landscapes have also survived only in small numbers. Otherwise not a single written document, such as a letter, in Ferdinand's hand survives from the journey, probably not because they were lost, but because they never existed.

The specimens collected form an additional source of information, but again the record is incomplete. Nothing seems to have survived of the zoological materials and very little of the geological and mineralogical specimens, among them a fragment collected by Hawkins on the island of Kimolos and now in the Museum für Naturkunde in Berlin. Very many botanical specimens survive in the Department of Plant Sciences, Oxford, with duplicate material in the Natural History Museum, the Naturhistorisches Museum, and possibly elsewhere, but locality information on labels is at best fragmentary and largely totally missing. In addition, not all specimens drawn by Ferdinand or mentioned by Sibthorp in his diary are correlated with herbarium material, which results in a patchy overall situation. Maybe all this is attributable to losses in transit, but it is more likely that a complete, correlated set of specimens, notes and drawings of plant material was never intended. Furthermore not a single participant of the journey had the vaguest notion of plant geography, then not yet formulated as a set of ideas, nor of the concept of endemics (organisms with restricted occurrence). Therefore no one had a reason to carefully record the provenance of the specimens collected and the notes taken, as would now be standard procedure to do.

The pair that left Vienna on 6 March 1786, 'the day cold – wind and snow', was very unequal indeed – the third Sherardian Professor at Oxford University, an academic and member of the social elite, well-off, single-minded, probably with immense confidence in his singular abilities, seeking prestige and fame, and the nondescript, but also extremely qualified young man from Feldsberg. Taking the slight difference of age – sixteen months – into account, Sibthorp and Ferdinand might have regarded themselves as equals in their respective fields, both brilliant young men. From a distance of more than two centuries, however, Ferdinand is regarded as the 'most perfect plant illustrator of all times', whereas Sibthorp is just one of many professors of botany in the late-eighteenth century. Although their qualifications were clearly complementary, their personal relationship seems to have developed in a complex, rather uneasy and ultimately strained manner. Most importantly, Sibthorp was either unable or unwilling (or both) to accept Ferdinand as his equal partner and seems to have treated him like a servant. In a sense this echoes Humphrey Sibthorp's treatment of Ehret. It is telling that in Sibthorp's letters to his father and his future travel companion, not even the name of his draughtsman is mentioned and in his diary he gives his first name only once, deliberately suppressing his surname. In fact, Sibthorp refers to him like he would refer to his horse. No doubt, the entirely different social background of Ferdinand was an important factor, but this can hardly provide a complete explanation: Hawkins, probably coming from an even richer family than Sibthorp, always had a high respect for Ferdinand, constantly referring to him by name, and treating him accordingly. In addition there was a language barrier: Sibthorp did not know German; Ferdinand did not know English. However, while Sibthorp was disinclined to learn German, an attitude typical of many in his country, Ferdinand did learn English, though only slowly and imperfectly. In short, Sibthorp expected the traditional subordination of the servant to his master, which must have led to tensions and resulted, eight years later, in an abrupt termination of the engagement. Several reasons seem to have worked together in bringing about this rift, but in any case Ferdinand was not prepared to join Sibthorp on his second journey to the Ottoman Empire.

Ferdinand and Sibthorp had left Vienna in a barouche and managed to reach 'Meerzuslag' (Mürzzuschlag) in the evening of the first day, a remarkable achievement considering the arduous passage over the Semmering pass in wind and snow. Passing through 'Graetz' (Graz), the capital of the archduchy of Styria, and 'Mahrburg' (Maribor in modern Slovenia), they quickly reached the eastern border of the duchy of Carniolia – Sibthorp's diary noting 'more snow, the country mountainous & wood'. On 9 March, when still in what is now Slovenia, Sibthorp started his botanical notes, which consisted merely of scientific names and running numbers, recording on this very day '*Helleborus niger* (1), *Primula veris* (2)'. This way of numbering records, in itself sensible, was highly inefficient since Sibthorp refrained from attaching any field labels to his specimens. As a consequence no correlation existed between his diary numbers, including the pertinent locality information and his herbarium specimens; another nightmare for anyone interested in plant distribution.

It is plausible to assume that on this occasion Ferdinand prepared a colour-coded pencil drawing of a plant he definitely knew from Feldsberg, the early blossoming common liverwort, *Hepatica nobilis*, making use of his new colour chart. However, he could also have done this later, for example when passing over the Apennines. Years later in Oxford, Ferdinand was to prepare on the basis of this drawing, which has not survived, a finished watercolour of this common and well-known plant. In this case, a specimen survives at Oxford, but no locality information is provided, which means that it cannot be correlated with any certainty to Ferdinand's watercolour. However, in contrast to his superior, Ferdinand at least regularly annotated his pencil drawings with locality information, though only in a very general way, such as 'Sizilien' (Sicily) or 'Cybrus' (Cyprus), which turned out to be extremely helpful when trying later to elucidate the provenance. However, as pointed out earlier, Ferdinand refrained from transferring his locality information from the pencil drawings to the finished watercolours.

The next day the pair reached the capital of the duchy, the small town of 'Laubach' (Ljubljana), where they paid a visit to Belsazar Hacquet, a surgeon, naturalist and alpinist of broad interests who had travelled widely and climbed many mountains, among them 'Tergolu' (Triglav), and was therefore of special interest to both. It goes without saying that Hacquet showed them his natural history cabinet, including crystals, a herbarium and 'a young Bear shot two Months' before, probably not yet free from odour. No doubt it was wise to have left Vienna at the end of winter, but further progress towards Venice turned out to be difficult. Again there had been heavy snowfall and at different posts two pairs of oxen had to be added to the pair of horses drawing Sibthorp's barouche on the mountainous roads leading to Trieste, where they arrived by midnight. In his diary Sibthorp, probably ignorant of Slavonic languages, had noted the 'peculiar language, the Carniolish or Windish' tongue, which he could only have heard for four days from 8 to 11 March 1786. Considering his Feldsberg background, Ferdinand was probably less surprised by listening to the sound of a Slavonic language, although he could hardly understand this one.

In Trieste Ferdinand must have seen the sea, which was to play such an important role in his later life, for the first time, but we have no information on the impression it made on him. For Sibthorp, this town, the rising port of the Hereditary Lands and an important gate to the East Mediterranean thanks to privileges from Emperor Charles VI and his daughter Maria Theresa, was the place where he engaged a 'Greek servant'. The latter remained anonymous until Sibthorp disclosed his name in his diary entry for 23 September 1787, the day of the servant's death – Joseph Ganzarowitz, definitely a name of Slavonic origin, which so very many were and still are in Trieste. It seems also plausible to assume that Sibthorp sold his barouche in this town, as it was of no more use for him by then. While Sibthorp noted only the seaweed bladder wrack, *Fucus vesiculosus*, in Trieste, Ferdinand is reported by Hawkins to have prepared in this town a drawing of rosemary, *Rosmarinus officinalis*, a culinary herb he may well have known from Feldsberg or Vienna. In this case, a specimen is preserved in Sibthorp's herbarium at Oxford, but from the island of Zakynthos in the Ionian Sea, which has nothing to do with Ferdinand's plant from Trieste. In short, there are three more or less independent lines of recording plants – as specimens, as illustrations and as field notes, which need not necessarily refer to the same locality, the same species or the same specimen.

When Ferdinand embarked in Trieste, leaving the Holy Roman Empire as well as the Hereditary Lands, he could not envisage that it would take twenty-eight years until he would return home. By then his mother Therese and Boccius were long dead, Jacquin was an old man of eighty-seven years and his son Joseph Franz his successor in the chair and as director of the Botanical Garden of Vienna University.

A 'sloop' brought Ferdinand, Sibthorp and Ganzarowitz to Venice, capital of the Serenissima Repubblica di San Marco – for all gentlemen of leisure a highlight and an absolute must on their Grand Tour, and at the same time the traditional main gate to the Levant, which, in Venice's better days, had partly belonged to the city's sphere of influence. By 1786 this was in the remote past, with many of the territories in 'oltramare', among them Crete and Peloponnese, long lost to the Ottoman Empire. At that time the conservative, corrupt Paolo Renier, formerly Venetian ambassador in Vienna and 'bailo' in Istanbul, was in power as the penultimate doge, described by Goethe in vivid detail only seven months later. Arriving in this most spectacular city on 16 March 1786, the first day 'wet & the atmosphere obscured by vapour', was spent on the market, mainly looking at fish. Maybe Ferdinand had already started making zoological drawings there, which in Oxford would be used as the basis for his accomplished watercolours of fishes, but we do not know, since again these are not accompanied by locality data. In any case many other opportunities would arise later, in particular during his long stay in Istanbul. Sibthorp records that he had let himself be rowed in a gondola to see the major sites of Venice, ending his diary entry by noting 'The Season of Mirth, the Carneval was now passed, the whole Town like the Gondola seemed to be in Mourning'.

After an unpleasant crossing by ship, the diary noting 'our cabin was full & and the Animal Effluvia mixing with the . . . Odour of the Vessel formed a Perfume highly unfriendly to the nerves', the small party entered the Papal States at a port near Ferrara and continued to Bologna in a hired carriage. By taking this route they had missed the chance to visit the Botanical Garden of Padua University, over centuries the most important place for the study of plants from 'oltramare' and the port of entry from many places, in particular from Crete. Instead, a visit to the Botanical Garden of Bologna University was on the programme, 'poor & scanty . . . scarcely containing half an acre . . . the Gardener very ignorant and [Gaetano

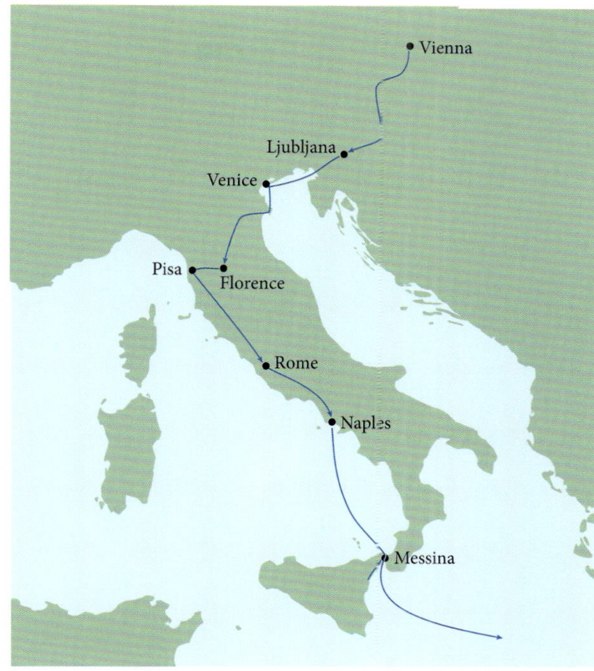

MAP 1. Ferdinand's route from Vienna to Messina in 1786.

Lorenzo] Monti so old that I did not visit him'. Sibthorp added ironically that 'Bologna is at present more celebrated for its Sausages than its Learning'. Missing Monti, then President of the Accademia delle scienze dell'Istituto in Bologna, meant also missing the chance to see the natural history cabinet of Ulisse Aldrovandi. This contained the oldest specimen and image in existence of an ornamental tree that Sibthorp and Ferdinand must have known from home – the horse-chestnut, *Aesculus hippocastanum* – a tree that Hawkins was the first to record from the wild in the Pindhos Mountains a few years later.[7] Instead Sibthorp, and probably also Ferdinand, paid a visit to Raffael's Ecstacy of St Cecilia, then in the church San Giovanni in Monte, later taken under Napoleon to Paris and now in the Pinacoteca Nazionale in Bologna. Continuing from this city towards Florence, the small party entered 'after much plague at the Customs' Habsburg territory again – the Grand Duchy of Tuscany, reigned by Grand Duke Leopold I, the younger brother and future successor of Emperor Joseph II. Magnificent 'views of the Apennine Mountains covered with Snow' presented themselves, and Sibthorp reported in lyrical words to his half-brother Humphrey 'The Environs of Florence are highly beautifull – the whole Country round it is in its full Force of Agriculture. The Olive wedded with the Vine, & Ceres smiling at their Feet. The grey Green of the Olive contrasted with . . .

the Hue of the Cypress . . . The Town proudly mantling with its Palaces, Convents, Churches; Villas meet the Eye in every Direction while the lofty Apeninnes form the Screen to the Whole'.

Like so many subsequent visitors, Sibthorp and Ferdinand took lodgings on the Arno river and toured the city with 'Abbé Fontana', almost certainly Felice Fontana, being his most important contact. The latter, a member of the Accademia dei Fisiocritici in Siena, was probably one of the most brilliant scientists Sibthorp ever met. Rightly regarded today as the father of modern toxicology, Fontana had studied a broad spectrum of topics, from viper venom to the properties of oxygen, from the anatomy of the human eye, describing for the first time the spatia anguli iridocornealis, to experimenting with curare. Ferdinand could not anticipate that years later his brother Franz would read Fontana's *Osservazioni sopra la ruggine del grano*, published in Lucca in 1767, when studying infections caused by cereal rust, *Puccinia graminis*, in cereals (see Chapter 8). It was Fontana who showed Sibthorp the natural history cabinet of the Grand Duke, then the Imperial Reggio Museo di Fisica e Storia Naturale, founded only eleven years before. It consisted of 'a hundred rooms of which fifty are already executed the Anatomical Preparations in Wax are of the greatest Beauty as are also those of the Succulent Plants, Fungi & c. the insects are particularly well preserved & ranged in Systematic Order according to Linnaeus & Fabricius, the Birds chiefly indigenous, a fine Collection of Fossils particularly from the Isle d'Elbe [Elba]'. Needless to say a gentleman like Sibthorp had to see Titian's lascivious 'Venus of Urbino', acquired by Guidobaldo II Delle Rovere, Duke of Urbino, in Venice, now in the Galleria degli Uffizi. Sibthorp knew a copy of the painting from the Anatomy School at Oxford, adding 'the Dean having judged it [the painting] too immoderate in the Library'. Surprisingly, the well-known Botanical Garden of Florence, like that in Padua one of the oldest in existence, is not mentioned in the diary.

The plants listed from Florence and surroundings numbered by Sibthorp 52 to 61 were possibly more relevant for Ferdinand. Among these was a '*Tulipa* nov: species', of which he made a pencil outline drawing,[8] which proves that his colour code was used right from the beginning of this journey. Later it was used as the basis for a finished watercolour documenting *Tulipa clusiana*, a tulip species native in Iran, Afghanistan and Pakistan[9] but naturalized on the Apennine peninsula. In this case Ferdinand had illustrated a plant that had not yet received a modern scientific name; his master had recognized it as new to science, although he neither proposed nor published a denomination. Very many plants new to science were to follow on this journey, nothing surprising for Ferdinand, apart from the fact that he must have so far observed few for the first time in the wild. His exceptional ability for precise recording is exemplified in this case by the brown wool protruding from the apex of the ovoid bulb in a short tuft and not omitted from his pencil drawing.

On 30 March 1786 Sibthorp recorded among others 'Cyclamen N. S. 73', another spring-flowering plant that was new to science, subsequently named *Cyclamen repandum*. Ferdinand must have prepared a pencil drawing that is no more extant; by contrast the watercolour based on it survived and was published decades later in London. In this case no specimen exists, making Ferdinand's watercolour of the new cyclamen the only record of the finding.[10]

The town of Pisa was Sibthorp's next destination, where he noted 'the Botanic Garden in much Disorder contains only a few Plants, tho' the Local is good & the Situation particularly favourable', indeed spectacular – then like today with the leaning tower, the cathedral, the baptistery and the campo santo within walking distance. At 'Leghorn' (Livorno) Sibthorp met John Graefer, gardener to the Queen of Naples, then Archduchess Maria Carolina, a daughter of Maria Theresa, and exchanged a few herbarium specimens with him. Sibthorp noted in his diary 'I gave Him a Specimen of the *Tulipa florentina* a new species discovered by me near Florence'. In accordance with the tradition of the time, herbarium specimens were exchanged in a liberal manner and without hesitation, which was another reason for gaps in Sibthorp's botanical collection. Returning to Pisa and passing through Siena, where Sibthorp and Ferdinand missed the Botanical Garden of Siena University then near the Ospidale delle Scale, the small party entered again the Papal States, reaching Rome on 6 April 1786.

ROME TO MESSINA

The oppressive prices for lodgings during the Holy Week seem to have been the main reason for a stay of only six days in the 'città eterna'. Ferdinand may have regretted his master's decision to continue so soon, since it meant that the encounter with his brother Joseph could only have been a brief one. Evidence for their meeting is almost nil, apart from the fact that the collection of Joseph's drawings in the Kupferstichkabinett in Vienna (see Chapter 4) contains a repeat of a view showing the Colosseum in Rome prepared by Ferdinand and now in the Sherardian Library in Oxford. The latter forms the first in a series of similar views (see below). Compared to his botanical work, they are rather conventional, only some approaching the quality of Caucig's 'vedute'. Though undated, they have been numbered chronologically in Ferdinand's hand, identifiable for example by the unconventional dot following the number 9. Years later the pencil sketches acted as the basis for the finished views painted almost certainly for Hawkins (see Chapter 9), which are referred to already in this chapter.

Otherwise it must have been a rather conventional visit to Rome and much shorter than Goethe's stays later that year and in 1787–1788. Only the Biblioteca Vaticana was the target of a vitriolic comment from Sibthorp – 'to see even a printed book it is necessary to have the permission of a Cardinal – this Gene to study renders the library a useless name'. A visit to the Botanical Garden of Rome University was also on the programme, then situated on the Gianicolo on a piece of land donated by Pope Alexander VII (Fabio Chigi). Sibthorp found it 'overrun by weeds', which led him to write 'From the little order that prevailed I could scarcely find the order of Tournefort was followed. The gardener a perfect ignoramus had never heard the name of Linnaeus. Indeed Italy in respect to gardening is greatly behind his neighbours & more than a century behind England'. This was indeed an appropriate comment on the situation, although in our context the reference to Tournefort is more relevant.

For Ferdinand, Rome was where he started his drawings of views and landscapes, which later resulted in three series: (1) the Oxford series with 141 monochrome watercolours, all numbered, annotated with locality information and in many cases signed by Ferdinand, kept in the Sherardian Library of Oxford University (with no. 55, 91, 95–97, 113–114, 116–118, 138 missing); (2) the Chichester series with six monochrome watercolours filling some of the gaps in the Oxford series, all numbered, annotated with locality information and signed by Ferdinand, now in the Cornwall Record Office in Truro, Cornwall; (3) the Göttingen series with 16 monochrome watercolours, all variants of the Oxford series, kept at the Niedersächsische Staats- und Universitätsbibliothek in Göttingen. Ferdinand's views are clearly not his forte and compare unfavourably with the records painted by many of his contemporaries on the Apennine Peninsula. However, in the later parts of the tour, Ferdinand's 'vedute' are sometimes the first pictorial record ever of a given locality or building and therefore of special interest.

Continuing southwards and crossing the Campagna, Sibthorp found his 'Head & Stomach very sensibly affected by the Marsh Miasmata', but neither he, his servant nor Ferdinand seem to have contracted malaria. Passing through Terracina, they left the Papal States and entered the Kingdom of Naples, then ruled over by Ferdinand IV (see Chapter 4), and reached the capital on 15 April 1786. These seem to have been idyllic days, with Sibthorp telling his father 'My Hours glide on smoothly. Early in the Morning I herborize with my Painter the Remainer of the Day we work he in drawing I in describing the Plants'. Full of confidence and pride he added 'I certainly on my Return shall have the first Collection of coloured Plants in Europe'. Interestingly, Sibthorp mentions the joint plant collecting, which may imply that Ferdinand was regarded as an acceptable companion in the field and when it came to spotting remarkable plants. He may also have acted as his master's assistant, changing drying paper or carrying the plant presses. Maybe there was also some separate collecting independently from each other, since James Edward Smith (see Chapter 9) later attributed some collections exclusively to Ferdinand. In a sense Sibthorp seems to have been impressed by Ferdinand's work, reporting home 'he has made more than a hundred Designs of different Plants found about Florence Rome & Naples . . . I never saw Beauty & Accuracy so fully combined together', meaning that within two months of travelling Ferdinand had produced a very remarkable number of drawings. The addition of 'at present we have only Time to sketch the Outline to finish a Flower & a leaf' confirms that Ferdinand continued to make use of his working method established in Feldsberg and Vienna (see Chapter 2). Surprisingly Sibthorp never mentioned the fact that Ferdinand supplemented his pencil outlines with the numbers of his colour chart. Was this a secret the latter wished to share with no one? Sibthorp also never recorded the number of herbarium specimens collected. In a sense this seems to mirror his preference of the pictorial record over the material record and may have led to the almost total lack of original labels accompanying the dried specimens.

The spring aspect of the Mediterranean flora apparently new to Sibthorp made him write 'The Ground exposed to the Sun was covered with Forest of rich Hesperion Fruit [i.e. *Citrus*] guarded by Foreign Troops of the Agave [*Agave americana*] & Cactus [*Opuntia ficus-indica*] while the Oriental Palm [? *Chamaerops humilis*] boldly rose above taking the Command at the Head of the Line. The Rocks were silvered over with Lotus Creticus 161. mixed with the Rosemary [*Rosmarinus officinalis*] & the Thymus communis 162. the Cistus salvifolius [*Cistus salviifolius*] now began to show its white Flowers on the Hills & to perfume the Air with its Fragrance. the Hedges that served as a Screen to the Olive Grounds & the Corn Fields were now empurpled with the Flower of Cercis [*Cercis siliquastrum*] mixing with the yellow of Spartium Spinosum [*Calicotome spinosa*].'

The stay in Naples lasted at least three weeks and included an excursion of five days to the island of 'Caprea' (Capri), then as now fashionable among tourists. Sibthorp contacted the British Ambassador in Naples, Sir William Hamilton, a Fellow of the Royal Society and recipient of the prestigious Copley Medal. He was a distinguished archaeologist, art historian and cicerone for many, and Sibthorp's main aim was to seek the recommendations to the 'different [British] Consuls in the Levant'. Only eleven days after Sibthorp and Ferdinand another visitor arrived in Naples – Amy Lyon, twenty-one years old, who was to become a few years later the famous Emma Lady Hamilton. Sibthorp might have met her, but she is not mentioned in his diary and letters. Interestingly he wrote from Naples to his father asking him to send £500 to Count Fries in Vienna, who would give credit to his contacts in Venice, Naples and the Ottoman Empire, adding 'this will set my Mind at Ease'. It is extremely likely that Sibthorp took pains to organize for himself, his servant and Ferdinand a passport to continue his travels within the Kingdom of Naples, then effectively including Sicily. In the following year Goethe had such documents issued in Rome and Naples, and these have survived in the Goethe und Schiller Archiv in Weimar.[11] No proper botanical garden existed at the time in Naples, but Sibthorp visited the small garden and the library of Domenico Cirillo and probably the herbarium of Terranus Imperati. After the end of the ephemeral Parthenopean Republic in 1799 Cirillo was executed and Imperati's plant collections largely lost, with a single volume surviving in the Biblioteca Nazionale di Napoli.[12]

Ferdinand prepared several drawings of landscapes in the surroundings of Naples, three of them documenting an excursion to Mount Vesuvius. Since he shows the 'grand aperture', he must have climbed the mountain, almost certainly accompanying his master and possibly also the British Ambassador, an authority in volcanology. In another view of Mount Vesuvius's summit three persons are shown, all with their walking sticks and wearing hats (Fig. 5.2.). An excursion to look at the cave of Pausilippo and Solvaterra was also undertaken, no doubt to satisfy Sibthorp's 'gusto greco' and to observe further manifestations of volcanism which had also been an important topic for Wutky (see Chapter 4). On 21 April 1786 the aforementioned excursion to the island of Capri followed, where Ferdinand is reported to have prepared a pencil drawing of Jupiter's beard, *Anthyllis barba-jovis*, a shrublet with dense inflorescences and flowers with lemon yellow corollas. The first of many Mediterranean islands to be visited seems to have disappointed Sibthorp, since he records in his diary 'The Rocks calcareous . . . the greater Part of the Island is entirely sterile . . . There are neither rivers nor Meadows nor Cows or Sheep a few Goats men upon the mountains, & the Inhabitants live Chiefly upon Fish & Vegetables'. Before Capri he seems to have had somewhat unrealistic expectations of bucolic life on the islands, and maybe he was even more disappointed on the dry and stony islands in the Aegean Sea (then Ege Denizi). Ferdinand's views of Capri corroborate Sibthorp's notes and show sterile hills, for example in the view from the island's harbour to Anticapri. At the time Mount Vesuvius must have been very active, since Sibthorp specifically records the 'continually beloking [billowing] forth its Flames, its sides empurpled with Lava'. Capri's flora is recorded by Sibthorp in a long list of names, including those of lichens and mosses, which are largely ignored in this text. Otherwise the tour to this island shows the limits of Sibthorp's knowledge of the Mediterranean flora, for he could determine several plants only to generic level. Again it was time for Sibthorp to write a letter to his father reporting from Naples 'I have taken a Passage aboard an English Ship bound for Smyrna [Izmir] called the *Levant* . . . in two or three days we set sail for Messina he [that is, the captain] promises to land me either at Canea [Hania] or some other Island near Candia [Crete] as Serigo [Kithera] or Argentiera [Kimolos] I expect the Firman . . . is waiting for me in Candia'.

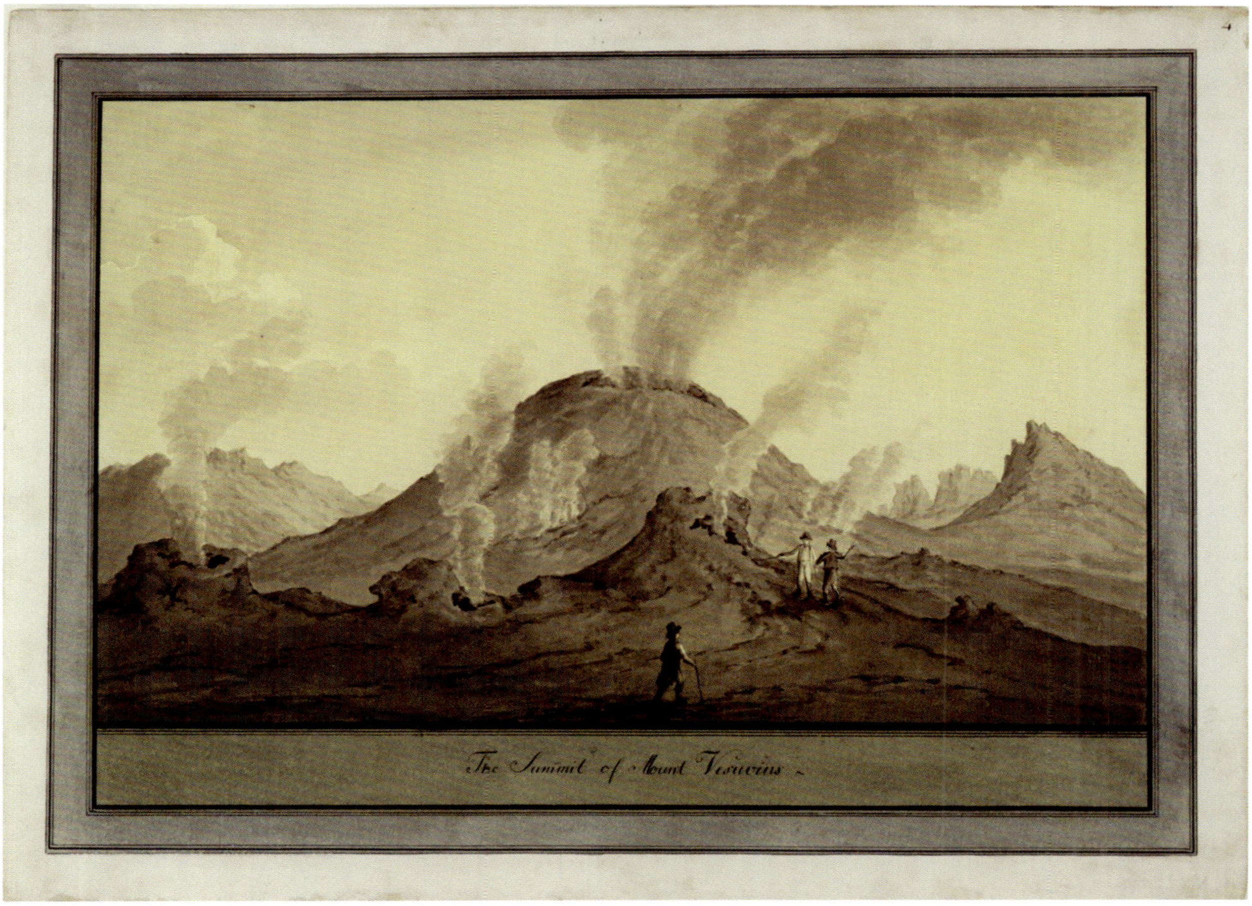

FIG. 5.2. On Mount Vesuvius. Watercolour by Ferdinand Bauer, after 1794. – Oxford, University of Oxford, Sherardian Library, MS Sherard 408, f. 4.

Sibthorp's diary now comes to an abrupt stop. However, it may be assumed that the small party left Naples in late May or early June, stopping for some days in Sicily and arriving on the island of Crete before 30 June, the day of Sibthorp's next letter home. The route taken is uncertain, but considering Ferdinand's views of the Aeolian Islands they must have passed Stromboli, Lipari and Salina at rather short distance. Four days after having left Naples the ship anchored in the harbour of Messina, where the small party observed another aspect of volcanism. Three years before, on 5 February 1783, a disastrous earthquake followed by a tsunami had destroyed major parts of the town, in particular the famous 'palazzata', a façade around 1.5 kilometres long along the harbour. In a letter to his sister-in-law, Sibthorp described the place as 'a melancholy Ruin, a Facade of noble Houses extended along the Quay of these only remain the Shells mournful Vestiges of its Ancient Grandeur which echo back the miserable Plaints of its wretched Inhabitants'. Interestingly, the following more general comments are added 'the Inhabitant of Messina regards his ruined Walls tight & looks no further. He has neither Force nor Courage to repair them Oppressed by their petty Princes by their Neapolitan viceroys & duped by their priests. these swarm over the whole Island & suggle [suckle] the People with Impunity, no less than three Miracles were worked during my Stay at Messina. but as I told an old Apothecary who invoked the Credit of his Saint that to me the greatest Miracle was that the Priests devoured their People without their crying out'. Clearly Sibthorp was a man of the enlightenment strongly opposed to the traditions and rituals of the Catholic Church in Sicily.

Unfavourable winds held the ship bound for Crete in the harbour and Sibthorp's party made use of the interval by examining the coastal mountains from which they were able to 'reap a rich Harvest of Sicilian Plants'. This was no exaggeration: on a single sheet of paper inscribed 'Sicilien' (Sicily), Ferdinand prepared colour-coded drawings of crown daisy, *Glebionis coronaria*, larkspur, *Delphinium halteratum*, toadflax, *Linaria purpurea*, evergreen rose, *Rosa sempervirens*,

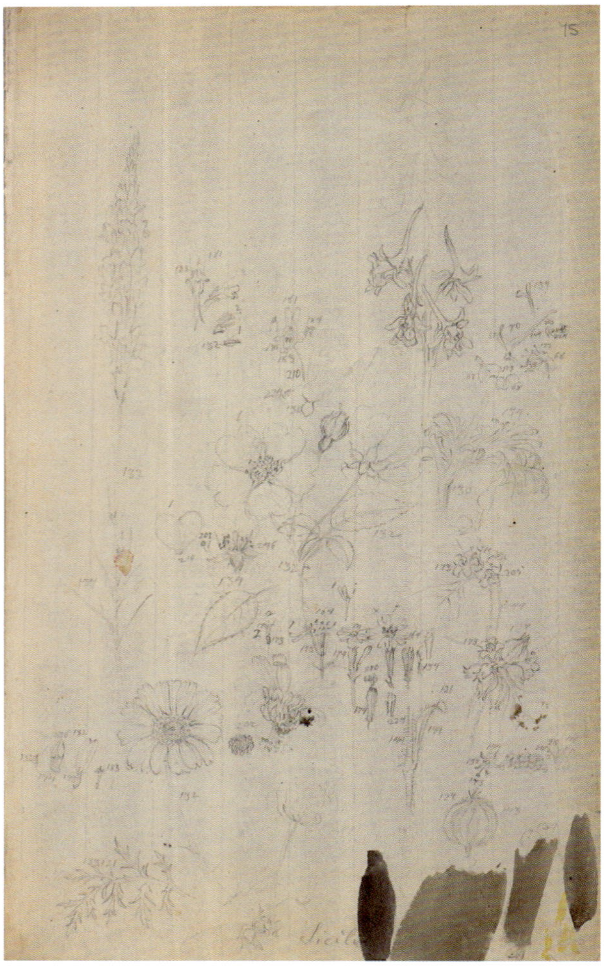

FIG. 5.3. Plants from the surroundings of Messina. Pencil drawing by Ferdinand Bauer, 1786. – Oxford, University of Oxford, Sherardian Library, MS Sherard 247, f. 75 r.

catchfly, *Silene colorata*, and *Withania somnifera*, the latter rare on this island (Fig. 5.3.). Although the flora of Sicily was relatively well known botanically, three of these species had not yet received modern scientific names. Several other plants flowering in May were also most probably documented by Ferdinand in Sicily, among them bear's breeches, *Acanthus mollis*, *Achyranthes aspera*, a pink-blossoming catchfly, *Silene coeli-rosa*, a deadnettle, *Lamium flexuosum*, *Lithodora rosmarinifolia* and *Urospermum dalechampii*, all western or central Mediterranean elements that Ferdinand could hardly have met further east. For the first time Sibthorp and Ferdinand also encountered a local endemic, a plant with an extremely small area of distribution, foreshadowing similar experiences later. Near Capo d'Ali south of Messina they found a species of sea lavender, *Limonium*, that had not yet received a scientific name and was later documented in a watercolour by Ferdinand. Today only known from a narrowly circumscribed maritime cliff, in danger of extinction[13] and in cultivation in the Botanic Garden of Messina University, *Limonium sibthorpianum* has rightly been dedicated to its discoverer.

A visit to Catania seems to have been intended, as Sibthorp noted in a letter 'Sr. Wm. Hamilton has furnished me with a Letter for the Prince Biscaris at Catania', but we do not know if this plan materialized. No doubt one of the reasons for this planned visit was the famous natural history cabinet brought together in Catania on the orders of Ignazio Paternò-Castello, 5th Prince Biscari, a fellow of the Society of Antiquarians in London. Ten years before it had been described in Domenico Sestini's *Descrizione del Museo d'Antiquaria e del Gabinetto d'Istoria Naturale di Sua Eccelenza il Sig.re Principe di Biscari I. Paterno Castello*. In any case Ferdinand recorded the view along the eastern coast of Sicily with Mount Etna, Europe's highest volcano, in the distance, possibly taken from the area of Taormina.

It is interesting to compare Goethe's stay in Sicily (he was to arrive in Palermo on 7 April 1787) with that of Sibthorp, who had arrived in Messina in early June 1786.[14] Both were accompanied by draughtsmen, Goethe by Christoph Heinrich Kniep, Sibthorp by Ferdinand. The arrangements may have been similar, stipulating in the case of Kniep that all drawings including 'alle Conture' (all outlines) would belong to Goethe. Both draughtsmen prepared pencil drawings, both with the aim of preparing later accomplished works, in the case of Kniep described as 'ad eseguire più tardi i disegni in bella' (to execute later the drawings in beauty). While Sibthorp and Kniep are not known to have prepared any botanical drawings in Sicily, Goethe probably sketched in Sicily an agave and an Indian fig opuntia, both drawings being kept at the Graphische Sammlungen Weimar. Goethe's interest in natural history is evidenced by his visit to the cabinet of the Biscari family, explicitly mentioned in *Italienische Reise*, his iconic autobiographic reflexions on the travels undertaken on the Apennine Peninsula. In early May 1787 Goethe had also been to Taormina and to Messina, and later vividly described the effects of the earthquake that had largely destroyed the town and significantly reduced its population.

The conventional part of Sibthorp's journey, echoing the classical Grand Tour so fashionable at the time for the upper echelons of society in England and elsewhere, now came to a close. Like Goethe the following year Sibthorp and Ferdinand had followed established routes, visited Venice, Bologna, Rome, Naples and Messina, and climbed Mount Vesuvius. The first three months after departing from Vienna had been used in order to get acquainted with the spring flora of the Apennine Peninsula and of Sicily, with Ferdinand preparing pencil drawings of about two hundred species. The party had visited several botanical gardens, natural history cabinets containing specimens, manuscripts and books, although they missed others, among them Michiel's *Codice Erbario*, at present in the Biblioteca Nazionale Marciana in Venice, and G. Dyckmann's watercolours of Cretan plants, at present in the Biblioteca Universitaria in Pisa. Now they were to start a tour on less-trodden paths, though certainly not a journey into the totally unknown. At least printed maps existed, albeit imprecise ones. Needless to say there were no more natural history cabinets or botanical gardens to visit on the second part of the journey now about to start.

We are unaware whether Sibthorp and Ferdinand knew about the living conditions in the Ottoman Empire that they were soon going to face. Asked 'How did you live?', Lord Byron's servant William Fletcher is reported to have answered after the death of his master in Missolonghi (Mesolongi) in 1824, 'Like dogs, on goat's flesh and rice, sitting on the floor of a hovel, all eating out of a dirty round dish: no knives, and only one or two horn spoons. They drink a stuff called wine, what but it tastes more of turps [turpentine] than grapes [clearly a description of retsina] and is carried about in stinking goatskins, and every-one drinks from the same bowl: then they have coffee which is pounded, and they drink it, dregs and all without sugar. They are all smoking when not sleeping; they sleep on the floor in all their cloths and shoes; they never undress or wash, except the ends of their fingers, and are covered with lice and fleas'. In a sense, it is a good thing when departing not to be fully aware of the discomforts and dangers lying ahead.

CRETE

There were good reasons why travelling the Ottoman Empire was unpopular at the time: meeting the plague or other recurrent epidemics was a danger, just like contracting malaria or falling into the hands of bandits, and even finding decent accommodation was not easy – with luck one could pass the night in a monastery or the house of a priest. If not, the travellers had to look out for a han, a public building, often empty and ruined. Furthermore communication was possible only with the help of interpreters, since Turkish was the principal language, with Greek, Italian and Albanian being other Levantine tongues a traveller would encounter in the western part of the Ottoman Empire. For these reasons Sibthorp must have hired an additional person described in a letter as 'my Servant an Armenian who spoke the Turkish Language in Perfection spoke a broken Greek scarcely intelligible', but we have no details as to when and where the interpreter was engaged and for how long he was a member of Sibthorp's party.

It seems plausible to assume that Sibthorp had provided himself with at least some travel information and equipment. Even so, the arrival in the Ottoman Empire must have been a shock for him and Ferdinand, though not for Ganzarowitz and the anonymous Armenian. Five days after embarking at Messina the coast of the Peloponnese came into view and Ferdinand must have prepared a drawing of the island of Cherigo (Kithera), then still belonging to the Serenissima Repubblica di San Marco. Sibthorp, his servant, his interpreter and Ferdinand subsequently landed on the island of Milo (Milos, then Milos Adasi), their first stop in Abdul Hamid I's Empire, where one of the most famous statues of Greek antiquity, the Venus de Milo, now in the Louvre, was still lying undiscovered in the ground.

Sibthorp's report on their arrival deserves quoting in full: 'firing a Gun a Pilot Boat put off brou[ht] us into the island of Milo. The Town was full two Mile[s] from the Beach. Sea Sickness, Confineme[nt] on board the Vessel, & the intuse [intense] Heat had so weakened me that I could scarcely crawl to it. On my Arrival I was panic struck half the Houses were already in Ruins the other half mouldering away. upon my asking the Cause I was told a contagious Disorder had carried off two third of the Inhabitants & that the living were scarcely in Number Sufficient to bury the dead . . . a few Shades glided along the ruined Walls with the Shrowds of Death. with Horror & dismay I returned to the Shore. The Ship was now sailed. my Servant . . . was upon his Knees praying to St. Anthony to deliver us from this Misery. Our first Object was to secure our Baggage a hollow Rock near the Shore seemed kindly to open its Bos[s]om to receive us & here we slept two Nights'.

With the sangfroid necessary under these circumstances Ferdinand made a drawing later to be used by him for the view of an islet next to Milos, while Sibthorp started to collect plants. Eventually a pilot was found to bring Sibthorp's party to the island of Argientera (Kimolos); the boat must have been a small one since when becalmed they continued with the help of oars.

We do not know when the group finally arrived in Crete (Kriti, then Gerit) nor when they left the island. Sibthorp's letter to his father was dated 31 [sic] June,[15] Ferdinand's views and his plant illustrations as well as the botanical specimens that were possibly collected in joint collaboration by both men are the only sources available. Canea was the first oriental town encountered by Sibthorp and Ferdinand, just as it had been for Tournefort, Gundelsheimer and Aubriet in the year 1700. With its mosques, minarets crowned by half-moons, fortifications, admirably recorded by Ferdinand, it possessed a Levantine, or rather oriental flair, stressed, as we have to assume, by veiled women and men wearing turbans in the streets. A realistic description of a botanical excursion to Cape Meliar (Akrotiri) is given in the letter cited above 'slept the first Night at the Convent of St. Trinité [Moni Tsagarolou, Moni Agia Triada]. The Greek Fathers gave us [sic] their best Accomodation such a miserable Chamber, Receptacle of Filth & Vermin. a Lamp during the Night cast a faint Glimmering on our Beds & a Host of Fleas, Lice, Bugs & Gnats presented themselves before us my Flesh yet tingles with their Nocturnal Revels'. This was one of the rare occasions when Sibthorp mentioned food, reporting to his father 'our Table was indeed somewhat better than our Beds Some Fowls roasted or rather burnt with rancid Oil, were sweetned by the best of Honey & the most delicate Lards the Honey was the Quintescance of Thyme & its aromatic Flavour rendered it gratful beyond any thing I had ever tasted'. Otherwise the stay in the monastery (Fig. 5.4.) seems to have been rather disappointing to Sibthorp, who notes 'Greek Fathers are miserable ignorant Friers, who are not only totally ignorant of every European Language but very deficient in the ancient Greek, nay I am assured there are some of their Bishops so ignorant as to be incapable of writing their Names'. Humphrey Sibthorp received in this letter some realistic background information on the living conditions in this monastery: 'these poor Greeks are in the most abject Slavery. Did the Turk confine Himself to refuse them Bells only they would have little Cause of Complaint but subject to daily Exaction oppressed by constant Tyranny. they drag out a most miserable Life. these convents are continually subject to the Visits of rambling Turks'.

Botanizing in unsafe country such as Crete made special arrangements necessary. Sibthorp reports being accompanied by the 'Janisary [an armed guard] of the French Consul [in Hania]', who belonged to his party together with Ferdinand, the interpreter, the servant, a muleteer, and the Consul's apothecary. Again Sibthorp is in a poetic mood and tells his father: 'on the Rocks which rose above us to freightful Height hung some of the rarest Plants. We gathered the Ebeny of Crete (Ebenus cretica) [*Ebenus cretica*], the Tree Pink (Dianthus arboreus) [*Dianthus fruticosus* subsp. *creticus*], the white Fleabane (Conyza candida) [*Inula candida*], the Immortal of the East (Gnaphalium orientale) [*Helichrysum orientale*] while the soft Dictammy [*Origanum dictamnus*] carpeted its Sides. Among the Rocks we found many other curious plants which the licentious Goats & the burning Sun had spared us. of these my Painter has drawn several which I hope some future Day will do me as much Credit as those of Aubriet have done Tournefort'. Indeed this is a most revealing passage, making clear what role Ferdinand was expected to play in the future plans of his master – he should serve as a vehicle to his further fame as the second Tournefort.

A lucky circumstance was reported to Cowley House that testified how quickly Sibthorp had adapted to Levantine realities: 'the Bashaw [pasha] of Canea who is a great Man here [in fact the head of the eyalet Girit] & a Bashaw of three Tails [an unusually high rank for this position] hearing I was a Physician sent to consult me. His Maladie was a Debility of the Stomach . . . I took Care to inform him that I was in Search of medicinal Plants. & those which were best adapted to his Disorder grew on the Mountains of Sfaccia [Lefka Ori]. But where it was imprudent for me to go without his Protection & Guards which I requested Him to furnish me with. These he promised us. After the Manner of the Orientals he gave us Coffee without Sugar & having played off the Charllatan for half an Hour I took my Leave . . . Tomorrow . . . I set off for the snowy Mountains of Sfaccia . . . the Janissaries, the old Apothecary I imagine are sufficient Protection I now botanize always with loaded Pistols in my Pockets. These Plants should be worth something'.

Long is the list of plants that Ferdinand must have documented during his stay in Crete, but again we are faced with the general lack of locality data. In the case

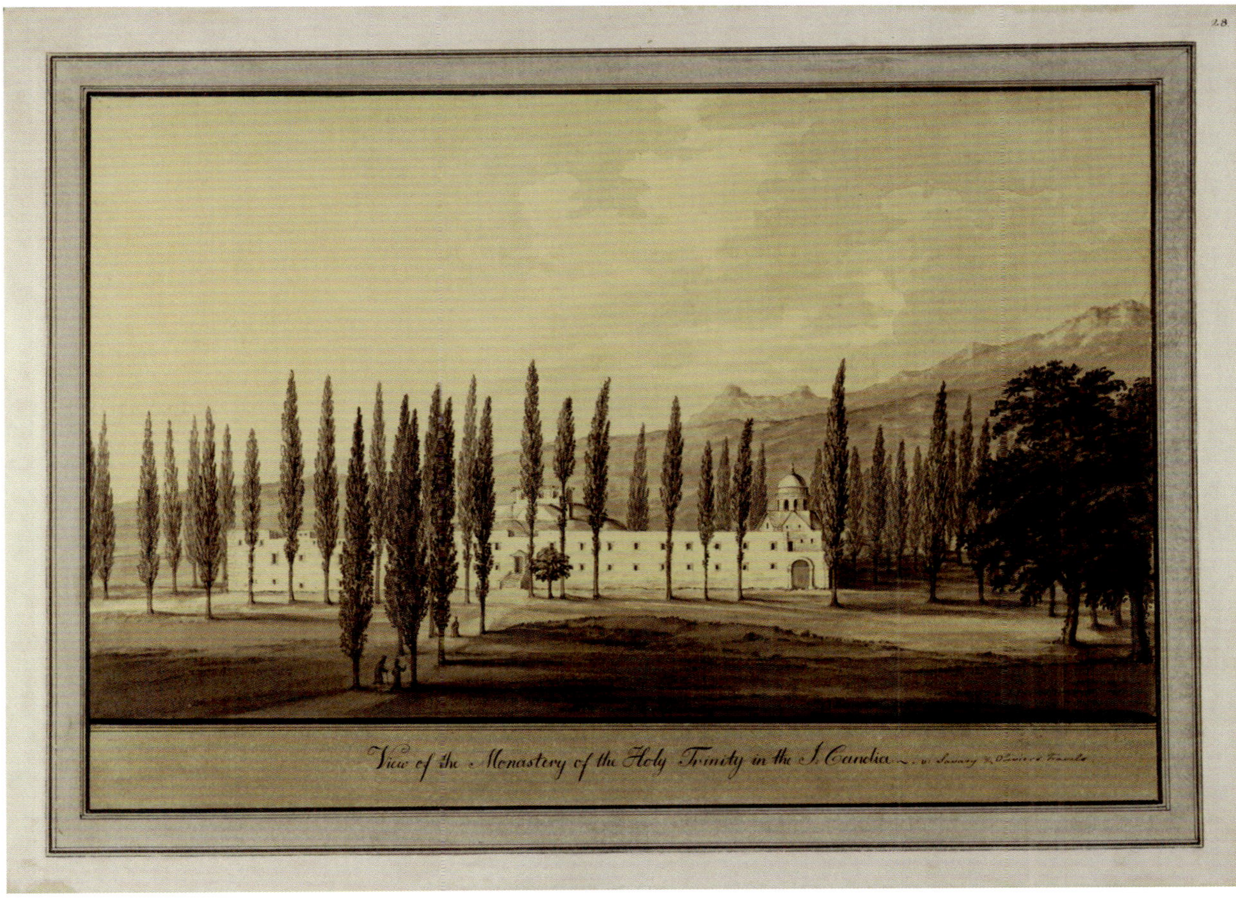

FIG. 5.4. View of the Monastery of the Holy Trinity (Moni Tsagarolou, Moni Aga Triada) in Crete. Watercolour by Ferdinand Bauer, after 1794. – Oxford, University of Oxford, Sherardian Library, MS Sherard 408, f. 28.

of a species occurring widely in the Mediterranean area that has been recorded by Ferdinand on Crete, this can only be ascertained when the pencil drawing has locality information written in his hand, otherwise it is only possible to make a guess. A good case for this argument is one of Ferdinand's pencil drawings marked by him 'Criti Candia' (Fig. 5.5., 5.6.), documenting ten colour-coded plants. Of these seven also occur outside this island, among them are widespread species like *Malva cretica*, which despite its epithet occurs also in southern Europe, western Anatolia and Cyprus. Ferdinand seems to have been very economic with his locality notes, marking the sheets of paper only in one place. Since he used both sides of a folded sheet, only one of the resulting two pages carried the locality information and only on one of its two sides. Ferdinand's straightforward logic became largely obscured when years later the half folded sheets were bound in a single volume in a chaotic way, which inevitably resulted in a laborious analysis of the foliation and the watermarks of the six different types of paper. However, this approach enabled the attribution of several pages carrying no locality information to specific phases of the journey. Among them is a page that carries Ferdinand's pencil drawing of the conspicuous, foul-smelling and rather sinister-looking inflorescence of dragon arum, *Dracunculus vulgaris*,[16] an aroid of widespread occurrence. It must have been prepared on Crete, because on one of the two sides of the folded sheet[17] it carries the annotation 'Candia'. This is corroborated rather nicely by the fact that we find drawings of two strict Cretan endemics together with dragon arum on the same side of the page: (1) *Verbascum spinosum*,[18] a distinctive spiny shrub forming hemispherical, thinly white-tomentose bushes characteristic of phrygana; and (2) *Sideritis syriaca* subsp. *syriaca*, a white-lanate perennial, growing on dry, rocky limestone slopes, which despite its epithet does not occur in Syria. These two taxa belong to the group of 'silvery plants' mentioned by Sibthorp.

Crete is famous for the considerable number of species restricted in distribution to this island, and a remarkably proportion of these were documented by

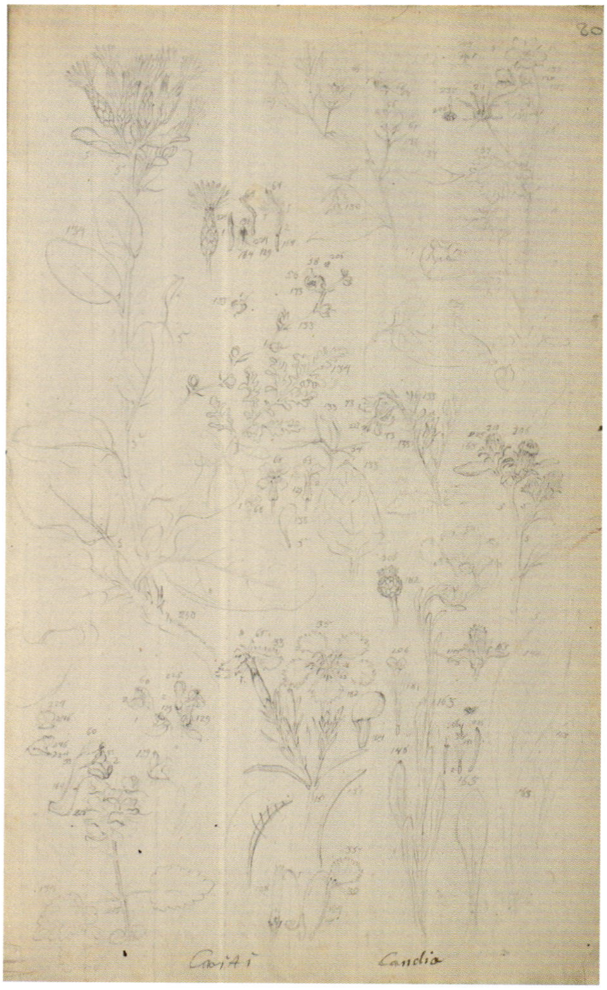

FIG. 5.5. Plants of Crete. Pencil drawing of Ferdinand Bauer, 1786. – Oxford, University of Oxford, Sherardian Library, MS Sherard 247, f. 80 r.

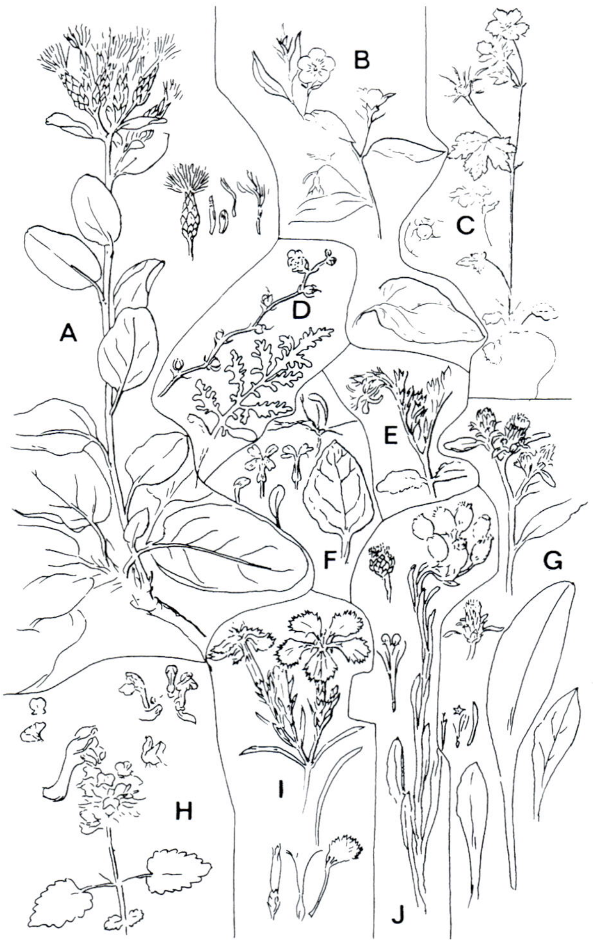

FIG. 5.6. Plants from Crete. Pen and ink drawing by M. Heilmeyer based on Ferdinand Bauer (cf. FIG. 5.2.), published in Ann. Naturhist. Mus. 98B (1996). – Berlin, Botanischer Garten und Botanisches Museum Berlin-Dahlem, Bibliothek.

Ferdinand. Among them were two species of woodruff, *Asperula pubescens* and *A. rigida*, both of which had not yet received a Linnaean name (a binomial). Furthermore, the pink-flowering Ebeny of Crete mentioned above, a much-branched bushy shrub of up to 1.5 metres, and two species of bedstraw, *Galium fruticosum* and *G. incurvum*, have been recorded. Ferdinand also prepared drawings of a skullcap, *Scutellaria hirta*, a plant of some horticultural interest, and *Staehelina petiolata*, a long-living woody chasmophyte growing in crevices of limestone rocks, both endemic to Crete.

Even more relevant in our context is the fact that some of the endemics recorded in pencil drawings occur only at considerable altitudes, proving that Sibthorp's party, like Tournefort, Gundelsheimer and Aubriet before, must have reached high altitudes in the mountains of Crete. The small *Phagnalon pygmaeum*, growing in crevices on limestone rocks and rocky slopes at 1600–2200 metres, and *Crepis sibthorpiana* of similar size from similar habitats at 1400–2400 metres are examples among the composites recorded by Ferdinand. Some of these Cretan endemics scarcely occurring below 1500 metres, for example the blue-blossoming *Anchusa cespitosa* restricted in distribution to the Lefka Ori, are locally abundant and were already known to the Venetian botanists many decades before Ferdinand documented them. Sibthorp's party must have also been on the southern side of the Lefka Ori massif, since Ferdinand was able to record *Teucrium cuneifolium*, a dwarf shrub densely tomentose with a greyish-white indumentum. This has never been found outside this clearly circumscribed area.

Ferdinand must have taken with him a powerful lens, since he records on a sheet of paper marked 'Candia'[19] in great precision the tiny flower of the extremely widespread grass *Panicum repens* strongly magnified – with its bracts, three anthers and bifid stigma – in a colour-coded pencil drawing. A lens was also necessary when he documented the tiny two-lipped blue corolla of *Solenopsis minuta*, a small acaulescent plant of wet habitats found in Crete, also shown on this very page.

Sibthorp seems to have been satisfied, reporting later to Sir Joseph Banks (see Chapter 8) in London 'the white Mountains of Spacia, retarded by the Snow & their great elevation, offered me a rich & ripe Harvest'. They brought Sibthorp 'the satisfaction of finding a greater part of the Cretan Plants mentioned by Tournefort in his Corollary [*Corollarium*], & even some which escaped his notice'. The reference to snow is no exaggeration – one of Ferdinand's views depicts a cave at high altitude filled with snow, probably Mile in the Lefka Ori,[20] located about half way between Volika and the summit of Mount Spathi. A second 'veduta' shows the small caravan in the mountains consisting of two men in European dress on horseback, possibly Sibthorp and Ferdinand, and three men in oriental dress on foot, plus three donkeys or mules carrying loads.

CRETE TO ISTANBUL

Information on the following months of the journey is extremely meagre. We neither know when nor from which harbour Sibthorp's party left Crete, which precise route they followed and when exactly they reached Istanbul, arriving there probably in early September. A fragmentary diary entry exists only for the first days of August; for the rest of the time Sibthorp's records have not survived or could not be located.

Ferdinand's views offer some clues: he recorded the islands of Hydra (Idra), Egina, Siphnos (Sifnos), Anti Paros (Andiparos), Amorgos and Samos (then Sisam). The numbers added by him to his landscapes are the only clue to the itinerary. Since these numbers follow a chronological sequence for the months on the Apennine Peninsula for which we have a diary, this is likely also to be the case for their first crossing of the Aegean Sea. As a consequence we may assume that the six islands listed

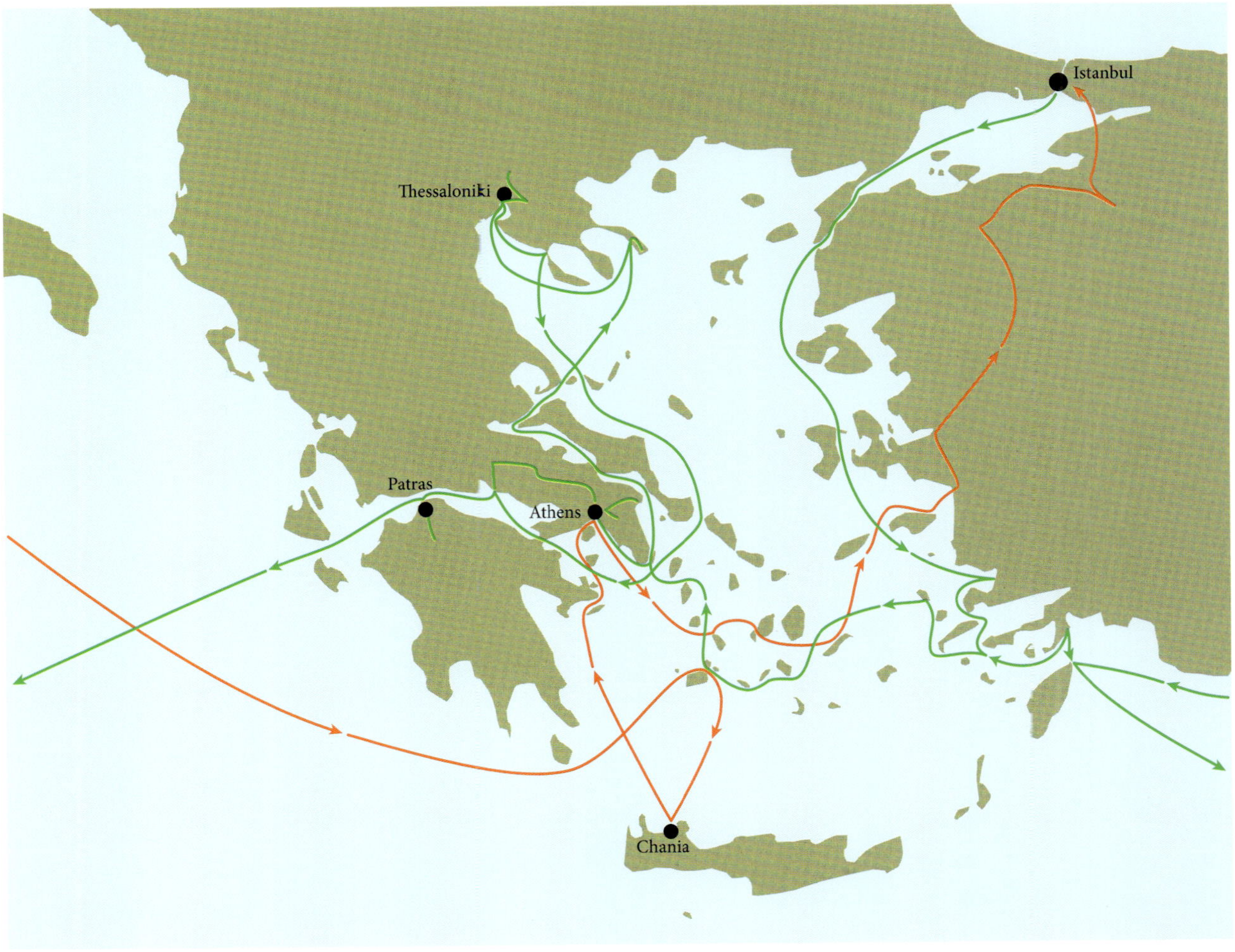

MAP 2. Ferdinand's route in the Ottoman Empire in 1786 (red) and 1787 (green) (based on Lack 1999).

above were visited in that sequence, which indicates a direction from west to east. Additional information comes from several sources: Smith reported, possibly using personal information received from his friend Sibthorp, that the latter and his company narrowly escaped shipwreck near Milos; there is reference to a visit to Attica in a letter by Sibthorp to Banks, corroborated by two passing references in passages of Sibthorp's diary describing events in the following year.

Strict endemics documented by Ferdinand offer further clues: one of them is the chamomile *Anthemis rosea* subsp. *rosea*, an appressed-pubescent annual of the composite family restricted in distribution to the island of Samos, where it occurs on rocky hills at 300–1200 metres. It was recorded by Ferdinand in a pencil drawing and collected by Sibthorp's party, with the specimen surviving at Oxford. In this case the accurate documentation of the pink colour of the ray flowers with the help of the colour chart is of importance to distinguish this subspecies from other annual species belonging to this genus and also occurring in the area. In Samos Ferdinand also documented the single column of a Greek temple, the Dipteros II, then standing. It still maintains its vertical position and consists of twelve column drums, precisely as shown in the view prepared in the summer of 1786 with two oriental figures, one wearing a fez, and a flock of sheep in the foreground.

The season for collecting plants was already advanced, and upon his arrival in Istanbul Sibthorp told Banks 'The Spring Plants had passed their flowers; & their Skeletons were too imperfect to ascertain their species: every thing was burnt up in the smaller islands'. So it had been decided to collect seeds, and small packets containing them were later sent from Istanbul to Sibthorp's botanical friends. One of them must have contained seeds of the marjoram *Origanum calcaratum*, a much-branched dwarf shrub long regarded as endemic to the island of Amorgos, but recently found also on a few other south-eastern Aegean islands. Tournefort had found it, Aubriet had illustrated it, now Ferdinand prepared a pencil drawing of this plant, and Sibthorp collected seeds, which quickly reached the Royal Garden at Kew, where plants were successfully raised. In his *Hortus Kewensis*, a catalogue of the plants cultivated at Kew, William Aiton the father gave a Linnaean name to the new acquisition, calling it *Origanum tournefortii*, adding 'Nat. of the Island of Amorgos. Introd[uced] 1788, by John Sibthorp'. Science, and plant taxonomy in particular, has always been highly competitive: Aiton, who had published his work after the 7 August 1789, was pre-empted by at least three days by Antoine-Laurent de Jussieu, who on the basis of the engraving and description given by Tournefort had published on 4 August 1789 the Amorgos plant under the name *Origanum calcaratum*, which is the correct scientific name. On this island Ferdinand had prepared two views, one of them showing the village of Chora with the steep cliff in its centre topped by the remains of a fortress from times when the island belonged to the Serenissima Repubblica di San Marco, with three windmills included on the right. Tournefort's footsteps were also followed on the island of Antiparos. Like Aubriet in 1700 it was now Ferdinand's turn to record the famous cave with its stalactites and stalagmites while a local guide with a torch in his hand showed the curiosity to the visitors (Fig. 5.7.). Another extremely local plant found during their first crossing of the Aegean Sea must have been *Campanula anchusiflora*, a bellflower new to science and restricted in distribution to Poros and Idra. It seems to have been collected on the latter island, recorded in a fine view by Ferdinand with the bay and harbour in the foreground surrounded by hills, topped again by many windmills.

According to the diary fragment, Sibthorp's party reached the western coast of Anatolia on 1 August 1786, possibly crossing over from Samos to Scala Nova (Kuşadası), and it was Ferdinand who prepared a drawing, later used as the basis for a view. To the right it shows the strongly fortified waterfront of this small town with two mosques and their minarets clearly discernible. Near this place an introduction was offered to the oriental style of travelling, and again Ferdinand acted as cameraman: two rings are shown, one of camels, one of horses, and in their centres are saddles and luggage; outside are groups of men in oriental dress, some of them armed, others near open fires. The view is appropriately named 'The Caravan at rest by the Coffeehouse near Scala nova in Asia' and shows a half-open wooden structure in the centre – probably the coffeehouse. Apart from Magnesia (Manisa), the locality information for the following days remains obscure; judging from information provided by Smith Sibthorp's party seems to have visited Bergas (Bergama), Lupada (Karacabey), Sousougheri (Susurlak), Thyatira (Akhisar) and Yalyam (Yalova) on their way to Istanbul. As before they repeated Tournefort's route, a traditional

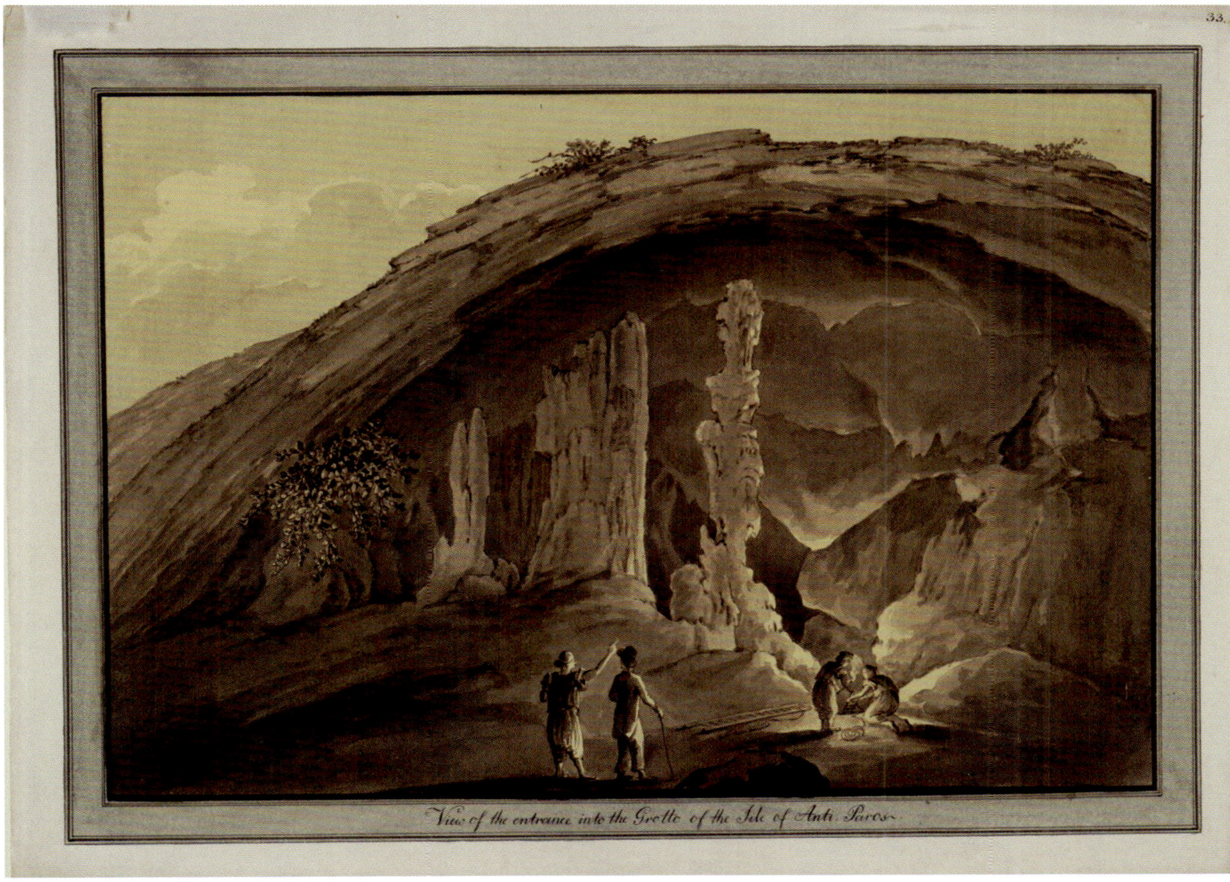

FIG. 5.7. The cavern on the island of Andiparos. Watercolour by Ferdinand Bauer, after 1794. – Oxford, University of Oxford, Sherardian Library, MS Sherard 408, f. 33.

caravan track linking Scala Nova with the capital of the Ottoman Empire.

Information is scarce on the ascent to Mount Olympus (Uludağ), the highest mountain in Western Anatolia, also known as the Bithynian or Mysian Olympus. Judging from Sibthorp's letter to Banks, the excursion lasted for three days with the travellers reaching 'superior heights', which may imply that they reached the summit at an altitude of 2542 metres. In any case they must have climbed at least to the upper montane zone, where, as always, in late summer the botanical harvest was more promising than at lower altitudes.

Ferdinand must have been busy preparing pencil drawings on Mount Olympus; maybe it reminded him of the Schneeberg near Vienna, a mountain he could have known. His harvest included a drawing of the rather widespread rusty foxglove, *Digitalis ferruginea*, then still in flower,[21] but also several much rarer species. One of them was *Scorzonera pygmaea*,[22] a dwarf cushion-forming plant, growing in rocky places at 1900–2500 metres. It was new to science and possibly reminded Ferdinand of the superficially similar *Crepis sibthorpiana* that he had documented on the Lefka Ori on the island of Crete just two months before. Another rare species then new to science documented by Ferdinand on the same side of the page marked by him 'M Olympus' was the stork's bill *Erodium sibthorpianum*, an endemic to this mountain massif growing on limestone at 2000–2400 metres. However, four more tiny perennial species from Uludağ recorded on the very same side of the page are widespread: the arctic-alpine fleabane *Erigeron uniflorus*, which extends to the high mountains of Anatolia, the very variable skullcap *Scutellaria orientalis*, the polymorphic white-flowering alpine rock-cress, *Arabis alpina*, and the speedwell *Veronica gentianoides* with its small blue to pinkish blossoms.

Considering Ferdinand's later itinerary and our current, rather detailed, knowledge of plant distribution in the eastern Mediterranean, several other plants could only have been documented by him on this high mountain, for example the tiny inconspicuous and late-flowering *Allium sibthorpianum*.[23] Another plant is the pink-flowering spike-heath, *Bruckenthalia spiculifolia*,[24] with the respective specimen collected by Sibthorp now kept in the Natural History Museum, corroborated by Ferdinand's pencil drawing on a sheet of paper marked 'Smirna'[25] (Izmir). Among the plants recorded by Ferdinand from Uludağ we find a few more endemics of this mountain. One of them is the bedstraw *Galium olympicum*,[26] another inconspicuous, cespitose perennial growing in rock crevices and screes at 2300–2500 metres, a second *Centaurea drabifolia*. However, there are also species with a wider distribution, among them the bellflower *Campanula spathulata*, albeit represented by a rather atypical, one-flowered specimen, and *Scorzonera cana*, the latter three documented by Ferdinand on a sheet of paper marked by him 'Olymbos'[27] (Fig. 5.8.). It was on this mountain that a plant of major horticultural importance was found – the avens, *Geum coccineum*, a rhizomatous perennial herb with a few-flowered inflorescence and flowers with bright orange-red to scarlet petals. It had already been collected by Tournefort, but was still without a Linnaean name, and Ferdinand documented it in a pencil drawing on a sheet of paper annotated 'Smyrna' (Izmir).[28] Admittedly the specimen, probably in its second flowering, is somewhat untypical, but this is in agreement with the relatively late collection date.

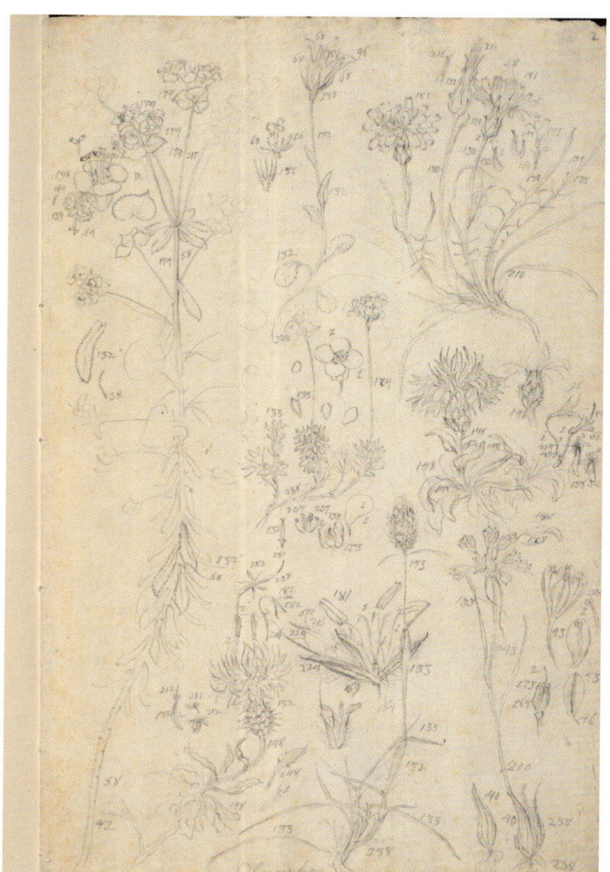

FIG. 5.8. Plants from Uludağ. Pencil drawings by Ferdinand Bauer, 1786. – Oxford, University of Oxford, Sherardian Library, MS Sherard 247, f. 2 r.

On the same side of the page we find a representation of *Asyneuma rigidum* subsp. *sibthorpianum*, another native of Uludağ, though not restricted to this massif. On this mountain Sibthorp and Ferdinand experienced for the first time the end of summer in the Mediterranean area, marked by the appearance of autumn-flowering bulbous plants. One of them is recorded in great detail in a pencil drawing[29] – the autumn crocus *Colchicum bivonae* with large flowers and conspicuously tessellated pinkish-purple petals, and leaves developing only in the following spring. Although this drawing has been previously misidentified as *C. parnassicum* and wrongly associated with the collecting tour in August 1787 to Mount Parnassos, Ferdinand's drawing is on a sheet of paper marked 'Olimbus' and therefore refers to Uludağ. In addition, Ferdinand recorded three views of the mountain they had just climbed together, two of them including the town of Bursa (Brusa), with its mosques and minarets.

After the weeks in the caravan and the mountain tour climbing Uludağ, to take lodgings in the British Embassy in Istanbul must have been a most welcome change, with Sibthorp reporting that they were treated 'in the full spirit of Oriental Hospitality' by Ainslie. For the following six months spent in and near the capital of the Ottoman Empire, no diary seems to have been kept by Sibthorp. No doubt Istanbul, with an estimated population of c.570,000 at the end of the eighteenth century was the largest city Sibthorp and Ferdinand had ever been to, much larger than London, Vienna or Naples. The capital of the Ottoman Empire was also the seat of the central power, i.e. the Sultan, then Abdul Hamid I, and the Grand Vizier (Vizier-i-Azam), effectively the Prime Minister. The latter position was then held by Koca Yusuf Pasha, a Georgian converted to Islam, residing in a famous building called the Sublime Porte (Bab-ı Ali). Hosted by the British Ambassador, Sibthorp indeed had a chance to see a military procession in Istanbul with both Abdul Hamid I and Koca Yusuf Pasha taking part, but the scene was not documented by Ferdinand.

Instead, Ferdinand continued to take views, later forming an impressive series of paintings, and to prepare pencil drawings of plants and also probably

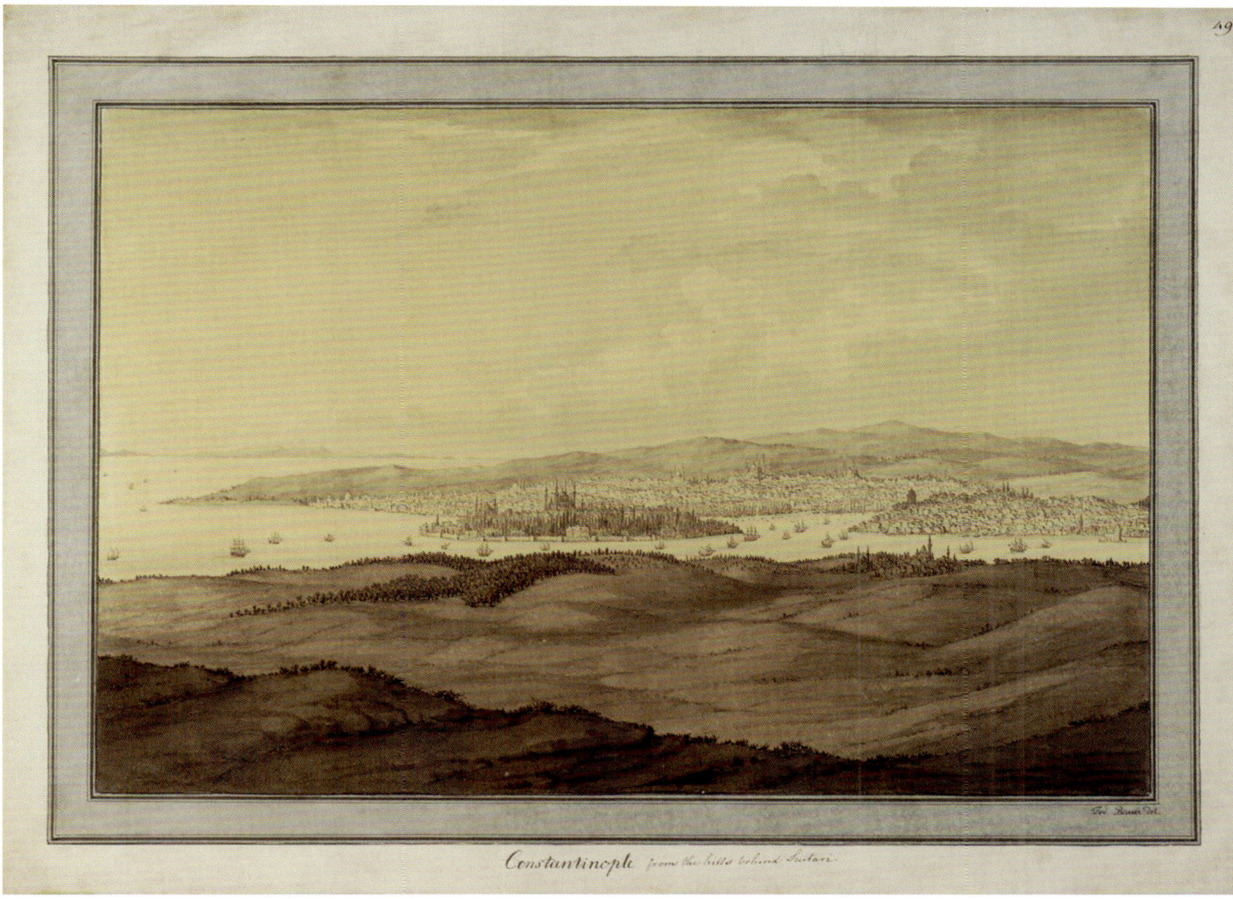

FIG. 5.9. **Panorama of Istanbul from the Asiatic Mainland. Watercolour by Ferdinand Bauer, after 1794. – Oxford, University of Oxford, Sherardian Library, MS Sherard 408, f. 49.**

fish. As so many artists before and after him Ferdinand seems to have been fascinated by this city, its buildings and in particular its unique position on the European and Asian shores of the Bosphorus. Some of his 'vedute' are taken from elevated points and at considerable distance aiming at the documentation of the panoramic views for which Istanbul has always been famous (Fig. 5.9.). One might expect the depiction of individual buildings of Istanbul – for example the Süleymaniye Mosque (Süleymaniye Camii) built by famous architect Misam Sinan, the Sultan Ahmet Mosque (Sultan Ahmet Camii) or the Hagia Sofia Mosque (Aya Sofiya Camii), the latter visited by his master – but this is not the case. Why Ferdinand refrained from this kind of documentation remains unknown; it was only several months later near Salonika (Thessaloniki) that he recorded a mosque at shorter distance. By contrast he prepared in great detail a view of a secular building, one of the residences of Abdul Hamid I, the Dolmabaçe Palace, at that time still a colossal wooden building on the shore of the Bosphorus.

Although the season for botanizing was now approaching its end, the surroundings of Istanbul continued to offer plants Sibthorp and Ferdinand had not yet come across. This is not surprising since Istanbul, because of its distinctly more northern geographical position and its more continental climate, harbours a flora different from that of Crete and the Aegean Islands. One of their trophies was a low-growing shrub with strikingly large yellow flowers and many names in English – rose of Sharon, Aaron's beard, great St-John's ward or Jerusalem star – *Hypericum calycinum*. It was already known to Linnaeus and reported later to have been found at Belgrade forest (Belgrad Ormanı), a mixed deciduous forest and classical botanical locality about twenty kilometres north of Istanbul. On the basis of a pencil drawing, later in Oxford Ferdinand produced a particularly fine watercolour, no doubt the most accomplished illustration then existing of the species, which is now widely cultivated as an ornamental. Other late-flowering plants almost certainly recorded from the vicinity of Istanbul include the bugseed, *Corispermum nitidum*, the bedstraw *Galium paschale* and the moor grass *Sesleria alba*. At Bujuchtar (Büyükdere), a village very near to Belgrade forest, a specimen of the autumn-flowering heath *Erica manipuliflora* was collected by Sibthorp, and in all probability a pencil drawing was prepared by Ferdinand.

In November Sibthorp moved to Chalcis [Heybeliada] in the Princes' Islands (Kizil Adalar), where he stayed with Ferdinand 'retired to a Convent . . . furnished . . . with Beds, Chairs, Tables' from the British Embassy, praising his new residence in an epic letter to his half-sister Lady Mary Elizabeth Sewell. Here he also mentions for the first time his specific interest in the local names of plants and animals, in particular fishes, writing 'I have made a Collection also of the different Fish & have been very attentive to the modern Greek names many of which correspond wonderfully well with the Ancient. I have at least upwards of Eighty upon my list'. This manuscript list may well have been the basis for a list of animal names without descriptions, including many fishes, published many years after the death of Sibthorp by Walpole. The Princes' Islands may have been also the place where Ferdinand produced a considerable number of pencil drawings documenting the ichthyology of the East Mediterranean. Here Sibthorp's report for the first time indicates his special interest for marine animal life 'A great variety of good fish is caught on the shores . . . We paid five piasters for drawing the net . . .Very fine lobsters are caught here . . .'. Ferdinand documented the Princes' Islands in five views – four of the island of Chalcis, one from the island of Principis (Pringipos/Büyükada) to the island of Chalcis and Istanbul. One of these 'vedute' inscribed 'View of the West part of the Village in the island Karki [Heybeliada]' is of particular interest, showing a rural scene with several buildings, to the left a group of women in long dresses and tall hats, and two musicians leaning under a tree with a tambourine and a fiddle. This is one of the few views by Ferdinand showing women, probably Greek Christians who enjoyed a more relaxed social life than their Muslim neighbours.

Sibthorp's epic letter from Chalcis to his half-sister is also relevant for understanding his relation to Ferdinand, with whom he was 'highly satisfied' since 'besides drawing my plants & incomparably well he delivers me of much Trouble in drying them & .' This was confirmed fourteen years later by Hawkins who told Smith 'Bauer . . . has a perfect knowledge of them [i.e. the plants], for he recollects their habitats & I believe the determinations of their species by Dr S -'.

Nothing is known for certain about how and where Sibthorp and Ferdinand spent the winter months. We may hypothesize that the former wandered through the famous bazaars of Istanbul collecting miscellaneous information on various vascular plants, leading to a manuscript published decades later by Walpole under the title *Natural history. Medicinal and economical uses of certain plants sold in the bazar, and herb-stalls at Constantinople*. On Christmas Day it was time for Sibthorp to send a letter to Banks, again containing relevant information on Ferdinand 'My Painter . . . has taken the outlines of above five hundred Plants & I shall at least have three hundred new species to add to the Linnean Nomenclature' adding 'I would have sent you at present some of the dried specimens . . . & my Painter at the same time pleading the necessity of several specimens to compleat his Figures, having taken only rough sketches in the course of our route'. This is another significant piece of information clarifying Ferdinand's method in preparing the finished watercolours: he was to base them both on his pencil drawings and on herbarium specimens, taking variability into account. Sibthorp's appreciation of Ferdinand's work is again expressed in his being 'particularly fortunate' with his illustrator, but does not bother to mention him by name. For Sibthorp Ferdinand was no more than 'my painter'.

Again it was time to make plans for the next year, and Sibthorp told Banks 'I expect about the latter end of January we shall set forward for Cyprus, from thence to Rhodes & Crete, whence I shall return again to Athens, making it my Head Quarters – attempts to penetrate on the one side into the interior parts of the Morea [Peloponnese] – visit Argos & Sparta [Sparti] & on the other side, if possible, pass thrò Thessaly to Mount Athos, as I wish to make my Flora Graeca as compleat as possible'. Taking the realities of travelling in the Levant in the late-eighteenth century into account, this was an extremely ambitious plan. As we shall see, it could be put into effect only in part the following year.

For several reasons Sibthorp's departure was delayed. One of them was his growing party: Captain Ninian Imrie or 'Emery of the Royals from Gibraltar', a man mainly interested in geology, topography and possibly spying, wished to join in, and Sibthorp was still waiting for his other projected travel companion Hawkins. On 30 January 1787 the latter arrived in the palace of the British Embassy in Istanbul after an exhausting winter journey from Vienna via Belgrade (then Dar-ül-Cihad) and Sofia (then Sofya), at the time both cities of the Ottoman Empire.

'A set of kitchen Utensils, a tent large enough for 9 persons a Casting net & besides Bedding' had been bought in Istanbul, as well as 'a stock of Sea provisions'. In addition to 'several Firmans . . . for various parts of the [Ottoman] Empire', Ainslie had provided the travellers with recommendations and 'procured others from the French Ambassador to the Consuls in the Levant'. As a consequence the party led by Sibthorp could rely on help from both the British and French representatives, not an unconventional situation before the outbreak of the French Revolution three years later. Finding a suitable passage turned out to be difficult, but in the end a 'large Venetian Merchantman' bound for the island of Cyprus (then Kibris), was selected.

ISTANBUL TO CYPRUS

On 13 March 1787, almost precisely one year after leaving Vienna, Sibthorp set off from Istanbul, this time accompanied by, in addition to his servant and Ferdinand, Hawkins, Imrie, their servants, and Hawkins's interpreter. Starting from this day until the end of the journey in Bristol, Sibthorp's diary entries are available, supplemented by information from letters and other sources.

'Sleety snow' was falling on Istanbul and a brisk wind was blowing when the *Bethlehem* left the harbour. They passed the island of Marmora (Marmara Adası), which was covered with snow, reaching the Dardanelles (Çanakkale Boğazı) on 16 March, first landing on the Asian shore near Abydos (Nagara) and then crossing to Sestos (near Eceabat) on the European shore. There they climbed a hill near the village, from where the party had a view over both sides of the straits and the island of Tenedos (Bozcaada), where Halil Hamid Pasha, a former Grand Vizier, had been executed the year before. For centuries the Dardanelles have been of strategic importance for the balance of power, and the British were particularly obsessed about their role. Imrie, then based in Gibraltar on the other end of the Mediterranean basin, must have been singularly interested in the fortifications, with the diary reporting 'we strayed into the precints of the Castle where we observed several large Canon[s] most of which were mounted only on stones, vast balls of Stone lay interspersed. This & the opposite Fort are of a very ancient construction totally unadapted to the purpose of modern Warfare but otherwise advantageously situated for blocking the passage of the Dardanelles'. Probably under the instructions of his master, Ferdinand produced several views of the fortifications taken from different angles, and one can hardly avoid the impression that this pictorial documentation served not only Sibthorp's curiosity.

For the botanist the first indications of spring were already visible, among them *Crocus flavus*, the common yellow crocus widely found in cultivation, and a specimen was documented by Ferdinand. Sibthorp and Ferdinand may have been the first to see it in the wild and to bring living material from this locality to England, since shortly after his return 'bulbs of liliaceous plants' are said to have been sent by him to William Aiton the father at Kew. Near Abydos Sibthorp shot 'a male and female *Sturnus Cinclus*' (possibly a dipper, *Cinclus cinclus*; but see Chapter 6), this being one of the few cases where detailed information is available for a bird record in the diary. A planned visit to the plain of Troy (Truva), needless to say then not yet excavated, had to be given up because of its considerable distance from a possible landing place, and on 19 March 1787 the *Bethlehem* finally passed the straits. Ferdinand was again asked to document the fortifications of the castle, maybe for intelligence reasons, the diary noting 'very considerable and of recent construction but unfortunately from the great breadth of the channel of little utility in preventing the entrance of hostile ships'.

Passing the island of Tenedos, Ferdinand took another view from the sea, which became very rough in the evening of the following day between the island of Scio (Hios, then Sakız) and the Asiatic mainland. The horrors of the following storm are described in detail in Sibthorp's diary: 'The wind had blown very fresh from the NE & had threatend us with a gale which by this time had blew[n] hard and raised the sea so high as to burst over the deck with sufficient frequency to alarm us and to spoil a small part of the corn with which the vessel was laden . . . In the Bogaz [Steno Folernou, between Samos and Ikaria] we were sheltered a little from the fury of the wind by the neighbouring headlands . . . The wind which abated a little at noon soon after rose upon us with greater fury than before . . . at the approach of the evening we found ourselves exposed to a wild sea thickly interspersed with isles. The Sun set when we were at some distance from a port on the Asiatic shore. The entrance to which so barricadoed with rocks as to be difficult & even dangerous. The Wind whistling wildly along the ropes, the confused cry of the mariners, the bellowing of the Steersman the Groaning of the Mast the irregular flashes of lightening with the white foam at the prow aggravated the uneasiness of our situation. At length after some moments of uncertain existence [the ship anchored in the bay of] Karabaglar [Karatoprak]'.

Due to poor weather conditions a few days were spent at this place, tortoises were observed, plants collected, fish caught, and Ferdinand prepared a view of the small harbour on the Asiatic mainland, which he did also at the next stop – Porto Cavalieri (near Akyar Burnu). Again it was time for botanizing, with Sibthorp reporting 'we were sufficiently amusd and occupied by the abundance & diversity of fine plants which grew within our reach & which were now mostly in flower. The Luxuriance of vegetation was even greater here than at Karabaglar and exceeded any thing we had before seen, the Conservatories of Kew contain not a richer display of vegetable beauty . . . Cap'n Imrie . . . reached the summit of the nearest mountain and brought back with him a new and elegant species of *Fritillaria*'. Both Imrie's specimen, Ferdinand's pencil drawing and his watercolour based on the latter survive at Oxford. All three document *Fritillaria sibthorpiana*, a rare fritillary which was found again in the wild only 180 years after Imrie and is known to occur locally in the extreme south-west of Anatolia and on the island of Simi. This case demonstrates the importance of Sibthorp's herbarium and Ferdinand's work and is at the same time one of the rare cases of a complete documentation of a specimen. It was mentioned in the diary with locality information, supported by a pencil drawing and an annotated herbarium specimen.

On 31 March the *Bethlehem* left the harbour and crossed over to the island of Rhodes (then Rodos) for a brief stop. Now a much-visited tourist site, Rhodes looked quite different in the year 1786, the diary recording: 'The streets appeared almost deserted and forlorn; the principal one, called the Knights' Street, offered a melancholy view of the remains of those houses defended with so much bravery by their various tenants. The arms of several of these knights are observable on the walls; and the ruins of the grand master's palace are covered with the golden Henbane'. This is clearly golden henbane, *Hyoscyamus aureus*, a poisonous plant of scattered occurrence in the eastern Mediterranean, still growing today on the walls in the town of Rhodes as well as on the walls of other fortifications, like those of the Venetian arsenal in Iraklion on the island of Crete. Ferdinand had prepared a pencil drawing of this plant,[30] producing later a watercolour showing the golden-yellow corolla of its flowers with a dark purple throat and exerted anthers. He also prepared a view of the harbour with its fortifications, windmills and sailing boats, the latter completely out of proportion.

The anchor was raised again on 1 April 1787 and the vessel headed for Cyprus with 'the high mountains of Lycia [on the Asiatic mainland] covered with Snow' on the portside. Birds were observed and shot, and fish caught, until the bay of Larnica (Larnaka) was reached on 8 April, where the party had to wait for several days for favourable winds to be able to enter the harbour.

Sibthorp's party had arrived in time with the glorious spring flora of Cyprus at its best and many plants as yet unknown to science waiting for their discovery. Tournefort, Gundelsheimer and Aubriet had not been to Aphrodite's island, and as a consequence the visit now beginning had all the ingredients of a botanical excursion into the unknown. Again we are faced with the well-known problem of locality data: if a widespread species was recorded in Cyprus, this can only be stated with certainty when Ferdinand had annotated his pencil drawing accordingly.

Lodgings were taken for a few days at the Salines, south-west of Larnica (Larnaca), and the diary records various observations on topography, geology and archaeology, such as 'considerable quantities of slags of an ancient Iron or Copper Furnace'. No doubt this was of special interest to Hawkins, whose brother owned copper and tin mines in Cornwall. Sibthorp seems to have been struck by the rich bird life, noting that hardly could a more favourable place be found in the Levant for observing migratory birds, and Ferdinand was commissioned to document some of those shot by his master.

On 12 April 1787 an excursion to the Mountain of the Holy Cross (Stavrovouni) started with Sibthorp noting, 'beautifull varieties of the ranunculus asiaticus, now in flower, we halted dined under a Carob tree, the air was soft & serene & the spot delightfull Several little rivulets cross'd the road, skirted with the Oleander [*Nerium oleander*], these were frequented by the beautifull Merops apiaster one of which we shot'. Watercolours prepared later by Ferdinand show the turban buttercup,

Ranunculus asiaticus, a species known for its wide range of petal colours (here shown in a dark red variant), *Nerium oleander* and the migratory bee-eater, *Merops apiaster* (Fig. 6.16), all at Oxford, unfortunately without any locality information. Since all three possess a wide distribution they cannot be associated with any degree of certainty to this excursion. Having lost their way, the party arrived late at a metochi belonging to the Stavrovouni monastery and ascended to the summit 'uncommonly steep & difficult'. However, the monastery itself was found deserted and 'locked up we attempted an entrance by force Our different instruments for herborizing & Fossil digging were employ'd without success when at length a peasant arrived with the keys & some milk and gave us admission. We found miserable accommodation for travellers. The Cells of the condemned Prisoners on Newgate [in London] could hardly be worse than the Cells of these Monks, the Church allthough built by the Empress Helena bore all the marks of poverty & neglect . . . we drew out some straw matrasses on the pavement . . . The wind howled mournfully along the dreary Aisles and shook the broken casements of the Chapel'. The botanical findings of this excursion comprise two endemics of Cyprus, both new to science – a thyme, *Thymus integer*, a low subshrub with procumbent stems, and an elegant gladiolus, *Gladiolus triphyllus*, smaller, with slender growth and paler flowers compared with the common sword-lily, *G. italicus*; both species were documented by Ferdinand.

The report of the descent testifies Sibthorp's ethnobotanical interests: for example, he mentions 'being thirsty a poor Peasant brought me a Pumpkin with water corked with a rush of the Poterium spinosum [*Sarcopoterium spinosum*] which served both as a Cover & a stainer [strainer] & prevented the entrance of flies and other insects. The stools on which we sat were made with the Ferula greca [*Ferula communis*], the stem cut into slips lay'd crossways & nail'd together'. The summer was approaching with swarms of locusts in their larval stage seen blacking the streets and threatening the harvest of wheat, now almost ripe.

After three days of rest at the Salines the party mounted mules again and set out for Famagusta (Ammochostos/ Gazimağusa), a town they found to be very disappointing, the diary recording 'the remains of a fine city now scarcely containing 200 inhabitants who are all Turks. The fortifications show the former strength of the place. The ditch is hewn out of the solid rock, some canon with the arms of Venice lay dismounted on the ramparts . . . Leaving the Bastion we pass'd through the Streets now melancholy pictures of Turkish Desolation. The town was of considerable extent & displayd still sufficient marks of its ancient magnificence. The Principal Church a considerable Gothic Structure is now a Mosque [today called Laia Mustafa Mosque], in the pavement I observed the Tombstone of a Knight Templar'.

On the way to Famagusta Sibthorp had mentioned in his diary a 'species of Arum unnoticed by Linnaeus' and Ferdinand had prepared a pencil drawing of it on a sheet of paper annotated by him 'Cybrus' (Cyprus)[31] showing *Arum dioscoridis*, a relative of Lords and ladies. Later in Oxford, Ferdinand was to prepare another of his masterpieces based on this drawing, showing the long spathe, the upper part of which is conspicuously spotted or blotched dark purple internally, and the clavate purple spadix (Fig. 6.11). Clearly the specimen must have been dug up and the underground part put into water in order to make it free from the adhering soil, so that Ferdinand could draw the subglobose tuber with its roots.

Leaving Famagusta, the diary mentions 'the Aloe vera, the Iris germanica and florentina' growing abundantly on the rubbish. Ferdinand's pencil drawings of bitter aloe, *Aloe vera,*[32] and common iris, *Iris germanica,*[33] both species of uncertain origin with a long history of cultivation, are on sheets of paper marked by him 'Cybrus', and may refer to Sibthorp's note. By contrast Ferdinand's pencil drawing of *Iris florentina*, possibly nothing more than a white-flowered variant of *I. germanica*, is on a sheet of paper annotated 'Asia' by him,[34] which almost certainly excludes Cyprus as the collecting site. In short, Ferdinand was very busy on this island, and possibly for this reason only prepared a single drawing of a Cypriot village – 'Upsera or Ipsa' (Gypsos), which the travellers had reached after a hard ride of five hours from Famagusta.

The convent of Antiphonitis (Antifonitis) in the northern part of Cyprus was the destination of a tour on 19 April, where a tree attracted the curiosity of Sibthorp, who described the monastery as 'most romantically situated in an alpine hollow & nearly embosom'd in trees . . . overlooks the whol Cilician Sea and commanding a prospect of the distant mountains of Caramania [on the Asiatic mainland]'. The published extract of Sibthorp's account reads 'The Eugoumenos [abbot] of the convent, a very old man, offered himself as my conductor, and leading me a few paces below the convent, into a garden, now covered with rubbish, pointed out a tree, which upon examination I found to be the Liquidambar Styraciflua. The trunk of it was much hacked. Different bits of it had been carried off by the curious and superstitious, as an ornament to their cabinets and churches . . . The name Xylon Effendi, and the tradition of the convent, testify the reputation in which this tree has long been held in the island. It was probably originally introduced by the Venetians during their possession of Cyprus. I could not discover, either from observation or enquiry, that it was to be found in any other part of the island; nor do I recollect that the Liquidambar styraciflua has been mentioned, by any botanist, as an oriental tree. I suppose it is from this that they extract a very fine perfumed oil, which, they say, as well as the wood, has the virtue of fortifying the heart and the brain. The common people here cut off the bark and wood together, toast it in the fire, and suck it, which they esteem a specific remedy in a fever, and seem to think it has a miraculous operation'. Again Sibthorp offers an enlightened interpretation, but his determination of the tree is incorrect. The tree planted next to a monastery and lost after the Turkish occupation of the northern part of Cyprus in 1974 was in all probability the oriental sweetgum, *Liquidambar orientalis*, a native of the island of Rhodes and the south-western part of the Asiatic mainland.[35] Furthermore, the tree had been mentioned by earlier travellers, among them Richard Pococke, formerly of Corpus Christi College, Oxford, who had found it in 1738 at this very monastery noting '. . . but [Cyprus] is most famous for the tree called by the natives Xylon Effendi and by the naturalists Lignum Cyprinum and Lignum Rhodium, because it grows on these two islands . . . It produces an excellent white turpentine; especially when any incisions are made in the bark'. The local name attached to this solitary tree 'Xylon Effendi' is simple to explain, meaning 'Wood, sir' and consists of the Greek word for wood (xylon) and the Turkish word for sir (effendi). Needless to say Ferdinand made a pencil drawing of a twig, including the male inflorescences arranged in racemes, the hanging female inflorescences with the protruding stigmas, and a ripe woody infructescence.[36] Surprisingly this image was later not used for the preparation of a finished watercolour for a simple reason – Ferdinand's master excluded plants from any further attention that he believed, wrongly in this case, to have been introduced into the Mediterranean area. Fortunately, the pertinent herbarium specimen, albeit without inflorescences and infrutescences, survives at Oxford.

The next morning the party descended to the northern coast of Cyprus, arriving late at Cerignes (Keryneia/Girne), 'a palty [paltry] seaport . . . the station of the passage boat that plies between the island & the main. The port is no other than a little bason which has been formed as much by art as by nature between the ledges of rocks on the beach; it is still so open to the swell that I observed the caikes [boats] moored to the rocks'. Upon arrival in this village a beautiful species of sage is recorded in the diary, recently interpreted as *Salvia hierosolymitana* or *S. pinnata*.[37]

'Having employed the morning in drawing and putting our plants in paper', which means pressing them, the party visited an old monastery dating from the Venetian period, but an attempt by Sibthorp and Imrie to climb the nearby mountains was abandoned for a simple reason – 'our guides . . . refused to satisfy it [our curiosity] by risking their mules in these mountainous tracts we therefore dismounted . . . The sun shone with uncommon force without the least breeze to mitigate the fervor of its rays, after a very hot & fruitless walk we return'd finding the spot too distant to reach & return before nights'.

On their way to Nicosia (Lefkosia/Lefkoşa) *Salvia veneris* seems to have been found, a tufted perennial herb, twenty to forty centimetres tall. Years later Smith added the note 'D[edit]. F. Bauer' (Ferdinand Bauer gave the specimen) to his account of this species, which indicates that Ferdinand had collected the specimen, possibly on a separate collecting tour. In any case he had made a pencil drawing of this very local species, another endemic of Cyprus. Approaching the capital of the elayet Kibris, Levantine life was recorded – 'a train of Camels, locust[s] . . . beginning to commit it[s] annual ravages myriads of them crushed under the feet of our mules'. The next day the travellers were received in great style by the pasha, who offered his firman for the whole island and a magnificent dinner giving 'a favourable idea of the Turkish Cookery & of Turkish Hospitality'. The pasha's palace had something special to offer – a gazelle, an extremely timid specimen that was required to be held while Ferdinand made a drawing of it. Although it was claimed to be an animal native in the Troodos Mountains, others said it was a present sent to the governor from the coast of Syria, a much more plausible explanation.

The party split in Nicosia, and Ferdinand returned with his master to the Salines, from where the pair continued to Limissol (Limassol/Limosol), then a small town on the southern coast of Cyprus. On this tour *Cephalaria syriaca*, a weed of agricultural lands widespread in the Near East, was recorded in the diary and collected, and a drawing of it was prepared by Ferdinand.

In Limissol Sibthorp and Ferdinand were reunited with Hawkins and Imrie and found lodgings with the British Vice-Consul. The next morning they started a tour to the Troodos Mountains heading for the 'Monastery of the Holy Cross' (Mesopotamos), which they reached at sunset. When the travellers ascended to the limit of the tree cover, the diary lists among other trees 'Quercus ilex'. In this Sibthorp was mistaken – he had almost certainly seen *Quercus alnifolia*, an oak endemic to Cyprus, which was to receive a Linnaean name only half a century later. The 'Cretan Maple' recorded in the same entry of Sibthorp's diary is *Acer obtusifolium*, a semi-evergreen shrub or small tree endemic to Cyprus, regarded by some as a geographical race of *Acer sempervirens*, a much more widespread species occurring also on Crete. The latter had been collected in 1700 on this island by Tournefort, who brought living material to the Jardin du Roi in Paris for cultivation. Sibthorp may well have seen this tree in the garden when he stayed in the French capital in c.1782. Today this very specimen continues to grow happily in the garden, renamed in the French Revolution Jardin des Plantes.

The next monastery in the Troodos Mountains to be visited was that dedicated to St John (Trooditissa Monastery), where the abbot asked Sibthorp for medical assistance. The diary noted 'a miserable old figure with one eye consumed by the Venereal disease . . . he was 60 & at 57 had been unfortunate in his amorous rencounters'. The following morning the party climbed on the back of mules and accompanied by a goatherd to the summit of the mountain range, where they found on the north-eastern side patches of snow. Hawkins observed 'the bearings of some important points but much regretted the want of a better instrument', while the geologically inclined members of the party studied rock samples. Sibthorp noted in his diary the 'Crocus vernus growing near the snow' and was again mistaken. Almost certainly he had observed another endemic to Cyprus, which had to wait for more than half a century until it was collected again by Theodor Kotschy and was soon after made known to science as *Crocus cyprius*. In this case Ferdinand did not prepare a pencil drawing, and Sibthorp did not collect a specimen, but considering the locality and the month of the collection *Crocus cyprius* is the only plausible assumption.

While descending the mountain the party was split for a second time. Sibthorp, accompanied by Ferdinand, was particularly well received in a Greek village with the diary reporting: 'The Papas himself was gone to the Harvest but the Elder of the house press'd me to sleep there that night. A fowl was kill'd for my supper and a bed prepared for me in a vacant part of a chamber devoted to the Silk worms. I was waited on by the daughter of the Papas the prettiest girl I had yet seen in the island. The roses of her cheeks were often heightened by her modest blushes. . . '.

On that day a tiny plant with a basal leaf rosette had been found in 'a little morass in the vale', noted in Sibthorp's diary as 'Pinguicula villosa' and recorded by Ferdinand in a pencil drawing annotated 'Wasertropfen' [water drops] in his microscopic hand. He must have carefully observed the slimy surface of the butterwort *Pinguicula crystallina*, an insectivorous plant, which indeed shows tiny droplets of water on its leaf margins. Based on this drawing Ferdinand later produces at Oxford a watercolour that was to serve as exemplar for a coloured copper engraving (see Chapter 6). The representation of the median lobe of the lower corolla-lip shown as bifid has been criticized as being uncharacteristic of Cyprus material, in which the central lobe has a rounded apex,[38] although there is some degree of variation.[39] In this context it has been suggested that Ferdinand was pressed for time and may have made this mistake, Sibthorp reporting in one of the published extracts of his diary 'My draughtsman stopping to sketch these plants [among them *Pinguicula crystallina*] was the cause for my losing my companion, who slept at a neighbouring monastery'. This interpretation may not be the whole story: when producing the watercolour Ferdinand could have been misled by the herbarium specimen, which comprises a single flower inadequately pressed, suggesting at first glance a bifid median lobe. This alternative interpretation for a minor error confirms that Ferdinand at least occasionally relied on additional herbarium specimens when preparing his finished works (see Chapter 6).

Meanwhile Hawkins and Imrie, who had lost Sibthorp and Ferdinand, met them again on the road to Nicosia. In the plains, summer had started and Sibthorp's reunited party seems to have been surprised

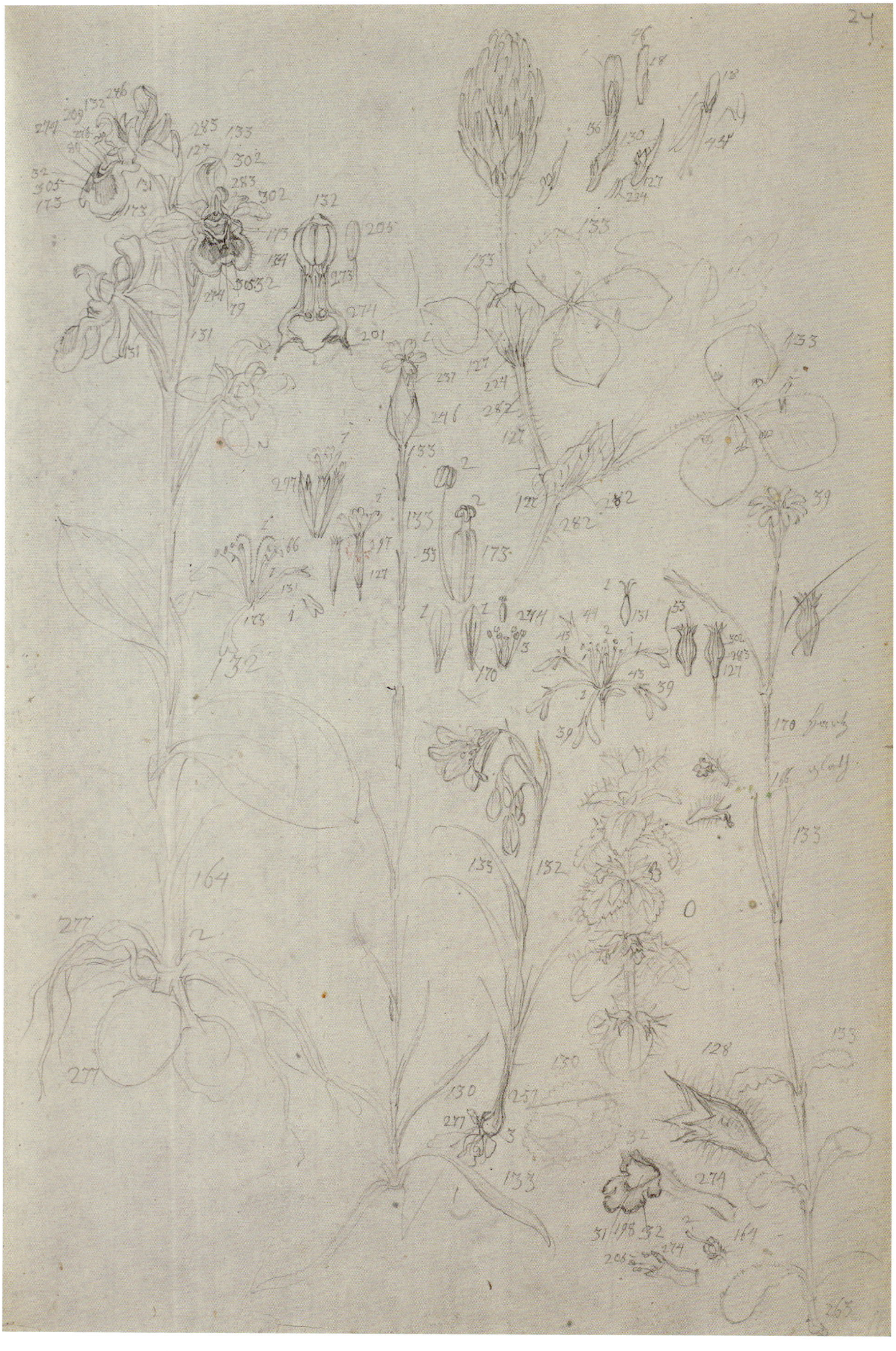

by the numerous herds of goats, the diary commenting 'there is not a shrub however bitter or hard for mastication on which this animal will refuse to browse even the most prickly ones are not exempted from its nibbling foracity . . . Each herd had usually some pastoral attendant who piped away his leisure hours and looked as wild and congenial with the landscape as a Painter could wish. . . . These Goats somewhat enlivened the scenery as we pas'd along for nothing could be more arid and chearless than the face of the Country'. After a short stay at the Salines, with Ferdinand probably busy preparing more pencil sketches, on 8 May 1787 the party was ready to embark on board a small vessel from Ragusa (Dubrovnik, in modern Croatia) heading for Baffo (Nea Pafos/Baf) on the south-western coast of Cyprus.

The village is described as a 'vast heap of ruins' and the classical minded members of Sibthorp's party started to take measurements of individual columns, studied the material of which they had been made and hypothesized that they had found 'the temple of Venus built by Augustus'. The list of plants recorded in Sibthorp's diary from the 'Turkish village called Iktinia [Ktima]' near Baffo exemplifies for another time the complexities of the expedition. Among others the following plants are recorded: 'Aloe vera', mentioned already in his diary at Famagusta, which implies that Ferdinand's pencil drawing annotated by him 'Cybrus' could also have been prepared near Baffo; 'Physalis somniferum' [*Withania somnifera*] first recorded by Ferdinand in Sicily; 'Sempervivum arboreum' [*Aeonium arboreum*], a native of Gran Canaria, cultivated in the Mediterranean area since classical times, possibly documented by Ferdinand at this locality for the first time; and 'prickly Chicory' [*Cichorium spinosum*], a widespread intricately branched, rigid subshrub forming spiny mats or cushions with few-flowered capitula, which Ferdinand could have recorded not only in Cyprus but also on many other occasions during the journey.

The sawfly orchid, *Ophrys tenthredinifera*, is a characteristic feature of the spring flora of garigue and phrygana, widespread in the Mediterranean area, but in this case we can safely assume that it was recorded in Cyprus, because the pencil drawing is on a sheet of paper marked by Ferdinand 'Cybrus'. Additionally a herbarium specimen, albeit damaged by insect attack, survives at Oxford. This finding is a great surprise: although the species has been recorded later several times for Cyprus, it is currently regarded as the rarest orchid species of the island. Until 2010 it was known from a single locality near Agios Minas on the Akamas Peninsula, an outstanding fact considering the number of collectors who have scanned Cyprus.[40] Found on that spot in 1992 and in 2002, its existence could not be confirmed in 2004 and 2005, and as a consequence the species has been listed among the data-deficient plants of Cyprus.[41] On 4 March 2011 G. N. Hadjikyriakou was able to photograph this very species at a second locality, near the village of Vasa Koilaniou (Kilanion), which is not far from the route of Sibthorp's party from Nicosia to the sea.[42] In his drawing Ferdinand shows himself the great connoisseur of orchidology: He had excavated the testicle-like underground parts of the plant, put them into water, and recorded them after the soil particles had been removed. In addition he also made a precise drawing of the flower, with particular attention to the lip with its clearly circumscribed pattern, annotating it with numbers referring to the colour chart (Fig. 5.10). Based on this colour-coded pencil drawing, at Oxford Ferdinand later produced one of his watercolours of superb precision (Fig. 6.10). Clearly the topic was not new to him – he or Joseph or Franz had already prepared several representations of *Ophrys* species for Boccius while they worked together in Feldsberg. But what Ferdinand produced in Cyprus was in a sense the apotheosis of colour-coding in orchidology. We can safely assume that neither Ferdinand nor his master had in their most vivid fantasies an idea of how the flowers of this plant deceive their pollinators. With a highly specific scent the sawfly orchid is able to attract male sawflies only, which recognize in the flower's lip, because of its shape and pilosity, a structure so similar to the abdomen of a female sawfly that they attempt copulation. Although of course unsuccessful, the males position their body in a characteristic way on the flower with their head touching the flower's pollinia. These then easily detach from the flower and become attached to the insect's head, again at a specific site. When the frustrated male, with usually a single pollinium attached to its head, tries to copulate with another flower, the pollen grains in the pollinium come into contact with the stigma of the second flower, resulting in cross-fertilization. In Sibthorp and Ferdinand's time no one had the faintest idea about this sublime form of deceit, now known as pseudo-copulation.

FIG. 5.10. Plants from Cyprus. Pencil drawing by Ferdinand Bauer, 1786. – Oxford, University of Oxford, Sherardian Library, MS Sherard 247, f. 27 r.

However, the great surprise is Sibthorp's manuscript name 'Cyclamen cypricum', which must refer to florist's cyclamen, *Cyclamen persicum*, the only spring-flowering cyclamen species native to Cyprus. This is confirmed by the pencil drawing that Ferdinand probably prepared there, although in this case no locality information is provided.[43] Since at that time this plant was only known in cultivation and presumed to originate from 'Persia', discovering it in the wild was another major discovery for Sibthorp. However, nothing indicates that he realized its importance. The species had received its epithet from Philip Miller on the basis of a plant grown in the Chelsea Physic Garden in London and now preserved in the Natural History Museum. Living material of *Cyclamen persicum* collected by Sibthorp was either sent or given to Aiton the father and subsequently cultivated in the Royal Garden at Kew. When the *Hortus Kewensis* was published the entry of *Cyclamen persicum* had the note 'Introduced by John Sibthorp, M. D. in 1788 Nat.[ive] of Cyprus' appended. Years later Ferdinand's watercolour (Fig. 6.2.), based on the pencil drawing prepared in Cyprus and painted at Oxford, was published as a coloured engraving in the *Flora Graeca* (see Chapter 6 and 9). The horticultural importance of this species can hardly be overestimated, for millions of these showy ornamentals are sold every year as pot plants all over the world. Today this cyclamen species continues to grow and flower copiously among the ruins of modern Kato Pafos.[44]

After some more botanizing and an unpleasant experience with Sibthorp's party being deliberately misled by local guides, it was high time to head back to the coast. Riding all night the travellers arrived at daybreak near Baffo 'when on firing a gun as a signal we were fetched on board by the ships boat'.

The visit to Cyprus had been an extremely successful one, in particular in the field of ornithology and botany. It seems appropriate to quote again from the published portions of Sibthorp's diary 'The naturalist, disappointed in finding so small a number of quadrupeds, is surprised on observing the great variety of birds which migrate to Cyprus . . . Great numbers of Grallae pass over in the spring from Egypt and Syria: these retreat farther, in proportion as the salt pools near Larnica are evaporated by the sun . . . The swallow, the martin, the swift, the Melba, the Pratincola, which frequent in numbers the pools of Larnica, visit also the island in spring . . . Those large birds which frequent the higher regions of the Troados [sic] . . . Of the rarer species . . . my draftsman [Ferdinand] has taken drawings'. Sibthorp's summary of plant life in Cyprus is equally correct in every detail: 'Cyprus, though possessing several of the Egyptian and Syrian plants, yet, from the scarcity of water, the great heat of the sun, and the thin surface which covers the upper regions of the mountains, can scarcely be considered rich in plants: and when compared to Crete must even appear poor'. Indeed, the number of vascular plant species of Cyprus is distinctly inferior to that of Crete.

Within a single month Sibthorp had managed to get an overview of the flora of Cyprus, listing 616 species, about a third of those known today. Because of the reasons set out above it is impossible to ascertain the exact number of pencil drawings prepared by Ferdinand, but it seems safe to assume that at least 200 plants were recorded, several of them endemic to the island and new to science (for example, the much-branched canescent shrublet *Onosma fruticosa*; see Chapter 6). To this figure the considerable number of zoological drawings, in particular of birds, has to be added. In short, spring time on Aphrodite's island had been a particularly intensive time for both naturalists.

CYPRUS TO PATMOS

The *Providence* with Sibthorp, Ferdinand, Hawkins, Imrie and their servants on board left the harbour of Baffo on 14 May 1787, their destination Rhodes. Due to calms and light breezes from different directions it took them much longer than expected. This meant losing time, with Sibthorp rightly complaining, 'Thus we were detained in the most mortifying state of inaction while the season for botanizing in these southern regions was rapidly passing away. Day after day we lay exposed to a schorching sun motionless or gently heaving on the smooth but swelling bossom of the Cilician sea'. On the other hand he could not resist describing the scenery and noted in his diary: 'To a Landscape Painter Compositions more striking could no where present themselves I have seen none which more truly possessed the character of sublimity. The Mountains . . . rose abruptly from shore . . . to the greatest degree of Alpine Elevation . . . awefull when veiled in twilight or wrapped in Thunder storms, serene and Majestic when gilded by the rays of the rising moon but Magnificent when the midday sun displayed their long glittering'.

The vessel anchored twice on the Asiatic mainland, first near Myra (near Kale), then at Finica (Finike), recorded by Ferdinand. Some botanizing was done and on one of the occasions a flowering specimen of the rather common mandrake, *Mandragora officinarum*, was found, dug out and documented by Ferdinand on a sheet of paper marked 'Asia'.[45] However, it cannot be excluded that this had been done several weeks earlier during a stop on the way from Istanbul to Cyprus. The mandrake's rootstock supposedly sometimes resembling a human shape had been associated since antiquity with magical properties. Ferdinand showed a distinctive male specimen exhibiting an erect, penis-like structure, possibly to amuse the person studying his representation. Later at Oxford he produced a fine watercolour (Fig. 6.3.), based on the colour-coded pencil drawing. For Sibthorp mandrake was of particular relevance: together with the opium poppy, *Papaver somniferum*, it was one of only a few reliable narcotics of eighteenth-century medicine, although still shrouded in myth.

The party penetrated a few kilometres inland and was able to watch scenery described in lyric words in the diary: 'Here The Pomegranate glowing with its scarlet flowers ornamented the hedges while the Vitis . . . overhung the river and perfumed the air with the most fragrant odour Orange & Lemon trees crowded in a wild luxuriance around the habitations of these untutord peasants & in vain solicited their art to prune & improve them. Fatigued with our walk & oppressed by the heat of the day we sought little repose under the shade of the spreading Plane tree [*Platanus orientalis*] on the margin of the clear & silent Lymirus and in front of a Caramanian cottage . . . We returned highly satisfied with our walk and richly laden with curious plants'. Again Sibthorp presents himself as the learned don from Oxford – he knew the classic name of the river and regards the local inhabitants as untutored. Ganzarowitz found out that the villagers kept a tame lynx, *Lynx lynx*, which was immediately recorded by Ferdinand, while Hawkins's servant Asfaroos Asaros, here mentioned for the first time by name, had bought a goat to be slaughtered.

After a 'tedious voyage' the *Providence* reached the harbour of Rhodes on 28 May 1787 and Sibthorp's party disembarked. Again their stay was a brief one, but Ferdinand seems to have prepared either on this occasion or during his first visit to the island a pencil drawing of the field bindweed, *Convolvulus scammonia*, made on a sheet of paper marked by him 'Rhodo'. The trailing, glabrous perennial with axillary long, pedunculate, cream-coloured flowers has a wide eastern Mediterranean distribution, but the locality given by Ferdinand is relevant. On the same sheet the umbellifer *Elaeoselinum asclepium* was documented, another species with a similar distribution, although apparently rarer in the Aegean region. The list of plants indicated for Rhodes on this occasion is a short note, including 'Juniperus Oxycedrus', very unlikely to be the species known today under this name but rather *Juniperus macrocarpa*, growing mainly on dunes and maritime cliffs. For once Ferdinand seems not to have prepared a drawing, nor is a specimen extant at Oxford.

After a visit to the French consul in Rhodes, obviously using the recommendations provided by Ainslie in Istanbul, and the usual delays with customs, Sibthorp and his party set out again for Crete, this time not by ship but in a small hired sailing boat, a caique, with five boatmen. As it turned out, they changed their plans and instead of a second visit to Crete they proceeded island hopping in a north-westerly direction, landing in the end at Piraeus. It was summer and the azure waters of the Aegean Sea lay before them. Their first stop was on the shore of the Asiatic Mainland near Porto Cavalieri, where they had been nine weeks before. By late May the vegetation had undergone a profound change, with Sibthorp noting in his diary 'a beautifull garden enammeld with some of the most curious plants of the east now presented a dry and wethered scene of desolation & furnished me only with the opportunity of collecting some seeds from the skeletons of those plants which I had before gathered in flower'. Indeed, skeleton is a very appropriate term for the desiccated, dead, brownish remains of annual plants which by that time of the year cover the lower altitudes in the Mediterranean area. Fish were caught, followed by a fish dinner on the beach.

Because of the prevailing weather conditions, the caique had to be rowed by the boatmen and Sibthorp's party passed three islands of the Dodecanese – Symi (Simi, then Sömbeki), Cos (Kos, then İstanköy) and Calimno (Kalimnos) – before entering the harbour of the fourth, i.e. Lero (Leros) on 31 May. Like in the plains of Cyprus it was harvest time and 'a Peasantry lively and active . . . furnish'd a pleasing contrast to the scenes of desolation which we had quitted on the Asiatic shore'. The hypothesis has been put forward that Ferdinand recorded the birthwort *Aristolochia hirta* after leaving Cyprus during a stop on one of the East Aegean Islands,

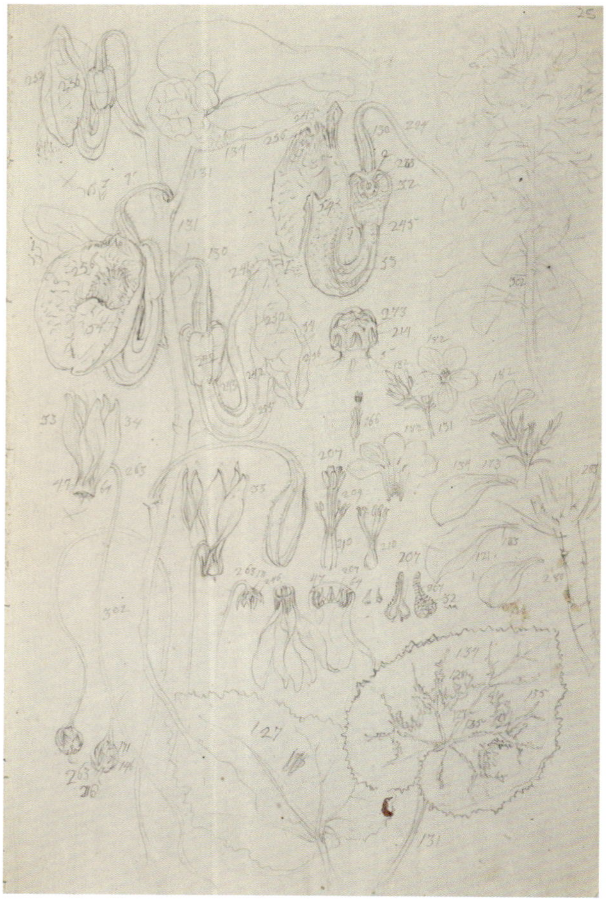

FIG. 5.11. Plants from Cyprus and the East Aegean Islands. Pencil drawing of Ferdinand Bauer, 1786. – Oxford, University of Oxford, Sherardian Library, MS Sherard 247, f. 25 r.

such as Leros, or on the Asiatic mainland. His most detailed pencil drawing[46] is on one of the few sheets of paper (Fig. 5.11.) left for good reason unmarked by him and placed on an unfolded sheet carrying the representation of *Cyclamen persicum*, which was definitely recorded in Cyprus. However, *Aristolochia hirta*, with its large, solitary, foul-smelling flowers and a distinctly curved flower tube abruptly expanded into a brownish-violet limb, has never been found in Cyprus, although it is frequent on the East Aegean Islands[47] and the adjacent Asiatic mainland.

Patmos (then Batmaz), the island of the Apocalypse, was reached in early June, presenting 'a brown arid surface blasting the hopes of the Botanist'. A woodcock, given in the diary as 'Scolopax glottis', was shot near a salt pool and two serpents killed, and maybe Ferdinand had to prepare pencil drawings of them. He definitely produced drawings later used for two views – one showing the harbour of this small island, the second the famous walled-in monastery of Apocalypse, though the latter was portrayed in a freely invented surrounding of hills. Like tourists and pilgrims today, Sibthorp's party was shown the cell where the evangelist had supposedly written his text. Later they admired the glorious view from the roof of the monastery of St John, then like today one of the holy places of Orthodox Christianity. Two days later the party walked up 'to the convent of St John to visit a learned monk whose name was Gregorio Zeno: 'He understood well the litteral Greek & possessed a large library for a man in his situation. Among the botanical books he showed me [i.e. Sibthorp] an Old Paulus Aeginatus & Matthiolius's Commentary on Dioscorides . . .'. This note again underlines Sibthorp's focus on the plants of Dioscorides and the visitor from Oxford received from the monk the 'Greek names and superstitious uses of several of the plants of the island'. Because of the advanced season, no botanizing seems to have been undertaken on this island, apart from the collection of seeds for later cultivation in the Physic Garden of Oxford University and elsewhere.

PATMOS TO ATHENS

Progress from Patmos was slow. Calms obliged the boatmen to row, island hopping continued and sometimes the night was passed in the caique. As a consequence the diary (now with major additions by Hawkins) concentrates on describing life at sea, giving full attention to the romance of sailing, for example by recording 'we were amply refresh'd by the coolness of the night. Our boatmen then resumed their labours, pushing on with vigour and timing the strokes of their oars by the modulation of a Greek song. The sky was so clear that we could still discern land and as we glided along were amused by observing now a rock & now an island growing gradually conspicuous & gradually obscure. Towards morning we spread our matrasses on the benches and were hull'd to sleep by the plunging of the oars and the melody of the rowers'. The sea offered the impression of perfect serenity, the diary continuing 'In such a state of the Air & Water as we experienced this & the preceeding [sic] day all the horrors of navigation vanish'd and were it not for the Rocks & Isles which rose here & there out of the glossy level and reminded us of an horizon our boat might have been conceived to hover in the air or like a balloon to float in a new Element as fancy dictated. The swell was scarcely perceivable and the surface was like glass save when gently rippled by a transitory Zephir which usually fainted away before it could reach our panting bosoms It was in vain that we spread the dropping sails, our boat lay as dead in the water that not a feather which dropped overboard parted company'. No comment is necessary to feel the mystique of those summer days in the Aegean Sea.

Stenousa (Shinoussa), a tiny island with an area of 8.14 kilometres squared, was reached in the caique, and again fish caught by the sailors was on the table for dinner. A deviating Scarus 'of a blood red colour with a golden spot a little above the caudal fin' attracted special attention, Hawkins continuing 'Mr. Bauer fortunately made a draught of it while the colours yet fresh and vivid'. No better description of the cameraman's role on this journey can be found. Leaving the island of Stenousa, the boat stopped at a little rock near the harbour, where Sibthorp is reported to have seen the cabbage *Brassica cretica*. The diary records that he collected 'several curious plants', among them *Catananche lutea*. This is an annual composite first recorded by Tournefort and now known to be widespread in the Mediterranean area, though of scattered occurrence. Ferdinand may have prepared on this occasion a drawing of a flowering specimen of this plant and of the *Brassica cretica* subsp. *aegaea*. The occurrence of intact plants on this rock is appropriately explained in the diary: 'This harvest of plants upon such a small rock must be attributed to their protection from the voracious Goats which wherever they are suffered to rang[e] totally blast the hopes of the Botanist'.

After a hot day without any protection from the sun apart from the flimsy sail the next landing was on Cosinissa (Koufonisi), possibly Kato Koufonisi, another island of the eastern Sporades, with an area of 5.77 kilometres squared even smaller than Stenousa. From there Sibthorp's party crossed over to Naxos (Naxia, then Nakşa adası), the largest of the Cyclades, to search for something essential on any voyage – fresh water. It was only at the third anchorage that this vital product was found in acceptable quality. As usual, seeds were collected, a serpent killed, a lizard caught. In the meantime Hawkins's servant had organized a kid and some goat's cheese, while the famous white marble of this island attracted the attention of Imrie, who would publish thirty years later in Edinburgh a *Catalogue of Specimens, illustrative of the Geology of Greece*, listing no less than nineteen specimens of this rock.

Navigation in the Aegean Sea continued to be unsafe. When leaving Naxos the Greek boatmen were alarmed by a 'two masted vessel that stood our course & was suspected to be a Pirate', but nothing happened. Sibthorp followed Tournefort's steps once more by selecting as the next place for dropping the anchor the island of Caloyero (Kalogeri) in the eastern Sporades, which he seems to have already visited the year before, possibly from Amorgos. On the first occasion he indicated that he had collected seeds of the Tournefort's 'Lunaria', now he was interested in flowering specimens of this plant, probably *Fibigia lunarioides*, a bright yellow flowering subshrub of the crucifer family endemic to the Aegean area, growing on coastal limestone rocks and cliffs. Ferdinand must have prepared a pencil drawing of this plant, but it is uncertain if it is based on a specimen from Caloyero. Sibthorp would have had no difficulty in identifying this plant in the field – Tournefort's *Voyage*, which he carried with him, contained a copper engraving showing this very shrub.

In the hope of finding some late-flowering plants on the eastern side of the island, a second landing was made, which could easily have ended in a catastrophe: the diary reports 'on our approach to the shore we were nearly overset by a partial gust or eddy of wind a more narrow escape not being conceivable'. Stoical botanizing was continued, though apparently with little success, and island hopping continued – on the same day to the southern extremity of Nio (Ios), and the following day to Sikino (Sikinos), where Sibthorp fell ill with fever and diarrhoea. He had his bed brought on shore, put it before the altar of a chapel on the beach dedicated to St John, and slept 'tolerably till midnight, when our Carrabechiras [kayikci, boatmen] pressed our departure for Candia'. Meanwhile Ferdinand was collecting seeds, a division of labour continued on the following days, when Sibthorp's party passed Policandro (Folegandros), touched at Milo (Milos) and finally on 11 June reached Argientiera (Kimolos), another island to which Sibthorp and Ferdinand had already been the year before.

Soon all members of the small group fell ill; the diary reported 'bilious complaints . . . which were aggrevated by the bad water which we procured from a well near the beach . . . faint & emaciated . . . as we were we still pursued with unremitted attention the several objects of our investigation . . .'. Argientiera was a paradise for Hawkins and Imrie, the mineralogists and geologists of the party, while Sibthorp added a few more botanical records into his diary. It was on this island that the plan to reach Crete was finally given up due to calms and contrary winds. Instead the decision was taken to continue to the coast of Attica.

Off the island of Siphanto (Sifnos) the travellers observed an eclipse of the sun. Nothing demonstrates more convincingly the incompletely known topography of the Ottoman Empire than the earlier proposal of the French ambassador at Istanbul to observe this rare, but projected astronomical phenomenon in the town of Rhodes – in order to rectify the longitude of the place. A creek on the south-eastern side of 'Thermia' (Kithnos), an island in the western Cyclades, offered the next site for anchorage. Sibthorp was disappointed, noting again 'so schorched that I collected but a few plants'.

Finally, the caique reached the coast of Attica and 'the columns of the Temple of Minerva on the promontory of Sunium [Akrotiri Sounio]' were already visible, but the currents carried the boat for some distance in another direction. The complete dependence of a small sailing boat on currents and wind is adequately portrayed in the diary: 'Before midday the wind became more favourable and allowed us to lay our course towards the Piraeus. At 5 a brisk gale sprung up from the Gulf of Corinth . . . It was a specimen of the old Athenian Wind Sciron we perceived the blackened appearance of the water at a great distance moving rapidly towards us Our boatmen prudently lowered the sails and put the boats head to the gale. Notwithstanding this precaution its violence was so great as to lay our broad side underwater and to occasion much alarm'. When the winds were again favourable, the time had come for the classicists, who could name each mountain and promontory while the boat was approaching the harbour, dropping the anchor at Phalerus (Faliron). A sailing tour of more than a month had come to a happy end, although by now all members of the party were 'sick of the Sea & tossed about in a little boat'. Proceeding to Athens meant high time for the philhellenes, with the diary reporting, 'Our hearts throbb'd high as we approach'd Athenes the view of this city was intercepted by some low hills but the Acropolis sufficiently marked its position and was an object of such pitturesque beauty independent of its classical interest as to encite in us the highest degree of admiration'.

The reality of the town, comprising then an estimated number of 6000–7000 inhabitants and recently plundered by marauding Albanians, differed neatly from the visionary dreams of the classicists. After all, Athens (then Atina) was a Turkish town surrounded by a newly erected 'low wall pierced with loop holes for the discharge of muskets'. Sibthorp's party waited on the Turkish governor, described as a particularly cruel man who had entered into personal negotiations with the son of the Desdár Aga (the commander of the fortress), the focus of discontent, and then ordered 'his hands to be tied behind his back he had him thrown headlong down a steep staircase where he broke his neck'. On this occasion the travellers had also been able to observe 'the pomp of an eastern procession' at close quarters when the governor returned to his residence.

We may suppose that Ferdinand was now fully occupied making drawings of the various views of Athens and of buildings dating from classical times, but this is by no means certain. He had already been in Athens the year before, and, in addition, the visit in late June 1787 was a brief one followed by a longer visit the following month. In any case Ferdinand paid particular attention to the new city walls, clearly discernible on his view of the town as seen from Mount Anchesmus (Likavittos), he also recorded three views of the 'Temple of Theseus', documented 'the monument of Philipappus', the 'theatre of Bacchus', the 'gate of Adrian' and 'Temple of Jupiter Olympus' (see below).

Mount Hymettos (then Imittos Dağı) in the vicinity of Athens was the destination of an excursion on 21 June. The view from its summit was rewarding and gave the travellers 'a compleat idea of the Topography of Attica besides which we had a view of many of the Cyclades the Saronic Gulf Egina the Morea [Peloponnese] as far as the Isthmus of Corinth [Korinthos]'. However, this famous panoramic view was not recorded by Ferdinand. From a botanical point of view the tour was unsuccessful, the plants 'parched & burnt up by the Sun & the few that remained were cropt by the Goats'. Grazing pressure near a human settlement was probably much stronger at that time and the effects omnipresent. Meanwhile Hawkins and Imrie may have studied the quarries where Hymettian marble had been extracted in antique times, with the diary reporting that 'the faces of the rocks had been hewn very evenly & still retained the marks of tools'.

MOUNT PARNASSUS

Sibthorp seems to have realized that in late June botanizing in the lowlands of Attica was as frustrating as it had been on the islands of the Aegean Sea. For this reason he decided to undertake an excursion to the high mountains northwest of Athens, in particular to the legendary Mount Parnassus (then Parnassus Dağı), to start in the evening of 24 June. This time the party was more numerous – in addition to Sibthorp, Ferdinand, Hawkins, Imrie and their servants it comprised Procopius Makri, the British Vice-Consul at Athens, who would later become the father of Teresa, Lord Byron's 'maid of Athens', as well as three Athenians in charge of the cattle. At that time this meant travelling into unsafe country, the area west of Athens having recently been 'the scene of a bloody slaughter of the Albanians by the Cap.n Pasha'. On the occasion the traditional 'pyramid of sculls as trophy of their victory' was erected, the standard memorial for insurrections suppressed by the Ottoman authorities.

After having passed through 'a venerable forest of Olives . . . stunn'd by the incessand din of the Cicada' casting 'a frequent look behind on the Acropolis which was now gilded by the rays of the setting sun and towered with unusual majesty above the plain' the first night was passed in Daphne (Dafni). The monastery was in ruins: it had been burnt and pillaged by Albanians, few monks had returned, and the travellers found a vacant unfurnished room to sleep in. Thebes (Thiva, then Thebai) was the next destination, in classical times the capital of Boeotia (Viotia), now consisting of 'about 500 Greek & Turk houses'. They had passed through agricultural land, greener and more fertile than Attica. Approaching the mountains, Sibthorp records in his diary, 'the grandeur of the Landscape . . . a flat of immense extent bounded by hills and mountains . . . Parnassus crowning the center of the view'. After a night spent in a tent, possibly the one bought in Istanbul, the party reached the little town of Livadia, where they had recommendations to a rich Greek living by the standards of the time in 'a Pallace . . . uncommonly magnificent, the Chambers have been fitted with the most costly furniture from Venice', again stressing the long-term influence of Venice on taste in the Levant.

During this stop at Livadia or, alternatively, on the journey back to Athens, a very local species was collected that had not yet received a Linnaean name. Ferdinand prepared a drawing of it: the bellflower *Campanula rupestris*, a plant that has never been found outside this small area of Sterea Ellas, growing near Livadia on vertical limestone rocks inside a ravine. In a sense this plant, now a popular garden plant, foreshadowed further new findings higher up in the mountains.

The famous 'grotto of Trophonius' in Livadia was an object of special attention. Ferdinand prepared a drawing of it and the diary commented 'a noble river boils up in three different places at the foot of the rock and turning immediately several mills tumbles down a deep bed through the town below Two of these three sources were probably the fountains of Lethe & Mneomosyne . . . The river was probably the Hircyne. The scenery . . . was exceedingly romantic Cliffs of the most craggy form rose abruptly on each side to the height of sixty fathoms a river running below the channel of which was pent up with their rude & accumulated ruins'. Again we find the philhellenic Sibthorp deep in thought, whereas the reality was different – Livadia was protected at the time by a fortified castle, 'mouldering into decay', but had acted as 'a secure refuge in times of danger and in a recent Albanian surprise had been the repository of the female inhabitants of the city and of its most valuable effects'.

On 28 June 1787 Ferdinand acted as a cameraman for a quite different object. In a village near Livadia a bas-relief in the 'Portico' of a cottage showing 'a rural figure of very ancient but not very elegant workmanship representing a Rustic, in one hand holding a staff, in the other offering a grasshopper to a dog which looks up ready to receive it' was found. Ferdinand's drawing survives in the Cornwall Record Office at Truro and another variant at the Niedersächsische Staats- und Universitätsbibliothek in Göttingen (Fig. 5.12.). The marble itself, known as the Alxenor Stele, was later presented to the Fitzwilliam Museum of Cambridge University and ended up in the National Archaeological Museum in Athens; it is reported to have been found at Orhomenos and is dated to the early fifth century BC.

Meeting a shepherd's boy near Livadia and going with him to collect plants made Sibthorp realize the many connections between past and present, the diary reporting 'my Pastoral Botanist surprised me not a little with his Nomenclature: I traced the names of Dioscorides & Theophr[astus] somewhat corrupted indeed by the pronunciation & by the long series annorum [i.e. of years] which had elapsed since the time of these Philosophers but many of them unmutilated and their virtues faithfully handed down in the oral tradition of the rural Sages of Betia [Viotia] my Shepherd's boy returned to his Fold, not less pleased & satisfied with some Paras that I had given him, than I was in finding in such a Rustic a Repository of ancient science'. In our context, one single aspect is of relevance:

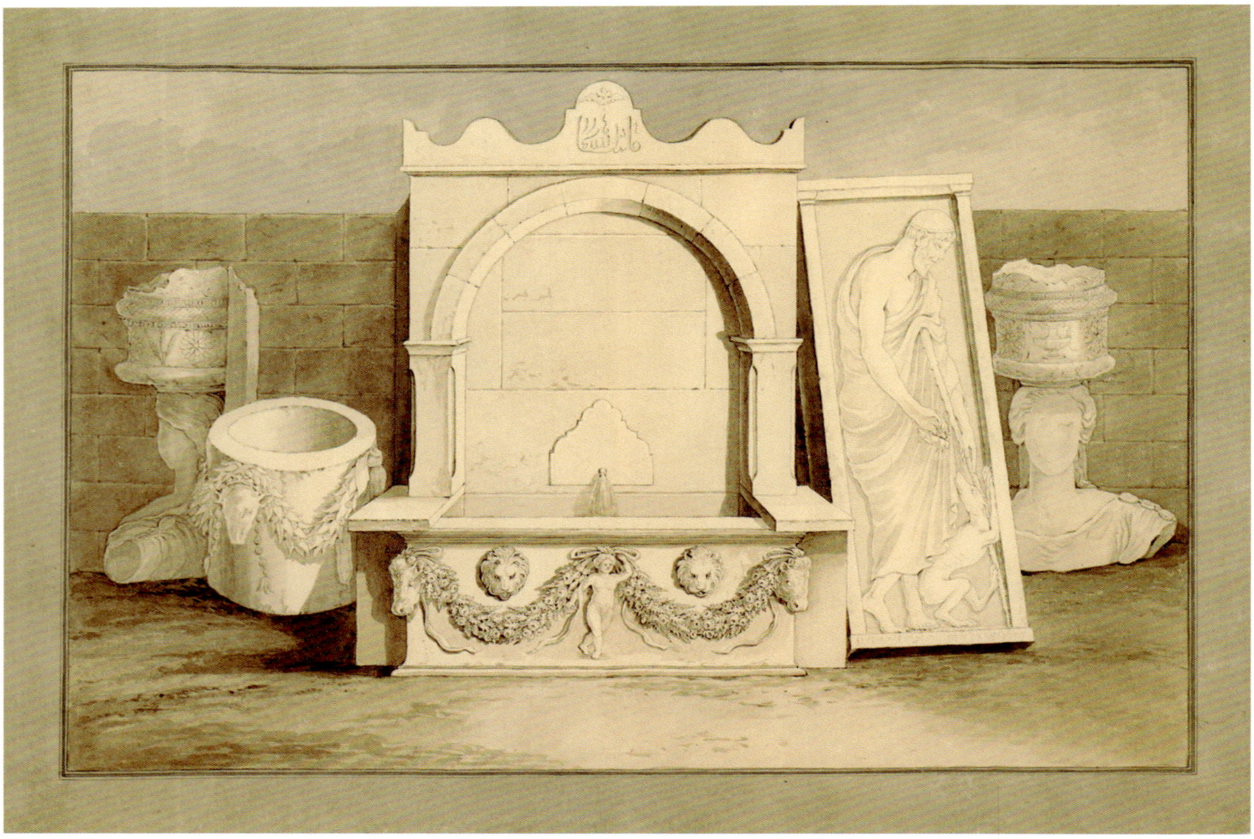

FIG. 5.12. A Turkish fountain and remains of Greek antiquity, to the right the Alxenor Stele. Watercolour by Ferdinand Bauer, undated. – Göttingen, Niedersächsische Staats- und Universitätsbibliothek, Handschriftensammlung, gr. 2° MS Histor. Nat. 94: XXI, f. 16.

what kind of sources for Greek plant names used in ancient times were available to Sibthorp at that time? The answer is straightforward – the copper engravings annotated by Kollár with the Greek plant names used in the two early Byzantine manuscripts kept in the Imperial Library in Vienna that Sibthorp had in his luggage. It seems plausible to assume that these manuscript names were the basis for Sibthorp's comparison between denominations in Byzantine times and those in 1787 near Livadia.

Sibthorp's linguistic observations were made under the harsh conditions of travelling off the beaten track in the Ottoman Empire, with the diary reporting 'I have a woefull recollection of the sufferings we endured from the bugs. The house . . . swarmed with these animals We were not free from their attacks by day but the outmost vigilance could not protect us from them by night so that we enjoyed but a few short intervals of repose. It may be usefull to any future traveller in Greece to know before hand what he has to expect from Gnats and Bugs. If he sleeps under tents he is assailed by the one, if in chambers he is beset by the other and if he can obtain a sound nap in either case good Lord! What must be the texture of his hide?'

Special precautions were felt necessary for the next part of the journey. Sibthorp's host in Livadia provided letters of recommendation to the local chiefs, arranged for his janissaries and servant to accompany the travellers and in the end the caravan set in motion. At a stop near a Turkish fountain Ferdinand prepared a drawing of the 'noble view of all Parnassus from its base to its summit' (Fig. 5.13.), which he was to use later for one of his frontispieces (see Chapter 6).

Albanian bandits had kept the small villages in the mountains in constant alarm, which resulted in Sibthorp's party being received in a somewhat unfriendly way, but the recommendations presented eased the situation. The night was spent in a guardhouse 'in the walls . . . loop holes to repel sudden attacks' and an additional guard of 'a dozen stout greeks with their fire arms' was thought necessary to protect the group in the ascent. Before setting out the following morning the guards 'crossed themselves with much devotion', hoping to return safe home from the projected mountain tour.

While no diary entries exist for the excursion to the Lefka Ori in Crete and Uludağ on the Asiatic mainland, we possess a detailed description of the ascent of Mount Parnassus (2455 metres), significantly lower than the Uludağ massif, a fact the travellers must have been unaware of. In addition, we do not know for certain if the peaks of the first two mountains were indeed reached, whereas this seems almost certainly to have been the case with Mount Parnassus. Since Sibthorp's party was in all likelihood the first group from outside the Ottoman Empire to climb this mountain, and since several plants new to science were collected for the first time on this excursion, the report merits being quoted in full, the diary reporting: 'We had now to climb up some dangerous cliffs and to scramble over accumulated craggs for a considerable span of time during which we were often obliged to recruit our exhausted strength by a little rest and to assist each other by mutual exertions. Then we had conquered . . . the most difficult part of our ascent we reach'd a patch of snow collected in the fissures of the rocks and soon after a Mandria or Sheep's stall where we drank some milk In half an hour after which we attained the summit. Our zeal in braving the difficulties of the ascent was now amply rewarded by a sight of allmost all Greece extended like a map beneath us In the immensity of view the eye found no repose but was dazzld by a brilliant display of lakes & mountains land and water capes inlets gulfs seas & islands as far as the utmost reach of the horizon. Our sky was not free from clouds which threw partial shaddows on the surface of Greece . . . and added to the confusion of their effect. Indeed A view like this would have baffled the labour of the correctest draughtsman. It was much too varied to be understood in all its parts during so short a stay & I may add that it possess'd too much classical interest not to produce an inebriating effect on our spirits & imaginations'.

Having reached the summit of Mount Parnassus, Sibthorp seems to have been enthusiastic – what had been the subject of his dreams in Vienna as reported to his father in Oxford had become true. We are unaware of Ferdinand's feelings, but high mountains and the panoramic views from their summits may have been less spectacular to him – we cannot exclude the possibility that he had climbed high mountains in the vicinity of Vienna.

FIG. 5.13. Mount Parnassus. Watercolour by Ferdinand Bauer, after 1794. – Oxford, University of Oxford, Sherardian Library, MS Sherard 408, f. 112.

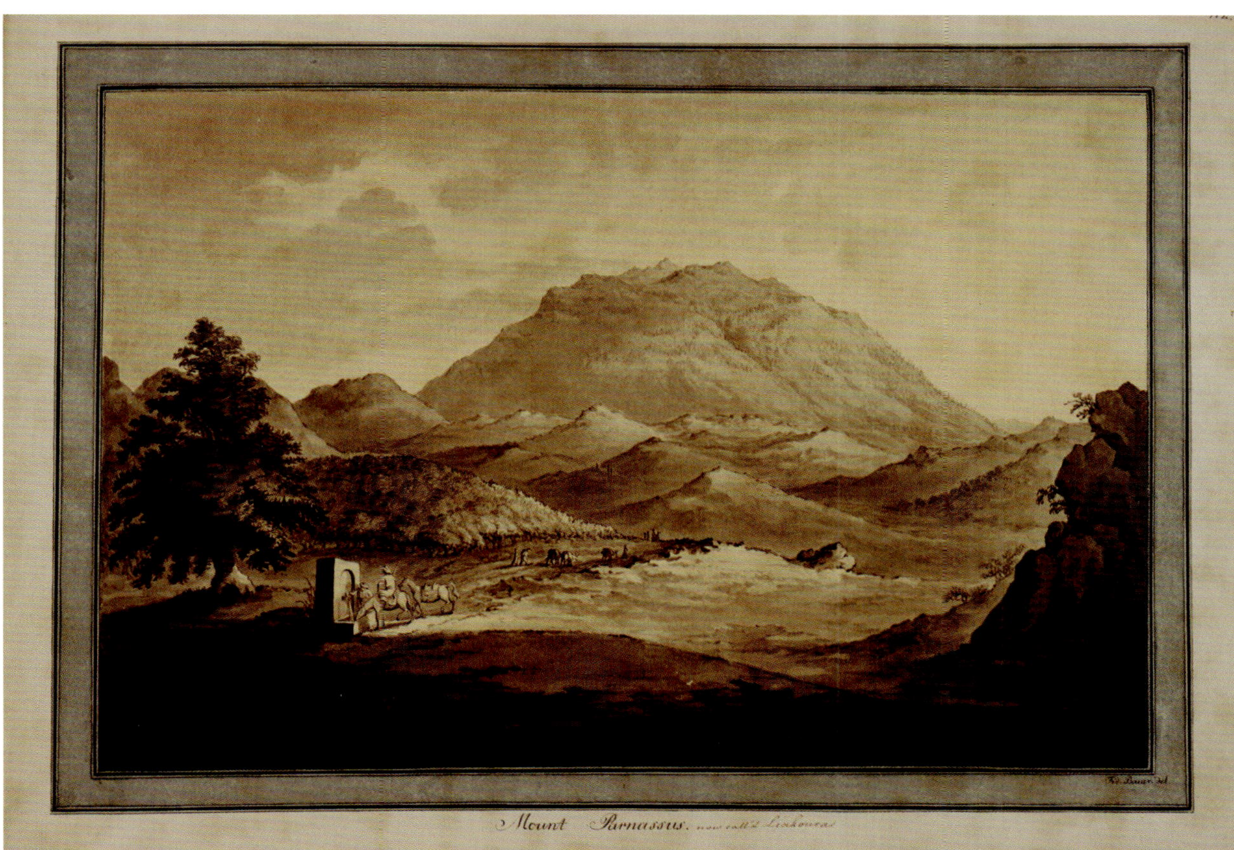

Lunch was prepared on the summit of Parnassus, in a style appropriate to the occasion. Sibthorp's party enjoyed roasted lamb and drank 'liquors tempered in the chrystalline snow the keen air of these empyreal regions sharpening our appetite', as the diary reports. Imrie took a rock specimen, proudly listed in his *Catalogue* as 'from the highest summit of Parnassus, the 30 June 1787', then the descent began. Contrary to expectations, Ferdinand had not recorded the magnificent panorama – he must have been extremely busy in recording plants and birds.

Again it is the diary, or rather Hawkins's additions to it, which offers us an insight into the findings: 'many new and rare plants were collected by Dr S whose botanical zeal was wrought up to the highest pitch of exulation [exultation] at the discovery of the magnificent Morina persica together with a new species of Salvia'. This time Sibthorp was not mistaken, indeed he had found *Morina persica* for the first time in Europe and identified it correctly, no doubt with the help of a copper engraving published in Tournefort's *Voyage*. Ferdinand prepared a drawing of this very spiny robust perennial[48] with flowers in dense whorls and corollas white turning pink. This formed the basis of a watercolour (Fig. 6.1.) painted by him the following year at Oxford. At the time the species was known only from the Asiatic mainland, and Sibthorp might have been one of the first botanists to realize the many connections between the flora of the Balkan Peninsula and that of Southwest Asia. In Europe *Morina persica* occurs in what is now Albania, Greece and Macedonia, needless to say regions then belonging to the Ottoman Empire. By contrast, the new sage was *Salvia ringens*, a species new to science that turned out to be restricted in distribution to the Balkan Peninsula. The list of plants that Ferdinand could only have recorded on Mount Parnassus is a long one, not because so many plants are endemics of this mountain, but because he had climbed only this one high peak near Athens in early July when many mountain plants were still in full flower. The woodruff *Asperula lutea*, a rather inconspicuous perennial with narrow spike-like inflorescences of three to six whorls, is a good case in point. It is an endemic of Sterea Ellas and the northern Peloponnese, where it occurs at 900–2000 metres, flowering from May to mid-July. Since Sibthorp and Ferdinand did not visit the high mountains in the northern Peloponnese at that time of the year, it is extremely likely that this plant was documented by Ferdinand on Mount Parnassus. The buckthorn *Rhamnus sibthorpianus*, a low shrub with adpressed branches and greenish-yellow tetramerous flowers, and *Silene auriculata*, a pulvinate catchfly with somewhat inflated calyces, show basically the same pattern of distribution. They must also have been recorded by Ferdinand on this excursion. Quite in contrast there is at least one plant depicted by Ferdinand that is definitely endemic to Mount Parnassus – *Geocaryum pumilum*, a tiny umbellifer with a strikingly globose root tuber. It must have been dug up from stony slopes or stabilized screes for this purpose. Sibthorp or Ferdinand must also have dug up here a late-flowering specimen of the fritillary *Fritillaria mutabilis*, which had to wait for no less than 204 years, no doubt a record, to receive a correct scientific name. Recorded on a sheet of paper annotated 'Parnassus'[49] by Ferdinand, there is no doubt on the provenance of the specimen. Several composites, among them *Staehelina uniflosculosa*, occurring also on more northern mountains in the Balkan Peninsula, an area Ferdinand never visited, were almost certainly recorded by him on Mount Parnassus.

The descent from this mountain was by the north-western side and led via the deserted village of Kallivia through agricultural land to a village called Castri or Delphos (Delphi), inhabited, as the diary explicitly mentions, by Albanians. This classical site is described in the following words 'We entered on the West by the Stadium coasted along the foot of the precipice, cross'd the Castalian rivulet and descending a little arrived at a Greek Convent where we lodged. The precipice of Delphi is divided from the summit to the base by a broad chasm which strikes deep into the body of the mountain. Thro' this chasm rushes a torrent in winter from the mountain above. At the foot of the rock on the right hand rises the Castalian fountain. Its waters are first collected in an ancient bath hewn out of the solid rock, they then join the waters of another fountain perhaps the Cassotis which gush out of the base of the opposite cliff and both forming a rivulet . . . have gradually worn out a deep bed and given a hollow theatrical form to the slope upon which are arranged on terraces the foundations of the ancient town. The inhabitants of the village make use of the bath & the fountains to season their wine casks. The water very cold & potable. Except these ancient terraces the foundations of the stadium a few inscriptions and some fragments of shafts & capitals we saw no antiquities'. This is hardly surprising – excavations in this part of the Ottoman Empire were to start only several decades later. Imrie's *Catalogue* reports 'the crumbled debris of Delphi . . . at present covered by the scattered huts of a miserable village' and had no scruples to take a specimen from 'one of the steps cut in the solid rock leading to the Bath of Pithia'. For some reason Ferdinand made no record of this place, which later became so popular among tourists interested in classical antiquity.

Sibthorp's party had dinner in the monastery at Castri and then continued towards the monastery of St Luke (Moni Osios Lukas), well known today, but severely criticized by Sibthorp. He noted 'styled as the Glory of Hellas in an age of barbarism and . . . still regarded by the tasteless Greeks as a magnificent structure but God Heavens how inferior to the Gothic Structures round in the same age in Western Europe'. Here Sibthorp shows himself as an opinionated expert of architecture focused on western stereotypes. In the lowlands the harvest time had started, and the monks were 'busyly employed in building an occasional refuge from the marauding Albanians and in securing their harvest', while the villagers because of the heat had started to sleep on the flat roofs of their houses.

Descending to the coast on the Gulf of Corinth, Sibthorp's party was attracted by the rich bird life. A boat was hired to carry them to the small islands of Didascolo (Staronisi) and Empelia (Ampelos) and the diary contains a miscellany of ornithological observations such as 'innumerable flights of the Melba almost darkened the Air . . . immense flights of Falcons which pursues the large owl S. Bubo with shrill piercing cries'. Ferdinand may have been commissioned to make drawings of the birds observed or shot, but since locality data are totally missing from his finished watercolours these cannot be attributed to this excursion. What is remarkable about Ferdinand's ornithological work is that he managed to record the species in their life-like positions – at a time when the concept of their 'jizz' (see Chapter 6) was virtually unheard of. Needless to say the necessary observations must have already been recorded in his pencil drawings made in the field, and not later at Oxford where he at best had access to a few preserved specimens, but not the live birds.

On 5 July the party climbed up to the monastery of Jerusalem (Ierusalim) on the eastern slope of Mount Parnassus and Sibthorp, accompanied by two monks from the convent, started a second excursion to climb the mountain but gave up the plan to reach the summit because of security problems. Two Greeks who had accompanied the travellers had been seized by an Albanian guard and are reported to have been 'maltreated under the pretence of conducting Corsairs into the country', but nothing of consequence happened. Sibthorp reports in his diary having come across '2 or 3 species of Rhamnus, the Fraxinus ornus & and a new and beautiful species of Prunus now in flower, which crept along the rocks'. The latter could be *Prunus prostrata*, a rigid shrublet with a cylindrical hypanthium and pink petals, which indeed had not yet received a Linnaean name. Ferdinand could have made a drawing on that occasion, though this is not certain, since the plant also occurs elsewhere, for example on the Lefka Ori in Crete. While the manna ash, *Fraxinus ornus*, is a readily identifiable tree, the '2 or 3 species of Rhamnus' are less clear, possibly *Rhamnus sibthorpianus* (see above) and *R. alpinus*.

Plant names in modern Greek seem to have been a constant preoccupation for Sibthorp. When 'a Monk of the Cloister a venerable Octogenian famous for his knowledge of Simples' arrived, Sibthorp successfully extracted from him several denominations 'barbarous Yet most of them significative, & some resting unaltered & uncorrupted the ancient names of Theophrastus & Dioscorides'. Again the proofs from Vienna may have played a role. Rewards had been offered by Sibthorp on his arrival at the monastery 'for procuring different birds'. This was the background for the acquisition of one of the more spectacular trophies of the journey. A young vulture frequent in the mountains was brought in, carefully measured and weighed with the results recorded in the diary. Again Ferdinand seems to have acted as the cameraman, although a specimen of the same species had also been shot earlier in Castri.

The return journey to Athens followed the same route. On 6 July 1787 Livadia was reached, obviously meanwhile plundered by 'Marauding Arnauts [Albanians] . . . ascribable . . . to the absurd policy of the Turkish Government in depriving them [the inhabitants of Livadia] of arms thus without the means of defence'. In Erimocastro (near Thespies) Sibthorp's party came across another reality: they were attracted by country music coming from a nearby village, but on arrival it became evident that the inhabitants were Albanians, who were unable to speak or understand Greek, while no member of Sibthorp's group was able to speak or understand Albanian. The Levant in general, and the Greek mainland in particular, was indeed of mixed ethnical composition.

ATHENS AGAIN

Soon after the arrival of the travellers on 9 July 1787, the end of Ramadan was approaching. Following a 'crowd of courtiers and dependants' Hawkins, perhaps with other members of the party, was received once again by the pasha of the town, with Asaros delivering a complimentary speech. Beyram, the feast to mark the end of Ramadan, was announced 'by the discharge of Cannon by rockets & by the sudden illumination of all the Mosques an Iman in the presence chamber [of the pasha] rechanted a prayer of some length . . . and each Turk threw himself into an attitude of profound devotion'.

Ferdinand may have attended the traditional festivities that followed, although he is not mentioned in the text: 'A Firebrand blazed in the middle of the court below around which a string of Turks & Albanians danced to the shrill melancholy music not unlike that of our bagpipe the martial Macedonian Dance each figure wore his usual Armour & most of them the Albanian Dress the picturesque effect of which was hightend by the catching reflection of the Blaze This Phalanx of warrior dancers was led by the Selektar of the Musselim'. An opium party was another aspect of the Bayram. Hawkins reports that rambling through the suburbs and hearing festive music he saw 'several black men and women at the door of a barn or whare house . . . filled with a crowd of the same complexion' and was 'conducted to the upper end of the appartment where . . . the opians male & female intermixed bid me welcome they were all squatted on their hams smoaking others wrangling & intoxicated but mostly surveying with looks of wild admiration and noisy marks of applause a man & woman who were dancing . . . The music consisted of the Turkish Pipe Cymbals & a drum'.

Permission was granted by the commander of the citadel for the party to visit the fortress, that is the acropolis, but we do not know if Ferdinand participated in this tour, since his views could also be recorded from other elevated points. The diary is again full of philhellenic sentiments, such as 'Here we had the melancholy pleasure of contemplating the remains of those venerable structures which since formed the Glory of Athens . . . of those monuments of Attic Art & of the Age of Pericles, which were to transmit to all succeeding Ages and to all Nations the most exquisite Models of Taste in Architecture and may I not add that Taste which accommodating itself to each purpose of Art and imprinting on each work its appropriate character combined Unity with Variety Simplicity with Ornament and Grandeur of effect with the utmost Nicety of Execution'.

The realities of Athens in 1787 were different. Hawkins reports that the aqueduct of Hadrian had suffered further from the construction of the new wall and that part of the pavement of the Peristyle of the Temple of Theseus had been removed by the pasha for his residence. This largely intact Doric peripteral temple is now interpreted as the Temple of Hephaestus, standing on the Agoraios Kolonos hill. Two particularly fine finished views by Ferdinand show this building, then used as a Greek Orthodox Church, with several oriental figures, male and female, the latter unveiled, in the foreground (Fig 5.14.). Belonging to the same series are views representing: (1) the interior of the Temple of Hephaestus; (2) the ruins of the Odeon of Herodes Atticus, a stone theatre located on the south-western slope of the acropolis; (3) the ruined Philipappos Monument, an ancient Greek mausoleum on Mouseion Hill; and (4) the ruined Temple of the Olympian Zeus. The latter building, one of the largest temples in the ancient world, had been pillaged in the 3rd century AD and for many centuries systematically quarried to provide building material. Interestingly, Ferdinand shows the ruins in a more complete state of preservation than they are in today, again with a group of oriental figures in the foreground. Another subject of Ferdinand's views is the Arch of Hadrian, a well-known structure from Roman times with its base still partially buried and forming an opening in the low, newly erected wall then surrounding Athens. The finished version of the view, admittedly with incorrect proportions, shows the Greek inscription in minute detail.

Hunting for antiquities was a favourite occupation of travellers in the Levant at the time, and Hawkins and Imrie were no exception to this trend. The *Catalogue* lists materials from several of the ruined buildings recorded by Ferdinand, and his finished views should be seen in this context. In addition to specimens taken from the 'Monument of Lysicrates, commonly called the Latern of Demosthenes', samples were taken from the hills surrounding Athens, measurements made, and Hawkins records 'The traveller [is] struck with The sharpness of the sculpture and the smooth immaculate surface of the blocks of marble [and] finds some difficulty in combining therewith ideas of such remote antiquity'. Hawkins is well aware of the reason for the excellent preservation

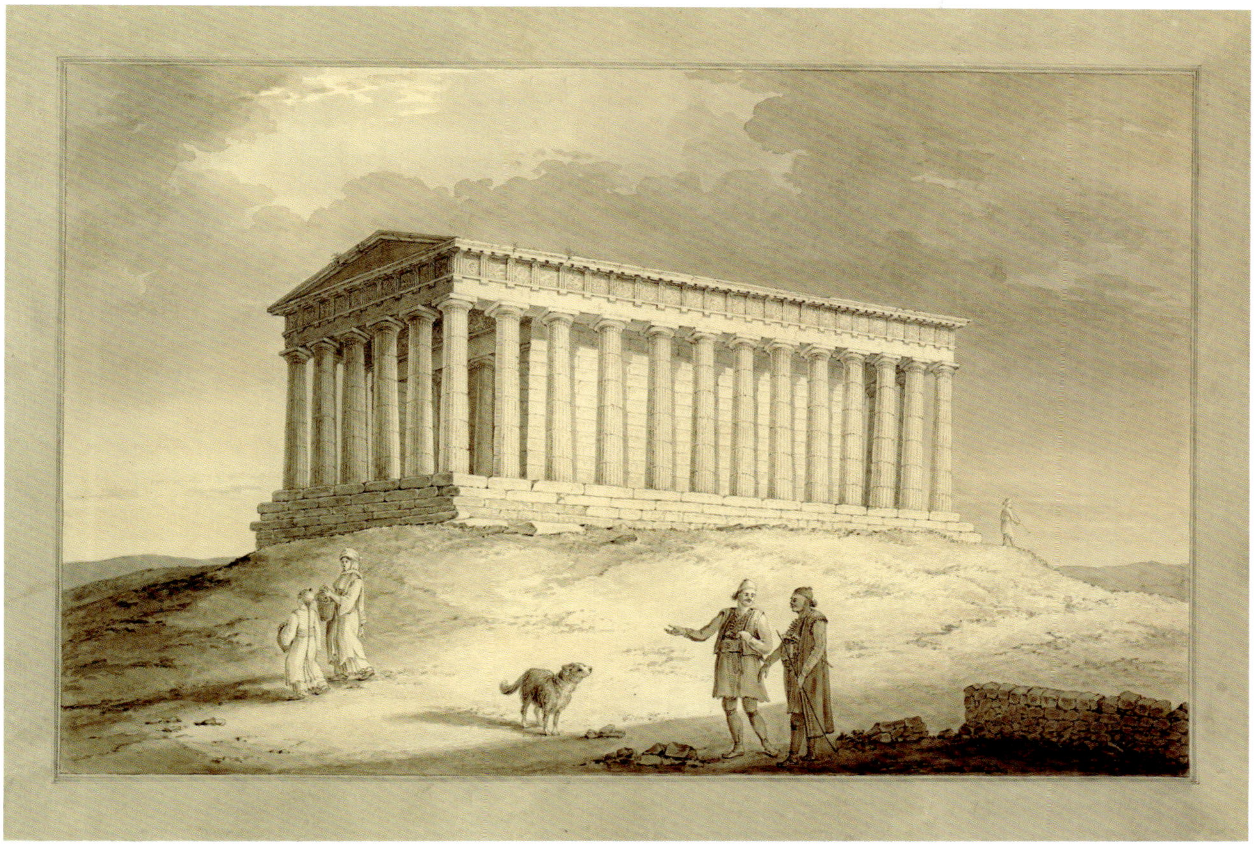

FIG. 5.14. Temple of Hephaestus, Athens. Watercolour by Ferdinand Bauer, undated. – Göttingen, Niedersächsische Staats- und Universitätsbibliothek, Handschriftensammlung, gr. 2° MS Histor. Nat. 94: XXI, f. 10.

of the sculpture that he encountered in Athens – the Mediterranean climate. His pertinent observation are very much to the point 'As for the native of Brittain he expect[s] to see in Greece the same effects of a damp and corroding atmosphere as in his own country. He figure[s] to himself the mouldring walls overhung with ivy, or cloath'd in moss or perhaps overspread with a thick incrustation of liverwort; but here every thing is the reverse; the white marble column still glitters in the sun after a lapse of two thousand years and looks allmost as fresh as when it first issued from the finishing hand of the sculptor, for even the marks of the chissel are not yet obliterated'.

Like Thomas Bruce, the 7th Lord Elgin, as ambassador of the United Kingdom in Istanbul a successor of Ainslie two decades later, Hawkins also looked out for a souvenir. He found a small column of porphyry in the acropolis 'for which I had bargain'd [to be] conveyed to the Piraeus by night'. However, this transaction came to nothing. Surprisingly Ferdinand seems not to have recorded the column, while we find a fountain with an inscription in Arabic letters depicted next to the Alxenor Stele among his finished views in the Niedersächsische Staats- und Universitätsbibliothek in Göttingen (Fig. 5.12.).

A joint excursion, including Makri, to Marathon (Marathonas) and Stamati (Stamata) ended with Ganzarowitz falling ill, interpreted by his master as due to the 'Marsh Effluvia'. Alarmed by the possible consequences the party headed back to Athens, where they remained until 23 July, Hawkins's additions to the diary recording, 'Dr S. chiefly at home in describing his specimens of new plants and animals and Mr Bauer in designing them.' Plans for an excursion to Thessaly and an attempt to climb Mount Olympus (Olymbos) were given up because of the plague said to prevail in Larissa and the danger from Albanian bandits. The following day 'a motley mixture of horses & asses & Mules . . . to the number of 20' drew up in front of the lodgings of the travellers and Sibthorp headed for the harbour in Piraeus with Bauer, while the others left for a geological excursion, in the end somewhat dissatisfying, to visit the 'ancient silver mines near Laurion'. The party was reunited near Cape Sunion (Akrotiri Sounio), and Ferdinand produced several drawings that later served for two impressive views of the famous ruined temple

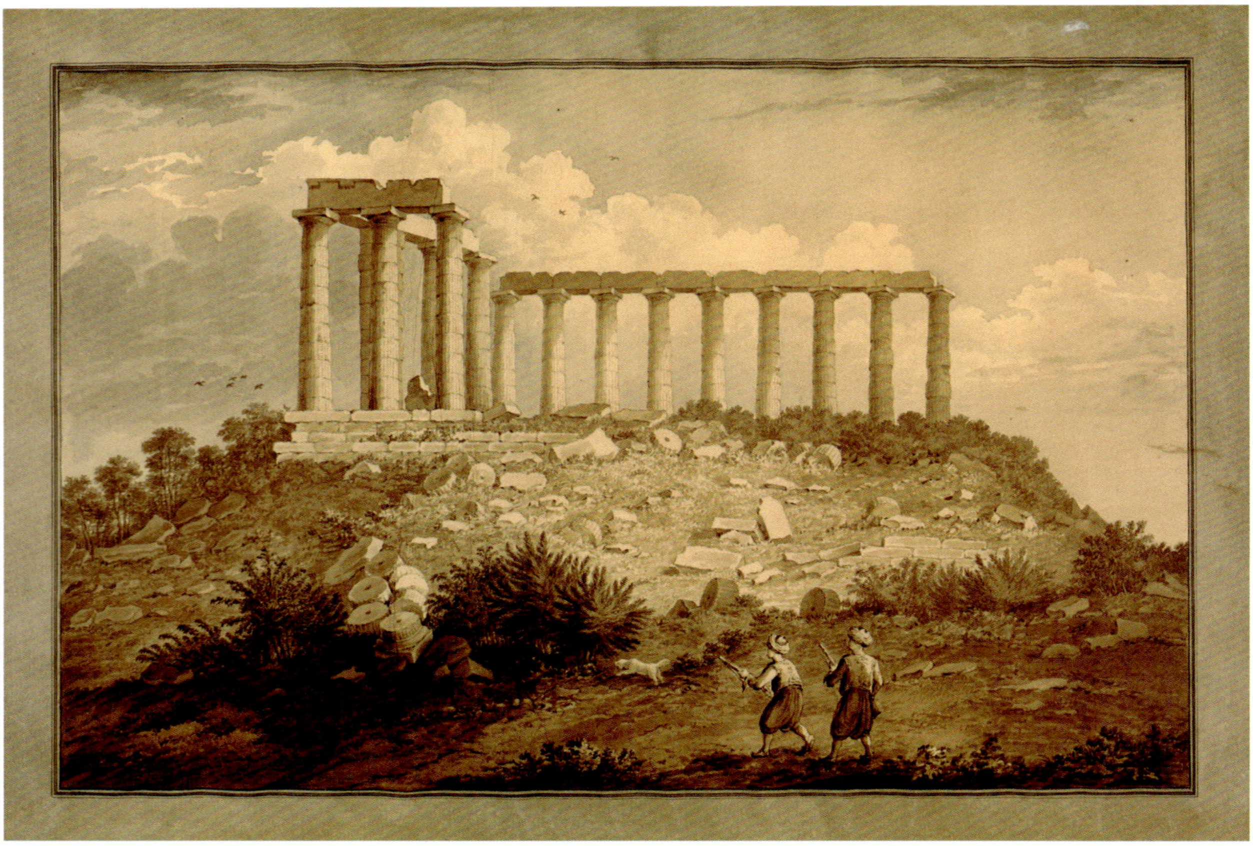

FIG. 5.15. Ruins of the Temple of Poseidon, Cape Sunion. Watercolour by Ferdinand Bauer, undated. – Göttingen, Niedersächsische Staats- und Universitätsbibliothek, Handschriftensammlung, gr. 2° MS Histor. Nat. 94: XXI, f. 1.

(Fig. 5.15.). Situated in a dominant position almost sixty metres above the sea on a promontory, the remains of a Temple of Poseidon, a hexastyle similar to the much better preserved Temple of Hephaestus just visited in Athens, continue to attract masses of tourists every day. By contrast, Sibthorp and his party were alone in this ancient sanctuary, where excavation had not even been envisaged, the diary noting 'neither inhabitants nor cultivation'.

MOUNT ATHOS

Another sailing tour in a northward direction had been contemplated, and on 25 July 1787 a single-masted caique with eight boatmen from the island of Poros brought the travellers to the harbour of Negropont (Halkida). Because of stormy weather progress was slow, necessitating stops at Madri (Lavrio), Macronesi (Makronissi) and Seraphina (Rafina). Meanwhile, birds were observed, seeds collected and the geology of the coast and the small islands noted.

Situated on both sides of the Euripos strait, a narrow channel of about forty metres' width separating the island of Euboea (Evvia, then Eğriboz Adası) from the mainland, the small town of Negropont offered a very special building – a wooden bridge spanning the strait – and Ferdinand took pains to have it recorded (Fig. 5.16.). Later he produced four finished views of the structure going back to a dike erected as early as 411 BC. Since Sibthorp intended to visit another mountain, Mount Delphis (Dirfis), 1745 metres high and thus significantly lower than Mount Parnassos, a visit had to be paid to the local aga. In fact he was the head of the Sanjak of Eğriboz belonging to the eyalet of the islands in the White Sea, and provided his firman for the village of Steni (Steni Dirfios). Arriving there 'the whole village' was at the disposal of Sibthorp's party and the Turkish authorities 'offered . . . as many Greek as we should have the occasion for, to carry our baggage, and accompany us up the mountain . . . The guards, having charged for the Greeks to procure us every possible accommodation, fired a salute and returned to their neighbouring village'. Sibthorp's party enjoined rural hospitality, with a 'basket of mulberries . . . brought, some chickens . . . caught and a lamb . . . seized, slaughtered and roasted', but it was quite evident where the power was concentrated.

On 3 August 1787 the ascent to Mount Delphis was made in bad weather. In pouring rain the travellers arrived at a mandra, where the shepherds 'kindled a fire and milked their flock for us'. At higher altitudes thick mist set in, the rain increased, but even under these adverse conditions rock samples were taken and Sibthorp recorded 'several of the more rare plants of Parnassos growing among the clefts of the rocks near the summit of the mountain, and some which I had not before observed in Greece'. Among these Ferdinand recorded the catchfly *Silene dirphya*,[50] a tiny mat-forming local endemic known only from the summit area of this mountain, as well as the pink-flowering composite *Crepis incana*, which also occurs on the high mountains of the mainland and the Peloponnese. Both pencil drawings are on a sheet of paper marked 'Negroponte' by Ferdinand.

With the storm increasing to a hurricane and hail driving 'furiously along the mountain', the diary reported 'on leaving the grotto [Ferdinand] was seized with a panic terror, and became almost motionless with fear . . . we were all shivering with cold and distress . . . A guide refused to wait for him [Ferdinand] and my servant dragged him by force down the mountain'. This episode underlines the relationship between Sibthorp and his drawing amanuensis, who after all was little more than his second servant. As might be expected, the following morning was used for drying the papers at Steni and then to return to Negropont, where Imrie continued to collect rock samples listed in his *Catalogue*.

Unimpressed by a warning of pirates, Sibthorp's party sailed in the small caique with a contrary wind to the island of Scopelo (Skopelos) and arrived on 10 August at the Athos peninsula, dropping anchor eastward of a bay called Daphne (Dafni). The first thing to be done was a visit to the Turkish aga resident in Caryes (Karies). There the travellers showed their 'passport', probably a recommendation from the 'chief Archon' they had met in Athens and effectively the equivalent to the modern diamonētērion, the special permit to visit Mount Athos (Agion Oros). It was also at this administrative centre that Sibthorp arranged for a guide to accompany the group during their visit on the Holy Mountain, then and now a semi-autonomous orthodox monastic community

FIG. 5.16. **The bridge crossing the Euripos strait. Watercolour by Ferdinand Bauer, after 1794. – Oxford, University of Oxford, Sherardian Library, MS Sherard 408: f. 122.**

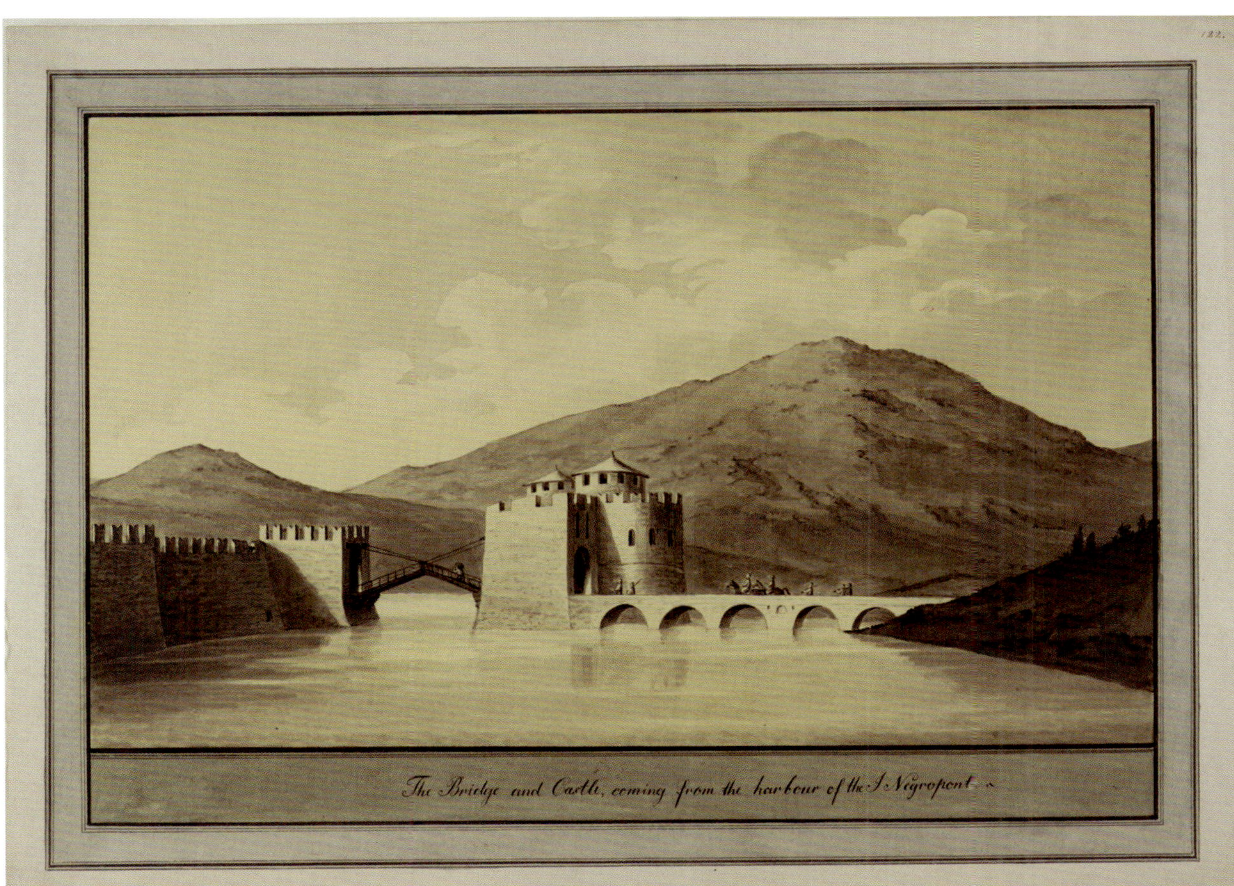

– off limits for women, beardless youth and female domestic animals.

The four days spent on this finger-like peninsula seem to have been a special highlight of the whole journey – Sibthorp gives a narrative of great detail and unexpected warmth in his diary, while Imrie remarks that the place was 'beautiful beyond description'. Many visitors have written accounts of their visits on the Holy Mountain, but Sibthorp's text again merits being quoted in full, giving a lively picture of his experiences in this very special place. In addition, he seems to be the first naturalist after an interval of several decades (that is, after the visit of Pococke) to give a report; in fact it is a report of literary quality. On 11 August Sibthorp's party 'gained . . . the monastery of Xeropotamo [Moni Xeropotamou], a large magnificent cloister of a quadrangular form. In the middle stands the church, a sumptuous building, paved with different colored marbles, and ornamented with daubings of strange bizarre figures of saints . . . Leaving the convent, we continued our walk to Caryes, shaded from the sun by some venerable old chestnuts, which composed here the principal part of the sylvan scenery. Nearly half way from Caryes, the prospect was uncommonly rich and beautiful: a vale stretched out into a verdant lawn, with cattle feeding on it . . . Caryes, though only a village, is the market of Mount Athos: here live the different artisans and mechanics who supply with their ware the numerous monasteries . . . The town appeared as it were pendant on the side of the mountain: some lofty pine trees towered above it: below, were hanging vineyards and fruit gardens contrasted with the monasteries and the somber hue of the cypress . . . Ivero [Moni Ivirou] appeared to us a Grecian Elysium: from our windows, we commanded a fine view of the sea; a short lawn intervened between it and the monastery; to the right, passing a torrent bed shaded with planes, were vineyards, that broke the roughness of the approaching mountain. The mules belonging to the convent were offered to us by the Hegoumenous and after riding for six hours over a mountainous rocky road near the coast, we arrived at the monastery of Laura [Moni Megistis Lavras] . . . It contains six hundred caloyers [monks], and has several cells or small convents dependent on it . . . Notwithstanding the grandeur of Laura, we met with indifferent fare: a strict fast of the Panagia, or the Virgin, proscribed at present to the monks of Monte Santo [Holy Mountain] the use of fresh fish: and the arid environs of Laura ill supplied so numerous a convent with vegetable products; and we with difficulty procured pickled sprats and olives. Here superstitious folly reigns in full force. The cocks crow only to wake the sleepy caloyer. Eggs are brought from Lemnos and other islands. Oxen graze in the valley: but neither cows, ewes, or she-goats are suffered to pollute this holy spot. In the library, which contains two small rooms, one for the printed books, the other for manuscripts, I was shown an ancient copy of Dioscorides, written in small ancient characters, with illuminated figures, but ill executed . . . mounting our mules [we] set forward for the summit of Athos. After riding for three hours we passed by the monastery of Keratia [Kerasia], on a rocky steep to the left far below us: our road led us over precipices difficult and dangerous to pass. In two hours . . . we arrived at a small chapel dedicated to the Panagia. Trees now became scarce; and the higher regions of the mountain rose naked above us; the summit being about an hour distant, and crowned with another chapel. After dinner, leaving our mules below, we ascended by a road formed by large pieces of marble . . . we earlier found no snow on the summit; nor did I observe any of those plants characteristic of an alpine region . . . we came . . . to the convent of St. Anne [skete of St. Anne] . . . The site of the convent is uncommonly picturesque and beautiful: hermitages excavated in the rock to the number of fifty or sixty, with their tops of argillaceous slate, glitter round it; several waterfalls, brought down from the upper region of the mountains by wooded conduits, irrigate numerous little vineyards. Above the convent, rises the mountain, breaking into rocks that pierce through the thick foliage, now of various hues; and high above these towers the naked summit of Athos: below, the sea extends itself in a wide expanse'.

On Mount Athos, as on other occasions, the members of Sibthorp's party differed in their predilections: Imrie took geological specimens, among them a piece of 'general rock' from the highest summit; Sibthorp collected seeds, among them of the foxglove *Digitalis lanata* subsp. *leucophaea*, which had not yet received a Linnaean name; Ferdinand recorded views, among them the monastery of Ivero, the skete of St Anne, and the hermitage in the summit area of Mount Athos. But there was much more to do for him – he had to make drawings of many plants, some fairly widespread and known already to Linnaeus, such as the bellflower *Campanula persicifolia*,[51] while others were local and new to science. The rarest of these is the spectacular

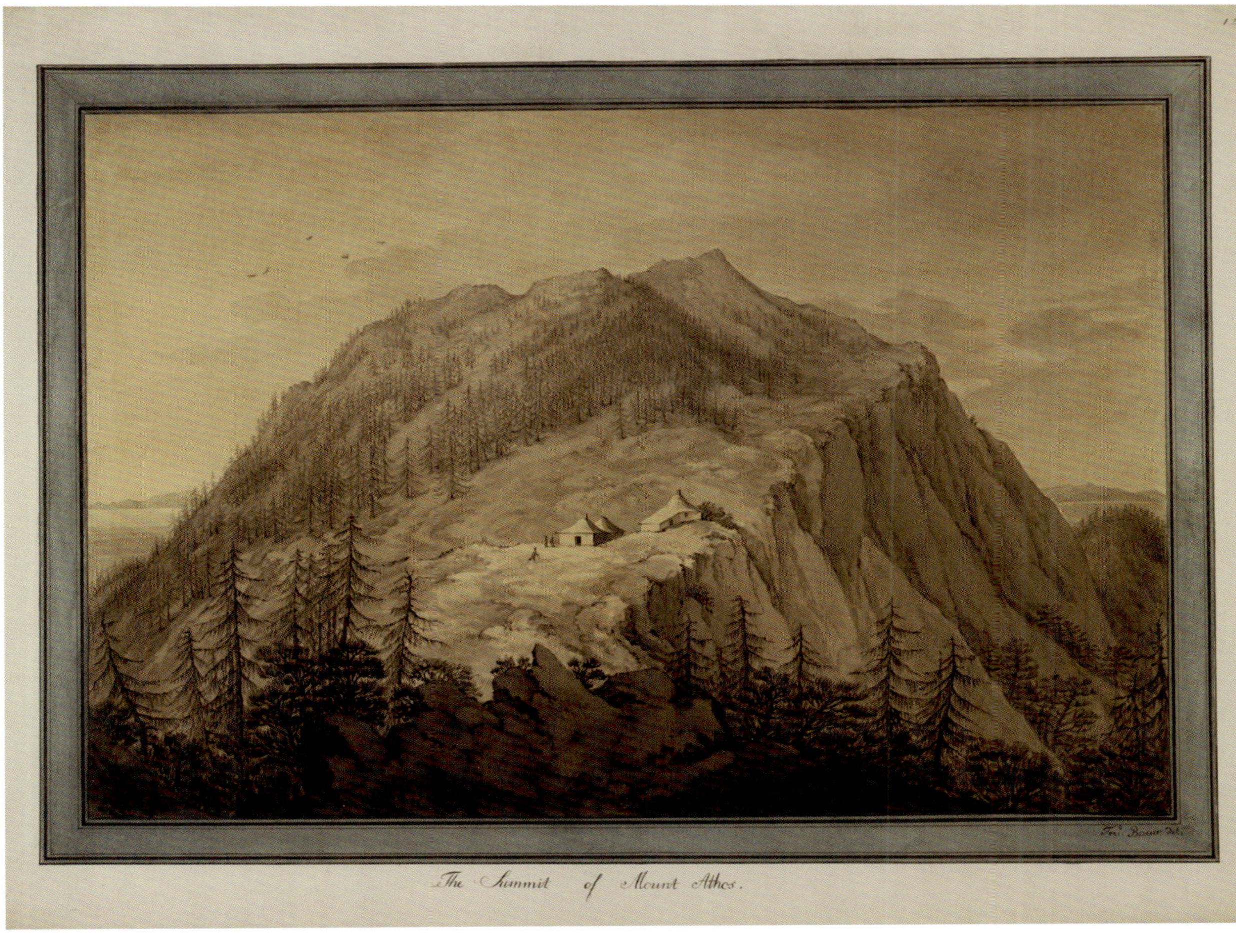

FIG. 5.17. The summit of Mount Athos with the skete of St Anne. Watercolour by Ferdinand Bauer, after 1794. – Oxford, University of Oxford, Sherardian Library, MS Sherard 408: f. 126.

endemic everlasting flower *Helichrysum sibthorpii*, a densely white- or greyish lanate cespitose herb belonging to the composites with white involucres and tiny white florets. It is only known from a population of maybe a thousand individuals. Growing in crevices of limestone rocks at 1800–2030 metres, it was new to science and proves that Sibthorp's party must have reached the summit area of the mountain at an altitude of 2033 metres (Fig. 5.17.). *Helichrysum sibthorpii* was recorded by Ferdinand as a pencil drawing together with one more strict endemic of Mount Athos on the same side of the page: the chamomile *Anthemis sibthorpii*, another composite with lower leaves 1–2-pinnatisect, glandular-punctate, recorded in minute detail. In addition to these two species restricted in distribution to Mount Athos, the white-flowering carnation *Dianthus petraeus* subsp. *orbelicus* is represented, a subspecies that has its most southern locality on Mount Athos but also grows in what is now north-eastern Greece and in modern south-western and central Bulgaria. By contrast, two juvenile specimens of liguliferous composites, almost certainly belonging to the genus *Crepis*, have later not been worked up by Ferdinand into finished watercolours.

There are a few more plants restricted in distribution to Mount Athos, and among these *Centaurea rutifolia* was also documented by Ferdinand. However, he also produced drawings of several less restricted plants, which because of their more northern range could only be recorded by him on Mount Athos, for example *Centaurea cuneifolia* and *Senecio othonnae*. The latter occurs very locally in a small wetland at an altitude of about seven hundred metres on Sibthorp's trail from the monastery of Megistis Lavras to the skete of St Anne. Another example of a plant certainly recorded by Ferdinand on Mount Athos is *Centaurea athoa*, with its golden yellow florets. It is a locally abundant plant at 1400–1800 metres, which at the time had no Linnaean name and is known today only from Mount Athos and mountains in north-western Anatolia, not visited by Sibthorp and Ferdinand.

Another find can be exclusively attributed to Ferdinand: the foxglove *Digitalis viridiflora*, a plant with both medicinal and horticultural value, which also had not yet received a Linnaean name and had to wait for more than three decades to get one. He must have made a pencil drawing of it, but no specimen survives at Oxford. Later Sibthorp did not wish to have a watercolour of it prepared, since he had not seen the plant with his own eyes. On the basis of the drawing Ferdinand later produced a finished watercolour, now kept in the archives of the Royal Horticultural Society in London, which was used as the exemplar for the coloured engraving of this very species included in John Lindley's *Digitalium monographia* (see Chapter 12). The seeds of *Digitalis lanata* subsp. *leucophaea* collected by Sibthorp on this occasion were to be passed on to Aiton the father at the Royal Garden at Kew, who raised from them living plants, very likely kept there for years in cultivation. Probably based on these specimens Ferdinand subsequently produced a watercolour showing this second species from Mount Athos, again in the archives of the Royal Horticultural Society, which was also used as exemplar for a printed illustration of this plant in Lindley's monograph (see Chapter 12). From a pharmaceutical point of view, finding these two foxgloves was the most significant discovery of the journey, though Sibthorp seems not to have been aware of their relevance for the treatment of heart failure. As on the island of Cyprus, plant hunting on Mount Athos resulted in many first records, among them several species new to science for a simple reason – Tournefort, Gundelsheimer and Aubriet had not visited the area. In consequence Sibthorp's party continues to be regarded today as the true discoverers of the floristic riches of this singular place.

The illuminated manuscript containing the writings of Dioscorides inspected by Sibthorp in the library of the monastery of Megistis Lavra merits a brief comment. Almost certainly he was shown Ω 75, about five centuries junior to the *Codex Vindobonensis* and possibly based on it or on its precursor. Needless to say, the reading room must have differed widely from the baroque splendor of the Imperial Library in Vienna, but Sibthorp was again in an atmosphere of concentrated study, no doubt familiar to him since his childhood from the many libraries at Oxford.

Four days after their arrival, Sibthorp's party descended 'by hanging vineyards and hermitages cut out of the rock to the beach', where the poriote boat was waiting to carry the travellers to Salonica (Thessaloniki, then Selanik). Sailing along the western coast of the peninsula with its shore 'bold and rocky, with ravines, each presenting the romantic site of a monastery', it was again time to record the scenery. In this case both Ferdinand and Imrie produced a drawing; the latter had it published in his *Catalogue*, listing in the legend all the monasteries to be seen along the western coast of Mount Athos.

THESSALONIKI TO PATRAS

Salonica was a sanjak (district) centre belonging to the Rumeli eyalet, a vast territory comprising the modern states of Albania, Bosnia and Herzegovina, Bulgaria, Macedonia, Serbia, Kosovo, and parts of Greece and Turkey, with Sofia as its capital. Like several towns in the Ottoman Empire, Salonica was a town of mixed ethnicity, with Muslims, Greek Orthodox Christians and Sephardic Jews living together. Ferdinand prepared a drawing of the town's panorama from the seaside, showing the fortifications and no less than sixteen minarets. Following established custom, Sibthorp's party took lodgings with the British Consul in this important harbour, but two days later the travellers proceeded to the country house of a Mr Abbot at Ourangick (near Asvestohori) 'escorted by most of the Frank [non-native] Merchants, forming a Calvacade [cavalcade] of upwards of thirty Persons'. The village was surrounded by agricultural land cultivated with corn, cotton and melons, and had a view of the gulf of Salonica. A visit to the low wooded Mount Courtiatch (Hortiatis) provided Sibthorp with an observation of another Levantine custom that he had already observed in Cyprus – 'ice being kept in pits . . . covered with dead leaves'. Ferdinand was commissioned to record the panorama from this elevation and another view of one of the nearby lakes later annotated 'Pechliger a village on the lake Petriski near Salonica' showing three water buffalos, a nearby rural mosque with a single minaret and for comparison of size two people in oriental dress (Fig. 5.18.). A Balkan endemic with a rather northern distribution must have been recorded in this area by Ferdinand: the milfoil *Achillea clypeolata*, then new to science but today a very common garden plant.

The marshy countryside to the north of Salonika was investigated on 23 to 25 August: fish were caught, birds were observed and shot, and plants were recorded,

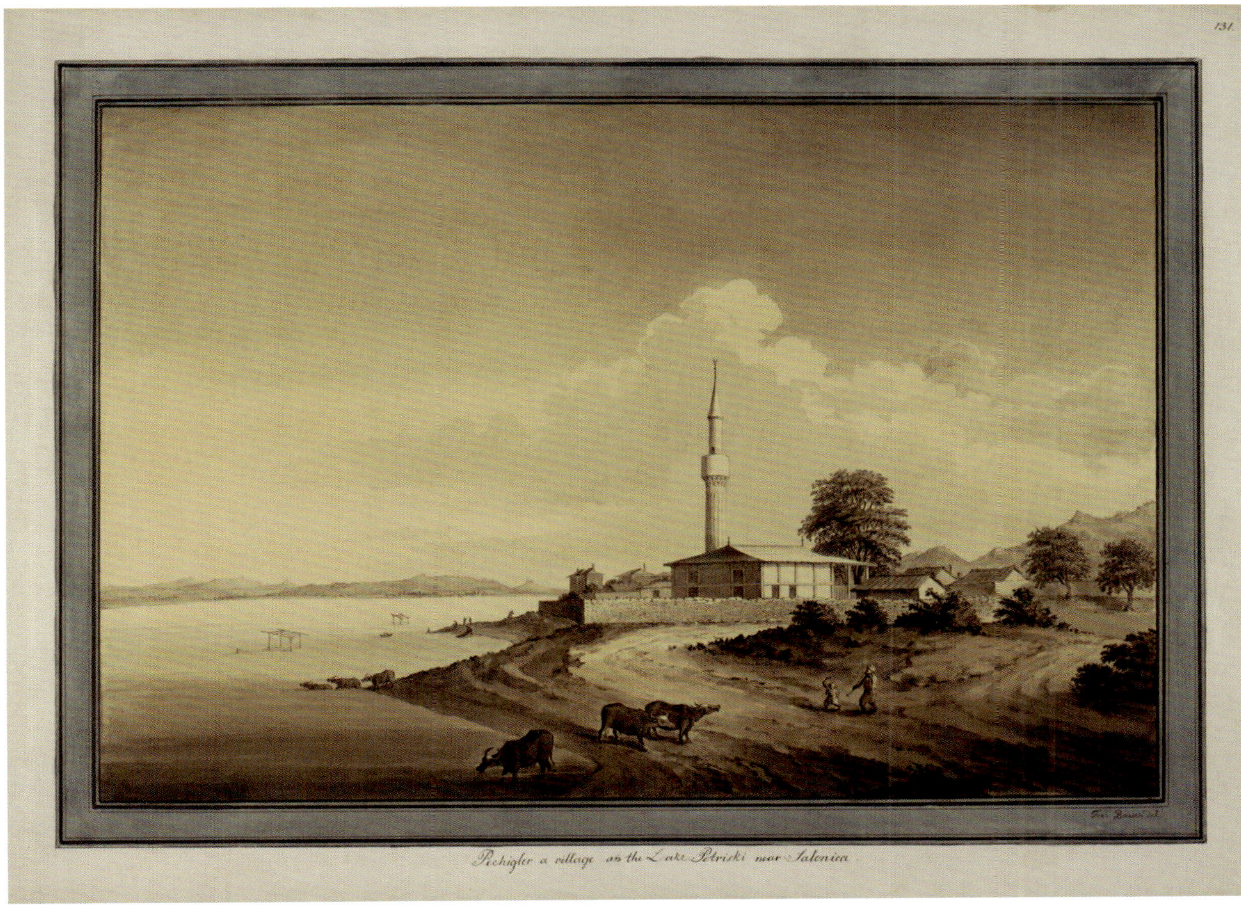

FIG. 5.18. **Mosque near Salonika. Watercolour by Ferdinand Bauer, after 1794. – Oxford, University of Oxford, Sherardian Library, MS Sherard 408: f. 131.**

among them elecampane, *Inula helenium*, a widespread Euro-Siberian species preferring wet shady localities, long cultivated as an ornamental and frequently naturalized. Although no specimen exists for this taxon in Sibthorp's herbarium, Ferdinand's pencil drawing has his unequivocal note 'Salonica'.[52] At this place Sibthorp's party may have been informed that Abdul Hamid I had declared war against Catherine the Great. The commencement of military conflict was located in the region near the mouth of the River Dnepr (now Ukraine), with General Alexander Vasilyevich Suvorov leading the Russian troops. Consequently it had no effect on Sibthorp's party during their last weeks on the western fringe of the Ottoman Empire.

On 30 August 1787 Sibthorp's party split up again. Hawkins, Ferdinand's later supporter and commissioner, stayed behind and continued his tour by visiting a silver mine at Siderocapse (Siderokastro) in the Paeonian Mountains (Mounts Belles, Vroudous, Orvilos). The rest of the group boarded again the caique from Poros with Cenchris (Kechries) in the west of the Saronic Gulf as the destination. This time sailing conditions were excellent, and within only five days the isthmus of Corinth (Korinthos) had been reached. There Sibthorp, Imrie and Ferdinand, having paid one hundred piasters, disembarked and entered the eyalet Mora comprising the Peloponnese with Tripoli (then Tripolis) as capital.

Again there seems to have been a feeling of disillusionment – 'a small hut near the port serves as custom house, the only remnant of the ancient Cenchreae'. A few days later Sibthorp's party crossed the isthmus and deposited the luggage at a warehouse near the harbour on the Golf of Corinth. Now there was time to explore the remnants of classical times, and Ferdinand recorded the ruins of the archaic Temple of Apollo, showing this building in a considerably more complete state of preservation than it is today, including walls that no longer exist. Much more accomplished are two panoramic views of the town, one of them showing the fortified town with a tall minaret in its centre and the fortress of Acrocorinth to the right, and the other depicting the view from the seaside with several sailing ships in the foreground.

In the harbour a small 'Zantiote boat', a caique from the island of Zante (Zakinthos, then Zante), then belonging to the Serenissima Repubblica di San Marco, was hired, and Sibthorp, Imrie and Ferdinand disembarked for Aspropiti, a place they had been to two months before. Three days later they began a third attempt to climb Mount Parnassus from the north-eastern side, but had to give up because of bad weather, and the party descended 'over steep precipices and turrent beds, covered with loose stones, with danger and difficulty, down the east side of the mountain'. From Aspropiti the caique brought Sibthorp, Imrie and Ferdinand to the town of Patras (then Balyabadra) on the Ionian Sea, where they again sought the services of the British Consul. His house became their base for the following weeks.

Autumn had started and botanizing had almost come to an end. Nevertheless, the decision was taken to make an 'excursion to [Mount] Olono [Erymanthos], the highest mountain of Morea'. This note testifies to their rudimentary knowledge of the topography of the Peloponnese, since the mountain ranges of Taygetos in the south are almost two hundred metres higher than the mountain Sibthorp's party intended to climb. Because of weather conditions, 'a thick mist' obscuring the view, the plan to reach the summit was given up and the travellers headed back to Patras.

Again, autumn-flowering bulbous plants were in flower and recorded by Ferdinand,[53] among them *Sternbergia colchiciflora*, which had not yet received a Linnaean name, a widespread tiny plant with a lemon-yellow perianth of six narrow segments. On the same page another *Sternbergia* is documented – the autumn dafodil, *Sternbergia lutea*, known already to Linnaeus, with a bright yellow perianth of six broadly elliptic lobes. Interestingly, Ferdinand had already come across the latter species the year before, most probably on Mount Olympus (Uludağ), since he recorded this very species on a sheet marked 'Asia' by him, but it was the drawing from Olono on which he based his watercolour later in Oxford.

Maybe exhausted from this last excursion, Sibthorp's servant Joseph Ganzarowitz died suddenly at Patras – 'seized with a Fit of Apoplexy his Puls ceased almost immediately'. All attempts to save his life, such as applying 'stimulants to his nostrils' and bleeding him, were in vain and on 24 September 1787 he was buried in the churchyard of the Europeans in Patras.

PATRAS TO OXFORD

It must have become clear to Sibthorp that further botanizing would be faced with limited success and that the approaching winter would be wet, cold and unfriendly. He may also have realized that procuring comfortable lodgings that could be adequately heated would not be easy in a provincial town like Patras. Therefore he had decided to return home together with Ferdinand, while Imrie remained behind; we are uninformed about the latter's movements over the following months, although he is likely to have returned to Gibraltar, where he published later on the geology of the mountain on which the settlement with its British garrison is situated.

Sibthorp had paid 40 guineas for the passage from Patras to Bristol on board the *Pomona*, but he had to procure his own provisions, the diary offering astonishing details. No less than 'a hundred Fowls & Ducks' were bought, as well as more conventional foodstuffs such as rice, coffee, sugar, biscuits, fruit and wine. Meanwhile Ferdinand must have prepared a view of the town with its massive fortifications from the seaside showing several vessels off the coast, with two mosques and several tall flagstaffs, possibly belonging to the consulates, clearly discernible. Late in the evening of 24 September 1787, Sibthorp and Ferdinand left Patras on board the *Pomona*, their tour of fifteen months through the Ottoman Empire having come to a conclusion.

What followed was a very long and dull voyage to Bristol, although definitely a comfortable one compared with Ferdinand's later voyage from Port Jackson (Sydney) via Cape Hoorn to Liverpool in the rotten *Investigator* (see Chapter 10). The island of Zante was passed two days after the departure, followed by Malta on 1 October, still ruled by the Knights of the Order of Malta, and Pentielero (Panteleria) on 2 October. Sibthorp's diary notes 'heavy Seas shipped much water on deck' and in late October Gibraltar, Imrie's home base, was passed. Two views of this place, which for centuries was of great strategic value for the British Empire, were taken by Ferdinand and later transformed into finished watercolours. By now Sibthorp and Ferdinand had not only left the Mediterranean Basin but also the Mediterranean zone with its benign climate. 'Dark cloudy weather' and strong gales are recorded in the diary; the coast of Ireland was spotted on 25 November, but it took more than one further week until the harbour of Bristol was reached on 5 December 1787.

It would have been standard procedure to submit Sibthorp and Ferdinand to quarantine, since they had arrived from the Ottoman Empire, known for contagious diseases such as the plague, which had already disappeared in Britain. However, for a member of the elite like Sibthorp and his illustrator, these regulations were waived and a few days later the unequal pair had arrived in Oxford in good time for Christmas.

ENDNOTES

1 Unless otherwise stated, this chapter is based on Lack (1999c) and Harris (2007). The scientific names of plants follow Strid (2009–2013).
2 Sir F. Cholmeley, pers. comm. (2002)
3 Russell (1954)
4 Krahnke (2001)
5 John Sibthorp to Humphry Sibthorp, 10 xi 1785 (3 SIB /1/9, Lincolnshire Archives)
6 Lack (2003)
7 Lack (2000b)
8 MS Sherard 247, f. 21 r (OXF)
9 Everett (2013)
10 Denney (2010)
11 GSA 25 / XXVII, Q
12 P. da Luca, pers. comm. (2013)
13 Giardina, Raimondo & Spadaro (2007)
14 Lack (2006a)
15 Lack (1996)
16 MS Sherard 247, f. 46 r (OXF)
17 MS Sherard 247, f.39 r (OXF)
18 MS Sherard 247, f. 45 r (OXF)
19 MS Sherard 247, f. 42 r (OXF)
20 N. Turland (pers. comm., 2013)
21 MS Sherard 247, f. 94 r (OXF)
22 MS Sherard 247, f. 115 r (OXF)
23 MS Sherard 247, f. 2 r (OXF)
24 Nelson (2007)
25 MS Sherard 247, f. 57 r (OXF)
26 MS Sherard 247, f. 57 r (OXF)
27 MS Sherard 247, f. 2 r (OXF)
28 MS Sherard 247, f. 60 r (OXF)
29 MS Sherard 247, f. 7 r (OXF)
30 MS Sherard 247, f. 48 r (OXF)
31 MS Sherard 247, f. 112 r (OXF)
32 MS Sherard 247, f. 49 r (OXF)
33 MS Sherard 247, f. 112 r (OXF)
34 MS Sherard 247, f. 107 r (OXF)
35 Hadjikyriakou (2007)
36 MS Sherard 247, f. 87 r (OXF)
37 Foley (2003)
38 Meikle (1985)
39 R. Hand, pers. comm. (2013)
40 Kreutz (2004), Karig (2010)
41 Tsintides et al. (2007)
42 http://www.flora-of-cyprus.eu (accessed 29. XI. 2013)
43 MS Sherard 247, f. 25 r (OXF)
44 R. Hand, pers. comm. (2013)
45 MS Sherard 247, f. 109 r (OXF)
46 MS Sherard 247, f. 25 r (OXF)
47 Panista & Tzanoudakis (2010)
48 MS Sherard 247, f. 92 r (OXF)
49 MS Sherard 247, f. 93 v (OXF)
50 MS Sherard 247, f. 69 r (OXF)
51 MS Sherard 247, f. 24 r (OXF)
52 MS Sherard 247, f. 104 r (OXF)
53 MS Sherard 247, f. 7 r (OXF)

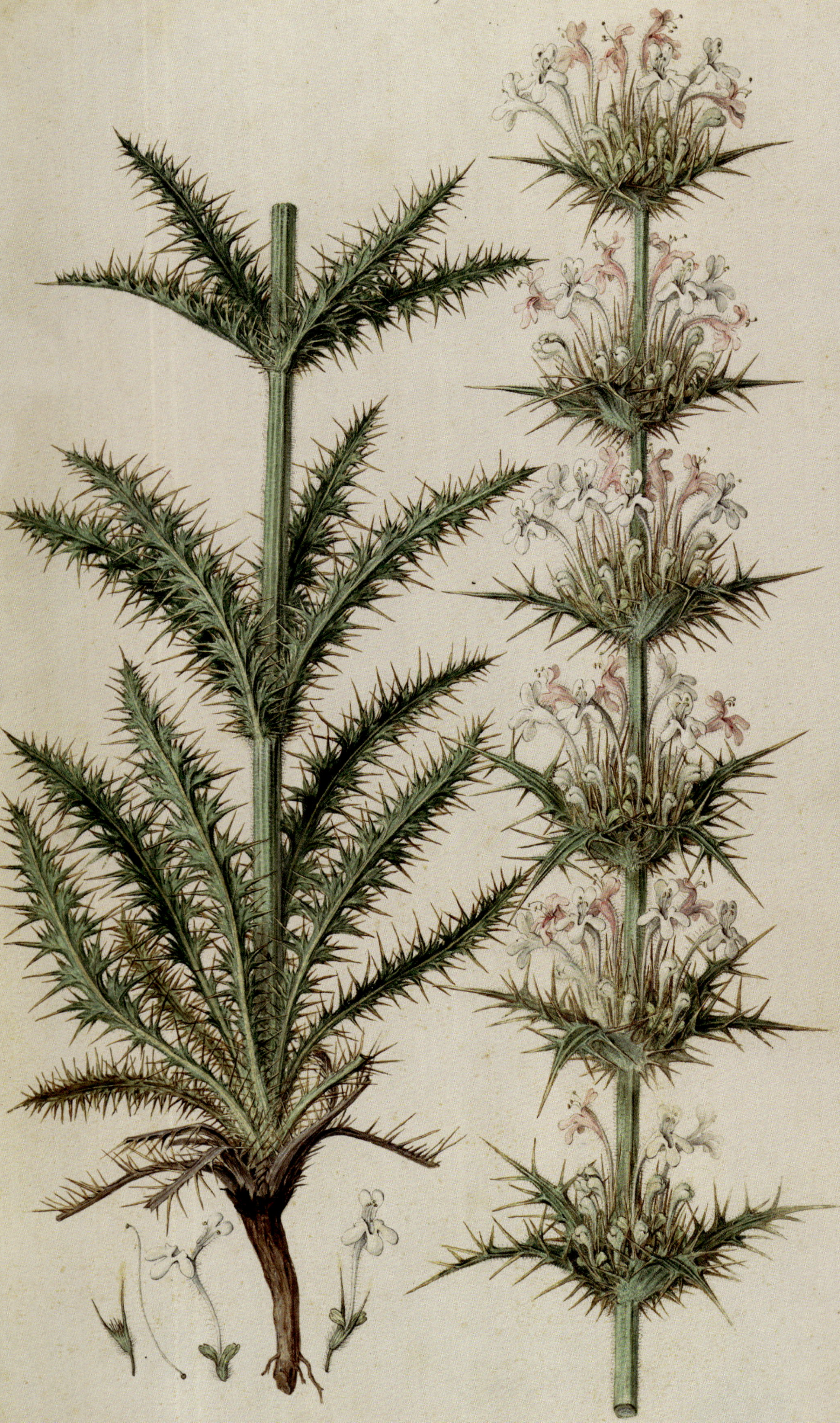

Bauer. del.

Morina persica

CHAPTER 6

Ferdinand Bauer in Oxford[1]

DISCREPANCY

THERE IS A STRANGE discrepancy between information gathered on expeditions and how it was later processed. Whereas a vast amount of primary information is available on many scientific expeditions and similar undertakings – among them the private journey of Sibthorp and Ferdinand to the Levant – albeit buried in diaries, letters, field notes and drawings prepared in the field, the collections themselves and the associated labels, much fewer data exist on how the materials and records were subsequently studied at home and later published. This means that the actual process of research is often very poorly documented, with only the end product, the printed publication and the appended printed illustrations, known. By contrast the various intermediate stages of scientific study, from scattered notes, drafts to preliminary versions of scientific texts, are mostly consigned to the dustbin. At best some manuscripts, drawings and watercolours submitted for publication have survived in the archives of learned societies and museums. In addition, history of science is very often focused not on the actual work of the scientist but rather on what is called progress in science, that funny mixture of fashion and trend, short-lived impact, genuine increase in knowledge, and the egos of the figures involved.

FIG. 6.1. ***Morina persica*****, Watercolour by Ferdinand Bauer, signed, c.1788. – Oxford, Sherardian Library, MS Sherard 243, f. 19.**

The iconic expedition of Alexander Baron Humboldt and Aimé Bonpland to tropical regions in Southern and Central America in 1799–1804 offers an excellent illustration of this point. Numerous publications have been dedicated, almost *ad nauseam*, to the travels of this unequal pair – to the diary, the letters, the field books, the collections, mainly botanical and archaeological, the topographical recordings, all brought back to Paris and Berlin – but scarce information is available on how their studies progressed over a period of many years until the results were finally published. In the case of the materials brought back by Sibthorp and Ferdinand, the situation is not entirely different, but in general there are more hard facts than in many other cases – several letters, among them the first extant letter written by Ferdinand, the first draft version of an incomplete manuscript by Sibthorp and, most importantly, an extensive series of watercolours of the plants and animals of the Levant prepared by Ferdinand. The scene of this chapter is Oxford.

JOHN SIBTHORP AT HOME

For Sibthorp, returning to Oxford after an absence of more than four years was a homecoming – to Cowley House, to his father Humphrey, then 76 years old, who was to survive him, and to his two widowed half-sisters Lady Mary Elizabeth Sewell and Sarah Cholmeley. It was also returning to his home country – Great Britain, and more particularly England, which since 1760 was under the reign of King George III. In the Seven Years' War Britain had defeated France and had become the dominant European power in North America and India. However, almost all American colonies had been subsequently lost in the American Revolutionary War, which had ended in 1783 with the Treaty of Paris and had resulted in the sovereignty of the United States. This profound change is seen today as the event defining the transition between the 'first' and the 'second' empire, in which Great Britain shifted its attention from the Americas to Asia, the Pacific and later Africa, all this having marked consequences on our story. It is not a coincidence that on 19 and 20 January 1788, just a few weeks after Sibthorp had returned to Oxford, the famous First Fleet reached Port Jackson (Sydney) in Australia, which opened up a new chapter in British activities overseas.

George succeeded to the throne aged twenty-two, after his grandfather King George II had died suddenly in 1760, and thereby became King of Great Britain, France and Ireland as well as Duke of Brunswick and Lüneburg, a Prince Elector of the Holy Roman Empire (see Chapter 2). The following year he married Charlotte, Princess of Mecklenburg-Strelitz, whom he first met on their wedding day; a fortnight later both were crowned at Westminster Abbey. They had fifteen children, nine sons and six daughters, all born between August 1762 and August 1783, with only two sons lost at an early age. Two of the surviving sons were to succeed their father on the throne – George to become King George IV and to reign from 1820 to 1830, and William to become King William IV and to reign from 1830 until 1837. Both held the titles of King of the United Kingdom of Great Britain and Ireland, and in addition that of King of Hanover. Their younger brother Ernest Augustus, later King of Hanover, and their niece Victoria, later Queen of the United Kingdom of Great Britain and Ireland, were to play an important role at the end of our story.

George III's reign, longer than that of any other British monarch before him, marked a period of great continuity and stability in his country, while much of the rest of Europe was in great turmoil due to the French Revolution and the subsequent Napoleonic Wars, which ended only in 1815 with the peace arrangements of the Congress of Vienna (see Chapter 11). When Sibthorp returned to Oxford, William Pitt the Younger was already Prime Minister and also Chancellor of the Exchequer, having been appointed to both positions at the tender age of twenty-four. He was to remain in office for a total of more than nineteen years. Today Pitt continues to be ranked highly among British Prime Ministers, 'if on no other ground than that he enabled the country to pass from the old order to the new without any violent upheaval … He understood Britain'.[2] His greatness was to come in the war with revolutionary France, with Great Britain's enemy setting the pace and Pitt being called 'the Atlas of the reeling globe'. Described as solitary and colourless, the Prime Minister was reported to have too often exuded his superiority to endear him to many. However, it was Pitt and the governments led by him that played a decisive role in conserving the established social stratification in Great Britain, with its remarkable stability and continuity in landed property. Later Prime Ministers under the reign of George III and the subsequent Regency era, among them Spencer Perceval and the long-serving Robert Banks Jenkinson, 2nd Earl of Liverpool, followed Pitt's general line of keeping Britain largely outside the long series of conflicts on the continent and aiming at restricting French influence. This was the political scenery in England to which the Levantine traveller returned.

Nearer to home and of more direct relevance to Sibthorp were three other personalities: Lord Robert Spencer, Sir William Dolben and Frederick North. Spencer was the third son of the immensely rich Charles 3rd Duke of Marlborough, who had been educated at Christ Church, Oxford. He was later elected Member of Parliament by the city of Oxford, a parliamentary constituency, holding his seat until 1790. Sir William Dolben, the 3rd Baronet, had also been educated at Christ Church and was elected MP by Oxford University, one of the few university constituencies, holding his seat until 1801. Frederick North, 2nd Earl of Guildford, known by his courtesy title Lord North, was at the time of Sibthorp's return the Chancellor of Oxford University, the key university officer. He had been educated at Trinity College, Oxford and served for twelve years as Prime Minister, today best known as the PM 'who lost America'.

Needless to say, John Sibthorp, seven months older than Pitt, was a man of a smaller calibre and engaged in an entirely different field, though definitely also a man with strong views on many subjects. Upon his return to Oxford in early December 1787 one of his first moves included the staging of his election to become a Fellow of the Royal Society. The pertinent document reads 'John Sibthorp Doctor of Physic and Professor of Botany in the university of Oxford who has for some years traveld in Asia Minor the Morae [Peloponnese] and the Grecian Islands, for the purpose of Comparing the natural productions thereof with the descriptions of Ancient Authors, and is now on the return to England being desirous of the honour of becoming a Fellow of this Society, we whose names are underwritten do on our personal knowledge declare, that we think him worthy of that honour and likely to be a usefull and valuable member' and is dated 6 December 1787. It had been signed by nine members and quickly resulted in Sibthorp's election on 6 March 1788. Smith (see Chapter 5) had been among the proponents, thirteen months younger than Sibthorp and years later the man to publish the first six of the ten volumes of the *Flora Graeca*, a work that the third Sherardian Professor had laid the foundations for, although he had not started it (see Chapter 9). However, today it is famous chiefly because of the coloured copper engravings based on the watercolours prepared by Ferdinand. In the year of Sibthorp's election to the Royal Society Smith founded the Linnean Society of London and became its long-serving first president, with Sibthorp among the first fellows to sign the roll book of this learned society,[3] which in the end purchased from Smith's estate Linnaeus's library, manuscripts and collections as well as Smith's library, manuscripts and collections. Later Smith was to recall 'In the spring of the year last mentioned [1788] the writer of this, with sir Joseph Banks [then President of the Royal Society] and Mr. [Jonas Carlsson] Dryander, passed a week at Oxford, which was devoted to a critical survey of the professor's [Sibthorp's] Grecian acquisitions; nor was the honey of Hymettos [Imittos], or the wine of Cyprus, wanting in this truly attic entertainment'. Almost certainly it was on this occasion that Banks got to know the pictorial record of plant and animal life of the Ottoman Empire prepared by Ferdinand and met him for the first time. Two years later Sibthorp received another prominent naturalist with strong botanical interests – Humboldt, aged twenty-one. On that occasion the latter must also have met Ferdinand, because on 21 February 1801 Humboldt referred in a letter sent from Havana to his botanical mentor Carl Ludwig Willdenow in Berlin to 'alle Bauer [Brüder]'[4] (all Bauer brothers), obviously including Franz and Josef as well. Humboldt may have seen Franz's works on his previous visit to Vienna in 1797 when he was hosted by Jacquin, elected Fellow of the Royal Society one month after Sibthorp.

Otherwise Sibthorp enjoyed the pleasant life of a young gentleman, which is evident from the following note sent to Hawkins 'it gave me much & real Pleasure to find you would come & shoot with me this September. Come then a most hearty welcome expects You . . . when tired with the Chase at Stanton Harcourt [Sibthorp's estate in Oxfordshire, bought in 1790, later bequeathed to Oxford University] here is the Flora & Fauna of Greece – we will comemmorate our Conquests – in a Display of the Exuvia of the Cyprian and Athensian Fields – the Parnassian Vultur shall appear & libations of rich Cyprus [wine] crown the Jubilee'.

Upon his return to Oxford Sibthorp was 29 years old and in comfortable circumstances: the income from his estate was supplemented by his salary as Sherardian Professor and his stipend from the Radcliffe Fellowship, which was to come to an end three years later. In 1793 his income was to increase further by a grant from the crown, which made him a Regius Professor; in addition he was listed among the physicians of the Radcliffe Infirmary at Oxford. In contrast to what might be expected, Sibthorp's main concern over the next years was not botany in general and the work on the projected *Flora Graeca* in particular, first mentioned in his letter from Istanbul to Banks (see Chapter 5), but the management, improvement and increase of his estates. This is explicitly mentioned by Smith, who noted Sibthorp's 'attention to his landed property and to agricultural pursuits, of which he was particularly fond'. However, some other commitments also had to be fulfilled. Unlike his father, Sibthorp gave regular undergraduate lectures, recorded in a 583-page manuscript in his own hand kept in the Sherardian Library at Oxford.[5] He focused on the use of plants, especially in agriculture and as foods and medicines, and seems to have illustrated his lectures with a wide spectrum of objects, among them books from his library, specimens from William Sherard's herbarium and living plants from the botanical garden that was under his supervision. In addition Sibthorp followed his exact equivalent in Vienna in making use of botanical illustrations – Franz had produced several for Jacquin's lectures (see Chapter 4), and now Ferdinand's watercolour of the olive tree, *Olea europaea*, definitely prepared in Oxford, was used in Sibthorp's academic teaching.

Needless to say, Sibthorp battled with the university authorities over funding for the botanic garden and praised it in his lectures 'Academic Gardens tho' greatly inferior in Magnificence & Splendour to those supported by Royal Expenditure may be considered as the more useful Schools of Botany. – not under the Restrictions, of royal and private Collections, they are at all Times open to the Public, & their Object is to inform as well amuse. Picturesque Beauty is not merely studied, but Method & Order as far as they conduct to a Systematick

Arangement must be preserved'. Of course this was an indirect criticism of Kew, at the time a Royal Garden and not a proper botanical garden open to the public. Like many others responsible for a botanical garden, Sibthorp had the planting arrangements changed in order to come closer to his ideal of teaching purposes. He also made sure that the living material collected during his journey in the Ottoman Empire was cultivated by his gardener. He could report to Hawkins 'I have been employed since my Return in raising a Plantation sacred to Greece and a quarter of my garden destined for that Purpose has received 800 Greek seeds – a deposit procured if you recollect cum multo Sudore [with a lot of sweat]'. This again echoes what Tournefort had done in the Jardin du Roi in Paris after his return from the Levant. Two years later Sibthorp confirmed the cultivation of this material by writing 'I have now a large Colony of Greek Plants & Time [thyme] of Hymettus [Imittos] enough to produce a Hive of Honey'. Apart from making the necessary arrangements for the cultivation of this material Sibthorp distributed herbarium specimens from his journey to the Levant, albeit as far as we know left unannotated, to Banks in London, Peter Simon Pallas in St Petersburg and Jacquin's elder son Franz, who came to see him in Oxford in late 1788 (see Chapter 7).

Sibthorp also started the proper identification of his material by comparing his herbarium collections from his tour of the Levant with those kept in the Sherardian herbarium in Oxford, the Linnaean herbarium in the possession of his friend Smith, then still in London, and the Banksian herbarium in London. In addition, Sibthorp referred to the rich collection of books and periodicals then available in Oxford and London. Furthermore, he extracted locality information scattered in his field notes and tried to correlate it with his unlabelled herbarium specimens. These data – scientific name, reference to the place of valid publication, reference to printed illustrations (including the proofs received from Jacquin), the name of the locality where Sibthorp thought the specimen had been collected or observed – were combined and included in a manuscript. However, this was left unfinished, since Sibthorp refrained from giving a single plant description. Because of the confused state of his field notes, the lack of any field labels accompanying the specimens and a somewhat unrealistic trust in his memory, Sibthorp was often totally wrong in his locality data. In retrospect Hawkins noted 'It is certainly a pity that Dr. Sibthorp did not mark his specimens… but trusted to his memory and dreamt of not dying'. Because of this state of affairs Sibthorp caused endless troubles for subsequent generations of botanists studying his collections and his locality data, often totally incorrect. In any case, his manuscript, effectively the skeleton for a draft version of the *Prodromus Florae Graecae*, remained a torso.

Nothing is known of Sibthorp studying the zoological material apart from a few pencil notes, possibly in his hand, on some of Ferdinand's pencil drawings of fish. In short, the plan of a *Flora Graeca* and *Fauna Graeca* seems to have faded away in Sibthorp's mind. In a letter written just a few days before Sibthorp's departure for his second journey to the Levant, Smith quite rightly called the third Sherardian Professor 'Florae Graecae ad calendas Graecas proditurae auctor' [the author of the indefinitely postponed *Flora Graeca*]. The only person working hard for the project during these years was clearly Ferdinand.

FERDINAND AT WORK

Coming to Oxford was quite a different experience for Ferdinand compared with that of his master. Ferdinand arrived in a country to which he had never been before, in a town where he knew no one and his knowledge of the English language was poor. It has been rightly said that the Oxford of his time was 'divided between town and gown, and the Austrian artist fitted comfortably into neither; he was an outsider. To the town he was a foreigner who spoke English poorly; to the gown he was neither a don nor student, merely Sibthorp's poorly educated draughtsman, his hired hand'. Sibthorp, no doubt a rather pompous young man, was not the person to smooth the delicate situation. The Oxford winters were less cold than those in Feldsberg, the summers less hot, there were many more rainy days, but most important for an illustrator working with colour there were many grey days, with a reduced quality of light, quite the opposite to what he had just experienced when crossing the Aegean sea in a sailing boat. Ferdinand had to adapt – as has been rightly said 'to the unfamiliar surroundings, the societal isolation and often trying . . . working conditions'. It was in this atmosphere that Ferdinand recreated the colours of the Mediterranean flora and fauna, very accurately and with all their splendour and diversity.

Although hard evidence is lacking, it may be assumed that Ferdinand lived in a way similar to most servants – under the roof of his master at Cowley House, or in Sibthorp's other place, the Professor's House opposite

Magdalen College, and sat at the servants' table. The 'plantation sacred to Greece' in the Botanical Garden was for him within walking distance, and so was the collection of herbarium specimens, animals dried and preserved in spirit, probably deposited in one or both places. If it were possible for Ferdinand to stroll today through the centre of Oxford, he would, as in the centre of Vienna, not need a map. The large, square Magdalen tower continues to be one of Oxford's landmarks on the eastern side of the city, the Great Quad and the New Building of Magdalen College remain unaltered, just like the triangular meadow bounded by the river Cherwell and the surrounding tree-lined path known as Addison's walk. Opposite this college and on its grounds Ferdinand would find, as in 1787, the Botanical Garden of Oxford University, founded in 1621 and the oldest in Britain, though significantly younger than those he had visited in Bologna, Pisa and Rome. Other landmarks, such as the University Church of St Mary the Virgin and the Radcliffe Camera, an iconic building designed by James Gibbs in the English Palladian style, have remained unaltered. The same applies to the Sheldonian Theatre, designed by Sir Christopher Wren, then and now the official ceremonial hall of Oxford University, where on 7 July 1791 Ferdinand's compatriot Haydn from Rohrau in the Archduchy of Austria below the River Enns was to receive his honorary degree. The same conservative attitude applies to the buildings of a large number of colleges, among them Christ Church, known simply as 'The House', where Banks had studied, Lincoln College, where Ferdinand's master had matriculated, Corpus Christi College, where Pococke had been educated, and St John's College, William Sherard's former college. The coats of arms of these colleges have not changed since then, and nor has Great Tom, the loudest of all Oxford bells, sounding one hundred and one times every night, which signifies the hundred original scholars of Christ Church plus one. The same continuity applies to the choirs of several colleges founded in late-medieval times, which continue to sing the church services during term. However, Ferdinand would be surprised to find female dons and female students as well as a Catholic Chaplaincy in Oxford today, since during his stay women were not allowed to enter academia and a catholic mass was read nowhere – not in Christ Church Cathedral, nor in any of the numerous churches or chapels of Oxford.

Providing himself with the necessary utensils was probably Ferdinand's first priority in Oxford. For his work he had to buy water colourist paper of the best quality, a selection of artists' brushes, in particular riggers, round brushes with long hairs and useful for fine lines, and watercolour paint. The latter included: (1) a rich array of pigments, mineral and organic; (2) gum arabic and sugars as binders to hold in suspension and fix the pigments to the paper; and (3) additives such as glycerine, ox gall and honey. Other ingredients included inert, white pigment like chalk for more opaque paint and egg yolk to produce shiny surfaces, for example for the eyes of Ferdinand's animals. At his time pigments could be purchased only at an apothecary or specialized 'colourman', among them William and Thomas Reeves at 80 Holborn Bridge, London. Although the provenance of Ferdinand's pigments has not yet been analyzed, one fact is clear – when purchasing his pigments in Oxford or London he must have considered the spectrum of hues in his colour chart prepared before he left Vienna.

Although nothing is on record, it is plausible to assume that Ferdinand made use of a drawing table, probably with a tilted surface in order to avoid chronic neck pain or spinal problems, a common problem with concentrated work on a horizontal surface. Additionally we may presume that drawers were attached to the drawing table or grooves made on the surface to keep tools and supplies, since looking away from his watercolour for a couple of seconds to find a brush, for example, could have broken the focus of his work. Anna Amalia Duchess of Saxe-Weimar-Eisenach's drawing table with its tilted surface and drawers for paper and pencils, dated c.1790 and preserved in the Stiftung Weimar Museum, serves as an example for this kind of furniture. We may also hypothesize that as in Feldsberg Ferdinand had a light box made for himself in order to backlight his colour-coded pencils with a sheet of watercolourist paper superimposed. Several fish pencil drawings (for example of the sword-fish, *Xiphias gladius*[6]) and plant pencil drawings (for example of the sawfly orchid, *Ophrys tenthredinifera*) and of *Erigeron uniflorus* (see Chapter 5), match the finished watercolours so perfectly in size that the use of such a utensil is the only plausible interpretation. However, the light source used in the light box remains unknown. In short, what Ferdinand produced in Oxford were watercolours on solid water colourist paper based on the pencil drawings on thin paper made in the field. This approach has been misunderstood by some, who thought that the field drawings were coloured, which was never ever the case.

FERDINAND'S PLANT ILLUSTRATIONS

Letters enable us to reconstruct the sequence of events on Ferdinand's hypothetical drawing table. As early as 1 July 1788 Sibthorp informed Hawkins, 'Bauer has just finished the Grasses which close the Triandria – nothing can be more beautiful & perfect than his Drawings – that I shall possess a unique work'. This statement indicates that Ferdinand followed the sequence of Linnaean orders and proves an important feature of his work – his speed; years later Banks very rightly was to call him 'indefatigueable'. In about six months he had produced a total of 102 finished watercolours of folio size showing plants recorded months ago in the Levant, among them the spiny *Morina persica* (Fig. 6.1.) from Mount Parnassus. And Ferdinand possessed the strength, discipline and determination to continue at the same speed, although it is difficult to imagine how he kept this work rate up during the short winter days. One year later, on 2 July 1789, Hawkins told his mother, Anne, 'Mr Bower [Bauer] has finish'd upward of 300 drawings of the plants which we found in Greece. These drawings surpass all that I have seen in this way even Ehrets which you may remember in my uncles room'. These watercolours included representations of florists' cyclamen, *Cyclamen persicum* (Fig. 6.2.), Sibthorp's trophy from the Baphos area in Cyprus, an image of the mandrake, *Mandragora officinarum* (Fig. 6.3.) from the Anatolian coast, and the flowering and fruiting specimen of the widespread umbellifer *Thapsia garganica*

FIG. 6.2. ***(left) Cyclamen persicum.*** **Watercolour by Ferdinand Bauer, c.1788. – Oxford, Sherardian Library, MS Sherard 244, f. 176.**

FIG. 6.3. ***(right) Mandragora officinarum.*** **Watercolour by Ferdinand Bauer, signed, c.1788. – Oxford, Sherardian Library, MS Sherard 244, f. 53.**

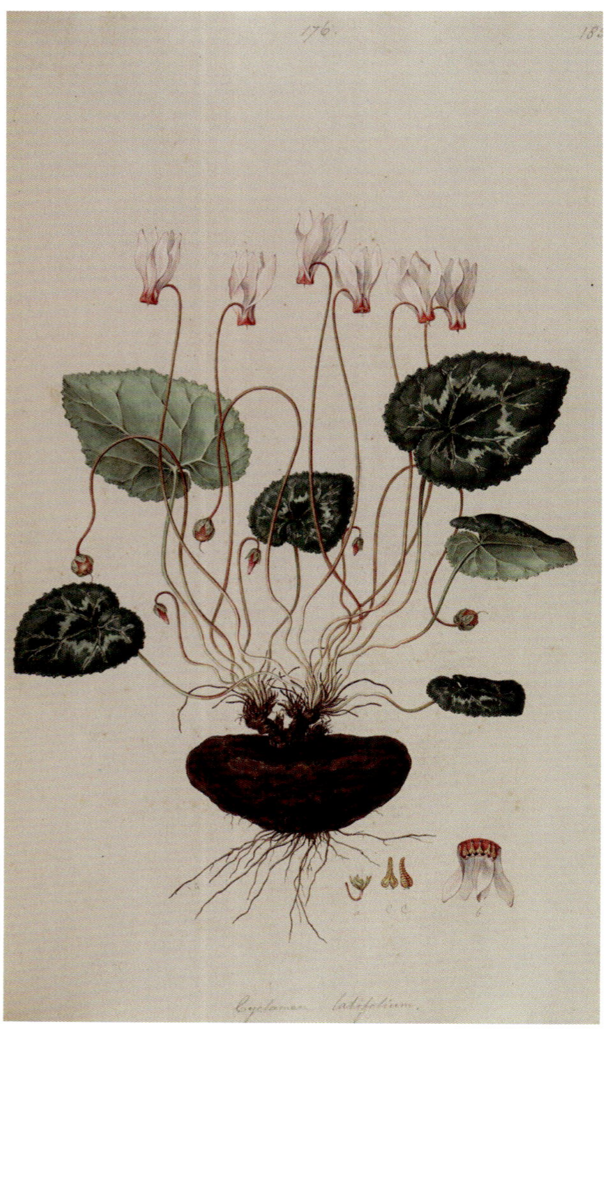

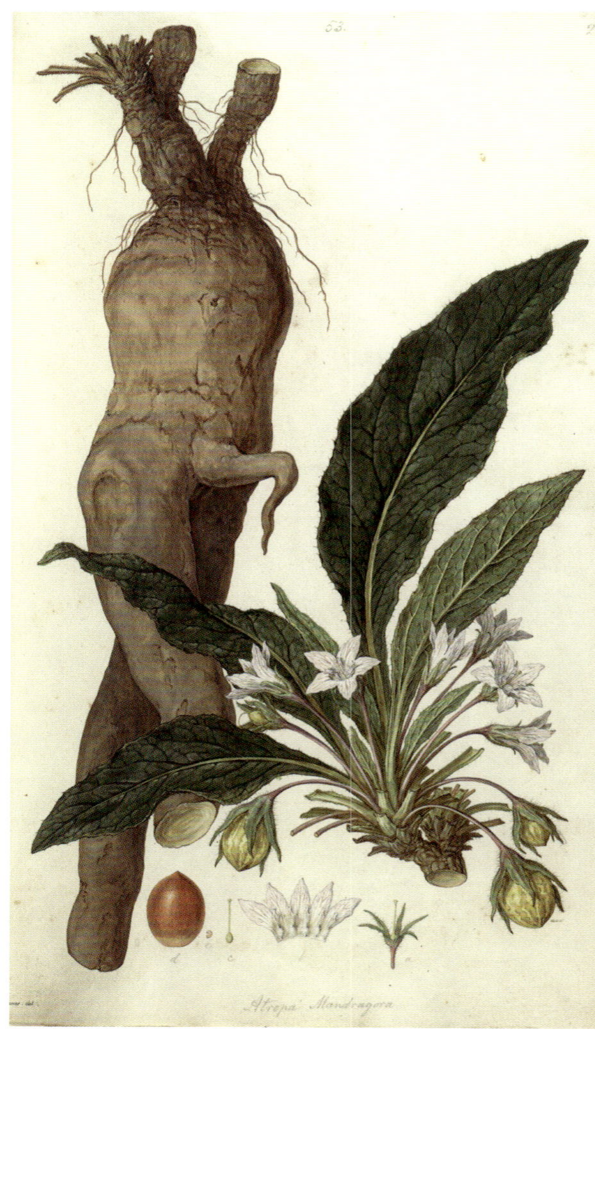

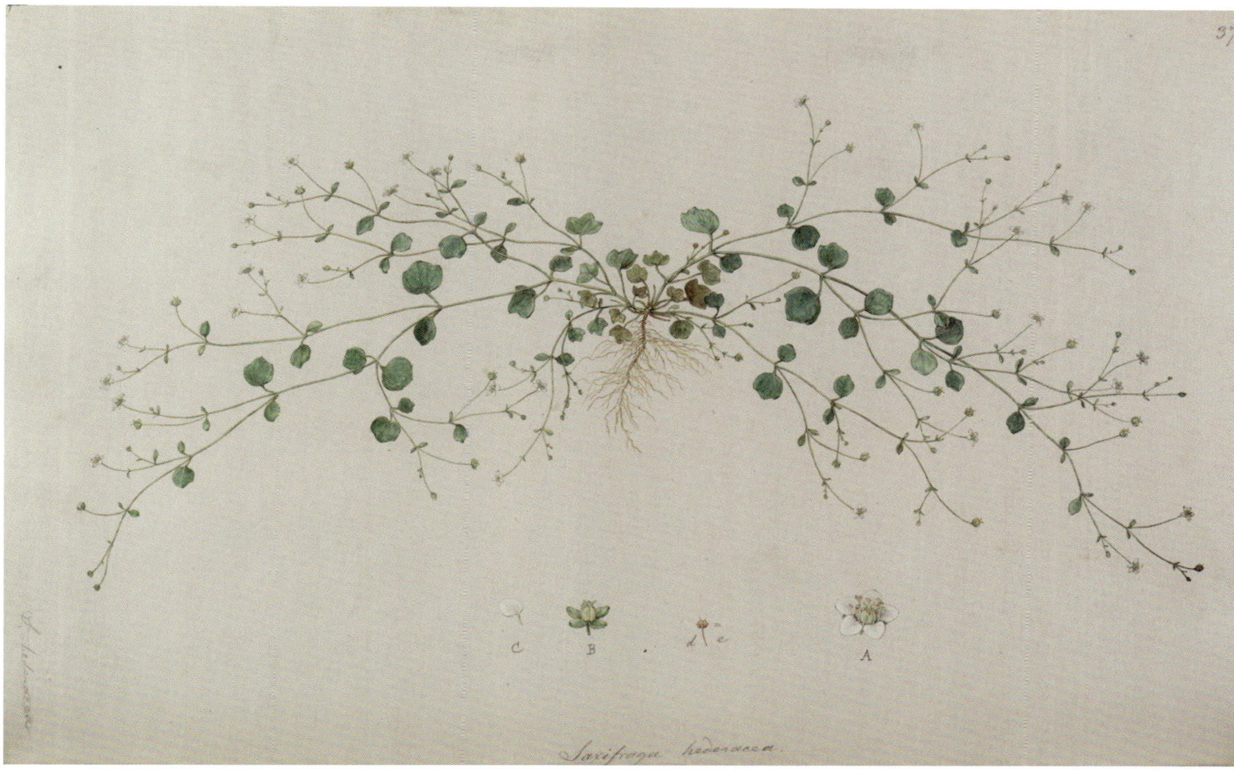

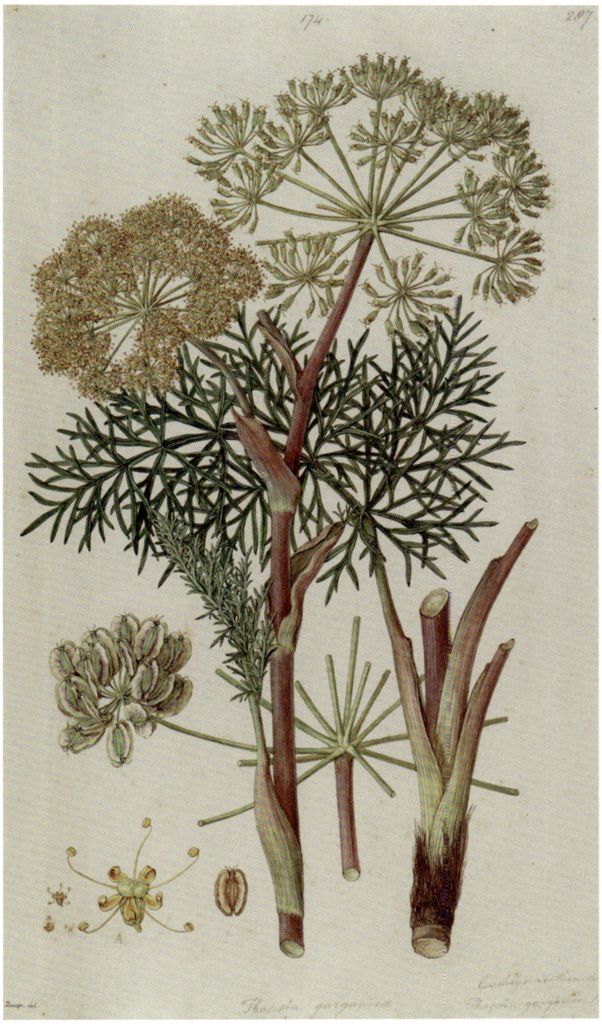

FIG. 6.4. (above) *Saxifraga hederacea*. Watercolour by Ferdinand Bauer, c.1789. – Oxford, Sherardian Library, MS Sherard 242, f. 145.

FIG. 6.5. (left) *Thapsia garganica*. Watercolour by Ferdinand Bauer, signed, c.1789. – Oxford, Sherardian Library, MS Sherard 242, f. 174.

(Fig. 6.5.). About two years later, on 11 June 1792, Hawkins informed Sibthorp 'I find that Bauer has completed the Flora & is about to begin the Fish'. Following the Linnaean system Ferdinand had produced among many others an illustration of the decumbent, small-flowered *Saxifraga hederacea* (Fig. 6.4.). Others include the spectacular *Dianthus fruticosus* subsp. *fruticosus* (Fig. 18.1.), confined to the crescent of islands from Andros and Kea to Folegandros, Sikinos and Anafi in the Aegean Sea, and the early-flowering *Helleborus orientalis* (Fig. 6.6.) from the Istanbul area. The later classes in the Linnaean order are also represented among Ferdinand's watercolours: for example the Polyadelphia with *Astragalus creticus* (Fig. 6.7.); the Syngenesia with *Ptilostemon afer* (Fig. 6.8.), probably collected on Mount Parnassus; the Gynandria with the orchid *Ophrys tenthrendinifera* (Fig. 6.9.); and the foul-smelling aroid *Arum dioscoridis* (Fig. 6.10.), the latter two from Cyprus. In this way a total of 966

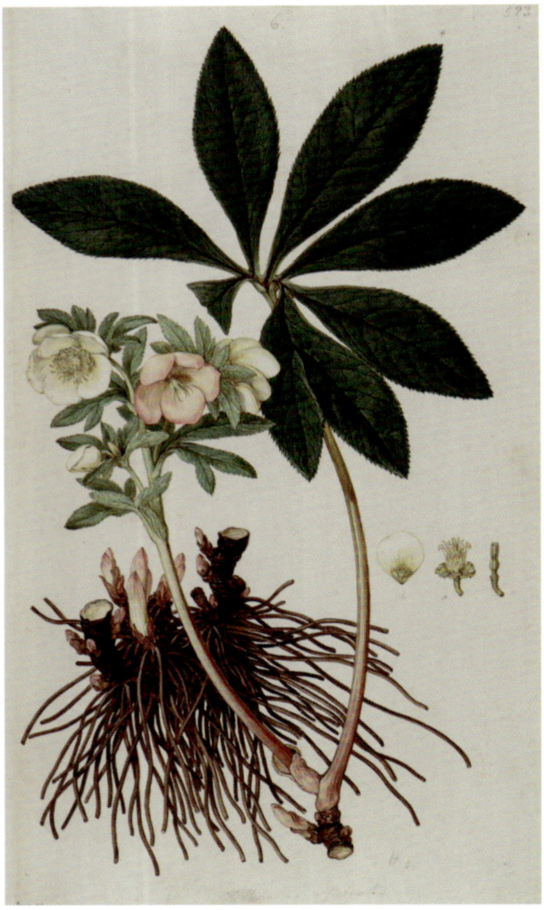

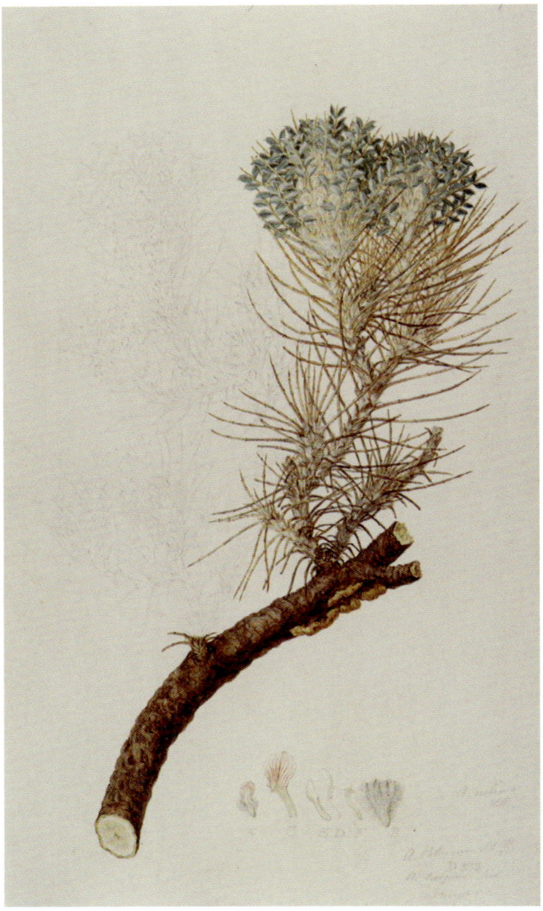

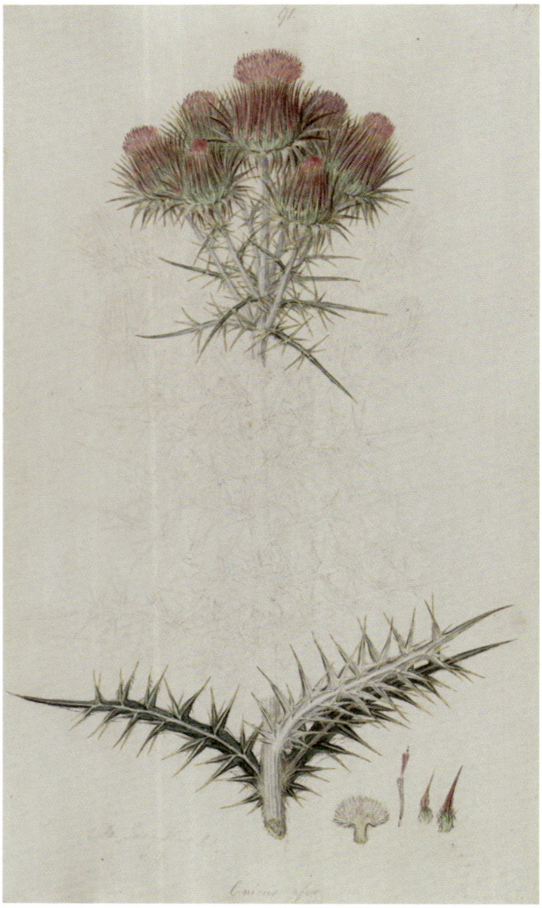

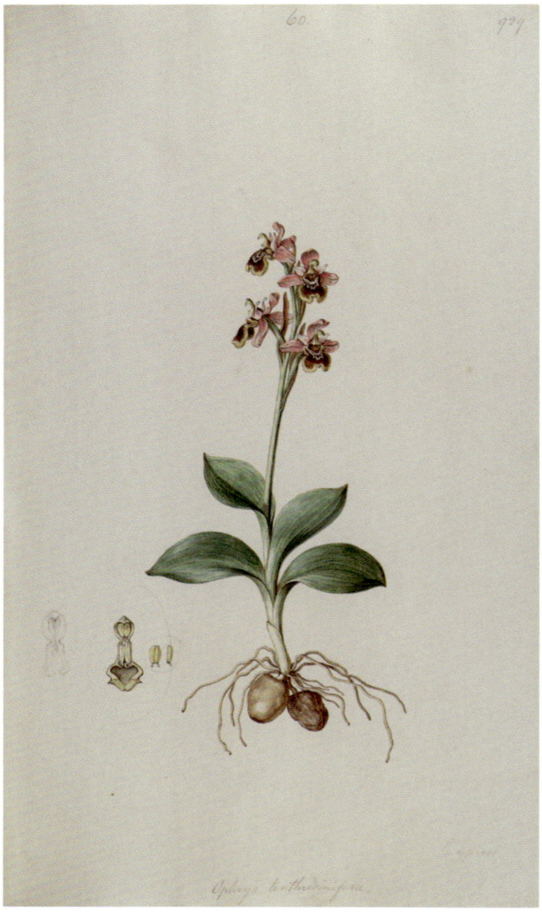

watercolours was produced in 53 months, or an average of 18 watercolours a month, a breath-taking speed and an outstanding achievement, even when considering that in some watercolours with repetitive plant parts, such as the flower heads in *Ptilostemon afer* (Fig. 6.8), only some were coloured whereas the contours of others were just sketched in pencil.

However, although the speed continues to impress, in particular when taking into account that Ferdinand worked in Oxford without an assistant and had to prepare his papers and grind his pigments himself, there are also other questions that puzzle the researcher. What formed the basis of Ferdinand's watercolours? Needless to say, he had his colour-coded pencil drawings to hand, all in small format and prepared in the field, as well as his indispensable colour chart. But how did he work when only a fragment of a branch had been recorded in the field and now he intended to paint a complete much-branched specimen? Did he rely on the preserved specimens in Sibthorp's herbarium? Did he possess a separate set of herbarium specimens that has not survived? Did he consider looking out for additional evidence in the 'Plantation sacred to Greece'? Did he possess additional large-format pencil drawings prepared in the field that have not been preserved? We simply do not know. However, when comparing Sibthorp's herbarium specimen of *Onosma fruticosa* (Fig. 6.12.) with the watercolour portraying the species (Fig. 6.13.), one cannot avoid the impression that Ferdinand at least in this case made use of the specimen. However, it is unclear how he managed to produce such perfect plant images, such as his *Cyclamen persicum* (Fig. 6.2.), on the basis of such incomplete, albeit colour-coded pencils, in particular since in this case no herbarium specimen exists. In addition, Ferdinand did not exclusively concentrate on plants found in the wild, which is what most conventional botanists would have done today, with their deep-seated dislike for everything cultivated. Instead, Ferdinand included several straightforward garden plants, among them bitter aloe, *Aloe vera*, and common iris, *Iris germanica*, both found in Cyprus (see Chapter 5), as well as a few domesticated crop plants such as chick pea, *Cicer arietum*,[7] and opium poppy, *Papaver somniferum*,[8] the source of another powerful painkiller and therefore of specific interest to Sibthorp.

Another aspect of Ferdinand's work merits attention: he never ever annotated a finished botanical (or zoological) illustration with a scientific or vernacular name. This job he clearly left to Sibthorp, who notably refrained from doing so. All determinations on the representations of plants are by Smith and added years after Sibthorp had died; for the scientific names and comments added to the images of the animals see below. This confirms a basic fact: Ferdinand, and as a matter of fact also Franz, were

FIG. 6.6. *(top left) Helleborus orientalis*. Watercolour by Ferdinand Bauer, signed, c.1790. – Oxford, Sherardian Library, MS Sherard 241, f. 6.

FIG. 6.7. *(top right) Astragalus creticus*. Unfinished watercolour by Ferdinand Bauer, c.1792. – Oxford, Sherardian Library, MS Sherard 242, f. 72

FIG. 6.8. *(bottom left) Ptilostemon afer*. Unfinished watercolour by Ferdinand Bauer, c.1792. – Oxford, Sherardian Library, MS Sherard 243, f. 91.

FIG. 6.9. *(bottom right) Ophrys tenthredinifera*. Watercolour by Ferdinand Bauer, c.1793. – Oxford, Sherardian Library, MS Sherard 245, f. 60.

FIG. 6.10. *Arum dioscoridis*. Watercolour by Ferdinand Bauer, c.1793. – Oxford, Sherardian Library, MS Sherard 245, f. 56.

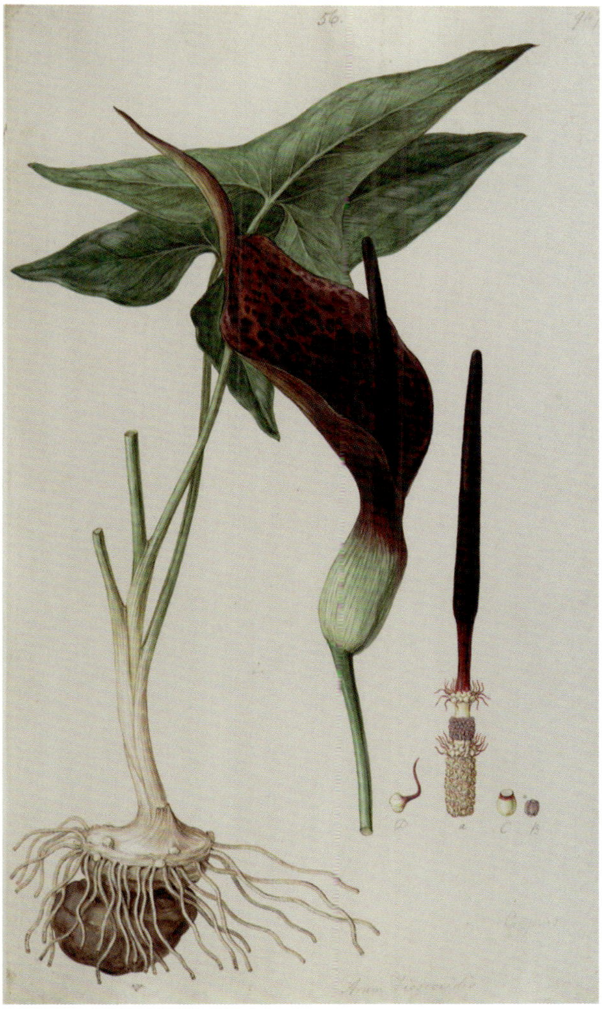

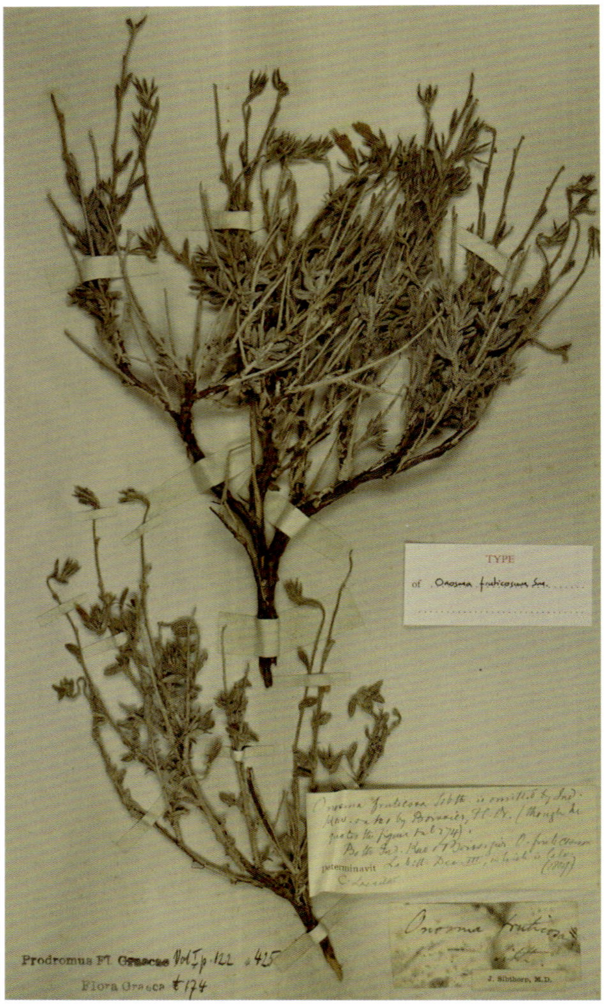

FIG. 6.11. ***Onosma fruticosa*. Herbarium specimen collected by J. Sibthorp in Cyprus, 1787. – Oxford, Department of Plant Sciences, Sibthorp Herbarium.**

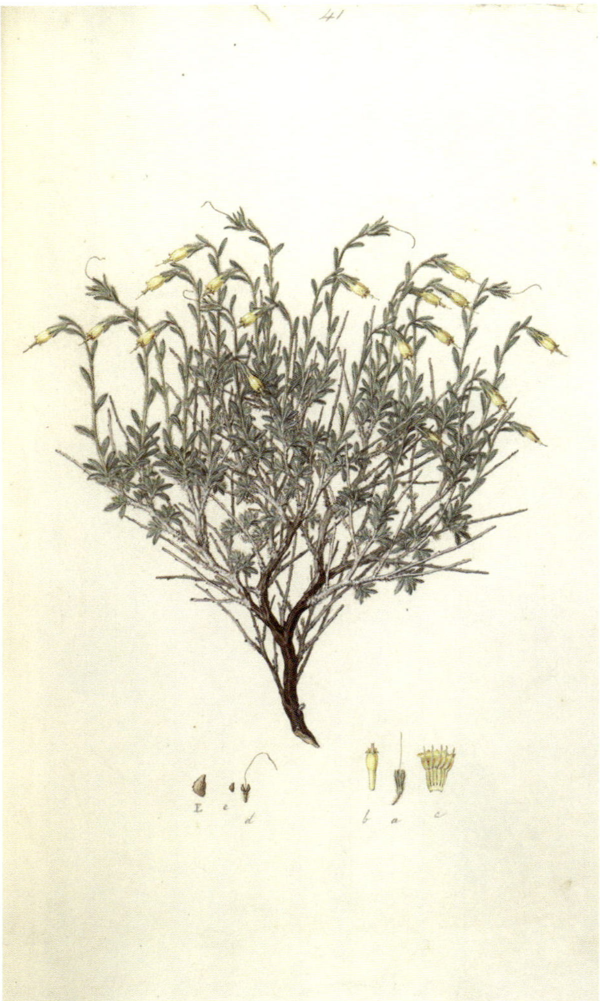

FIG. 6.12. ***Onosma fruticosa*. Watercolour by Ferdinand Bauer, signed, c.1788. – Oxford, Sherardian Library, MS Sherard 244: f. 41.**

illustrators, not taxonomists. The universal questions of cataloguing the diversity of life, in particular the naming of all its forms 'driven with the petty rivalries and transient egos of its practitioners, just like in any other science or area of academic pursuit', clearly lay outside their focus. Instead their field was the accurate painted record of nature, and more particular of the diversity of life.

FERDINAND'S ANIMAL ILLUSTRATIONS

Having finished the plant illustrations, Ferdinand continued with the representations of animal life in the Levant – exclusively vertebrates – a work completed in January 1794, when Sibthorp told Hawkins 'Bauer has just finished all his Drawings of Natural History'. Another tour de force had come to an end. Compared to the *Flora*, much less is known about Ferdinand's *Fauna*, totalling 293 folio sheets and comprising mammals, birds, reptiles, amphibians and fish. In contrast to the plants, all of the animal watercolours were completed by Ferdinand, because of the much smaller number of repetitive structures in animals. One of the reasons we know so little is that very few colour-coded pencil drawings survive, all in large format. Another reason is the total lack of any locality data on the pencil drawings of animals. As a consequence, the animal illustrations can only be correlated with a hypothetical locality when the travel diary explicitly und unambiguously mentions the occurrence or observation of a particular animal species at a specific site. A further reason for the general neglect of the animal illustrations is very easily explained – they remained unpublished and consequently were of relevance only to the zoologists at Oxford who happened to have heard of their existence. These were: famous ornithologist John Gould, author of *The Birds of Europe*, *The Birds of Australia*, *The Birds of Great Britain* and many more works, as well as the identifier of the birds now nicknamed 'Darwin's finches', which played an important role in the inception of Darwin's concept of evolution by natural selection; Frederick Holmes, a tutor at Corpus Christi College, Oxford; geologist Hugh Edwin Strickland, who had been educated at Oriel College, Oxford and was the author of *The Dodo and its Kindred*; ichthyologist and ornithologist William Yarrell, author of *A history of British fishes* and of *History of British birds*, an original member of the Zoological Society of London. The first three were interested in birds only and left behind annotations on Ferdinand's watercolours, marked 'G.', 'F. Holmes' and 'H.E.S.', respectively; Yarell added his determinations to many representations of mammals, reptiles, amphibians and fish without giving his name or initials.

Among the mammals we find an excellent watercolour of a dromedary, *Camelus dromedarius* (Fig. 6.13.), probably drawn when Sibthorp and Ferdinand joined the caravan from Scala Nuova to Istanbul, or later on the island of Cyprus. The Eurasian lynx, *Lynx lynx*, mentioned in Sibthorp's diary from one of the stops on the Anatolian coast, is documented in its summer coat, while the illustration of a water buffalo, *Bubalus bubalis*, seems to have been echoed in Ferdinand's view of a mosque near Salonica (Fig. 5.18.). He also depicted a hedgehog, *Erinaceus europaeus*, and four turtle species.

Ferdinand's birds merit special attention. It has rightly been said that had they been published he would have been admired for the previous two centuries as a famous ornithological illustrator, belonging to the same category as John James Audubon or, a century later, J. Gould (see below). No bird depicted could not be found today along some part of the route taken by Ferdinand, a few of them meriting special attention: the ring ouzel, *Merula torquata*, is rare along the route followed; the white-winged snowfinch, *Montifringilla nivalis*, and the alpine chough, *Fregilus graculus*, must have been observed at high altitudes, for example on the excursion to Mount Parnassus. The great surprise is the Cyprus warbler, *Sylvia melanothorax*, a passerine bird wintering in what is now Israel, Jordan and Egypt but breeding in Cyprus. The point of interest is that the Cyprus warbler was not recognized as a separate species until 1872, eighty-five years after having been first recorded by Ferdinand. Neither Strickland nor Gould recognized Ferdinand's watercolour as a representation

FIG. 6.13. ***Camelus dromedarius*. Watercolour by Ferdinand Bauer, signed, c.1792. – Oxford, Sherardian Library, MS Sherard 238: f. 9.**

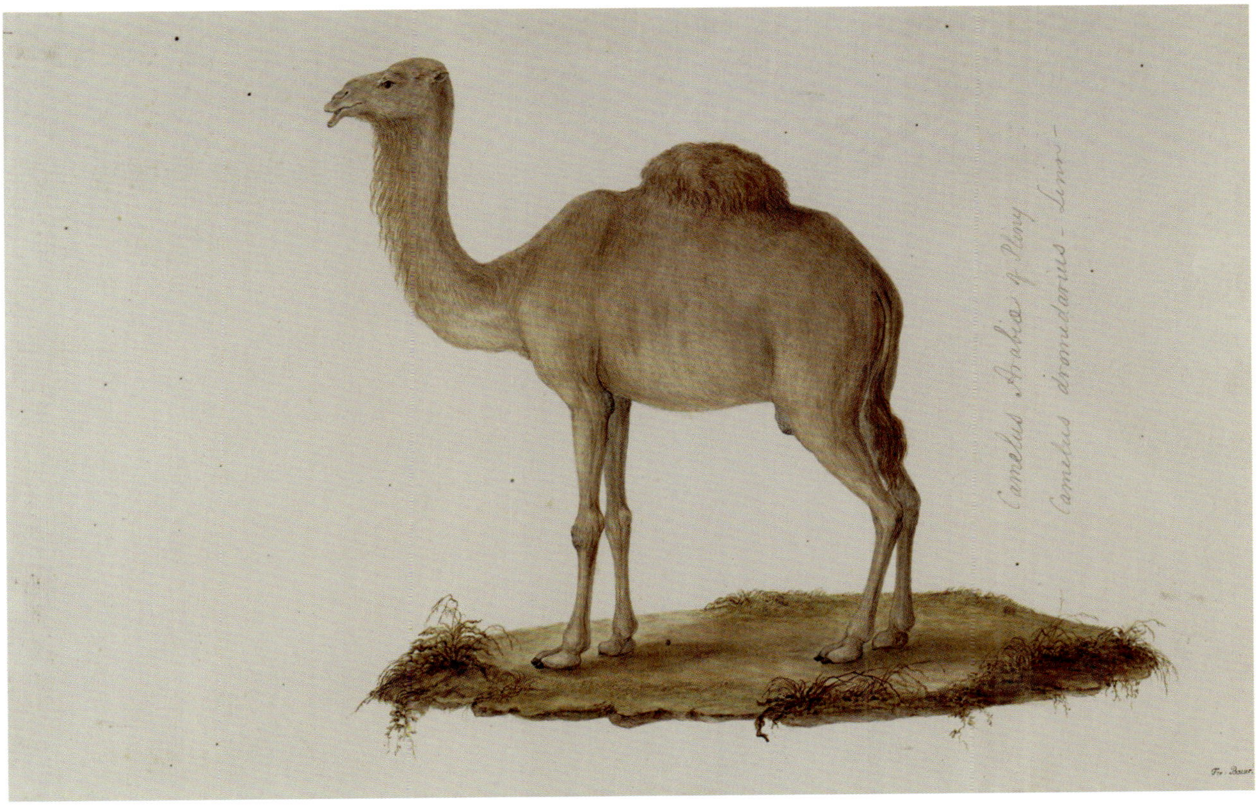

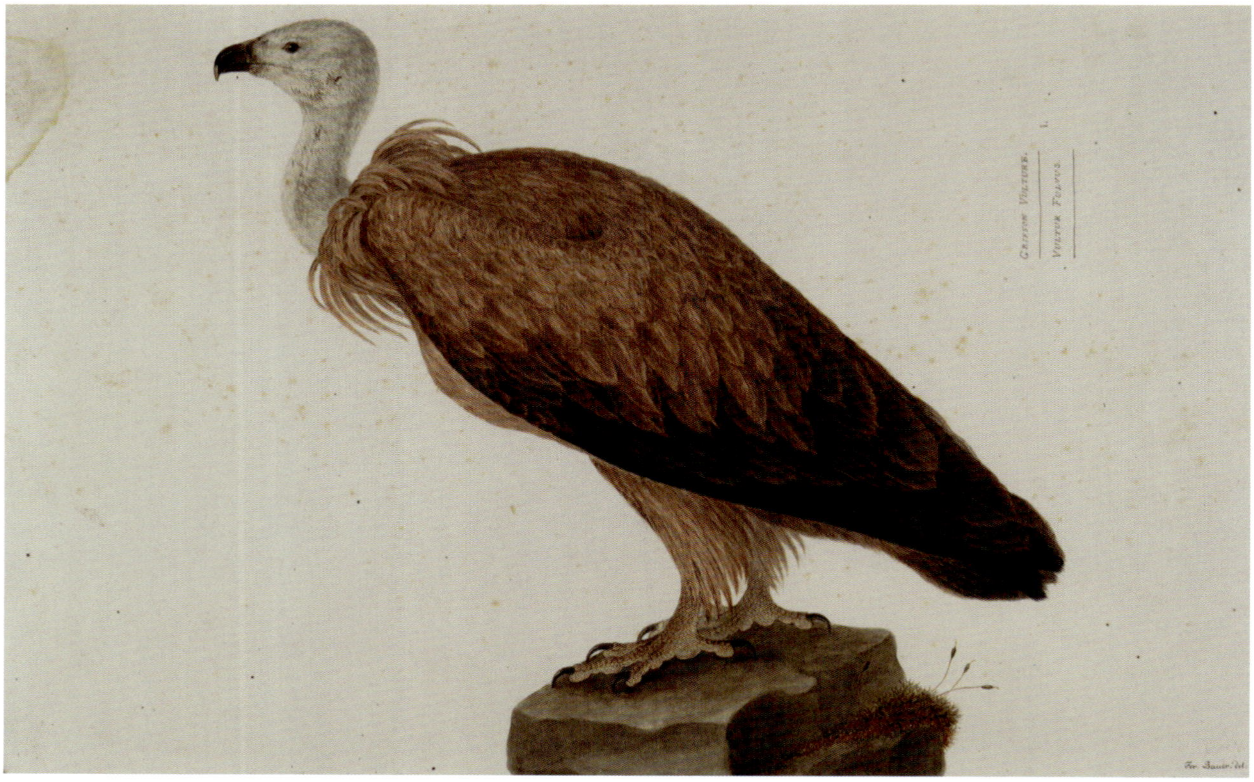

FIG. 6.14. ***Gyps fulvus*. Watercolour by Ferdinand Bauer, signed, c.1793. – Oxford, Sherardian Library, MS Sherard 240: f. 1.**

of a species new to science. The masked shrike, *Lanius nubicus*, a striking bird with black-and-white plumage and orange flanks, is a rare vagrant in summer, though possibly not in Cyprus. A striking male black-headed wagtail, *Motacilla flava feldegg*, another migrant, was probably a summer visitor to the area. By contrast to these rarities, many birds documented by Ferdinand are common and widespread. These include the Griffon vulture, *Gyps fulvus* (Fig. 6.14.), Sibthorp's 'Parnassian vultur', a scavenger, feeding mostly from carcasses of dead animals, which it finds by soaring over open areas, often moving in flocks. Among the distinctly smaller birds we find the richly coloured European bee-eater, *Merops apiaster* (Fig. 6.15.), a gregarious bird nesting colonially, and the white-headed duck, *Oxyura leucocephala* (Fig. 6.16.), known to dive and swim underwater and reluctant to fly, preferring to swim for cover. The former is mentioned in Sibthorp's diary in Cyprus, whereas the latter may have been observed and drawn during Ferdinand's extended stay on the Princes' Islands and in Istanbul. The European bee-eater may well have already been known to Ferdinand, though hardly to his master, from the vineyard terraces near Feldsberg, where the bird continues to occur.

When Ferdinand was preparing the finished watercolours of the birds, more than five years after the end of the journey, he ensured that the waterfowl were accompanied by clearly recognizable seaweeds, thus paying attention to their natural environment. For the same reason other birds were painted sitting or standing on ground covered with mosses, lichens, or, with a single, well-considered exception, phanerogams with inconspicuous flowers. In this, Ferdinand followed precisely the models he had almost certainly seen years ago in the Imperial Library in Vienna – the representations of birds on velin in the *Receuil d'Oyseaux* (see Chapter 2).

The relevance of Ferdinand's animal illustrations is best explained by a simple fact. Without a single exception all zoological specimens collected by Sibthorp in the Ottoman Empire have been lost.[9] In addition, the names used in Sibthorp's diary – in English, Greek (in Greek script) and anglicized Greek – are often ambiguous; for example, his 'bald-buzzard' refers to both the marsh harrier, *Circus aeruginosus*, and the osprey, *Pandion haliaetus*. To make things even more difficult, a considerable number of European bird species did not yet have accepted scientific names, and Sibthorp tried in his diary to force Linnaean names onto species Linnaeus had not known. Even some of the identifications in Sibthorp's diary that should be unequivocal may not be. He refers, for example, to 'Sturnus cinclus' [*Cinclus*

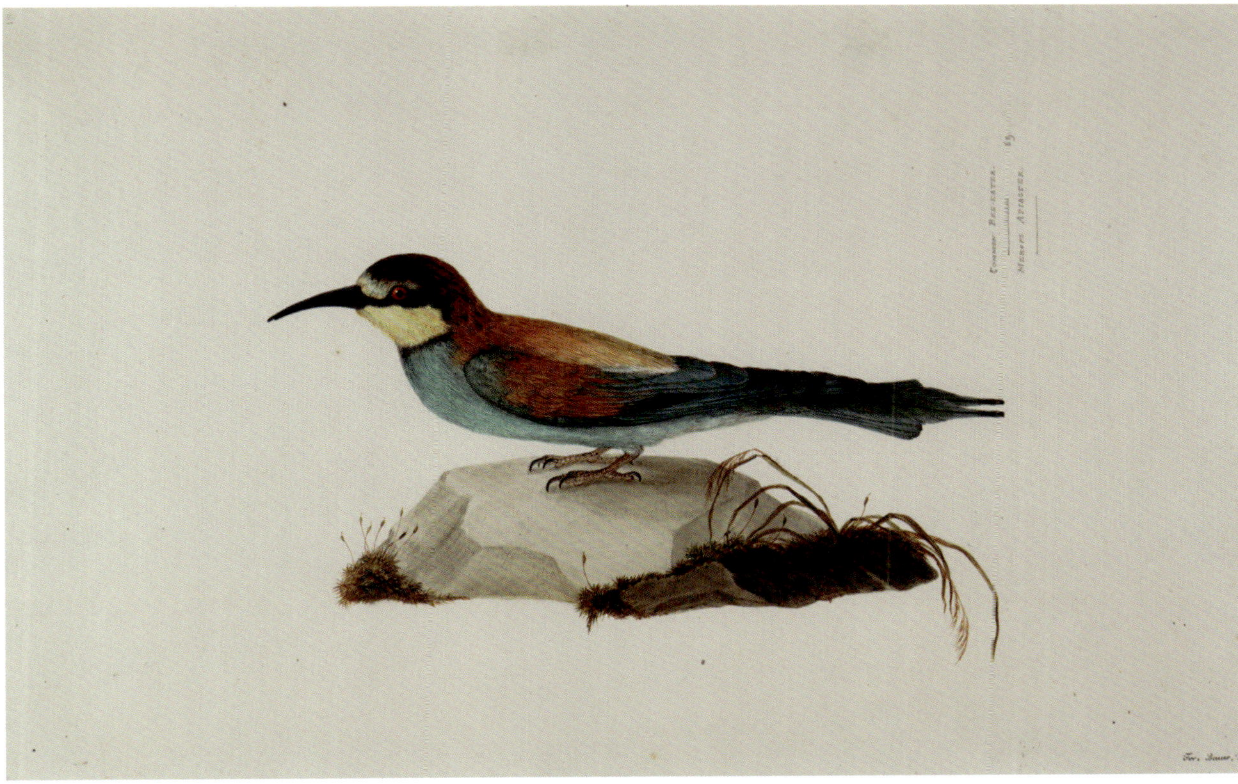

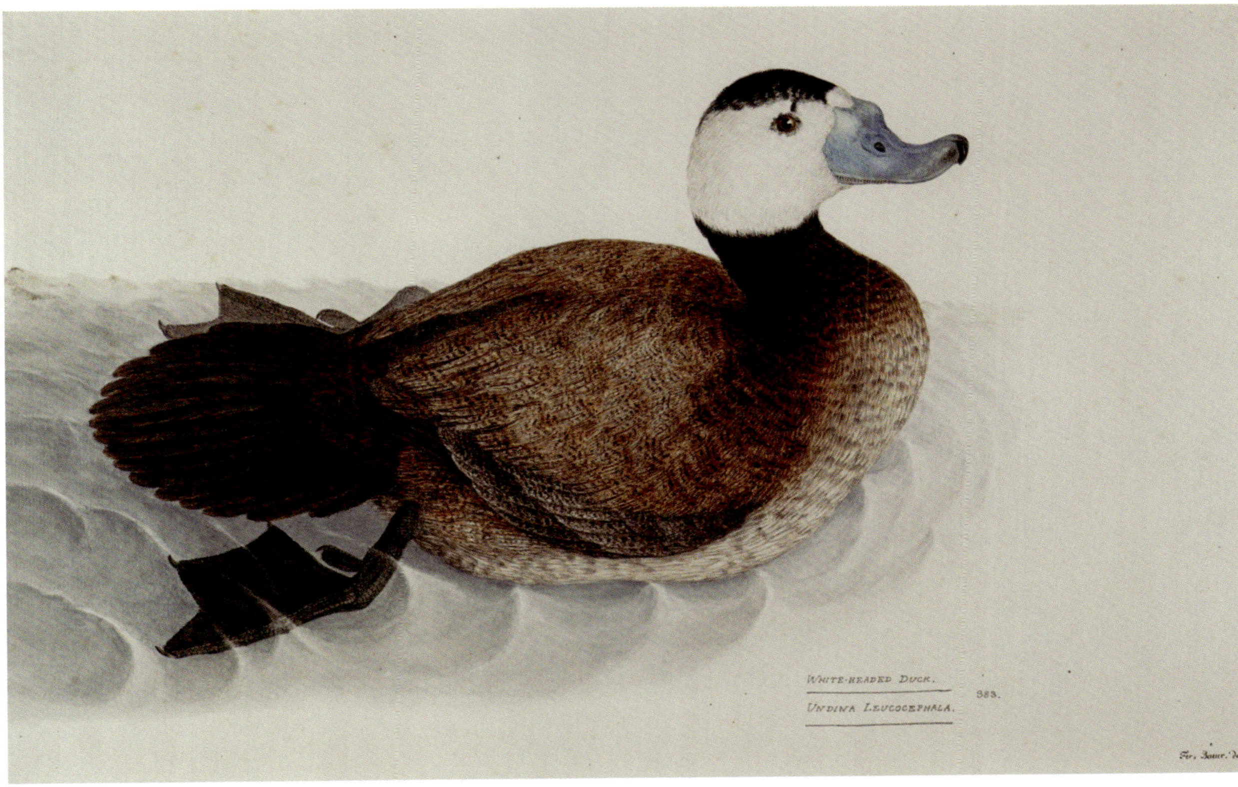

FIG. 6.15. *(top) Merops apiaster*. Watercolour by Ferdinand Bauer, signed, c.1793. – Oxford, Sherardian Library, MS Sherard 240: f. 18.

FIG. 6.16. *(bottom) Oxyura leucocephala*. Watercolour by Ferdinand Bauer, signed, c.1793. – Oxford, Sherardian Library, MS Sherard 240: f. 101.

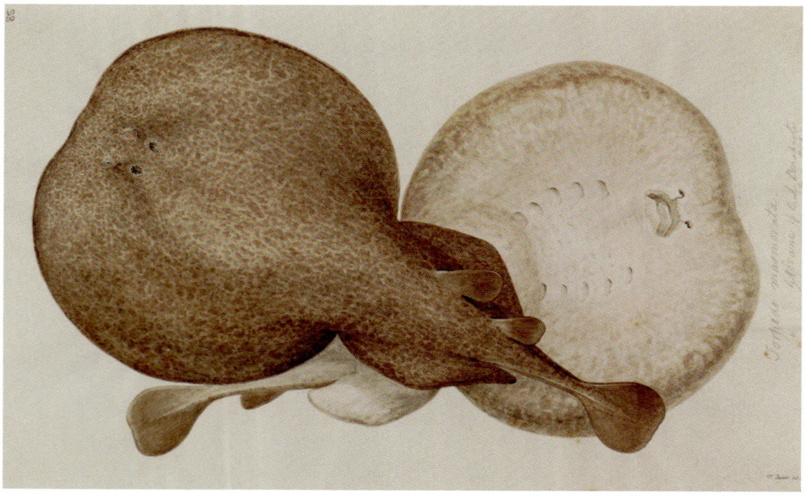

FIG. 6.17. *Torpedo marmorata.* Watercolour by Ferdinand Bauer, signed, c.1793. – Oxford, Sherardian Library, MS Sherard 239: f. 85.

cinclus] several times, the epithet never having been applied to any bird but the dipper, a species that stays close to running water, and one which must have been familiar to Sibthorp. However, in his diary he says that he shot a male and female of this species 'in a grove of Planes in the outskirts of Abydos' (see Chapter 5), certainly not the sort of place in which one would expect to find a dipper.

As the loss of Sibthorp's zoological material was complete, neither specimens of mammals, birds or reptiles (for example of the common European green lizard, *Lacerta viridis*), nor of the very many fish (for example of the marbled electric ray, *Torpedo marmorata*, a nocturnal and solitary species often found lying on the sea floor; Fig. 6.17.) exist any longer. In short, Ferdinand's painted record is the only reliable source of information on the animal life observed on the journey to the Levant that has survived.

FERDINAND'S ACCURACY

It has been said that a botanical or zoological illustration is a detailed, scientifically accurate representation of a plant or animal that gives particular attention to the features that distinguish it from other plants and animals. It has also been stated that a botanical or zoological illustration may be idealized as an amalgamation of elements from numerous specimens, or it may represent a specific individual, as a rule the statistical 'standard' within a range of variants (see Chapter 2). However, all this is only part of the story. The botanical or zoological illustration must also depict all the features the specialist needs to see and to study in order to be able to identify a plant or animal, which implies that either the commissioner of the illustration or the illustrator himself must be perfectly aware of what these characters are. A key criterion for quality is therefore the percentage of illustrations that can be determined unequivocally by the specialist. In this Ferdinand ranks extremely highly: the identity of 965 out of his 966 botanical illustrations prepared for Sibthorp has very recently been stated to be accurate without any doubt. Similar figures exist for mammals, birds and amphibians, whereas several images of reptiles and, in particular, fish remained undetermined by Yarell. This fact is somewhat difficult to interpret, since the pertinent illustrations are not known to have been recently studied by specialists. In any case several undetermined fish represent juvenile specimens, which are always difficult to identify.

The life-like feature of Ferdinand's zoological work is outstanding, especially in what birders call the 'jizz' – the overall impression or appearance of a bird garnered from shape, posture, flying style or other habitual movements, size and coloration. In Ferdinand's time, the concept of 'jizz' did not even exist, which makes his animals, in particular his mammals and birds, so remarkable. In three cases, the white-headed duck (Fig. 6.16.), the cinereous shearwater, *Puffinus cinereus,*[10] and the Manx shearwater, *P. puffinus,*[11] he actually portrayed the birds swimming. This possibly prevented Ferdinand from committing the standard artist's error of the time in drawing the shearwaters standing upright, as on land they sit on their bellies and shuffle along.

However, occasionally even the most diligent illustrator like Ferdinand made incorrect observations: the bill of the Egyptian vulture, *Neophron percnopterus*, is somewhat too short and thick;[12] the beak in the barred warbler, *Sylvia nisoria*, is also too short;[13] and the black-legged kittiwake, *Rissa tridactyla*, does not have a hind claw, whereas Ferdinand's illustration shows one.[14] The tail of the European pond turtle, *Emys lutaria*, is deficient in length[15] and the dorsal fin of the Atlantic bonito, *Sarda sarda*, is missing.[16] In a single case incorrect proportions have been recorded: the gull-billed tern, *Geiochelidon nilotica*, stands on legs that are rather too long.[17]

Dissections are totally missing in Ferdinand's animal illustrations, but he carefully takes the ventral side of many fish into account, for example in the marbled electric ray (Fig. 6.17.); enlargements are rare, for example the foot and beak of the Griffon vulture. An accurate depiction of the comb-like structures on the claws of the grey heron, *Ardea cinerea*, used in combing his plumage is given,[18] and Ferdinand also correctly depicted the lesser kestrel, *Falco naumanni*, as having pale claws,[19] a character that distinguishes it from the common kestrel, *F. tinnunculus*, which has blackish claws.[20] Needless to say the illustrator has to be very close to the bird in order to be able to observe the differences or study a specimen that had been shot dead. Significantly, Ferdinand showed both sexes for both species of kestrel. In this work he evidently followed the traces of his father Lucas, then long dead, who must have accompanied His Grace, the Reigning Prince of Liechtenstein, on his hunting excursions on his Feldsberg estate (see Chapter 2).

In contrast to animals, which can fly, run or swim away, plants do not move, and consequently there are no special features to be summarized as 'jizz', although each and every plant species exhibits its particular 'architecture', preserved also in herbarium material. To this Ferdinand paid attention, for example in his representations of grasses, praised by his master. Enlargements, in botany conventionally called analyses or details, are infinitely more frequently shown and much more important for identification in vascular plants than in vertebrates, and very often several are given on a single botanical illustration. The Bauer brothers had started this approach in Feldsberg and later, under the guidance of Jacquin in Vienna, they seem to have developed it further. When starting out on his journey to the Levant, Ferdinand was presumably already aware of what kind of floral and fruit parts needed to be recorded in particular plant families in the field for an analysis (see Chapter 5). Therefore the hypothesis is put forward that only occasionally a hint was received by Sibthorp. After all, Ferdinand was an experienced illustrator with excellent training, fully qualified to work and observe independently from his master. In order to achieve utmost accuracy Ferdinand indicated in his analyses in watercolour which plant parts he had recorded in natural size and which he had depicted enlarged – by adding in pencil letters in lower case and upper case respectively. For example in the florists' cyclamen, *Cyclamen persicum* (Fig. 6.2.), he documented a corolla cut open with the five stamens attached to it in natural size (marked b), the sepals with the gynoecium in natural size (marked a), and two separate stamens weakly enlarged (marked C). Cross sections and longitudinal sections of fruits are only very rarely found, for example a cross section of the fruit of *Peucedanum obtusifolium*[21] and a longitudinal section of the gynoecium of the widely cultivated pomegranate, *Punica granatum*.[22]

Ferdinand seems to have been aware of the importance of what we now call intraspecific variation, in particular in annual plants, which show markedly different size, branching and number of flowers depending on the time of the year, the habitat, moist or dry, and the altitude. In several cases he portrayed more than one specimen: for example, two specimens of rocket, *Eruca vesicaria*, an annual crucifer commonly cultivated as salad, differing in the number of basal leaves,[23] or four specimens of *Scrophularia lucida*, a biennial species of figwort, differing in the extent of branching.[24] Similarly Ferdinand took intraspecific variation depending on age into account by documenting, for example, a juvenile, flowering and fruiting specimen of *Matthiola fruticulosa*.[25] He did the same when he recorded two specimens of *Urospermum picroides*, an annual species of the tribe Cichorieae, in different stages of development, maybe collected at an interval of a few weeks[26] – even though it seems doubtful that he fully understood the facts. Ferdinand's observational power enables the specialist to discern taxonomic entities within a species: a good example of this is the two subspecies *Nigella arvensis* subsp. *aristata*, a Greek endemic, and *N. arvensis* subsp. *glauca*, a native of Greece and Turkey, documented by Ferdinand. The illustration of the former subspecies includes an analysis of the flower parts, comprising no less than ten elements, labelled a, b, c, C, D, e, f, g, G, H: almost a record in Ferdinand's work for Sibthorp. The apotheosis of this kind of analysis was to be reached several years later, when he prepared the watercolours documenting the plants observed and collected in Australia in Banks's house in London (see Chapter 11). Another example is Ferdinand's recording of eligulate and ligulate forms of *Anthemis rigida*, an annual composite widespread in the Mediterranean region. Again it seems questionable whether he or his master understood the relevance of this aspect of intraspecific variation.

In contrast to his zoological material, Ferdinand's botanical illustrations were published as coloured copper engravings in the *Flora Graeca* between the years 1806 and 1840, a topic not dealt with here since neither he nor his elder brother Franz was involved in the later stages of this project (see Chapter 9). Since many of the resulting prints formed part of protologues (that is, pertained to the material used by the original author when validly publishing a plant name new to science), interest in them was much more intense than in the zoological material. Very recently this resulted in an in-depth analysis and re-issue of all 966 plant illustrations in reduced size.[27] Very rightly Ferdinand is called in the introduction of this work 'the true hero of the *Flora Graeca*' and his work is described as the 'Herculean task of preparing the watercolours'. Even when all his botanical work is rigorously submitted to the most severe analysis, little remains to be criticized.

As with the animals there are very occasionally minor observational mistakes; for example, the patent hairs on the styles of *Fimbristylis bisumbellata*, visible at higher magnifications, were overlooked by Ferdinand.[28] The lateral horns on the indurate stigmata crowning the siliquae in *Matthiola fruticulosa* are depicted as unusually well developed.[29] Of course Ferdinand also recorded a few atypical, depauperate or fragmentary specimens. One of these is a specimen of tree flax, *Linum aboreum*, a South Aegean endemic, with unusually elongated branches densely covered with leaves, possibly from a cultivated specimen;[30] another is the poorly developed, few-flowered specimen of the yellow star-of-Jerusalem, *Gagea* spec., of uncertain identity;[31] a third one is an incomplete, dubious specimen of a mullein, *Verbascum* spec.[32] The latter two examples are of unknown geographical origin. In a sense this situation is too well known to every plant taxonomist – herbaria also contain atypical, depauperate and fragmentary specimens. Among the 966 plant illustrations the only truly enigmatic watercolour illustrates an annual species of *Senecio*, either *S. lividus* or a poorly developed specimen of *S. vernalis*. Never could Ferdinand be convicted of producing an illustration of a 'specimen' composed of two species, although this has recently, but wrongly, been claimed for *Valantia aprica*.[33] And there are all those innumerable details that make the true connoisseur happy: the short leaves of *Bruckenthalia spiculifolia* ending in a long stout hair,[34] the corolla of *Scutellaria orientalis* split open to see the inside[35] and the villous hispid stems of *Papaver pilosum* with hundreds of painted hairs of equal length correctly represented,[36] all three from Mount Olympus (Ulu Dağ). Added to this we find what could be called plant architecture documented very true to nature: the trailing, sometimes twining stems of *Convolvulus althaeoides*,[37] the very dense, rigid, intensely branched cushion of Greek spiny spurge, *Euphorbia acanthothamnos*, an important constituent of phrygana, and the divaricately branched, subglabrous shrublet *Alhagi graecorum*.[38] Ferdinand is also very good in presenting plant surfaces: for example the somewhat shiny unripe fruit of the mandrake (Fig. 6.2.), the dark red fruit surface with conical papillae of the strawberry tree, *Arbutus unedo*,[39] a common constituent of macchie, and the strongly rugose achene of *Reichardia tingitana*.[40] Convincingly, Ferdinand records the endless gradations of prickliness – the rhachis in *Astragalus creticus* protracted in a yellow spine (Fig. 6.7.), the pinnatifid leaves in *Ptilostemon afer* with thickened veins produced into sharp marginal spines (Fig. 6.8.), the ovate phyllaries in cardoon, *Cynara cardunculus*, abruptly contracted into a spinose apex,[41] or the sharp prickles on the stems of the rough bindweed, *Smilax aspera*, an evergreen climber common in the Mediterranean area.[42] The observer can almost feel the different forms of pain when touching the images of the different kinds of spines, in particular as they appear three-dimensional on paper.

For good reasons Ferdinand was praised for his accurate hues perfectly true to nature, clearly one of his great specialities. When leafing through the carnations (for example, Fig. 18.1.) and legumes (for example, Fig. 6.7.) his sublime sense for all the hues of the rainbow is confirmed over and over again. However, there is the curious fact that the deep blue corollas of *Anchusa cespitosa*,[43] because of this character one of the most striking endemic plants of Crete, have not been adequately documented, their hue being far too light and rather grey-bluish. Several explanations present themselves – perhaps Ferdinand did not have adequate pigments available, or the paint has faded away in the last two centuries. The latter interpretation is possibly an inappropriate hypothesis, since the watercolour was never exposed to light for any extended time. Another explanation is that this is an example of one of Ferdinand's rare observational errors. Several months after having preparing this watercolour, Ferdinand painted dark blue petals in *Lupinus pilosus*[44] – did he manage to improve his array of pigments? In another case Ferdinand's recent critics were wrong. They had stated that the corolla colour in one of the watercolours documenting *Echium angustifolium* should be blue instead of red,[45] but in fact they were wrong in the identification of plant depicted in the watercolour, having made the assumption that it represents *E. arenarium*, which indeed is a species with blue corollas.

Ferdinand is also to be praised for the exceptional care he took in recording underground plant parts in a precise manner, with convincing examples from the Cichorieae. In the annual *Urospermum picroides* he paints a thin taproot,[46] in the perennial *Scorzonera cretica* a massive, thick, ellipsoid rootstock,[47] in *Leontodon tuberosus* a cluster of root tubers,[48] in *Sonchus bulbosus* a thin rhizome bearing small, irregularly shaped tubers,[49] and in *Crepis sibthorpiana*, another Cretan endemic, a woody base reaching deep into the limestone rocks and slopes, on which this species grows.[50] It may come as a surprise that Ferdinand did not record the large, irregularly shaped and roughly fissured root tuber of *Corydalis rutifolia*,[51] an endemic of Cyprus, which then had not yet received a scientific name. Two explanations present themselves: the tuber is deeply buried in the soil,[52] which consequently could have escaped the attention of Sibthorp and Ferdinand. Secondly the plant grows on the stony mountain summits of the island flowering in spring. Indeed, their tour with Hawkins to the peak of the Troodos Mountains was undertaken at the end of April 1787 when the substrate may still have been frozen and the tuber difficult to dig out. Most diligently, Ferdinand also recorded the annual legume *Onobrychis caput-galli* with the pod from which the plant germinated, as in other annual species of this genus, persisting at the base of the stem.[53]

Ferdinand's animal and plant illustrations differ in an important point: whereas only a few animals were new to science when Ferdinand completed this part of his commission in 1793, about a tenth of the plants had not yet received a Linnaean name, and several more had never been properly recorded in watercolour. These included not only strict endemics, such as *Anchusa cespitosa* from Crete, *Gladiolus triphyllus* from Cyprus and *Helichrysum sibthorpii* from Mount Athos, but several widespread species, among them *Cyclamen repandum* and three species of milfoil (*Achillea* sp.), the latter four today of major horticultural importance and today widely cultivated as ornamental plants.

FERDINAND'S FRONTISPIECES[54]

When in Feldsberg in late 1785 or early 1786 Sibthorp must have seen the first six volumes of the *Codex Liechtenstein* already completed (see Chapter 2). No doubt he marvelled at the six elaborated title pages, which were effectively flower pieces of exceptional elegance and refinement (for example, Fig. 2.16.) and the landscapes at the beginning of volumes five and six. Eight years later he seems to have remembered these materials and commissioned Ferdinand to prepare watercolours that would serve as patterns for frontispieces (full-page copper engravings with reference to the contents of the work facing the title pages and repeating the title) for the projected *Flora Graeca*. Ferdinand combined elements from his previous work for Boccius, resulting in seven watercolours. This work must have started in early 1794, since Sibthorp told Hawkins, then overwintering on the island of Zakynthos, on 31 January 1794 that 'Bauer . . . is now forming the Vignettes of Grecian landscapes as Frontispieces'. In fact he combined landscapes from the Ottoman Empire with flower garlands, as he had done for the first time in the view from Skalky towards the Pollau Mountains combined with fruits of three solanaceous genera in the fifth volume of the *Codex Liechtenstein* (see Chapter 2).

As is evident from the skeleton of the manuscript of the *Prodromus Florae Graecae* (see Chapter 9), the main work, the projected *Flora Graeca*, was to follow the Linnaean system, and it seems also to have been clear that the eventual publication was planned in volumes, each containing one hundred copper engravings. The botanical component of the patterns for the frontispieces commissioned by Sibthorp was to reflect the contents of the respective volume. The first of them was to contain the beginning of the Linnaean system, the Diandria and part of the Triandria, and consequently the flower garland in the frontispiece pertaining to this volume should depict flowering specimens dealt with in this very volume. It is not known who conceived the idea of this iconographic programme – Sibthorp, Ferdinand or both – but significantly it was followed with minor variations in all seven patterns, with the seventh dealing with the second part of the Didynamia to the first part of the Diadelphia. There is a single exception: the genus *Aristolochia* is included among the Hexandria, not among the Gynandria of Linnaeus, since the two Mediterranean species illustrated by Ferdinand possess six stamens as a rule. As in several volumes of the *Codex Liechtenstein*, the flower garland was painted as arranged on an oval structure. By contrast, the topographical component of the frontispieces (the landscapes) does not reflect the contents of the respective volume. The botanical component of the pattern for the frontispiece of the intended sixth volume (Fig. 6.18.), for example, contains several widespread plants, among them

FIG. 6.18. Frontispiece with a view of Athens. Watercolour by Ferdinand Bauer, signed, c.1794. – Oxford, Sherardian Library, MS Sherard 241: i.

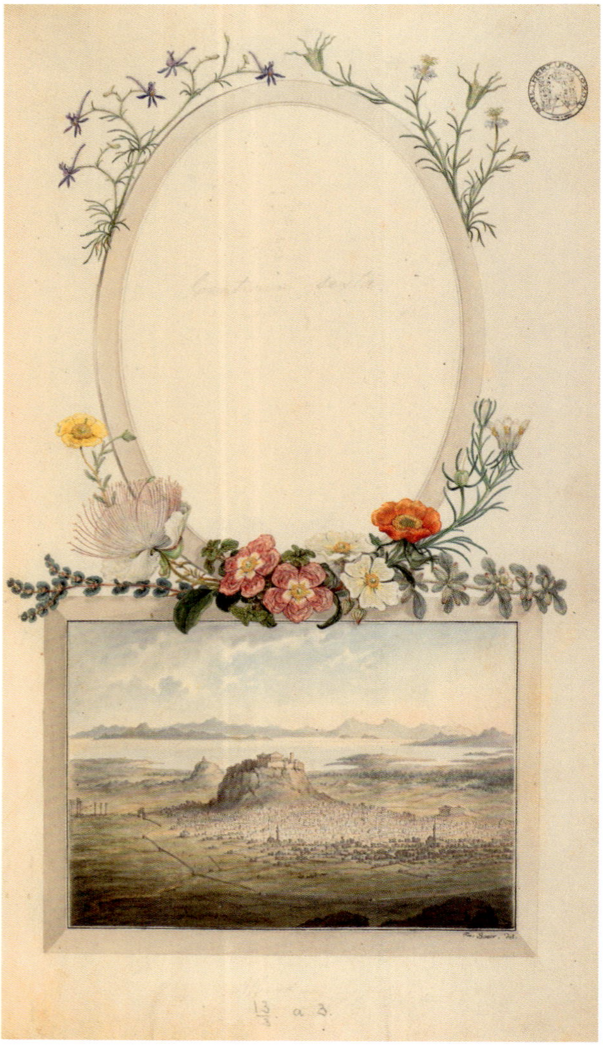

two species of rock rose (*Cistus monspeliensis* and *C. parviflorus*) as well as the caper bush, *Capparis spinosa*, and also avens, *Geum coccineum*, a native of the high mountains of the Balkans and western Anatolia, never occurring at sea level. The topographical component of this pattern, however, shows the town of Athens, where the latter species was never found.

Of the final choice, three views show high mountains – Mount Parnassus, Mount Athos and Mount Olympus (Ulu Dağ) – which have all been particularly profitable botanizing territory for species new to science, and this is likely to have been the reason why they were selected to act as frontispieces for the first three volumes of the projected *Flora Graeca*. The remaining views integrated into the patterns document the city centre of Istanbul from the north, the northern mouth of the Bosphorus seen from the south, the town of Athens (Fig. 6.20.) surrounded by the newly erected walls (see Chapter 5), and the town of Corinth. All seven 'vedute' are in colour, painted in small format, surrounded by a greyish frame and signed; they seem to be the last coloured landscapes Ferdinand ever produced.

Why Ferdinand prepared patterns for only seven frontispieces and not ten remains an enigma, in particular since he had prepared more than two hundred watercolours documenting species belonging to the final Linnaean classes, Syngenesia, Gynandria and Cryptogamia. Does this hint at an abrupt end to his engagement by Sibthorp? Did Ferdinand possess further pencil drawings of plants of the final Linnaean orders to be used as the basis for more watercolours and therefore he refrained from preparing patterns for the frontispieces accompanying the subsequent volumes? We simply do not know. A single point is clear: the patterns for the frontispieces, among the finest works by Ferdinand and after their publication, admittedly, the best-known frontispieces in botanical literature, were clearly Sibthorp's last commission.

FERDINAND'S LEVANTINE COSTUMES, DEVONSHIRE VIEWS AND FLOWER PIECES

Following his great role model Tournefort, who had asked Aubriet to record costumes on his journey in the Ottoman Empire later published in his master's travel account *Voyage du Levant*, Sibthorp had commissioned

Ferdinand to do the same. Little is known about this work undertaken in the field, apart from the fact that in a manuscript list 'Contents and Index of Mr. [Franz] Bauer's Portfolios' kept in the Niedersächsische Staats- und Universitätsbibliothek in Göttingen, '26 plates Pencil sketches of . . . Costumes in Turkey taken on the spot in 1784–1785 [recte 1786–1787] in and near Constantinople by Ferdinand Bauer' are listed. These drawings belonged to the material that George Williams, John Sibthorp's successor in the Sherardian chair, had intended to buy for the Bodleian Library, but in the end failed to do so. After Franz's death these sketches were auctioned in London, and their current whereabouts are unknown. Nothing indicates that Ferdinand produced coloured repeats based on these pencil drawings, although this cannot be excluded.

During his engagement by Sibthorp, Ferdinand seems to have visited Instow, his master's estate near Bideford in Devon. There he took a number of views, which he later worked into pen and ink drawings with grey washes on paper, documenting the farm belonging to the manor with its buildings, the local church and the countryside near the sea. This material is kept today in the archives of Tapeley Estate at Instow[55] and is very similar to Ferdinand's 'Mediterranean Scenes' kept in the Sherardian Library (see Chapters 5 and 9). These Devonshire views document the rural life so well known to Ferdinand from his youth in Feldsberg – with its horses, cows, geese, an occasional dog and cat, cottages, barns, a dovecot and, no doubt new to him, a tidemill. Two further views by Ferdinand have recently been auctioned – the first inscribed 'From the Ropery, Instow, towards Bideford' on 12 October 2007 by Mallams, Oxford, the second described as 'Cattle at pasture in a wooded coastal landscape' on 28 March 2002 by Christie's, London. A third 'veduta' entitled 'Marino [sic] – Instow belonging to Col. Sibthorpe' must have been painted after Sibthorp's death, when the Instow estate had already passed to his elder half-brother Humphrey; it is kept in the Print Room of the British Museum.[56]

A pair of flower pieces signed 'Ferd Bauer del.' belongs to a totally different world – that of the still lifes. Painted on paper the two watercolours show plants native in England arranged in an artistic way. Auctioned on 22 March 1985 at Christie's, London[57] the two works are described as 'A Thistle with Honeysuckle' and 'A Thistle with Forget-me-not [recte *Veronica* spec.] and other flowers'. Among the later we find the common Eggs and Bacon, *Lotus corniculatus*. Measuring 25.4 × 20.3 cm, the pair are rather small in size and associated with one of the country houses of the Sibthorps, the auction catalogue noting the provenance as 'by descent to Miss Sibthorp, Canwick Hall, Lincoln'. The two flower pieces are unique in documenting common plants which Ferdinand, after his move from Feldsberg to Vienna, otherwise avoided as subjects. The whereabouts of the two watercolours are unknown.

ARGUMENTS OVER MONEY

The personal relationship between Ferdinand and Sibthorp remained strained, possibly since their travels. According to a letter sent by Hawkins to Sibthorp, probably in July 1791, Ferdinand felt that he was being 'treated with little respect by his master', 'pressed by work' and 'deceived to the terms of the engagement having no idea that it could continue so long otherwise he [Ferdinand] should have stipulated for much more his present income not being more than he had earned in Vienna'. In addition, Hawkins informed Sibthorp about Ferdinand's concern relative to his vision – one of his eyes is reported to have deteriorated. Hawkins also told his friend in Oxford that Banks had expressed a desire to hire Ferdinand after his engagement with Sibthorp had come to an end – with the prospect of sending Ferdinand to the West Indies. Clearly Hawkins was on Ferdinand's side and supported his expectation of a higher salary by telling Sibthorp in this letter 'First I presume that I have the fullest conviction of the smallness of his present emolument when compared to his first rate merit as an Artist & the necessity of offering terms greatly superior'.

This resulted in a harsh reply 'When he [Ferdinand] set out with me from Vienna his Terms were eighty Sequins with all the Travelling Expenses paid. – he did not express any Discontent at this Price, till his brother [Franz] came to England. after an Interview with his Brother he became discontented this Discontent was however removed on my Promise of an additionall twenty Sequins [sic] p. Ann. Since this Time he has gone tranquilly with his work – I am highly satisfied with his Drawings & sensible of their Beauties. but like a prudent Man with a beautiful Mistress I must not suffer myself to be intoxicated with the Beauty. & run through lengths which Prudence cannot approve. More than I pay Him at present neither can I afford nor am I willing to pay'.

The positions of all three were perfectly clear: Ferdinand knew that Franz received an annuity of £300 for life from Banks, probably without being pressurized, whereas he was paid £100 per year for the duration of the project. Sibthorp was not prepared to pay more than a third of his stipend from the Radcliffe fellowship to Ferdinand. Hawkins felt the income too low for an illustrator of such extraordinary talents. The apparent conflict seems to have remained unresolved, and no indication exists that Ferdinand succeeded in raising his income any further.

About a year later Hawkins and Ferdinand discussed another project – engraving, printing, colouring and publishing the birds at the latter's risk. Sibthorp was informed about this plan by Hawkins, who added 'I know it was not your intention to have the sketches of the birds made out as they are well known species. I should therefore think that you could have no objection to accorde to Bauers proposal especially as he is sanguine enough to suppose that he can get something by it, and upon which I could give him no opinion . . . You will do right to encourage him by giving him up the use of the materials and even parting with the penciled outlines under the express'd conditions of publishing in a proper scientific way'. This proposal was immediately rejected by Sibthorp, his answer again being harsh, but interesting because of his description of Ferdinand's working method: 'Now I confess I lay more Stress upon the Animals than the Plants because Plants may be revived by their Seeds but from a dead Fish of which Spiritus Vini has extracted all the Colours & Lustres which it had living in its native Element; what Artist can possibly divine the Colour or make out the numberless Shades of those Tints which correspond to the Numbers of his [i.e. Ferdinand's] Sketches. I therefore shall certainly look to the finishing of these as well as the Plants'. Needless to say, nothing came out of Ferdinand's plan. Sibthorp was possessive of Ferdinand's talents, in the same way as he would be of the beauty of his 'mistress'. In addition, there were deep societal differences: for Sibthorp, the wealthy landowner, the professor at Oxford University, the Fellow of the Royal Society, in short the member of the elite, Ferdinand was no more than a servant with special qualifications to be dismissed at short notice and at any time. Clearly there was a clash of characters – both seem to have possessed considerable faith in their own abilities, a fact Sibthorp seemingly found difficult to swallow.

Very appropriately it had been said that 'It would be too easy to place Sibthorp in the position of sinner and Bauer in that of saint, but their outlooks and stances need to be placed in the context of an English society where slavery was condoned, rural labourers were in poverty, and the political class and aristocracy were suffused with corruption and beholden to the whims of a dysfunctional monarchy. Also you do not get remembered for being nice, you get remembered for being good at what you do, and Sibthorp clearly had an eye on posterity' – expecting to become famous due to Ferdinand's work. Clearly Boccius and Jacquin, a professor in comfortable financial circumstances like Sibthorp and, in contrast to him, already ennobled, had treated Ferdinand differently.

FERDINAND DISCONTINUES WORKING FOR SIBTHORP

The plan to visit the Ottoman Empire for a second time was formed early, when Ferdinand was still deeply immersed in his botanical illustrations. It was Sibthorp who told Hawkins in an undated letter, possibly written in May 1790, 'How happy should I be – could we once more visit together the Attic shore', to which the answer was 'no plan would give me more real pleasure and it is a journey so practicable & now rendered so easy to us by past experience that I see no objection to its speedy accomplishments'. However, it took two years before Sibthorp received the approval of his family for a second journey to the Levant, and almost two more years had passed, until he finally set out from London on 20 March 1794. One of the reasons for this delay was Sibthorp's purchase of another estate.[58] In July 1791 Hawkins, acting in a sense as intermediary between Ferdinand and Sibthorp, had told the Sherardian Professor that Ferdinand could be induced to join them on a second journey when the finishing of the Flora & Fauna was to be done in Hawkins's house in London instead of Oxford. Clearly this proposal did not make Ferdinand's complex situation at Cowley House easier. On 30 June 1793 the possible continuation of Ferdinand's engagement was still considered by Sibthorp, a man of immense dedication, but also of remarkable single-mindedness, who wrote to his past and future travel companion 'I am so fully satisfied with his Drawings as an Artist I wish we could have his Pencil to render the Flora & Fauna of Greece more complete' and 'In respect to Bauer – my wish to keep it a perfect Secret from my Family is the Reason for not mentioning to him at present my Intentions. I know my Father & my Sister are impatient he should finish his Drawings & I would not commit myself by a Discovery of any Intention of taking Him with me in a further Tour. I do not wish to put out his Eyes by calling upon Him to represent very diminutive Objects but I would not cho[o]se by a long List of Negatives to confine my Requests'. Considering his poor health – Sibthorp probably suffered from tuberculosis – it was not a wise move to even consider a second journey with its privations and discomforts. But we have to regard Sibthorp by now a driven man, chaotic in his planning, supercilious, focused strongly on the prestige and recognition his ego demanded.

In the end it was Hawkins who invited Ferdinand a few weeks later to join Sibthorp and himself on a second journey. This letter is lost, but Ferdinand's reply, in fact his first extant letter dated 31 July 1793, survives in the Cornwall Record Office and is worth quoting, because of its content and because of its language. He wrote, and this is a verbatim quotation, 'I think Dr Sibthorp will beging His tour quite in the oldstyle never come to a determination before the last day. then at ones all in haste which is most unblessent, for mypart I conclude from all the obstacles and dissagreaments by Dr Sibthorp that He himself has not much inglination for compleating his work . . . all Drs delay and obstacles will attribut to some dissinclination, if I should take that engagement, I which it would be whit autany opposition, in case if it is, I must big pardon if I shall Esgus myself to acconpeng Dr Sibthorp'.

Five years later, and more than a year after his master's early death, Ferdinand bitterly wrote to Hawkins, then again in Salonica, 'I am sorry that I must recollect my disappointmend at the end of my work from Greece by Dr. Sibthorp'. Although a confrontation between Sibthorp and Ferdinand in late 1793 or early 1794 is plausible with possibly things said 'which Bauer never forgot or forgave',[59] we have no record in the archive. In any case either Ferdinand was dismissed or he decided not to continue. The fact remains that he did not accompany Sibthorp on his second journey, with the consequence that no pictorial record of this tour exists (see Chapter 9). Nothing indicates that Ferdinand continued to work for Sibthorp in Oxford after the latter's departure for Vienna and Istanbul. Clearly the prospects of Ferdinand, then aged 34, were not in Oxford, but elsewhere. He had worked extremely hard, had dedicated eight years of his life to Sibthorp and created what Hawkins called 'works of art . . . superior to any thing of the kind existing'. Clearly Ferdinand had set standards. But what steps should he take next?

For a continuation of Ferdinand's life see, Chapter 9.

ENDNOTES

1 Unless otherwise indicated, this chapter is based on Lack (1999c) and Harris (2007).
2 Petrie (1959)
3 Page & Stearn (1988)
4 Moheit (1993)
5 Harris (2011)
6 MS Sherard 238: f. 47 (OXF)
7 MS Sherard 242: f. 39 (OXF)
8 MS Sherard 241:f. 24 (OXF)
9 J. Pickering (pers. comm., 22 v 1994)
10 MS Sherard 240: f. 113 (OXF)
11 MS Sherard 240: f. 114 (OXF)
12 MS Sherard 240: f. 3 (OXF)
13 MS Sherard 240: f. 106 (OXF)
14 MS Sherard 240: f. 112 (OXF)
15 MS Sherard 238: f. 14 (OXF)
16 MS Sherard 239: f. 39 (OXF)
17 MS Sherard 240: f. 106 (OXF)
18 MS Sherard 240: f. 78 (OXF)
19 MS Sherard 240: f. 10,11 (OXF)
20 MS Sherard 240: f. 8,9 (OXF)
21 MS Sherard 242: f. 175 (OXF)
22 MS Sherard 242: 121 (OXF)
23 MS Sherard 241: f. 68, 69 (OXF)
24 MS Sherard 244: f. 83, 84, 85, 86 (OXF)
25 MS Sherard 241: f. 57, 58 (OXF)
26 MS Sherard 243: f. 146, 147 (OXF)
27 Strid (2009, 2010, 2012, 2013)
28 MS Sherard 245: f. 111 (OXF)
29 MS Sherard 241: f. 58 (OXF)
30 MS Sherard 241: f. 162 (OXF)
31 MS Sherard 245: f. 96 (OXF)
32 MS Sherard 244: f. 58 (OXF)
33 MS Sherard 242: f. 191 (OXF)
34 MS Sherard 243: f. 188 (OXF)
35 MS Sherard 244: f. 140 (OXF)
36 MS Sherard 241: f. 25 (OXF)
37 MS Sherard 244: f.13 (OXF)
38 MS Sherard 242: f. 56 (OXF)
39 MS Sherard 243: f. 191 (OXF)
40 MS Sherard 243: f. 164 (OXF)
41 MS Sherard 243: f. 88 (OXF)
42 MS Sherard 245: f. 110 (OXF)
43 MS Sherard 244: f. 35 (OXF)
44 MS Sherard 242: f. 20 (OXF)
45 MS Sherard 233: f. 49 (OXF)
46 MS Sherard 243: f. 147 (OXF)
47 MS Sherard 243: f. 138 (OXF)
48 MS Sherard 243: f. 142 (OXF)
49 MS Sherard 243: f. 156 (OXF)
50 MS Sherard 243: f. 158 (OXF)
51 MS Sherard 241: f. 31 (OXF)
52 Meikle (1977)
53 MS Sherard 242: f. 59 (OXF)
54 Lack (1997b)
55 Brain (1986), Grant (1999)
56 Moore (1998)
57 Anon. (1985)
58 London, University College London, MS Add 254, Lord Odo Russel Collection. - Unpublished letter Franz Bauer to Joseph Franz Edler von Jacquin, 21 v 1792.
59 Bruce (1970)

Tab. 4.

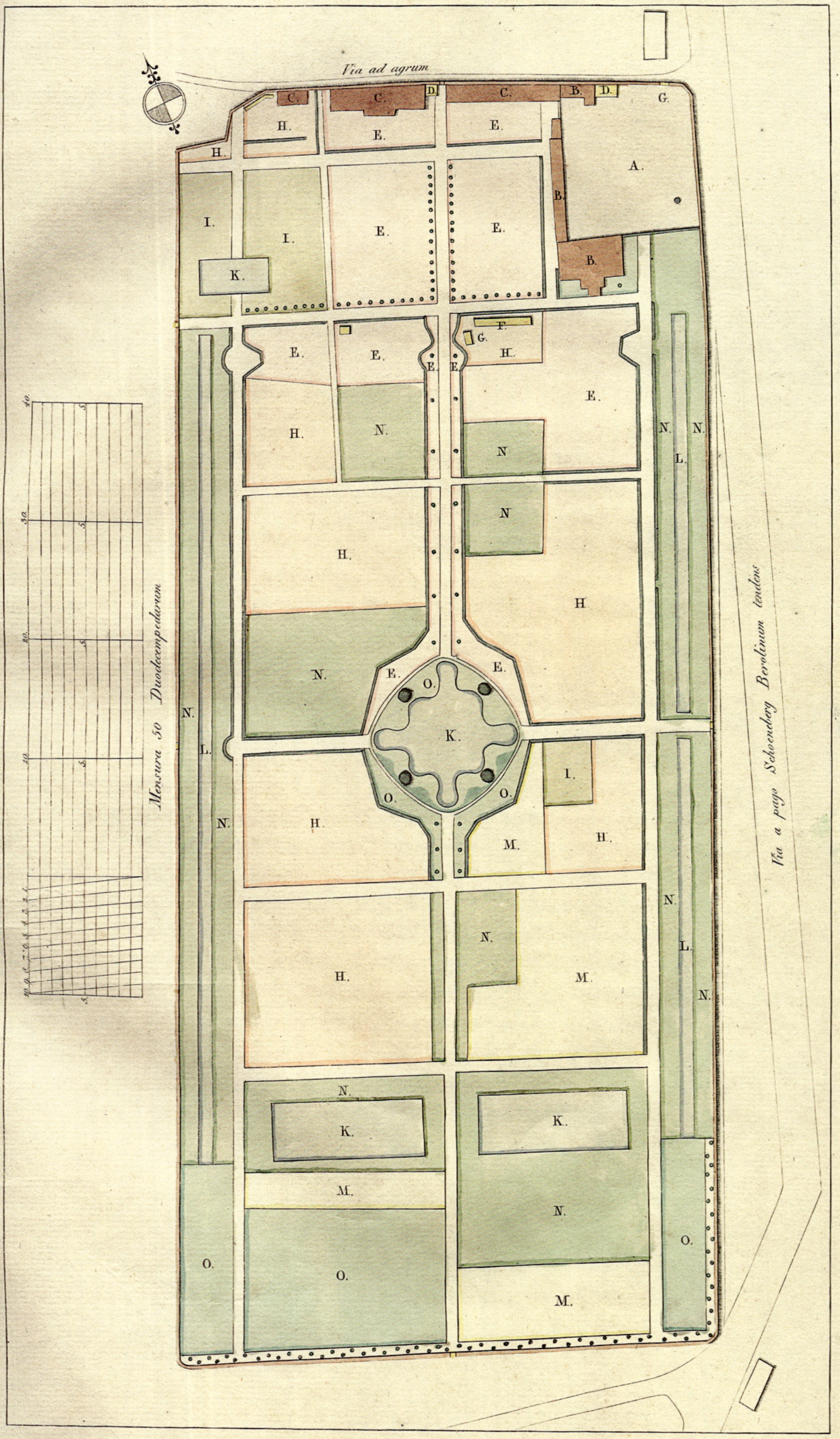
Via ad agrum
Mensura 50 Duodecempedarum
Via a pago Schoenberg Berolinum tendens
A.
B.
C.
D.
E.
F.
G.
H.
I.
K.
L.
M.
N.
O.

CHAPTER 7

Franz Bauer and Joseph Franz Edler von Jacquin tour Europe[1]

NETWORKING IN VIENNA

At an early age the Bauer brothers had produced samples of their outstanding talents, with Franz's drawing of a flowering specimen of *Pulsatilla pratensis* subsp. *nigricans*, published when he was 13 years old, as a convincing example (see Chapter 2). However, they were not the only examples of wunderkind at that time. Eight years after Franz and only some one hundred and fifty kilometres to the east of Feldsberg another Franz was born who showed an unusually early development of his intellect. This was Joseph Franz Jacquin, the eldest child and first son of Jacquin and his wife Katharina Josepha, née Schreibers (see Chapter 3).

Joseph Franz was born on 7 February 1766 in Selmecbánya, where his father had been appointed professor of chemistry and metallurgy. Two years later when his father accepted Swieten's offer (see Chapter 2) to take over the combined Chair of Botany and Chemistry at Vienna University,[2] the family moved to the imperial capital. Thanks to the double salary for the two chairs, plus 1000 guilder per year from the Board of Minting and Mining and Katharina's substantial dowry, the Jacquin family lived in comfortable circumstances. Joseph Franz seems to have been brought up speaking French and Dutch, the languages of his father, as well as German, the language of his mother, daughter of Johann Heinrich Schreibers,[3] a secretary in the Government of the Archduchy of Austria below the River Enns. Later Joseph Franz was tutored by Franz Molitor,[4] a student of medicine at Vienna University and subsequently Professor of Chemistry, Pharmacy and Materia medica at Mainz University. The circles of the learned in Vienna were closely interlinked – in 1780 Molitor was to translate an English text into German, originally published by Ingen Housz, his former pupil's uncle and the father of photosynthesis, and had it printed in Leipzig.[5]

From his early youth Joseph Franz is reported to have accompanied his father on his botanical excursions into the surroundings of Vienna. On one of such excursion to the Schneeberg, a limestone massif southwest of the city reaching an altitude of 2075 metres, Joseph Franz, then already Edler von Jacquin (henceforth called Jacquin the son), was lucky enough to observe a common lizard, *Zooteca vivipara vivipara*, giving birth to live young. The boy, then eleven years old, produced a written report on this phenomenon and had it sent in 1778 to one Respinger in Basel. Several years later it appeared in a Latin translation in the journal *Nova Acta Helvetica Physico-Mathematico-Anatomico-Medica*[6] in this Swiss city, accompanied by a copper engraving showing an adult and a newborn common lizard. When he was thirteen, Jacquin the son sent another manuscript to Basel entitled 'Tria genera nova ex horto viennensi', which contained fully fledged Latin descriptions of three plant species and was belatedly published in the same volume of the *Nova Acta Helvetica*, testifying both

FIG. 7.1. Plan of the Royal Botanic Garden in Schöneberg near Berlin. Anonymous copper engraving, coloured. – C.L. Willdenow, Hortus Berolinesis, Berlin, 1806, Plan (A). – Berlin, Botanischer Garten und Botanisches Museum Berlin-Dahlem, Bibliothek.

linguistic skill and considerable knowledge of botanical terminology. One of the three species was *Lachenalia aloides*, a geophyte restricted in distribution to three localities on granite hillsides on the Cape of Good Hope, introduced by Francis Masson in 1774 at Kew and cultivated in the botanical garden at Vienna University.[7]

Well aware of the necessity of networking in Vienna, Jacquin (henceforth called Jacquin the father) had wisely dedicated his first substantial publication *Selectarum stirpium americanarum historia* to Emperor Francis I Stephen, who had financed his expedition to tropical America (see Chapter 3). Maria Theresa, who had founded the botanical garden of Vienna University now placed under his care, was the dedicatee of his second major work *Hortus Botanicus Vindobonensis*. It was undoubtedly Jacquin the father, probably as ambitious to promote his son like Leopold Mozart, who arranged for the publication of a mathematical tract authored by his son when twelve years old. A public disputation on Jacquin the son's theorem had been held in the main hall of Vienna University on 25 January 1778. In the same year *Sätze aus der Mathematik* appeared in print, dedicated as the title continues 'to the august name of Maria Theresa', then 61 years old. While still at school Jacquin the son translated from Dutch into German a treatise on shoes by the extremely versatile physician and naturalist Petrus Camper and had it published as a separate treatise,[8] needless to say in Vienna.

However, Jacquin the son's first and major contribution to science was of a different calibre. In 1784, when just 18 years old, he published *Beyträge zur Geschichte der Vögel*, an ornithological text of 45 pages with 19 coloured copper engravings appended, an independent work containing the descriptions of 33 birds. One of these was new to science – the northern crested caracara, *Caracara cheriway*, a scavenger mainly feeding on carrion and occurring in a range from the southern parts of today's United States, including Florida, to what is now northern Peru and northern Brazil. This publication is partly based on his father's ornithological notes taken during his expedition to the islands in the Caribbean Sea and the northern coast of the South American mainland (see Chapter 3), plus two descriptions provided by Franz Xaver Freiherr von Wulfen, a former Jesuit then based in Klagenfurt in the Duchy of Carinthia. In addition, Jacquin the son had observed and described several birds kept in the aviary at Schönbrunn, partly introduced by his father, partly of Old World or of completely unknown provenance. No doubt *Beyträge zur Geschichte der Vögel* is not a masterpiece of the ornithological literature, but it provided Jacquin the son with an early introduction to the realities of publishing a scientific text and of collaborating with an illustrator, whose identity remains, however, unknown. Having completed his school education Jacquin the son enrolled at Vienna University and began to study medicine following established family tradition.

The desire to make one's own son the only possible successor in the job is understandable. In eighteenth- and nineteenth-century academia such a succession in the chair was not an exotic phenomenon, with the third Sherardian Professor of Botany at Oxford University being a son of the second Sherardian Professor as an example (see Chapter 5). In a parallel manner Thomas Martyn, Professor of Botany at Cambridge University at that time, was the son of his predecessor John Martyn,[9] The former is reported to have said: 'I succeeded because I canvassed with my father's resignation in my pocket before anybody was aware of it'. Long before his resignation Jacquin the father used his excellent network of contacts at court to make his elder son the only possible candidate for the two combined chairs – by successfully petitioning Emperor Joseph II to pay for an official journey, a kind of scientific Grand Tour across Europe, to be undertaken by the promising young man.

The journey's primary source of information is a series of letters, all in German, sent by Jacquin the son to his parents and his brother Gottfried (see Chapter 3), but the current whereabouts of these is largely unknown. Only five letters from Znaim, Leipzig, Bonn, Rotterdam and London respectively are kept in the Österreichische Nationalbibliothek in Vienna.[10] However, two series of extracts taken from the complete run of letters have been published – the first on mineralogical, geological, paleontological and chemical subjects, the second on botanical and horticultural matters – with the rest of the contents remaining a mystery. These extracts form the backbone of this chapter. Unfortunately, the letters sent by Jacquin to his son are also unknown; they must have existed, as indicated by occasional quotations taken from them, naturally in French, in the letters of his son; the letters sent by Jacquin the son's mother and brother are also untraceable. Unfortunately, the administrative aspects of the journey have also not yet been studied in any detail, but we know for certain that the expenses totalling no less than 8650 guilders were to be paid out by the representatives of the Emperor outside the Hereditary States,[11] which implied that Jacquin the son had to keep the necessary contacts. In addition, he is reported to have had at his disposal a carriage for his private use, which may imply that he had access to the carriage park of the imperial representations.

The petition to the Emperor entitled 'Plan of a journey to be undertaken by my son Joseph von Jacquin under the supreme orders of his Imperial Royal Majesty [Joseph II]' had been drafted by Jacquin the father. Again, only published extracts are known, from which the following quotations are taken: 'Since the principal purpose of my son's journey is making him more completely knowledgeable in those fields of science, which he has selected as his main topics, i.e. natural sciences, chemistry and botany, it is necessary to choose the itinerary in such a way, that all cities and places are to be passed through, where famous naturalists, chemists and botanists are resident, or mines, factories, gardens and public institutions relevant for him are to be found. In addition, he must attempt to spend at least part of the summer in those regions where the most famous botanical gardens are located . . . Finally, he has to remain in correspondence with the Imperial Royal gardener in Schönbrunn [Richard van der Schot], in order to provide him with those plants, which may not yet be present in the imperial garden of Schönbrunn'. The message was clear: neither botanizing nor making chemical experiments was the purpose of the tour, but networking. What we do not know is the timing of Jacquin's initiative in the appropriate circles at the imperial court, but it seems feasible to assume that the safe arrival of Sibthorp and Ferdinand in Oxford in early December 1787 had been quickly reported to Vienna and acted as a trigger for a plan of action possibly outlined beforehand. Clearly Sibthorp was Jacquin the father's exact equivalent and Oxford, as a traditional centre of learning, was an important place for his son to visit. Jacquin the father had previously sent herbarium material to Sir Joseph Banks, then already President of the Royal Society and based in London. In addition, Banks must have also been considered right from the beginning as an important contact for Joseph Franz. On the same line of improving scientific exchange was Jacquin the father's election to Fellowship of the Royal Society effectuated on 3 April 1788, just a few weeks before the beginning of the trip (see Chapter 6).

It would have been logical for the Jacquin family to attach an assistant to their elder son, who was about to leave Vienna for an extended journey across Europe of at least two years; in fact Joseph Franz returned to Vienna only in the summer of 1791 (see Chapter 9). Since botany was to play such a prominent, though not exclusive, role – in particular the visits to botanical gardens – Jacquin the father may have considered a qualified illustrator a sensible choice, particularly as the latter could record plants of special interest in drawings very true to nature. The Jacquin family may have had a preference for a young man, though slightly older than their son, then aged 22, ideally already known to them and with a trustworthy personality. This may have been the reasoning for the selection of Franz, eight years older than their son, as a travel companion and assistant. However, we have no archival record for this interpretation. Neither do we know the financial and organizational aspects of the arrangement: Was Franz paid for by the Jacquin family? Were his travel expenses included in the government funds available to Jacquin the son? We are also ignorant about Franz's motives to join the son of his commissioner on this special form of a Grand Tour. Wanderlust, the desire for a change from the routine of documenting plants cultivated in the Botanical Garden of Vienna University, or the wish to meet or join his younger brother Ferdinand in Oxford? To Franz and Jacquin the son a third party was attached, about whom we know almost nothing – Abbé Leonhard Gruber 'a priest well-known in religious education circles in Vienna . . . middle-aged, he was already well-travelled and was seen, presumably, as a source of guidance, moral if necessary, for the two young men'. Since Gruber, who had been for a short period a professor of philosophy at Salzburg University, is nowhere mentioned in the extracts of Jacquin the son's letters and seems to have played no role in this journey, he is not further dealt with in this chapter and the text is written as if the two naturalists were travelling without the accompanying priest.

Decision taken, the time had come for Franz to say farewell to the persons who had shaped his early years – to his mother Therese and to Boccius, both now 57 years of age and living in Feldsberg. Franz will also have taken leave from his brother Joseph, just back from Rome and since 1 May 1788 employed by Aloys I Joseph in Vienna and Feldsberg,[12] from his eldest brother Johann, about whom we know so little (see Chapter 16), and from Jacquin the father. In early May 1788 a carriage left the 'Kaiserliche Haupt- und Residenzstadt' with Franz and Jacquin the son on board. When the unequal pair, though probably less unequal than Sibthorp and Ferdinand, crossed the border between the Archduchy of Austria and Moravia near the town of Retz, the first part of Franz's life had come to its conclusion. He was 30 years of age and certainly not aware that he would neither see Boccius, his mother nor his two elder brothers again, nor ever return to Feldsberg or Vienna.

For the reasons set out above, the record of the journey is fragmentary, the itinerary taken by the two young men in part unclear; in addition, the published extracts are partly chaotic, often redundant and sometimes difficult to understand. Like Sibthorp two years earlier, Jacquin the son only occasionally writes about his travel companion, but Franz is always mentioned by his name, indicating a different sort of relationship. During the year 1788 Franz is not known to have written a single letter, and only two of his plant illustrations prepared during this period have survived; a third one is known only from the copper engraving that was based on it. However, the journey provided Franz with new contacts and important impressions relative to his work as an illustrator.

VIENNA TO KARLSBAD

The first stop of Jacquin the son and Franz seems to have been Znaim on the River Thaya, from where a first letter was sent to Gottfried in Vienna, asking him to send a manuscript list of Austrian plants forgotten and left behind on Joseph Franz's writing table. Meanwhile Franz may have recollected his early impressions of wildlife in the riverine marshes and woods near his home town of Feldsberg, only some fifty kilometres to the east of their location on 5 May 1788. Prague, hailed as Golden Prague (Zlatá Praha), was the first destination, and the two travellers seem to have followed the standard route via Iglau (Jihlava), Deutschbrod (Havlíčkův Brod) and Tschaslau (Časlav). Mozart, who had been to Prague just a few months before to conduct the first, rapturously received performance of his *Don Giovanni*, may have taken the same road. For the two naturalists a stop at Sedlitz (Sedlec), now a suburb of Kuttenberg (Kutná Hora), a mining town in Central Bohemia, was a must. They were not primarily interested in the Sedlec ossuary, estimated to contain the skeletons of between 40,000 and 70,000 people, known for its enormous chandelier of bones with garlands of skulls draping the vault of the nave, but in a particular lime tree (*Tilia* spec.) in the Sedlec cemetery. Called 'Kapuzinerlinde' (lime tree of the order of Capuchin Friars Minor), this extremely old specimen was known to possess hooded leaves and had become a topic in legends and myth.[13] On 8 May 1788 Joseph Franz informed his mother that only a tiny fraction of the leaves, about every hundredth, showed this anomaly, that he had collected a dozen herbarium specimens of this singular tree, written a report on it and that Franz had depicted one of the twigs. The respective drawing could not be traced. However, twelve years later Jacquin the father published a description of the peculiar lime tree in the Sedlec cemetery together with a coloured copper engraving showing this phenomenon,[14] mentioning that the illustration had been prepared ten years earlier.

Soon afterwards Jacquin the son and Franz arrived in Prague, one of the most spectacular cities in central Europe. Just four years earlier the four municipalities – Hradschin (Hradčany), Old Town (Staré Město), New Town (Nové Město) and Little Quarter (Malá Strana) – had been merged into a single entity. Although we know very little about the visit of the two naturalists, it is very likely that they went to at least three of these four municipalities – for a simple reason. At that time a single bridge, the famous Karlsbrücke (Karlův Most), a late-medieval construction and not yet adorned with as many statues as today, crossed the Vltava River, separating Old Town and New Town on the right bank from the other two quarters on the left bank. It is inconceivable that the guests from Vienna would not have seen Hradschin, the castle district, with St Vitus Cathedral (Katedrála sv. Víta), St George Basilica (Bazilika sv. Jiří), the Royal Palace (Starý královský palác) and several pompous city residences of noble families, e.g. of the Czernins (Černínský palác) and Schwarzenbergs (Schwarzenbergský palác). In a sense they must have felt at home: parts of the Royal Palace, just to mention one example, had been worked over by Nikolaus von Pacassi, the architect who had also been in charge of major construction works at the Imperial Library in Vienna and the summer residence of the Habsburgs in Schönbrunn. Jacquin the son and Franz must have also been to the Old Town district, since the former reported to his brother Gottfried that they had visited a 'splendid library' where he had met the 'librarian Ungar'. This was clearly the Imperial Royal Public and University Library

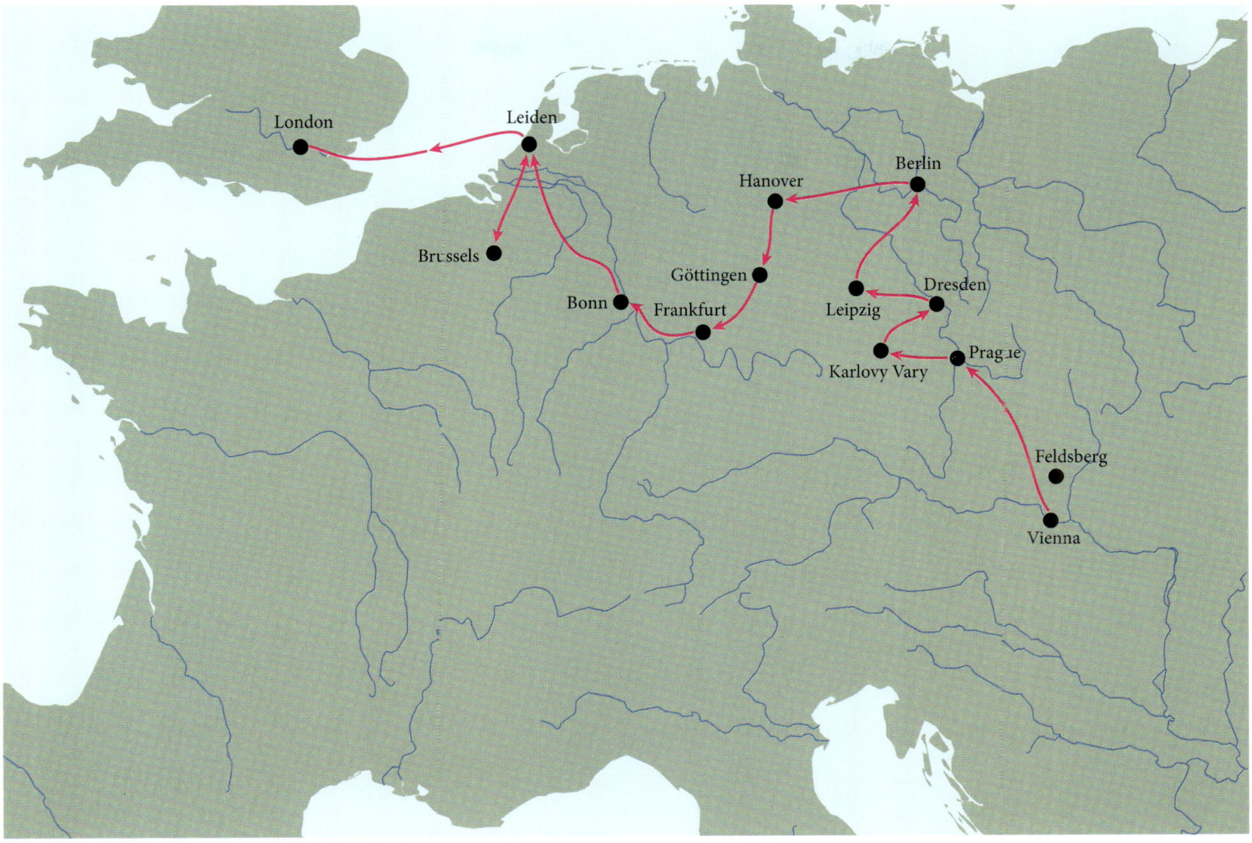

MAP 3. **Franz's route from Vienna to London in 1788.**

(Cisařska – královska veřejna univerzitní bibliotéka), a denomination given to it by Maria Theresa in her position as Queen of Bohemia. She had arranged for its premises in the Klementinum, an extensive building previously owned by the Jesuits and located in the Old Town. Since 1780 Karel Rafael Ungar was director of this library, which had been very generously supported by Franz de Paula Ulrich 3rd Prince Kinsky of Wchinitz and Tettau. The former complained to Jacquin the son that table 27 (in fact t. 26) of his father's *Florae Austriacae icones* showing bastard balm, *Mellitis mellissophyllum*, had been cut out – not an uncommon problem with lavishly illustrated large-format botanical works. On the Old Town Square (Staroměstské náměstí) nearby, Franz could have contemplated the date 21 June 1621, the day when 27 protestant rebels were executed on precisely this spot – under the presidency of Karl 1st Reigning Prince of Liechtenstein (see Chapter 2). That act of loyalty had endeared Karl I so much to Emperor Ferdinand II that he soon afterwards became viceroy of Bohemia. In addition he was given the duchy of Jägerndorf (Krnov) in Silesia and, in exchange for nullifying the Emperor's debts, gained much confiscated 'rebel property' thereby laying the foundations of the family's future wealth. Admission to the Knights of the Golden Fleece followed in due course (see Chapter 2). However, we may doubt that Jacquin the son and Franz had such deep historical insights into the consequences of the famous battle of White Mountain (Bílá Hora), fought in the vicinity of Prague on 8 November 1620.

The two naturalists were also shown the Botanical Garden of Prague University, an institution founded by Maria Theresa in 1775 and situated in the village of Smichov on the left bank of the Vltava River, way out of the city centre,[15] no more extant and built over in the late-nineteenth century by the Národní dům na Smíchově. Boccius's contact in Prague, Mikan, the first professor of botany and chemistry at Prague University and director of its botanical garden, acted as guide to the son of his colleague in Vienna and the latter's assistant. Jacquin the son was not impressed: he found the grounds large, but only a quarter of it planted out, the rest a meadow of clover. Judging from Mikan's *Catalogus plantarum*, which had been published in Prague twelve years before, the labelling of the plants in the garden followed the Vienna model and offered nothing new (see Chapter 3). In order to have reached Smichov, the two travellers must have been to the Little Quarter, where many noble families possessed their fine city residences with coats of arms above the entrances that both would have had little difficulty in recognizing – among them those of the Hrzán, Liechtenstein, Lobkowicz, Pálffy and Waldstein families. Of these, three also possessed city residences in Vienna (see Chapter 3). Three years after the visit to Prague Aloys I Joseph had his residence in the Little Quarter renovated to be able to receive in style Emperor Leopold II after the latter's coronation as King of Bohemia; ten years later Ludwig van Beethoven played his second piano concerto (op. 19) in the Prague residence of one of his most important sponsors – Franz Joseph Maximilian Prince Lobkowicz. Because of their architectural qualities the Liechtenstein Palace (Lichtenštejnský palác, now mostly called Ledeburský palác after its later owners), and the Palais Lobkowicz (Lobkovický palác) continue to attract many visitors. Considering Jacquin the son's elevated status as traveller 'under supreme orders', he should have had contact with members of the higher strata of Prague society, but again nothing is on record.

Very briefly a visit to 'Dr. Mayer Herbarium' is mentioned, referring to Johann Mayer, until 1786 secretary of the Böhmische Gesellschaft der Wissenschaften (Bohemian Society for Science), which would soon have the epithet 'königlich' (royal) added to its name, effectively the first Bohemian Academy of Sciences.[16] In this collection a specimen of sweet galingale, *Cyperus longus*, attracted the attention of Joseph Franz and he hypothesized that it had come from the same source as the specimen in Vienna (that is, from Schwencke's garden in The Hague; see below). It is interesting to note that two of the naturalists they had met later became Vice-Rectors of Prague University – Ungar in 1790, and Mikan in 1799.

The following weeks were mostly dedicated to geology and mineralogy. Again, the reason for this is simple to explain: Jacquin the father had wished that his son was to receive further qualification not only in botany but also in chemistry in the broad sense, since he wished to pass both chairs to Joseph Franz – and as a matter of fact succeeded in the end. From Prague the two travellers moved westwards to Karlsbad (Karlový Varý), a spa town famous for its hot alkaline springs. Three years before their arrival Goethe had stayed there for the first time. Many years later he was to praise exuberantly *The genus Pinus* with its copper engravings based on the works of Ferdinand and Franz (see Preface). Jacquin the son sent a report to his father and brother, giving details of his activities and announcing a small box with feldspars and granites and later a second box with more specimens, including petrified wood, all from Karlsbad – some bought from a local stone cutter, the rest collected by himself. Many more boxes from this journey were to follow. Excursions led both to Schlackenwald (Horní Slavkov) southwest of Karlsbad, where ores containing tin, silver, lead, tungsten and uranium were extracted and processed, and to Mies (Stříbro) on the River Mies (Mže), the Czech name of the town meaning 'silver'. Reports are given on cassiterite from Schlackenwald, on baryt from Mies, on the primitive process of melting tin ores and on the ruined state of the production site. On 22 May 1788 the two naturalists visited the Einheitsschacht, probably a mine near Joachimsthal (Jáchymov), where small pieces of 'krystallisiertes Rotgulden' were extracted with a few fine specimens broken from the shaft by Franz. This seems to have been ruby silver ore, but the identity of this mineral was only clarified in subsequent years, notably by Abraham Gottlob Werner of the Mining Academy in Freiberg. During this tour Jacquin the son and Franz were accompanied by one Graf Czernin, possibly Johann Rudolf Graf Czernin von und zu Chudenitz, owner of extensive possessions in Bohemia among them Schönhof (Krásný Dvůr), at a distance of around thirty kilometres from Joachimsthal. He was a well-known art collector, amateur musician and subscriber to Mozart's concerts given in Vienna in 1784. Jacquin the son also mentions in his letter that one 'Bergrat Peithner', probably Joseph Wenzel Peithner Ritter von Lichtenfels, then inspector of mines in Joachimsthal,[17] had given him a few small pieces of 'Opal im Bleiglanz', probably opal in galena, from Bleistadt (Oloví), 'really among the rarest things in existence'.

KARLSBAD TO BERLIN

The next letter to Vienna was dated 26 May 1788 and came from Johanngeorgenstadt in the Erzgebirge (Ore Mountains), a small mining town north of Karlsbad. By now Jacquin the son and Franz had left the Hereditary Lands of the Habsburgs and entered the Electorate of Saxony that was reigned over at the time by Prince Elector Friedrich Augustus III of the House of Wettin. Mineralogy continued to dominate the succeeding weeks, as is evident in reports from Freiberg, dated 10 June

1788, and from Dresden, the capital of the Electorate, dated 23 June. Judging from the published extracts, visits to mineralogical cabinets were the most important topic on the agenda – in Freiberg and in Schneeberg, infamous because of the Schneeberg disease, a form of lung cancer frequently observed in the miners triggered by the radioactive gas radon omnipresent in the shafts, described as early as 1567 by Paracelsus. Jacquin the son mentions having seen specimens of 'Rothgülden' 'of the kind I shall possibly never see again' in the cabinet of one 'Bergmeister Schmidt', possibly in Freiberg. In addition Joseph Franz was given by his host crystallized 'Kobaltblüte', now called erythrine, a cobalt arsenate mineral, described as late as 1832 with the type specimen from a mine in Schneeberg. In Freiberg Jacquin the son must have seen the mineralogical cabinets of Christian Ehregott Gellert, a metallurgist, and of Carl Eugenius Pabst von Ohain, the latter collection regarded as 'beyond all imagination'. This was no exaggeration: Pabst von Ohain, a curator of the Mining Academy in Freiberg founded three years after that in Selmecbanyá,[18] had brought together a collection of c.7500 items, after his death catalogued by the famous Werner, who published in Freiberg in the years 1791–1793 *Ausfuerliches und sistematisches Verzeichnis des Mineralien-Kabinets des weiland kurfuerstlichen saechsischen Berghauptmanns Herrn Karl Eugen Pabst von Ohain*. This collection was acquired by Portuguese diplomat António de Araújo e Azevedo, Primer Conde da Barca, sent to Portugal and later to Brazil, where it is now kept at the Museu Nacional da Universidade Federal do Rio de Janeiro. Surprisingly no meeting with Werner is recorded, who two years before had been the mineralogical teacher of Hawkins.[19]

In Schneeberg Jacquin the son had called upon Adolf Beyer, commemorated in 1943 in the naming of the mineral beyerite, a bismuth carbonate, and marvelled at his rich collection of crystallized 'Hornsteine', possibly quartz. However, Joseph Franz told his father that he possessed among his collection in Vienna one of the best specimens of this mineral. The only shaft visited seems to have been the Himmelsfürst to the west of Brand-Erbisdorf, famous for its rich layers of silver ore and a particularly productive mine – a record of sixty-eight kilograms of silver is reported to have been extracted on a single day (12 August 1749). When the two travellers from Vienna were present, large specimens were brought up and Jacquin the son bought a few of them for his father at a high price. What is remarkable is the subsequent note in the letter of 10 June 1788, informing him that if he felt the price too high he should pass the specimen on to the mineralogical cabinet of the Emperor. This testifies to the excellent connections of the Jacquin family at court – or alternatively to send it on to England, where Jacquin the son was confident to be able to sell it at a good price.

In late 1785 Sibthorp had visited the Grünes Gewölbe in Dresden (see Chapter 5), admittedly the largest collection of treasures in Europe founded in 1547 as 'Geheime Verwahrung' (Privy Repository) by Moritz Elector of Saxony. The two travellers arrived in mid-June 1788, when refurbishments took place and only a small selection of items could be inspected by visitors. For a connoisseur in mineralogy like Jacquin the son it would have been natural to comment on the most famous objects: (1) the Dresden Green, a forty-one-carat (8.2-gram) natural green diamond, one of the greatest diamonds in the world, its colour a consequence of exposure to radioactive materials, probably originating from a mine in India; (2) the moor with emerald cluster, manufactured in c.1724 by goldsmith Johann Melchior Dinglinger, a statue 63.8 centimetres high richly decorated with jewels, including the famous emerald cluster from a mine in present-day Colombia, a gift of Emperor Rudolf II; (3) the jewel-studded 2.28-metre-high obelisk *Obeliscus Augustalis*, a cabinet piece conceived as an indoor monument, made by the same artist for Friedrich August I (Frederick Augustus I), Elector of Saxony, who was also Friedrich Augustus II King of Poland and Grand Duke of Lithuania. However, Jacquin the son neither mentions these three most spectacular objects, nor the no less spectacular jewel garnitures or the 648-carat (129.6-gram) sapphire, a present from Peter I, Emperor and Autocrat of all Russias. Therefore we have to assume that these representative objects could not have been on show. Instead Jacquin the son described for his father 'Hornerze' and large pieces of silver ore and of green fluorite, a halide mineral, all from the cabinet of the Elector. Joseph Friedrich Freiherr von Racknitz, in 1788 chamberlain at the court of Friedrich Augustus III, was owner of another fine mineralogical cabinet, in which Joseph Franz was fascinated by the specimens of 'Wermsdorfer Weltauge', also called 'Oculus mundi'. These were very rare opals found in Wermsdorf in the northern part of the Electorate, capable of absorbing water, thereby becoming transparent and then diffracting light in a special way, resulting in transient iridescent colours.

The stay in Dresden, which lasted for about two weeks, also offered the chance to visit several gardens, four of which are explicitly mentioned in the letters sent to the father, the mother and the brother: (1) the 'Herzogengarten', recte Herzogin Garten, in the very centre of Dresden and now fallow land; (2) the Großer Garten (Large Garden) to the south of it; (3) the garden near the Japanese Palace in the Neustadt quarter of Dresden; and (4) the gardens at Pillnitz, way out of the city centre. The latter two are situated on the right bank of the Elbe River. Jacquin the son was impressed by the 'botanical' orientation of several gardens (the focus on plant diversity) and had a simple explanation to offer: the Elector is a 'passionate botanist and the books of Papa are permanently in his cabinet'. This may indeed have been the case, since Friedrich Augustus III is known to have possessed a profound interest in botany. It is on record that starting from c.1785 he had rare plants recorded by 'botanical court painters', with Johann Christian Jacob Friedrich the first of these, which resulted in the famous *Pillnitzer Centurienbände*, now in the Sächsische Landesbibliothek – Staats- und Universitätsbibliothek Dresden.[20] These volumes form the Dresden equivalent to the *Codex Liechtenstein* in Feldsberg, which had been continued by inferior illustrators after the departure of Joseph, Ferdinand and Franz. The Elector of Saxony had also made arrangements for a botanical library to be brought together and a herbarium to be started. However, the published extracts do not contain a mention of any personal contact between the two travellers and Friedrich Augustus III. Friedrich's fully illustrated *Anweisung zum Zeichnen und Blumenmalen* (see Chapter 2) published in 1786 in Friedrichstadt, now a part of Dresden, would have been of great interest to Franz, but this fine treatise is also referred to nowhere in the extracts. Otherwise mentioned in the report to Vienna are rose mallow, *Hibiscus rosa-sinensis*, and hedges of bridewort, *Spiraea salicifolia*, just like the breadfruit tree, *Artocarpus altilis*, Indian fig opuntia, *Opuntia ficus-indica*, spotted laurel, *Aucuba japonica*, and gardenias, *Gardenia* spec., cultivated in a conservatory in the Herzogin Garten. In agreement with his father's instructions Jacquin the son reported to Vienna that he had made an arrangement with Richard van der Schot: two juvenile specimens of the banana, *Musa sapientum*, were to be sent from Dresden to Schönbrunn by carriage – in exchange for a specimen of the sacred fig, *Ficus religiosa*, and of the shooting star, *Dodecatheon* spec., plus, if possible, seeds of *Gloxinia perennis* and fruits of Turkey oak, *Quercus cerris*, to be sent from Schönbrunn to Dresden. There were, however, also some more astonishing plants – among them in the garden of Pillnitz a specimen of what Jacquin the son called *Veronica decussata*. This is a great surprise – the name had been validly published in 1785 by Conrad Moench for a plant that had been originally collected on the Falkland Islands and was subsequently cultivated in a garden near Weissenstein in the vicinity of Kassel. The name was almost never used again and was therefore most recently rejected, to be replaced by *Hebe elliptica*.[21] However, Jacquin the son explicitly used *Veronica decussata*, and correctly associated the name with a specimen he claims to have received from the Magellan Straits, collected by Philibert Commerson – indeed *Hebe elliptica* occurs in New Zealand, where it had been collected by the Forsters (see below), Chile, Argentine and the Falkland Islands. A fifth garden in Dresden was recorded only by Franz, who prepared a watercolour documenting the showy lady's slipper, *Cypripedium reginae* (Fig. 7.2.), an orchid native in the north-east of the United States and the south-east of modern Canada. The illustration is annotated 'Prince Antony's garden at Dresden, 1788' in a later hand and kept in the Natural History Museum. Prince Anton was the youngest brother of Frederick Augustus III to succeed him decades later as Anton I, King of Saxony. Franz's illustration with its dissections of the floral parts and the fruit clearly demonstrates the lessons learned from Jacquin the father in Vienna. Since Jacquin the son was an official traveller it is almost certain that he called upon Franz Anton von Hartig, imperial envoy in Dresden, chamberlain to Emperor Joseph II and privy councillor, before setting out with Franz for their next destination – Leipzig.

In 1785 Sibthorp had been to Leipzig and had paid a visit to Johann Hedwig, then the physician in charge of the military hospital in this important commercial city, well known for its trade fairs,[22] and so did Jacquin the son and Franz in July 1788. Meanwhile their host, aged 57, had been elected Fellow of the Royal Society and appointed Professor Extraordinary at the medical faculty of Leipzig University, which was the fifth oldest university in the Holy Roman Empire with an uninterrupted tradition. Only the universities of Prague, Vienna, Erfurt and Heidelberg were senior. However, Hedwig's speciality was not medicine, but mosses and liverworts. At a time when very little was known on the sexual reproduction

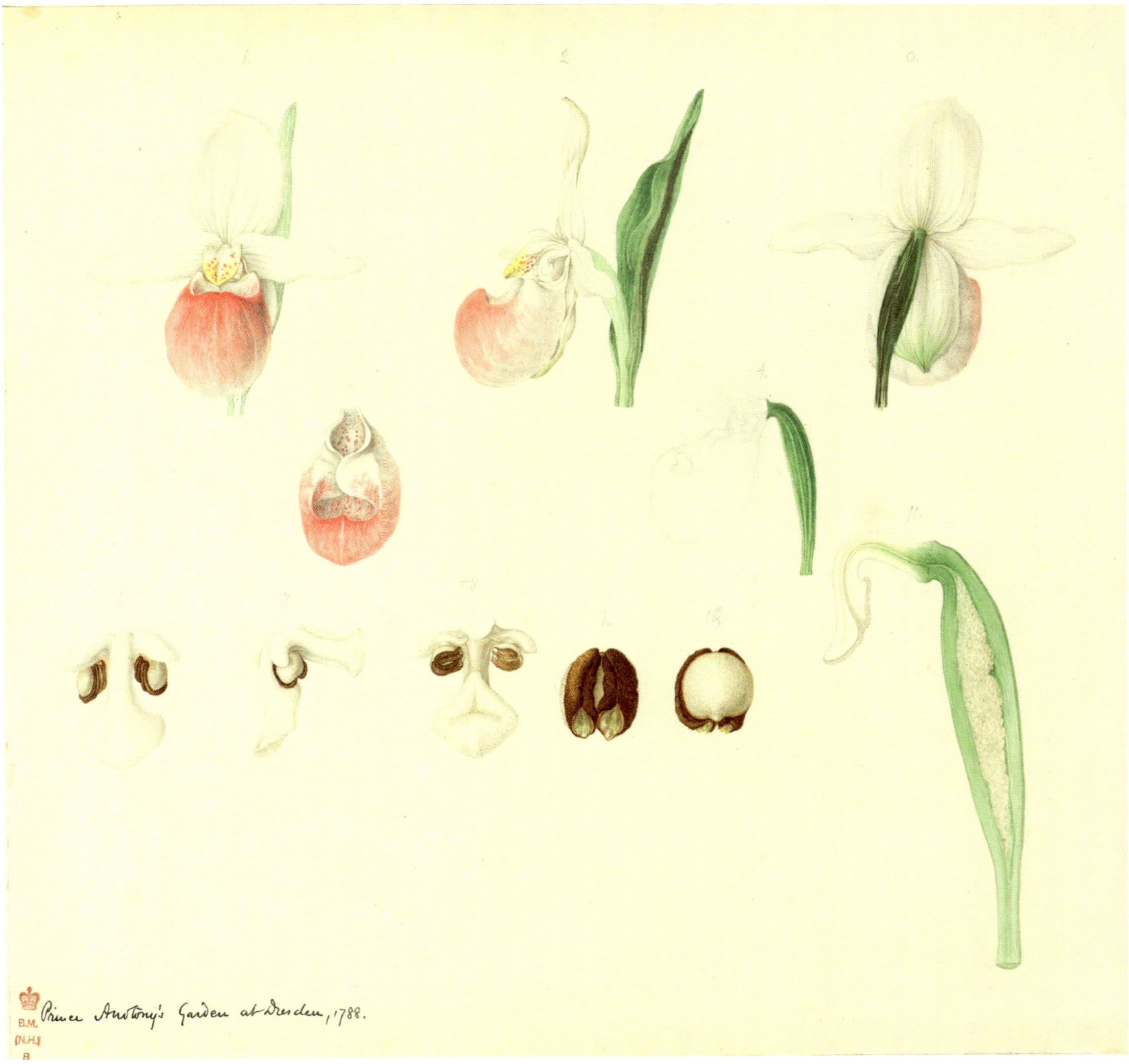

FIG. 7.2. *Cypripedium reginae*. Watercolour by Franz Bauer, 1788. – London, Natural History Museum, Library and Archives, Unfinished Drawings of Plants vol. 4, no. 219.

of the so-called lower plants – hence Linnaeus's name for them 'cryptogamia' (plants of hidden marriage) – Hedwig had made two important observations and published them in his very first paper: firstly that the inconspicuous antheridia (which he termed anthers) were the true male organs, and secondly that the minute cells emitted from them are the male gametes that fertilize the archegonia (the pistilla of Hedwig). As we now know, this was the correct interpretation, quite contrary to the prevailing view at his time, which regarded the spores of mosses as male organs and the spore capsule the equivalent of the anther. Hedwig had also observed the germination of moss spores and the growth from them of the filamentous protonema (the cotyledon of Hedwig), which is the juvenile growth form of the moss plant. In addition, he clearly distinguished mosses from liverworts. Hedwig also demonstrated the value, in the classification of mosses, of the peristome, the minute, elaborately constructed teeth that surround the mouth of the spore capsule. Hedwig's subsequent work on the taxonomy of mosses, which was already in progress when Jacquin the son and Franz paid him a visit, continues to be of outstanding relevance for modern science. As a consequence the International Code of Nomenclature for Algae, Fungi and Plants fixes the valid publication of names in mosses (except *Sphagnaceae*) as beginning with 1 January 1801, the date of publication of Hedwig's posthumously published *Species muscorum frondosorum*. In flowering plants Hedwig is regarded as the first to have described stomata, observed their opening and closing, and had a fair understanding of their function.

In a broad sense, Hedwig was a compatriot of his guests. He had been born in Kronstadt (Brașov, now in Romania), then belonging to the Kingdom of Hungary, where Emperor Joseph II ruled as King II. József. He had attended school in Vienna, Pozsony (Bratislava, now Slovakia) and Zittau, before being enrolled at Leipzig University. Maybe this had eased contacts with Jacquin the son and Franz. The former reports only briefly to his brother Gottfried that Hedwig had demonstrated the genitals of mosses as well as 'evaporation holes' (stomata), and spiral vessels in flowering plants adding with admiration 'who has not seen this, is unable to have an idea of it'. A modern account of Hedwig's achievements contains the following statement: 'Distinguished above all for exact and patiently repeated observations, Hedwig's work depended on the skilful use of dissection and the compound microscope. At first he used a simple lens magnifying 6X, but by successive improvements to the compound microscope . . . he was eventually able to use magnifications of up to 290X. In his *Fundamentum [historiae naturalis muscorum frondosorum]* he explains his method of dissecting mosses with the aid of needles and small knives, mounting the preparation in drops of water on glass slides – essentially the technique used today. He recorded his observations in accurate drawings. Although he did not teach himself to draw until the age of forty, the figures of his *Descriptio (aut adumbratio microscopico – analytica)* are among the most accurate and beautiful illustrations of mosses of his own and any other period'.

Franz may have been more impressed than Jacquin the son. After all, he was better qualified to judge the exceptional accuracy of Hedwig's microscopical drawings and his mounting of the pertinent preparations. In 1794 at the latest, Franz, then already settled at Kew, was to start his studies on pollen,[23] which made him a pioneer in this field very much in the same way as Hedwig was in mosses; however, unlike Hedwig, Franz left his findings unpublished. Eleven years later Franz was the first to accurately observe and draw the germination of pollen cells on the stigma (see Chapter 14), much like Hedwig had done when observing the germination of spore cells in mosses, but again he did not publish his findings. Needles, scalpels and, most importantly, microscopes, were to become standard utensils for Franz. In fact the visit to Hedwig may have had a long-lasting effect on Franz, who today is regarded as 'one of the greatest of The Royal Society's [early] Victorian microscopical observers'.[24]

Leipzig had provided a contact to an outstanding microscopist who has rightly been called the Linnaeus of bryophytes. Halle, the next destination, offered a meeting with another rare animal in the world of science – the circumnavigating naturalist and ethnographer. The only specimen in central Europe at that time was Johann Reinhold Forster, recipient of the honorary doctorate of Doctor of Civil Law from Oxford University and now professor of natural history and mineralogy at Halle University. The distance between Leipzig and Halle was a rather short one, but it meant entering the Electorate of Brandenburg, reigned over at the time by the Elector Friedrich Wilhelm III of the House of Hohenzollern, who was as Friedrich Wilhelm II (Frederick William II) also King of Prussia, a territory outside the Holy Roman Empire. Forster[25] had taken part together with his son George in the second voyage of James Cook on the *Resolution* – in lieu of Sir Joseph Banks and his party (see Chapter 8). As a matter of fact, this had been the first Antarctic expedition ever – reaching extreme southern latitudes, with landfalls including the Society Islands, Easter Island, Vanuatu, Norfolk Island and New Caledonia; in the latter two places those persons landing were the first westerners ever to do so. However, in 1788 Forster, now aged 58, one of the most widely travelled naturalists on Earth, was deeply involved in university feuds, in particular a fight against the hated, degrading 'Oberschulkollegium', imposed on the professors in the Kingdom of Prussia by the ministry in Berlin. Always a bold, difficult and embittered man, possibly his own worst enemy, he had been 'relieved' from his obligations for the museum of natural history at Halle University in that year. This was a surprising decision since Forster could claim to be the most famous of all its professors, a member of more than twenty academies of sciences and erudite societies, who had been received by five kings – among them Frederick II, the uncle of the now reigning monarch. It is hardly astonishing that Forster was – according to Jacquin the son in a letter to his mother – 'very much dissatisfied' and had expressed an interest in the position at Pavia University previously held by Professor Giovanni Antonio Scopoli, who had died two months earlier. Forster is also reported to have donated fifty herbarium specimens and 'Tahitische Zeug', possibly ethnographical objects from the Friendly Islands, to Jacquin the son, and had promised more herbarium duplicates. However, in the same letter a report is sent to Vienna that the circumnavigator, a bibliomaniac and therefore always short of money, although one of the best-paid professors of Halle University, was ready to sell his remaining herbarium for 3000 (thalers). According to Jacquin the son this collection had been offered previously at the same price to Friedrich Wilhelm II, King of Prussia, in Berlin, but the negotiations had broken down. Forster's mineralogical cabinet was judged by Jacquin the son to be 'quite good', adding in a condesending manner that he had received the best specimens 'from us' (the Jacquin family). Not unexpectedly, a long list of Forster's mineralogical desiderata is sent to Vienna. The conversations with Forster must have been interesting, although none of those taking part could know that thirteen years later Franz's brother Ferdinand would embark on the *Investigator* for another circumnavigation of the globe and spend more than five months on Norfolk Island, the tiny island in the Pacific that had been discovered on the voyage of the *Resolution*.

The recently enlarged botanical garden of Halle University had remained nominally under the care of Forster, but the actual work was supervised by the newly appointed Professor of Botany Philipp Caspar Junghanss.[26] Rather unsurprisingly, the latter is reported to have expressed the hope of receiving seeds from Jacquin the father, asking his son for advocacy. Again a box is announced to be sent to Vienna containing, among other things, roots of badan, *Bergenia crassifolia*, goat's horn, *Astragalus tragacantha*, and sensitive fern, *Onoclea sensibilis*, all three species already known to Linnaeus.

Instead of proceeding from Halle directly to Berlin, the two travellers chose a minor detour to visit Wörlitz, thereby entering for a short interval the Principality of Anhalt-Dessau, then reigned over by Leopold III Friedrich Franz, Prince of Anhalt-Dessau, known as 'father Franz'. He was an anglophile and a strong supporter of the Enlightenment, with a special interest in the education of the population of his principality in science and nature. Friedrich Wilhelm von Erdmannsdorff had erected Wörlitz Palace for him, the first neoclassical building in Central Europe; later the first and largest of the English parks in the Holy Roman Empire, today called Dessau-Wörlitz Garden Realm, was created on the prince's orders. The gymnosperms planted out in this park attracted the attention of Jacquin the son and Franz, the former reporting to Vienna on the existence of plantations of balsam fir, *Abies balsamea*, Weymouth pine, *Pinus strobus*, loblolly pine, *Pinus taeda*, and Atlantic white cedar, *Chamaecyparis thyoides*, all from North America, and of several species of juniper, *Juniperus* spec. Receiving cones seems to have been the main purpose of the visit. Franz did not anticipate that only a few years later he would be involved together with his younger brother in an in-depth study of gymnosperms, resulting in the famous monograph *The genus Pinus* (see Chapter 9).

Jacquin the son sent four letters to his parents and brother from the next destination of the tour – Berlin, the capital of the Electorate of Brandenburg and of the Kingdom of Prussia. The published extracts do not mention the city, which was then a town of c.150,000 inhabitants, with around twenty per cent belonging to the military, and therefore called by Sibthorp in 1785 'too military to please me much' (see Chapter 5). Three years later construction works for the Brandenburg Gate, a neoclassical triumphal arch designed by Carl Gotthard Langhans and modelled on the Propylaea in Athens, which he knew only from an engraving, had just started to replace a former city gate. Many decades later the iconic building became the very symbol of Berlin, Prussia and subsequently of Germany. In 1788 the architectural centre of the city was clearly the so-called Forum Fridericianum, consisting of the Royal Opera House, St Hedwig's Church, the Prince Heinrich Palace and the Royal Library, the latter erected on a plan bought from the architect Joseph Emanuel Fischer von Erlach based in Vienna (see Chapter 3). These buildings were arranged around a large square, the latter transversed by the famous boulevard Unter den Linden (Under the Lime Trees), its western end marked by the future Brandenburg Gate. It is extremely likely that Jacquin the son and Franz walked Unter den Linden, since the former reported that he had met Carl Ludwig Willdenow, then aged 22, the son of an apothecary, who owned a house on this boulevard. Willdenow was a medical student of Halle University, soon to be examined by Forster, with a strong interest in botany. Later he was to become a member of the Royal Academy of Sciences in Berlin, Director of the Royal Botanic Garden in Schöneberg near Berlin and eventually the first Professor of Botany at the newly founded Berlin University. It was Willdenow – 'a very amiable young man' in the words of Jacquin the son – who showed the two visitors round the botanical garden, which is reported to be extensive and in an excellent position (Fig. 7.1. p. 166). Some curious plants are mentioned, among them the smoketree surge, *Euphorbia cotinifolia*, from tropical America, and *Sibthorpia prostrata*, a prostrate perennial, its generic name a dedication from Linnaeus to his equivalent at Oxford, Humphrey Sibthorp. After this tour Jacquin the son and Franz went to the cemetery of the Dorotheenstadt quarter of Berlin to pay their respects to Johann Gottlieb Gleditsch. He had been the former director of the Royal Botanic Garden and a great admirer of Jacquin the father. The grave is reported to have been planted with a tree whose generic name referred to Gleditsch – the honey locust, *Gleditsia triacanthos*. Long is the list of plant species that Jacquin the son reports would be sent from Berlin to Schönbrunn, 'once the great heat is over'. Among them we find the Arctic raspberry, *Rubus arcticus*, winter aconite, *Eranthis hyemalis*, *Hebe elliptica*, and a species of *Lachenalia*. The latter was no doubt of personal interest to Jacquin the son, because he had formed the generic name, although it had been validly published by Johann Andreas Murray, a pupil of Linnaeus (see below). In Berlin the two guests from Vienna were also asked to write entries into Willdenow's poetry album, a kind of friendship book, though much more intimate than in other countries, containing quotations, poetry and worldly wisdom as well as signatures of friends and guests. The two entries in this *album amicorum*, today still in private hands,[27] differed widely – Joseph Franz, a professor's first son and a professor-to-be, chose a quotation in Latin and added a dedication in Latin; Franz wrote German prose in his neat hand and signed 'your true friend Franz Bauer from Vienna Berlin 24 July 1788'. This entry seems to be the oldest sample of Franz's handwriting.

Among the names mentioned in the extracts of the letters from Berlin we find one Nicolai, possibly Christoph Friedrich Nicolai, a key figure of the Enlightenment in the Prussian capital, a critic and publisher who had been asked to send four copies of George Forster's *De plantis esculentis* to Vienna. A second name is Holmskjold, almost certainly the mycologist Johan Theodor Holmskjold, Director General of the Danish Postal Services and of the Danish Porcelain Factory, who is reported to have sold his mineralogical cabinet for 20,000 thaler to Christian VII, King of Denmark and Norway, Duke of Holstein, in the latter position a prince of the Holy Roman Empire. The most outstanding figure mentioned, however, is Klaproth, clearly Martin Heinrich Klaproth, apothecary, chemist and member of the Royal Academy of Sciences in Berlin, who is reported to have found out that apathite is 'limestone with phosphoric acid'. Only one year after the visit of the two travellers from Vienna, Klaproth managed to extract a new element, which he named uranium, from pitchblende found at Wittigsthal near Johanngeorgenstadt. On 24 September 1789 he was to announce his finding at a meeting of the academy, no doubt a major breakthrough in chemistry and mineralogy, opening new gates to science. However, Klaproth was wrong in one detail – the black powder he had isolated was uranite rather than the element uranium, a metal, which took several more decades to be found and characterized. Before leaving Berlin, Jacquin the son and his travel companion are likely to have called upon the imperial envoy Heinrich XIV Prince Reuß of Greiz, the younger brother of Heinrich XIII Reigning Prince Reuß of Greiz, belonging to a family that had received the special privilege of having all their sons named Heinrich, thereby creating endless troubles for the historian. Nothing is on record on this meeting, since the bills of the official journey kept in the Kamerale in the Österreichisches Staatsarchiv in Vienna have not yet been analyzed. When crossing the Forum Friedericianum for the last time, Franz could not anticipate that the Royal Library would become the repository of his younger brother's swan-song, the so-called Berlin passion flowers (see Chapter 12).

BERLIN TO BONN

Much less is known about the subsequent journey to Bonn – in particular the precise itinerary is not completely clear. The first destination seems to have been Braunschweig (Brunswick), the capital of the Electorate of Braunschweig-Lüneburg. Only visits to mineralogical cabinets are mentioned, including the ducal cabinet (that pertaining to George III, the Duke of Braunschweig-Lüneburg, who was also King of Great Britain), containing extraordinary calcite druses. Since one Brückmann, probably Urban Friedrich Benedict Brückmann, a professor of anatomy and surgery at Braunschweig University, is mentioned, it is possible that his mineralogical collection was also inspected.

Continuing in a southerly direction, the next stop seems to have been Goslar, Imperial and Free City, where specimens of anthophylite were acquired. An ascent to the 'Blocksberg' (Brocken) followed, a 1141-metre-high peak in the Harz Mountains west of the town of Wernigerode, the capital of the County of Wernigerode, then ruled by Christian Friedrich Count of Stolberg-Wernigerode. The two travellers, both obviously familiar with the flora of the Alps to the west of Vienna, seem to have been disappointed. They had expected many alpine plants but found only a very few, among them alpine pasqueflower, *Pulsatilla alpina*, and alpine hawkweed, *Hieracium alpinum*. Bog rosemary, *Andromeda polifolia*, and a species of peat moss, *Sphagnum palustre*, were encountered on the boggy summit, where peat was dug in turbaries. The Brocken is known for its very rough climate, similar to that of Iceland, characterized by strong winds, high precipitation and an average temperature of 2.9°C. Clausthal, a nearby mining place in the Electorate of Braunschweig-Lüneburg, offered a contact to one Stölzner, who was interested in Hungarian and Austrian minerals in exchange with minerals from the Harz Mountains, 'of which no one has any idea in Vienna', as stated by Jacquin the son.

After this excursion to the Harz a visit to Göttingen followed, a Mecca of learning like Oxford or Cambridge, famous for the university that had been founded by George II, Elector of Braunschweig-Lüneburg, the grandfather of George III. For Franz, a visit to the university library must have offered a special moment – among its treasures was kept a copy of the luxury edition of the *Selectarum stirpium americanarum historia*. It is still there and has a title page either painted by Franz or by Ferdinand, although not signed (see Chapter 3). From the published extracts of Jacquin the son's letter it remains unclear whether the copy had actually been inspected, but it is reported to have been kept in a locked cupboard. Two more publications of Jacquin the father are reported to have been in that library – the *Hortus Botanicus Vindobonensis* and the *Florae Austriacae icones*, with sheets of thin paper between the coloured copper engravings to protect them. Franz could hardly have imagined that some six decades later this very library would become the repository of a considerable collection of his drawings, notes and manuscripts – of that part of his estate which was to be acquired at the auction of his belongings in London in 1841 by Ernest Augustus, King of Hanover, and donated by him to Göttingen University (see Chapter 17).

The Botanical Garden of Göttingen University, founded by Albrecht von Haller in 1736 and one of the few botanical gardens in central Europe that subsequently had never changed site, was found to be in good order. A banana and a species of *Murraya*, probably the orange jessamine, *Murraya paniculata*, were seen in flower. The generic name of the latter was dedicated to Murray, then the professor of botany at Göttingen University and a full member of the Royal Society for Sciences, also founded by George II.

In 1788 parts of central Europe formed a mosaic of semi-independent territories under the increasingly weak umbrella of the Holy Roman Empire. In order to reach the small university town of Marburg an der Lahn the two travellers had to enter the Landgraviate of Hesse-Kassel, then ruled over by the landgrave Wilhelm IX. He was a wealthy man, known to have hired Mayr Anselm Rothschild, the founder of the Rothschild family dynasty, to supervise the operation of his properties and tax-gathering. Marburg University is the second oldest protestant university in the Holy Roman Empire – after the ephemeral Liegnitz University (Legnica, now Poland); however, its botanical garden had been founded only two years before Jacquin the son and Franz arrived in mid-August 1788. It was situated west of the Elisabethkirche, one of the earliest purely Gothic churches in central Europe and held by some as the model for Cologne Cathedral. Conrad Moench, the author of the name *Veronica decussata*, was in charge. He had moved the botanical garden from the Collegium Carolinum in Kassel to Marburg, when the former institution was amalgamated with Marburg University. It goes without saying that there was nothing worthwhile to report to Vienna on the nascent botanical garden, apart from the fact that Moench intended to exchange minerals for seeds with Jacquin the father. The letter written in Aschaffenburg on 31 August 1788 and directed to Gottfried mentions a visit to Frankfurt, another Imperial and Free city, the place where following established tradition the election and coronation of the Emperor took place (see Chapter 3). Jacquin the son and Franz may have visited the respective sites – St Bartholomew Church and the famous Römer, a medieval building, where the coronation banquets were organized. In a family close to the court like the Jacquins all this must have been common knowledge. However, in the published extracts of the letters to Vienna only a meeting with one Salzwedel, no doubt Peter Salzwedel, the owner of the pharmacy 'To the White Swan' in Frankfurt, is mentioned. He possessed a mineralogical cabinet and a fine private botanical garden on the Main River, where Jacquin the son noted arrowhead rattlebox, *Crotalaria sagittalis*, *Dicliptera paniculata* of the Acanthaceae family, and a species of the leguminous genus *Pterocarpus*. A part of this garden still exists, while the other part has been lost to the construction of a particularly fine villa in the neoclassical style, the Villa Metzer erected in 1803, now housing the Museum Angewandte Kunst under the posh Frankfurt address Schaumain 15.

The next stop on the itinerary seems to have been Mainz on the left bank of the River Rhine, the capital of the Electorate of Mainz. In 1788 Friedrich Karl Joseph Reichsfreiherr von Erthal was archbishop of Mainz and ex officio Elector of Mainz, holding at the same time the office of Archchancellor of the Holy Roman Empire, with his deputy Rudolf Joseph Reichsgraf von Colloredo-Waldsee, the penultimate Vicearchchancellor of the Holy Roman Empire, based in Vienna (see Chapter 3). We cannot exclude the possibility that Jacquin the son, travelling under supreme orders of the Emperor, called upon Erthal, although this does not seem likely. The penultimate Elector of Mainz, who was also prince-bishop of Worms, opposed the dynastic tendencies of Emperor Joseph II and had even joined the mostly Protestant 'Fürstenbund', a coalition of princes opposing the Emperor's plan to exchange the Habsburg Low Countries for the Electorate of Bavaria. Clearly Erthal was an enlightened prince who had even a hymnal published in the vernacular. In addition he had convoked with others, notably the Elector of Trier, then Clemens Wenzeslaus Prince of Saxony, the Elector of Cologne (see below) and the archbishop of Salzburg and *Primas Germaniae*, the Congress of Ems, where twenty-three antipapal articles were drawn up and signed. Needless to say, the long reigning Pope Pius VI (see Chapter 4) rejected the idea of lowering his authority to a merely honorary primacy. According to Ehrtal's enlightened concepts the two universities of the Electorate – Erfurt and Mainz – were reformed and a few months after the visit of the two naturalists from Vienna he engaged George Forster (see above), a protestant, as his librarian. At Mainz University Molitor held a professorship, and this may have been a motivation for the two guests to visit the city. Otherwise one Fibig, no doubt Johann Fibig, Professor of Natural History,[28] and one Lau, certainly Bernhard Johann von Lau, Professor Extraordinary for Economics, held positions at Mainz University. The latter two were interested in mineralogy and are mentioned in a letter to Vienna.

From Mainz the travellers most probably continued down the river Rhine through the Upper Middle Rhine Valley to the city of Koblenz, a region usually called the Rhine Gorge. At the northern gateway to this area and on the right bank of the river the mighty fortress of Ehrenbreitstein, then belonging to the Electorate of Trier, would have attracted their attention, for two reasons. Firstly the Holy Tunic, a robe said to have been worn by Jesus and one of the top relics of the Roman Catholic Church, was kept in safe custody in this place belonging to the Archbishop of Trier, who was also ex officio the Elector of Trier. Since this object, also called 'Chiton of the Lord', was highly popular at the time and shown to the faithful only at large intervals, the two enlightened travellers could have heard about the place where it was conserved. Secondly Lucas Baur, Franz's father, had been born in Ehrenbreitstein, now part of the city of Koblenz (see Chapter 2), but we are unaware if Franz knew about this, since he had lost his father so early in his life and oral tradition may have been interrupted.

On 7 September 1788 at the latest the two naturalists from Vienna had reached Bonn on the left bank of the river Rhine, the residence of the Archbishop of Cologne and ex officio Elector of Cologne. Both these positions were held by Archduke Maximilian Franz, the youngest son of Emperor Francis I Stephen and Maria Theresa, brother of Emperor Joseph II and another enlightened prince like Erthal. This had resulted in problems with the conservative, non-enlightened Cologne University and led to the foundation of Bonn University in 1786, which received the necessary sanction from the Emperor. Needless to say, no botanical garden existed two years later, and again botany is not mentioned in the published extract of the pertinent letter sent to Vienna. A meeting of Jacquin the son with the Elector of Cologne seems more plausible than with the Elector of Mainz, and there is a special reason for this: Ingen Housz, Jacquin the son's uncle, had saved the Archduke's life by inoculating him against smallpox when he had been a boy of eleven years living in Vienna (see Chapter 3). The successful vaccination of Archduke Max Franz, his elder brother Ferdinand and their niece Maria Theresa, the daughter of their eldest brother Joseph, opened the gates for mass inoculations in the Hereditary States of the Habsburgs, a fact which must have been known to Ingen Housz's nephew who was now passing through Bonn. He may even have heard of the awards received by his uncle for the successful treatment of the imperial children – an annuity for life of 5500 ducats, the continuation of Ingen Housz's court lodgings, servants and equipage in Vienna, worth a considerable sum of money, an immediate payment of 3000 ducats, a diamond ring worth 1000 ducats, and from Emperor Joseph II a golden snuffbox with his portrait.

During their stay in Bonn Jacquin the son and Franz could also have met a young man who had just returned from Vienna – Ludwig van Beethoven, aged seventeen. Archduke Maximilian Franz, an admirer of Mozart like his eldest brother, had arranged the journey to the imperial city with the aim of having Beethoven, born in Bonn and therefore his subject, tutored by Mozart. However, nothing came out of this scheme. Instead young Beethoven had some lessons with Haydn, Franz's compatriot, born like him in the Archduchy of Austria below the River Enns.

BONN TO LEIDEN

Jacquin the father had been born in Leiden, where he had received elementary education. Later he had attended secondary school in Antwerp and after an interval at Leuven University had returned home in order to begin his medical studies. Among Jacquin the father's professors at Leiden University we find Pieter van Musschenbroeck and Bernhard Siegfried Albinus, who gave lectures in physics and surgery respectively, both Fellows of the Royal Society. Jacquin the father also had followed chemistry classes by Hieronymus David Gaub (Gaubius). However, the botanical lectures by Adriaan van Royen, Professor of Medicine and Botany, Director of the Botanic Garden at Leiden University and another Fellow of the Royal Society, seem to have made the strongest impression on him. Although Jacquin the father had left soon afterwards for Rouen, Paris and finally Vienna without completing his medical studies, he maintained links with Leiden, where his very first publication written in Vienna was printed. In a sense it would have been natural for him to recommend to his son a visit to the Republic of the Seven United Netherlands, the first state outside the Holy Roman Empire on his journey. Published extracts of four letters written in Rotterdam, Leiden and Utrecht, exist. The chronology proposed is tentative, since the localities mentioned are often adjoining and, furthermore, the two naturalists seem to have returned to Leiden before continuing to London. In addition, the extracts include two letters written on 20 December 1788, one written in Leiden, the other in London, which cannot be correct.

Maastricht, then an exclave of Brabant of the States within the Holy Roman Empire, seems to have been the first destination, where fossils and flint were studied. Jacquin the son and Franz had their first impression of the sea at Scheveningen, today a suburb of The Hague, but were driven away because of bad weather, noting sea-buckthorn, *Hippophae rhamnoides*, sea holly, *Eryngium maritimum*, and common stinkhorn, *Phallus impudicus*, occurring in masses. At The Hague, then the residence of William V Prince Orange-Nassau (Willem Batavus), the Stadtholder of the Republic of the Seven United Netherlands, henceforth the Dutch Republic, they may have called upon Karl von Merode Westerloo, the imperial ambassador accredited at the Stadtholder. However, this visit is not mentioned. Instead there is proof for a visit to 'Schenkischer Garten' (Schenk's garden) in The Hague, described as 'completely at its end' and – apart from a tulip tree, *Liriodendron tulipifera*, a large specimen of bald cypress, *Taxodium distichum*, and spruge-laurel, *Daphne laureola* – 'empty'. Why spend time in such a garden? The answer was not easy to find. Its owner, Martin Wilhelm Schwencke [sic], a physician and amateur botanist at The Hague, had died[29] and the contents of the garden had been auctioned, with the bulk of the living specimens bought by Emperor Joseph II for Schönbrunn. It seems that Jacquin the son tried to find out if anything of interest had been left behind, which obviously was not the case, noting for example that the large magnolia sold for 130 guilders had 'just made the testament'.

In Leiden Jacquin the son and Franz were hosted by Sebald Justinus Brugmans, Professor of Botany at Leiden University, who showed them the botanical garden and the natural history cabinet under his care. In the conservatory a *Heliconia* attracted the attention of the party; they reported back to Vienna that it was 'larger than ours, because it is a descendant of this one' testifying to the contacts between the two botanical gardens. However, Jacquin the son was not totally convinced about the botanical knowledge of Brugmans. In the Botanical Garden of Leiden University a meeting is recorded with a gardener named Meerboerg, who claimed to know Jacquin the father from the years when he had been an apprentice of Adriaan van

Steckhoven, who had subsequently moved from Leiden to Schönbrunn. Meerboerg is also reported to have demonstrated to Jacquin the son herbarium specimens sent by the latter's father during his first year in Vienna to Jan Frederik Gronovius in Leiden. Mention was also made of several trees in the Botanical Garden of Leiden University dating from the time of the great Boerhaave, regarded as the founder of modern clinical teaching as well as of the modern academic hospital who counted Haller among his pupils.

Jacquin the son's letter to his father written on 28 October 1788 in Utrecht contained no information about this city, but mainly on Amsterdam, in particular on what is called the 'Athäneum', a city-sponsored school, often considered the predecessor of Amsterdam University, though it was then not yet possible to earn a degree. It was the botanical garden attached to the Athenaeum that attracted the two guests from Vienna, who were given a tour by Nicolaas Laurens Burman, formerly a pupil of Linnaeus at Uppsala University and now a teacher at this school, under whose care the garden had been placed. Jacquin the son regarded the collection of living plants as the largest and most valuable in the Dutch Republic, which is mirrored by his note that he had visited this garden three times. His only criticism was the heavy pruning of the trees because of lack of space, a sign of 'corrupt' taste. A rather long list of remarkable species follows – among them the 'Dutch' *Heliconia*, the silver tree, *Leucadendron argenteum*, from the Table Mountain area of the Cape of Good Hope, the East Cape dwarf cycad, *Encephalartos caffer*, from the eastern province on the Cape of Good Hope, ericas, *Erica* sp., also from the Cape of Good Hope, as well as a cucumber tree, *Averrhoa bilimbi*, from Malesia. All these specimens exemplified the close connections of the Athenaeum garden with the Dutch possessions in Africa and Asia. Franz recorded one of the rarities in a pencil drawing kept in the Naturhistorisches Museum in Vienna. It is annotated 'Pekea Caryocas Amstelodomi 1788 Bauer del:', is not colour coded and shows an opened-up flower of the souari tree, *Caryocar* spec., from tropical America with special attention to the very numerous fine stamens exceeding the petals. Burman, who had shown Jacquin the son and Franz around this garden, was the author of *Flora Indica*, published in Amsterdam twenty years before they had arrived in this city. This work contained the second oldest printed illustration of plants collected in Australia (*Acacia truncata* and *Synaphea spinulosa*) though totally misinterpreted by Burman, who thought they were ferns from the East Indies.[30] At that moment Franz could not have imagined that his brother Ferdinand would soon afterwards make such an important contribution to botanical illustration in Australia when circumnavigating the fifth continent on the *Investigator*.

Amsterdam was also the place to pay a visit to the natural history cabinet of physician Martens Houttuyn, who had published the encyclopaedia *Natuurlijke Historie of uitvoerige Beschryving der Dieren, Planten en Mineraalen* in thirty-seven volumes. However, the cabinet had already been dissolved; only the herbarium was still waiting to be sold. Jacquin the son noted 'extraordinary magnificent things', among them superb specimens of a pitcher plant, *Nepenthes distillatoria*, from the East Indies. Other contacts in Amsterdam included physician, physicist and chemist Johan Rudolph Deiman, who demonstrated to a party of naturalists including Jacquin the son the production of artificial cold by mixing aqueous salt solutions. A chemist and mineralogist named Engelbronner, who traded in a wide range of chemicals, including acids, selling them to pharmacies, was also found to be interesting because of his mineralogical cabinet. It is reported to have included a magnificent collection of precious stones, agates and jasper from India as well as large borax crystals. As before in Leiden, Jacquin the son managed to locate a specimen sent by his father to Amsterdam – a specimen of ore from Idria (now Idrija in modern Slovenia) with a label in his father's hand, bought at auction from one Gronovius. Months later Haarlem, a town west of Amsterdam where two orchid species were seen, no doubt in a conservatory, is mentioned in a letter sent from London. These specimens could have come from De Hartecamp, George Clifford's estate near Haarlem, where Linnaeus had been engaged. The property was sold by the former's heirs in the year when the two travellers from Vienna were touring central Europe.[31]

For reasons unknown, Jacquin the son and Franz interrupted their stay in the Dutch Republic and returned to the Holy Roman Empire, more precisely to the Habsburg Low Countries. They definitely visited Brussels, where Archduchess Maria Christina, the favourite daughter of Maria Theresa (see Chapter 4), represented her brother as vice-regent, and Leuven. Nothing is known of the stay in Brussels, and very little on that in Leuven, where only the old colleges and the

Botanical Garden of Leuven University, 'better than expected', are mentioned in passing in the published extracts. After this interval the two travellers seem to have returned to Leiden, from where the final letter from the Dutch Republic was sent to Vienna.

Back in Leiden, this university town seems to have become for a second time the travellers' place of temporary residence, where Brugmans is reported to have accompanied his guests to the garden formerly owned by Boerhaave – at the latter's country house Oud Poelgeest in Oegstgeest, today a suburb of Leiden. It was a small world – Boerhaave had been the academic teacher and mentor of Swieten (see Chapter 3). A monumental tulip tree with pendant branches planted by Boerhaave is mentioned, it continues to blossom there even today. Further treasures shown to the guests from Vienna included three famous herbaria – those of Leonhard Rauwolf, Jakob Breyne and Paul Hermann, the latter herbarium from Suriname – all then in the keeping of the library of Leiden University. More to Franz's taste were probably two collections of plant illustrations also kept in this library: those brought together by Rumphius on the island of Ambon (Pulau Ambon in Indonesia), forming the basis for the illustrations posthumously published in his *Herbarium Amboinense*; and the famous *Codex Clusius,*[32] published in part in Clusius's *Rariorum plantarum historia*. The latter contained watercolours documenting macrofungi, a few with their vernacular names in German and Hungarian added; the material had been assembled when Carolus Clusius lived in Vienna. The specimens are reported to have been collected in 'Pannonia', the region stretching from what is now western Slovakia to what is now the northern part of Croatia. Franz would later also document a few macrofungi, among them the common stinkhorn, but he would get also deeply involved in the documentation of the ravages of microfungi on wheat (see Chapter 13).

Another academic contact was Floris Jacobus Voltelen, professor extraordinary at the medical faculty of Leiden University, who demonstrated chemical experiments on the fermentation of milk involving, surprisingly, mercury. He also showed Boerhaave's former chemical laboratory to his guests, where about half a century earlier Jacquin the father had followed the lectures by Gaub. In a sense, the stay in the Dutch Republic meant for Jacquin the son tracing the steps of his famous father and of the relatives of other Dutch expats now in Vienna. Among these was the brother of the gardener Richard van der Schot, of whom, as Jacquin the son reported, no one had heard in Delft, but who was known in The Hague. Other members of the Jacquin and Swieten families were either dead or lived in Vienna, like Jacquin the son's aunt Agatha Ingen-Housz, or Gottfried van Swieten, the man who was later to translate the texts used by Haydn in his *Die Schöpfung* (The Creation) and *Die Jahreszeiten* (The Seasons) from English into German.

For Franz the stay in the Dutch Republic may have made him aware of something quite different: the masses of unknown plants that had arrived from the Dutch overseas possessions in botanical gardens. Much more of this was to be seen during the next stop of their trip.

LEIDEN TO LONDON

Nothing is known of when and from which harbour the two travel companions left the Dutch Republic, and neither do we know when and in which harbour they arrived in England. The first, though somewhat doubtful, report (see above) is in a letter from London dated 4 December 1788. Twenty-one more letters sent by Jacquin the son were to follow, all directed to his father or his brother, all written in London. Not a single letter by Franz is known from the year 1789 and only two watercolours as well as two copper engravings based on his drawings are known.

Travelling seems to have now almost come to an end for Jacquin the son and Franz. The reason for this temporary intermission becomes evident in the letter of 16 December 1788, Jacquin the son writing 'Incontestably there is no place for botany where you can see and learn more than here in London. The large number of excellent facilities of all kinds available to everyone on the one hand and the indescribable zeal with which our science is pursued here on the other hand make London in this respect the first place in the world'. He praises the 'extraordinary number of herbaria of famous botanists of all nations' kept in the British Museum, those at Oxford, the Linnaean herbarium in the possession of Smith, and the 'richest herbarium in the world' kept by Banks (see Chapter 8). He continues 'Add to this the large number of excellent gardens and the enthusiasm for botany which is so general here. What an opportunity!'. In short, Jacquin the son feels he has entered paradise, and this may well have also applied to Franz.

Even before writing this letter Joseph Franz and Abbé Gruber had been invited to take part as guests in an evening meeting of the Royal Society on 27 November 1788. It is on record that the former subsequently attended fifteen such meetings before the summer recess of 1789. Clearly the son of a Fellow of the Royal Society and professor-to-be at Vienna University was regarded as a suitable invitee. About a year later, on 5 November 1789, Ingen Housz, a fellow like Jacquin the father, invited his nephew and Franz to take part in the evening meeting of the Royal Society in the pompous rooms in Somerset House, no doubt a memorable event for the country lad from Feldsberg. Banks also invited Joseph Franz, his drawing amanuensis Franz and Gruber to 32 Soho Square (see Chapter 8).

The published extracts from the twenty-two letters from London, biased as they are in their focus on natural history, deal with a plethora of topics, mainly botanical, while mineralogy and chemistry are now only of marginal relevance. Names mentioned in this context include John Hawkins, who had travelled with Sibthorp and Ferdinand in 1787 in the Levant (see Chapter 5), and one Rashleigh, probably Philip Rashleigh III, Member of Parliament, Fellow of the Royal Society and eminent mineralogist. The production of mercury oxide in the Apothecaries' hall in London is described. The occurrence of elaterite in Derbyshire, an elastic bitumen, of which specimens had been acquired at a good price by Jacquin the son, is reported back to Vienna. The discussions in the circle of the London cognoscenti on the electrolysis of water resulting in 'inflammable air' (hydrogen) and 'dephlogisticated air' (oxygen) found also an echo in the letters sent home.

Nothing transpires about where the two guests from Vienna had taken their lodgings in London, in which circles of society they moved or what sort of more general impressions they received during their stay, which was to last almost fourteen months for Jacquin the son and for Franz for the rest of his life, though he could not know this upon arrival. Both visitors remained in contact with the imperial ambassador at the court of St James – Carl Emmerich Alexander Baron Reviczky von Revisnye (Reviczky Károly Imre Sándor), followed in 1790 by Johann Philipp Count Stadion Warthausen, later as minister of foreign affairs the direct predecessor of Metternich and as minister of finance the founder of the Austrian National Bank. Keeping up good relations was necessary not only to receive the funds for the prolonged stay in London but also to be able to make use of a safe and speedy channel of communication with Vienna. Diplomatic bags were used to send letters, but possibly also boxes of seeds and books directed to members of the Jacquin family and others. Publications played a significant role in the correspondence, with Jacquin the son telling his father on 16 December 1788: 'It is unbelievable how many botanical works, good and bad, have appeared here over the last years. I would like to know, what I should buy and what not, and, in case I should make the Imperial Library more complete, I have to know, which botanical books from England are already there'. This note again hints to networking in Vienna: Gottfried van Swieten, the son of Jacquin the father's protector in Vienna (see Chapter 3), had in the meantime become prefect of the Imperial Library. Clear evidence exists that Jacquin the son acted also as London agent for the selling of the expensive, fully illustrated botanical publications of his father, to which Scheidl, and later Franz and Ferdinand Bauer, had contributed (see Chapter 3). Needless to say, there was plenty to write about the financial arrangements with booksellers and about the numerous other botanical works appearing at the time all over Europe, often in fascicles creating endless problems for the precise dating of scientific names with its well-known consequences on priority.

However, the letters from London are dominated by topics that continue to interest botanists most – the identity and naming of plants, in particular those new to science that flooded the herbaria and gardens in London from all four corners of the world. In addition, a lot had to be written about the masses of specimens both living and permanently preserved that were already in London, in particular in the Linnaean herbarium and in the collections owned by Banks. Several letters contain annotated lists of specimens, living and permanently preserved, received or sent out, synonymies and queries concerning botanical literature and individual herbarium specimens – after all, Jacquin the son had told his father 'In case you continue to be dubious about a plant, please kindly write to me, because this is the place [London], where one can find certainty about every subject'. Indeed, botany was held in high esteem at the time and so were the current publications – on 6 February 1790 Jacquin the son informs his father from Paris that he had been unable to study an illustration in the *Icones plantarum rariorum* because Banks had given his copy of this work on loan to Queen Charlotte.

In our context a single aspect needs special consideration – communication between London and Vienna and vice versa took place not only via texts (plant names and plant descriptions) and plant specimens (living material or herbarium specimens) but also via images (botanical illustrations), either drawings of a specimen living or permanently preserved, tracings of pressed herbarium specimens, or copper engravings. Exchange of information with the help of illustrations sent by mail was definitely easier, safer and cheaper than with the help of brittle and bulky herbarium specimens sent by mail. Sending potted plants was not quite an alternative because of the excessive cost and the repeated losses in transit. Sending seeds, bulbs or corms was more attractive but needed great patience – many years had sometimes to pass until the plant raised from the seed or corm finally flowered and became determinable. This state of affairs again emphasizes the advantages of being accompanied by a competent botanical illustrator. As a matter of fact Jacquin the father had earlier used the same method of communication for exchanging botanical information when he corresponded with Linnaeus in Uppsala, as evidenced from the former's letters with appended botanical illustrations now kept in the archives of the Linnean Society of London. Numbering specimens also helped communication between London and Vienna, including numbering of seeds, or rather seed packages, in order to be able to communicate more easily about the plants raised. A good example to illustrate this point is the letter of 3 November 1789 containing a lengthy discussion of an illustration of a *Lachenalia* species sent from Vienna to London. Jacquin the son offers a tentative determination, adds a critique of the leaves represented and compares the species depicted with its relatives by sole reference to other published illustrations. And on 7 August 1789 it was announced that a drawing of a species of cape pondweed, *Aponogeton* spec., prepared by Franz would be sent in the other direction, from London to Vienna.

Several pages have been dedicated to the various aspects of plant and seed transport. It is the speed of seed distribution that is almost unbelievable. The first indication of seeds received in London from Botany Bay is found in Jacquin the son's letter to Gottfried dated 25 May 1789, and on 12 June 1789 Jacquin the son confirms that Banks had received very many seeds from this locality (see Chapter 8), adding that the latter had promised him material from all species to be sent to van der Schot in Vienna. This is extremely early – only weeks after the first two ships, among them the *Prince of Wales,*[33] sent out from Portsmouth on 13 May 1787 as part of the First Fleet, had returned to England.

As were so many visitors, the two guests were impressed by the mild climate, noting that magnolias, rhododendrons and kalmias were cultivated in gardens without shelter, adding that 'all English wish to die when the thermometer is at minus 2°'. A great surprise is the comparatively small number of gardens mentioned in the published extracts of Jacquin the son's letters – and not a single garden is described. This is somewhat difficult to explain, particularly since the guests from Vienna spent almost fourteen months together in London – did Jacquin the son simply focus on the few strictly speaking botanical gardens that were important for him? While ignoring gardens that were less botanical, was he more interested in herbarium material than in living specimens, which is hard to believe? Was he so focused on plant diversity that he excluded any idea of a garden in the conventional sense? We simply do not know. Another hypothesis is of course the suppression of any 'non-scientific' texts by the editor of Jacquin the son's letters.

The most frequent references are made to the Royal Garden at Kew, then a royal estate and, though containing fabulous botanical collections, not yet a botanical garden in the strict sense (that is, open to the public on a regular basis, see Chapter 6). Further details about this garden, now called the Royal Botanic Gardens Kew, are given in the following chapter. However, in the extracts of Jacquin the son's letters more is found on the books about the plants cultivated in Kew, in particular on William Aiton's *Hortus Kewensis* (see below), than about the actual garden. The Chelsea Physic Garden, the second oldest botanical garden in England, is also only mentioned but not described. A small number of nurseries in the villages near London are listed as sources of plant material, among them such famous companies as those run by James Lee jun., William Curtis and Joachim Conrad Loddiges. The latter two were Jacquin the son's main suppliers of the Venus flytrap, *Dionaea muscipula*, a insectivorous plant from marshlands in south-eastern North America then fashionable in cultivation and very much sought after in Vienna. Syon House, a famous country house and garden then belonging to Hugh Percy, 2nd Duke of Northumberland, on the left bank of the Thames opposite the Royal Garden at Kew is mentioned only in passing. When Jacquin the son informed his father on 11 December 1789 about an invitation from Mary Watson-Wentworth, the Marchioness of Rockingham, widow of a former Prime Minister, to come and see her in Hillingdon House, all he comments about her garden is a brief note that it is 'very nice'. However, he explicitly notes having seen in this garden *Portlandia grandiflora* and *Oreopanax capitatus* in flower. The garden of Amelia Egerton, Lady Hume, the wife of Sir Abraham Hume, 2nd Baronet, another Fellow of the Royal Society, is mentioned, but only as the place where *Sterculia balanghas* flowered. In short, we cannot exclude the possibility that Jacquin the son was a conventional plant taxonomist who loved, studied and described the diversity of plant life in all its forms but had no genuine interest in the idea of a garden.

Following his father's instructions, Jacquin the son engaged himself in the transfer of living material from London to Vienna – of seeds and, more rarely, of bulbs and potted plants. As usual, acquisition was by way of purchase, exchange and gift. Clearly novelties and rarities had the greatest appeal; of these only three species announced by Jacquin the son in his letter of 10 July 1789 as to be sent to Vienna are mentioned here: the wintersweet, *Chimonanthus praecox*, a deciduous shrub native in China with strongly scented pendant flowers opening in the dormant season; the Madeira giant bellflower, *Musschia aurea*, an endemic of Madeira with distinctly erect, yellow flowers; and the great swamp orchid, *Phaius tankervilleae*, a terrestrial orchid native in tropical south-eastern Asia, said to have been first introduced from China by John Fothergill, a Quaker physician and Fellow of the Royal Society. Novelties and rarities from Vienna were requested by Jacquin the son in his letters of 30 March and 10 July 1789 to be sent to London to reciprocate, of those listed three are given as examples here: Zoys's bellflower, *Campanula zoysii*, a tiny light-blue bellflower restricted in distribution to the high limestone mountains of the south-eastern Alps, described as a species new to science by Wulfen; the tartar bread plant, *Crambe tataria*, a crucifer known to the Bauer brothers from their work for Boccius in Feldsberg (see Chapter 2); and the Wulfenia, *Wulfenia carinthiaca*, a local endemic of the Carnic Alps whose generic name coined by Jacquin the father commemorates its discoverer Wulfen. The latter three species were probably sent as seeds to London, whereas according to Jacquin the son the great swamp orchid, for example, would be prepared for packing up for transport when the plant had shed its leaves. Apart from supplying alpine and pontic plants, Jacquin the father is also asked to send Banks fruits of the dwarf palmetto, *Sabal minor*, a palm native of tropical America.

For reasons set out in the following chapter, only two watercolours by Franz, both kept in the Natural History Museum, are to be unequivocally datable to 1789. One of them depicts a species of *Gladiolus* sect. *Hebea* of unknown origin inscribed in black ink on the verso 'Franc: Bauer pinxit 1789', the other shows a flowering twig of the wintersweet cultivated in the Royal Garden at Kew (Fig. 7.3.) with three analyses added – a flower with the tepals removed, a flower with the tepals and stamens removed and a longitudinal section of the flower showing its thick cup-like receptacle. In the same year a copper engraving based on this very watercolour was published as plate 10 of *Hortus Kewensis*, a catalogue of the plants then being cultivated at Kew and a work that ranks prominently in Jacquin the son's letters to his family in Vienna. Collaboration between botanists in London and Vienna took many forms. On 10 July 1789 Jacquin the son tells his father that van der Schot will receive a specimen of lemongrass, *Cymbopogon citratus*, which had not yet produced flowers in England, hoping that the gardeners in Schönbrunn would be more successful. This plant from south-eastern Asia had to wait for decades until it received a Linnaean name and until it was again recorded – after Rumphius – in an accurate illustration.

While in England the two guests seem to have travelled little. A trip to Edinburgh was mandated by Jacquin the father, but we are unaware what came of it. In the letter of 12 June 1789 a visit to Oxford is briefly mentioned, but we have few details. Five days were spent in this city in order to become acquainted with the various institutions of Oxford University. Only the drawings and watercolours for the *Flora Graeca* could be inspected, but not the specimens, since Sibthorp was bedridden with a fever. This may have been the first opportunity for Franz to meet Ferdinand after an interval of two years, during which so much had happened. In the report to Vienna mention is made of rare minerals, such as zeoliths from Scotland, given by William Thomson, a Fellow of the Royal Society, physician of the Radcliffe Infirmary at Oxford and founding member of the Linnean Society.[34]

The main point of reference for Jacquin the son and Franz was clearly London, and more particularly Banks and his librarian Dryander, who both treated their guests

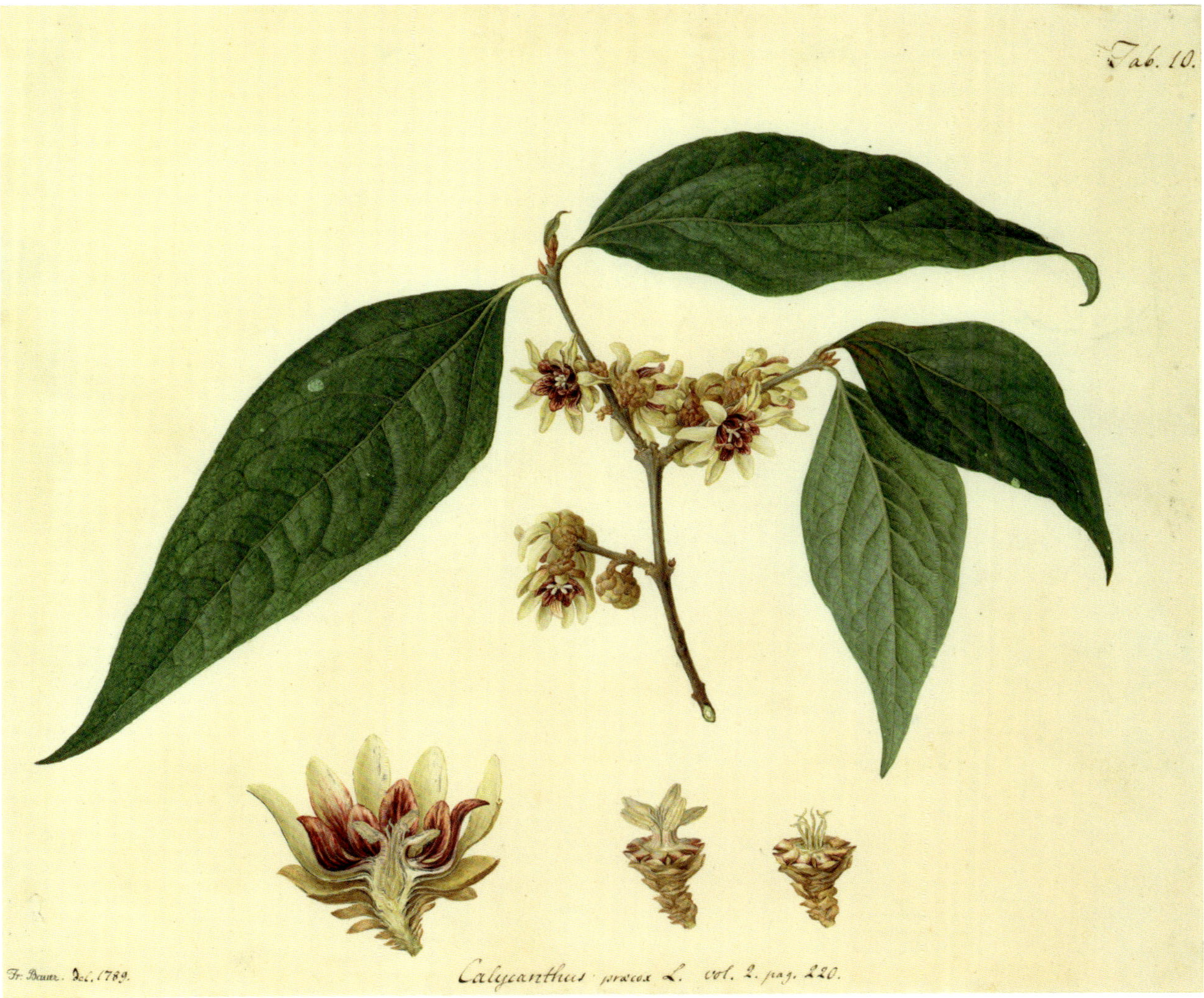

FIG. 7.3. ***Chimonanthus praecox.*** **Watercolour by Franz Bauer, 1789. – London, Natural History Museum, Library and Archives, Original Drawings for Hortus Kewensis, t.10.**

in the most liberal manner – by permitting not only access to the library and herbarium but also in donating copious herbarium specimens. On their journey across central Europe the two travellers had seen many natural history cabinets, but the 'crown of all', as very appropriately worded by Jacquin the son, was to go to 32 Soho Square. Here is his description: 'Banks lives in a very large, nice house in Soho Square. His library and herbarium have been placed in three large rooms, two for the library, one for the herbarium. His library consists of natural sciences only, travel books and all journals and serials of academies and societies; in botany it is as compleat as possible. The plant collection [i.e. the herbarium] is kept in wooden cupboards all made of mahogany, covering all walls of the room. The inner arrangement is roughly like in my [herbarium cupboards], however, all specimens are mounted. On each herbarium sheet the [scientific] name is noted and from whom he [Banks] got it [the specimen]. In addition there is a manuscript [catalogue] of about fifty volumes in octavo belonging to the herbarium, started by [Daniel] Solander. There is a sheet of paper for every known plant, on which all more recent synonyms are noted down, as well as corrections and notes, collected from whatever source. Above the chimney there is a portrait of Linnaeus by [Alexander] Roslin, to the right a box containing plants in spirit and to the left a collection of drawings. Every foreigner, who has been introduced to him [Banks] and who has been given permission, can spend his time from 9.30 a.m. until 4 p.m. [there] and make use of everything just as if it were his property, even if neither he [Banks] nor Dryander are present. In the centre of the room three to four tables are placed, on each of them ink, pens, pencil, paper etc. All newspapers are also there. Until 10 a.m. there is breakfast for everyone, and up to that moment Mylady [Lady

Dorothea Banks] and Miss Banks [Sarah Sophia Banks] remain there. The herbarium is rich beyond expectation; whatever genus is studied one is amazed!' In short, 32 Soho Square was not only the city residence of the Banks family, but also a private botanical institute with a fine library, a superb herbarium and working facilities for Banks's librarian and curator as well as his visitors.

On 3 November 1789, about eleven months after his arrival in London, Jacquin the son had written to his brother 'Compliments from [Franz] Bauer. I believe it is high time to bring him away from here, otherwise he is filtered from me. Banks has his eyes very much on him'. This indeed was the case. Banks, then aged forty-six, made an extraordinary proposal to Franz – an annuity of £300 for life, an enormous amount of money, under the sole condition that he would remain in England and work for him. An offer from a man like Banks could not be rejected, since it effectively meant a permanent position as a scientific illustrator under very generous conditions. Further details are given in a letter by Sir James Everard Home writing shortly after the death of Franz, summarising the latter's reminiscences: 'He [Franz] came over with Jacquin then on their way to Paris, passing through London. He [Franz] was to remain in Paris 3 years. He was to go [together with Jacquin the son to Paris] not withstanding the Revolution which was then in motion. His passage was engaged and place taken by the coach on which he was to go the next morning to Dover, on the last day of his stay in London . . . he [Franz] had dined with Sir Joseph Banks, who after Tea took him aside (Into the Voyage Room) and made his proposals to him to remain in England, which he at last accepted'.[35] This is how it came to be, that Jacquin the son left London for Paris in late January 1790, whereas Franz remained behind. Nothing is on record about the former's feelings on losing Franz, but maybe he regretted having told his father on 7 August 1789 'When you write to him [Banks] I would like to ask you to thank him for all his almost fatherly kindness with which he has treated me'.

Surprisingly, no document of the arrangement between Banks and Franz exists, apart from a note signed 23 years later by Franz stating, 'I the undersigned do hereby declare that all the sketches and drawings which I have made since the year 1790 (as well as those I have already delivered, as those which I still have under my care in my house) are the property of, and belong to the Right Honble Sir Joseph Banks, K. B., I having during that time received from him an annual salary. I likewise declare that all the sketches and drawings which I shall make thereafter, as long as I shall continue to be engaged for that purpose by the Right Honbl Sir Joseph Banks, K. B., and any of them which might be found in my house at the time of my death (should that happen during my continuance of my so being engaged) shall be the property of, and belong to the Right Honbl Sir Joseph Banks'.[36] This statement indicates that Franz's engagement started in 1790; it was to last until his death. As will be shown later (see Chapter 16), the document signed by Franz was to have wide-ranging consequences after Banks's death in 1820 and subsequently after Franz's death in 1840, when the future of his extremely diverse pictorial oeuvre had to be decided upon (see Chapter 17).

Banks's intentions for recruiting Franz in 1790 on a permanent basis become perfectly clear in the second codicil to his will dated 3 March 1820, signed just three months before his death: 'I still continue hold and believe to be founded in truth, that the establishment of a botanic garden cannot be compleat unless a resident artist to be constantly employed in making sketches and finished drawings of all new plants that perfect their flowers or fruits in it be part therefore'. Furthermore Banks had expressed in this codicil the hope that 'the charge of maintaining Mr. Bauer would . . . be . . . transferred from me and placed on the establishment of the Garden', but this had never become true nor was it to become true after Banks's death. In short, Banks acted in late 1789 for the King in order to promote botany and science in general by recruiting Franz as illustrator, or in the words of Banks as 'resident artist'. Indeed he selected an extremely many-sided and experienced young man, who had visited more botanical gardens and, possibly, mines than he had himself.

Having accepted Banks's offer, Franz, aged thirty-one, must have felt as if he were in a dream – like his two elder brothers he had gained a permanent position, but on much more favourable terms. Joseph, back from Rome, had found in Vienna a permanent job on the payroll of the Reigning Prince of Liechtenstein (see Chapter 12), and Johann on the estates of this very Prince, although we have only the most fragmentary information on the engagement of the elder brother (see Chapter 16). Furthermore Franz's salary was indeed grandiose, in the same order of magnitude as a professorial emolument at Oxford at the time. For comparison: Captain Cook's salary for his second circumnavigation of the globe lasting for more than three years had been merely £500, while Franz was to receive £300 per annum – for life. In retrospect it has been rightly said that Banks placed Franz in 1789 'in a state of moderate independence, enabling him to pursue through life the bent of his Genius unshackled by the caprice of the Public, and independent of booksellers and critics'.[37] Arrangements between Banks and Franz seem to have been informal right from the beginning: Franz enjoyed complete financial security and wide freedom to choose the subjects of his studies. In addition there was no exclusivity: Franz was permitted to work also for other commissioners, and this indeed he did in subsequent years (see Chapters 15 and 16). Right from the beginning Banks seems to have had a high opinion of Franz. In the year of the latter's death one of Franz's neighbours in Kew wrote of Banks's 'estimation conceived from the beginning of their acquaintance and friendship for Mr. Bauer'.

But what kind of personality was Sir Joseph Banks? What were his motives in hiring Franz as a botanical illustrator on a permanent basis under such generous terms? How could he afford to pay out such a noble salary?

ENDNOTES

1 Unless otherwise indicated, this chapter is based on Kronfeld (1909, 1921).
2 Lack (2000c)
3 Madriñán (2013)
4 Fitzinger (1840)
5 Beale (2011)
6 Jacquin, J. F. (1787)
7 Duncan (2012)
8 Camper (1782)
9 Lack (1999c)
10 Autogr. 14/1, 141/87.
11 Petz-Grabenbauer (1996)
12 Vienna, Fürstliche Sammlungen, Archiv. - Majorat-Kassa Rechnung 1788, No. 524.
13 Kronfeld (1919)
14 Jacquin (1800), p. 19, t. 11.
15 Jirásek (1962)
16 Maiwald (1904)
17 Anon. (1836)
18 e. g. Fettweis (2001)
19 Russell (1954)
20 Helm (1965)
21 Bayly & Kellow (2004)
22 Heltmann (2002)
23 Stewart & Stearn (1993)
24 Ford (2001)
25 Hoare (1975)
26 Kaiser & Piechocki (1968)
27 Currently deposited in the Library of the Botanischer Garten und Botanisches Museum Berlin-Dahlem in Berlin.
28 Pertlik (2013)
29 Heine (1963)
30 Nelson (1994)
31 J. Compton (pers. comm., 2014)
32 Leiden, Universiteitsbibliotheek Codex BPL 303; Clusius (1983)
33 Cavanagh (1990)
34 Torrens (2004)
35 London, Natural History Museum, Library and Archives, R. Brown correspondence 1: f. 214. – Sir James Everard Home to R. Brown, 17 xii 1840, unpublished letter; cf. Meynell (1983)
36 Desmond (2007)
37 Scheer (1840)

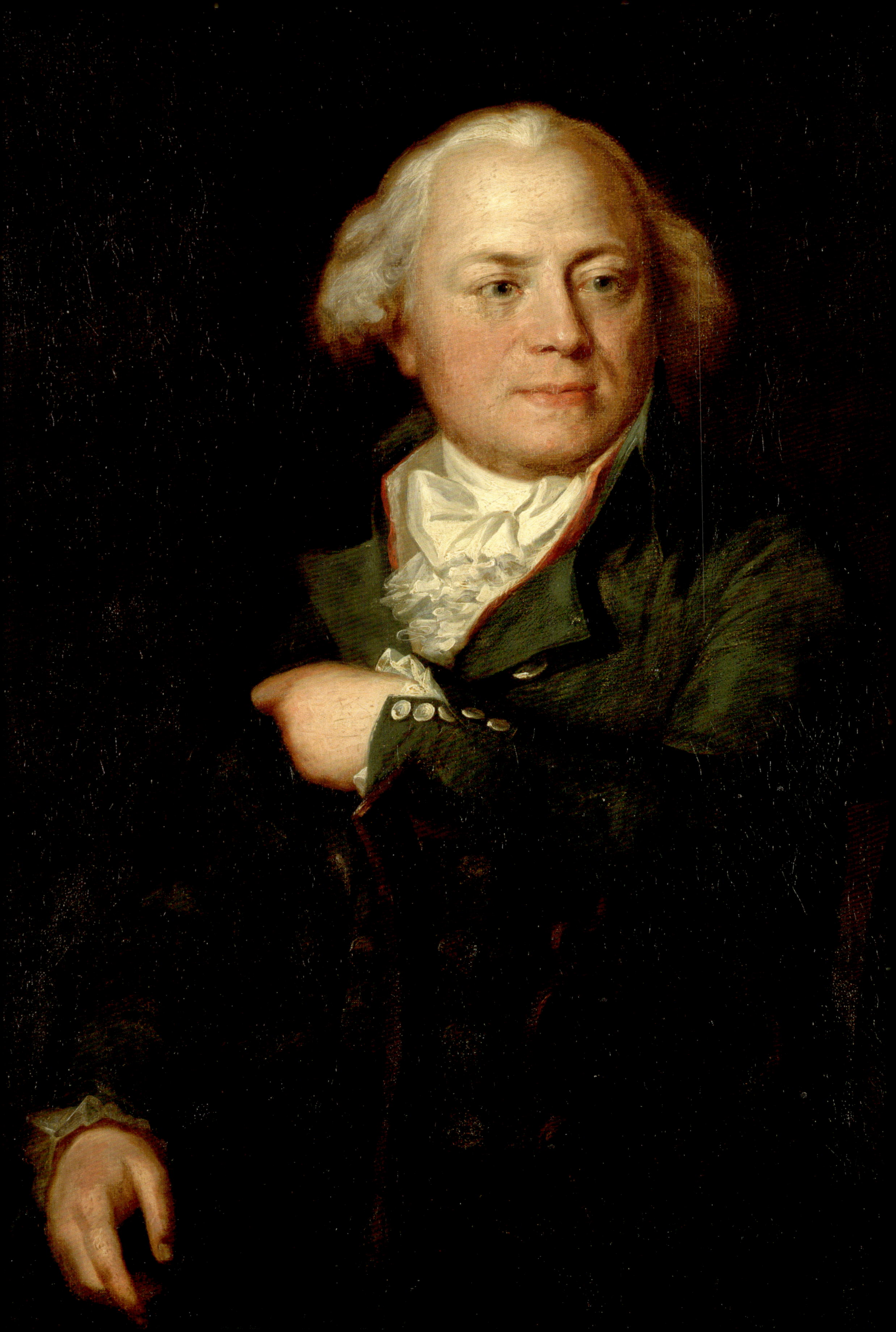

CHAPTER 8

Franz Bauer in the circle of Sir Joseph Banks in London and Kew[1]

SIR JOSEPH BANKS, BT, PC, GCB, DLC, PRS, FSA

JOSEPH BANKS WAS ONE of the great figures in Georgian England, not as a scientist in his own right, but as a great promoter of science, in particular of the natural sciences. His outstanding achievements were duly acknowledged – he was knighted, created Baronet, nominated Privy Councillor, received the red ribbon of the Order of the Bath and, more importantly, was elected Fellow of the Royal Society at an early age. Later he became its longest-sitting president, holding this position for over forty years – from 1778 until his death in 1820. From his home in 32 Soho Square, London, Banks was involved in many projects relevant to an extremely broad spectrum of topics. Placed in the centre of progress in the natural sciences and an early networker on a global scale, he had his finger in 'every scientific pie', as rightly observed by one of his several biographers. William Eden 1st Baron Auckland rightly called Banks 'His Majesty's Ministre des Affaires Philosophiques', which may be translated with less polish into modern terms as 'His Majesty's Minister of Science and Technology'. This unofficial position went with great influence, particularly as he filled it for more than four decades. Banks never held a government position, nor was he ever a member of Parliament, but as a man with a huge private income he dedicated his life to the service of his country. The reasons for choosing his seal may be regarded as the motto for this great philanthropist: 'I have taken the Lizard, an animal said to be endowed by nature with an instinctive Love of Mankind, as my Devise, and have caused it to be engraved on my Seal, as a Perpetual Remembrance that a man is never so Well Employed, as when he is labouring for the advantage of the Public; without the Expectation, the hope or Ever a wish to Desire advantage of any kind, from the Result of his Exertions'.

FIG. 8.1. Franz Bauer, c.1795. Anonymous portrait. Oil on canvas. – Kew, Royal Botanic Gardens.

In our context three aspects of Banks's personality merit special attention. He possessed a lifelong passion for botany in all its aspects and ramifications, energetically promoting horticulture and agriculture and acting among his many duties as the unofficial superintendent of the Royal Garden at Kew. Secondly he was fully aware of the importance of an accurate graphical record in science in general, and in botany in particular. And, thirdly, he was almost obsessed by the thought of Britain being outrivalled by France in the struggle for supremacy on a global scale. At the same time, Banks was an utterly practical man who never lost contact with the realities of life, in particular its financial aspects, and possessed an excellent feeling for the political and economic potential of new developments. He played a leading role in the colonization of Australia by Europeans, in the voyages to the South Seas in order to transfer the breadfruit tree to the British possessions in the Caribbean, in the importation and improvement of the Spanish merino sheep that in the long run became the basis for the Australian economy, and, back home, in the foundation of the Ordnance Survey.

While today any contact with artists or knowledge of the arts is almost anathema to many natural scientists, this was not the case for Banks. He was a genuine connoisseur and member of the Society of the Dilettanti, a London dining club, which became the prime mover in the establishment of the Royal Academy of Arts. Sir Joshua Reynolds, its first president and a member of the Dilettanti, was the man to paint one of the more famous portraits of Sir Joseph Banks. Needless to say, Banks had great respect for the talents and qualifications of artists, which was to play such an important role when hiring Franz for life (see Chapter 7) and providing long-term support for Ferdinand via the Admiralty (see Chapters 10 and 11). Since these arrangements started in 1789 and 1801, respectively, and lasted until Banks's death, Franz and Ferdinand experienced the established President of the Royal Society (Fig. 8.2.), later the autocrat of the 'philosophers' and domineering old baronet confined to his chair by crippling gout. They did not meet the romantic young man of wealth who stooped to sail with Cook onboard the *Endeavour*, which is the stereotype of Banks for the less initiated.

FIG. 8.2. Sir Joseph Banks, 1809. Portrait by Thomas Phillips. Oil on canvas. – London, Royal Society, RS 9544.

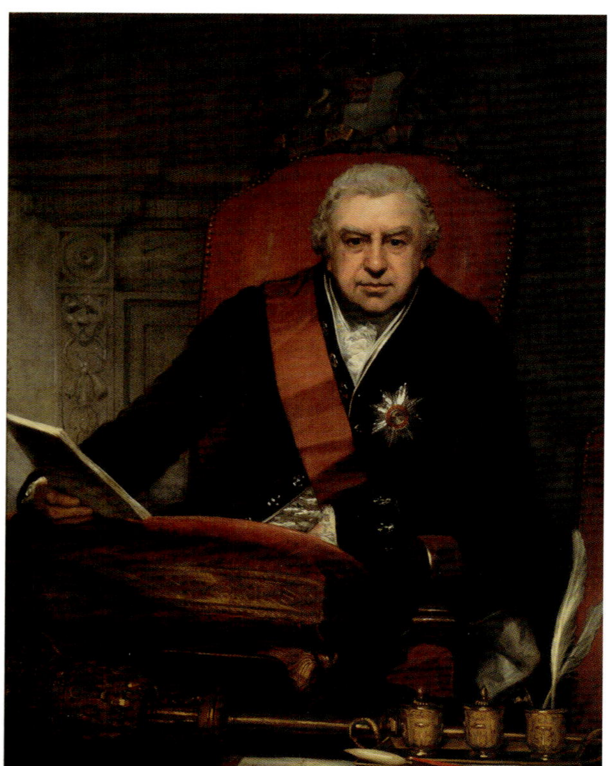

Joseph Banks was born on 13 February 1743 in London as the only son of William Banks, MP, and his wife Sarah, née Bate, and grew up with his sister Sarah Sophia at Revesby Abbey, one of the family's country houses near Boston in Lincolnshire. As with so many of his class he was first sent to Harrow and later, surprisingly, also to Eton, where Banks's lifetime engagement with the study of nature, and the plant world in particular, began. This found visible expression in his first collection of plants, insects and shells gathered along the River Thames, augmented by ramblings among the fens of Lincolnshire. Christ Church, Oxford was the next step in his education, but little is on record about these years. Considering the inertia of Humphrey Sibthorp, the second Sherardian Professor of Botany (see Chapter 5), it seems that Banks was largely self-educated in the natural sciences. Before his first university year was out, the premature death of his father took him back to Lincolnshire and to his very substantial inheritance, placing him among the happy few in England, though not in the same class as the titled nobility. In these last years of his minority he seems to have served a sort of apprenticeship in estate management under the tutelage of his uncle, which was extremely valuable for him in later years. Quite naturally Banks, with his obsession with natural history, remained an oddity in the Junior Common Room at Christ Church, as he was to remain an exception in the council rooms and with his academic friends. But he was always a respected oddity. Banks's lifelong association with the British Museum, then in Montagu House, started with his first reader's ticket issued in 1764, just five years from its opening to public use. It is plausible to assume that it was not long before he entered a working relationship and lasting friendship with Daniel Solander, a pupil of Linnaeus and then an assistant of the Department of Natural and Artificial Productions at the British Museum. Networking in London started early for Banks – at the age of twenty-three he was elected Fellow of the Society of Antiquaries and Fellow of the Royal Society, owing clearly more to the cluster of relationships formed in and about the British Museum than to anything else.

After seven years of increasingly academic discipline there followed three voyages that offered ample opportunities for investigative field work, with the second of these making Banks world-famous. The first voyage on the *Niger*, bound on fisheries protection duty, led Banks, accompanied by a servant, to the coasts of Newfoundland and Labrador. It demonstrated to him the realities of sailing and working in adverse conditions. Here is a brief extract of his report: '. . . we had a very hard Gale . . . which has almost ruind me in the Course of it we shipp'd a Sea which Stove in our Quarter & almost Filld the Cabbin with water in an instant where it washd backward and forward with such rapidity that it broke in Pieces Every chair & table . . . among other things that sufferd my Poor Box of Seeds was one which was intirely demolish'd as was my Box of Earth with Plants in it which Stood upon deck . . .'. This voyage, lasting a total of ten months, turned out to be a practical experience in the hazards as well as the opportunities of field work in natural history and was thoroughly profitable for the years to come. Back in London, and typically for Banks, he had a selection of his trophies documented in watercolours very true to nature – the plants by the ageing Ehret (see Chapter 5), the birds, fishes and insects largely by the nascent Sydney Parkinson. However, the voyage on the *Niger* was just a prelude for a much longer and more significant undertaking.

On 19 November 1767 the Committee of the Royal Society had recommended that the Government be petitioned to send a ship with observers to follow the transit of Venus from islands in the South Sea. This was an exceedingly rare astronomical phenomenon, wrongly thought to be of direct relevance for improving navigation at high seas. Not deterred by his rough crossings of the Atlantic, the twenty-six-year-old Banks did everything to join the expedition, which was to become the triumph of British seamanship – the voyage of the *Endeavour* commanded by Lieutenant James Cook. After lengthy deliberations the Admiralty agreed to accept 'Joseph Banks Esq. and his Suite consisting of eight persons with their Baggage, bearing them as Supernumeraries for Victuals only'. This suite was paid for by Banks and comprised his friend Solander, who was to be the botanist on the voyage, four servants, a secretary and, most importantly for our story, Parkinson, responsible for the painted record of plant and animal life.

The voyage that started on 25 August 1768 from Plymouth, mulled over and romanticized ever since, with all its moments of triumph and disaster, has been told so often that neither a summary nor another analysis is necessary here. What matters in our context is the rich pictorial record of plant and animal species prepared by Parkinson, supplementing the botanical and zoological collections. Because of the difficulties in preserving biological specimens under tropical conditions, the material record is incomplete, in particular for animals, whereas the pictorial record is rich and has survived the vicissitudes of the *Endeavour* voyage. However, Parkinson was so overwhelmed by the material to be documented during the second half of the voyage, in particular during the landfalls on the eastern coast of Australia newly discovered by westerners on that occasion, that he prepared only incompletely coloured drawings. In short, what arrived on 12 July 1770 in the Downs was merely the basis for a pictorial record of plant and animal life, which still had to be completed – a total of 673 unfinished and 269 finished botanical drawings all by Parkinson, plus 298 finished zoological drawings, largely by Parkinson.[2] What mattered more were the huge losses in the ship's company, among them Parkinson: of the original ninety-four men only forty-one survived, a mortality rate of fifty-six per cent. It has rightly been said that the expedition on the *Endeavour* remains unique among British voyages of exploration, for no other was conceived or conducted on this special pattern – a collaboration of civilian science under royal patronage with royal finance executed under Admiralty management, joined with private enterprise funded from a county rent roll. Of course the latter refers to Banks's income from his estates, which was not as gigantic as that of the Duke of Marlborough or the Reigning Prince of Liechtenstein, but still very considerable.

Back in London, Banks was quite naturally a subject of gossip and social interest. He was quick to pay his respects to the Royal Society; a few days later he was presented to King George III at the Court of St James, and shortly after that, together with Solander and accompanied by the President of the Royal Society, to the King and the Queen at Kew. This contact with the royal couple was of long duration, lasting until George III went insane, but was maintained by Queen Charlotte until her death. Indeed, Banks served as a sort of informal advisor to the royal couple on a broad spectrum of matters, which gave him considerable political power and influence. A plan soon emerged for a second voyage to the South Seas, to be undertaken in the opposite direction. As a consequence the Admiralty ordered the Navy Board to purchase two vessels, renamed *Resolution* and *Adventure*, for service to remote lands. It was clear that Banks would again take part in this voyage, although under a somewhat different regime. If any incentive beyond exploration was needed, it came from Paris. An expedition sent out by Louis XV, King of France, on *La Boudeuse* and *L'Étoile* had preceded the *Endeavour* in the Pacific, although it missed Australia. However, the travelogue was speedily published as *Voyage autour du monde* in the French capital, in contrast to the much slower publication of the official account of the *Endeavour* voyage. The race between Britain and France for the South Seas had started.

Because of disagreements with the Admiralty, Banks's plan came to nothing and in his place Johann Reinhold and George Forster (see Chapter 7), together with Cook, explored the distant shores of the Pacific. They circumnavigated the globe in an easterly direction and penetrated deep into southern latitudes. Banks, still a youthful man of adventure, had organized a party of no less than sixteen participants to join Cook, among them four illustrators, which - instead of disbanding them – he now took on a northern voyage under his sole leadership. On the same day that Cook drew out of Plymouth harbour with the *Resolution* and *Adventure*, Banks left Gravesend with his party, including Solander and the brothers James and John Frederick Miller as illustrators, on the chartered *Sir Lawrence* for the Hebrides, Iceland and the Orkneys. The voyage, lasting five months, was less spectacular than the much longer circumnavigation of the globe, and the pictorial record brought back was definitely less extensive and less interesting, comprising both botanical and zoological drawings; they ended up in 14 New Burlington Street, Banks's London address at the time.

After a tour to the Dutch Republic the years of travel had come to an end for Banks and he became sedentary, or rather he did not leave England again but remained constantly on the move between London and his country houses and estates. He began to consider the publication of the vast amount of material from the *Endeavour* voyage and the much smaller collections from the *Sir Lawrence* voyage. Banks was, however, a busy man with many interests and obligations, in comfortable circumstances, under no pressure to publish and possibly also easily diverted by new topics on his agenda. As a consequence there were numerous delays. At least a start was made to complete the works of Parkinson by engaging the Miller brothers; they were sons of Johann Sebastian Müller of Nuremberg, himself a botanical illustrator and engraver, and had anglicized their names. Completing did not mean the completion of the unfinished botanical drawings prepared in Australia and later in Batavia, but producing very accurate copies of the incompletely coloured illustrations and adding full colour to them. Preparing copper engravings was not part of Banks's commission for the Miller brothers.

Much less is known about what happened with the painted record of the *Sir Lawrence* voyage, apart from the fact that John Fredrick Miller decided to publish some of the animal drawings. This he did without a proper title page or any descriptions, together with a miscellany of botanical drawings, the latter without any relationship to the northern voyage. Miller's work dating from 1776 to 1782 included sheets containing the legends and has often been cited as *Various subjects of natural history* or as *Icones animalium et plantarum*. The copper engravings were issued in fascicles in London and later also included in George Shaw's *Cimelia physica*, with several illustrations representing new and rare plant species. Among them we find, for example, prints documenting the large-leaved kōwhai, *Sophora tetraptera*, from New Zealand cultivated in the Chelsea Physic Garden, and of *Fothergilla major* from eastern North America. Both illustrations included a scientific name and analyses, which means that the two scientific names were validly published. The copper engravings of bird-of-paradise, *Strelitzia reginae,*[3] and of the silky stewartia, *Stewartia malacodendron*, both from the Royal Garden at Kew but without such analyses, are the oldest printed illustrations of the two species in existence. All this was done without Banks's permission, which quite naturally resulted in the end of the association of John Frederick Miller with his commissioner. In the end Parkinson's drawings from Australia and Java were completed by Fredrick Polydore Nodder. The latter was also engaged together with several others, among them Daniel MacKenzie, in the engraving of the copper plates based on the finished repeats of Parkinson's illustrations. Banks's engagement for the accurate graphic record of plant and animal life is testified to by a simple fact – he paid for the immense labours of his natural history illustrators and engravers for about fifteen years, and they were busy until c.1784 with the results of the *Endeavour* voyage.

Solander's death in 1782 must have been a major contributing factor in the eventual failure to publish the botanical results of the voyage on the *Endeavour*. But there were also other aspects, among them the economic consequences of the American War of Independence. Since Banks's income was particularly affected by arrears of rent from his agricultural estates and since a period of continuing depression held little prospect of successful sales of a luxury publication, the wise, though deplorable decision was taken not to continue. At what point this was eventually decided by Banks remains unclear.

It would, however, be inappropriate not also to consider a shift in Banks's mind – away from the botany of the circumnavigation of the globe in the *Endeavour* and towards a much stronger engagement in public affairs in London. His election as President of the Royal Society was one reason, his growing attention to the Royal Garden at Kew and plants cultivated in gardens in general another. Two engravings showing *Phaius tankervilleae*, the orchid referred to in Jacquin the son's letter to his father (see Chapter 7), and *Strelitzia reginae*, respectively, both paid for by Banks, may be regarded as indicative for this change in focus. Both are based on watercolours by James Sowerby, who then worked for several aristocratic commissioners, among whom were William Legge, 2nd Earl of Dartmouth and the Marchioness Rockingham[4] (see Chapter 7). Both illustrations with analyses were engraved by MacKenzie and have been very recently dated to around June 1788.

This is the more general background for Banks's decision to engage Franz in late 1789. More proximate reasons may have been the lack of a competent illustrator in his circle, since both Nodder and MacKenzie were rather more engravers than plant illustrators. Banks was probably also influenced by his experiences with two visitors from Paris. Like Jacquin the son and Franz in 1788, Charles-Louis L'Héritier de Brutelle, a high-ranking magistrate and amateur botanist based in Paris, had come to England in 1786 for a stay, which eventually lasted for fifteen months. He was followed in April 1787 by the illustrator Pierre Joseph Redouté,[5] later to become famous and in a sense the equivalent of Franz in the French capital (see Chapter 16). In early January 1789, a few weeks after the arrival of the two guests from Vienna, the *Sertum Anglicum* had started to appear in the French capital, containing descriptions by L'Héritier and copper engravings based on drawings largely by Redouté and Sowerby, showing plants either rare or new to science. Among these were specimens cultivated in the Royal Garden at Kew, and in other gardens in and around London as well as material from the herbarium at 32 Soho Square. For various reasons, Banks could not have been pleased – feelings of national pride may have played a role, but more important was the fact that the nascent *Hortus Kewensis* had in a sense been pre-empted. Banks may also not have been amused when realizing that the first instalment of the *Sertum Anglicum* contained the first description of a tree restricted in distribution to Australia that had first been collected during the landfall of the *Adventure* in Tasmania – the Australian oak, *Eucalyptus obliqua*. The first illustration of this widespread tree, up to thirty metres tall, was to appear in the third fascicle in April 1792. In addition some of the few copper engravings illustrating the *Hortus Kewensis* compared rather unfavourably with those to be published in the *Sertum Anglicum*, although this was not yet apparent. Furthermore, Banks would have realized by 1789 that finding a competent botanical illustrator was more difficult than finding a competent engraver, and far more difficult when it came to hiring an experienced botanical illustrator capable of working independently from his commissioner. Banks may also have been frustrated by the ever-changing group of botanical illustrators that were working for him at his new London residence, 32 Soho Square.

In Franz Banks encountered precisely this exceedingly rare species – a dedicated young man, with an experience in botanical illustration gleaned over at least fifteen years, who had been trained in an exemplary manner by Boccius and Jacquin. In addition, he promised to work in the same dedicated manner as his brother Ferdinand did for Sibthorp in Oxford (see Chapter 6). Furthermore, Franz was likely to offer the same kind of blend of naturalist and artist that Sibthorp had praised in his first letter from Vienna about Franz's younger brother (see Chapter 5). As a matter of fact it is likely that Franz had studied more botanical diversity than Banks, in Feldsberg, Vienna and in the many botanic gardens that he had visited on his travels with Jacquin the son. In addition Franz was a foreigner who had not yet formed a network of contacts in England and therefore would not have other commissioners in the first years of his engagement. With an experience of twenty years in supervising botanical illustrators and with his connoisseur's appreciation and respect for a highly talented and fully qualified artist, Banks will have known that this was a moment not to miss. He was not to regret his decision. Seventeen years later, in a memorandum dated 25 January 1806, Banks was to write about Franz and his more adventurous brother Ferdinand as 'the most skilful painters of natural history in this Kingdom, & in my poor opinion are not equald in any part of Europe'.

FRANZ SETTLES DOWN IN KEW

The archival record nowhere indicates a binding obligation for Franz to settle down in Kew, but it was the logical place for him to do so. 32 Soho Square had a room assigned to the engravers hired by Banks, but being a city residence it did not comprise a garden or a conservatory. Since Franz was expected to prepare mainly representations of living plants, just as he had done before in Feldsberg and Vienna, his place was near the garden informally supervised by Banks – the Royal Garden at Kew. No document survives indicating when he moved to Kew, but it seems safe to assume that he took lodgings there in 1790.

Today a western suburb of London and a fine residential area, Kew is situated near the very busy Heathrow airport and right below the point where the planes' engines are switched back. Furthermore, Kew is connected with the city centre by tube and roads that are often crowded with heavy traffic. In 1790 this was not the case: when Franz moved to Kew it was a peaceful small village of approximately hundred-fifty houses on the River Thames way outside of London, the place where several members of the aristocracy, including the Royal Family, had their summer houses and gardens. Erasmus Darwin had hailed the place in his poem 'The Botanic Garden' published just after Franz's arrival at Kew as 'Imperial Kew by Thames glittering side' and a few years later Josiah J. Boydell was to write 'The river now proudly flows between the spreading lawns of Sion [House], and the Royal Gardens of Richmond, which together, form a scene of superior beauty'.

Although at first glance bucolic, Kew had recently seen a catastrophe: in the winter months of 1788–1789, just when Jacquin the son and Franz arrived in London, King George III went insane. Not being able to tell a tree from an ambassador, so it is reported, he had to be confined to the cold and uncomfortable rooms of Dutch House, now called Kew Palace, near the river Thames. Then, as now, St. Anne's Church formed the centre of Kew, which had been promoted to a parish as late as 1769. Just before the arrival of Franz two famous artists, both founding members of the Royal Academy of Arts, had been buried in the small churchyard of St. Anne's – Thomas Gainsborough, one of Britain's greatest painters, andTübingen born Jeremiah Meyer, miniaturist to King George III and Queen Charlotte. There were many fine houses and gardens at Kew, among them Dutch House, where the sons of the royal couple used to spend their summer, the Richmond Gardens, and the Royal Garden at Kew, the latter open to the public during summer on Thursdays. This garden had been adorned by several garden follies, among them a pagoda, an Alhambra, and a mosque, all built by Sir William Chambers between 1758 and 1762. In short, both gardens were private, and several years later the established privilege of public access on certain days was to be withdrawn by King William IV, the third son of George III.

All this could have hardly made a strong impression on Franz. The village of Kew comprised markedly fewer houses than Feldsberg, although definitely much finer ones. The Thames was less impressive than the Danube in Vienna. St Anne's Church, markedly smaller than today, was a tiny building compared with the majestic church where Franz had been baptized. Dutch House and White House, the latter still in use but demolished in 1802, were far smaller and distinctly inferior in splendour to the summer residences of the Liechtensteins in Feldsberg and that of the Habsburgs in Schönbrunn, which Franz must have known. The gardens and parks at Kew, and Schönbrunn were comparable in size, in the number of follies and conservatories, the latter definitely more spectacular because of the relief, the large waterworks, and the dominating Gloriette, a large building erected in the final years of Maria Theresa marking a victory against her arch enemy Frederick II. At that time Schönbrunn also compared well with Kew in the number of species cultivated. This may be regarded as a surprise, but is reflected by Banks's remark in a later letter to William T. Aiton, His Majesty's gardener at Kew, that Schönbrunn is 'the only rival to Kew that I am acquainted with'.[6] Like in Schönbrunn, the botanical garden in the strict sense, then called the 'Physic and Exotic Garden' at Kew, covered only a small area of the estate. As in Schönbrunn this garden was situated next to the Great Stove. At variance with the situation in Schönbrunn, the orangery at Kew, the only conservatory surviving intact from the time of Franz's arrival, was also near the stove and next to a Temple of the Sun, the latter destroyed by a cedar of Lebanon, *Cedrus libani*, blown over by a storm in 1916.

There is a unique document of Franz's early years in Kew: his name scratched into a piece of window glass kept in Denmark House, today 356–358 Kew Road at the junction with Mortlake Road, Franz's first address. This house has maintained the atmosphere of the late-eighteenth and early-nineteenth centuries. Today a grade II* listed building,[7] Denmark House has three storeys and a basement, built for the Engelhardt family, which had come from Hanover to England with King George I. The first floor reception rooms of nos 356 and 358 still have a plastered ceiling by Francis Engelhardt, the former with agricultural motifs, the latter with corner urns, putti and anthemion motifs. However, Franz is not likely to have rented these rooms, but rather lodgings on the second floor. Thomas Engelhardt, the son of Francis and brother of George Engleheart, who had anglicized his name and was one of the greatest English miniaturists working on ivory, could have been Franz's landlord at the time.

Having settled down in Kew, Franz wisely concluded that it would be best to adapt to the English climate – milder in winter and cooler and wetter in summer than in Feldsberg – to adjust to the English way of life and to learn English, which he seems to have mastered soon. Very appropriately he had already scratched 'Francis' in the window glass and subsequently used the English form of his Christian name in his English letters and Franz in his German letters. Later he totally switched to English, even when writing to Jacquin the son in Vienna, which indicates that he had integrated himself largely into the society in Kew. However, neither Franz nor Ferdinand anglicized their family name, though they were often referred to as 'Bower'. Another document of Franz's early years at Kew is an anonymous, undated portrait of Franz in oil (Fig. 8.1. p. 190), now in the possession of the Royal Botanic Gardens Kew. It shows a young, fair-haired man dressed in a jacket with a high collar typical of the late rococo period, but without any attribute or reference to his profession. Thirteen years after Franz's death and upon the opening of The Economic Botany Museum No. 1 at Kew, proposed by King George IV, but built only in the early years of the reign of Queen Victoria, this painting was permanently exhibited there and has recently received an even more prominent place (see Chapter 20).

Apart from his extensive painted record (see below), very little is known about Franz as an individual. He must have been in contact with his brother Ferdinand in Oxford (see Chapter 7) and his brother Joseph in Vienna, but the pertinent letters could not be traced, although they must have existed. Franz also remained in contact with the Imperial Ambassador in London, first Johann Philipp Count Stadion (see Chapter 7), followed in 1793 by Ludwig Count (later Prince) Starhemberg. His family owned not only the spectacular Starhemberg Palace opposite the Liechtenstein Palace in Bankgasse in the very centre of Vienna but also a property in Wieden, then a suburb of Vienna, near what is now Operngasse. In a primitive theatre building erected on this piece of land the first performance of Mozart's opera *Die Zauberflöte* (The Magic Flute) took place on 30 September 1791 – just three weeks after Franz's first extant letter from Kew to Vienna.[8] The availability of an arrangement of this famous opera 'as Quartetts for 2 V: Tenor & Bas' at a London bookshop near the Admiralty was noted by Franz, albeit without giving a precise date (see Chapter 3).

The reasons for maintaining the ties with the Imperial Embassy in London were the same as those for Jacquin the son – letters could be sent in diplomatic bags to Vienna, which is explicitly mentioned in Franz's next letter, dated 21 May 1792.[9] On that day a packet of seeds was included in the diplomatic mail destined for Vienna – the seeds had been collected in Botany Bay in Australia, sent to Banks and handed over to Franz for speedy transport via courier to Vienna. Both letters sent from 'Kew-green' contain almost no information on Franz's work but offer a vignette of the circle surrounding Banks to which Franz now belonged. Dryander, Solander's successor as curator and librarian in 32 Soho Square, is characterized: 'he eats, - drinks – grows fat – is always in a Hurry – and – dos [sic] nothing'. Building works in Banks's house are reported to be planned with the aim of getting more light and space into the library - resulting in chaos with herbarium specimens and books scattered all over the place, making Dryander unhappy. There are also private matters dealt with in this letter: Therese Bauer had suddenly died of a stroke on 19 October 1790 in Feldsberg[10] (in the house no. 8 next to the church where all her seven children had been baptized), and a payment by Franz and Ferdinand to Jacquin the son is arranged, which was subsequently to be transferred to their brother Joseph in Vienna. Fifty guilders, a small sum of money, was sent, destined for their aunt, who had lived with the defunct mother and was now in desperate need of financial support. Almost simultaneously with the Bauer brothers, Jacquin the son had also lost his mother Katherina Josepha and Franz sent his condolences. The latter had received this information from Ingen Housz, which demonstrates his good contacts at the Jacquin family, with whom he kept in touch until the end of his life. Ingen Housz had also used the diplomatic bags of the imperial representative for urgent, safe and gratis transport to Vienna[11] – in April 1791 *A treatise of the plague, containing an historical journal and medical account of the plague at Aleppo* by Patrick Russell, a Fellow of the Royal Society like Ingen Housz, was sent to Gottfried van Swieten (see Chapter 3). In the covering letter Ingen Housz intimated that the book, rightly arguing that this pestilence was a highly contagious disease, should be made available to Emperor Leopold II, 'because of the danger of the spread of this terrible disease while the war with the Turks has . . . come almost to Vienna itself'.

Effectively Franz had become illustrator-in-residence at the Royal Garden at Kew, in fact the first ever. However, on a European level, this was not a new concept – a 'peintre du Roi' had already existed at the Jardin de Roi in Paris for more than a century, for example, with the position held by such luminaries as Nicolas Robert, Jean Jobert and others. The former's works may have been known to Franz via the *Receuil des Plantes* in the Imperial Court Library in Vienna (see Chapter 2). This may have been in the back of Banks's mind when engaging Franz. Exactly why the latter was shortly thereafter called 'Botanick Painter to His Majesty', for example on the title page of the *Delineations of exotick plants*, remains unclear (see below). In any case Franz was never a painter to King George III, but received an annuity for life from Banks's pocket.

In the following three decades ending with the death of Banks in 1820 Franz was faced with several challenges, unexpected as well as expected ones. He may have been surprised to become 'drawing master for botanical subjects' to Princess Elizabeth, the third daughter of George III, a young lady of twenty years of age when Franz arrived at Kew. Nothing is known of when tuition started, how it proceeded and when it ended. Princess Elizabeth is known for her drawings, a selection of which was put on show in 2010 in an exhibition entitled 'Eliza, Arbeiten auf Papier. Das graphische Werk der Landgräfin Elisabeth 1770 – 1840' in the Gothisches Haus in Bad Homburg.[12] A fine flower piece with a bird's nest by Princess Elizabeth, a watercolour and bodycolour on vellum copied from a painting by Margaret Mee, is kept in the Royal Collection.[13] At the age of thirty-eight married to Fredrick Prince (later Landgrave) of Hesse–Homburg, she became upon the death of her father-in-law Elisabeth Landgravine von Hesse-Homburg. According to an unpublished letter sent to Robert Brown (see Chapter 10) 'all the remuneration he [Franz] ever received [from Princess Elizabeth] was half a stick of Indian Ink, which she broke in half and gave him'.[14] In 1814 Banks asked Franz to prepare two congratulatory cards to be sent to Princess Elizabeth, which are kept in the Fitzwilliam Museum in Cambridge,[15] while the pertinent pencil sketches are in the Natural History Museum, London.[16] The surrounds are by Franz, and one has the following oleaginous text written in the reserve in Banks's best handwriting 'March 30 1814 May the much honord Flower of Strelitz, the budding Myrtle of Affection, the Elegant Emblem of Remembrance & the Sacred Symbol of Recollection Perpetualy Encircle the Reverd Name of ELIZA. May Her Royal Highness, under the Mild & chearing influence of our Guiding Star, Ever Continue to Enjoy that Happiness, which Superior Talents unexampled affability and Angelic Virtue, never Fail & bestow; is the heartfelt wish of one who in Token of his Sincerity, has subjoind his name, his Seal & his flower. Jos: Banks' (Fig. 8.3.). The iconographical programme, probably devised by Banks, Franz or both, consists of bird-of-paradise, *Strelitzia reginae*, scarlet banksia, *Banksia coccinea*, myrtle, *Myrtus communis*, heartsease, *Viola tricolor*, forget-me-not, *Myosotis sylvatica*, or, alternatively, Siberian bugloss, *Brunnera macrophylla*, and a pink-flowered rose, so far unidentified. A depiction of Banks's seal showing the lizard, his motto, is added next to his signature, and a star crowns the surround. The scientific name of the first mentioned plant refers of course to Princess Elizabeth's mother, Queen Charlotte, née Sophia Charlotte Princess of Mecklenburg-Strelitz, then a small duchy in the Holy Roman Empire; however, in contrast to widespread opinion, Queen Charlotte was not born in Strelitz but in the tiny town of Mirow. The companion piece has a similar iconographical programme, probably again devised by Banks, Franz or both, consisting of heartsease, forget-me-not or, alternatively, Siberian bugloss, a blue-flowered composite, probably cardoon, *Cynara cardunculus*, and the pink-flowered rose; again the surround is crowned by a star. The flowery text in the rectangular reserve is in the hand of Sarah Sophia Banks and reads 'Deeply Engraved upon my Heart, are those Sentiments of Respectful Attachment, which Your Royal Highness has permitted me to Cherish withen it, never never while its warm Pulses continue to beat, shall they be sullied or impaired, ever ever may they be increased and improved S. S. Banks March 30th 1814'. Nothing is known about the date, identical on both cards. It may refer rather to a special occasion in the life of the addressee than to the battle of Paris, which was fought on this very day, ending the war of the sixth coalition in which Great Britain participated with a victory over France and its allies. At the same time the two iconographical programmes mirror the plant life in the British Empire – bird-of-paradise representing Africa, the scarlet banksia Australia, all others Eurasia; significantly America is missing. Princess Elizabeth was also in contact with Lady Dorothea Banks; the former

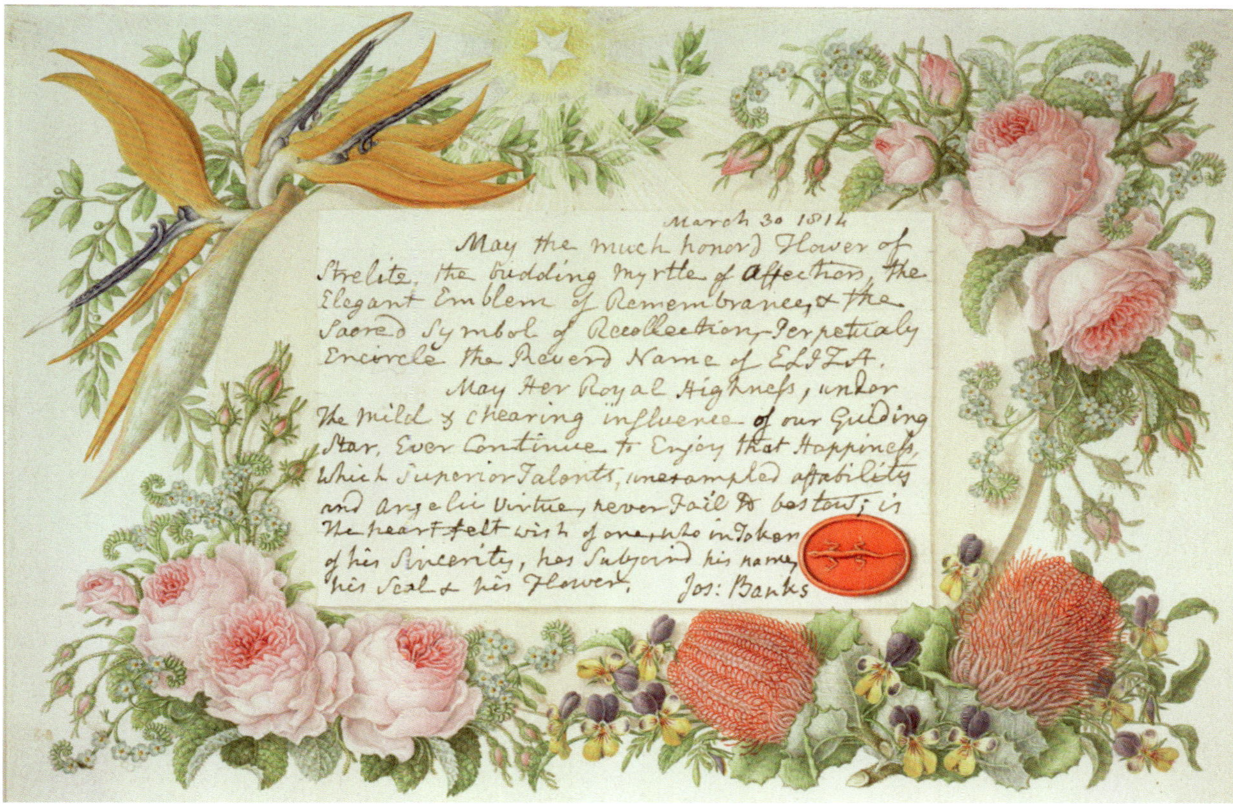

FIG. 8.3. A surround of mixed flowers by Franz Bauer, 1814. Paper, pencil, watercolour. – Cambridge, Fitzwilliam Museum, PD. 174 – 1973.

presented the latter with a small pocket-book closed with a lockable flap containing about forty silhouettes cut out by Princess Elizabeth and Theodore Tharp. Acquired by Mary, Queen Consort of George V, King of the United Kingdom, the booklet ended up in the Royal Collection.[17]

It is likely that Franz at least occasionally also met Queen Charlotte and her second daughter Princess Augusta Sophia, who was to remain unmarried. The former could speak German to Franz, although her Mecklenburg intonation would have differed markedly from his rural accent from the Weinviertel. According to the letter directed to Brown cited above both ladies 'used to colour some of the plates of the Heaths [i.e. *Delineations of exotick plants*] of which M[r]. Bauer had made the drawings. the plates so coloured were sold at the sale of Queen Charlottes effects'. Again, no details are available, but the colouring could not have started before the prints were available, which was hardly before c.1794, when Queen Charlotte was fifty years old and the Princess was twenty-six. Decades later Frederick Scheer (see Chapter 16) reported that Franz 'was in frequent attendance on the King and Royal Familiy', but no further details are given.[18]

Henriette Charlotte Countess Itzenplitz was another young lady who was tutored by Franz in drawing.[19] She was the daughter of Adrian Heinrich von Borcke, Prussian envoy in Dresden and Stockholm, and was visiting England on her honeymoon together with her husband Peter Alexander Count Itzenplitz, the owner of estates in Brandenburg. In December 1792 they had dinner with Banks, who became their most important contact in London, with Smith, Franz and others attending. On that occasion Henriette Charlotte received from Lady Dorothea Banks copper engravings showing *Strelitzia reginae* and *Phaius tankervilleae*[20] (see above). Engraved by MacKenzie, they had been published in the *Hortus Kewensis*. When pregnant Henriette Charlotte moved from London to Kew in spring 1793, living in the same house as Franz. In letters to her mother she praised the accuracy of Franz's botanical work, whom she repeatedly met over tea, and offers the only detailed description of him as a person 'small in size, somewhat above forty years old [in fact he was then 35], who unites a considerable amount of good nature, principles, education, a mind of delicacy, modesty with a self-reviving mood and satire . . . noble in his behaviour . . . precise in his thinking'.[21] On 10 May 1793 Henriette mentions having been with her husband and Franz in London to see an exhibition of the Royal Academy of Arts and of the Orléans collection.[22]

The latter was a collection of German, Dutch and Flemish paintings previously owned by Louis Philipp II, Duke of Orléans of the Bourbon family, who was to be guillotined in Paris in November of that same year. This seems to be the only explicit reference to Franz's interest in art exhibitions and the arts in general. Like Franz, Jacquin the son and L'Heritier before her, Henriette had received a privilege from Banks - to become 'a true fellow' of the garden, which meant that she had unlimited access to the Royal Garden at Kew and its conservatories and was permitted to ask the staff for assistance. She is not known to have continued drawing later but maintained an interest in botany. Henriette is known for having hosted in her country house in Kunersdorf east of Berlin in summer 1813 the famous French refugee Adelbert von Chamisso. The latter was a poet and a plant taxonomist, writing on that occasion the iconic novel *Peter Schlemihls wundersame Geschichte* for Henriette 's children.

Franz had been engaged by Banks for the documentation of all kinds of objects. One of the more unconventional commissions was to copy accurately the Chinese characters from a china cup to a piece of paper. This was sent on 28 February 1807 in a letter from Kew to 32 Soho Square,[23] together with the phonetic equivalent in the Latin alphabet as collected from one Hy Hee from Canton. Lady Dorothea Banks was known for her interest in china, of which she possessed a fine collection at Spring Grove, the country house of the Banks family in Isleworth, Middlesex in the vicinity of Kew. Along the same lines of interest are Franz's illustrations added to Sir Joseph Banks's manuscript 'Collections on the Subject of Old China and Japan Wares with some Remarks on these interesting Manufactures made in Lady Banks's Diary at Spring Grove in 1807', now in the Kent History and Library Centre at Maidstone.[24] The text is both a catalogue of his wife's collection and a history of china production in the Far East, perhaps the earliest of its kind existing in the west. Interestingly, most of the illustrated Chinese ceramics in Banks's catalogue are not vessels but figurines of the type made at Dehua, which were later called *blanc de chine* by Europeans.[25] To this manuscript Franz added images of dragons, of the qilin or kirin, of the foo dogs and of other emblematical animals taken from printed books for comparison with the actual figures in Lady Dorothea's collection.

An example of what could also be expected by Franz from the new arrangement with Banks was a commission received in 1791 at the latest. In that year a 'vegetable monstrosity' had been sent from Switzerland to Banks, which Franz was asked to draw. Thirteen years later this 'plant' became the subject of a paper authored by Charles Konig (Karl Dietrich Eberhard König),[26] the assistant to Dryander. The drawing could not be traced but on the basis of it a copper engraving was produced by MacKenzie, carrying nine printed signatures of witnesses, among them Banks, Dryander, Smith and of other Fellows of the Linnean Society. A copy is kept in the archive of the Linnean Society in London, inscribed on the recto 'The Linnean Soc.y May 3, 1791 By S.r J. Banks + HM.LS [Honorary Member of the Linnean Society]',[27] in addition the copper engraving was included in Konig's paper. This is his description: 'its hairy stalk, and the form of the leaves, left no doubts of its being either [creeping crowfoot] *Ranunculus repens* or [*Ranunculus*] *polyanthemos*, both very common in Switzerland. Intermixed with its own flowers, and on its own peduncles, it bore several other flowers, that every botanist allowed to be those of [the common daisy] *Bellis perennis*'. Yet when Banks soaked it in boiling water 'the flowers of the daisy almost immediately detached themselves from the stalk, and floated in the liquid'. In this way the enigma of the 'daisy-flowering crowfoot' (*Ranunculus bellidiflorus*), clearly a fraud, was clarified, with Franz having been selected as the illustrator to document the artefact. Why this object was sent to Banks remains unknown, just as it is unclear why a similar fake had been described in 1753 by Johannes Gessner, Professor of Mathematics and Physics at the Collegium Carolinum in Zürich, in his *Dissertatio de Ranunculo bellidifloro et plantis degeneribus*. However, Gessner could indeed have been misled in his observation,[28] since critical observation as mandated by the Enlightenment did not start in a synchronous way among the learned.

What other kinds of conventional challenges the new position would bring became evident to Franz on a snowy day at Kew, possibly during his second winter in England. Fifty years later he remembered it in a conversation that Sir James Everard Home, the son of the more famous Sir Everard Home (see Chapter 15), recorded in a letter 'The Peonia [sic] Mouton [*Paeonia suffruticosa*] was at first placed in the stove [at Kew]. One day in the depth of a severe winter as M[r]. Bauer was walking around the Melon ground [in Kew] in company with M[r]. Aiton he observed something red peeping through the snow, he poked it out with his stick. Saying 'here is something alive' M[r]. Aiton exclaimed 'It is my Moutan' and took possession of it. The leaves having fallen off the gardener who had the care of the stove thought it was dead, and as it was a plant of much interest fearing blame thought it best to throw it away and say nothing about the matter. In this way it was saved and was at the same time proved to be Hardy'. This is a plausible story: the much sought-after tree peony or Mu-Tan, a native of China and in cultivation since the Han Dynasty, has been reported to have first arrived in Europe in 1787 sent by one Dr Duncan of the East India Company to Banks. The latter almost certainly had it cultivated at Kew, where the incident happened two or three years later. It is likely that 'M[r.] Aiton' refers in this context to William Aiton, not to his son and successor in the job William Townsend Aiton, who was to outlive Franz (see Chapter 16). Among Franz's finished plant illustrations painted in utmost refinement we find an undated watercolour showing this very species (Fig. 8.4.), which had to wait for several more years until it received its scientific name.

FIG. 8.4. Tree peony, *Paeonia suffruticosa*, by Franz Bauer, undated. Paper, watercolour. – London, Natural History Museum, Library and Archives, Finished Drawings of Plants, pl. 2.

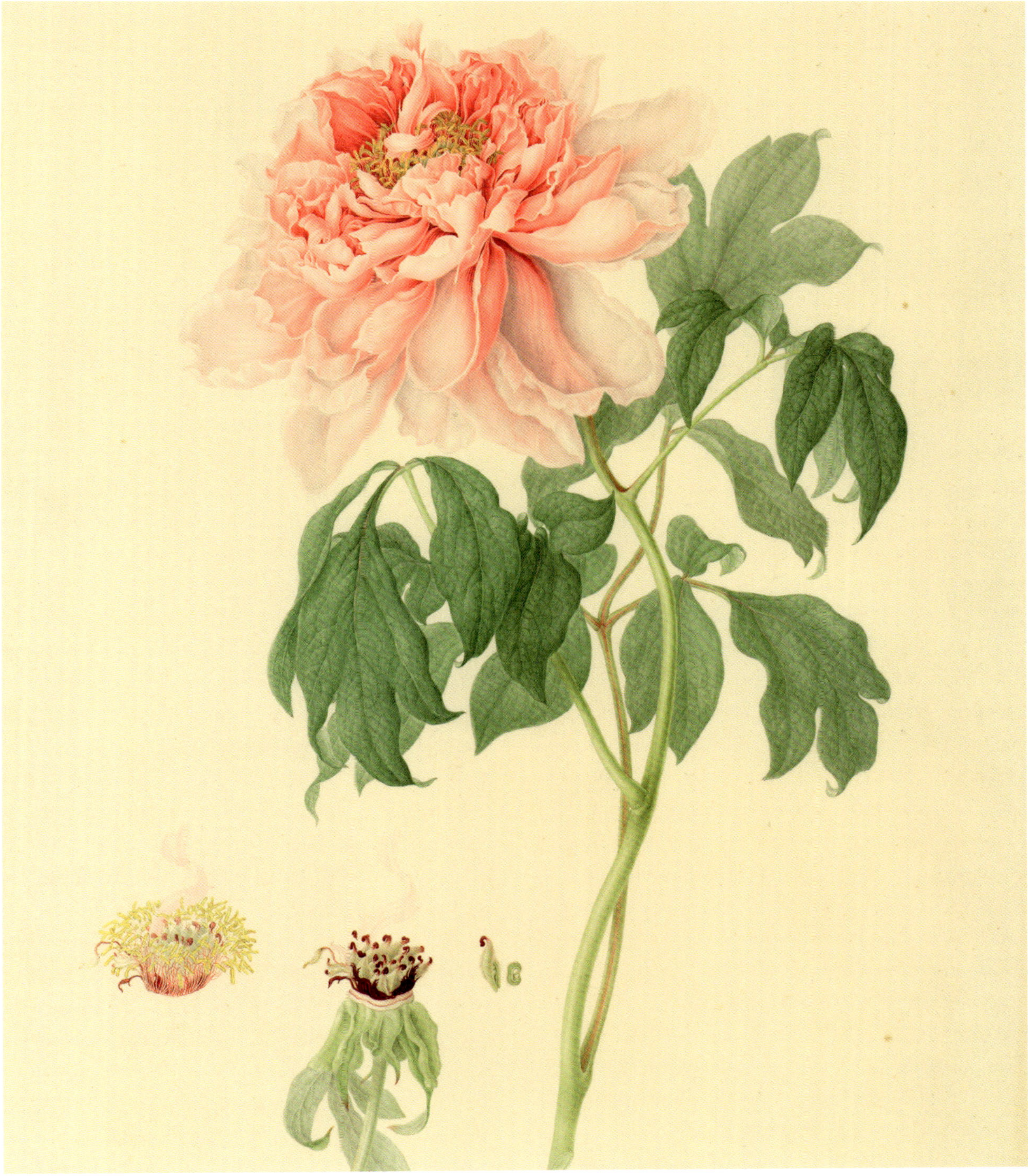

THE PAINTED RECORD OF THE ROYAL GARDEN AT KEW

Starting with his engagement by Banks, Franz produced over the next three decades a painted record of the more remarkable plants cultivated at the Royal Garden at Kew – in a sense continuing his work for Jacquin in Vienna interrupted in early 1788 and, to a certain extent, that done previously for Boccius in Feldsberg. The steady flow of highly accurate botanical illustrations leaves the scientist and the connoisseur equally breathless; both quality and quantity are absolutely stunning. Clearly this almost endless series of plant portraits is unique in England and ranks Franz in line with the greatest names among the French botanical illustrators, who were documenting the rarities cultivated in Paris for the successive kings of France (see above).

However, this extremely valuable material, clearly Franz's most important achievement documenting plants cultivated in gardens, is very difficult to interpret and therefore has not yet been fully appreciated by the scientific community and the aficionados of botanical illustration. Two sets of problems exist, practical as well as contextual ones. Firstly Franz's painted record of plant life covering the period from 1790 to 1820 is kept in four institutions: the major part in the Natural History Museum; smaller parts in the Niedersächsische Staats- und Universitätsbibliothek in Göttingen and the Naturhistorisches Museum in Vienna; and at least one watercolour of an orchid, the early purple orchid, *Orchis mascula*,[29] at the Royal Botanic Gardens Kew. Generally speaking, London owns the finished and the unfinished watercolours, but only very few pencil drawings, Göttingen a few unfinished watercolours and other preliminary materials, and Vienna selected pencil drawings only. Furthermore the three institutions differ in their curatorial practice. In Vienna the pencil drawings are kept loose in boxes and have not been foliated, in Göttingen the drawings have been foliated and bound into volumes - as a rule glued on supporting paper, which makes an inspection of the watermarks impossible. In London the drawings received the well-known 'B. M. (N. H.)' stamp with a little crown always placed in one of the corners, they were foliated and bound into several volumes according to systematic order - with the respective content given on a handwritten title page, for example 'C1 British Orchids Francis Bauer 1792–1817' or 'D4 Unfinished drawings Francis Bauer. Monocotyledons I. Orchideae I.' Again the inspection of watermarks is impossible. Needless to say, a few drawings by other hands were found intermixed in London, among them a few signed or ascribed to Ferdinand, testifying to the close collaboration between the two brothers (see Chapter 9). Unfortunately, several drawings kept in London were brutally trimmed in the binding process. Since annotations (see below) had often been placed on the very margin of the drawings, information had to be copied by a librarian or amanuensis in ink on the watercolours – often in quotes, with the identity of the original hand lost. Fairly recently several of these volumes seem to have been unbound; the individual drawings were mounted, framed and are now kept in Solander boxes, with the original sequence reflecting systematic order now lost for ever. In addition, the remaining bound volumes and the unbound material in the Solander boxes are kept separately, the former in the Botany Library, the latter largely in a strong room, which is not very practical when it comes to comparing drawings. All this hinders reference to a particular drawing.

Furthermore, and in contrast to Ferdinand's watercolours for the *Flora Graeca* and *Fauna Graeca* (see Chapter 6), this painted record of plant life is not available on the internet, which necessitated repeated visits to London, Göttingen and Vienna in order to analyze the complexities of this material. As a consequence the reconstruction of Franz's working method was more tedious than for Ferdinand, since all the latter's work done for Sibthorp is kept in a single place (see Chapter 6).

The contextual problems are even more difficult to resolve. Firstly Franz, like Ferdinand, largely refrained from signing his works – only those watercolours regarded by him as ready to be sent to the engraver or lithographer, for example his documentation of the life cycle of wheat, *Triticum aestivum*, and its diseases (see below), were consistently signed by him, as a rule on the recto. Secondly, Franz only occasionally dated his watercolours, which impedes the establishment of any reliable chronology; when given, dates appear on the recto or the verso. Between 1790 and 1820 there has been no extended period of time when cultivated plants were not portrayed. Thirdly, Franz often refrained from giving further data, such as an identity of the plant depicted or the provenance of the specimen documented – that is, he only rarely indicated where the plant had come from originally, who had collected it in the wild and where it had been subsequently cultivated. Many drawings are therefore like herbarium specimens of plants cultivated in a garden devoid of any further information. In addition, it would not be surprising if some of the plants depicted, such as terrestrial orchids native to the British Islands, were not cultivated but collected in the wild and brought in pots to Franz's studio. Fourthly, generations of librarians and botanists have added occasional comments to these drawings, in particular determinations. However, contrary to modern standards, only very occasionally these annotations were signed and dated. In order to distinguish these later additions from Franz's own, mostly fragmentary notes the expertise of a graphologist is necessary. Often the note in ink 'Kew plants', 'Bauer Kew plants' or 'Bauer's Exotick Plants' is found on the drawings followed by a number; this is never in Franz's hand, but by an anonymous librarian.[30] In addition, notes from the verso have often been copied by a later hand on the recto of the drawing. Fifthly, a surprisingly high number of illustrations remained unfinished. Why this is so remains an enigma. Was Franz overwhelmed by the sheer quantity of the plant material that had to be documented? Was it as a consequence of any lack of direction by Banks? Did Franz intend to finish his drawings at a later stage and ran out of time? We simply do not know. Sixthly, Franz's plant illustrations are mostly not correlated with herbarium specimens, which would be the standard practice today. All these obstacles have a straightforward consequence: Franz's drawings of plants cultivated at Kew, and there are very many, have been rarely commented on or cited. Although often in quality similar to those of his younger brother from Australia (see Chapter 11), these illustrations have never received equal applause. One of the reasons may have been that they are not connected with the romance of sailing to distant lands and the adventure into the unknown, but merely plants cultivated in the Royal Garden at Kew. Of course these are stereotypical concepts totally unrelated to the realities of the circumnavigation on the *Investigator* (see Chapter 10), but they seem to have opinionated historians of science and connoisseurs of botanical illustration alike.

We may hypothesize that as a rule Franz followed a working pattern different from that of his younger brother. He seems to have first made a pencil sketch of the plant, sometimes on a rather thin sheet of paper. In a second step this image was transferred in pencil on solid water colourist's paper, maybe with the help of a light box, and worked out in greater detail. As a third step paint was applied to the contours, making use of an extremely fine brush, resulting in thin lines surrounding the image, as seen in an image of the crape myrtle, *Lagerstroemia indica* (Fig. 8.5.). Only in the final step were the contours fully filled with paint. Nothing is known of how Franz managed to erase the fine pencil lines, but rubbers were already commercially available in his time. A good case to illustrate this point is the pencil sketch of a flowering branch of African linden,

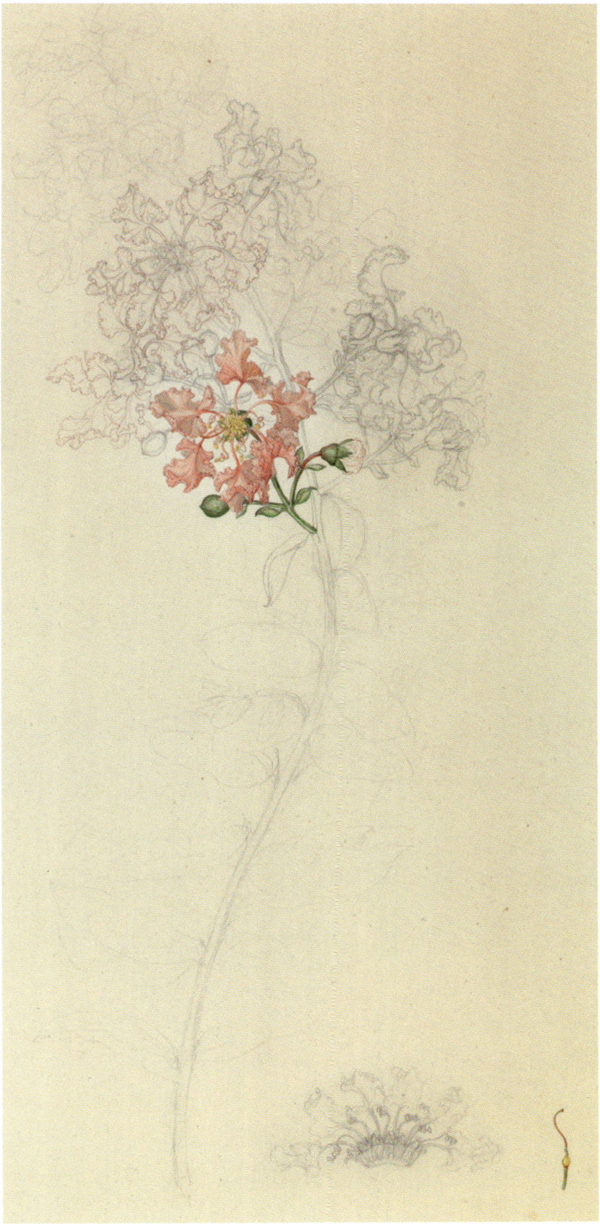

FIG. 8.5. Crape myrtle, *Lagerstroemia indica*, by Franz Bauer, undated. Paper, pencil, colour pencil, watercolour. – London, Natural History Museum, Library and Archives, Unfinished Drawings of Plants, vol. 1, no. 125.

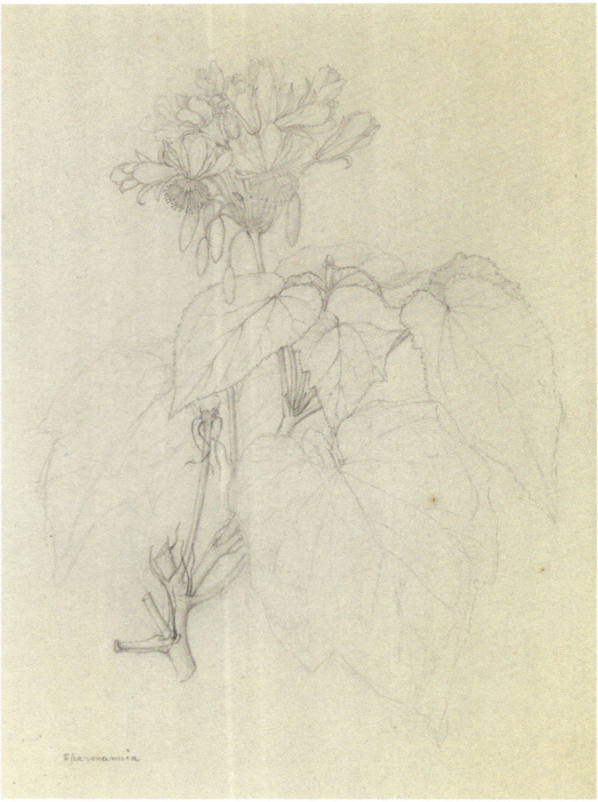

FIG. 8.6. African linden, *Sparmannia africana*, by Franz Bauer, undated. Paper, pencil. – London, Natural History Museum, Library and Archives, Unfinished Illustrations of Plants, vol. 1, no. 68.

Sparmannia africana (Fig. 8.6.), and the incompletely coloured repeat (Fig. 8.7.) based on it. Refraining from the use of the colour code may have had a simple reason – Franz was a resident botanical draughtsman. As a rule the subjects to be depicted would have been brought to him and he could work on them right from the beginning in his studio at leisure, probably at Denmark House. By contrast, Ferdinand had served Sibthorp as his travelling illustrator and consistently made use of his colour code when recording plants and animals in the field, very often being pressed for time (see Chapter 5). Even during the more sedentary phases of his life Ferdinand very often used his colour code, for example when preparing the pencil drawings for *The genus Pinus* (see Chapter 9). Only very occasionally Ferdinand did not follow his routine, possibly when working right from the beginning in the studio, and adopted his brother's approach – for example, when documenting some of his passion flowers (see Chapter 12). Considering the close collaboration of the two brothers, it is extremely likely that Franz was perfectly au fait with his brother's method and occasionally also relied on it.

Although generally very incompletely annotated, a few of Franz's drawings offer glimpses of his private life in Kew and the circle of his contacts. He cultivated plants at least occasionally in his own garden, since his watercolour of the heath spotted orchid, *Dactylorhiza maculata*, bears the note 'My own garden No. 1. April 7 1803 No. 2 May 10. 1803' and includes an image of a rhizome dug out with a longitudinal and transverse section of the tuber added. There are occasional remarks such as 'Lady [Dorothea] Banks' Garden 1798', 'R. H. S.'s garden Joseph Banks 1817' or 'From the Duke of Northumberland [then Hugh Percy 3rd Duke of Northumberland]'s Garden Sion House October 23 1819', but the bulk of the material seems to have come from 'K: G:' [Kew Garden] – although this is very rarely explicitly mentioned.

As President of the Royal Society, founding member of the Horticultural Society of London[31] and one of the first Honorary Members of the Linnean Society of London,[32] Banks was a man with very many contacts. A few names of fellows of these three learned societies appear occasionally on Franz's watercolours of plants in cultivation prepared during the three decades leading up to the death of Banks. Examples to illustrate this point are shown on two representations of the vanilla orchid, *Vanilla planifolia*, which bear the names 'Greville' (see Chapter 13) and 'Hibert', respectively. Both were Fellows of the Royal Society, George Hibbert an eminent merchant, wealthy slave- and ship-owner with sugar plantations in Jamaica, and a member of Parliament from 1806 until 1812.

Surprisingly, on Franz's watercolours only a single reference to another garden owned by a member of the Royal Family is found – Claremont House. The associated note 'Prince's Leopold's Garden' enables the determination of a *terminus post quem*. This country house near Esher in Surrey was bought in 1816 by the English nation by an Act of Parliament to become a wedding present. Charlotte Augusta Princess of Wales, the only legitimate heir of George, then the Prince Regent and later King George IV, and her husband, Leopold Prince of Saxe-Coburg-Saalfeld, became the first owners of this country house. While the princess died the following year in childbirth,

FIG. 8.7. African linden, *Sparmannia africana*, by Franz Bauer, undated. Paper, pencil, watercolour. – London, Natural History Museum, Library and Archives, Unfinished Illustrations of Plants, vol. 1, no. 67.

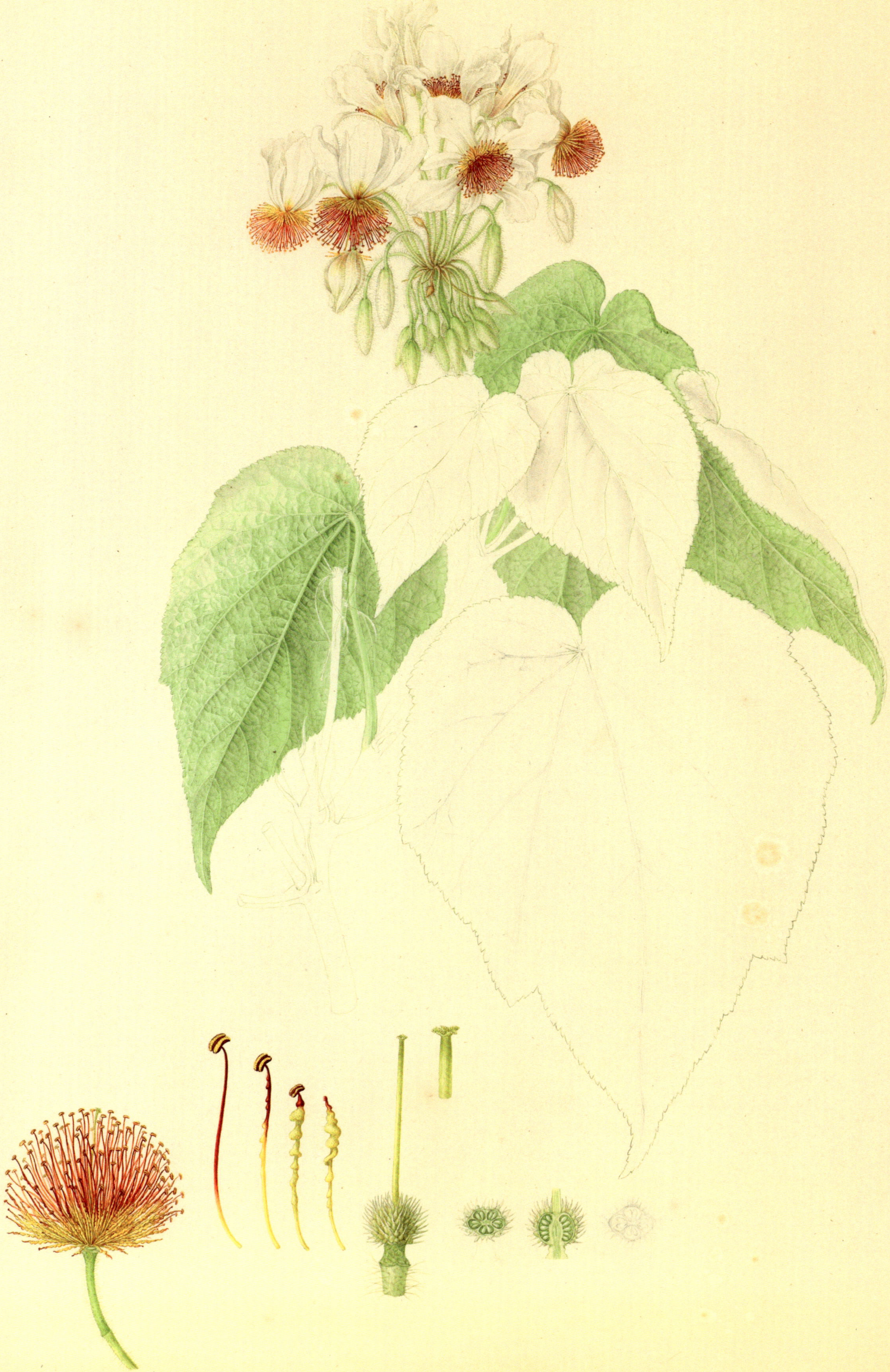

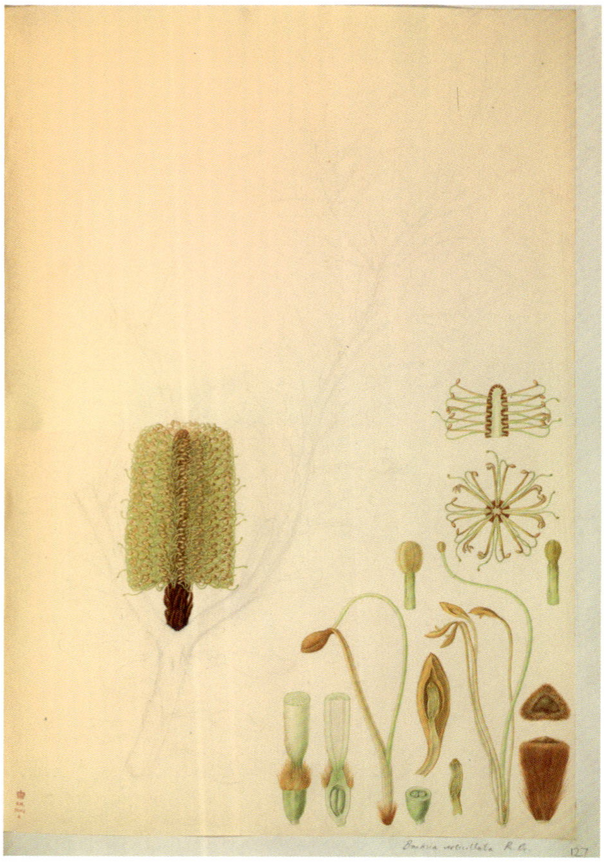

FIG. 8.8. Banksia, *Banksia* spec., by Franz Bauer, undated. Paper, pencil, watercolour. – London, Natural History Museum, Library and Archives, Unfinished Drawings of Plants, vol. 2, no. 127.

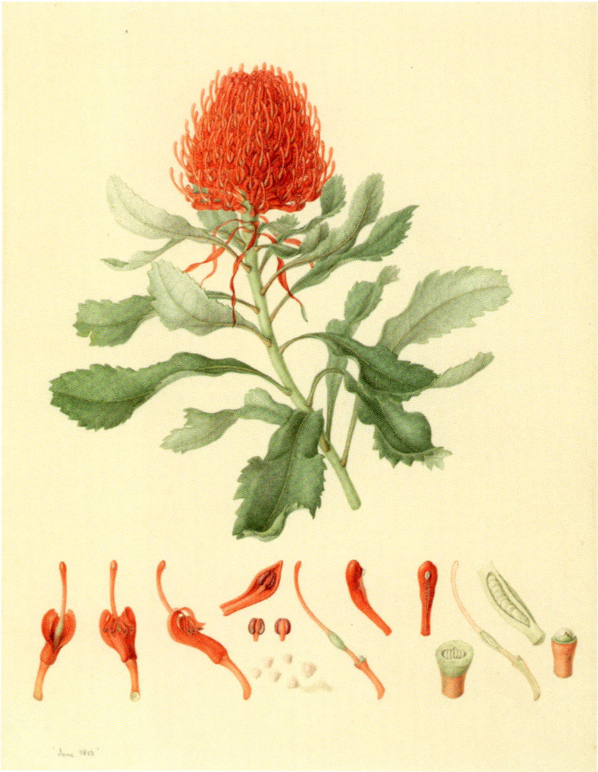

FIG. 8.9. Waratah, *Telopea speciosissima*, by Franz Bauer, undated. Paper, pencil, watercolour. – London, Natural History Museum, Library and Archives, Finished Drawings of Plants, pl. 84.

Leopold became the first King of the Belgians in 1831. Her death left George III, then a completely blind old man on whom permanent madness had descended, a father of many children but not of a single legitimate grandchild. The absence of an heir to George III's three elder sons was the principal reason for his fourth son Prince Edward Augustus, Duke of Kent and Strathearn, aged fifty, to marry the widowed Marie Luise Victoria Princess Leiningen, née Princess of Saxe-Coburg-Saalfeld, an elder sister of Leopold. The wedding had been celebrated in Dutch House at Kew, but the ensuing marriage was a very short one due to the early death of Prince Edward Augustus. However, the resulting child – Princess Alexandrina Victoria – was to play an important role at the very end of our story. She was to succeed her uncle King William IV in 1837 and became the long-reigning Victoria, Queen of the United Kingdom of Great Britain and Ireland. It was to her that Franz directed his famous request, which she accepted (see Chapter 17).

Albeit legally owned by Banks, Franz's graphical record remained for years largely in his care in Kew. Most importantly, only a minute fraction was ever published (see below), which is deplorable though perfectly understandable – Banks, who would have possessed the necessary financial means was overcommitted with numerous other projects and duties to fulfill, and Franz, living on an annuity for life, was under no pressure to publish and preferred to record the novelties arriving at Kew. In a way it was work similar to that in Feldsberg, differing in being more selective and much more refined in the analyses recorded. Franz was clearly the specialist for the new arrivals at Kew, not the common, well-established plants. We may well imagine Franz walking regularly through the garden and the conservatories, or the gardeners, in particular William Townsend Aiton, another founding member of the Horticultural Society, calling on him with novelties or rarities to be depicted.

The recent introductions from Australia were a special challenge, since Franz could have seen only a very few of them elsewhere before starting to work in Kew. He recorded *Banksia* species (Fig. 8.8.) and species of cone flower, *Isopogon* spec., as well as the waratah, *Telopea speciosissima* (Fig. 8.9.), with its bright-red inflorescence, and the tall kangaroo paw, *Anigozanthos flavidus* (Fig. 8.10.). A few of these Australian plants had been raised from seeds collected by Peter Good, the gardener on the *Investigator* voyage (see Chapter 10), and cultivated at Kew, for example the watershield, *Brasenia schreberi*. It is illuminating to compare Franz's finished watercolour of this aquatic plant cultivated at Kew with Ferdinand's finished watercolour of the

FIG. 8.10. Tall kangaroo paw, *Anigozanthos flavidus*, by Franz Bauer, undated. Paper, pencil, watercolour. – London, Natural History Museum, Library and Archives, Unfinished Drawings of Plants, vol. 5, no. 50.

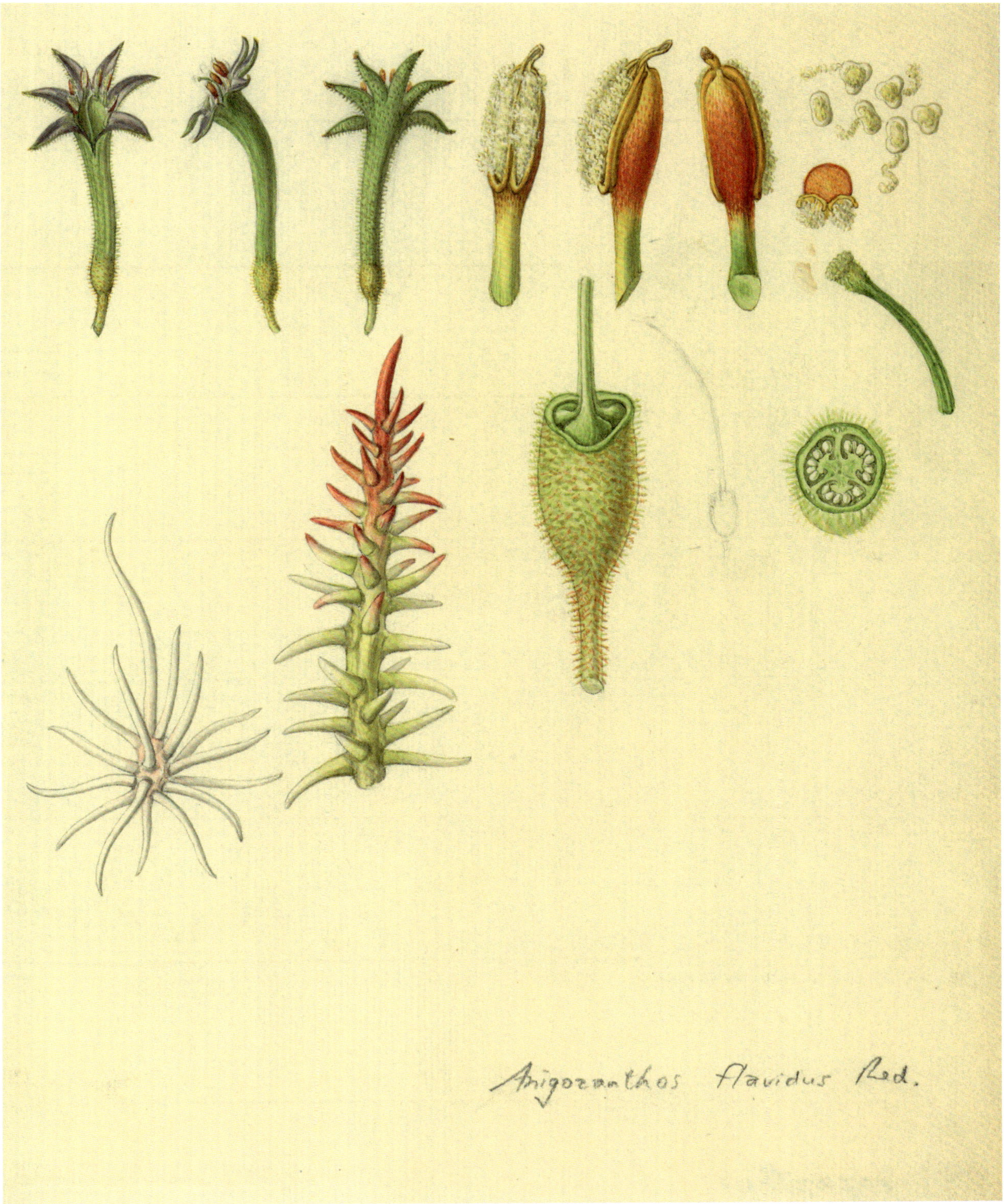

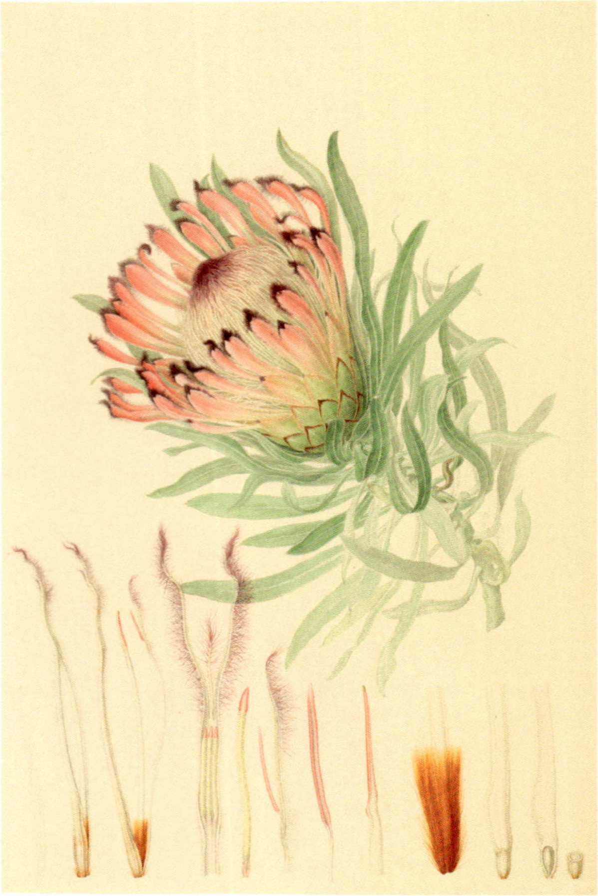

FIG. 8.11. *Protea burchellii* by Franz Bauer, undated. Paper, watercolour. – London, Natural History Museum, Library and Archives, Finished Drawings of Plants, pl. 86.

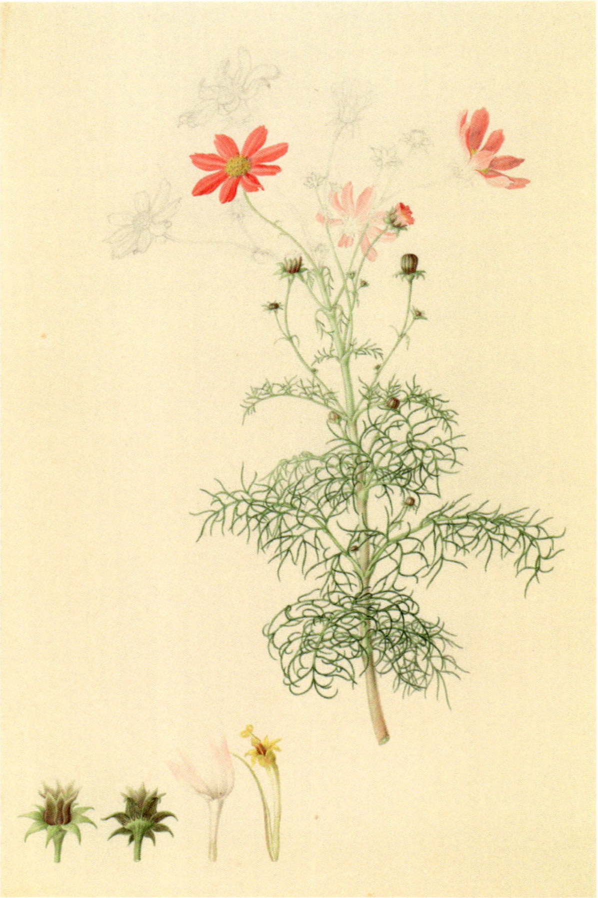

FIG. 8.12. Garden cosmos, *Cosmos bipinnatus*, by Franz Bauer, 1792. Paper, watercolour. – London, Natural History Museum, Library and Archives, Finished Drawings of Plants, pl. 21.

same species based on a colour-coded pencil of a plant in the wild prepared by him in the field.[33] Many genes in common as well as a joint training by Boccius and Jacquin resulted in an almost unbelievable similarity of the two end products – the two representations of *Brasenia schreberi*, which were painted in Kew and London, respectively.

Many of the plants documented by Franz with utmost precision have nowadays become widely known to the general public, but in the late-eighteenth century and early-nineteenth century this was not the case. These include the African linden, *Sparmannia africana*, the Cape gardenia, *Rothmannia capensis*, and *Protea burchellii* (Fig. 8.11.), all three from the Cape of Good Hope. The last of these had to wait for more than a century until it received a scientific name. Franz dedicated six drawings to the papyrus sedge, *Cyperus papyrus*, a native of tropical and subtropical Africa long used by humans, notably in ancient Egypt, as the source for papyrus paper. The living plant may have originated from James Bruce, a Fellow of the Royal Society, who had also sent specimens of aromatic resin and of balsam of Mecca, *Commiphora gileadensis*, to Banks in 1789.[34] Introductions from South and Central America recorded by Franz include garden cosmos, *Cosmos bipinnatus* (Fig. 8.12.), *Gloxinia perennis* (Fig. 8.13.) and the tiger flower, *Tigridia pavonia*, a pencil drawing of the latter plant has recently been traced in the Naturhistorisches Museum in Vienna. The watercolour of the tiger flower demonstrates the perfect botanical illustrator documenting all those flower parts enlarged and in utmost detail relevant for the scientist – Franz depicts not only a branch with the fully opened flower but also no less than fifteen figures, from left to right: (1) the gynoecium and (2) a longitudinal section of it, (3) an unripe stamen and (4) a ripe stamen with pollen protruding from it, (5) pollen cells (see Chapter 14), (6, 7, 8) transverse sections of the ovary, (9) the tip of the style with the stigmas, and, to the right, (10) a longitudinal and (11, 12) two transverse sections of the capsule, as well as (13) a seed, plus (14) a longitudinal and (15) a transverse section of it. This is another masterpiece of a brilliant observer and accurate recorder of plant life. Several other representations of petaloid monocots show a similar refinement in the details recorded, among them the bizarre-looking black flag, *Ferraria nuda* (Fig. 8.14), a malodorous native of South Africa, and the garden hyacinth, *Hyacinthus orientalis*, a highly decorative, spring-blooming plant with very fragrant purple flowers, native in what is now Turkey, Syria, Lebanon and Israel. As in the work done for Boccius, Franz very carefully records the underground parts of both species – culms in the case of the black flag, bulbs in the case of the garden hyacinth, adding in the latter case a transverse section of the bulb and an instructive longitudinal section of the complete plant. However, there were also plants to be portrayed at Kew that Franz knew already from Vienna – among them *Schotia afra*, denominated *Schotia speciosa* by Jacquin and incorrectly described by him as a species new to science on the basis of a living plant received from South Africa, just before Franz's departure from Vienna with Jacquin the son (see Chapter 7).

While working for Boccius and Jacquin, Franz and Ferdinand seem never to have documented gymnosperms. In 1804 in Kew Franz recorded common juniper, *Juniperus communis*, with its dark blue berries, which contain small cones, correctly depicted by him as enclosing seeds. A few years before that Franz had embarked together with Ferdinand on a project to document various gymnosperms, in particular species of the genus *Pinus* (see Chapter 9). In addition, a few macrofungi were recorded, among them the obscene-looking common stinkhorn, *Phallus impudicus*, with its foul-smelling, sticky spore mass born on a cap-like structure on top of an erect stalk, which Franz had observed with Jacquin the son near Scheveningen (see Chapter 7).

FIG. 8.13. *Gloxinia perennis* by Franz Bauer, undated. Paper, watercolour. – London, Natural History Museum, Library and Archives, Finished Drawings of Plants, pl. 79.

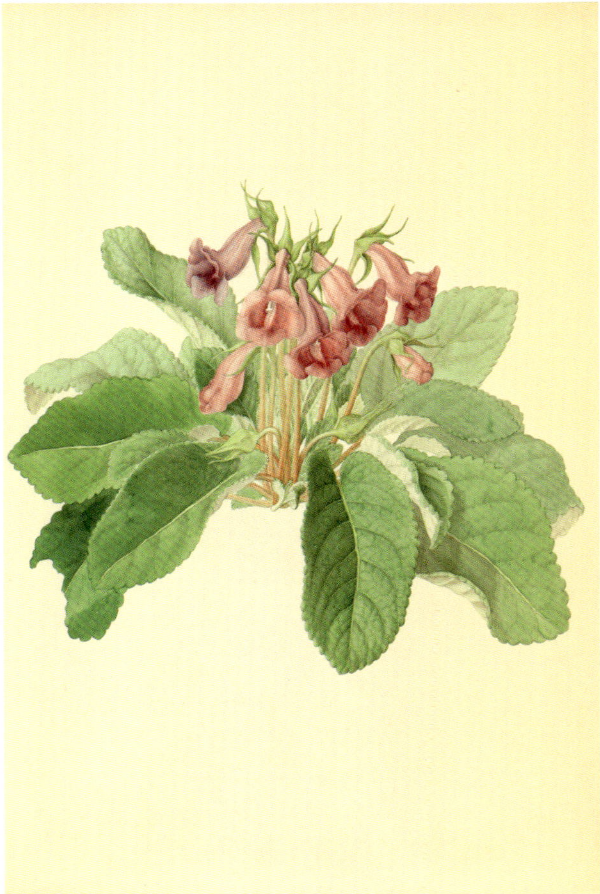

FIG. 8.14. Black flag, *Ferraria nuda*, by Franz Bauer, undated. Paper, watercolour. – London, Natural History Museum, Library and Archives, Finished Drawings of Plants, pl. 153.

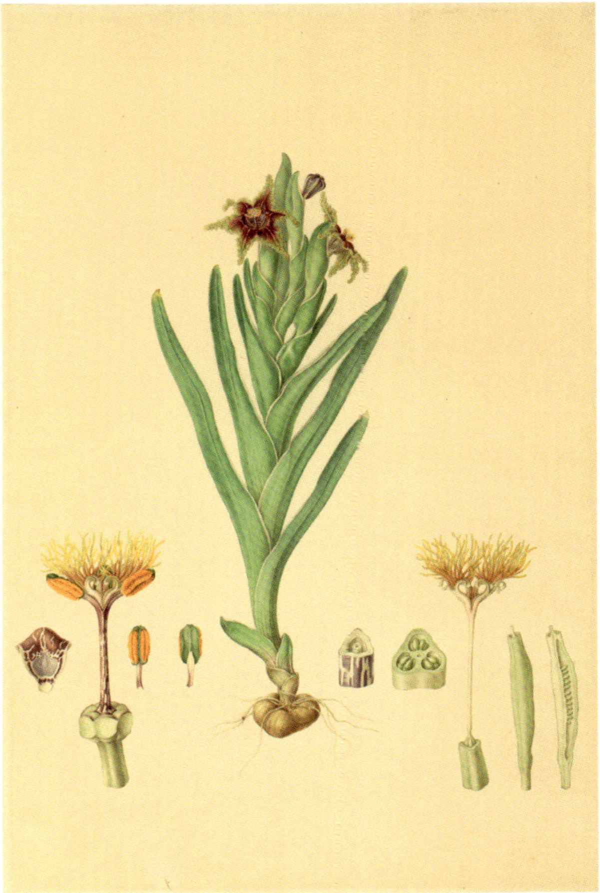

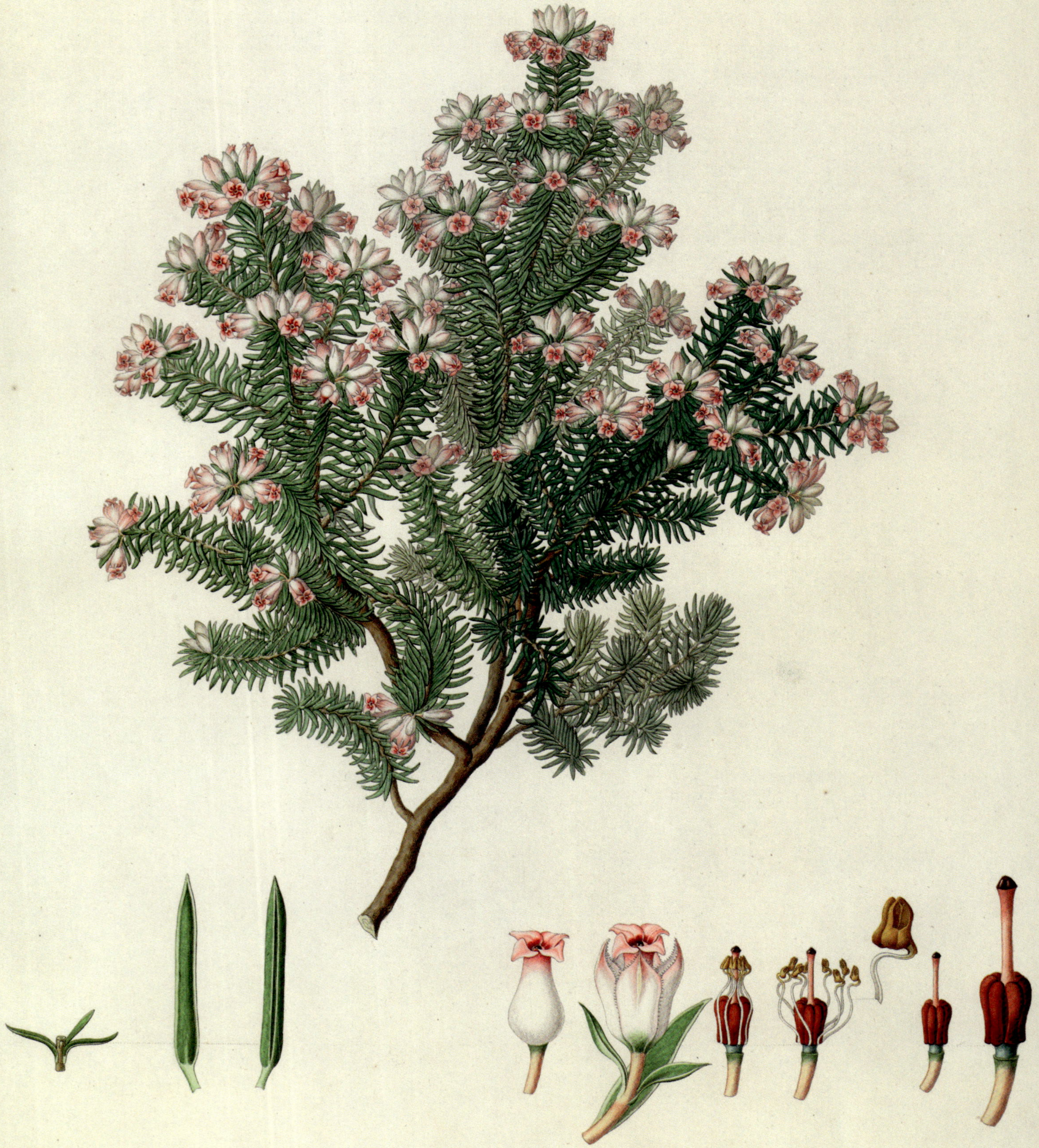

Erica taxifolia.

Mackenzie

DELINEATIONS OF EXOTICK PLANTS AND *STRELITZIA DEPICTA*

On 17 April 1790 Jacquin had written to Banks, 'The degree of joy my heart was filled with at the receipt of your last letter cannot be felt by him who is [not] a father. And so my son has behaved well in England, so as he ought, and you my dear Sir, who are so excellent a judge in this matter, testify it? My son then has answered all my wishes . . . I wish Mr Bauer continues giving you satisfaction by his works . . . All that I fear is, that your greatest trouble will by [be] finding mens [sic] for illuminate your works'[35] Indeed, publication, in particular the engraving of the watercolours prepared by Franz, was to turn out to be a problem on the long run not to be surmounted.

When Franz was engaged by Banks in late 1789, MacKenzie had already worked for several years as the latter's engraver in 32 Soho Square. It was MacKenzie who would engrave twenty-seven of Franz's watercolours, surprisingly all heaths (*Erica* spec.) from the Cape of Good Hope,[36] almost all sent by Francis Masson, one of the plant collectors engaged by Banks and paid for by King George III. These heaths were subsequently cultivated in the Royal Garden at Kew. Two more heaths were engraved by James Basire and one by Ferdinand, which was the first engraving he ever produced (see Chapter 9). The thirty engravings are all dated: t. 1–20, 22 and 30 1 January 1793, t. 21, 23, 25–28 1 January 1800, t. 24 and t. 29 1 January 1801. However, there is conflicting evidence: it has also been stated in writing that the three fascicles were published in the years 1796, 1797 and 1803, respectively. The engraver's work is reported by Franz to his earlier travel companion in a letter of 21 May 1792, mentioning that he himself had already drawn sixty heaths. Publication took place in fascicles of ten engravings, the print run decreased from ninety for the first to fifty for the last fascicle.[37] In a conversation with Countess Itzenplitz in early 1793, Franz mentioned that the process of engraving took place under his supervision in his home at Kew, and that specially trained children coloured the plates, then a widespread procedure. A title page giving Franz as 'Botanick Painter to His Majesty' was published as well as a foreword by Banks, but there was no text. From a modern point of view this would be regarded as an eccentricity, but it was in line with Banks's concept of publishing sophisticated plant illustrations with details and a scientific name but no text, as he had done previously when publishing the engravings showing *Phaius tankervilleae* and *Strelitzia reginae* (see above). Here is his reasoning: 'It will appear singular, at first sight, that engravings of plants should be published without the addition of botanical descriptions of their generic and specific characters; but it is hoped, that every Botanist will agree, when he has examined the plates with attention, that it would have been an useless task to have compiled, and a superfluous expense to have printed, any kind of explanation concerning them; each figure is intended to answer itself every question a Botanist can wish to ask, respecting the structure of the plant it represents; the situation of the leaves and flowers are carefully imitated, and the shape of each is given in a magnified, as well as in a natural size. The internal structure of the flower, respecting the shape and the comparative size of its component parts, is also, in all cases, carefully displayed'. All this is praise for Franz's work and no exaggeration. In a most meticulous way Franz had consistently prepared analyses for all thirty species of *Erica* – documenting for example in *E. taxifolia* (Fig. 8.15.), from left to right, (1) a piece of stem with three leaves, (2) a single leaf in two views, (3) a single flower, (4) a branch with a single flower, (5) a flower without corolla and with stamens in early development, (6) a flower without corolla and with stamens in late development, (7) an anther enlarged and (8) the gynoecium in two views. Indeed, all plates are so very precise that the identity of the species documented was never in doubt. In addition, several scientific names were validated by Franz on these copper engravings, for example *Erica viscaria*, *E. halicacaba* and *E. longifolia*, because they contained analyses. Jacquin had acted in a similar way when publishing his *Icones plantarum rariorum* in Vienna.[38] However, there is another possible explanation for the absence of a text in the *Delineations*. Under the copyright of the time in England, books of engravings without text were exempt of the requirements to present copies to the nine copyright libraries for free, a number which was increased to eleven in 1801.

FIG. 8.15. *Erica taxifolia*. Coloured engraving by D. MacKenzie based on a watercolour by Franz Bauer. Franz Bauer, *Delineations of exotick plants cultivated in the Royal Garden at Kew*, t. 19, London, 1793. – London, Natural History Museum, Library and Archives.

The preface of the *Delineations of exotick plants cultivated at the Royal Garden at Kew* also merits special attention in another sense, since it clearly indicated that more was planned than the three fascicles that were published, Banks writing 'Under circumstances so commonly fortunate to a gardener . . . it has been thought proper, to make preparations for publishing a new edition of the Hortus Kewensis; and, in the mean time, to present the public with the work to which this is prefixed, intended to be annually continued, provided it meets with a favourable reception, as long as the excellent draughtsman, who has undertaken the execution of it, can be prevailed upon to remain in England, for that purpose'. This is a laurel wreath for Franz from his commissioner, whose attention to the Royal Garden at Kew was indeed global in approach, stating in the foreword 'Mr. Masson, who had travelled as a Botanist, at his Majesty's expense, for twenty-five years, was never so fortunate in furnishing abundance of fresh seeds and living plants, as during his last mission on the Cape of Good Hope. The settlement of a colony on the coast of New South Wales [Port Jackson, today's Sydney] has opened to us a fresh source of botanical wealth, in a climate nearly congenial to our own; and of this a large share has been transmitted to Kew, by Arthur Philip. Esq. the Governor. William Bligh, Esq. Commander of his Majesty's ship *Providence*, who was sent to the South Seas for the purpose of carrying the Bread-fruit from thence to the British colonies in the West Indies . . . has also been enabled to place in Kew Gardens some hundreds of species, natives of the East and the West Indies, which had never before been seen in Europe . . . Commanders of ships have employed the leisure of their homeward bound passages in taking care of the vegetable produce of the climates they visited, anxious to furnish, on their return, any degree of increase to her Majesty [Queen Charlotte]'s amusements. The Directors of the East India and the Sierra Leone Companies have forwarded to Kew such plants as their servants abroad were able to procure for them; and the Government of Jamaica has sent a public present to his Majesty [George III] of plants collected on that Island'. Concurrently the Royal Garden at Kew had been enabled by its owner, George III, to keep pace. Banks notes 'new houses [conservatories] have been built, upon a most ample scale, and of a most convenient construction: the garden itself has lately received a considerable addition of space; and the whole has been laid out . . . in a manner so commodious, that it is not only capable of containing the whole of the present Collection with ease, but of receiving all such additions as may probably be made to it, for several years to come'. This is the description of the Royal Garden at Kew during a particularly lucky period of its existence. It lay just opposite Denmark House, where Franz had taken his lodgings; he had only to pass St Anne's Church and cross Kew Green in order to arrive at the main entrance. In 1804 Franz moved across the road to 23 The Green, a late-eighteenth-century pair of houses with three storeys with dormers in a slated attic and a brown brick parapeted front, today a grade II listed building.[39] Within a few minutes on foot he would have entered the Royal Garden.

For Franz the publication of the *Delineations*, a work dedicated to Queen Charlotte, brought fame and recognition only a few years after his arrival in England. It was a luxury publication, destined for the happy few who could afford such a costly work or were fortunate enough to receive it as a gift. Several plates of this work were donated in 1795 to Maria Feodorovna, Grand Duchess of Russia, who had visited Vienna in 1781 and had subscribed to a luxury copy of the *Selectarum stirpium americanarum historia* (see Chapter 3). She was shortly after that to become the Empress Consort of Pavel I, Emperor and Autocrat of all the Russias in St Petersburg. Almost certainly Banks had his hand in this present to the Russian court. However, in retrospect Franz had to tell Jacquin the son on 19 August 1814, 'my situation and occupation is much the same it has been for so many years. Of the Kew Plants [i.e. *Delineations*] are only three numbers published; the work being on too large a scale, and requiring too much labour, it could not be sold but with considerable loss, and my Engraver diing [dying], the work was dropt nearly twelve years ago [i.e. in 1802]'.[40] Indeed, MacKenzie had died in 1800 and Banks had not found a suitable replacement. For unknown reasons Franz's watercolours forming the basis of the *Delineations* have been split between the Natural History Museum, and the Royal Botanic Gardens, Kew, with a few apparently lost; preliminary material is at Göttingen, repeats of a few drawings probably by Franz himself are at Kew and a single one is in the Hunt Institute for Botanical Documentation in Pittsburgh, Pennsylvania. As with his other drawings of plants cultivated in the Royal Garden at Kew, Franz refrained from adding scientific names; this was done by Dryander.[41] Why Franz had not mentioned in his letter to Vienna the work done in collaboration with his brother Ferdinand on gymnosperms for Aylmer Bourke Lambert remains unknown (for further details see Chapter 9).

Four years after sending this letter to Vienna Franz published eleven of his watercolours documenting two more species restricted to the Cape of Good Hope. He made use of a brand-new printing technique, which seems never to have been used for botanical illustration before – lithography. Just as in the *Delineations* the format was very impressive, and again there was a title page but no text. The print run is unknown, but was probably even smaller than for the *Delineations*: only four copies seem to be known today from public libraries in England (the British Library, London; the Natural History Museum; the Royal Botanic Gardens, Kew; the Bodleian Library, Oxford) and only three from public libraries on the continent, not all of which are complete. One of them was acquired for that great bibliophile and connoisseur in Vienna, Franz I, Emperor of Austria (see Chapter 17), a nephew of Emperor Joseph II. The complete print run was coloured by hand. It has been hypothesized that this was done by Franz, which is a plausible assumption, but no evidence for this could be traced; we are also uninformed as to whether the work appeared in instalments. The title *Strelitzia depicta or coloured figures of the known species of the genus Strelitzia from the drawings in the Banksian Library* bears no reference to the Royal Garden at Kew, although the plants were almost certainly cultivated there. Instead of an explanatory text a complete graphical account is given of two species of the genus *Strelitzia*: the bird-of-paradise, *Strelitzia reginae* (Fig. 8.16.) and an arborescent, white-flowering species, *S. alba*, both endemics of small areas in Southern Africa, the former today a common plant in the florist's shop. Nothing is known for certain regarding when precisely the work was published; on the title page 1818 appears, which is repeated in the standard bibliographies. In fact this was the year in which the life of Queen Charlotte came to an end. Her death was very much a Kew story – it took place in the presence of her eldest son, the Prince Regent and later King George IV, in Dutch House, the brick building very near to Franz's lodgings. In her final years she is reported to have repeatedly called upon Banks at Spring Grove, from where Franz also had received plant specimens to be recorded. Maybe the publication of *Strelitzia depicta*, no doubt initiated and shaped by Banks, was his final tribute to her.

Banks's arrangement with Franz was not an exclusive one. As a consequence the latter was permitted to accept commissions from others, and this he did. Many of them were fellows of the Royal Society or scientists soon to be elected fellows of the society. One of the first was Richard Anthony Salisbury né Markham, who had asked Franz to prepare an illustration of *Bauera rubioides*, no doubt a flattering job for the country lad from Feldsberg to do. The species is said by Salisbury to have been first collected by Banks and Solander on the eastern coast of Australia when circumnavigating the world in the *Endeavour*.[42] Coined by Banks, the generic name had been dedicated to both Franz and Ferdinand; it was validated in Henry Charles Andrews's *Botanical Repository* in 1801, together with a description and an illustration showing a flowering twig from a specimen cultivated in the conservatory of Andrews's father-in-law, John Kennedy, in Hammersmith.[43] The coloured copper engraving contains four analyses. In 1804 John Sims, the first editor of *Botanical Magazine* after the death of its founder William Curtis, published a second coloured engraving

FIG. 8.16. Bird-of-paradise, *Strelitzia reginae*. – Coloured lithograph based on a watercolour by Franz Bauer, Franz Bauer, Strelitzia depicta, t. 1, London, 1818. – London, Natural History Museum, Library and Archives.

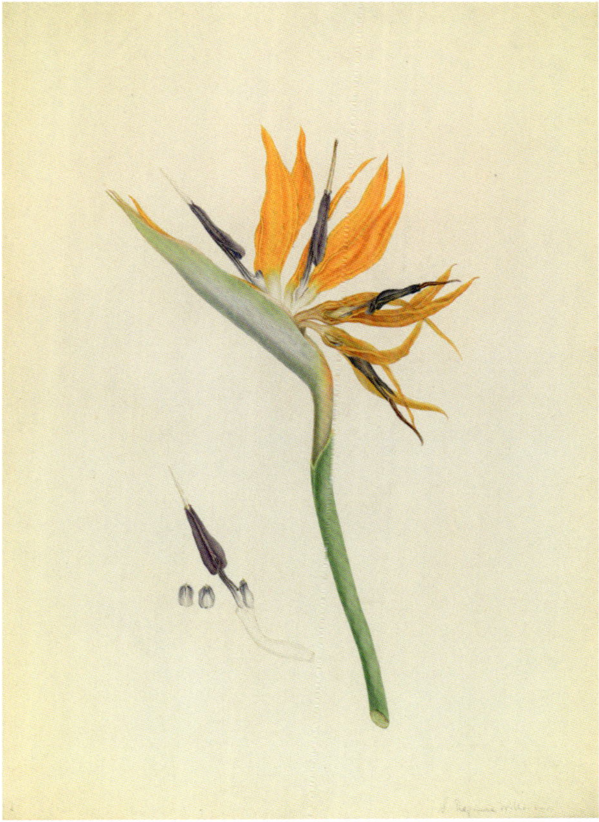

of *Bauera rubioides* in his journal based on a specimen cultivated in the garden of Messrs. Grimwood & Wyke in Kensington;[44] it included no analysis. In the following year the paper by Salisbury, another founding member of the Horticultural Society, appeared. It gave a much more detailed description, based on a specimen cultivated on his estate Mill Hill and including an uncoloured engraving based on a much more sophisticated drawing by Franz, which could not be localized. Salisbury's text is full of praise: 'The works of the older of these brothers [i.e. Franz] are indeed most admirable, far excelling all that have appeared before, and hardly to be rivalled in future ages . . . The great Linné [Linnaeus], in naming plants after celebrated men, has often traced with much ingenuity some resemblance between their manners

FIG. 8.17. ***Bauera rubioides.*** **– Anonymous copper engraving based on a drawing by Franz Bauer. – Annals of Botany 1: t. 10, London, 1805. – London, Natural History Museum, Library and Archives.**

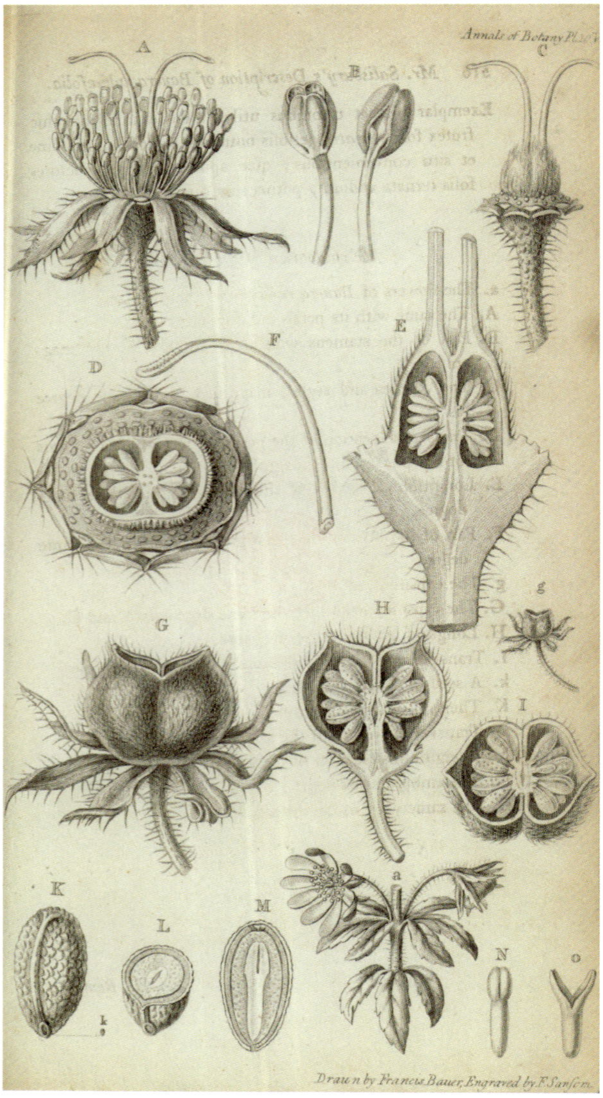

and habits, and those of the vegetable. Thus the stem of *Bauera*, firm and inflexible, pushes rapidly forward, in spite of every obstacle, into full flower and fruit, like the transcendant [sic] pencil of these painters whose name it bears; stipules or bracts it has none, disdaining as it were these vulgar supports in arriving at perfection: moreover, almost every part is produced in pairs; the cotyledons in pairs, the branches in pairs, the leaves in pairs, the flowers in pairs, the styles in pairs, and, lastly, the cells of the fruits [carpels] in pairs'. This is a rather far-fetched analogy, but probably based on personal acquaintance with the two brothers; however, at the time of publication of Salisbury's paper Ferdinand was still on Norfolk Island in the Pacific (see Chapter 10). The copper engraving of *Bauera rubioides* (Fig. 8.17.) published by Salisbury is a considerable advance over the slightly older illustrations published by Andrews and Sims. It consists of no less then fourteen figures, labelled A–O. As with Ferdinand's botanical drawings prepared for Sibthorp, Franz used capital letters for enlarged, and lower case letters for non-enlarged flower parts. More significant are several details: Franz represents an anther in dorsal and ventral view, the slightly enlarged tip of the style, and the embryo prepared free of the endosperm in which it is embedded, all characters that were never documented in an illustration of *Bauera rubioides* before. This engraving exemplifies a particular forte of Franz – the preparation of minute botanical subjects, almost certainly with the help of a sharp scalpel and a needle, and the accurate recording of anatomical details. At the same time the three engravings of *Bauera rubioides* show how quickly imports from Australia were passed from one garden to another in the early years of the nineteenth century.

Robert Brown, back in London after his circumnavigation on the *Investigator* with Ferdinand (see Chapter 10), was another commissioner for Franz, relying on the latter's experience in the representation of anatomical details. Why Brown had not asked Ferdinand, who was equally sophisticated in this kind of work (see Chapter 11), to do the job remains unclear. Brown, one of the greatest botanists of all times and soon to become an iconic figure because of his observations of the stochastic movements of particles in fluids, subsequently called Brownian motion, was a man of many interests. One of them was the fern genus *Woodsia*, which in 1813 he had recognized as new to science because of 'the peculiar structures of their [its] involucrum [i.e. indusium]'.[45] Apparently this character had not yet been observed and was to be documented by Franz. A herbarium specimen of alpine woodsia, *Woodsia alpina*, a fern widespread in northern latitudes occurring in Eurasia and America, was portrayed in thirteen figures, arranged on a plate that is now kept in the archives of the Linnean Society in London. It shows: (1) parts of fronds, (2) the sori (groups of spore-containing sporangia), (3) the indusium ending in long, bent hairs, (4) intact sporangia in different views, (5) dehisced sporangia and (6) an individual spore – all in different magnifications. Brown is not explicit on the provenance of the specimen, apart from the note that one of the thirteen figures to form the exemplar for the copper engraving was based on cultivated material.

Even less is known about the contact between Franz and John Bellenden Ker, né Gawler, the first editor of the *Botanical Register* founded in 1812 in opposition to the *Botanical Magazine*, then edited by Sims (see above). Interested in petaloid monocots, Ker published a substantial paper on the Amaryllis family;[46] one of his plates, admittedly of rather poor quality, contains a single figure based on a watercolour by Franz, kept in the Natural History Museum.

Nothing is known so far about who commissioned the extraordinary flower piece signed 'Francis Bauer MDCCIIC' kept in the Fitzwilliam Museum, Cambridge (Fig. 8.18.).[47] Because this still life seems to be the only such work painted by Franz while in England, it merits special attention. Measuring 477 × 352 mm and painted on vellum, a particularly fine support, it has been rightly described as 'On a marble ledge, a peeled orange and a decorated brown pot containing flowers that include roses, pelargoniums, tropical orchid, herbaceous Clematis and a Red Admiral butterfly'. The orchid referred to is clearly *Phaius tankervilleae*, which Franz had recorded several times; the roses include the pink cultivar very similar to that shown on the two congratulation cards prepared for Banks; and we also find the forget-me-not or, alternatively, the Siberian bugloss, and heartsease. However, there are several more plants in the flower-piece – at least three species of *Erica* from the Cape of Good Hope, among them *E. cerinthoides*, hummingbird fuchsia, *Fuchsia magellanica*, tiger lily and a species of iris, a white rose as well as several plants not yet determined to species level, among them jasmine (*Jasminum* spec.). As on the title page of the luxury edition of the *Selectarum stirpium americanarum historia* painted for Jacquin (see Chapter 3), we find a memento mori in the form of a house fly sitting on the

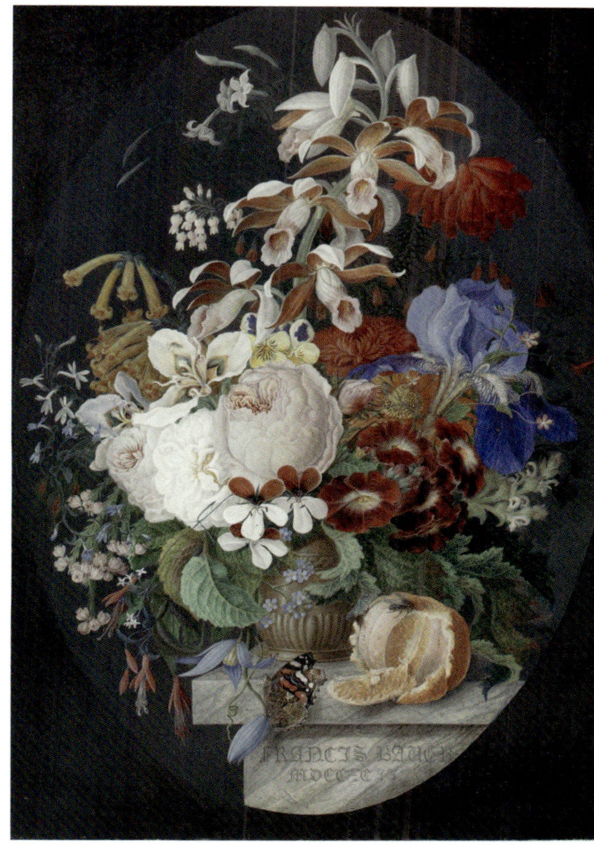

FIG. 8.18. **Flower piece by Franz Bauer. Vellum, bodycolour, signed, 1798. – Cambridge, Fitzwilliam Museum, PD. 58 – 1975.**

peeled orange and there is also the logo from the Jacquin workshop – the red admiral butterfly, *Vanessa atalanta*, the larvae of which feed exclusively on the stinging nettle, *Urtica dioica*. The 'aphid-like insect' also referred to in the catalogue cited above still needs to be more closely studied. The great surprise, however, is a reference to Franz's youth in Feldsberg – *Salvia aethiopis*, a rarity from the Pollau Mountains very near to Nikolsburg and previously included in the *Codex Liechtenstein* (see Chapter 2). Considering the iconographical programme and taking the support material, unconventional for Franz, into consideration, a member of the Royal Family could have been the commissioner of this flower piece, maybe Princess Elizabeth or Queen Charlotte. As has been rightly said, this is 'a fine and finished example of Bauer's technique at its most spectacular'.

When Banks died in Spring Grove, two years after the queen, the idea of a journal dedicated to the rarities and novelties of the Royal Garden at Kew was long dead and only a tiny fraction of Franz's painted record of plant life at Kew prepared over thirty years had been published. A small number of Franz's respective drawings had been previously deposited at 32 Soho Square and some of them have been cited in the *Botanical Magazine*, for example as '[Franz] Bauer's Sketches v.[olume] 1. Bibl[iotheca] Banks sine titulo [Banks Library, without a denomination]', indicating that they were at least occasionally consulted by the botanical community. Otherwise all other material, hundreds and hundreds of drawings, finished and unfinished, plus sketches and a few manuscript notes, remained with Franz at Kew, forming a unique painted record of the Royal Garden. Several reasons may have contributed to this state of affairs: firstly, both Franz and Banks seem to have been perfectionists, not prepared to publish plant illustrations except of the very best quality and in large format, which always remained cost-intensive; secondly, both seem to have felt unable to accept a compromise on inferior standards or larger print-runs; thirdly, the market for luxury botanical publication in England was always limited, even when supported by a foundation as in the case of the *Flora Graeca* (see Chapter 9); fourthly, Banks felt unable or was unwilling to organize the demands for big money from the crown necessary for grandiose projects, like those supported by other Maecenas, such as Joséphine, Empress of the French, the first wife of Napoleon; fifthly, Banks as a person was not, as rightly observed by Countess Itzenplitz in 1793, sufficiently focused on the idea of publishing Franz's works - with too much competition from other fields of interest in his mind.[48] The result is that Franz's vast amount of work was right from the beginning destined largely to be shelved, or as worded somewhat bluntly by Countess Itzenplitz 'would rot for the greater part in some library'.

THE LIFE CYCLE OF WHEAT

The same happened to very many other watercolours by Franz. In his letter of 19 August 1814 to Jacquin the son he had noted in retrospect: 'I have lately finished a very interesting work to which I gave the title of "Ilustrations [sic] of the Germination and Vegetation of the Wheat, and the several Diseases in Corn". I was near seven years occupied with that work; Sir Joseph [Banks] wished to have it published, but we cannot find any engraver here to execute the plate'. The finished illustrations survive and have recently been described: 'The forty-eight drawings in the British Museum (Natural History) are all finished work, far removed from sketches, and measure *ca* 320 mm high by 210 mm across. They are shelved in two Solander cases, each of which has a notebook in handwriting resembling that at Kew [see below]. The first notebook has a formal title page, as well as a heading above the next sheet, which reads "Illustration of the germination and vegetation of wheat, and of several diseases in corn.", and has pp. 1 – 22 describing Tab. I – XXIII which show normal wheat. The notebook in the second case has no title page but above the text is the heading "Illustration of several diseases in corn", followed by pages numbered 23 – 75 describing Tab. XXIV – XLVIII, all concerned with infected wheat. The notes on each drawing amount to short essays with references to other authors (e.g. pp. 70 – 75 are devoted to ergot). The second notebook ends in the same style as that at Kew: "Kew March 20 th 1813 Francis Bauer" '.[49] In addition to the 48 plates with appended legends and the two notebooks, a copy in Franz's hand of the second notebook entitled "Memoranda and observations of several diseases in corn" in quarto is kept in the archive of the Royal Botanic Gardens, Kew. Furthermore many preliminary drawings and sketches pertinent to this subject together with copious manuscript notes fill several boxes in Göttingen.[50] Because of Banks's will and the codicils (see Chapter 17) a notebook in small format containing several watercolours by Franz and of particular relevance for this topic ended up in the Kent History and Library Centre in Maidstone.[51] The description cited above is most adequate: the forty-eight plates in the Library and Archives of the Natural History Museum are of superb quality and show Franz for the first time as the microscopist and nascent researcher. However, in order to properly understand their value, it is necessary to provide some general background information.

With an income of more than £20,000 per year in the year of his death, Banks was one of the large landholders and more wealthy county esquires in England. Rightly described as 'a stalwart Georgian country gentleman of independent means and a truly independent cast of mind', in 1820, he received an annual income of c.£16,000 from a total of 523 tenant families. Even when considering his substantial expenses for 32 Soho Square, Spring Grove, Revesby Abbey and his other country houses – taxes, servants, carriages and horses, buildings and repairs, farming and gardening expenses, furniture, coal and woods, liquor, etc. – a substantial sum of money remained for the private use of Banks and his family.

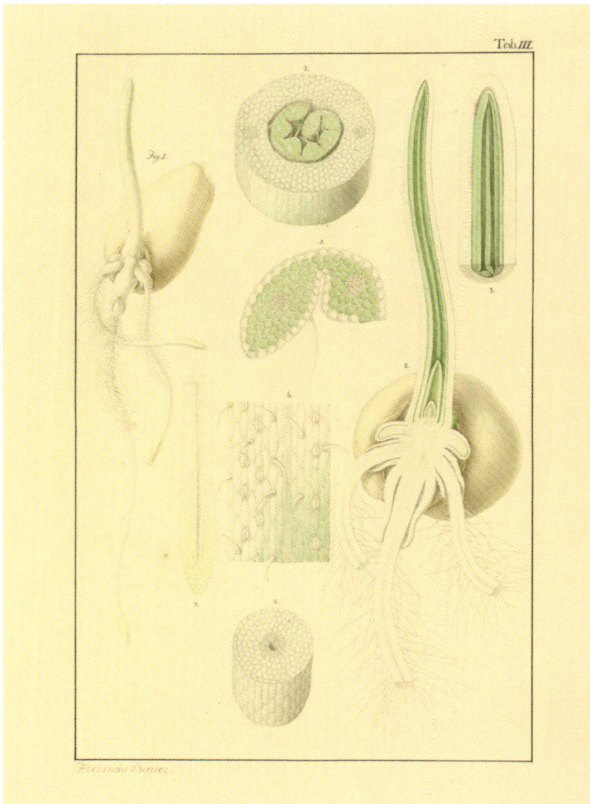

FIG. 8.19. Wheat, *Triticum aestivum*, by Franz Bauer, c.1805. Paper, watercolour, signed. – London, Natural History Museum, Library and Archives, Germination and Vegetation of Wheat, pl. 3.

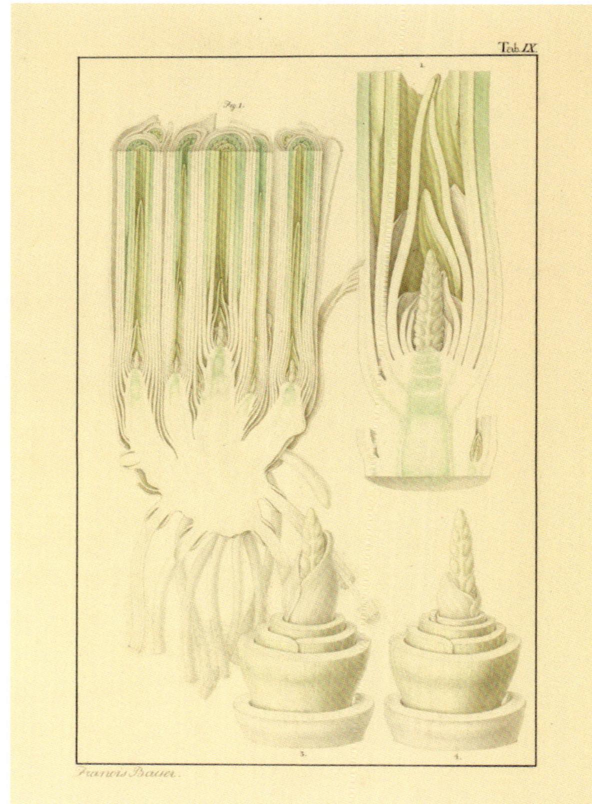

FIG. 8.20. Wheat, *Triticum aestivum*, by Franz Bauer, c.1805. Longitudinal section of basal portion of juvenile plants. Paper, watercolour, signed. – London, Natural History Museum, Library and Archives, Germination and Vegetation of Wheat, pl. 9.

As a man well versed in financial matters, he knew perfectly well that the economic basis of his tenants was their income from agriculture and animal husbandry, with horticulture and fishing providing some additional revenue. Clearly wheat was a very important crop for most of them and therefore this was the plant on which to focus his particular attention.

A native of the Fertile Crescent in the Near East and for many millions of people the basis for nutrition for more than fifteen millennia, wheat has always attracted the interest of scientists. However, the first in-depth documentation of the life cycle of wheat was prepared by Franz, no doubt under orders from Banks. No proof of this is available, since communication between commissioner and illustrator was by word of mouth or by letters, which Franz did not keep. Franz prepared a painted record of utmost precision clearly intended for publication – all 202 figures are numbered and carefully arranged into a total of 23 plates (nos 1–23), and all these compositions are surrounded by a frame in black ink, though markedly thinner than in the *Codex Liechtenstein* (see Chapter 2). In calligraphy and again in black ink Franz had added running numbers to the plates in the upper right-hand corner and in pencil his signature in the lower left-hand corner, in both cases outside the frame; no dates are given. The sequence of the plates strictly follows the chronology of the wheat plant's life cycle. In addition, Franz provided the text for all the legends, forming a bound manuscript comprising twenty pages kept together with the 23 plates.

The first plate selected for this subchapter (Fig. 8.19.) shows: (1) a juvenile wheat plant seven days old, (2) a longitudinal section of it, (3) a longitudinal section of the tip of it, (4) a transverse section of the first leaf enveloped by the coleoptile, (5) a transverse section of this leaf, (6) the epidermis of the leaf, (7) the tip of the root and (8) a transverse section of the root. Like all other plates of this sequence this is a feast for the observer – the three-dimensionality of the young plant is accurately presented, including the very many root hairs emerging from the root and the faint yellowish-green colour of the first leaf. Longitudinal sections of the juvenile plant

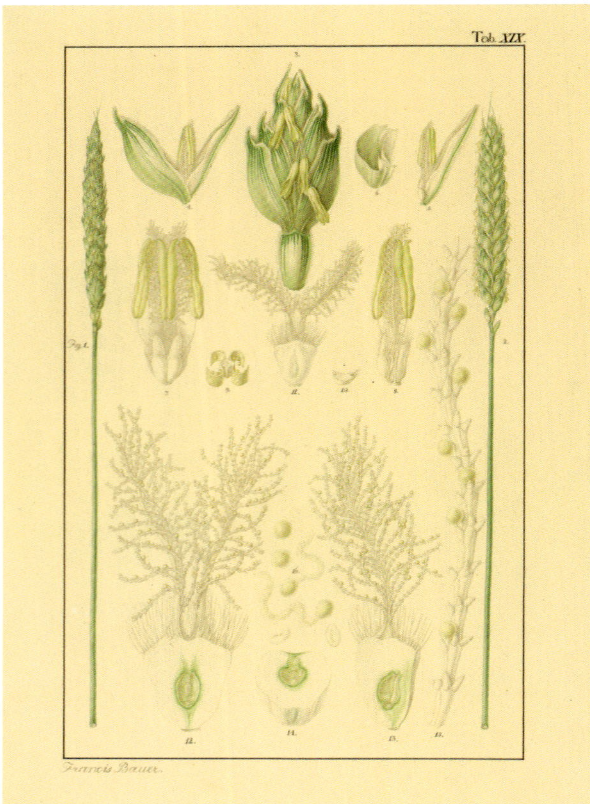

FIG. 8.21. Wheat, *Triticum aestivum*, by Franz Bauer, c.1805. Flowering ear. Paper, watercolour, signed. – London, Natural History Museum, Library and Archives, Germination and Vegetation of Wheat, pl. 19.

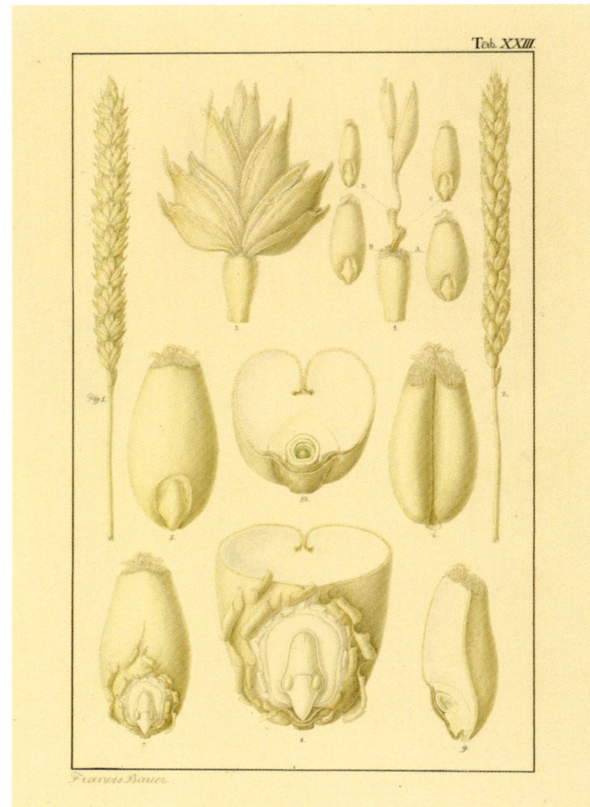

FIG. 8.22. Wheat, *Triticum aestivum*, by Franz Bauer, c.1805. Fruiting ear. Paper, watercolour, signed. – London, Natural History Museum, Library and Archives, Germination and Vegetation of Wheat, pl. 23.

(Fig. 8.20.) follow with an accurate representation of the minute ear primordia surrounded by the bases of the leaves not yet containing chlorophyll. The plate that documents the flowering of wheat (Fig. 8.21.) forms the apotheosis of the whole series – no less than sixteen figures represent, among others: (1) the ear in flower, (2) a spikelet magnified, (3) a flower from the spikelet with its bracts, (4) a flower with stamens discharging pollen, (5) a transverse section of the anther, (6) an ovary with its two feathery stigmas, (7) longitudinal sections of the ovary with the two feathery stigmas, (8) a transverse section of the ovary, (9) a single hair of the stigma with the pollen cells attached, and (10) individual pollen cells. The life cycle comes to an end in a plate (Fig. 8.22.) representing (1) the ripe ear, (2) a ripe spikelet, (3) a ripe spikelet with its four fruits, i.e. grains, removed, (4) a single grain, and (5) dissected grains showing the embryo. In several respects this is botanical illustration at its best: firstly, because of the documentation of a plant's entire life cycle; secondly, because of the use of different magnifications from natural size to considerable enlargement, with the pollen cells visible only in the microscope; thirdly, because of the depiction of often more than one structure, for example more than one single flower, in order to demonstrate variation within a single specimen; fourthly, because of the representation of different views, for example of grains in a dorsal and a ventral view and of longitudinal as well as transverse sections. Added to this, the detailed legends written by Franz in the accompanying notebook, this is simply a brilliant work worthy of the highest praise. It was rightly regarded in an anonymous obituary as 'perhaps the most splendid and important monument of microscopic observation'.[52]

THE ENIGMATIC BARBERRY BUSH

Wheat is prone to attack from several parasitic fungi, and mass infection can completely ruin a year's harvest, with immediate consequences to farmers' income and far-reaching consequences to the economy of a country as a whole. This was the more general background for another commission for Franz, no doubt coming again from Banks – to document 'Diseases in corn'. The end product consisted of finished plates (nos 24–48) ready for publication and accompanied by the text for all legends, forming a bound manuscript, with extensive additional comments and notes, all in Franz's hand. Both the plates and the manuscript are in the Library and Archives of the Natural History Museum.

The stem rust, *Puccinia graminis*, is one of the most pernicious fungal infections in wheat, causing very major concern due to the number of people dependent on wheat for sustenance. For this reason Banks must have asked Franz to study this pest in detail. With his rural background, with information he may have received about the failure of wheat crops during his early days in Feldsberg, and also his experience with working and drawing with the help of a microscope, he was best suited for this job. The considerable losses suffered by farmers in England from stem rust in 1803 and 1804 together with the unreliability of grain imports because of the strained political situation in Europe seem to have been an additional motivation for Banks to commission precisely this type of research, in 1804 at the latest.

The resulting notes are more extensive than on the life cycle of wheat. The material includes, apart from the four finished plates (nos 24–27) in London, preliminary drawings and manuscript notes in boxes in Göttingen and the notebook in Maidstone. Franz, the botanical illustrator, transmuted into Franz the nascent researcher. He extracted the pertinent literature, documented his observations in drawing and writing and summarized his findings in text. However, Franz refrained from publishing the results of his observations, but rather communicated them to Banks. It was Banks who in 1805 published a treatise entitled *A short account of the cause of the disease in corn called by farmers the blight, the mildew and the rust*, but significantly he called his text simply 'Mr. Bauer's exploration', which in fact it largely was. The preface is dated 30 January 1805. It should also be noted that in his text Banks referred to a single wheat disease covered by three different English names – blight, mildew and rust. The two copper engravings appended to Banks's *Short account* are solely based on the graphical record by Franz. They show longitudinal and transverse sections of culms infected with the fungus and various developmental stages of the teleospores. In the reprinted version of the *Short account* and latter editions, which due to Banks's prominent position in science reached a very considerable circulation among the land-owning gentry and professional men, the contents of the two plates are amalgamated into a single copper engraving. Because of the considerable print-run, Willdenow's slightly earlier publication on stem rust, which had appeared in Ratisbon, was totally eclipsed (see below). It is illuminating to look more closely into this research done at Kew in order to discern the input from Franz in relation to the input from Banks.

As we now understand, more than two centuries after Banks and Franz, the pathology of the stem rust is most extraordinary and most complex. Spores called uredospores land on the green wheat plant, germinate, develop a mycelium and form pustules that invade the outer layers of the stalk. Infested plants produce fewer tillers and set fewer fruits and the whole plant may die in cases of severe infection. Within a few weeks an apparently healthy wheat plant can be transformed into a black tangle of broken stems and shrivelled grains. The infection is visible with the naked eye by the presence of so-called uredinia on the plant: orange to brownish, elongated pustules. These most frequently occur on the leaf sheaths, but are also found on the stems, leaves and glumes. On the leaf sheath and glumes these pustules rupture the epidermis and release more uredospores, which in summer repeat the infection again and again, often with disastrous consequences. Towards the end of the growing season the mycelium of the stem rust forms black teliospores on the wheat plant, which is the reason why this pest has also been called black rust. These teliospores hibernate and in spring produce a third type of spore called basidiospores, which instead of infecting the wheat plant infect the leaves of the barberry bush, *Berberis vulgaris*. There they germinate, forming a mycelium that produces pycnia, small yellowish clusters on the upper surface of the leaves. These produce a fourth type of spore called pycnospores in a sticky honeydew. These, sometimes transferred by flies, sooner or later infect other leaves of the barberry bush, where they germinate, form another mycelium and develop into cup-like structures called aecidia, breaking through the lower surface of the leaves. The aecidia are filled with orange-yellow powdery aecidiospores; only the latter are capable of infecting the young wheat plant in spring. The developing mycelium produces uredospores and the whole cycle starts again. In short, the stem rust is an obligate parasite with a complex biology. It is heteroecious, requiring two host organisms, the cereal host and the alternate host. At the same time this refined pest is macrocyclic, exhibiting five different spore types.

Needless to say, only a small part of this complex life cycle was known to Banks and Franz, and even less of it was really understood. In a draft manuscript kept in Göttingen, Franz summarized his observations in retrospect: 'In the month of August 1805 [1804?], I received, specimens of wheat plants very strongly infected with this disease . . . & was desired to ascertain by microscopic investigation, whether the black eruption with which the wheat, and almost all gramineous plants were that year, strongly affected, was an organized, or unorganized substance? On attentive observation I soon ascertained, that the supposed black powder, consisted of innumerable very minute, but perfectly organized parasitic fungi (which I subsequently found, Persoon had recorded (in his Synopsis fungorum) as Puccinia graminis . . . The above ascertained facts I illustrated by detailed microscopic figures, and communicated it, as the result of my investigations; the correctness of which was soon confirmed by a small treatise onthat [sic] disease by Felice Fontana, under the title Osservazioni sopra la Rugine del Grano, published in Lucca in 1767. This Treatise was obtained after I had compleated my researches; it gives a very detailed and correct account of the disease, and is also illustrated with very good microscopic figures. Fontana must therefore be considered the first author who discovred that this disease is occasioned by parasitic fungi. . .'[53]

FIG. 8.23. Stem rust, *Puccinia graminis*, on wheat, *Triticum aestivum*, by Franz Bauer. Paper, ink, watercolour, 1805. – Göttingen, Niedersächsische Staats- und Universitätsbibliothek, Handschriftenabteilung, gr. 2° MS Histor. Nat. 94, V: f. 8.

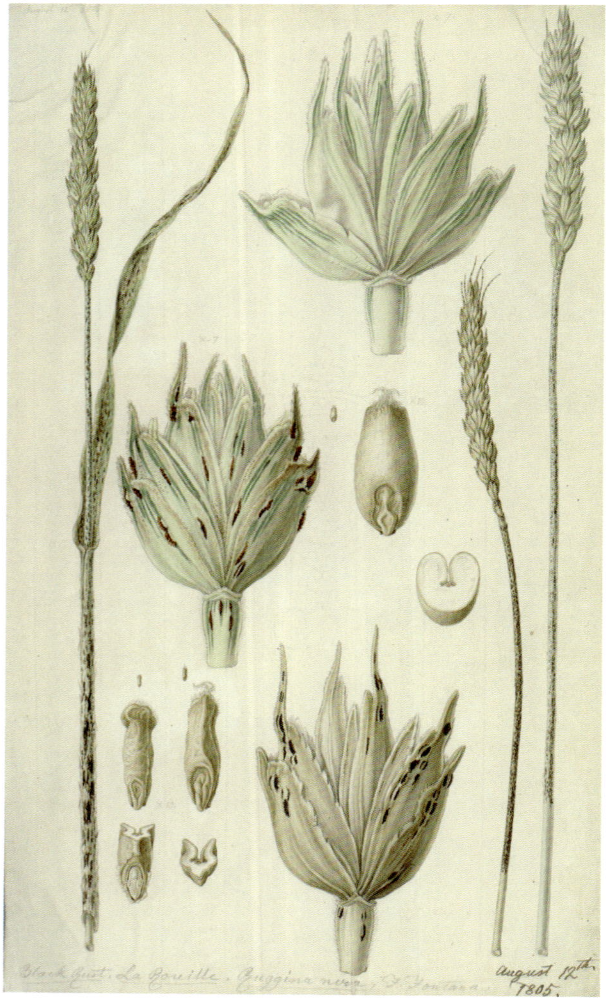

This is a very correct scientific report. Franz had observed the teliospores, noting 'the black eruptions' on the wheat plant and these he had documented in several figures in Göttingen (Fig. 8.23.) and London. In addition, he had correctly identified the pathogen as *Puccinia graminis* and found the extremely rare treatise by Fontana, no doubt in the library of 32 Soho Square, where a copy was recorded by Dryander. Franz had also realized that Fontana had observed the same fungal parasite on wheat. Interestingly, his younger brother Ferdinand had met Fontana in Florence in the spring of 1786 on his way to Rome (see Chapter 5), but they may not have discussed 'la ruggine', the Italian denomination for stem rust. Several figures taken from Franz's unpublished plates documenting this pest (Fig. 8.24.) were used as exemplars for the copper engraving accompanying the *Short account* (Fig. 8.25.). The handwritten legends to go with the two plates testify to the close collaboration between Franz and Banks – they are in the hand of Franz, with Banks adding his corrections. However, when comparing the copper engraving published by Fontana with that published almost four decades later by Banks, no increase of knowledge appears.

FIG. 8.24. *(opposite top)* Stem rust, *Puccinia graminis*, on wheat, *Triticum aestivum*, by Franz Bauer. Paper, watercolour, signed. – London, Natural History Museum, Library and Archives, Unfinished Drawings of Plants, vol. 6, no. 77.

FIG. 8.25. *(opposite bottom)* Stem rust, *Puccinia graminis*, on wheat, *Triticum aestivum*. Anonymous copper engraving based on a drawing by Franz Bauer. – Joseph Banks, A short account, London, 1805, f. 1. – London, British Library.

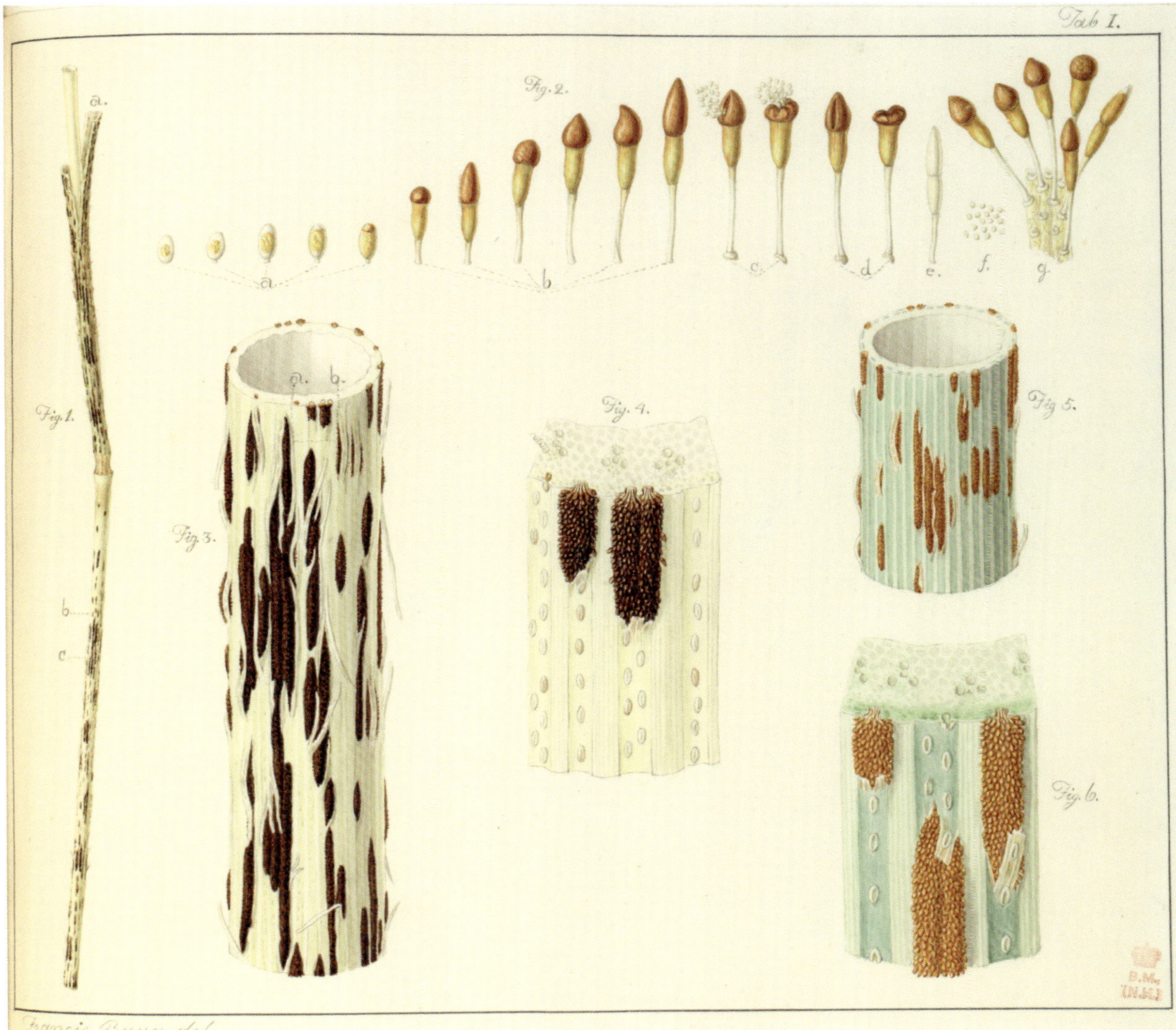
Tab I.
Fig. 1.
Fig. 2.
Fig. 3.
Fig. 4.
Fig 5.
Fig. 6.
a.
b.
c.
d.
e.
f.
g.
Francis Bauer del.
B.M. (N.H.)

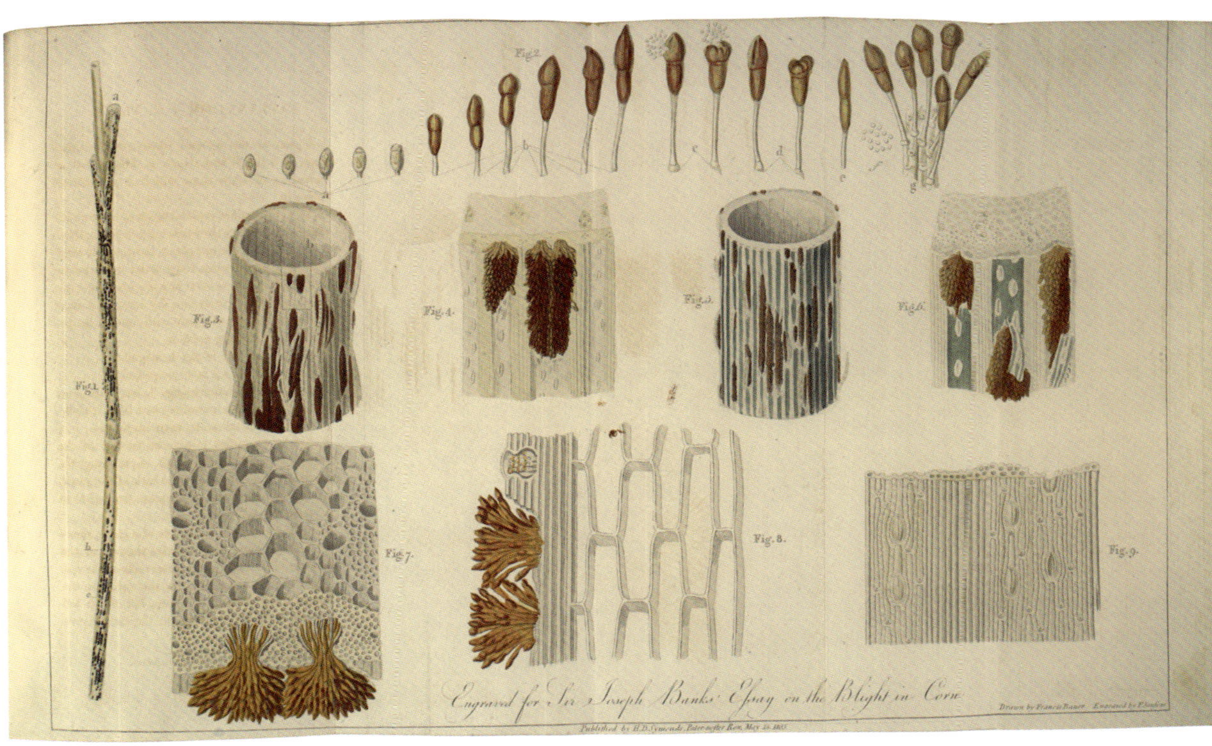
Fig.1.
Fig.2.
Fig.3.
Fig.4.
Fig.5.
Fig.6.
Fig.7.
Fig.8.
Fig.9.
Engraved for Sir Joseph Banks' Essay on the Blight in Corn

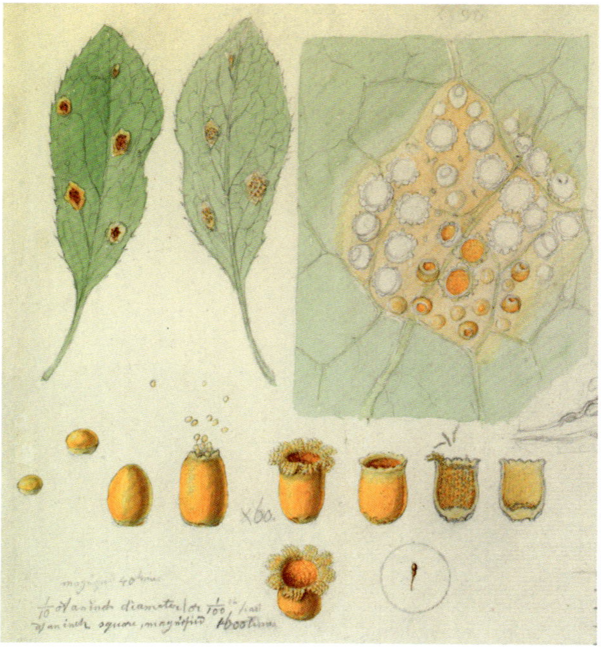

FIG. 8.26. Stem rust, *Puccinia graminis*, on common barberry, *Berberis vulgaris*, by Franz Bauer, 1805. Paper, pencil, ink, watercolour. – London, Natural History Museum, Library and Archives, Unfinished Drawings of Plants, vol. 6, no. 81, f.22.

How could it happen that the *Short account* became famous in the history of plant pathology? It was the following hypothesis put forward in the text: 'It has been long admitted by farmers, though scarcely credited by botanists, that wheat in the neighbourhood of a barberry bush seldom escapes the Blight [*Puccinia graminis*]. The village of Rollesby in Norfolk, where barberries abound, and wheat seldom succeeds, is called by the opprobrious appellation of Mildew Rollesby [i.e. Stem Rust Rollesby]. Some observing men have of late attributed this very perplexing effect to the farina of the flowers of the barberry, which is in truth yellow, and resembles in some degree the appearance of the rust, of what is presumed to be the Blight in an early stage. It is, however, notorious to all botanical observers, that the leaves of the barberry are very subject to the attack of a yellow parasitic fungus, larger, but otherwise much resembling, the rust in corn. Is it not more than possible, that the parasitic fungus of the barberry and that of the wheat are one and the same species, and that the seed transferred from the barberry to the corn is one cause of the disease? Misletoe [sic], the parasitic plant with which we are best acquainted, delights most to grow on the apple and the hawthorn, but it flourishes occasionally on trees widely differing in their nature from both of these; in the Home Park, at Windsor, misletoe may be seen in abundance on the lime trees planted there in avenues. If this conjecture is founded, another year will not pass without its being confirmed by the observations of inquisitive and sagacious farmers'. A short comment on Banks's text is necessary: by farina he almost certainly meant pollen and wrongly equated the farina with the 'blight in an early stage' (that is, the uredospores); by the 'yellow parasitic fungus on the barberry' Banks meant either the pycnia or the aecidia or both; and by 'seed transferred from the barberry to the corn' he referred to the aecidiospores infecting the wheat. No doubt this was a congenial hypothesis, although Banks was pre-emptied by Willdenow, whom Franz had met in 1788 in Berlin, years before they both became interested in stem rust. In his estimate of the time necessary to find confirming evidence for his hypothesis Banks was far too optimistic. The complexities of the reproductive cycle and the obligate host change were unravelled experimentally only decades later by Heinrich Adolf de Bary, then at Freiburg University.

There is, however, a strange fact concerning Franz and the fungus. Franz had recorded the orange aecidia on the leaves of the barberry bush, he had even documented aecidiospores in the moment of dehiscing from the aecidia together with an enlarged teliospore in a drawing in the Natural History Museum (Fig. 8.26.). It is dated 14 July 1805 and was most probably available before the *Short account* was reissued. Furthermore, the notebook in Maidstone shows whole clusters of aecidia recorded by Franz, again with a single enlarged teliospore. Twenty-eight years later, in 1833, and in retrospect Franz wrote 'the most obstinately defended opinion was [in c.1805], that the black rust is occasioned, or atleast [sic] promoted, by the existence of Barberry bushes in the vicinity of wheat fields; and it is a fact that a highly respected, and very extensive and influential landowner, actually effected the exstirpation of nearly all the Barberry bushes in a whole county, and thus propagated and confirmed those absurd and injurious prejudices'. It is probable that Franz was referring to Banks, the influential landed gentleman, and continued to feel unable to concur with the latter's hypothesis on the role of the barberry bush. By contrast, Banks had proposed, admittedly on the basis of the most fragmentary evidence, a fruitful hypothesis, while Franz, who had prepared the drawings, did not have the necessary imagination or courage to understand the concept of obligate heteroecism.

After 1805 Franz continued his observations on stem rust and, in addition, he started experiments. He noted, again in retrospect, '. . . I repeated my investigation the next season 1806 . . . and to be fully convinced of the fact I selected in my small experimental ground, some strong and perfectly sound plants which had as yet escaped the infection, and bruises [sic] and bent some of the leaves in the middle, but without rupturing the epidermis . . . and after about fortyeight hours I observed that the bruised and bent ends of the leaves were also dropping; and after 9 to 10 days were covered with Black rust . . .'. Franz probably experimented with a far too small a number of plants and we do not know what kind of spores he tested, but in a sense this was a beginning of experimental plant pathology. He was not alone. In 1804 Thomas Andrew Knight, elected Fellow of the Royal Society in the following year, had written to Banks '. . . I sowed, in the autumn of 1804, a row of wheat round a plant of that kind [the barberry], which grew in my garden . . . and I also sowed several small portions of seed of that same kind in a meadow . . . situated at a considerable distance from it. All the plants continued perfectly healthy till the beginning of July, when those near the barberry bush showed evident symptoms of disease . . . The increase of the disease was from this period extremely rapid, and in the course of a single week the straw was covered with yellow spots exactly as they were delineated by Mr. Bauer . . . The plants of wheat, which grew at a distance from the barberry bush, remained free from the disease, I carried a branch of the barberry bush, with the diseased fruits [the aecidia] upon it, to one of them, and wetting it with water, I brushed the wheat plants with it, repeating this operation three successive days. In the course of ten days the plants of wheat, which I endeavoured to infect by means of the barberry branches and fruit, became covered with the disease . . .'. This is the first account of a successful experiment, but Knight was not convinced. A few lines later in his account he claimed that water could also cause stem rust in wheat, without considering the possibility that it might have contained spores capable of infection.

In short, the concept of a fungal parasite living on two totally different hosts and forming several different spore forms was too novel to be considered seriously. This also becomes evident in an anonymous review of the *Short account* published in *The Farmers' Magazine* stating 'We are free to say, that, if the author were not a man high in rank, and at the head of the Royal Society, his publication would have been disregarded by every farmer in Great Britain'. In addition, we cannot be certain that Franz considered the orange to brown uredospores to belong to the same organism that produces the black teliospores, since in the legends he calls the first disease 'red rust' (Tables 24–25), the second 'black rust' (Tables 26–27). Polymorphism in parasitic fungi continues even today to be an arcane subject to all except the cognoscenti and seems to have remained obscure to Franz. There is thus a second lesson to be learned: even a perfect observer like Franz is not automatically the best interpreter of his observations.

Following the example of Knight, Franz started an additional series of experiments trying to infect wheat with uredospores, but the results are only found among his scattered manuscript notes in the Niedersächsische Staats- und Universitätsbibliothek Göttingen. It was a test making use of no less than 800 wheat plants cultivated on different soils, a beginning of experimental agriculture in the early years of the nineteenth century, though not continued.

OTHER CEREAL KILLERS

Stem rust is not the only pest in wheat, and wheat was not the only cereal of economic relevance for Banks. A large field of research had opened up for Franz, and judging from the number of his pertinent drawings it is plausible that he was indeed busy with mycological studies for several years, as he had told Jacquin the son in Vienna on 19 October 1814.

One of the other infections of wheat is caused by a parasitic fungus called the common bunt or the stinking smut, *Tilletia tritici*, a contagious disease that can lead to complete crop loss due to the grain being unsaleable because of its discolouration and a revolting smell. As a matter of fact such crop losses were frequent in the 1780s, and are believed by some to have been factors contributing to food shortages in France prior to the outbreak of the French Revolution. The life cycle of this pathogen is markedly simpler than that in stem rust. As we now know, the spores on the surface of the grain germinate along with the grain, each producing a short fungal thread terminating in a cluster of elongated cells, which produce a second type of spore, called ballistospores. These infect the coleoptile of the growing wheat plant before the emergence of the first leaves and develop a mycelium growing internally within the developing ear, which becomes infected. Affected plants develop apparently normally until the ear emerges, when it can be seen that some grain sites have been replaced by so-called smut balls, seed-like structures each containing millions of greasy, black, foul-smelling spores. In severe cases, the whole field may smell of rotting fish, the stink due to trimethylamine. In bad weather conditions the ears may appear to be covered in a black ink-like substance as the spores are released in masses and run out of the protective glumes onto the ear and stem. As each smut ball contains a vast number of spores, the capacity for contamination in a field is enormous.

In 32 figures arranged in two plates (nos 32–33) Franz documented common bunt, showing affected ears, individual infected spikelets, infected flowers with sterile anthers, longitudinal and transverse sections of the infected pistil, diseased grains in both views, a transverse section of diseased grains and clusters of spores. His manuscript description in the notebook dated 1805 and referred to above covers several pages and contains the following statements: 'At their first intrusion the parasites vegetate [form a mycelium] and nourish themselves upon the cellular or medullary substance of the unimpregnated [not yet fertilized] germen, and when that is consumed upon the Tunica or seed-coat which in the Pepper Brand [*Tilletia tritici*] generally continues growing the same length of time and to the same size as in the sound plants . . . the fungi of the Pepper Brand after they have entered the germen, by their rapid propagation and absorption of the nourishment seem to effect within the grain a partial diminuation [sic] of vitality which incapacitates the parts of the fructification and thus impregnation never takes place, and neither album nor embryo exist in branded grains'. This is an exact description by Franz, but his ideas about the causes of the fungal infection are totally wrong, since he wrote 'the predisposing cause [of the infection] is a rapid and luxuriant vegetation of the plants arising from an increased degree of vitality'. Clearly we find here a situation well known to every historian of science – an amalgam of perfect observations with an interpretation that in hindsight makes no sense. However, it should not be forgotten that Franz had studied a wheat disease of great economic importance at the right time – just two years before Isaac-Bénédict Prévost published a more competent interpretation in his *Mémoire sur la cause immédiate de la carie ou charbon des blés*, an exceedingly rare treatise printed in Paris in 1807. Unsurprisingly neither the mycological studies by Franz nor those of his French colleague had an immediate effect on the development of plant pathology.

Franz was a man to follow ideas and observations over several years, and this is also the case for his studies on the common bunt. Almost three decades after their start and thirteen years after the death of Banks Franz decided to summarize his results in a short paper. This came out in 1833 in a popular journal with a large circulation – the *Penny Magazine*, published in London by the Society for the Diffusion of Useful Knowledge, definitely less elitist than the *Philosophical Transactions*. This was a considered move – Franz, elected on 8 February 1821 to Fellow of the Royal Society (see Chapter 13) could have also chosen the *Philosophical Transactions*. The title of the paper in the *Penny Magazine* reads 'The smut balls or the pepper brand';[54] the first part refers only to the common bunt, the second only to loose smut, *Ustilago nuda* (see below); unfortunately the printing quality of the illustrations is extremely poor. Franz must have studied the anatomy of many ears infected by common bunt, since he reports 'I found some ears having one side infected, while the opposite was perfectly sound. Sometimes five or six perfectly sound grains are found in an infected ear, and a few thoroughly infected grains are found in an otherwise sound ear. The infected grains are always in the last spicket [spikelet] at the apex of the ear'.

For both fungal infections, common bunt and loose smut, Franz proposed a sensible prevention – steeping the grains. For the first of these he noted that the grain should be put in 'properly prepared limewater, leaving it to be soaked at least twelve hours, and then to dry it in the air before sowing it', although he, like his contemporaries, remained unconvinced about the success of the method. When writing his text for the *Penny Magazine* Franz, like his contemporaries, apparently did not know that a much more effective method against spores of the common bunt had already been found by Prévost in France – highly diluted solutions of copper sulphate. This information was buried in the latter's *Mémoire sur la cause immediate de la carie* and remained without any impact until its rediscovery seven years after Franz's death.

As rightly observed by Franz, loose smut is less host-specific and attacks barley, *Hordeum vulgare*, and oats, *Avena sativa*, as well as wheat. Franz studied all three systems and had them documented in 58 figures arranged in five plates (nos 34–38). He was also very well aware of the differences between loose smut and the common bunt: in the second part of his paper in the *Penny Magazine* he noted that the former has no smell, much smaller spores and is far more destructive than the latter. Spore sizes are given in an unconventional way: Franz tells his reader that 2.56 million spores of the common bunt cover one square inch, while 7.84 million spores of the loose smut fit into the same space. In addition, Franz depicts the maturation and germination of spores. Concerning loose smut, he observed in the *Penny Magazine* 'I have ascertained, by repeated experiments of inoculation, that the seeds [spores] of the fungi [*Ustilago nuda*] . . . is absorbed by the roots of the germinating seed-corn and, being so extremely minute, is mixed and propelled by the circulating sap, and deposited in about every part, even in the cellular tissue of the plant, where these seeds [spores] continue to vegetate and multiply rapidly . . . The whole ear is often found entirely destroyed many weeks before even the individual florets are quite developed, or the sound ear emerges from the hose'. No doubt this is an adequate description of the devastating effects of the mycelium of loose smut growing within the cereal plant.

More relevant than describing in similar detail the two other parasitic fungi studied by Franz – fusarium head blight, *Gibrella zeae*, and powdery mildew, *Blumeria graminis*, both wheat pathogens (documented on four plates, nos 28–31), – is another fact. While Franz was absolutely certain about the often catastrophic effects of fungal infections, many of his contemporaries thought otherwise. As he rightly pointed out some considered insects the true cause of these clearly circumscribed diseases in cereals, others blamed 'blasts of wind' and yet others 'the corruption of the sap of the plants'; no doubt the peak of nonsense was the concept that these diseases were caused by the effect of meteorites.[55] Only a few, such as the congenial Prévost, held views similar to Franz.

Parasitic fungi were and still are a constant threat to crops. The same applies to many animal pests, of which only two are dealt with here – ear-cockle nematodes, *Anguina tritici*, an endoparasite in wheat and rye, *Secale cereale*, and woolly apple aphid, *Eriosoma lanigerum*, a common pest on apple trees sucking sap from woody stems. Both were studied by Franz: he published two papers[56] on the ear-cockle nematodes, the first based on his Croonian Lecture delivered on 5 December 1822 before the Royal Society and appearing in the *Philosophical Transactions*, the second in the *Penny Magazine*. In addition, Franz contributed the graphic record of the woolly apple aphid to another paper.[57]

According to Franz's second paper in the *Penny Magazine*, specimens of wheat infected with the nematodes were first received by him in July 1807 from Kent. He adds that the odd name ear-cockles or brown purples came from 'the distorted shapes and dark-brown colour of the diseased wheat-grains, which bear some resemblance to the [seeds of a] weed generally known in corn fields, and vulgarly called Corn-Cockles, or Purples, the Agrostemma Githago [*Agrostemma githago*]'. In these distorted grains Franz reports to have found a white fibrous substance consisting of 'real organized animals' and wondered how these were introduced. As with the common bunt and loose smut Franz performed infection experiments, in placing the eggs of nematodes extracted from diseased wheat grains on the posterior side of other wheat grains and placing them into the soil. As expected, after several days small nematodes appeared 'between radicle and the plumula', which answered the question Franz had asked himself. The finished watercolours documenting ear-cockle nematodes form part of the series 'Diseases in corn' (nos 39–41) described above.

Among all his other interests and activities Banks was a man who also collected information on agricultural and horticultural pests, often over an extended period of time. As witnessed by the notebook in Maidstone, he was assisted in this by Franz, who in addition to parasitic fungi (for example *Roestelia cancellatum* on the leaves of the pear tree, *Pyrus communis*, mentioned in a paper by Knight[58] published in the *Transactions of the Horticultural Society of London*) also recorded all kinds of insects, as a rule in their larval form. Only very rarely did Banks make his own records available in printed form to the interested public; one of the exceptions is his account 'Notes relative the first appearance of the *Aphis lanigera*, or the apple tree insect, in this country', published in the *Transactions of the Horticultural Society* in 1817. The content reads like an extract from Banks's diary and deals with the woolly apple aphid (Fig. 8.27.), believed by him to have been first found in England in Kennington in 1787 as an introduction from France. Over seven years, from 1790 to 1797, Banks assembled information from the community of gardeners in and about London, and had his notes arranged in chronological order and published. Franz had been asked to record the beast, or rather the colonies of the beast, sitting on the twig of an

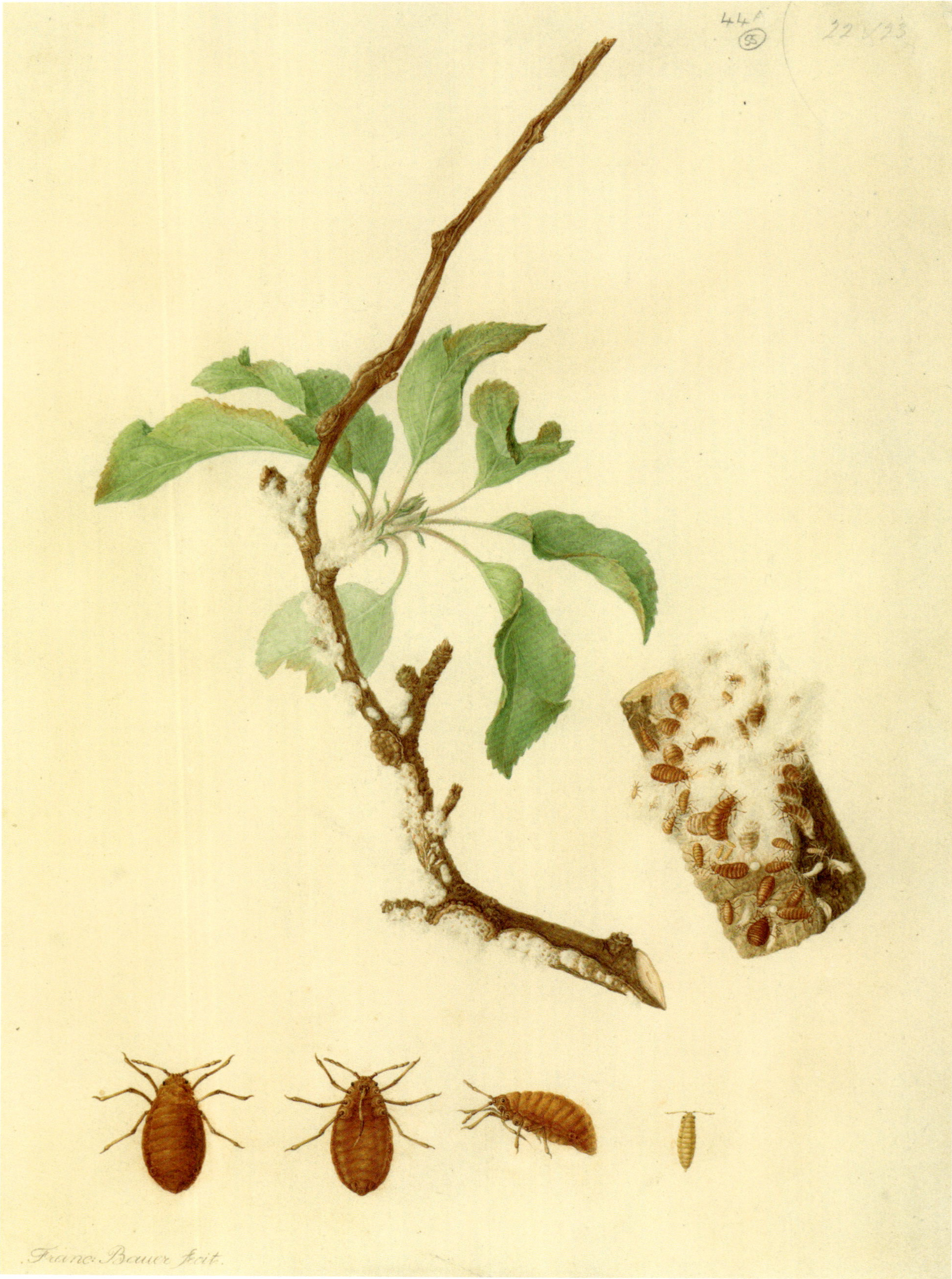

FIG. 8.27. Woolly apple aphid, *Eriosoma lanigerum*, on apple tree, *Malus domestica*. Watercolour by Franz Bauer, undated. – London, Natural History Museum, Library and Archives, Australian Zoological Drawings, pl. 55.

apple tree sucking sap. Nothing is known about when and where the twig had been collected; the published image comprises an additional winged male. The text mirrors the circle in which Banks and, in a sense also Franz, moved – visits are mentioned to: gardeners James Lee, father and son, at Hammersmith, and William Aiton, father and son, at Kew; to Charles Francis Greville (see Chapter 13) at Paddington; to William Forsyth, like Greville a founding member of the Horticultural Society and His Majesty's gardener at Kensington; to the garden of Queen Charlotte at Frogmore near Windsor, and to Banks's own garden at Spring Grove. Sooner or later most of these places harboured the rather harmless aphid. The account ends with a description of how Banks, for his final years reduced to a wheel chair, had his apple trees protected by having their stems 'painted over with a mixture of one pound of brimstone mixed with four of soft soap'. Three years after this paper was published, Banks died on 19 June 1820 at his country house in Spring Grove. The drawing of the woolly apple aphid by Franz, now in the Natural History Museum had been their last collaborative project.

The death of his employer marked a change for Franz, but not a turning point in his life. For more than thirty years he had lived on a very generous salary from Banks. In the same way as after the death of Maria Theresa Ingen Housz could have been certain that he would receive his annuity from her legal heirs, the Emperors Joseph II, Leopold II and Franz II, Franz could be certain to receive his annuity from the legal heirs of Banks. Aged 62, Franz was in an extremely privileged position, of which very many could only dream. He was to survive twenty more years, to remain active and in good health, and to get engaged in a surprising number of challenging new projects (see Chapters 13–16).

ENDNOTES

1 Unless otherwise stated, this chapter is based on Carter (1988), Desmond (2007) and Lack (2008b).
2 Diment, Humphries, Newington & Shaugnessy (1984), Diment, Humphries, Press, Shaugnessy & Newington (1987), Wheeler (1986)
3 Mabberley (2011)
4 Simpkins (1974)
5 Stafleu (1963)
6 Chambers (2000)
7 http://www.britishlistedbuildings.co.uk/en-205525-356-and-358-kew-road-greater-london-auth (accessed 26 iii 2014)
8 Pittsburgh, Hunt Institute for Botanical Documentation, Archive. – Franz Bauer to Joseph Franz Edler von Jacquin, 12 ix 1791, unpublished letter.
9 London, University College London, Lord Odo Russell Collection. – Franz Bauer to Josepf Franz Jacquin, 21 v 1792, unpublished letter.
10 Brno, Moravský zemský archiv, Matriky Valtice, Sign. 65/3531, f. 65.
11 Beale (2011)
12 Grzechca-Mohr (2011)
13 Inv. No. RCIN 45291
14 London, Natural History Museum, Library and Archives, Brown Correspondence 1: f. 214. – Sir James Edward Home to Robert Brown, 17 xii 1840, unpublished letter.
15 PD. 173 – 1973, PD. 174 – 1973; Scrase (1997).
16 London, Natural History Museum, Library and Archives – Original Drawings of Francis Bauer f. 23. 25.
17 Inv. No. RCIN 1047678
18 Scheer (1840)
19 S. Lindemann (pers. comm.)
20 Potsdam, Brandenburger Landeshauptarchiv, Rep. 37, (Alt-) Friedland, No. 382. – H. C. Itzenplitz to H. C. von Friedland, 20 xii 1792, unpublished letter.
21 Potsdam, Brandenburger Landeshauptarchiv, Rep. 37, (Alt) Friedland, No. 382. – H. C. von Itzenplitz to H. C. von Friedland, 4 iv 1793, unpublished letter.
22 Potsdam, Brandenburger Landeshauptarchiv, Rep. 37, (Alt) Friedland, No. 382. – H. C. von Itzenplitz to H. C. von Friedland, 10 v 1793, unpublished letter.
23 Sacramento, California State Library, Sutro Branch. – Franz Bauer to J. Banks, 28 ii 1807, unpublished letter.
24 Kerr (1989)
25 Pierson (2007)
26 Konig (1804)
27 Lack (2004)
28 Jäggi (1893)
29 Sherwood & Rix (2008)
30 C. Jarvis (pers. comm., 2008)
31 Elliott (2004)
32 Gage & Stearn (1988)
33 Mabberley (1999a)
34 O'Brien (1987)
35 London, University College London, MS Add 254, Lord Odo Russell Collection. – Nikolaus Joseph Edler von Jacquin to Sir Joseph Banks, 17 iv 1790, unpublished letter.
36 Britten (1899)
37 Henrey (1975)
38 Dolezal (1973)
39 www.britishlistedbuildings.co.uk/en-205482-23-and-25-kew-green-see-details-for-furt
40 London, University College London, MS Add 254, Lord Odo Russell Collection. – Franz Bauer to Joseph Franz Freiherr von Jacquin, 19 viii 1814, unpublished letter.
41 Britten (1901)
42 Salisbury (1805)
43 Andrews (1801)
44 Sims (1804)
45 Brown (1813)
46 Ker (1817)
47 PD. 58 – 1975; Scrase (1983)
48 Potsdam, Brandenburger Landeshauptarchiv, Rep. 37, (Alt) Friedland. No. 382. – H. C. von Itzenplitz to H. C. von Friedland, 21 viii 1793, unpublished letter.
49 Meynell (1981)
50 Göttingen, Niedersächsische Staats- und Universitätsbibliothek, Handschriftenabteilung. – gr 2° MS. Histor. Nat. 94, I – V; cf. Anon. (1893)
51 Inv. No. U 951. Z 33
52 e.g. Anon. (1841 a)
53 Lack (2011b)
54 Bauer (1833a)
55 Money (2007)
56 Bauer (1823, 1833b)
57 Banks (1817)
58 Knight (1818)

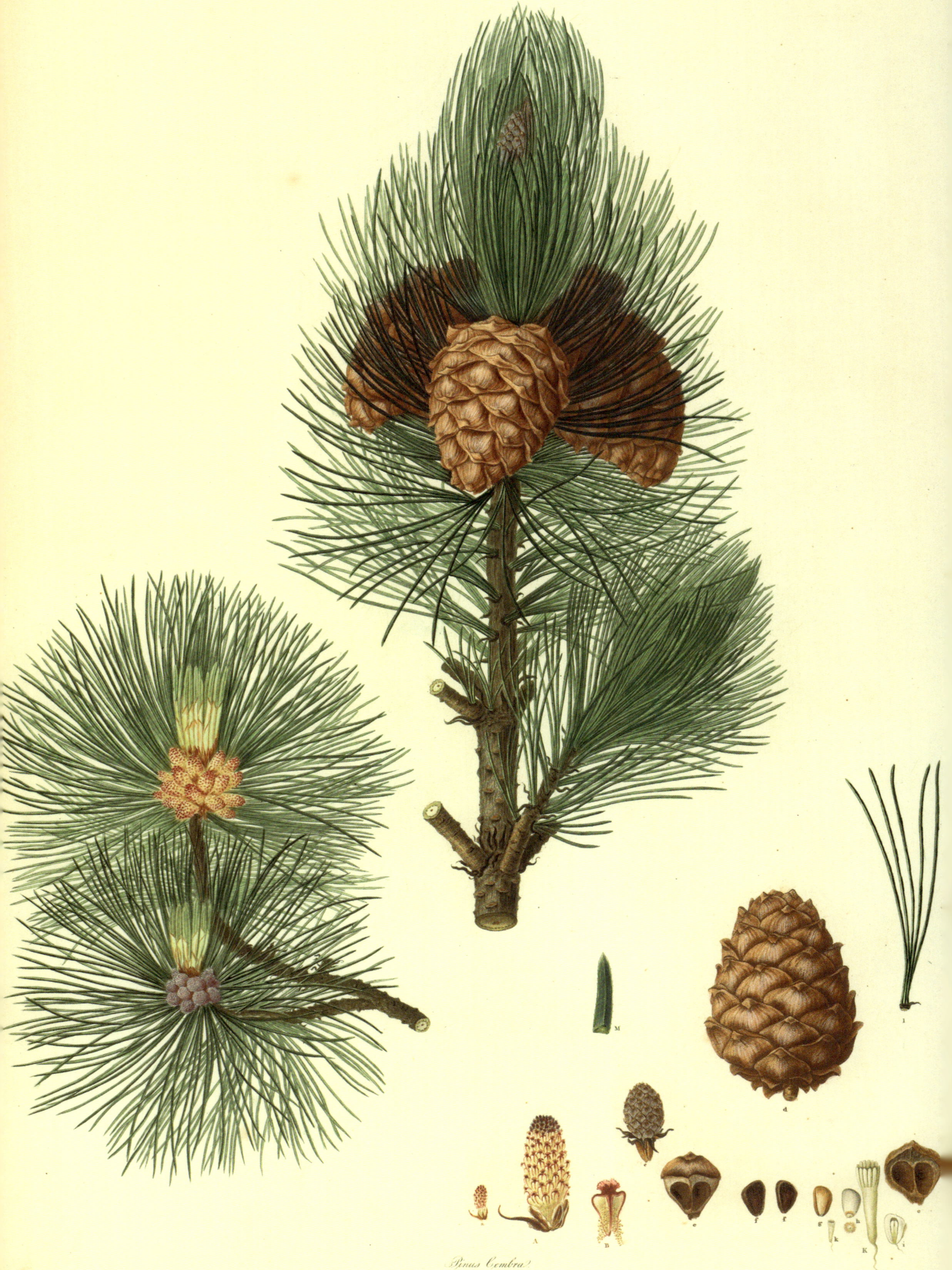

Pinus Cembra.

CHAPTER 9

Ferdinand Bauer, the years between[1]

WATERMARKS

Because of the paucity of sources available this chapter is short. It deals with the period in Ferdinand's life beginning with the moment when he discontinued work for Sibthorp in early 1794 and ending with his engagement as botanical draughtsman on the voyage of the *Investigator* in July 1801. Watermarks form an important guide for these years, although they offer no assistance for the precise chronology of events. A single letter by Ferdinand survives from these years, giving his address as 'London No. 53 Holburn' and containing the statement 'From the time I left Oxford, I have been the greatest part in London, however I have not had any particular engagement I have found always something to keep myself employed, now I have begun a small Botanical work, which I shall finish abaut [sic] that time when you proposed our meeting in England, after which I shall determine if I remain longer or not in this Contry [sic]'.[2] The date is given as 'Jully 25 the 1797', the addressee is Ferdinand's former travel companion John Hawkins, who had been elected a Fellow of the Royal Society on 5 May 1791 and was in 1797 on his second journey in the Ottoman Empire and staying at that moment in Selanik [Thessaloniki]. Ferdinand's London address is confirmed in a short note referring to an exhibition of the Royal Academy of Arts in 1794.[3] The title of the work is given as 'Flowers with Touraco Bird, from nature', which is untraceable. Turacos are birds from sub-Saharan Africa belonging to the Musophagidae group, and Ferdinand may have studied either a living bird in one of the private or public aviaries in London or a stuffed specimen in one of the many natural history cabinets. Judging from the regulations of the Royal Academy of Arts at that time, it is plausible to assume that this was an oil painting, the first of the extremely few Ferdinand is recorded to have prepared, of which a single one survives (see Chapter 11). However, he remained basically a draughtsman with pencil, brush and watercolours as his main utensils.

The watermark '1794 Whatman' is discernible on some of the landscapes documenting Ferdinand's past journey in the Levant with Sibthorp in 1786–1787 (see Chapter 5), proving that these views, often signed 'Ferd. Bauer' were prepared at the earliest in that year, but maybe considerably later. They are definitely the result of studio work. Ferdinand used solid card, sometimes two sheets of paper glued on top of each other, making an inspection of the watermarks impossible. No illustrator would have taken such heavy paper on a journey and worked on such large solid sheets. Who commissioned the series comprising landscapes numbered chronologically 1–141 with only a few missing? The most plausible candidate is Hawkins. All views, kept largely in the Sherardian Library at Oxford with a few more in the Cornwall Record Office at Truro, have been annotated with locality data by Ferdinand in his clumsy, often orthographically incorrect English with letters crossed out, added or erased. However, further corrections and additions are

FIG. 9.1. Arolla pine, *Pinus cembra*. Coloured copper engraving by Warner based on a watercolour by Ferdinand Bauer. A. B. Lambert, *A description of the genus Pinus*, London, 1803, t. 23. – London, Natural History Museum, Library and Archives.

all in Hawkins's hand. Furthermore, some of the sketches of landscapes taken in the Levant survive among the Hawkins Papers in Truro, but have not yet been studied in detail. There is another fact: the landscapes in the Sherardian Library were acquired by the University of Oxford in the 1930s after the death of George Claridge Druce, a Fellow of the Royal Society and Fielding Curator in the Department of Botany at Oxford University, who had purchased them from Wheldon & Wesley in 1923.[4] It has been suggested that this series had been inherited by Thomas Platt (see below) from Hawkins and that it later passed to the well-known dealer of natural history books and manuscripts, which is a possibility. By contrast, all other materials in the Sherardian library dealt with in this text had become the University's property soon after the death of Sibthorp in 1796 as stipulated in his will (see below). Several landscapes have small holes in the corners, which indicate that they had been nailed on a wall, possibly in Hawkins's Grand Tour Cabinet in one of his country houses, such as Bignor Park, Sussex. By contrast, nothing is known about for whom the small set of views from the tour with Sibthorp, kept in Göttingen, was painted (see Chapter 5).

Hawkins, in marked contrast to Sibthorp, had always maintained good relations with Ferdinand, and it is very likely that he gave the commission to prepare the views. This is reflected by Ferdinand writing in the letter cited above 'Your unexpected letter . . . has much revived my recollection of our past acquaintance & time in which I had the pleasure to be in your Company, from which I remember many marks of Frindship [sic] from you, for which I always shall feel the greatest regard . . . '. This is likely to be the general background for the preparation of the 'Views of Mediterranean Scenes', which, however, could in principle also have been prepared after Ferdinand's return from his circumnavigation of the globe in the *Investigator*.

The views have often been considered to be sepia drawings, but close inspection reveals that they are not. Ferdinand started by making a drawing with pencil or pen and black ink; the pencil lines are only rarely visible with the help of a lens and were mostly covered by paint. Later he painted his landscapes in shades of grey wash and finally laid a pale thin wash over the whole painting. For this process he often made use of a transparent brown, light red or yellow. The final step seems to have been the preparation of the frame around the painting (usually consisting of three parallel lines in black ink) and adding some written information on the subject (see above). Though it is almost impossible to find mistakes in Ferdinand's watercolours of plants and animals (see Chapter 6), this is definitely not the case with his landscapes. Most importantly, in several views the proportions are quite wrong – for example the man on horseback passing through the Arch of Hadrian in Athens is presented far too small, and sailing boats of a curious size, often devoid of crew, are repeatedly shown, possibly used only to indicate proportions; in this Ferdinand often failed. When sketching a town from a distance he was not always correct; for example the position of the Parthenon is wrong in one of his views of Athens. Some of the archaeological details, like the partly destroyed metopes of the Thesseion in Athens, were 'improved' by Ferdinand. It is unclear how a hand, which had produced such an accurate representation of the view on Feldsberg or the vista towards Nikolsburg (see Chapter 2) could have made such mistakes.

Along the same lines are five views preserved in the Ashmolean Museum at Oxford, 'View of the East Bridge at Oxford', 'View of the great walk to Madalene [sic] College Oxford', 'View from the walk of Christ Church Meadow to Magdalene College in Oxford' (Fig. 9.2.), 'View from the great walk to Christ Church Oxford' and 'View of the Church at Southley in Oxfs-'. The first of these is prepared on paper watermarked '1804' and must have been prepared after Ferdinand's return from his circumnavigation of the globe in the *Investigator*. All five views are very similar in material, style and technique to the set of Mediterranean scenes, but a precise date cannot be attributed to the remaining four. Two further views by Ferdinand came up for auction: the first, briefly described as 'Magdalen College', passed hands when Franz's belongings were auctioned at Christie's in London in late 1841[5] (see Chapter 17), the second at Christie's in London on 4 November 2010, when the drawing was described as 'An extensive view of Oxford from Christchurch Meadows [sic], showing from left to right Christchurch College, the spire of St Mary's Church, the Radcliffe Camera, All Soul's College and Magdalen College and Bridge, pen and ink and grey wash, on paper, 12 ½ × 20 ½ inches', with no dates given. A view of Magdalen Tower and Bridge in pencil, annotated 'Magdalen Colledg, July 1794' and kept in the Bodleian Library at Oxford with the papers of Edward Dayes, could also be by Ferdinand, with the odd spelling 'Colledg' reminiscent of his orthography when adding

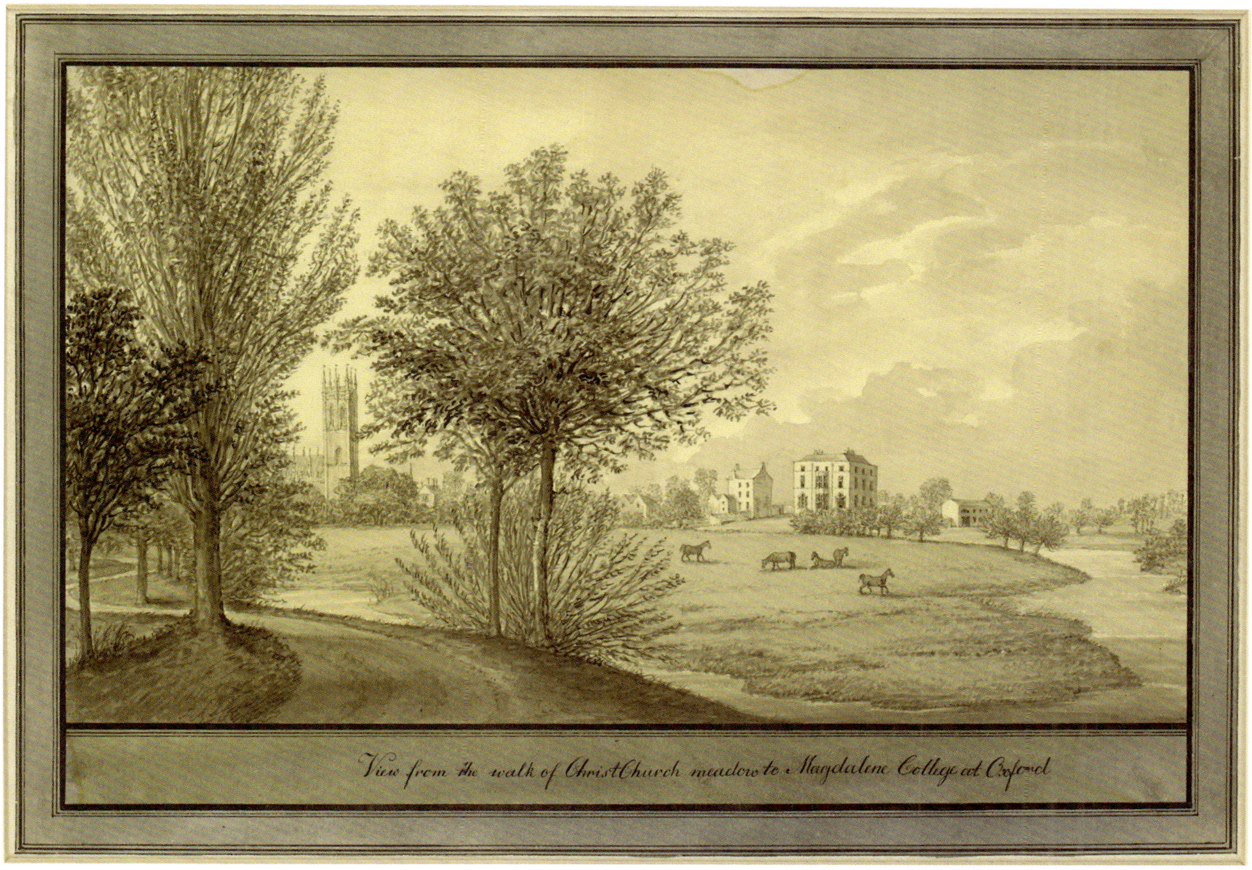

FIG. 9.2. View from the walk to Christ Church Meadow towards Magdalen College in Oxford by Ferdinand Bauer, c.1795. Paper, pen, grey wash. – Oxford, Ashmolean Museum, Department of Prints and Drawings.

locality information to the Mediterranean scenes. It has been suggested that this pencil drawing was intended for publication in the Oxford Almanacks, calendars attached to an illustration of local significance and published by the Delegates of the University Press. The Oxford Almanack for 1797 shows Magdalen College and Bridge, but since the pencil described above has not been signed, it is not possible to prove this assumption, although a plausible one. A further picture entitled 'A farm yard' by Ferdinand, possibly referring to Instow (see Chapter 6), was likewise auctioned at Christie's in late 1841, when fourteen 'pencil sketches of views in England' were also acquired by an unknown collector.

The watermark 'J. Whatman 1794' or 'J. Whatman 1797' offers the only clue to a very curious subject, so far totally unknown – a folio sized print entitled *The life, death and burial of Cock Robin: dulce est desipere in loco*, of which only a single copy could be traced in the Princeton University Library. It is a score for unaccompanied mixed chorus, with the title page reported to have the printed note 'Ferd. Bawer [sic] delt., Mno. Bovi sculp.'. The Latin verse in the title is taken from Horace, Odes 4, 12, 28 and translates 'it is sweet on occasions to play the fool'. The single traceable copy has a bookplate of Anne Rushout, a daughter of John Rushout, 1st Baron Northwick of Northwick Park, MP and, like Banks, a Fellow of the Society of Antiquaries. A second copy kept in the University of Glasgow Library is reported as missing. Some background information on this curiosity is found in the auction catalogue of Franz's belongings, where this item is listed as 'Life, Death and Burial of Cock Robin, edited by the late Duke of Marlborough, frontispiece by Ferdinand Bauer, only 50 copies printed'. The text paraphrases the English nursery rhyme 'Who killed Cock Robin', which has been much used as a murder archetype; the earliest record of the rhyme is believed to have been published in c.1744. Clearly Ferdinand must have received a commission to do this work. The connection between him and the 'late Duke of Marlborough' is perplexing. Since the auction catalogue was printed in 1841, the reference may refer either to George Spencer, 4th Duke of Marlborough, who died on 29 January 1817, or to his son George Spencer-Churchill,

5th Duke of Marlborough, who died on 5 March 1840. The first, a Knight of the Garter, Member of the House of Lords and previously Lord Privy Seal, seems to be the more plausible of the two, since he was also, like Banks, a Privy Councillor and a Fellow of the Royal Society. His principal country residence was Blenheim Palace near Oxford, then like now one of the most spectacular stately homes in England. How Ferdinand managed to attract this commission from a member of the top stratum of English aristocracy, extremely rich and very near to the throne, the man who had borne the sceptre at the coronation of King George III and was comparable in status to Aloys I Joseph in Vienna, remains an enigma. Maybe a link with Banks had existed.

A PUZZLE TO EVERYBODY

Aylmer Bourke Lambert, a Fellow of the Royal Society, the Linnean Society and the Society of Antiquaries, was a member of the landed gentry like Banks. Like the latter he was an only son and had lost one parent early, with sugar plantations in Jamaica and extensive lands in Ireland descending on his shoulders. In his boyhood he had also been a passionate collector, especially in the field of botany and had formed a small private museum at Boyton House, his father's country house in Wiltshire.[6] Lambert matriculated at St Mary Hall, Oxford (now incorporated into Oriel College), where he became acquainted with Banks and John Sibthorp, and like Banks left without having taken a degree. He led the privileged life of a gentleman of leisure, definitely less engaged in public affairs than Banks, although his inclination towards botany possessed a similar profile – Lambert was an avid collector of botanical collections, a bibliophile with an extensive library and a true patron ready to support the works of others. Before 1788 Lambert established a lasting friendship with Smith, who later appointed him one of the vice-presidents of the Linnean Society, a position he held until his death. Like 32 Soho Square for Banks, Boyton House became for Lambert the receptacle of his collection of botanical specimens, books and manuscripts, freely open to men of science. Many years later and after the death of his wife in 1828 Lambert left his country seat and took as his home 10 Cumberland Place at Kew, now 300 Kew Road, preferring the proximity of the Royal Garden. In his old age he thus lived in close proximity to Franz, the latter three years his senior, and may well have belonged to Franz's botanical circle (see Chapter 16); Lambert outlived him by just thirteen months.

How Ferdinand attracted a commission from Lambert is unknown; Banks may have been a factor, but also Franz in Kew. Ferdinand's first job for Lambert was somewhat unconventional – he was asked to prepare drawings of a few herbarium specimens kept in the collections at 32 Soho Square, Boyton House and possibly elsewhere. Since the natural colours had long been lost, Ferdinand painted in monochrome, the resulting watercolours being in a sense reminiscent of his landscapes since he used grey wash. All eight drawings survive in the Natural History Museum. Although Lambert chose the title *A description of the genus Cinchona* for a work to include prints based on Ferdinand's monochromes, only two species treated by him in the text belong to this genus as currently circumscribed.[7] One specimen depicted by Ferdinand is of special interest: collected by Forster the father and the son (see Chapter 7) on Tongatapu, the main island of today's Kingdom of Tonga, during Cook's second voyage, it belongs to *Badusa corymbifera* (Fig. 9.3.).

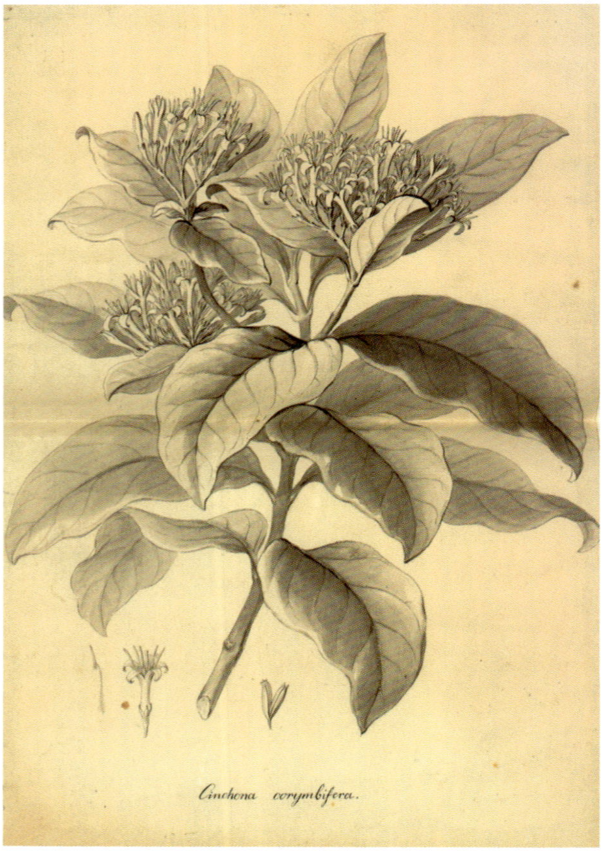

FIG. 9.3. *Badusa corymbifera* by Ferdinand Bauer, c.1795. Paper, grey wash. Bound into copy of A. B. Lambert, The genus Cinchona, London, 1797. – London, Natural History Museum, Library and Archives.

George Forster had previously prepared an illustration of this species, a copper plate had been engraved and a few 'pulls' taken, but then the whole project came to a standstill.[8] Later Banks bought these illustrations, which are now also in the Natural History Museum. In a sense Ferdinand repeated the work of George Forster, who had died on 10 January 1794 in poverty in Paris, while The Terror at the onset of the French Revolution was about to culminate, with George's father, a professor at Halle University (see Chapter 7), outliving the son.

Rather unexpectedly, Lambert's book, which was published in 1797, also contains the description of a genus totally unrelated to *Cinchona* and new to science – *Hyaenanche*, a member of the Picrodendraceae family. A native of South Africa, its name refers to the fact that the fruits were used locally to prepare a powder, which when put into the carcasses of lambs had the potential of poisoning hyenas. A female specimen of *Hyaenanche globosa* growing in the conservatory of Charles Bennet, 4th Earl of Tankerville, at his country house Mount Felix in Walton-on-Thames was documented by Ferdinand in 1795 and published in Lambert's book as a mezzotint with engraved outlines. Charles Bennet's wife Emma had been the lady selected by Banks as the dedicatee of the orchid *Phaius tankervilleae*, repeatedly painted by Franz (see Chapter 7). Dedications were very much in vogue at the time, and an almost endless list of generic names and specific epithets commemorate figures of major and minor importance in botany, in science in general, in politics, and in other fields of life. For obvious reasons Banks, to whom Lambert had dedicated *The description of the genus Cinchona*, was particular successful in this respect.

The second commission from Lambert was more conventional – documenting the diversity of the gymnosperms by producing watercolours of living specimens cultivated at the time in various gardens and arboreta in England, with a few additional figures based on herbarium material. This became a joint undertaking: Ferdinand produced the bulk of the work, and four images each were prepared by Franz and Sowerby (see Chapter 8). Nothing is known of where Ferdinand had his drawing desk for this project, but 32 Soho Square, Denmark House or 53 Holburn seem reasonable guesses. At any rate he must have become familiar with the Royal Garden at Kew, which was to become an important place of reference after his return from the voyage on the *Investigator* (see Chapter 11).

As usual the finished work and some pencil drawings became the property of the commissioner and ended up in his library. At the auction of Lambert's effects in 1842 this material was acquired by Edward Stanley, 13th Earl of Derby, a Knight of the Garter, Member of Parliament and later President of the Linnean Society. It remained for generations in the family and is currently owned by Edward Richard William Stanley, 19th Earl of Derby. The collection comprises sixty drawings, ranging from 'pencil studies of individual cones to complete illustrations' and is housed in the library of Knowsley Hall near Liverpool, Merseyside.[9] This gem has never been studied in any detail, although it is known that four of these drawings were signed by Franz and one is dated 1801. Several more colour-coded pencil drawings referring to Lambert's *opus magnum* ended up in the Natural History Museum, of which only very few have been published so far, among them one with a coloured detail.[10] The background of the acquisition of this material is complex: Ferdinand seems to have left this material with his brother Franz; when the latter died, his belongings were auctioned and the pencil drawings relative to Ferdinand's work for Lambert, forming lot 157, were acquired by Brown for 18 shillings (see Chapter 17). When Brown died these drawings were acquired by the British Museum and are now in the Library and Archives of the Natural History Museum. Even Ferdinand's drawing of mataī, *Prumnopitys taxifolia*, thought lost, has recently been found – bound into a copy of Ferdinand's *Illustrationes florae novae Hollandiae* (see Chapter 11) in private hands in Australia.[11] The respective pencil drawing is kept in the Library and Archives of the Natural History Museum and has been published recently. The mataī belongs to the *Podocarpus* family and is restricted in distribution to New Zealand, where it occurs on both islands, as well as the smaller Rakiura or Stewart Island.

On the basis of these watercolours and with ample funding Lambert had a work produced entitled *A description of the genus Pinus*. He chose a misleading title, since it refers not only to the genus *Pinus* as then understood, but also to several other quite unrelated genera. Work seems to have started in about 1798 or earlier, with Banks offering advice, George Jackson, then in charge of Lambert's herbarium, writing the Latin descriptions, and Smith assisting Lambert as well as correcting the text. Although this folio book measuring 57 × 45 cm, dedicated to Banks, has received high praise for its contents and the quality of the copper engravings, it is a bibliographical monstrosity. In a letter from Henry G. Bohn, the publisher of Lambert's work, to John Kidd, Radcliffe Library, Oxford, dated 30 March 1835 and glued into the Bodleian copy, is a statement appropriate to this book: 'Most of the sets in Public Libraries are very deficient and . . . the Radcliffe copy is so. It would be advisable for you to send me & I will inform you the cost of the completion. I may say that Mr. Lambert's book is a puzzle to every body, as being published in single plates, without any printed list, + some of the text inlaid from the 8to, it can only be understood by those who have arranged it with the assistance of Mr. Lambert himself. There are not 12 complete + properly coloured copies of the folio in existence'.[12] Furthermore many copies contain an amalgam of uncoloured, partly coloured and completely coloured engravings, the text includes a labyrinth of notes, additions and corrections. Bohn referred to the first volume only, of which complete copies comprise plates 1–38, 21*, 34*, 37*, 38*, plus an appendix with the plates 39–43; Ferdinand and Franz were not involved in the second volume. Sowerby contributed the plates 37*, 38*, 42 and 43; Franz 9, 16, 20 and 38; all others are by Ferdinand – with the exception of plate 5, which is based on one of Ehret's drawings dated 1744, then in Banks's library at 32 Soho Square, and of plate 41, which is based on a drawing by Parkinson from Cook's first voyage showing rimu, *Dacrydium cupressinum*, a large tree endemic to New Zealand and the neighbouring Stewart Island.[13]

Goethe had no occasion to inspect the watercolours, but only the coloured engravings in the Großherzogliche Bibliothek in Weimar; he was enchanted (see preface). Since this copy was incomplete, he wrote on 9 March 1817 to Carl Franz Anton Ritter von Schreibers, director of the Imperial Royal Natural History Cabinets in Vienna (see Chapter 17), to ask him to have three watercolours of the missing plates made and sent to Weimar. The prestige of *A description of the genus Pinus* was such that Goethe knew who possessed copies in Vienna and mentioned that either the copy in the library of Franz I, Emperor of Austria, or the copy in the library of Prince Albert of Saxony, Duke of Teschen, a son-in-law of the former's grandmother Maria Theresa (see Chapter 7), could be used for this purpose. While the watercolours in Knowsley Hall, being in private hands, have become known to the cognoscenti only very recently, the list of those praising Lambert's book is a long one. It started with Banks, who told the publisher in the presence of the author that it was 'the finest work ever published', by which Lambert no doubt must have felt satisfied and related this remark to his friend Smith. One eulogy is being particularly poignant and to the point: 'What is marvellous in the *Genus Pinus* is the play of the pine needles, wherein one soon comes to understand how it was that the Chinese *literati* evolved a whole school of artists who took for their subject nothing of more substance than the movement of bamboos. It is, also, the incredible variety of tasselling in the branches; the diversity of the pine cones; even, the salubrity and pungency of the pinewoods with their therapeutic properties, as in the calm, clean air of the mountains in a hundred different lands, with the noise of the wind among the pine needles, but returning, always, to this incredible skill in their delineation where every individual needle leads its own independent life and has been given the importance of, it could be said in slight exaggeration, a sword blade by one of the great sword masters of the Far Orient. Ferdinand Bauer, and his brother Francis, are almost unique among painters in this capacity to invest a single plant form, or part of one, with prime importance'.[14]

A work regarded 'in terms of aesthetics [a] marvellous instance of controlled intoxication' nevertheless needs a careful analysis of its contents and the role played by Franz and Ferdinand. As is so often the case, chronology is a problem – neither the pencil drawings nor the finished watercolours are dated. We know that the publication of the first volume (except the appendix) came to a conclusion in 1803 when Ferdinand was still in Australia, but we do not know when the individual plates and text pages were available to the public. In any case it is extremely likely that at least Ferdinand finished his work for Lambert before joining Flinders and Brown on the voyage in the *Investigator* in 1801. The appendix of *A description of the genus Pinus* was published in 1807 at the latest and contained only five copper-engravings, three of these based on watercolours by Ferdinand. In theory, these could have been made after his return from his circumnavigation. A print-run of 200 has been reported, and we know that one Warner produced the engravings.[15] The delicate, but monotonous, task of colouring is said to have been committed to William Hooker, who regarded himself in 1805 a 'pupil of Franz Bauer'[16] and years later was to predecease him. However, this assumption is not corroborated by an inscription in one of the copies in the Library and Archives of the Natural History Museum reading 'One of the twenty-five original copies coloured by Mr. Bauer, of Kew', which is more likely to refer to Franz than to Ferdinand. Whatever the percentage of the print-run that was coloured, it must have been tedious work lasting for many months, which could have hardly been accomplished by either Hooker alone or even by both Hooker and 'Mr. Bauer', but only with the help of qualified assistants. As so often the success of the plates was due to the collaboration of several highly qualified people – Ferdinand and Franz, Warner and Hooker – with the necessary funds coming from Lambert's county rent roll. Needless to say, the balance of this slightly esoteric project ended in the red, but this could not have been a serious worry to a man like Lambert. On 23 November 1803 Lambert had written to Smith 'I shall have only 175 copies to sell which I shall clear by eight hundred pounds, to repay me fourteen hundred pound; then I have the Drawings and the Copper plates, which I reckon at two hundred pound[s], so that I shall be about four hundred pounds minus'.

Gymnosperms from all continents except Africa were included in Lambert's work – common ones in Europe, both in the wild and planted, such as Scots pine, *Pinus sylvestris*, or Norway spruce, *Picea abies*, as well as species then only very rarely in cultivation outside its native range such as Masson's pine, *Pinus massoniana*, one of the two most widespread species of pine in China occurring from Sichuan and Yunnan to Taiwan (see below). There are species Franz and Ferdinand may well have known already from their earlier excursions to the Alps, for example the larch, *Larix decidua*, the mountain pine, *Pinus mugo*, or the arolla pine, *Pinus cembra* (Fig. 9.1. p. 228). Ferdinand must have seen the stone pine, *Pinus pinea*, on the Apennine Peninsula when heading for Naples with Sibthorp. Now he depicted material from a flourishing tree in a garden in Clapham belonging to Henry Cavendish, a Fellow of the Royal Society famous for his analysis of atmospheric gases and the discoverer of hydrogen. Ferdinand painted an illustration of the Aleppo pine, *Pinus halepensis*, a tree he had already recorded for Sibthorp on their tour to the Levant. Franz had seen the plantations of Weymouth pine, *Pinus strobus*, at Wörlitz on their way from Halle to Berlin (Chapter 7); now his brother was documenting 'a fine specimen growing in the Royal Garden at Kew' (Fig. 9.4.). This is a tree with an extremely disjunct distribution – with by far its largest range in eastern North America, but with other populations far to the south in Mexico and Guatemala. However, there were also several other species the two brothers could not have observed or painted before, such as the jack pine, *Pinus banksiana*, the only pine species in northern North America occurring from the Atlantic almost to the Pacific Ocean. This was one of the species that had its scientific name validated in Lambert's work. For his watercolour Ferdinand used a specimen cultivated at Pain's Hill (now Painshill), today one of the finest remaining examples of an eighteenth-century English landscape park, then owned by Henry Luttrell, 2nd Earl of Carhampton, a Member of Parliament, who had led the British oppression of the United Irishmen Rebellion. Herbarium material collected by Archibald Menzies on the north-western coast of America and kept at 32 Soho Square was used by Ferdinand to prepare a drawing of Douglas fir, *Pseudotsuga menziesii*. Its English name refers to David Douglas, who had recognized the enormous potential of this species for forestry, while the Latin epithet commemorates Menzies, who had taken part in the famous expedition commanded by George Vancouver circumnavigating the world in the *Discovery*. By contrast, the material of the tamarack, *Larix laricina*,

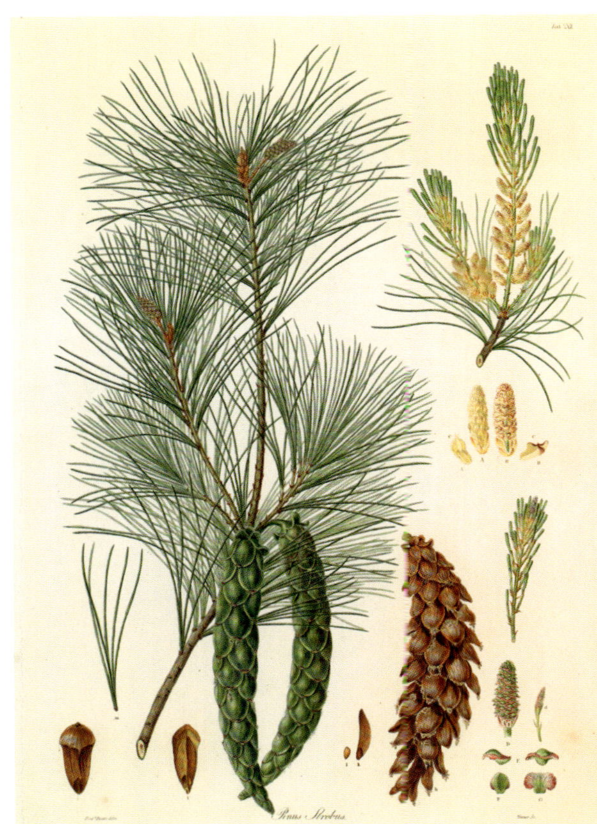

FIG. 9.4. Weymouth pine, *Pinus strobus*. Coloured copper engraving by Warner based on a watercolour by Ferdinand Bauer. A. B. Lambert, *A description of the genus Pinus*, London, 1803, t. 22. – London, Natural History Museum, Library and Archives.

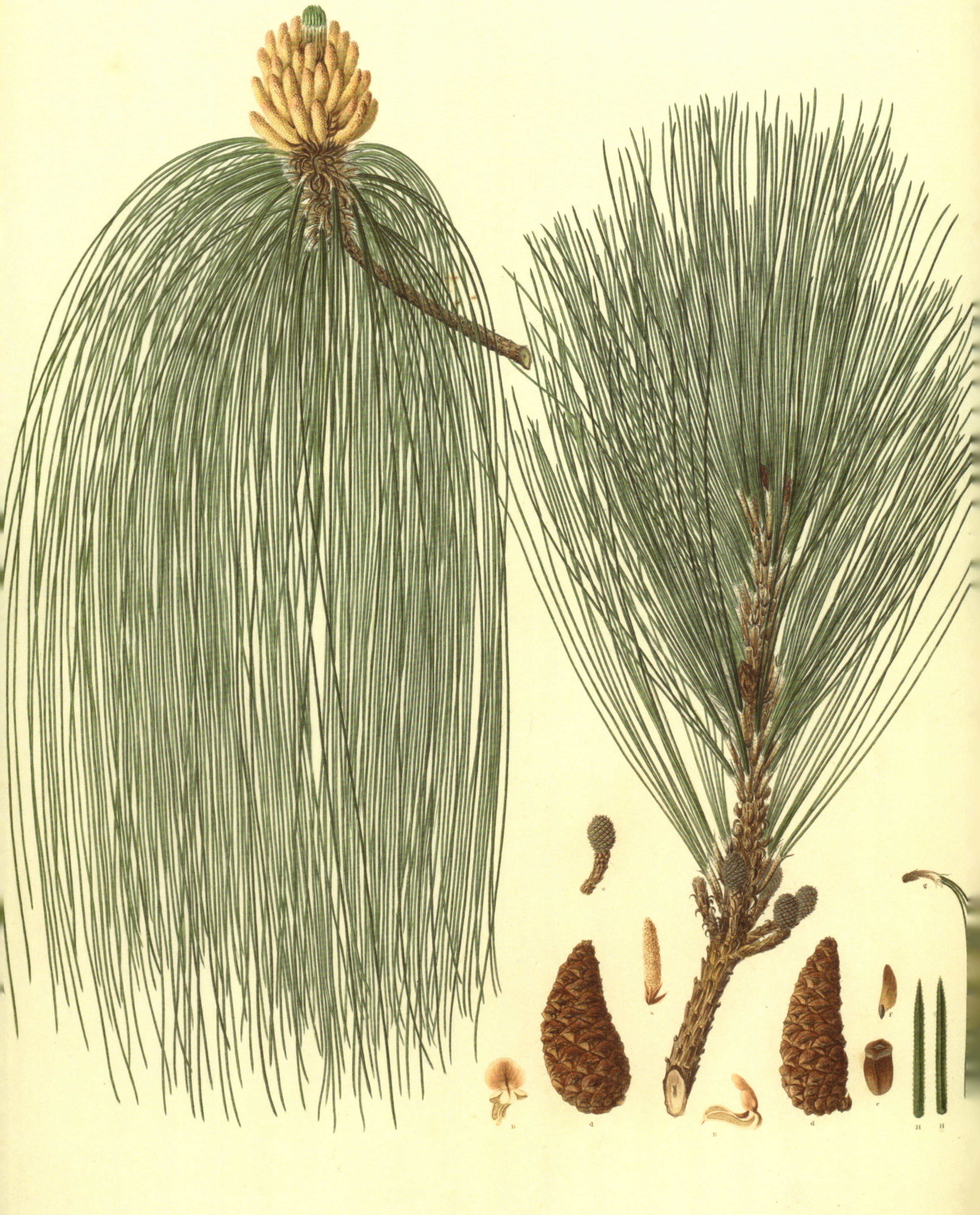

Ferd.ⁿ Bauer delin. *Pinus longifolia.* Warner sculp.

FIG. 9. 5. Chir pine, *Pinus roxburghii*. Coloured copper engraving by Warner based on a watercolour by Ferdinand Bauer. A. B. Lambert, *A description of the genus Pinus*, London, 1803, t. 21. – London, Natural History Museum, Library and Archives.

came from a large tree cultivated at Whitton Place, an estate then belonging to John Campbell, 5th Duke of Argyll, another Member of Parliament and field marshal. Lambert described the specimen as having a 'remarkable beautiful appearance in the summer, being covered with a great number of bright purple cones'. It had been planted by John Campbell, 4th Duke of Argyll, a Knight of the Thistle, Member of Parliament and the father of the then owner. A very cold-resistant tree tolerating temperatures to at least – 65°C, the tamarack is one of the boreal conifers of North America with a continent-wide distribution ranging from Alaska to the area of present-day Quebec. From Asia the chir pine, *Pinus roxburghii* (Fig. 9.5.), is described by Lambert on the basis of a manuscript by William Roxburgh giving the mountains of Nepal as provenance, which is very correct since this species occurs along the greater length of the Himalayas; in this case no information is available on which material Ferdinand based his watercolour. For Masson's pine only herbarium material kept at 32 Soho Square of a tree cultivated on the Cape of Good Hope 'raised from seeds which had been sent from China' was available to Ferdinand; *Pinus massoniana*, the scientific name for this species, was also validated in Lambert's work. The Library and Archives of the Natural History Museum keeps a colour-coded pencil documenting the cedar of Lebanon, *Cedrus libani* (Fig. 9.6.); however, the copper engraving has the note 'Sowerby del.', probably in error, since Sowerby is not known to have ever used a colour code. This ornamental tree is widely cultivated in gardens and parks worldwide, but is restricted in its natural distribution to what is now Lebanon, Syria and Turkey, with a small somewhat deviating population

FIG. 9. 6. Cedar of Lebanon, *Cedrus libani*. Paper, pencil by Ferdinand or Franz Bauer, c.1800. – London, Natural History Museum, Botany Library, Bauer Unit, Sketches of Coniferae, f. 1r.

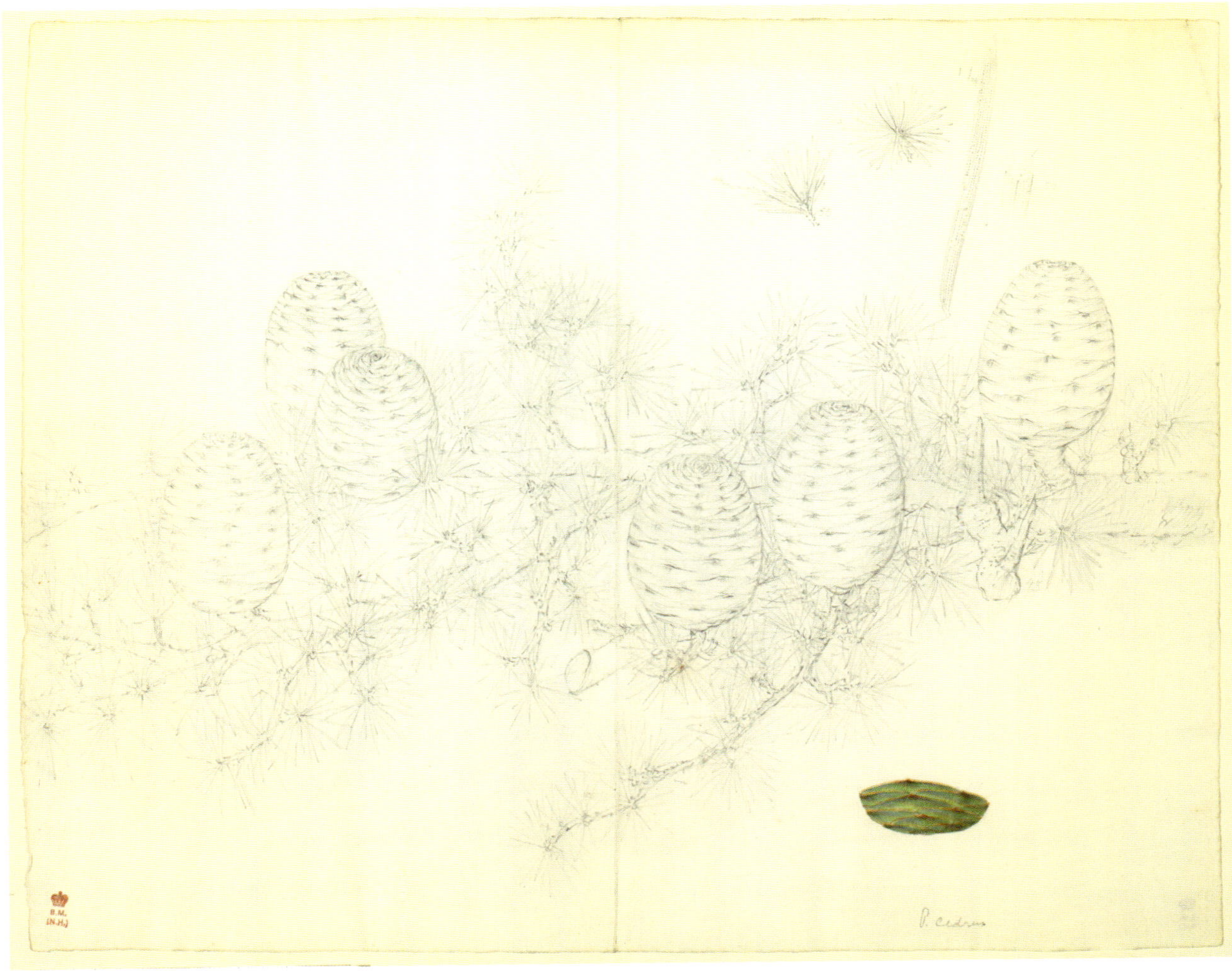

in Cyprus – where this species was certainly observed, though not documented, by Ferdinand in 1787.

The gymnosperms from Oceania are a surprise. One of them is the Cook pine, *Araucaria columnaris*, a distinctive, narrowly conical tree up to sixty metres high, restricted in distribution to New Caledonia. It was first recorded during Cook's second voyage and noted by Cook, Johann Reinhold Forster and William Wales, the astronomer, in their diaries. Subsequently the scientific name *Cupressus columnaris* was given to the plant by Forster the father (see Chapter 7). In this case the provenance of the specimen documented by Ferdinand is precisely indicated - 'a most beautiful specimen and the largest in this kingdom is now in the Royal Gardens at Kew, to which it was first introduced by Governor [Arthur] Philipp' (Fig. 9.7.). The second gymnosperm from Oceania is the Norfolk Island pine, *Araucaria heterophylla*, an endemic of Norfolk Island, a tiny spot of thirty-four square kilometres in the Pacific Ocean, where Ferdinand was to spend six months in 1804–1805 (see Chapter 10). The tree reaches a height of fifty to sixty-five metres and has a straight vertical trunk, even in the face of incessant onshore winds. To Cook, who discovered Norfolk Island only a few days after having discovered New Caledonia, this tree appeared suitable for making masts and yards for sailing ships. However, it turned out that the wood is not sufficiently resilient for these uses. Today this tree is known worldwide as an ornamental tree, grown either in pots indoors or planted in warmer climates in gardens, parks or along roads. When the different figures were combined to form table 39 (Fig. 9.7.), Lambert seems to have been confused, since the legends only partly refer correctly to the plant parts represented. Fig. a, b, g, h, i and k belong to the Cook pine, whereas Fig. c, d, e, f and F belong to the Norfolk Island pine. As a rule, the printed plates of *A description of the genus Pinus* are delicate compositions consisting of several figures, documenting twigs, needles, male and female inflorescences, cones, scales of cones, seeds and embryos, including various cross-sections. They all are highly diagnostic, meaning that the identity of the gymnosperms documented was never in doubt.

In addition to documenting the gymnosperms of the world, Lambert also had at least one further, though minor commission for Ferdinand. He asked him to depict a flowering specimen of wild rice, *Zizania aquatica*, which had been cultivated in a pond at Spring Grove[17]. The seeds 'put up in jars of water' were received via the

FIG. 9.7. Cook pine, *Araucaria columnaris*, and Norfolk Island pine, *A. heterophylla*. Coloured copper engraving by Warner based on watercolours by Ferdinand Bauer. A. B. Lambert, *A description of the genus Pinus*, London, 1807, t. 39. – London, Natural History Museum, Library and Archives.

Banks network: they had come from Canada sent by the physician John Mervin Nooth, a Fellow of the Royal Society and resident of Quebec. Interest was mainly economic, with Lambert writing: 'I am persuaded to think it [the Wild Rice] may be sown with some advantage where no other grain will grow, in many shallow places of water in Great Britain and Ireland, especially on the latter country, where I have seen several extensive lakes which appear well suited for the purpose'. A copper engraving based on Ferdinand's drawing was appended to a short note by Lambert on this subject published in the *Transactions of the Linnean Society*. The paper had been read on 6 December 1803, when Ferdinand was at Sydney Cove, and appeared in print on 8 November 1804[18] when he was on Norfolk Island (see Chapter 10).

Preparing a watercolour of a species of *Lambertia*, the name commemorating this patron of botanical

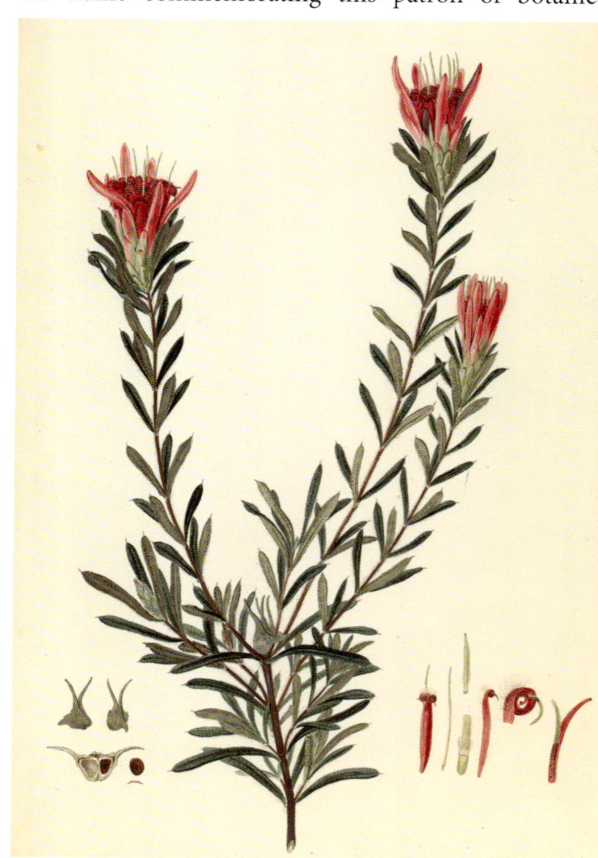

FIG. 9.8. Mountain devil, *Lambertia formosa*. Watercolour by Ferdinand Bauer, c.1797. Paper, watercolour. – London, Linnean Society, Archive.

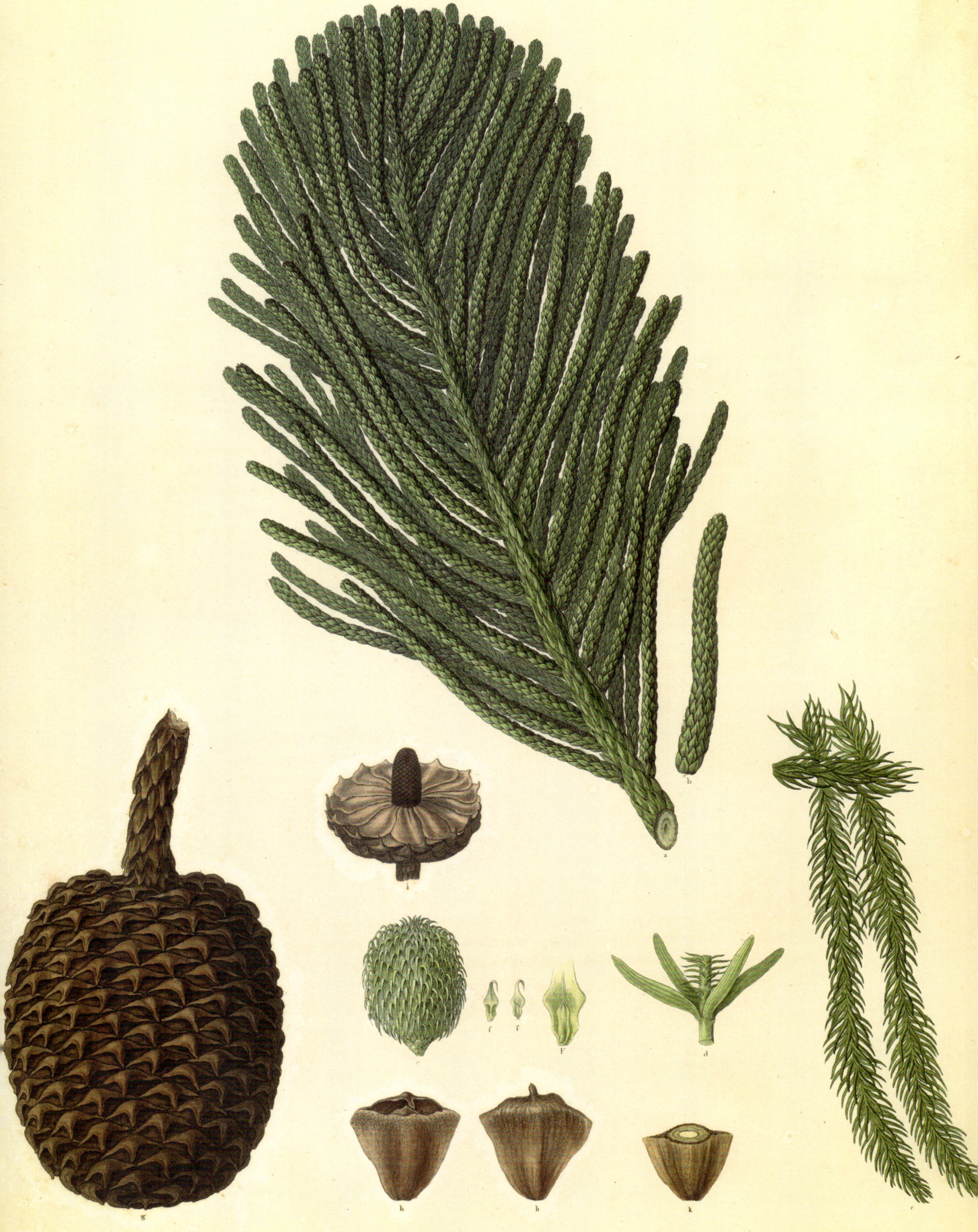

Dombeya excelsa.

illustration in London, was a minor commission for Ferdinand, probably coming from Smith. The resulting plant portrait (Fig. 9.8.) is kept in the archive of the Linnean Society. As for the *Delineations*, MacKenzie prepared the copper engraving showing the mountain devil, *Lambertia formosa*, which formed the only plate to accompany a paper by Smith entitled 'The characters of twenty new genera of plants', published in the *Transactions of the Linnean Society* in 1798. Smith explicitly stated 'engraved from a drawing made in New South Wales and corrected from wild specimens'. This has recently been interpreted as indicating a re-drawn illustration on the basis of the watercolour of the mountain devil prepared by Sydney Parkinson on the *Endeavour* voyage or the repeat by John Frederick Miller based on it. However, this is incorrect, since Parkinson showed a totally different specimen and flower heads without attention to the involucral leaves arranged in several rows. As a matter of fact Ferdinand used as the basis for his work the specimens of this shrub kept in the herbarium of Smith, which show precisely all those characters recorded in the watercolour. John White, the first Surgeon-General of New South Wales, is given as collector of these specimens; he had arrived with the First Fleet in Australia in 1788 and returned to England in 1794. The two specimens,[19] now in the Herbarium Smith kept by the Linnean Society, were collected in 1791, almost certainly in the surroundings of what is now Sydney; fragments[20] of these are conserved in the Botanisches Museum Berlin-Dahlem and probably elsewhere. The only reference Ferdinand could have taken from Parkinson's incomplete drawing or the repeat was something extremely important to him – the colour of the flowers and the inner involucral bracts; they are identical in both. Ferdinand could never have divined their tint from the herbarium specimens, which are devoid of any annotations concerning colour.

SIBTHORP'S LEGACY

Europe was at war when Sibthorp set out from London for his second journey to the Levant on 20 March 1794. He was accompanied by Francesco Borone to act as his servant and assistant, since Ferdinand was not prepared to come with him a second time (see Chapter 6). As a consequence there is no graphical record of this journey. In order to avoid becoming a French prisoner of war Sibthorp had to take an easterly route to Istanbul, which was his first destination. The departure of the newly appointed British Ambassador to the Porte, Sir Robert Liston, offered a favourable opportunity to reach the capital of the Ottoman Empire safely, quickly and cheaply. However, Sibthorp was foolhardy to leave England without being in good physical condition, for he soon became seriously ill, writing to Hawkins, 'I felt myself very unwell . . . with a considerable Degree of Fever & a pleuritic Complaint that gave me some Alarm'. Being a physician himself and having already travelled in the Ottoman Empire, he ought to have been aware of the necessity of being in good health when starting such a strenuous undertaking.

The journey from Ostend [Oostende], where they entered the Holy Roman Empire, to Istanbul was undertaken by the ambassadorial party at great speed and with little comfort using the route via Brussels, Bonn, Frankfurt and Würzburg. In the vicinity of this town Sibthorp saw Archduke Franz, on his way to Frankfurt, where he was to be crowned Emperor of the Holy Roman Empire, in fact the very last one in an extremely long line. Sibthorp would have been surprised if someone had told him that he had just seen the only monarch who – deeply interested in botany – would later subscribe to his posthumous *Flora Graeca* with the copy now in the Österreichische Nationalbibliothek in Vienna. Over roads 'broken up by the armies' the three coaches advanced slowly via Nuremberg, Regensburg to Schärding, where they entered the Hereditary Lands of the Habsburgs. During the brief stop in Vienna, Sibthorp met Jacquin the son back from his tour across Europe, which had included stays in Paris, Turin, Florence, Naples and Venice,[21] and the latter's future son-in-law Carl Anton Ritter von Schreibers (see Chapter 12). From the imperial capital the ambassadorial suite continued in an easterly direction, entering the Kingdom of Hungary near Kittsee [then Köpcsény]. The three coaches followed the River Danube downstream, crossed the river at Buda (now Budapest) and continued through Ketskemet (Kecskemét) and Tsegedin (Szeged) to Hermannstadt (Sibiu). On 29 April they left the Kingdom of Hungary and entered the Ottoman Empire near Rosenturm (Turnul Roșu), from where the group of English travellers was accompanied by guards or couriers, as protocol dictated. After the stay in Bucharest, Sibthorp fell seriously ill, an attack of malaria being the possible reason. Having crossed the Danube a second time at Georgiova (Giurgiu), the ambassadorial party proceeded southwards to Provadia (Provadiya) in what is now Bulgaria, with Sibthorp noting, 'A most violent bilious cholic prevented my enjoyment of this scenes [sic] and confined my botanic enquiry.' On 19 May 1794 the travellers had reached the palace of the British Ambassador in Pera (part of Istanbul). By then, Sibthorp was already in a poor condition, the last sentence of his diary indicating his feelings 'The cemitries [sic] and sepulchral gardens with the accompaniment of the funeral cyprus [cypress] impressed the mind with a kind of horror – which the glittering domes and minarets of the place could scarcely chase away'. With some luck Sibthorp survived, possibly being treated with the *Cinchona* bark, which would have been the standard treatment at that time.

On 8 August 1794 Sibthorp met Hawkins at the British Embassy in Istanbul and after another excursion to the 'Bythinian Olympus' [Ulu Dağ], similar to that undertaken with Ferdinand in 1786 (see Chapter 5), the two travellers with their servants and Hawkins's interpreter set out for a sailing tour to Athens. The season for botanizing was almost over, and during the following weeks Sibthorp's harvest consisted only of seeds that would later be sown in the Physic Garden at Oxford. Unfavourable winds, pressure of time and concerns over security, in particular pirates, but also over the health of Sibthorp and the two servants, determined a programme different from that originally intended. In the end it was island hopping, similar to seven years earlier, with a stay of eleven days on Mount Athos before the party continued to Skiathos. The finding of provisions turned out to be difficult, the decayed mast of the sailing boat had to be repaired at Egripo (Halkida) and after some more delays the anchor was dropped in the harbour of Piraeus on 15 October. Five days later Borone fell from a window in Athens and 'was bruised in such a manner that he lived no more than thirty minutes' as reported in the diary of Hawkins's servant. When Sibthorp and his party left Athens in mid-November, the rainy season had started. Their destination was the island of Zakynthos, then still belonging to the Serenissima Repubblica di San Marco, where the winter months were spent under 'the hospitable roof of Consul [Spiridon] Foresti being as convenient and comfortable as any we have been accustomed to in England'.

In late February of 1795, Sibthorp together with Hawkins and their servants returned to the Ottoman Empire for a tour of the Peloponnese, which turned out to be the most tedious and exhausting experience of their joint travels. Several factors made the following weeks extremely uncomfortable: heavy rains, snow and hail; nightmarish accommodation in the local roadside inns; and the constant fear of being robbed. The only reward was the satisfaction of visiting more or less virgin territories, where only very few travellers had been before and where the spring flora, now at its best, was totally unknown. The botanical harvest of these weeks consisted of poorly annotated herbarium specimens scattered over the Sibthorpian Herbarium at Oxford, but devoid of a graphical record, since Ferdinand was not part of the team. On 29 April 1795 Sibthorp's party reached Zakynthos, but instead of taking a few weeks of rest on the island, Sibthorp decided to rush back to Oxford. Furthermore, he did not wait for a ship bound for England but decided to sail to the eastern coast of the Apennine Peninsula in a small open boat. What followed was a disaster: after stops on the island of Santa Maura (Lefkada) and Preveza on the mainland, where he caught 'a bad Cold & Cough', Sibthorp landed at Corfu and provided himself with a health certificate for entering the Kingdom of Naples. Leaving the island the small boat had to seek shelter near the island of Fanno (Othoni), a few miles from the north-western tip of Corfu, and approaching the shore was stranded among the rocks. Because of an 'imperious Mistral' the travellers became prisoners of the tiny island, while the storm nursed Sibthorp's fever and cough. After finally being able to leave Fanno and cross the Adriatic Sea there were more delays, since Sibthorp's bill of health had to be first sent to Naples and then forwarded to Othonna (Ortona). Only after having passed quarantine in this town was he permitted to proceed. It was from this harbour that Sibthorp wrote his last letter to Hawkins stating 'When I consider my sufferings both in going to & coming from Greece I think I have as much Merit as any botanist had . . . tho I am not ambitious to become canonized in any Calendar You . . . will I know kindly add to my Greek Laurels by the by'. Very little is on record of Sibthorp's travel from this Adriatic harbour to Oxford; Ancona, Venice, Verona, Erlangen and Göttingen are briefly mentioned, and it must have been in late summer when Sibthorp finally reached home. Already critically ill, he wrote of feeling 'since my Return . . . very much indisposed and in a very infirm State of Health a nasty low Fever – with a Cough that alarms me from the affection of the Lungs'.

Taking his poor health into account, it seems unlikely that Sibthorp was able to resume his responsibilities as Sherardian Professor, as Keeper of the Physic Garden and as honorary physician. There is no evidence that he continued to work on the skeleton of a manuscript for the *Prodromus* or that he started the projected *Flora Graeca*. By now very ill, Sibthorp decided to try the mild climate on the west coast of England, but was advised to go to Bath. It was in this famous spa that he died of his unconquerable disease, probably tuberculosis in combination with malaria, on 8 February 1796, and was buried in Bath. He had only reached the age of thirty-seven years.

Although chaotic and inconsiderate in many of his dealings and no doubt a difficult man, Sibthorp had drawn up a crystal-clear, though unconventional, testament, which made him famous. His extensive landed property was to go to his father and creditor, then eighty-three years old – with the exception of a single estate. This was the subject of a special clause here quoted verbatim because of its special relevance for our story. 'All my Freehold Estate situated in the Manor of Sutton in the Parish of Stanton Harcourt in the County of Oxford . . . to the Chancellor Masters & Scholars of the University of Oxford for the purposes herein after-mentioned viz: that the Profits or Rents of such Estate be applied in the following manner first the Publication of a Work for which I have collected the Materials to be entitled *Flora Graeciae* [sic] which is to consist of ten Folio Volumes each Volume to consist of one Hundred Plates also a small octavo Edition without Plates entitles *Prodromus Florae Graeciae* [sic] & I do request of my Executors to find out & nominate some Person well skilled in Botany & Natural History for that Purpose & under whose direction the Plates may be engraved & coloured from the Drawings & MS Notes which shall be found in my Journals or loose Papers & have before him all my Collections of dried Plants & Natural History. These two Works being completed & published the Rents of the said Estate to be applied as following viz. the complete sum of two Hundred Pounds to be paid annually to a Professor of Rural Economy . . . And I further give & bequeath unto said University, all my Books of Natural History, Botany & Agriculture & all my Drawings or coloured Figures of the Said subjects & all my dried Plants & Collections of animals preserved either dry or in spirits to be deposited in the Library of the Professor of Botany provided that the collections be not injured & if subject to Injury than [sic] to be placed in some dry department or placed somewhere where it shall be considered by the University as most expedient.' In the general part of his will, Sibthorp had very wisely appointed three friends as his executors – Hawkins, Thomas Platt, a prominent solicitor who had been principal clerk to three Chief Justices, and Francis Wenman, Regius Professor of Law at Oxford University.

In modern terms, Sibthorp had created a foundation – the estate in Stanton Harcourt had passed to the University of Oxford, which was bound to use the profits first for the publication of a *Flora Graeca* and a *Prodromus* and 'these two Works being completed & published' for a Professorship in Rural Economy. Thus Sibthorp had laid down a precise sequence of actions to be taken, which was binding for the University and for his executors, of which Wenman was one. The latter was to die soon after Sibthorp, drowned in the River Cherwell. At the same time, the testator had put a certain pressure of time on the project, since the University of Oxford obviously had a strong interest in the stipend to be paid to the Professor of Rural Economy, but had to wait until the two books were published. Certain clauses in Sibthorp's will are worded in such sublime legal terms that the advice of a competent lawyer, maybe Wenman, is likely. When more than four decades later Franz applied to the Crown to rule over a clause in Banks's will, he must also have collected similar legal advice (see Chapter 17).

Hawkins, back from his second journey to the Levant, and Platt seem to have had their first meeting on the subject of this foundation in late 1798 or early 1799, possibly in London. No details survive, but some form of agreement seems to have been reached on the division of work, with Hawkins dealing with the organization and Platt with the legal and financial matters. The next thing to be agreed on was 'to find some Person well skilled in Botany & Natural History for that Purpose & under whose direction the Plates may be engraved & coloured from the Drawings [i.e. Ferdinand's drawings]'. There was an obvious candidate – James Edward Smith, an extremely able, industrious and skilled botanist, a Fellow of the Royal Society and first President of the Linnean Society, one year younger than Sibthorp had been when he died and, like him, a doctor of medicine. Furthermore, Smith was prepared to do the job, with the emolument from the foundation no doubt being an additional fillip. A letter to the Reverend Dr [Michael] Marlow, Vice-Chancellor of the

University of Oxford, signed by Hawkins and Platt and dated 16 May 1799 reads, 'We request you to deliver to Dr. Smith the bearer hereof all the materials which he requires for the completion & publication of D.r Sibthorp's Flora Graeca'. In the following weeks Smith had this material transported to Norwich, where he lived from 1797 onward with his wife Pleasance. This is how Ferdinand's pencil drawings and watercolours, all the herbarium specimens of the journey, and Sibthorp's diaries and field-notes, as well as the skeleton of his manuscript for the *Prodromus*, arrived on Smith's desk.

For the production of the engravings and their subsequent colouring there was also an obvious candidate – Ferdinand. At an early moment he had not only been informed by Sibthorp the father about the will of the latter's son but was even shown the testament. On 25 July 1797 Ferdinand had told Hawkins 'all [relative to Sibthorp's will] was to remain to your return to England, particular after the death of D:r Wenman, who I heard that he made inquirys for executing this work [the *Flora Graeca*] which now shall be left to be settled by you, to your plan'. Choosing Ferdinand may well have been one of Hawkins's options even when still en route in the Levant. Before 15 July 1799 Hawkins had a meeting with Ferdinand in London, which is described in a letter to Smith 'Bauer wishes to have the busyness settled as soon as it conveniently may be, in order to make those arrangements which would be necessary for so great an enterprize [sic]. In the first place, he says, he must look about assistants, and secondly it will be expedient for him to remove into more airy & roomy apartments. He repeated his wish to undertake both the engraving and the colouring having reason to think, he says, that he could execute the former better & cheaper than Mackensie [Daniel MacKenzie] and wishes that it might be decided by a trial'.

Ferdinand's position was quite clear, but the whole matter was in fact more complex than he thought, the critical point being the most precious thing of all – time. Sooner or later Ferdinand must have realized that he could not be 'certain of procuring assistants and that it would probably be a work of time to train up engravers in this line'. Under these circumstances it was a considerate decision of Hawkins to insist that 'the trials about to be made must serve to determine the quality of the work, the expense & the time and from these data must be digested our plan'. Indeed it was an enormous project – to have about a thousand copper plates of folio size engraved, impressions taken, and the prints coloured. Hawkins mentions an edition of fifty copies, which would mean c.50,000 engravings to be coloured by hand.

The letters from Hawkins to Smith over the following months are the only source of what happened. In November 1799 the trial seems not to have taken place yet and Hawkins summarized his view in the following words, 'There is certainly an advantage in having the whole work executed by one person as you observe because it is more easy to overlook one artist than two Besides if [Ferdinand] Bauer is competent to the engraving there can be no doubt but the plates will be executed with superior exactness after the original drawings. But here the following objections obviously present themselves. First, it is clear that all the plates cannot be engraved and even the limited number of 50 impressions coloured by Bauer alone Secondly, that if he has to form a school of Artists for the purpose, much time will necessarily be lost before they are able to lend any material assistance Perhaps both plans may be combined without inconvenience and with one manifest advantage, that of superior expedition. Sufficient employment may be given to Mackenzie if he executed only a part while the other is committed to the industry & ingenuity of Bauer. By these means too a sort of competition will be excited leading [to] the improvement of the work I think neither of these ingenious Artists can reasonably be satisfied with a partial execution of this work for the enterprize is so vast that with all the expedition it is possible to use I fear it will not be brought to a conclusion within twenty years. . .'. In fact, even this estimate was quite wrong; from the first fascicle published in 1806 it took thirty-four years until the last fascicle appeared in 1840, just a few months before Hawkins died on 4 July 1841. In his letter to Smith the former continued: 'The importance of Bauers zealous cooperation will more forcibly strike you when it is considered how often you will need his assistance in referring the drawings to the appropriate specimens, the seeds to their parent plants, how many new drawings must be made from dry specimens collected on the last tour and how many unfinished drawings completed'. As a matter of fact the last two options were subsequently given up totally.

Further letters were exchanged and possibly meetings held and on 23 December 1799 Hawkins wrote again to Smith. 'I have allways been of opinion that the oeconomy of time is a far superior consideration with us to that of expense and that for this reason also we cannot commit this work to Bauer without associating with him Mackenzie, which indeed was what Sir Joseph [Banks] observed at the beginning & which induced him strongly to recommend A division of the work among these Artists. Bauer I believe contends that the engraving is a trifling part of the labour . . . On the other hand it may be said that if so much time is required for colouring the less can be spared by him for engraving and the colouring alone of fifty impressions of each plate would occupy him so long that he would soon perceive it impossible to get thro' the work . . . and if Bauer objects to working in concert with Mackenzie let each engrave & colour separately his own portion of the work I am moreover of the opinion that the immensity of the work requires us to enlist in it other artists of merit such as Sowerby for I must again observe that one man whatever be his industry will never be able to get thro' it . . . Have you given drawings to Mackenzie for engraving? If not let it be done without delay, by which means we preclude any farther loss of time, the same with Bauer'.

Indeed, Smith seems to have done this and Ferdinand received a payment of £7.7s for the engraving of 'specimens of two Plates', whereas no payment is recorded for MacKenzie, who possibly did not take part in the long discussed trial – he died before mid-February 1800 in London (see Chapter 8). The two copper plates could not be traced; it is likely that they ended up like almost all copper plates used for printing - by being melted down. Another meeting was held with Ferdinand, and Hawkins reported on 13 February 1800, 'I called upon Bauer and found him less desirous than I expected of engaging in the great work He had even accepted in part some other commissions. In short I believe if he cannot execute the Flora Greca in the way he likes & on the terms he has proposed he will absolutely decline it, and then Heaven knows how we shall get it done! Bauer mentioned the sacrifize [sic] of his liberty in so tedious a work as an objection which he could not easily get over. He certainly will not execute it in conjunction with any artist and will not bind himself to any time for he says it is impossible to say when he may be able [to] form good assistants, for those whom his brother [Franz] has had have proved unsteady and even inimical. He had thought he said if he engaged in this work of taking for his assistants Girls . . . Thus the work in all probability would proceed very slowly for the first few years until these assistants were properly trained'.

Hawkins's letter is revelatory: on the one hand it shows Ferdinand as a hard-headed perfectionist, unwilling to divide the work for a gigantic project with a second graphical artist, a man with limited understanding of the value of time, even of his own working time. On the other hand it is understandable that Ferdinand was not prepared to bind himself for life on a single, highly monotonous graphical project, which he must have considered in every respect a markedly inferior challenge – in particular, when compared to recording the diversity of plants and animals by making drawings true to nature. In addition, the project would have fixed him permanently to London, which may not have been to his liking, as evident from his earlier letter to Hawkins. At the same time Ferdinand had a realistic grasp for the complexities of being bound to collaborate with other graphical artists of inferior skills – in a project to which he had laid the foundations by making the exquisite watercolours and in danger to be 'spoiled' by reproductions prepared by lesser hands. From the arguments discussed at length between the parties concerned two critical points transpire. Firstly, Sibthorp had insisted in his will on 'Plates . . . engraved & coloured', which explicitly excluded the publication of an edition with uncoloured engravings, not only easier to accomplish but also much cheaper. Secondly, two of the three co-executors were principal figures in the top stratum of the British legal system. When contemplating a possible intervention into the nascent *Flora Graeca*-project the lawyers of Oxford University must have known that they would risk losing a case against Platt, as several years later the Trustees of the British Museum would do.

In the end it was the third candidate who was commissioned by the two surviving co-executors to prepare the coloured copper engravings – James Sowerby, who had already worked for Smith before Hawkins's second return from the Levant. No details on the decision are known, but one point clearly emerges: right from the start the successful candidate had a team of collaborators at hand – in fact he ran a family business, which neither Ferdinand nor MacKenzie had, although perhaps they were more brilliant. This is how it came that Ferdinand's hopes concerning the *Flora Graeca*, unrealistic as they seem to us from the beginning, came to an end. As a consequence he was not in any way involved in the publication of the great work. At the same time the decision taken by Hawkins had the effect that the plants collected by Sibthorp on his second journey to the Levant were not depicted in the *Flora Graeca*.

What other commissions Ferdinand had taken on in early 1800 remains unknown: maybe the work for Lambert jointly undertaken with Franz (see above) had started late, maybe the work on foxgloves in the end published by John Lindley had started early (see Chapter 12), maybe Ferdinand produced additional flower-pieces (see Chapter 11). No doubt, producing copies of existing engravings could also have been an option for Ferdinand – in the age before the camera this was always a field in which one might earn some money. As a matter of fact, Franz had produced repeats of the plant illustrations contained in the luxury edition of Jacquin the father's *Selectarum stirpium americanarum historia* (see Chapter 3) – on paper water-marked 1807.[22] However, Ferdinand could also have become engaged in projects outside science, such as producing engravings – his London address being given as 'Paas Engraver No 53 . . . Holburn'. In any case a very major commission was waiting for Ferdinand in 1801, and again Banks would be the great mover behind and on the scene.

ENDNOTES

1 Unless otherwise stated, this chapter is based on Mabberley (1999a) and Lack (1999c).
2 Lack (1997 b)
3 Graves (1905)
4 http://www.bloomsburyauctions.com/detail/720/6.0 (accessed 1 v 2014)
5 Anon. (1841f), Lack (2003)
6 Renkema & Rendle (1930), Miller (1970)
7 Andersson (1998)
8 Nicolson & Fosberg (2004)
9 Edmondon (2002 a, b)
10 Knapp (2003)
11 Mabberley & Moore (1999)
12 Lack (2004)
13 Farjon & Filer (2013)
14 Sitwell (1956)
15 Savage (1930)
16 Salisbury (1805)
17 Lambert (1804)
18 Raphael (1970)
19 LINN – HS 159.1, LINN – HS 159.2
20 B – W 2469
21 Pittsburgh, Hunt Institute for Botanical Documentation, archive. – Franz Bauer to Joseph Franz Edler von Jacquin, 12 ix 1791, unpublished letter.
22 Lack (2003)

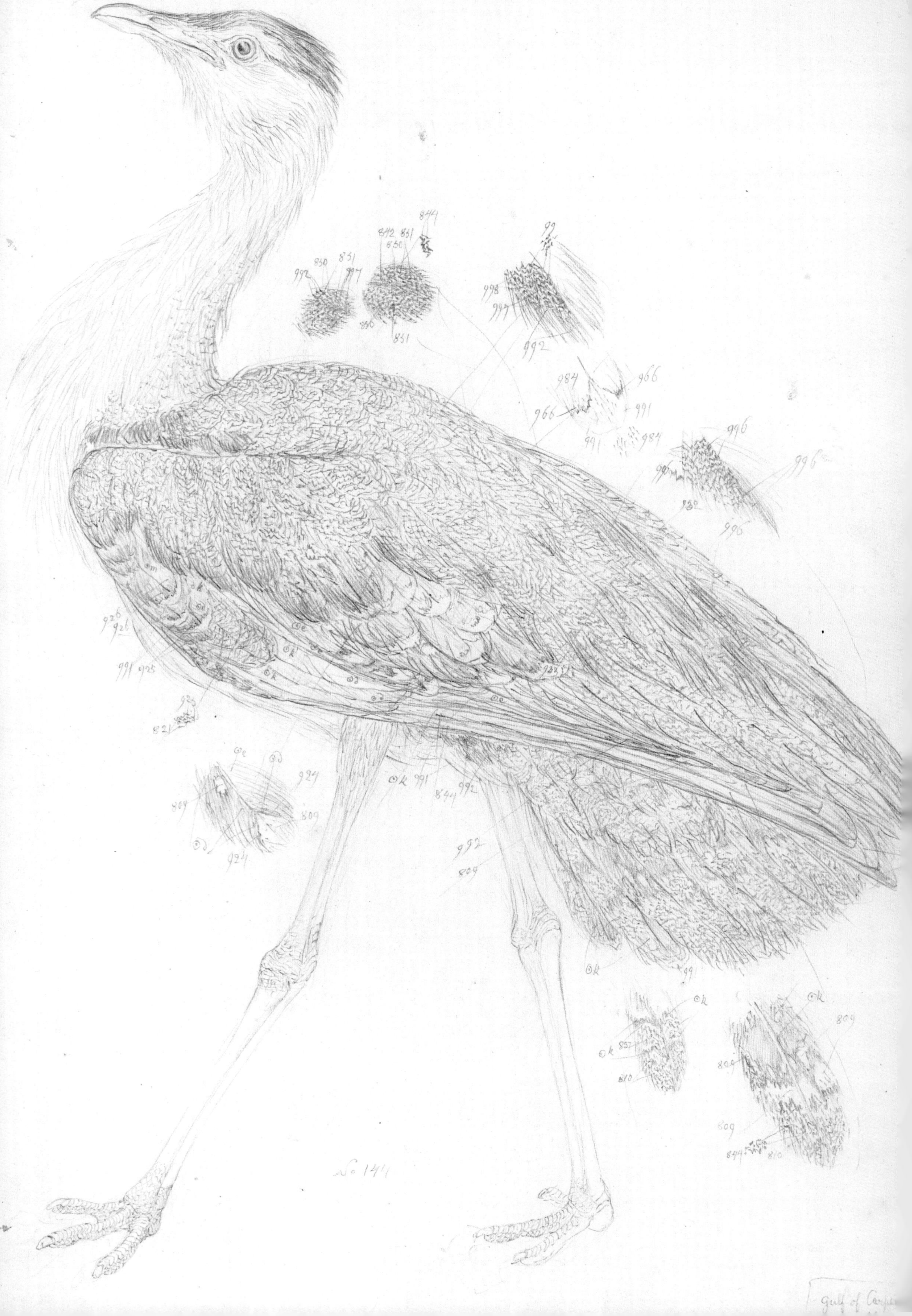
No 144
Gulf of

CHAPTER 10

Ferdinand Bauer and Robert Brown explore Australia[1]

THE BACKGROUND FOR A MAJOR EXPEDITION

FERDINAND HAD ARRIVED in England with Sibthorp in December 1787; Franz followed him with Jacquin the son in December 1788. Only a few months later a political, economic and social crisis that was first focused on Paris and Versailles rapidly grew into a major conflict of truly European dimension – the French Revolution. Right from the beginning Britain was deeply involved in the resulting political and military struggle against revolutionary France, later Napoleonic France, which ended only with the famous Congress of Vienna, concluded in 1815 (see Chapter 12).

The Declaration of Pillnitz in 1791 by Emperor Leopold II and Friedrich Wilhelm II, King of Prussia, was followed by the execution of Louis XVI, King of France, and Marie Antoinette, Queen of France and sister of the Emperor, in Paris. This led to the so-called War of the First Coalition, intended to contain Revolutionary France. When French troops occupied the most westerly parts of the Holy Roman Empire, this was perceived to be a threat to the British Navy, and George III as King of Great Britain and Ireland and Elector of Hanover joined the First Coalition in 1793. From the British point of view this was basically a naval conflict, involving the Mediterranean, the Atlantic and the Caribbean. It resulted in the British occupation of several formally Dutch possessions, among them the Cape of Good Hope and Ceylon. An attempted invasion of Ireland by French troops remained unsuccessful, and the British Navy fought victories against the Spanish fleet at Cap St Vincent and against the fleet of the Batavian Republic, a French client state, near Camperdown. However, on land the tide turned against the First Coalition, and the conclusion of the Peace of Campo Formio in October 1797, which brought both losses and gains to Emperor Franz II, the son of Emperor Leopold II, left Britain as the only one in the field fighting against France.

The War of the Second Coalition was a second attempt by the established powers, led by Emperor Franz II , King George III and by Pavel I, Emperor and Autocrat of all Russias, to contain or eliminate revolutionary France. With successes in Egypt, on the Apennine Peninsula and in Central Europe it started well for the coalition. The battle of Copenhagen on 2 April 1801 with Vice Admiral Horatio Nelson as one of the British commanders defeated the Danish fleet, although the political importance of the victory was reduced by the murder of Paul I in St Petersburg. This was the general political background when Banks, as President of the Royal Society, received a letter from Louis-Guillaume Otto, later Comte de Mosley, on 9 June 1800. He was the French commissioner for the exchange of prisoners of war in London (and soon later the French Ambassador to the Court of St James), and asked for a British passport to be issued for a French scientific expedition. This was speedily organized by Banks and on 19 October

FIG. 10.1. Australian bustard, *Ardeotis australis*. Pencil drawing by Ferdinand Bauer, 1802. Paper, pencil. – Vienna, Naturhistorisches Museum, Archiv.

1800 *Le Géographe*, commanded by Nicolas Baudin, and *Le Naturaliste*, headed by Jacques Félix Emmanuel Hamelin, left for Australia with no less than twenty-four scientists, illustrators and gardeners on board. Clearly the expedition had been sanctioned by Napoleon, by then already Premier Consul and soon to become the key figure in European politics. In December 1800 Banks, always worried about Britain being eclipsed by France in its struggle for supremacy on a global scale, sat down to write a letter to a Fellow of the Royal Society – George John Spencer, 2nd Earl Spencer, Knight of the Garter, Privy Councillor and First Lord of the Admiralty.[2] In it he sketched the course of a British expedition to Australia, with an emphasis on the careful survey 'of the Coast from Basses [Bass] Street to King Geo[rge] 3ds Harbor . . . to explore the Gulph of Carpentaria, & those parts to the westward of it between the 130th and 140th degrees of East Longitude, where the Coast of New Holland [Australia] is not Laid down as Continuous . . . Next to this, a Carefull investigation & an accurate survey of Torre's Strait & [of] the Coast of New South Wales as low down as the Isles of Direction [South Direction Islands, North Direction Islands, Lizard Islands] . . . appears the most interesting'. This was just a first draft, but Banks had no difficulty in convincing Spencer of the necessity of such an undertaking to outbalance the competing French expedition.

As a result of many factors, the answer of the Admiralty was dictated by parsimony – a single ship, the *Investigator*, was employed, the 'scientific gentlemen' comprised only five persons, and the departure was delayed until July 1801. Although this rather poor and belated response to French scientific enterprise has always been deplored by the scientific community in Britain, it was actually a most sensible move by the Admiralty. With Russia a somewhat hesitant political and military partner in the Second Coalition, in particular concerning its naval forces, the British Navy remained the only major power bound to oppose the combined French and Spanish naval forces. The all-decisive Battle of Trafalgar, which was directed by Nelson on the British side and annihilated the French and Spanish combined navies, was still to be fought. The three vessels selected were rather similar in size and age. The *Investigator* had been built in 1795 as a collier of three hundred and forty-four tons, was extra copper sheathed and repaired before setting out on the voyage.[3] *Le Géographe* and *Le Naturaliste*, with a displacement of three hundred and fifty and four hundred tons, respectively, had been launched in 1800 and in 1795. All three ships had been renamed for the purpose of the two competing expeditions, the *Investigator*'s earlier names being *Fram* and *Xenophon*. Clearly one of the purposes of the French expedition was to collect intelligence about the British settlement established in 1788 at Sydney Cove, which – as specified by François Péron and laid down in a subsequent secret report[4] – could easily and successfully be attacked from the sea. There would also be a competition in the field of science – the French travel account *Voyage de Découvertes aux terres australes executé par l'ordre de Sa Majesté l'Empereur et Roi*, which appeared in 1807 in Paris, was to predate *A Voyage to Terra Australis* published belatedly in 1814 in London. The same struggle for priority applies to the creation of several toponyms and the validation of many scientific names based on the numerous plant and animal specimens and illustrations brought back from the two expeditions.

Taking his own experiences into consideration on the *Endeavour* voyage with Cook, as well as the frictions that had existed between Johann Reinhold Forster and Cook on the voyages of the *Resolution* and *Adventure*, Banks felt it necessary to draft a Memorandum of Agreement to be signed 'by all the persons employed as scientific assistants on board H. M. Ship the *Investigator* for the purpose of exploring the country of New Holland'. Having determined who the scientific personnel should be, Banks now paid scrupulous attention to every detail, revealing his awareness of the potential problems and conflicts that might arise for such a party on a Royal Navy vessel. Although formal in terminology, it is a nicely worded document setting the general scene on the *Investigator* for the years to come. The memorandum starts by stating that 'Their Lordships have been pleased to Issue the following instructions and commands to be obeyed by all Persons so employed. It is expected that every Person so employed do sign his name in testimony of his acquiescence in the terms on which their Lordships are pleased to employ them. 1st. Their Lordships require every Person employed as a Scientific Assistant on board the *Investigator* to render Voluntary Obedience to the Commander of the Ship in all Orders he shall from time to time issue for the direction of the conduct of his Crew, or any part thereof. 2dly. Their Lordships require that all Persons so employed do on all occasions conduct themselves peaceably, quietly and civily to each other, each readily Assisting the other in his respective department to the utmost of his ability,

[and] in such manner as will best promote the success of the Public Service in which they are jointly engaged, and unite their individual endeavours into one general result . . . 4thly Their Lordships consider the Salary allotted to each Person employed as full compensation for the whole of his time. They expect therefore all Journals, remarks, Memorandums, Drawings, Scetches, Collections of Natural History, and of Habits, Arms, Utensils, Ornaments & c. of every Kind be delivered up on the return of the Ship to such Persons as their Lordships shall direct to receive them'. The pecking order on the *Investigator* is made clear in another statement: '3rdly. Their Lordships require the Draughtsman employed for Natural History to pay due attention to the directions he shall receive from the Naturalist, and the Draughtsman employed for Landscape and Figures to pay regard to the Opinion of the Commander in choice of Objects most fitting to be delineated; and their Lordships moreover require the Gardener and the Miner to pay obedience to the Naturalist in all such Orders as he shall think fit to give them'. The salaries reflect the pecking order – Brown was to receive four hundred guineas per year; Ferdinand was to receive three hundred guineas per year, more than triple the emolument received from Sibthorp; whereas both Good and Allen were to receive only one hundred guineas per year. Most importantly, Ferdinand and Brown were basically colleagues who received their salaries from the same employer during the voyage on the *Investigator*, a situation quite different to Ferdinand's total dependency on Sibthorp (see Chapter 5). Both were also to enjoy their own cabins on the *Investigator*, which was not the case for the gardener and the miner. Furthermore Brown and Ferdinand were accompanied by their servants; the name of Ferdinand's valet has been given as Powel.[5]

For our story the final, rather involute, clause is particularly relevant: '7thly. Their Lordships moreover declare that after such descriptions, Drawings and Scetches as shall be found necessary for the Illustration and Embellishment of the intended Publication shall have been Selected by such Persons as their Lordships shall be pleased to appoint, and such Specimens of Natural History, Arms, Implements, Habits, Ornaments, &c. as their Lordships think fitting shall have been applied to such purpose as their Lordships shall approve, the remainder of the descriptions of Plants and animals &c., and the Scetches of all Kinds, shall be at the disposal of the Persons who have made them for the purpose of being Published by them whenever it is thought proper, at their own risque, and for their own advantage, provided, however, that all such Drawings as shall be finished during the Voyage, and such Scetches as their Lordships shall Order to be finished after the return of the Ship, shall be considered as the Property of the Public, and lodged in the Depot of the Admiralty when required so to be, and that the remainder of the Collections of Natural History, Arms, Habits, Implements, Ornaments &c. shall be at the disposal of the Persons who have collected them. All this, however, on condition that each Person shall during the Voyage have behaved himself with propriety to the rest, Their Lordships reserving to themselves the Power of punishing all deviations from good humor and perfect harmony among the parties by witholding from the Persons offending such parts of the benefits above described as they shall think proper'. As shall be shown later (see Chapter 11), this arrangement effectively gave Ferdinand the permission to take with him all his pencil drawings as well as his natural history collections from the voyage on the *Investigator* when he decided to return to Vienna in 1814.

In the presence of Banks this Memorandum was signed on 29 April 1801, probably at 32 Soho Square, by the five 'scientific assistants' in the following sequence – Robert Brown, the naturalist; Ferdinand Bauer, the botanical draughtsman; William Westall, the landscape & figure draughtsman; Peter Good, the gardener; John Allen, the miner. It is significant that the document was also signed by Banks, the true promoter of the *Investigator* voyage, who had selected the 'scientific assistants'. No doubt a good deal of diplomacy must have been involved, since only a few weeks before signing the memorandum a new Prime Minister had come into office – Henry Addington, 1st Viscount Sidmouth – and there was also a new First Lord of the Admiralty, John Jervis, 1st Earl of St Vincent. Surprisingly, neither Brown nor Ferdinand and Westall were Banks's first choices. Mungo Park, a Scottish explorer who had travelled in 1795–1797 as the first westerner into the central portion of the Niger River Valley, had declined the offer to join the *Investigator* voyage as naturalist because of an amorous attachment. William Alexander, who in 1792–1794 had accompanied Sir George Leonhard Staunton, 1st Baronet, as a draughtsman on a diplomatic mission to China, had rejected Banks's proposal because of his wife's ill-health. Julius Caesar Ibbetson and William Daniell, both landscape artists, who had previously exhibited at the Royal Academy of Arts, had also turned down Banks's offer. In the end, Banks had recruited five people of outstanding qualifications – possibly with the exception of Allen, of whom Brown, who could be very acerbic, told Banks that he is 'really of very little use'.[6]

On 22 June 1801 the First Lord of the Admiralty issued instructions to Lieutenant Matthew Flinders, the Captain of the *Investigator* and commander of the expedition. Again it is worth quoting an extract in order to appreciate the difference between what was intended and what was the final route of the expedition and the end result of the project. The document starts by stating 'Whereas the sloop under your Command has been fitted and stored for a Voyage to remote Parts; And whereas it is our intention that you shall proceed in her to the Coast of New Holland for the purpose of making a complete Examination and survey of the said Coast, on the Eastern side of which His Majesty's Colony of New South Wales is situated; You are hereby required and directed to put to sea the first favorable opportunity of Wind and Weather, and proceed with as little delay as possible in execution of the service abovementioned, repairing in the first place to Madeira and the Cape of Good Hope in order to take on board such supplies of Water and Live Stock as you maybe in want of. Having so done you are to make the best of your way to the Coast of New Holland, running down the said Coast from 130 Degrees of East Longitude to Basses Straits . . . use your best endeavours to discover such Harbours as may be in those Parts; and in case you should discover any Creek or opening likely to lead to an inland Sea or Strait, you are at liberty either to examine it, or not . . . When it shall appear to you necessary, you are to repair to Sydney Cove for the purpose of refreshing your people, refitting the sloop under your Command, and consulting with the Governor of New South Wales upon the best means of carrying on the survey of the said Coast . . . and taken under your command the Lady Nelson Tender which you may expect to find at Sydney Cove, you are to recommence your survey . . . You are to be very diligent in your examination of the said Coast . . . fixing in all cases, when in your power, the true positions both in Latitude and Longitude of remarkable Head Lands, bays and harbours, by astronomical observations . . . When you shall have completely examined the whole of the Coast from Basse's Straits to King George the third's Harbour you are . . . to proceed to and explore the North West Coast of New Holland . . . Having performed this service you are carefully to examine the Gulph of Carpentaria, and the parts of the Westward thereof, between the 130th and 139th Degrees of Eastern Longitude . . . you are to proceed to a careful investigation and accurate survey of Torres Straits; and when that shall have been completed, you are to examine and survey the whole of the remainder of the North, the West, and the Northwest Coasts of New Holland . . . So soon as you shall have completed the whole of these Surveys & Examinations . . . you are to proceed to, and examine very carefully the East Coast of New Holland seen by Captain Cooke [sic] . . . moving the Investigator onwards from one Harbour to another as they shall be discovered in order that the Naturalists may have time to range about, and collect the produce of the Earth, and the Painters allowed time to finish as many of their works as they possibly can on the spot . . . And when you shall have completed the whole of the surveys and Examinations as abovementioned, you are to lose no time in returning with the sloop under your Command to England for further orders, touching on your way, if necessary, at the Cape of Good Hope . . .'. Essentially the instructions stipulated a survey of the coasts of Australia in a clockwise direction starting from Port Jackson (Sydney) and a return voyage to England in a westerly direction. It is a remarkable fact that the instructions also contain a clause relative to botany, which also merits being quoted: 'And whereas you have been furnished with a Plant Cabbin [sic] for the purpose of depositing therein such Plants, Trees, Shrubs &c . . . you are, when you arrive at Sydney Cove, to cause the said Plant Cabbin to be fitted up by the Carpenter on the Quarter Deck of the sloop you command . . . and you are to cause Boxes for containing Earth to be made and placed therein . . . You are to place the said Plant Cabbin with the Boxes of Earth contained in it, under the charge and care of the Naturalist and Gardener, and to cause to be planted therein during the survey such Plants, Trees, Shrubs &c, as they may think suitable for the Royal Gardens at Kew . . .'. It is hard to avoid the impression that this part of the instructions was masterminded, if not drafted, by Banks.

The survey of the largely unknown coastline of Australia was clearly the first priority of the expedition, which resulted in endless, highly repetitive and monotonous work for the captain and his officers, with work for the scientific assistants largely concentrated during the landfalls. Charting the coastlines of Australia was like a basso continuo in many pieces of baroque music, with the activities of the scientific assistants forming the continuously alternating melody. For the botanically inclined members of the team, notably Ferdinand, Brown and their servants, another basso continuo was the repeated changing of the paper in the plant presses to absorb moisture from the newly gathered plant specimens. At the change a fresh supply of paper was introduced and the earlier, by now moist sheets were dried for reuse, a practice to be continued until the respective specimens were completely dry.

Since the flora and fauna of the Australian coasts were almost totally unknown, the spectrum of objects that were collected and recorded on the *Investigator* voyage is extremely varied, more or less everything being new. In short, all five scientific assistants must have constantly been busy. Only one of them had earlier taken part in a genuine voyage of exploration – Ferdinand, from his earlier work in the Levant, perfectly au fait with the routine of collecting and preserving plant and animal material. At the same time he was both the oldest man on board, aged forty-one upon departure, and the only non-British member of the team – in a sense an outsider, particularly as his command of English was still incomplete. For him the voyage on the *Investigator* must have offered striking differences to his earlier tour with Sibthorp in the Levant, but also several similarities.

DIFFERENCES AND SIMILARITIES

These differences and similarities merit some comments. Firstly the *Investigator* voyage was clearly an expedition in the strict sense, with written instructions, a clearly set-out sequence of actions planned long in advance, not a leisurely tour by an Oxford don and his Cambridge friend to satisfy their personal curiosity. In addition, there was a straightforward division of work between the five scientific assistants, which focused Ferdinand's attention on the accurate representation of flora and fauna, leaving the recording of the coasts and the landscapes to Westall,[7] and the collecting of living plant material to Good, whereas before he had been obliged to act as personal assistant – or rather, slave – for all sorts of duties for the egotistical Sibthorp (see Chapter 5). Furthermore, Ferdinand enjoyed the luxury of having for the first time his own servant (see above).

Secondly, the voyage on the *Investigator* covered significantly larger distances than the tour with Sibthorp. Although intended as the first circumnavigation of Australia, it also became a circumnavigation of the globe, which meant for Ferdinand and Brown crossing the immense spaces of the southern Pacific and passing the Straits of Magellan, which among their London contacts only Banks had experienced before when with Cook on the *Endeavour* (see Chapter 8). The tour with Sibthorp had started on 6 March 1786 in Vienna and ended on 5 December 1787 in Bristol, lasting for 21 months. The *Investigator* sailed on 18 July 1801 from Spithead and was back in Liverpool on 13 October 1805. Although the expedition effectively came to an end in July 1803, it meant for Ferdinand and Brown an absence of more than 50 months from London.

Thirdly, printed maps, albeit inaccurate ones, were available for the Levant when Sibthorp and Ferdinand set out, whereas the making of a first map of Australia covering the complete coastline, though not the interior of the continent, was the main purpose of the *Investigator* voyage. The more common and widespread species of the fauna and flora of the Levant were already known to the scientific community; by contrast, only a tiny part of the fauna and flora of Australia had ever been collected by westerners and an even smaller fraction were studied and named when the *Investigator* expedition started. Very rightly Smith, who had never been to Australia, had stated in his *A specimen of botany of New Holland* 'When a botanist first enters on the investigation of so remote a country as New Holland, he finds himself as if he were in a new world . . . Whole tribes of plants, which at first sight seem familiar to his acquaintance, as occupying links in Nature's chain, on which he has been accustomed to depend, prove, on a nearer examination, total strangers, with other configurations, oeconomy, and other qualities, not only all the species that present themselves are new, but most of the genera, and even natural orders . . . '. In short, the probability of collecting and observing a plant or animal new to science was much higher during the *Investigator* voyage. This fact subsequently ignited much more interest in this project, both among the scientists and the historians of science, than in Sibthorp's tour in the Levant. Naturally all this resulted in a rich body of knowledge on the *Investigator* voyage, scattered over numerous publications. In addition, the exploration of Australia was always nearer to the heart of the Anglo-American public than the Ottoman Empire, which never was to such a degree in the focus of its political and economic interest.

For Ferdinand, there were also several similarities between the tour in the Levant and the expedition starting at Spithead. Firstly, according to the signed memorandum of understanding, he again had a superior to whom he was expected to report – Brown. Although we know very little about their relationship on board the *Investigator*, during the landfalls and later during their joint exploration in the hinterland of Port Jackson, it seems never to have been difficult or strained like it had been between Ferdinand and Sibthorp. Even more important is the fact that Brown and Ferdinand, and as a matter of fact also Brown and Franz (see Chapters 13 and 14), continued to collaborate for years after the end of the voyage, with Brown writing after the death of the two brothers, 'I may be permitted to add how fortunate I consider myself in having so long enjoyed the friendship and so often have been indebted for the important assistance of these two distinguished men, whose merits in the branch of art they cultivated have never been equalled'.[8] When the estate of Franz was auctioned in London in 1841 it was also Brown who acquired pencil drawings by Ferdinand documenting the vegetation of Norfolk Island that he had not been able to see himself (see Chapter 17). We have no indication that Brown pressurized Ferdinand in the way Sibthorp had done, although he was clearly the leader of the team. It rather seems that in the field, when Brown and Ferdinand were together, the two worked in loose cooperation or even independently from each other. The link between a pencil drawing by Ferdinand, a note in Brown's diary, a piece of written information in the famous 'Brown Slips' (see below) and on the herbarium labels is not perfect. As a consequence it would be utterly wrong to automatically assume that the plant specimen that was recorded by Ferdinand was the very same plant specimen described by Brown, although a few cases are known when this was definitely the case. It would also be wrong to assume that all the plant collecting was done only by Brown. Four persons were involved – Brown, Ferdinand, Good and, at least occasionally, Flinders – with uncertain percentages for the three main figures. It should also be noted that Ferdinand worked for many months independently from Brown: after the latter had left for an extended stay in Tasmania and during Ferdinand's exploration of Norfolk Island. In any case Ferdinand possessed a rich collection of herbarium specimens from the *Investigator* voyage when he returned to Vienna in 1814 (all unlabelled, however). Not every specimen may have been collected by himself, but for this we have no evidence. The same can be said of the animals – it would be wrong to assume that an animal specimen recorded by Ferdinand in a pencil drawing was necessarily the animal caught or shot by him, as others were also involved, such as Flinders in the case of the young turtle caught in the Atlantic (see below). At any rate Ferdinand also owned a considerable collection of preserved zoological specimens from the expedition in the *Investigator* when he moved back to Vienna. The fate of this precious plant and animal material will be elucidated later (see Chapter 17).

Secondly, an additional similarity with the voyage to the Levant existed: the division of work between the naturalist and the natural history illustrator was basically the same. Like Sibthorp before him, Brown noted preliminary plant and animal names in his diary, as a rule together with field descriptions, sometimes brief or missing. Meanwhile Ferdinand prepared the colour-coded pencil drawings, a work that he must have accomplished at great speed – just noting dozens and dozens of numbers on a single pencil drawing takes its time. Most importantly, Ferdinand is not known to have ever described a plant or animal specimen. The concept of text or self-expression in words written down simply did not appeal to him.

Thirdly, Ferdinand must have been very well aware of the realities of the voyage ahead, having no doubt a more realistic grasp of the cramped and difficult working conditions on board a vessel than any other scientific assistant recruited by Banks. As a consequence it was clear to him that he would have to again follow his well-established method of preparing colour-coded pencil drawings in the field, which would enable him to produce watercolours at a later moment in a studio. Humidity on board and the constant movements of the vessel would exclude the preparation of sensitive watercolours. Ferdinand therefore took with him as his basic utensils drawing paper, which soon became a worry for him because of the humid storage conditions on board the *Investigator*, and pencils.

Fourthly, by signing the memorandum of agreement Ferdinand had been indirectly informed that after the end of the *Investigator* voyage he would be expected to prepare finished watercolours, which was roughly the same procedure as arranged by Sibthorp. What Ferdinand did not know was that back in London only a relatively small fraction of the pencil drawings was to be selected in order to be worked up by him in colour. Whereas only a tiny fraction of the pencil drawings of plants from the tour of the Levant extant at Oxford was not used for finished watercolours (see Chapter 6), out of the 1543 pencil drawings of plants from mainland Australia (see below) only 236 became the basis of watercolours. This stands in marked contrast to the 966 watercolours of plants prepared for Sibthorp at Oxford. Similarly out of the 176 pencil drawings of animals from mainland Australia (see below) only 52 were worked up as watercolours.[9] This also stands in marked contrast to the 293 watercolours of animals prepared for Sibthorp. In short the whole project remained – as far as the graphical record is concerned – a torso, with the ratio of pencil drawings versus watercolours somewhat better for the animal drawings than for the plant drawings.

In addition, there is a complication well known from the works by Franz (see Chapter 8): the pencil drawings in the Naturhistorisches Museum in Vienna are separate today from the watercolours in the Natural History Museum. In Vienna the pencil drawings had a complex fate (see Chapter 19), with a considerable part of them lost in May 1945 at one of the evacuation sites of the Naturhistorisches Museum in Oberhöflein. Currently they are in no systematic order, apart from being divided into plant and animal pencil drawings. In addition, for a variety of reasons the pencil drawings in Vienna have never attracted the same attention as the finished watercolours in London and are not as thoroughly studied as they should be. As a result, the overall picture of Ferdinand's contribution to the *Investigator* voyage is strongly biased towards his watercolours in London.

Another important fact is that not a single pencil drawing of a plant or animal from the other British possessions, i.e. Cape of Good Hope and Norfolk Island, or from the Portuguese respectively Dutch territories, i.e. Madeira and Timor, was used as the basis of a watercolour. In addition, we are faced with further Bauerian complications: date and locality information is noted inconsistently on the pencil drawings by Ferdinand and often totally missing; sometimes just Roman numerals arranged in chronological order for the different landfalls are given, in several cases Arabic numbers are given which may be interpreted as conventional collection numbers, possibly also arranged in chronological order. Such information is almost always lacking in his watercolours. Often Brown's scattered notes, called 'Brown Slips', provide an indication of a date for a collection, which may not necessarily be the date when the pencil drawing was made, but at least offers some guidance. However, one fact remains: in cases where Ferdinand did not give a date, number or a locality to a pencil drawing and there is no reference in the *Voyage* (see below), the diaries by Brown and Good, the 'Brown Slips' or related evidence, it is impossible to determine exactly where it had been prepared – unless the species recorded is extremely local. This is exactly the same situation faced by every student studying the pencil drawings from the Levant.

When back in England, Ferdinand recorded in colour-coded, and often dated, pencil drawings additional plant material from Australia cultivated in the Royal Garden at Kew, which he integrated into some of his finished watercolours, thereby creating an amalgam that is difficult to analyze and interpret (see Chapter 11). One further amalgam was created later when Ferdinand, back in Vienna, and possibly in collaboration with others, produced pen-and-ink tracings of many pencil drawings from the *Investigator* voyage, often rearranging the figures (see Chapter 12). Several of these tracings were published after Ferdinand's death (see Chapter 18). Many of the pen-and-ink tracings have been colour-coded in pencil by Ferdinand; other pen-and-ink tracings are not colour-coded, and some of this group contain substantial mistakes, probably because less capable hands were involved (see Chapter 12). Today the extant pen-and-ink tracings are intermixed with the extant pencil drawings from the *Investigator* voyage in the archives of the Naturhistorisches Museum (see Chapter 19).

Apart from these differences and similarities relevant to Ferdinand and his work, there are several more general aspects that merit attention. Every researcher of the *Investigator* voyage is faced with an oversupply of sources: the official, somewhat sanitized account, Flinders's *A voyage to Terra Australis*, has appeared in print; the private diaries by both Brown and Good have been published; several letters from the more important figures involved (notably Banks, Brown, Flinders and several others) survive together with a considerable amount of official reports. In this text preference is given to the evidently unpolished diaries and personal letters. All these documents have quite naturally led to a substantial amount of secondary literature of varying levels of sophistication – exactly what one might expect for such a major accomplishment of British seamanship and exploration. In all this Ferdinand is not among the central figures. However, in contrast to Sibthorp's diary Ferdinand is regularly mentioned by name in the diaries of Good and Brown, and the same applies for Flinders's report. In the latter reference is often made to the 'botanical gentlemen', which as a rule must have included Ferdinand. Often, however, the relevant entries are rather laconic, for example with Good noting for 19 January 1803, 'Anchored to South of a small Island [Burney Island] & land[ed] with Captain [Flinders], Mr Brown & Bawer [sic] & collected a few plants', or Brown noting on 2 March 1803, 'Capn Flinders accompanied by Mr Bauer & Surgeon [Hugh Bell] left the Ship & rowd in the whale boat intending to examine the head & other unexplored parts of the bay [Arnheim Bay]'. In brief, reconstructing the day-to-day routine of the expedition is like solving an enormous puzzle of information scattered over many, in some cases conflicting, sources, with the collection dates, albeit fragmentary, and the meagre annotations on the pencil drawings particularly valuable.

A few, rather technical letters written by Ferdinand to Franz in Kew and Joseph in Vienna have survived, as have his letters to Banks, but the main source of information about the former's activities are the pencil drawings, clearly his greatest achievements. Ferdinand seems to have felt this too: back in London he produced a 'List of Sketches of Plants and Animals made during the Voyage of his Majesties Ship Investigator and subsequently at Port Jackson & Norfolk Island by Ferdinand Bauer', now in the Natural History Museum.[10] Here is his total breakdown 1762 pencil drawings of plants (of these 79 from the Cape of Good Hope, 1543 from mainland Australia, 60 from Timor, 80 from Norfolk Island) and 216 pencil drawings of animals (176 from Australia, 40 from Norfolk Island). However, this list is definitely not complete, since Ferdinand refrained from listing his pencil drawings of animals from the Cape of Good Hope and from high seas. Overall these are impressive figures indeed, though not as impressive as those for the much shorter tour in the Levant. Ferdinand arranged this list according to taxonomic groups, relying for plants on the Linnaean system, which was the standard at his time (for example, the Triandria are numbered 57–87). Clearly these numbers have nothing to do with the chronology of the voyage and are not always given on the pencil drawings. The numbers of Ferdinand's list are not to be confused with the inventory numbers given decades after the receipt of the pencil drawings by the staff of the Naturhistorisches Museum. More importantly in our context, we find a significant advance over Ferdinand's earlier work for Sibthorp – the meticulous attention given to anatomical detail, the inclusion of numerous accurate dissections, enlargements and further anatomical details (see Chapter 11). In order to be able to prepare these, scalpels, needles, tweezers must have been part of Ferdinand's equipment.

In addition Ferdinand divined a new, even more sophisticated colour code for the *Investigator* voyage, ranging from 2–994 with various abbreviated notes added, a further improvement over the colour code that he had used in the Levant. As was the case with the one used for the tour with Sibthorp, no colour chart for the expedition to Australia survives, but it may have never existed or rather existed only in Ferdinand's mind. Of course, this is just a hypothesis, but in the same way as the refined ear of a conductor is able to discern several hundred tones and sounds produced by the dozens of individual musical instruments forming an orchestra, Ferdinand's brilliant eye may have been able to do the same: to discern several hundred hues of the rainbow and equate them with numbers, noting them down on his drawings. Based on the numbers attached to the pencil drawings documenting plants from what is now Western Australia, an attempt has been made to understand and reconstruct Ferdinand's 'Australian' code.[11] Code numbers 1–100 stand for red and dark red, 101–200 for purple and pink, 201–300 for pink and mauve, 301–400 for lilac, violet and blue, 401–500 for pale sage green and green, 501–600 for dark green and yellowish green, 601–700 for yellow, 701–800 for orange, 801–900 for brown and 901–999 for white and black. Compared to the colour-code used when working for Boccius, there are similarities and differences; both codes begin with the reds; then the yellows, blues and greens follow in what may be called the Feldsberg code, whereas in the Australian code the sequence is blues, greens and yellows. To the uninitiated the sheer number of different hues distinguished by Ferdinand is impressive, but on second thought it is not. When Horace-Bénédict de Saussure in the late 1780s devised his cyanometer, now conserved in the Musée d'histoire des sciences in Geneva, to measure the different intensities of the sky's blue at alpine heights, he distinguished gradations from 1–52.[12] By contrast Ferdinand made use of the numbers 301–400 for lilac, violet and blue, which are different hues of the rainbow plus different intensities. In short, quantifying colours and colour intensities was definitely not an isolated phenomenon at the time. Although it has been stated in the literature, Ferdinand never ever used Roman numerals to indicate the various hues; these numerals always stand for collection sites.

The abbreviated notes that Ferdinand occasionally added show him to be bilingual. 'G' or 'gel' stands for yellow (German gelb) and 'l gl' for light shiny (German leicht glänzend). In other cases the abbreviations in the two languages are so similar that it is impossible to attribute them either to English or German – for example, 'Pur' for purple (German purpur). Other annotations are definitely in English; for example, we find terms such as 'fine glossy', 'milky', 'transparent' or 'Blak green' on the pencil drawings from Norfolk Island. With the exception of his first draft representations, which lack any references to colours, Ferdinand consistently and persistently colour-coded his 'complete' pencil-drawings from the voyage in the *Investigator*. So far three exceptions to this rule has been found – a pencil drawing showing a Quoll (*Dasyurus* spec.), a carnivorous marsupial with a nocturnal lifestyle, said to be exclusively annotated with colour terms such as 'Brown-Blak', 'BrownGray' and 'grayRed'[13] and two further animal drawings, both taken on Norfolk Island (see below).

THE SCIENTIFIC ASSISTANTS

Of the five scientific assistants, only one had received formal recognition before setting out in the *Investigator* – Westall, born on 12 October 1781 in Hertford, had been admitted to the Royal Academy School in London at the age of eighteen. Little is known about Good, who had been working as a kitchen gardener at Wemyss Castle in Scotland before being recruited by Banks; even his date and place of birth remain unknown. Earlier he had been a foreman at the Royal Garden at Kew and assisted in the transport of living plants from England to Calcutta and back, but it is doubtful if he collected plants under the conditions of a genuine expedition.[14] Allen, born on 7 May 1775 in Ashover, is reported to have started working at a mine on one of Banks's estates at the tender age of twelve. During the voyage in the *Investigator* he played a rather minor role, but he was the first scientific assistant to be back in London to give Banks a report on the ill fate of the expedition.

By contrast much more is known about Brown. Born on 21 December 1773 in Montrose in Scotland, he had started reading medicine at Edinburgh University but was strongly attracted to botany. Without completing his studies he decided to join the Fifeshire regiment of Fencible Infantry as an assistant surgeon and went to Ireland, where he maintained his interest in botany. Soon he came into contact with specialists in England, among them James Dickson, one of the eight original members of the Linnean Society, who published on cryptogamic plants from the British Islands. On 12 December 1800 Banks, who was to open his library at 32 Soho Square to Brown, sent out a letter to him, then at the Fife Regiment based in Kilcock. It began by stating 'a ship is this orderd to be fitted out for the purpose of exploring the natural history (& many other things) of New Holland, & it is resolved that a naturalist & a Botanic Painter shall be sent in her, The salary of the naturalist will be £ 400 a year . . . if you choose to accept the appointment I will certainly recommend you '. Five days later Brown replied to Banks accepting his offer, and on 24 December 1800 he was in London, his diary reporting 'Wait on Sir Jo Banks – dine with him, my situation more fully explaind . . . attend Sir Joseph's Library every forenoon in order to form a herbarium of New Holland plants from the specimens brought by different collectors from Botany Bay . . . I then employ my time in extracting information relative to New Holland plants from the different publications in which they occur . . . Dind on the 20 Jan. with Mr. Flinders who is to command the vessel Mr. [Ferdinand] Bauer the botanical & Mr. [William] Daniel[l, who was subsequently to resign from his engagement] the Landscape painter of the expedition'.

While we possess ample evidence for Brown's preparations for the voyage ahead in the first half year of 1801, no information could be found to reconstruct Ferdinand's activities (see Chapter 9). However, he must have arranged for the stock of drawing paper to be taken on board, writing on 8 April 1803 from Timor to Franz at Kew, 'The paper which I took with me on this cruise has gone mouldy because of the dampness and warmth of the cabin and is covered with spots of mould', occasionally still visible today. Ferdinand may have also taken with him a drawing aid – a camera lucida, a very special prism then already available from William Hyde Wollaston, one of the last great natural history polymaths, a pioneer in developing optical instruments and a Fellow of the Royal Society.[15] The essential principle is that the light from the object is reflected twice within the prism and that the illustrator observes the reflexion through half of his eye's pupil and the drawing paper with the other half. The tiny prism, measuring about one cubic centimetre, could easily have been taken on board the *Investigator*, known for its excellent instruments. However, no description of the use of the prism by Ferdinand throughout the voyage is known to exist. Still it cannot be excluded that occasionally Ferdinand may have used a camera lucida (see below).

After the unavoidable delays, among them waiting for a French passport for the vessel, Ferdinand set out from London with Brown and Westall on the afternoon of 14 June 'to join His Majesty's Sloop Investigator lying at Spithead [southwest of Portsmouth]'.

FROM SPITHEAD TO KING GEORGE III SOUND

The first month was spent waiting at Spithead for the *Investigator* to sail. The time was used to compare the readings of the different thermometers, among them one by Ferdinand, who is also reported to have brought on board a 'few Submarine plants found thrown on the beach', which testifies to his broad spectrum of interests. On the 18 July 1801 favourable winds finally permitted the expedition to start, and two weeks later we find a first record of Ferdinand preparing a drawing in sight of Madeira – of 'a small Turtle [caught by Flinders] asleep on the water', a pencil drawing now in the Naturhistorisches Museum. Many similar notes follow in Brown's diary; for example, on 3 September 1801 in the North Atlantic Ocean, '[I] describd the Albecore [*Thunnus alalunga*] & Mr Bauer made a drawing of it'. In the afternoon of 16 October the *Investigator* anchored in Simon's Bay near the Cape of Good Hope, then under British control, now belonging to the Republic of South Africa.

On 21 October 1801 Ferdinand wrote his first letter to Franz at Kew, of course in German and in an orthography reminding the attentive reader that he had grown up in the rural parts of the Archduchy of Austria below the River Enns, but here quoted from the English translation provided by Marlene Norst (see Preface): 'I must admit that I have enjoyed very good health . . . and, to my great surprise, I found that from the time I first set foot on the ship till this moment I have not felt the slightest touch of any sea sickness'. However, Ferdinand is considerably worried about the state of the *Investigator* by telling his brother 'I am sorry to say that throughout the entire voyage so far we seldom had a dry cabin because water was coming in everywhere through the sides of the upper-middle deck despite the fact that during this time we were not exposed to any great storm'. Indeed, the Admiralty's choice had not been a good one and the rotting, structurally unsound and cramped sloop remained a constant worry for the following four years, not only for the captain but for all on board. Focused on plants Ferdinand continues by telling Franz, 'The mountains [near the Cape of Good Hope], however, which are for the most part large sand hills are covered with the most beautiful plants so that our first excursion yielded us a great number of the most beautiful blooms, of these I wish I could send to you in England all the orchid types which I have seen so far . . . this is the place and great opportunity for collecting'. Indeed Ferdinand and Brown had arrived at the best season of the year. On 3 November 1801 Ferdinand sent a second letter to Franz deploring the imminent departure for Australia 'because a great number of beautiful things which could be found here if one had more time to look for them', mentioning the 'Orchites [orchids], Proteax [proteas] and much else'. This is all the information his brother received on the 79 pencil drawings of plants recorded in Ferdinand's list for the Cape of Good Hope (see above). Among them was an image of the terrestrial orchid *Bartholina burmanniana*, sometimes called spider orchid, because of the numerous narrow petal lobes resembling a spider. It is a winter-blossoming, summer-dormant geophyte that was already known to Linnaeus, which turned out to be Ferdinand's only illustration from this region ever published – in fact even twice, in 1818 in London by Ker, né Gawler[16] (see Chapter 8), and in 1838 in Vienna, twelve years after Ferdinand's death, by Stephan Ladislaus Endlicher (see Chapter 17). Considering Franz's focus on orchids (see Chapter 13) this pencil drawing would have been a feast for him. Judging from the note, 'each [member of the party] carried provisions and a large tin box for specimens etc.' in Good's diary for 24 October 1801 it appears that plant specimens were not pressed in the field, but kept in tight-fitting tin boxes until the return to the *Investigator*, when they would be pressed and dried.

In his two letters to Franz from the Cape of Good Hope no mention was made of the birds recorded by his younger brother – among them, drawn on a single sheet of paper, the male of the southern double-collared sunbird, *Cinnyris chalybeus*, and the male of the Cape sugar bird, *Promerops cafer*,[17] needless to say both colour coded (Fig. 10.2.). Particular attention was given to the tail feathers of the latter species and specifically, as shown in a detail, their tips. Ferdinand not only had a superb eye for distinguishing the hues of the rainbow, he was also a keen bird-watcher. Both species, though

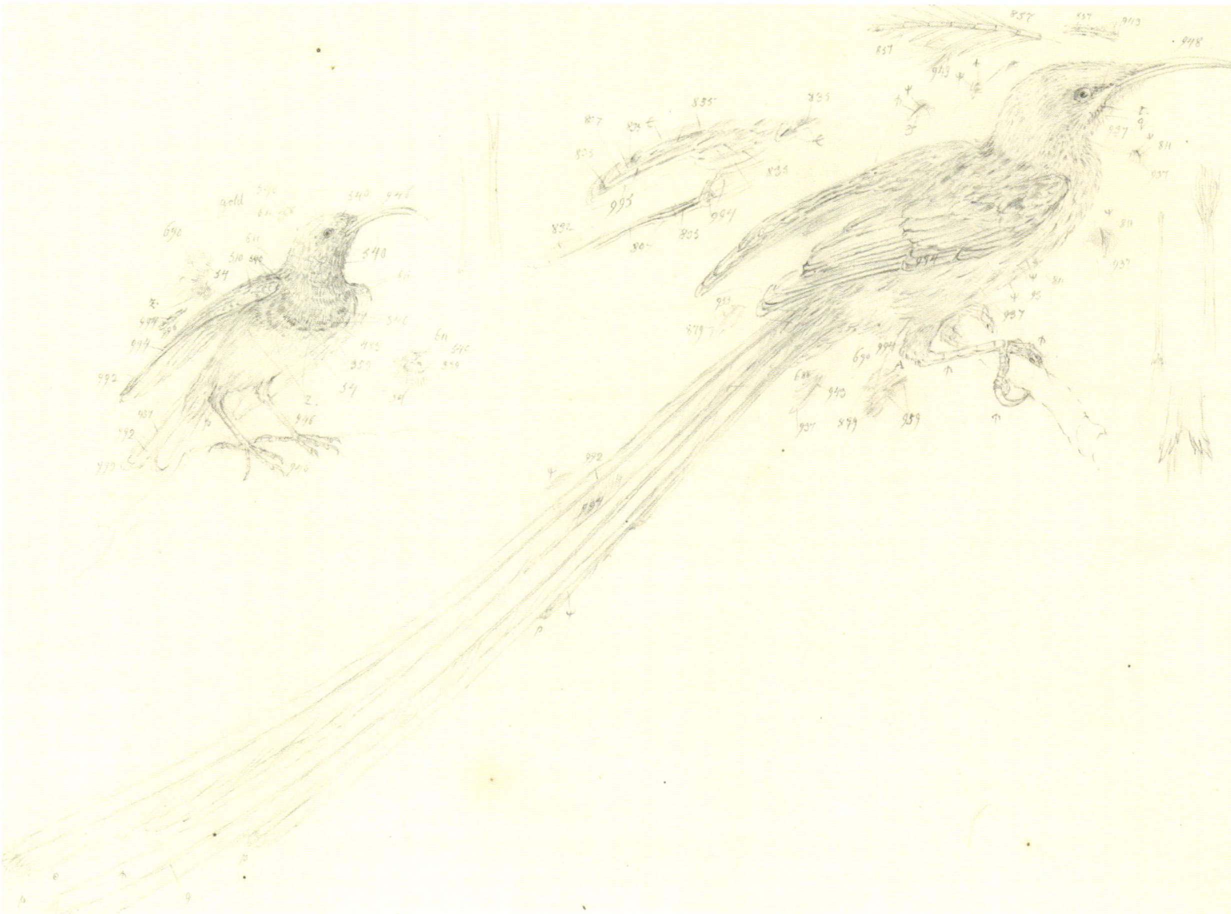

FIG 10.2. **Left: male of the southern double-collared sunbird, *Cinnyris chalybeus*. Right: male of the Cape sugar bird, *Promerops cafer*. Pencil drawing by Ferdinand Bauer, 1801. Paper, pencil. – Vienna, Naturhistorisches Museum, Archiv.**

not rare, are endemic to what is now South Africa and Namibia, but they were already known to Linnaeus. This would scarcely ever occur in the later part of the voyage. Ferdinand could also have mentioned to his brother that he had seen and collected material of the genus *Erica*, the subject of Franz's earlier *Delineations*. Two specimens survive as herbarium material annotated 'C. B. Sp. Ferd. Bauer' in the Naturhistorisches Museum.

On 4 November 1801 the *Investigator* sailed from Simon's Bay, Cape of Good Hope to Australia, with Cape Leeuwin coming in sight on 6 December and arriving at King George III Sound two days later. On 9 November 1801 Good reported, 'after breakfast Mr Brown & party was landed on South shore where we collected many specimens and Some seeds'. The party possibly included Ferdinand; on that occasion the sticky tailflower, *Anthocercis viscosa* subsp. *viscosa*, a shrub so called because of its viscosity and endemic between what is now Denmark and Bremer Bay, may have been among the plants encountered. Ferdinand produced two colour-coded pencil drawings of this species, including details, such as two cut-up corollas and two stamens (Fig. 10.3.), which he worked up later in London into a finished watercolour (see Chapter 11). On 14 December 1801 Brown records his first meeting with Aborigines, writing 'After breakfast went on shore & observing smoke at the head of the harbour walkd to it, our party consisting of Mess[rs] Westal[l] Good & Allen with Bauer & [Thomas ?] White [and] servants. When we reachd the place where the fire had been we found it just extinquishd. No shells or any other mark of the natives having been there. But in a few minutes we saw on the beach one man who advan[ce]d towards us loudly hollowing at short intervals in a very peculiar manner'. Good recorded more details 'he . . . began to call to us with great force and seeing us to persist in advancing to him he brandished his spear & struck off from the Beach among the bushes calling to us in a threatning manner & making signs with his hand for us not to advance . . . he then begun to set fire to the bushes to prevent us from following him and gain time for his family to escape'.

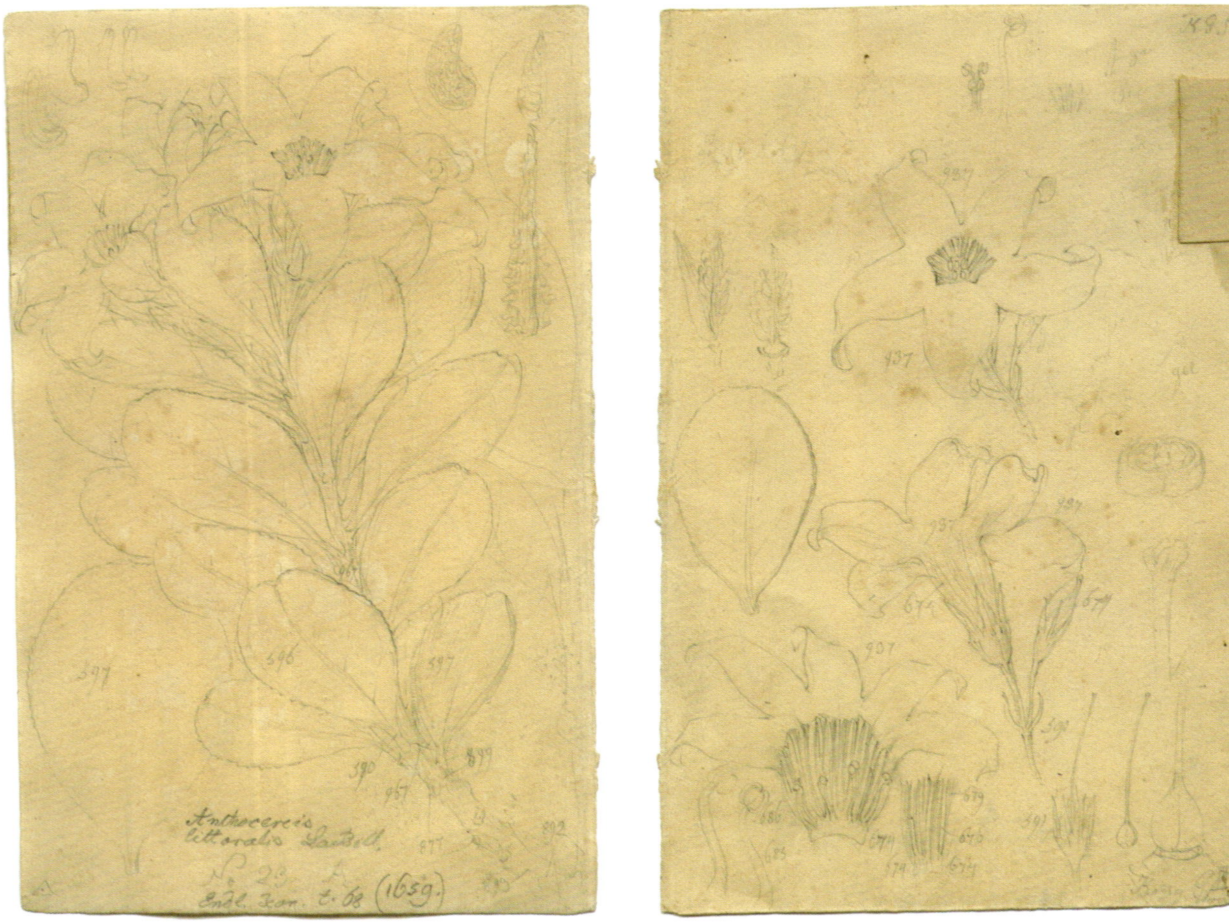

FIG. 10.3. Sticky tailflower, *Anthoceris viscosa* subsp. *viscosa*. Pencil drawings by Ferdinand Bauer, 1801. Paper, pencil. – Vienna, Naturhistorisches Museum, Archiv.

Not accustomed to the oppressive heat and extreme dryness of the high summer in the Southwest of Australia, Ferdinand became critically dehydrated during an excursion to Torboy Inlet on Christmas Eve 1801. Brown reported in his diary: 'Mr Bauer, quite exhausted, lay down & seemd unable to proceed. After in vain attempting to get him along with the party, Mr Good, myself & Wright, a seaman, remaind with him. He felt the strongest inclination to sleep every time he stopt, wch was nearly every 40 paces. In this way we movd on slowly resting every little way, Mr Bauer & soon afterwards Mr Good in great distress for [lack of] water, Mr Bauer often insisting on being left to his fate & after dark, we were uncertain of our direction. We at last not having found water arrivd on the beach and . . . about 12 oClock [i.e. midnight], at the tents'. This seems to have been the most critical moment for Ferdinand on the whole voyage – he clearly suffered from pronounced dehydration, showing severe inhibiting symptoms. Quite naturally no electrolytes for intravenous substitution were available at that time, let alone on the remote coast near the south-western tip of Australia. However, oral rehydration must have helped Ferdinand to survive the crisis, which could easily have determined his life. In the days before that near-catastrophe he had already received his first impressions of the very diverse flora and fauna of this part of the fifth continent: on 16 December 1801 he seems to have prepared a pencil drawing of Brown's leatherjacket, *Acanthaluteres brownii*, a fish new to science, which had his oldest scientific name published in 1846 exclusively based on the finished watercolour by Ferdinand. On 22 December he had probably been busy with a pencil drawing of the King's skink, *Egernia kingii*, an omnivorous and viviparous reptile new to science, which received its scientific name more than three decades later. Ferdinand seems to have recovered quickly: two species of the genus *Banksia* were waiting for him to be recorded as pencil drawings, both lost in 1945 in the Oberhöflein fire (see Chapter 19). However, on the basis of one of them, he produced a watercolour, now at the Natural History Museum, which was used as the exemplar for a coloured engraving published in Ferdinand's *Illustrationes Florae Novae Hollandiae* (see Chapter 11). It recorded the scarlet banksia, *Banksia coccinea*, another species new to science, with its name validated by Brown in a paper published in the *Transactions of the Linnean Society*. It is a shrub or small tree endemic to the south-west of what is now Western Australia, growing on white or grey sand in shrubland, heath and open woodland, and today the most important species of *Banksia* in the cut-flower industry.

In retrospect we have to regret that nobody on board possessed the skills of Sibthorp and Hawkins when it came to recording the more general impressions gathered during this first contact with the fifth continent. The extremely intense light characteristic of the South-Australian summer, even stronger than in the Aegean Sea, is not mentioned, nor is the unbelievable clearness of the sky at night. Neither Ferdinand nor Good nor Brown, always dry and stoic, seem to have had the necessary qualifications and preparedness to venture into these more poetic aspects of travel records, in which both Sibthorp and Hawkins had been highly successful. Even the strangeness of the plants and animals did not elicit a more general note of wonder or surprise. Ferdinand must have been confronted with the strikingly vivid and bright colours of the Australian flora and fauna, but again there is no written record by him. All this makes the extracts from the diaries, letters and reports a much more arid, even monotonous matter to read when compared to the various records from the tour of the Levant.

For 2 January 1802 we find the following entry in Brown's diary: 'Mr Good went in search of the pitcher plant wch Messrs Bauer & Westall had found yesterday in flower. He returned with it in flower'. By pitcher plant Brown referred to the Western Australian pitcher plant, *Cephalotus follicularis*, another endemic of the southern coastal districts of what is now Western Australia, growing there on moist patchy sands, but cultivated today worldwide. In his *Voyage* Flinders remembered 'Among the plants collected by Mr. Brown and his associates [including Ferdinand], was a small one of a novel kind, which we commonly called the pitcher plant. Around the root leaves are several little vases lined with spiny hairs, and these were generally found to contain a sweetish water, and also a number of dead ants. It cannot be asserted that the ants were attracted by the water, and prevented by the spiny hairs from making their escape; but it seemed improbable, that this was a contrivance of nature to obtain the means necessary either to the nourishment or preservation of the plant'. In 1802 further specimens of the Western Australian pitcher plant were collected by Jean Baptiste Leschenault de la Tour, a botanist on the French expedition to Australia, with the first description of the plant new to science being published in 1806 by Jacques Labillardière in his *Novae Hollandiae plantarum specimen* in Paris. This is one of the several cases where Ferdinand's and Brown's first finding and Ferdinand's first graphical record, published in 1814 in Flinders's *Voyage*, was eclipsed by others. This resulted from the late return of Brown and Ferdinand to England and the lack of determination and focus among those with money and responsibility in London in making the botanical results of the *Investigator* voyage speedily available to the scientific community. Another member of the Protea family, the lanolin bush, *Franklandia fucifolia*, must also have been recorded by Ferdinand during this first landfall in Australia in a pencil drawing, which suffered the same fate as the drawing of the scarlet Banksia in Oberhöflein. Again there is a finished watercolour by Ferdinand in the Natural History Museum, which was published as a copper engraving, again in Flinders's *Voyage*.

For the following day Flinders remembers, '. . . it was the 3rd of January in the afternoon, before we steered out of Princess-Royal Harbour. It was not my intention to proceed immediately to sea; and therefore took the opportunity of standing backward and forward in the sound, with the dredge and trawl over board; and a variety of small fish were brought up. These were of little use, as food; but with the shells, sea weeds, and corals, they furnished amusement and occupation to the naturalist and draughtsman, and a pretty kind of hippocampus, which was not scarce, was generally admired'. This finding is neither mentioned in Brown's nor Good's diary, but there is a record by Ferdinand – two specimens of the weedy seadragon, *Phyllopteryx taeniolatus*, later painted in Vienna in watercolours, one of his most famous works. It is the exact representation of the fleeting colours that continue to fascinate ichthyologists. This small fish, reaching about forty-five centimetres in length, has small, but conspicuous leaf-like appendages that provide camouflage and a number of short spines for protection. Restricted to the southern coast of Australia and the coasts of Tasmania, the weedy seadragon was new to science. Of striking reddish colour with yellow and purple markings, this fish was also collected by the competing French expedition, with its oldest scientific name *Syngnathus taeniolatus* published as early as 1804, when the *Investigator* was still in Sydney Cove. Naturalist Bernard Germain Comte de Lacépède in Paris, Membre de l'Institut de France and later also a Fellow of the Royal

Society, was its author. Surprisingly, males brood the eggs externally on a patch of skin beneath the tail, a detail accurately observed by Ferdinand (Fig. 11.14.). In later phases of the voyage submarine life seems to have been only occasionally recorded, with its rocky reefs, seaweed beds and seagrass meadows. In the late morning of 5 January 1802 the stay in King George III Sound had come to an end, the anchor was raised and the *Investigator* started to sail along the southern coast of Australia in an easterly direction.

ALONG THE SOUTHERN COAST OF AUSTRALIA

Flinders felt free to re-interpret the instructions received from the First Lord of the Admiralty, who had commissioned him to proceed now directly to Sydney Cove for repairs and refurbishment and start the survey of the southern coast from this harbour. The Captain found this impractical, and as a consequence the circumnavigation of the fifth continent was commenced in an anticlockwise direction from King George III Sound. In short, a long and tedious work of navigation in largely unknown waters began, following basically the pattern as set out during the first landfall in Australia, which meant 'fixing the true positions of various points' as first priority and giving the scientific assistants ample possibilities to collect the various materials for their special interests as the second priority. What is now called the Recherche Archipelago offered both unpleasant and pleasant surprises. On 10 January 1802 Brown noted in his diary, 'Eat about 16 nuts of the Zamia, partly raw & part roasted. Felt no bad effects. . . . Mr Bauer felt only a Rise of Curd at his stomach all day

MAP 4. Route of the Investigator circumnavigating Australia in 1801 (blue), 1802 (red), 1803 (green), 1805 (pink) (based on Vallance, Moore & Groves 2001).

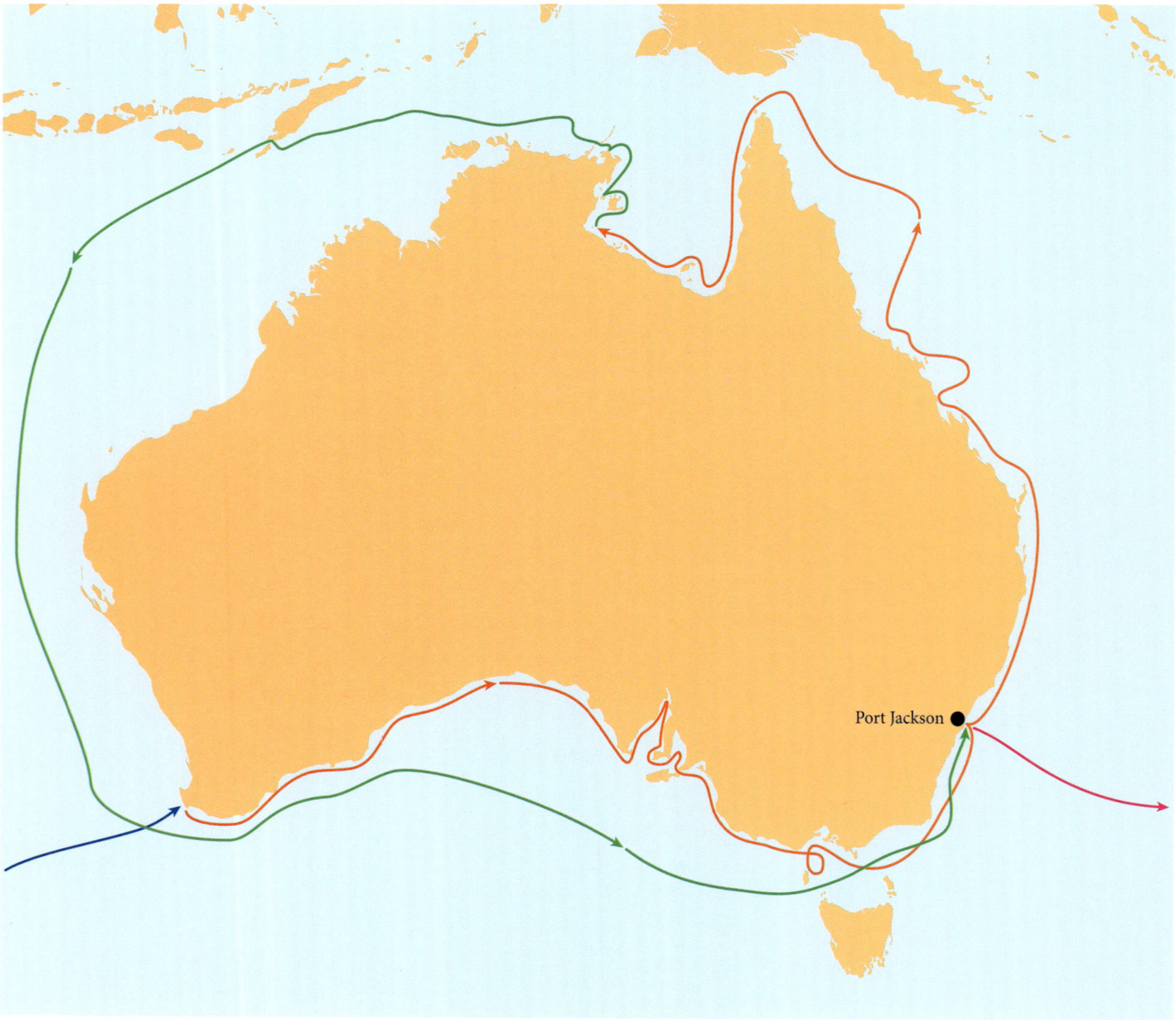

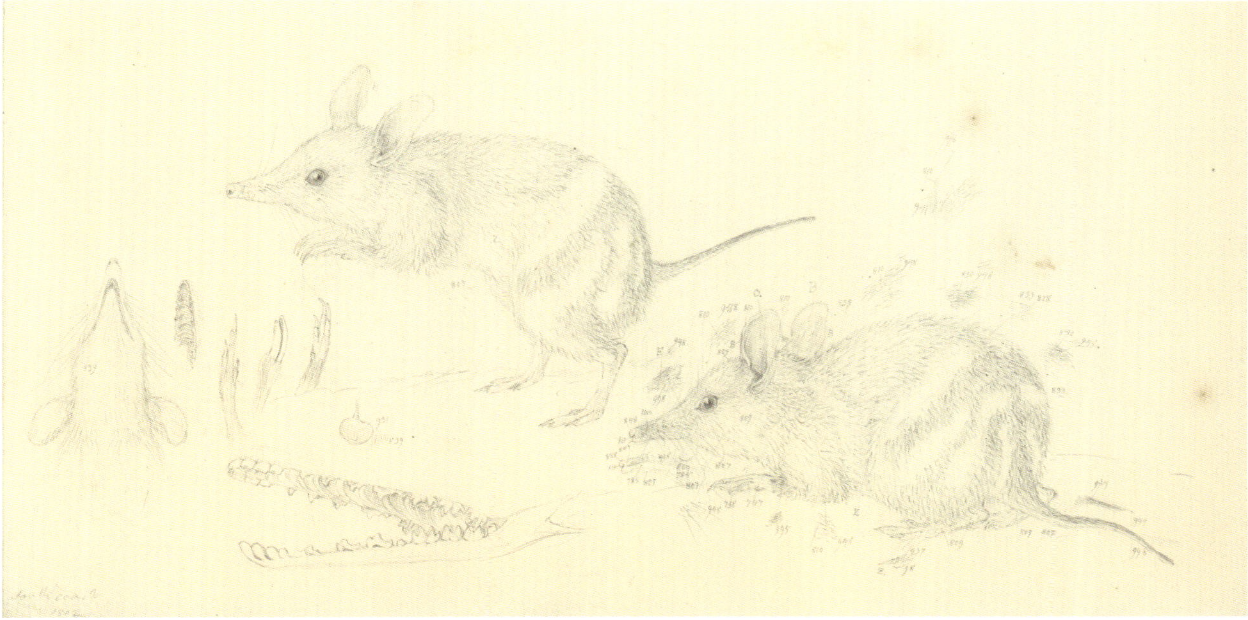

FIG. 10.4. **Western barred bandicoot, *Perameles bougainville*. Pencil drawing by Ferdinand Bauer, 1802. Paper, pencil. – Vienna, Naturhistorisches Museum, Archiv.**

but about 10 at night was attacked with sickness'. The cause of the problem, probably *Macrozamia dyeri* or *M. riedlei*, does not seem to have been recorded in a pencil drawing, whereas 'a small kind of Kangaroo' in Brown's words mentioned two days later attracted Ferdinand's attention and must have been shot. This was probably a specimen of the Recherche rock-wallaby, *Petrogale lateralis hacketti*, a subspecies restricted in distribution to the inaccessible islands of the Recherche Archipelago. With the original specimen apparently either not preserved or lost, this is a particularly relevant finding, since the species did not receive its scientific name until several decades later. Sharks were the main topic on the agenda on the following day, and Ferdinand documented a specimen of a spotted wobbegong, *Orectolobus maculatus*, in a dorsal and ventral view in a pencil drawing, which never became the basis of a watercolour and was only published two hundred years after Ferdinand's return to Vienna in 2014.[18] Since wobbegongs are known to quickly lose their camouflage, consisting of bold markings, when drying,[19] it is likely that Ferdinand prepared his drawing of a dead specimen still in water, for example in a container. No doubt this must have been a challenge – carpet sharks, found in shallow waters in the eastern Indian Ocean from Western Australia to southern Queensland, reach a length of three metres. It is unlikely that anyone on board had heard or understood the name wobbegong, meaning 'shaggy beard' in one of the Australian aboriginal languages. This referred to the growths around the mouth of these nocturnal sharks, nicely documented by Ferdinand, who recorded in a detail also the jaws and the dentition of the beast.

In the *Voyage* Flinders remembered, 'Several seals were procured on this and the preceding day, and some fish were caught along-side the ship; but our success was much impeded by three monstrous sharks, in whose presence no other fish dared to appear. After some attempts we succeeded in taking one of them; but to get it on board required as much preparation as for hoisting in the launch. The length of it, however, was no more than twelve feet three inches (c.3.7 metres], but the circumference of the body was eight feet [2.4 metres]. Amongst the vast quantity of substances contained in the stomach, was a tolerably large seal, bitten in two, and swallowed with half of the spear sticking in it with which it had probably been killed by the natives. The stench of this ravenous monster was great, even before it was dead; and when the stomach was opened, it became intolerable'.

For the 13 January 1802 Flinders recollects, 'An observation of the latitude and a set of angles were there taken, and they brought back . . . a few small kanguroos of a species different from any I had seen before', an entry which may also refer to the same species and/or specimen recorded previously by Ferdinand, demonstrating how difficult it is to unequivocally interpret the written sources. The western barred bandicoot, *Perameles bougainville*, another marsupial, was another trophy. Ferdinand annotated his pencil drawing 'South coast / 1802 III', the Roman numeral standing for Fowlers Bay[20] (Fig. 10.4.). Brown's diary entry for 29 January 1802, 'On the whole this [is] the most barren spot we have seen. No water . . . Caught a small quadruped genus unknown', is likely to refer also to the western barred bandicoot. The term 'barred' refers to the alternating pale and dark bands on the rump of this small crepuscular hunter, measuring about forty-six centimetres in length. This was another marsupial new to science, which had to wait for more than two decades to receive a scientific name. Good noted for this day, 'This is the most miserable part of the Coast we had landed on – not a drop of fresh water and little variety of either animal or vegetables – we saw in some places the prints of naked feet in the mud and sand but no recent many very recent prints of the feet of Dogs – This was the only part of the main [land] where none of the bushes had been set on fire'.

Submarine invertebrates seem to have only occasionally attracted the attention of the captain, but for 18 January 1802 Flinders recorded in the *Voyage*, 'The afternoon we passed a number of pale red medusas . . . and which, on being touched, produce a sensation like the stinging of a nettle. There was also a red scum on the water, and some of it was taken up to be examined by Mr. Brown in a microscope. It consisted of minute particles more than half a line in length . . . '. Significantly, this observation was not recorded by Ferdinand in a drawing; there seems to be little doubt, that his brother at Kew, by then an experienced microscopist (see Chapter 14), would not have missed such a chance.

On 5 February 1802 Flinders marked three points on the Australian mainland coast near Streaky Bay with the letters F, G and H in his log. In his *Voyage*, the first, at c. 32° 33' S, 134° 04' E became Cape Brown, the second, at c. 32° 43' S, 134° 04' E Point Westall, while the third 32° 43' S, 134° 04' E became Cape Bauer, thereby immortalizing three of the scientific assistants. Needless to say, in 1802 none of them was aware of the honour that Flinders would create by later bestowing such toponyms. Naming Thistle Island to commemorate John Thistle, the First Mate on the *Investigator*, had a more sombre background – together with Midshipman William Taylor and six crew members, he perished on 21 February 1802 when sent out to seek an anchorage where something essential might be obtained – fresh water. In his letter of 22 May 1802 from Port Jackson Ferdinand tells Franz (in German) 'We would have considered ourselves fortunate to have completed this part of the journey which in several places, especially between the islands, is very dangerous, if we had not the misfortune to lose the master of our ship, Mr. Thistle, a midshipman, Mr. Taylor, and 6 sailors in a boat which went to shore and from there back to the ship'. On the day following the catastrophe, Brown noted in his diary 'the wreck of the boat, bottom up, discovered . . . without masts & and with none of the people saw several pieces of the boat & one oar . . . Mr [Robert Merritt] Fowler found the boat['s] compass floating in the water, but nothing of the people'.

Ferdinand also produced a pencil drawing of an Australian sea lion, *Neophoca cinerea* (Fig. 10.5.), known to breed only on the south and west coasts of Australia. Since the *Investigator* did not anchor on the west coast, a specimen from the south coast must have been documented, which is substantiated by Ferdinand's note 'South Coast Islands'. Typical for Ferdinand he records the 'jizz' of the animal by documenting the sea lion on land in an upright position and swimming in the sea. Brown seems to have also mentioned the Australian sea lion on the Waldegrave Islands by noting on 11 February 1802 'In our walk along the beach we found several families of Seals' and on 22 March on Kangaroo Island 'The common seal was numerous', while Flinders has 'a few hair seals were killed upon the shore' on the day of Brown's first entry. A colour-coded pencil drawing of the crimson rosella, *Platycercus elegans*, was simply annotated 'Kangoroo Island'. In his diary entry for 23 March 1802, when the *Investigator* was off this island, which is in fact Australia's third largest, Brown has 'Psittacus Pentantii' in his diary; two Brown slips in the Natural History Museum, carry this name from different anchorages at Kangaroo Island, one of them dated 23 March 1802 and coinciding with the diary. Native in eastern and south-eastern Australia, this is a common parrot with striking colours, very accurately documented by Ferdinand – the representation of the showy tail feathers is annotated with no less than seventeen numbers. No watercolour is known to have been based on this pencil drawing. In the case that Ferdinand's image would be 'revived' (see below) it would be possible to determine the subspecies to which the specimen had belonged. Deeply involved in his ornithological work on the southern coast of Australia, Ferdinand would probably not have reflected about a simple fact – he was doing precisely the same work his father Lucas, of whom he could not possess a recollection, had done for the Reigning Prince of Liechtenstein half a century earlier (see Chapter 2).

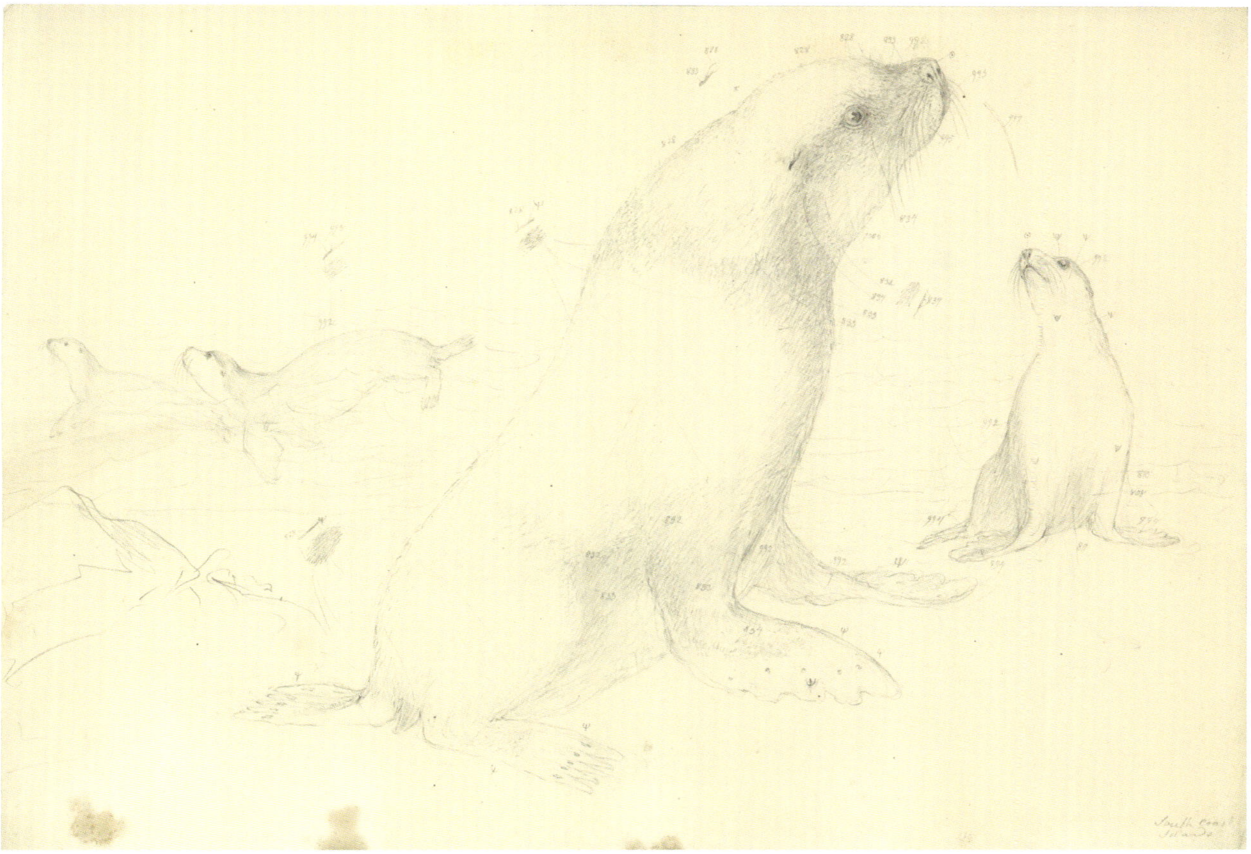

FIG. 10.5. Australian sea lion, *Neophoca cinerea*. Pencil drawing by Ferdinand Bauer, 1802, Paper, pencil. – Vienna, Naturhistorisches Museum, Archiv.

On 23 February 1802 Brown, then at Memory Cove, noted in his diary, 'Shot 3 specimens of a new species of Parroquet, the same as Mr Bell[']s'. This entry has been rightly interpreted as referring to the Australian ringneck, *Barnardius zonarius zonarius*, since Brown's slip indicates 'Feby 23 Memory Cove, desc ss, March 1802' with the addition of 'Fig. Pict. Ferd. Bauer Nos. 20 et 21'. This medium-sized, basically green parrot with characteristic yellow rings around the hind neck is endemic to Australia and was one of the several bird species that were not yet known to science. A gregarious, day-active bird, feeding on a wide range of foods including nectar, insects, seeds and fruits, it could not have been difficult to Brown and Hugh Bell, surgeon on the *Investigator*, to shoot a specimen. As evident from Brown's annotation a specimen was recorded by Ferdinand, who based on this pencil drawing a watercolour of this parrot. Nowhere did Flinders and his scientific assistants venture from the coast into the interior, with marine life continuing to offer interesting objects to the 'scientific gentleman'. One of them was the blue swimmer crab, *Portunus pelagicus*, native in the intertidal estuaries of the Indian and Pacific Oceans and on some of the coasts of the Mediterranean. It is abundantly recorded in Australia; consequently Ferdinand must have had ample opportunity to collect and record this beast. However, his pencil drawing of a male specimen, showing both a dorsal and a ventral view (Fig. 10.6.), is clearly annotated 'South Coast XIIII', standing for St Vincent's Gulf, on which shore the city of Adelaide is located today. Neither mentioned by Brown, nor by Good or Flinders, it is this annotation on the pencil drawing that offers the only clue to localize this graphic record. Today this is a commercially important crab, considered a delicacy throughout Asia. Ferdinand will not have been aware of this, nor of the fact that because of its fast growth and ease of larviculture it is an ideal species for aquaculture. In our context another aspect merits attention: the pencil drawing is of breathtaking precision. Many measurements of the bilateral crab taken show a degree of accuracy that is significantly above what could be expected – less than 0.2 millimetres difference between the right- and the left-hand extremities. This may indicate that Ferdinand at least in this case could have tested a camera lucida, although hard evidence is lacking for this hypothesis. Another pencil drawing carries the same locality information and documented

XIII

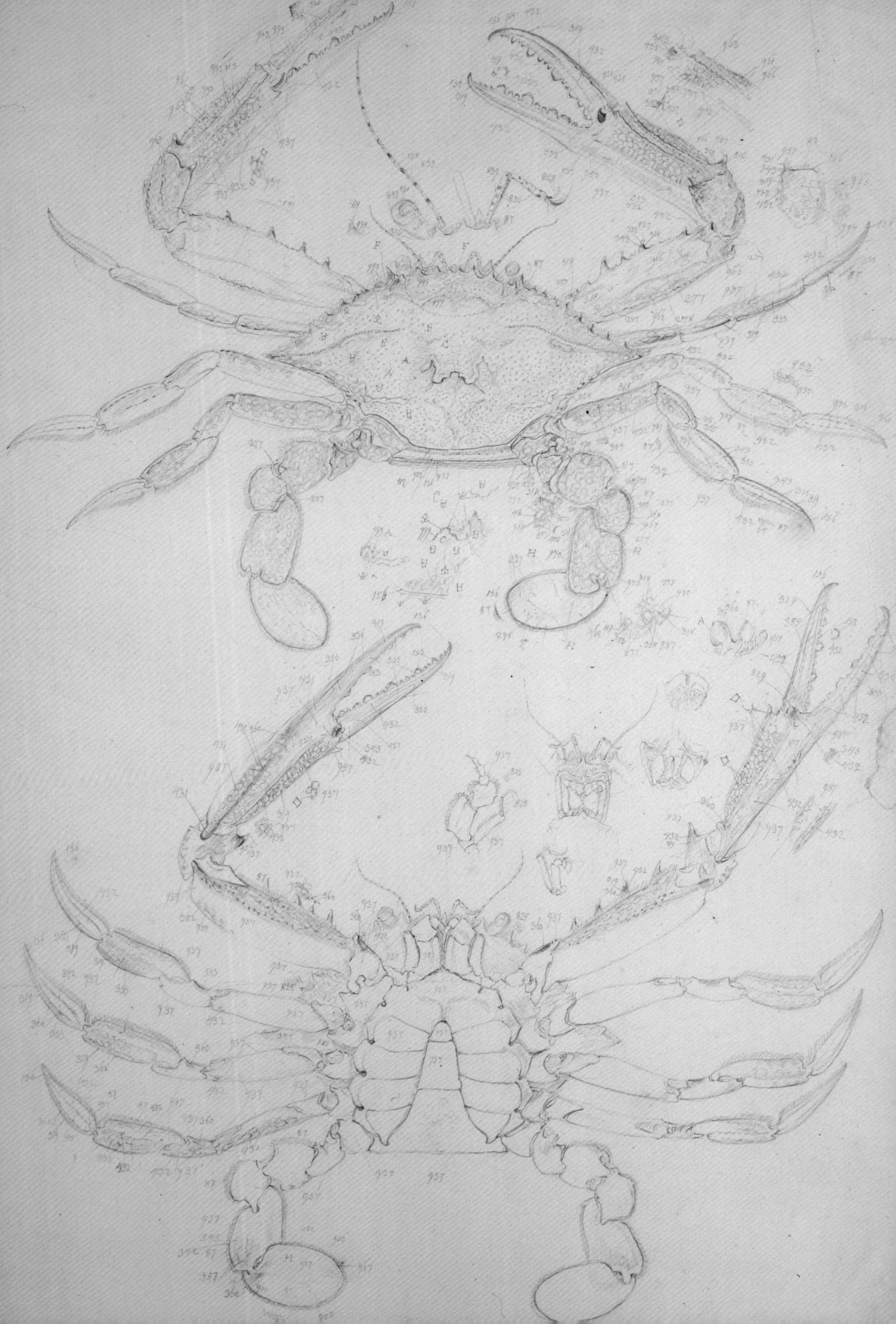

the white-bellied sea eagle, *Haliaetus leucogaster*, a large diurnal bird of prey. With its white head, breast and under-wing coverts this is a distinctive bird; the upper parts are grey and the underwing flight feathers black. Resident from India through South East Asia to Australia along the coasts and waterways, Ferdinand seems to have shot a specimen in a typical habitat and near the southern fringe of its distribution area. An opportunistic carnivore, the white-bellied sea eagle hunts mainly aquatic animals, but it also takes birds and feeds on carrion. Ferdinand's obsession with the very correct colour representation is evident from the fact that in the pencil drawing the iris of the white-bellied sea eagle's eye is annotated with no less than five numbers.

A few days later, on 8 and 9 April 1802, an unexpected and politically interesting meeting took place between Flinders on *Investigator* and Baudin on *Le Géographe* – in a bay now commemorating the event by having been named Encounter Bay. At this moment the Peace of Amiens had already been concluded between the French Republic and the United Kingdom, but this neither Flinders nor Baudin could know, both then charting the southern coast of Australia and being cut off from communication with Europe. Significantly both expeditions carried passports from the other competing naval power. Here is an extract from Brown's diary entry for the two days: 'In the afternoon . . . A sail in sight . . . She appeard to be standing towards us. We bore down, beat to quarters & hoisted our pennant & union Jack. She soon after hoisted english colours, having also french vanes flying at the main . . . provd to be the french Frigate Le Geographe . . . I accompanied C Flinders on board . . . In the morning . . . again went on board Le Geographe . . . We then parted company with her'. A considerable body of information exists on this famous encounter, which in the end was beneficial for both sides: the two captains were able and prepared to share their views and experiences on the coast lines just surveyed. Brown was important, since he knew some French, whereas Flinders did not. Rather unsurprisingly, Ferdinand played no role whatsoever in this meeting.

A few pen-and-ink tracings of Ferdinand's pencil drawings (see Chapter 12) later ended up in a place he knew well, now the Royal Botanic Gardens, Kew. One of them carries the locality information 'South Coast p. Bay XVI', which stands for Port Philip Bay, a large bay covering c.1930 square kilometres in what was to become the state of Victoria; on its northern shore the city of Melbourne is now located. This image shows the orchid *Eriochilus cucullatus*. The scientific name *Epipactis cucullata*, validated by Labillardière for this species, predates the name *Eriochilus autumnalis* coined by Brown for the same species. As with the Western Australian pitcher plant, the French botanist had eclipsed the work of his English competitor. On 27 April 1802, the day on which this tiny plant is assumed to have been recorded, Brown noted in his diary, 'Mr Bauer found the short club of a native. It resembles those of the natives about Port Jackson [Sydney] wch they call Waddies. It was markd in circles of zig zags equidistant & pretty regular. The surface where grasped was made rough'. This ethnographical object had its name from the Dharuk Aborigines and was in use in hand-to-hand combat, capable of splitting a shield and killing prey, or employed as a projectile. Ferdinand must have taken the stick with him to Vienna; in the inventory of his estate it appears as '1 Neuholländischer Stoßkolben';[21] a search in the Weltmuseum in Vienna, the most likely place for ethnographical objects, proved unsuccessful. On the same occasion Brown mentions in his diary 'Psittacus Banksii', an unpublished name with the epithet commemorating Banks, his protector and later employer (Chapter 11). This clearly indicated that Brown thought the bird new to science, which in fact it was. Judging from one of Ferdinand's pencil drawings labelled 'Bay XVI', Brown's entry referred almost certainly to the glossy black cockatoo, *Calyptorhynchus lathami*, also known as the Casuarina black cockatoo after one of their preferred food items. However, the locality is significantly to the west of the bird's current distribution range. The glossy black cockatoo shows considerable sexual dimorphism. Although the pencil drawing was not selected for being used as a basis for a finished watercolour, we are able to tell from it the sex of the specimen – because of Ferdinand's colour code. His numbers for the tail feathers indicate bands of bright red, hues characteristic of a male. With individual feathers colour coded, including patterns on them, this is Ferdinand at his best - comparable to his earlier documentation of the clearly circumscribed pattern on the lip of the sawfly orchid (Fig. 5.10.), which he had depicted when he was with Sibthorp and Hawkins on the island of Cyprus.

FIG. 10.6. Male blue swimmer crab, *Portunus pelagicus*. Pencil drawing by Ferdinand Bauer, 1802. Paper, pencil. – Vienna, Naturhistorisches Museum. Archiv.

Three days after Brown's entry on Ferdinand's waddy we find another bird name 'Psittacus Haematodes' in his diary. This may indicate that Ferdinand made at that locality (Port Philip Bay), and around that date, a pencil drawing that later in Vienna formed the basis for one of his most famous bird watercolours – showing the rainbow lorikeet, *Trichoglossus haematodus moluccanus*, as two birds on a branch, one facing to the left and the other to the right (see Chapter 11). This bird species feeds on pollen and nectar, with the generic name *Trichoglossus* referring to its brush-like tip of the tongue. On 3 May 1802 the *Investigator* left Port Philip Bay with the tide. The survey of the southern coast had come to an end, and six days later they reached Port Jackson or Sydney Cove on the eastern coat, where Flinders found *Le Naturaliste* in the harbour.

ALONG THE EASTERN COAST OF AUSTRALIA

The *Investigator* had last been repaired at the Cape of Good Hope; now a stop of ten weeks was felt necessary for refurbishments and preparations for the voyage northwards. By that time, fourteen years after the foundation of this settlement in 1788, Sydney Cove had a population of about two thousand inhabitants, mainly convicts, sailors and soldiers with their families. The district, mainly inland in the area of Parramatta, supported a similar number.

It was time to write letters. Flinders was the first to give his report to Banks in a letter dated 20 May 1802, noting 'I say nothing of our scientific gentlemen, Mr Brown being so much better qualified to tell his own story than I am. It is fortunate for science that two men of such assiduity and abilities as Mr. Brown and Mr. Bauer have been selected; their application is beyond what I have been accustomed to see'. This is more than respect and recognition, it is almost a eulogy. However, in our context it is more important to realize that Flinders regards the two on a par as to dedication and qualification. Ferdinand, in his typical matter-of-fact style, reports from Sydney Cove two days later to Franz (in German): 'The land which we saw on this [southern] coast looks very barren for the most part and consists largely of sand hills which are partly covered with shrubs while others are quite bare. Further inland, however, it is heavily timbered and in many places the trees have been burned down. We found very few inhabitants on this land and none at all on the islands if we are not going to count the kangaroos as such, of which we found very many on the islands. Although the season for plants in bloom had almost passed in New Holland, I nevertheless made 350 sketches of plants besides 100 sketches of the animal kingdom during this time . . . As far as I am concerned, although we were exposed to great heat and great fatigue in the course of our land excursions, I am nevertheless in good health and spirits . . .'. Although the story has been repeatedly told that Ferdinand was the first naturalist from the Hereditary States to visit Australia, this is not the case: from 12 March to 11 April 1792 Haenke (see Chapter 3), born in Kreibitz (Chřibská, now in the Czech Republic) and a participant of the circumnavigation of the Pacific ocean led by Alessandro Malaspina on the *Atrevida* and the *Descubiera* flying Spanish colours, had been in Sydney Cove.[22] Haenke possessed at that time the early colour chart of the Bauer brothers on board and was to send a letter to the same addressee as Brown did – Banks. It is a small world indeed: when young, Ferdinand and Haenke had enjoyed the support of the same patron in Vienna - Jacquin.

A week later Brown wrote to Banks about Ferdinand, 'Mr. Bauer desires me to add that he has made 350 drawings of plants and 100 of animals. He has indeed been indefatigable, and has bestowed infinite pains on the dissections of the parts of frucification of plants'. Although nothing has been reported by Brown or Flinders, speed must have been another aspect of Ferdinand's graphical record prepared in the field: many plant specimens as well as their flower and fruit parts often quickly fade away in the intense heat of the Australian summer, so he had to be quick. It has been said that Brown induced Ferdinand to pay special attention to these details, and it is likely that he indeed exerted some influence. However, the Bauer brothers had already very accurately documented details, such as in a cross section of a fruit capsule of henbane, *Hyoscyamus niger* (Fig. 2.15.), at a time when Brown was merely six years old. Brown's comments to Banks are high praise indeed; by contrast, Ferdinand is not mentioned by Brown in his letters to Dryander and Charles Francis Greville (see Chapter 13), written on the same day in Sydney.

A second letter was written by Ferdinand in Sydney Cove on the day before the departure of the *Investigator* and directed to Franz at Kew stating (in German): 'I have so far found much to occupy me. We made several short excursions inland, for example to Botany Bay, Parramatta and Hawkesbury and from there to the river Grose which brought us to the foot of the Blue Mountains after which I returned to Sydney with the Lady Nelson on the Hawkesbury River. During this time I made considerable additions to my collection and all my sketches which now number 700 I have left behind in the Governor's house'. Two aspects of this letter merit attention: firstly Ferdinand mentions his collection, which proves that he already started to amass plant and animal specimens, which, many years later ended up in the Naturhistorisches Museum (see Chapter 17); secondly he was wise enough to leave his pencil drawings in safe storage with Philip Gidley King, then Governor of the nascent colony of New South Wales, where also all the living plants collected on the south coast had been deposited. Unfortunately we have no record of the preparation of Ferdinand's collection. Whereas this is a rather straightforward process for plant specimens, the preparation of vertebrates shot or caught is much more complex. Contrary to what has been stated in the literature, he did not preserve specimens in alcohol. The specimen of the spider genus *Araneus*, allegedly collected by him in Australia and now in Vienna, is of Neotropical origin.

Ferdinand may have prepared on 16 July 1802 at Sydney Cove a pencil drawing of, admittedly, the most curious mammal endemic to Australia – the duck-billed platypus, *Ornithorhynchus anatinus*. It is one of the five extant species of the monotremes, the only mammals that lay eggs instead of giving birth. Evidence for the date is meagre – there are no entries for that day in the diaries of Brown and Good, and in his much later report Flinders does not mention the recording of the strange semi-aquatic animal. The only evidence is found in the collection of unbound Brown slips with the shelf mark Z 1, 'Descriptions of animals observed on the *Investigator* voyage', where we find two dates – on f. 149 'Port Jackson July 16 1802', and at f. 169 'September 7 1803'. It is more plausible to assume that this happened on the first date, since on 8 September 1803, another day without an entry in Brown's diary, Ferdinand was collecting plants between Parramatta and Hawkesbury. The platypus's unconventional appearance – duck-billed, otter-footed, beaver-tailed, the male having a spur in the hind foot that delivers a venom – had baffled naturalists before, with some thinking it to be a hoax. However, when Ferdinand prepared the pencil drawing and probably also the specimen (Fig. 10.7.) now at the Naturhistorisches Museum, it was already known among the initiated. One of them was George Shaw, co-founder of the Linnean Society, Fellow of the Royal Society and assistant keeper of the Department of Natural and Artificial Products at the British Museum. He had published the first scientific description of the platypus in 1799,[23] even before the voyage of the *Investigator* had started. In the following years Sir Everard Home, another Fellow of the Royal Society who was later to become an employer of Franz (see Chapter 15), studied the odd anatomy of the strange beast and had his results published in the *Philosophical Transactions*[24] with William Clift

FIG. 10.7. Male platypus, *Ornithorhynchus anatinus*, collected by Ferdinand Bauer in Australia, c.1802. – Vienna, Naturhistorisches Museum, 1. Zoologische Abteilung, Säugetier-Sammlung, Inv. Nr. 26.513.

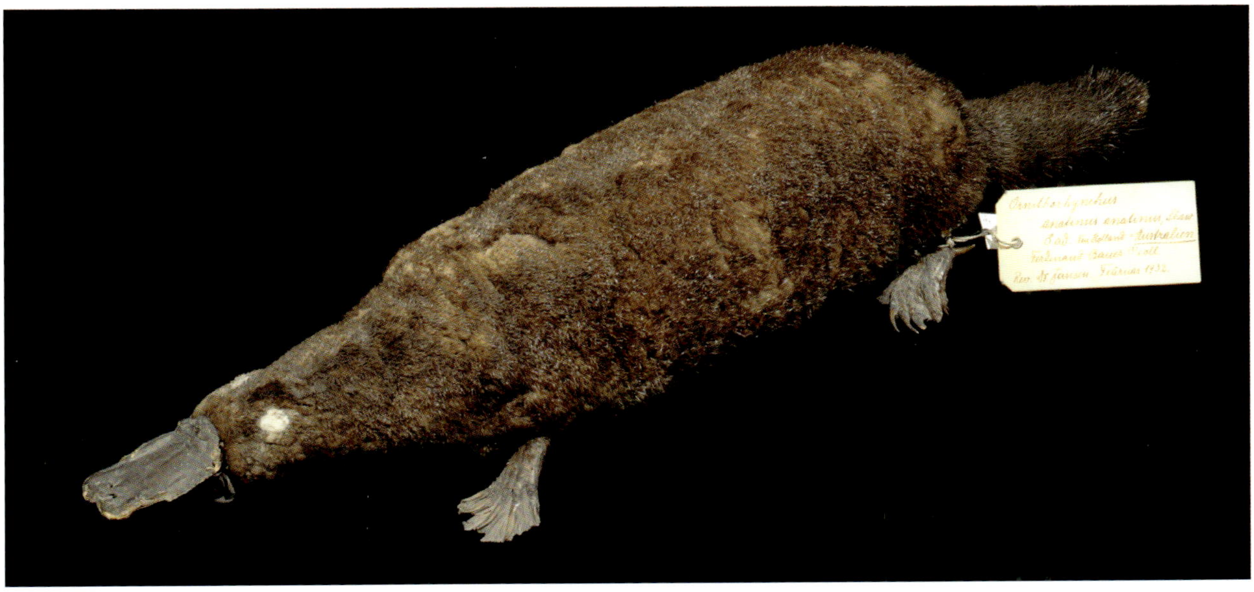

preparing some of the illustrations. Even before that Johann Blumenbach, professor at Göttingen University and Fellow of the Royal Societies in London and Göttingen, validated the generic name *Ornithorhynchus*. Banks, with his lively interest in Antipodean novelties, had provided both Home and Blumenbach with a specimen of the platypus. However, Ferdinand was the first to base his drawing on a relatively fresh specimen and produced a very accurate documentation of the elusive 'duck-billed mole', as it was then called by the colonists. Unfortunately nothing is known of how Ferdinand managed to catch his specimen of the platypus, an animal with crepuscular habits and secluded, timid behaviour. Otherwise very little is known about Ferdinand's first stay at Sydney Cove; there are gaps in the diary of both Brown and Good, and the excursion to Parramatta mentioned above is only briefly alluded to. It took place on 17 June 1802, and the party included Ferdinand, Good, Westall and Brown, who explicitly mentioned Government House, now called Old Government House and belonging to the National Trust.

On 21 July 1802 the *Investigator*, accompanied by the tender *Lady Nelson* with John Murray in charge, sailed out of Sydney Cove and turned northwards to start the exploration of the eastern coast, very much along the same lines as in the first part of the circumnavigation. Flinders must by now have made the necessary arrangements to set up on deck the conservatory for living plants. In his *Voyage*, and more particularly in the section referring to July 1802, he reminisced 'Amongst our employments was that of fitting up a green house on the quarter deck, and sawing plank to make boxes for the reception of such plants as might be found by the naturalist, and thought worthy of being transported to His Majesty's botanic garden at Kew. This green house had been received at Sheerness, and stowed away in pieces; but I saw that when filled with boxes of earth, the upper works of the ship, naturally very weak, would be incapable of supporting the weight; and that in bad weather, we would be obliged to throw it over board for the safety of the ship. I therefore proposed its reduction to two-thirds of size; and Mr. Brown being of opinion it would then contain all the plants likely to be collected in any one absence from Port Jackson, it was reduced accordingly . . . This arrangement required an alteration in the tiller, and a short one, with two arms, was fitted to the after part of the rudder head; with which expedient, and leading the main braces forward, the green house was not likely to cause much inconvenience to the working of the ship'.

Like for the first part of the voyage information on Ferdinand is also poor for the second part. He definitely prepared a pencil drawing lost in the Oberhöflein disaster, of the plant mentioned by Good on 7 August 1802 as 'an elegant Embothrium' and by Brown on the same day as 'Embothr[ium] banksianum'. Eight years later Brown published the name *Grevillea banksii* for this plant new to science; after a further three years Ferdinand published a coloured copper engraving of this plant in his *Illustrationes* (see Chapter 11). Several kilometres north at Keppel Bay, Ferdinand prepared another pencil drawing documenting *Turraea pubescens*. Although a wide-spread species described from China and common in Malesia and Australia, the specimen recorded by him probably comes from this Bay, since, in this case Brown gives in his manuscript notes the date 10 August 1802 with the addition 'Bauer dr[awing]'.

Five days later Good recorded in his diary an incident that shows the realities of exploring the unchartered coastline in the tropical part of Australia: 'one of the Petty officers & a seaman were missing they had been last seen near Cape Keppel their absence as night approached caused considerable uneasiness'. On the next morning 'the people were seen coming along the Beach with a number of Natives – The Boat went ashore immediately and brought them off when they made a ludicrous figure they had been wading the greater part of the night to the middle in mud in mangrove Swamps – their cloths all rags without Shoes or Stockings having all Stuck in the mud – Mr [Thomas] Evans the midshipman reported – that they were seen in the morning (while still bewildered with Mangroves & Swamps) by a large party of natives one of which came to them & presented him with a bough of a tree making Signals for him to return another which he did as a signal of peace – he then took our two forlorn people to his party & offered them roasted Ducks & fish & afterwards led them to the Beach which they were very near although they did not think so till they saw it & believed they would not have reached it for several hours but for the friendly assistance of the Natives, Mr Evans being very Thirsty made Signal as if to drink the Salt water when they made a dolefull noise very expressive of sorrow & immediately led him to fresh water which was very near & then followed him to the Beach'.

According to the annotation on one of his pencil drawings, Ferdinand documented the blue-winged kookaburra, *Dacelo leachii leachii*, at 'Shoalwater Bay 1802'[25] in what is now Queensland, visited by the *Investigator* from 29 August to 2 September 1802 (Fig.

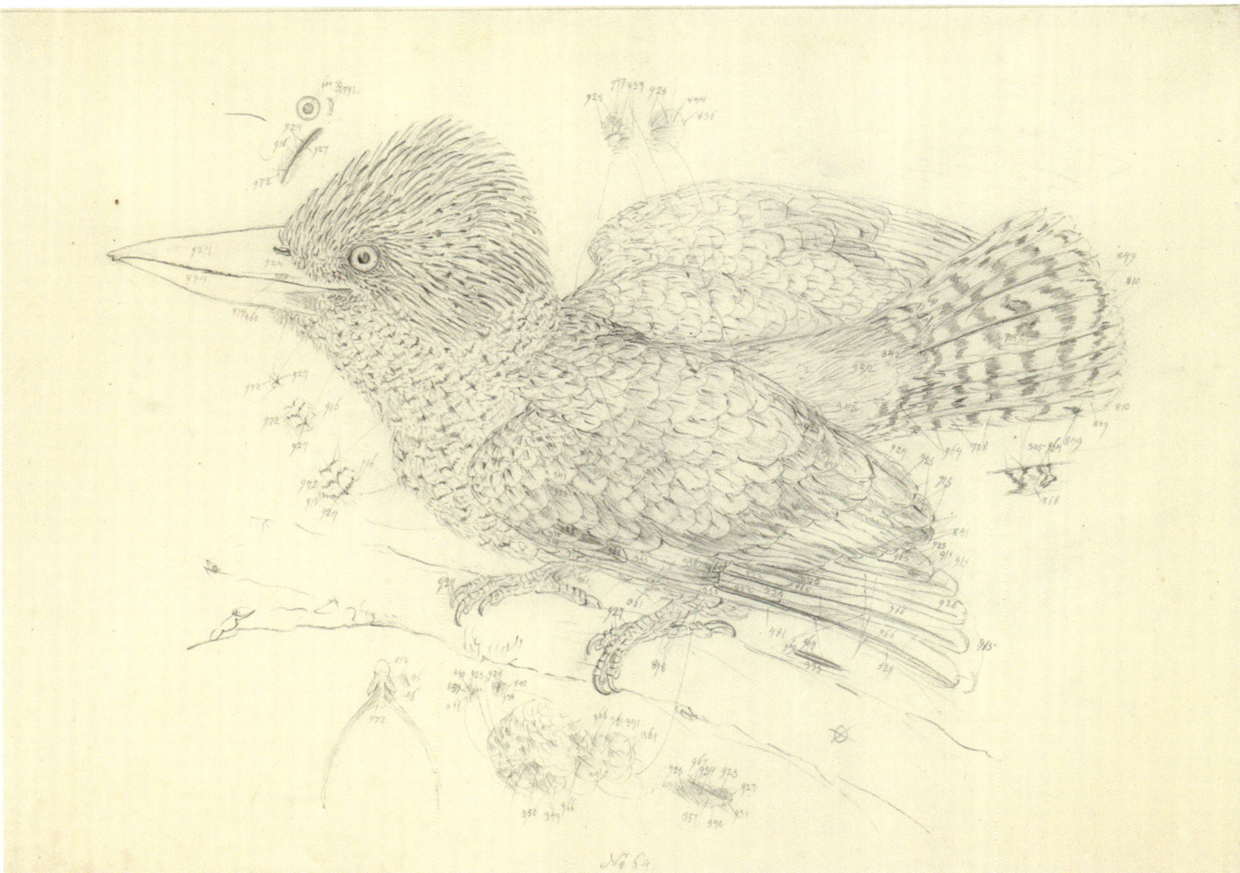

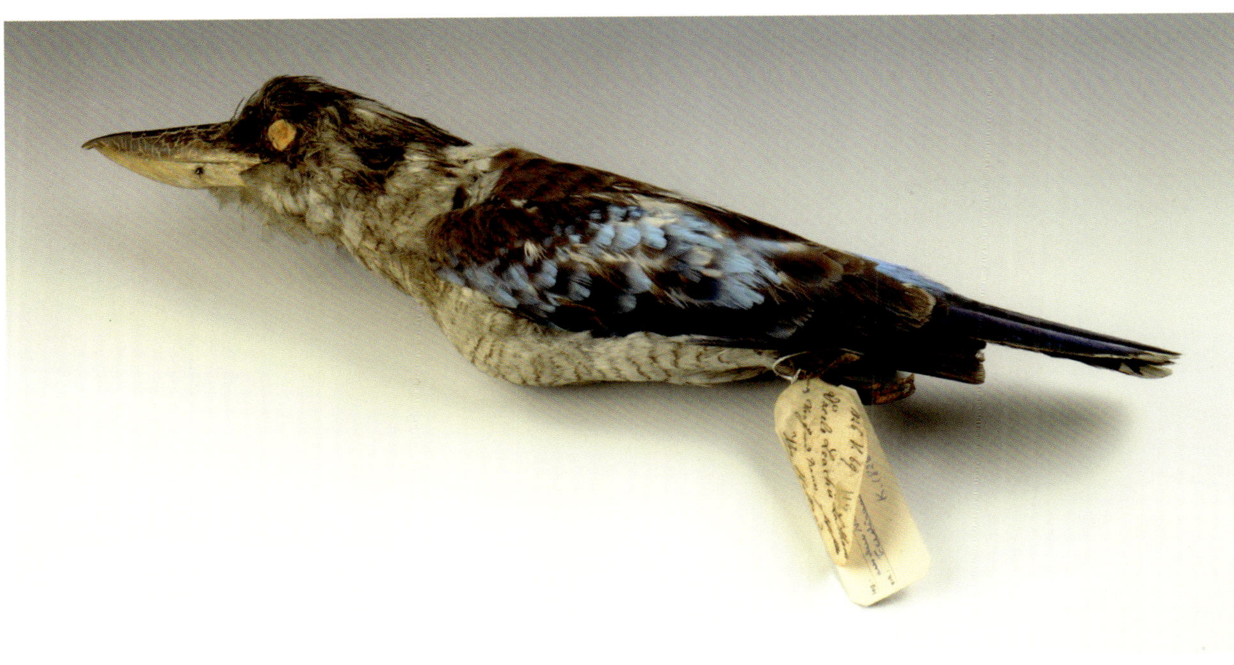

10.8.). An attentive bird-watcher, he recorded both a male and a female specimen, only the former with the locality information given above. No watercolours are known to have been prepared by Ferdinand based on these drawings, and only the specimen of the male bird survives in the Naturhistorisches Museum (Fig. 10.9.).

FIG. 10.8. *(top)* Male blue-winged kookaburra, *Dacelo leachii leachii*. Pencil drawings by Ferdinand Bauer, 1802. Paper, pencil. – Vienna, Naturhistorisches Museum, Archiv.

FIG. 10.9. *(bottom)* Male blue-winged kookaburra, *Dacelo leachii leachii*. – Vienna, Naturhistorisches Museum, Vögelsammlung Inv. Nr. 50.545, 1826.V.69.

Apart from getting lost or dehydrated, there were also other minor dangers such as getting bitten or stung. On 9 September 1802 on West Hill Island both Ferdinand and Brown suffered the same fate, from ants, probably the green tree ant, *Oecophylla smaragdina*, and also plants. Brown reported in his diary, 'On the hill I had observed a single plant of Morus incendiarius . . . & having incautiously touch'd it was stung in the middle finger of the left hand. . . Mr Bauer was also stung by Tragioides pruriens'. The stinging tree, *Dendrocnide moroides*, which is reported to be potent enough to kill humans, or the less dangerous shining-leaf stinging tree, *D. photinophylla*, were the plants to be blamed in the case of Brown, whereas *Tragia novae-hollandiae*, already depicted by Parkinson, was the culprit in the case of Ferdinand.

On 30 September 1802 Brown went ashore on Middle Island and 'observed a few plants not seen before by us, among others Caryota oleracea'. There is no reason to doubt that this refers to the cabbage palm, *Ptychosperma elegans*, of which Ferdinand must have made a pencil drawing, now lost for the same well-known reasons. On the basis of this he prepared in London three watercolours, which – hard to believe but true – were published as late as 1999. However, before that two of these were redrawn and integrated into an image of a tropical landscape and published five years after Ferdinand's death – in Carl Friedrich Philipp von Martius's famous *Historia naturalis palmarum* (see Chapter 18). The cabbage palm was new to science; Brown had validated the name *Seaforthia elegans* for it, which was taken up by Martius, who correctly noted 'secundum Ferd. Baueri iconem' (based on Ferdinand Bauer's image). It is no surprise that such a monumental palm also appeared in two of the landscapes prepared by Westall, now in the National Library of Australia in Canberra. A few days before Flinders found the so-called Flinders Passage, a way through the Great Barrier Reef, Brown had visited Calder Island, where he was surprised, as he put it in his diary, by 'Cycas circinalis [seen] for the first time, but it had been seen in Keppel Bay by C Flinders, I observd it with fruit & and with young male flowers'. This entry both testifies to Flinders's interest in botany and efficient communication between the Captain and the scientific assistants on board the *Investigator* – in contrast to the tensions between Cook and the Forsters on the *Resolution*. The plants observed, both males and females, and recorded by Ferdinand in pencil drawings now lost, belonged to the ricketty bush, *Cycas media*. This palm-like tree is a widespread in the northern parts of Australia, but the locality recorded on the pertinent Brown's slip is clearly Calder Island. Near this locality Ferdinand prepared a colour-coded pencil drawing of a megabat, annotated by him 'Fliegender Hund Northumberland Islands 1802', kept in the Naturhistorisches Museum. It seems not to have ever been studied by a bat specialist and shows Ferdinand again as a man with both German and English in his mind. In the *Voyage* Flinders mentioned this finding and wrote 'I returned on board in the evening, without having discovered any traces of the lost cutter or seen any thing worthy of particular notice; unless it were three of the large bats, called flying foxes at Port Jackson: when on the wing and at a distance, these animals might be taken for crows'. On 26 August 1802 the Captain mentioned the loss of the boat – 'the cutter had been upset from the rapidity of the tides, which ran above four knots, the man in her was thrown out, and the boat went adrift. The man was taken up by the Lady Nelson; but the boatswain, who with two men in a small gig had gone after the cutter, was not heard of till the next morning, when he returned without any intelligence of his object, having been bewildered in the dark by the rapid tides in a strange place, and in danger of losing himself . . . The opening through which he had come was named Strong-tide Passage'. No doubt charting the coast of what is now Queensland was not an easy task. It has been presumed that a specimen of butterfly cod, *Pterois* spee., has been caught near that passage and documented by Ferdinand in a pencil drawing, which later became the basis of a finished watercolour. This is the only animal which could not be precisely identified, as the finray counts do not much match any known species (see Chapter 11).

By then the *Investigator* must have entered tropical waters. A month later a serious problem presented itself to Flinders, which he recorded in the *Voyage* in an entry referring to 25 September 1802: 'I found, to my no less surprise than regret, that the time keepers had again been let down; and no more than one day's rates had been since obtained . . . it appeared that lieutenant [Samuel] Flinders [Mathew Flinders's brother] being intent upon these, had forgotten to wind up the time keepers on 22d at noon. This fresh difficulty was very embarrassing. To go away for Torres' Strait and the Gulph of Carpentaria without good rates, was to cripple the accuracy of all our longitudes; and on the other hand, the expected approach of the contrary monsoon on the North Coast admitted of no longer delay in Broad Sound'.

Ferdinand does not seem to have recorded underwater life in what Good described on 5 October 1802 as 'a continual labyrinth of Coral reefs Shoals and Breakers with waters at time as smooth as a mill pond at other times looking like a whirlpool'. This coincides with Brown's note for the period from 6 to 15 October 1802: 'Among the reefs. In vain attempting a passage. Sometimes in very dangerous situations, chiefly arising from the depth & foul ground of our anchorages & the rapidity of the tide in many places'. Flinders's recollections of 9 October 1802 show his amazement of what presented itself to his eyes: 'In the afternoon, I went upon the reef with a party of the gentlemen [possibly including Ferdinand]; and the water being very clear round the edges, a new creation, as it was to us . . . was there presented to our view. We had wheat sheaves, mushrooms, stags horns, cabbage leaves, and a variety of other forms, glowing under the water with vivid tints of every shade betwixt green, purple, brown, and white; equalling in beauty and excelling in grandeur the most favourite parterre of the curious florist. These were different species of coral and fungus, growing, as it were, out of the solid rock, and each had its peculiar form and shade of colouring; but whilst contemplating the richness of the scene, we could not long forget with what destruction it was pregnant . . . The edges of the reef, but particularly on the outside where the sea broke, were the highest parts; within, there were pools and holes containing live corals, sponges, and sea eggs and cucumbers; and many enormous cockles (chama gigas) [*Tridacna gigas*] were scattered upon different parts of the reef. At low water, this cockle seems most commonly to lie half open; but frequently closes with much noise; and the water within the shells then sprouts up in a stream, three or four feet high: it was from this noise and the sprouting of the water, that we discovered them, for in other respects they were scarcely to be distinguished from the coral rock. A number of these cockles were taken on board the ship, and stewed in the coppers; but they were too rank to be agreeable food, and were eaten by few. One of them weighed 47 ½ lbs. [c. 22 kg] as taken up, and contained 3 lbs. 2 oz. [c. 1, 5 kg] of meat'.

On 18 October 1802 a sudden opportunity presented itself to Ferdinand to write to his brother at Kew (in German): 'I am taking this opportunity just to let you know that the Investigator has come along this coast as far as the Cumberlands Islands and between the Coral Reefs, where this night we lay at anchor and where I heard that the brig Lady Nelson was going back to Port Jackson. About myself I can only report that I am still well as are all on our ship and that so far there has been no particular accident to the boat and the anchor although the coast on this side is extremely dangerous. I have made good additions to our collection during this time'. It took almost to the day one year until the letter arrived in Franz's hand at Kew.

It would be quite wrong to think that the *Investigator* would have entered no-man's land. Good describes a meeting with Aborigines on 28 October 1802: 'Saw many Natives on the Beach & in the water – immediately on Anchoring the Natives launched 3 Canoes which soon reached the ship one after the other – they were at first shy but being shown some Hatchets at length one plunged off the Canoe & swam to the Ship bringing a piece of Bamboo filled with very bad water & some plantains – after which various articles were exchanged & some Cocoa [Coco] nuts plantains Bows arrows & shell ornaments were exchanged'. Two days later Good continues by noting, 'They were all well made of a good stature & in good condition very active & expert swimmers they were naked but wore various ornaments, most all had the ears Cut in several places & pieces of a kind of pearl shell cut & fixed in the ears of different shapes – they had various tassells made from fibres of the Bark of some Tree neatly plait & fixed some round the neck others the Arms some round the middle & some the legs . . . some had a piece of a large Shell with string fastned to to [sic] lye round the middle which completely covered the privy parts but few wore it . . . at 8 AM we begun to weigh Anchor & when they were given to understand they were in the way they readily retired & at desire sat down on the Deck at the Stern of the Ship and when at least they were desired to leave the Ship they readily did so & leapt into the Sea Swam to their canoes Set Sail and Stood for the Shore . . . we soon came in sight of Taits [Zuizin] Island . . . Anchored to the Lee of a small low Island covered with trees & Captain [Flinders] Messr Brown Bawer & Westall landed & collected a few things'. One of the plants was given as 'Pedalioides tribulis Sol' in Brown's diary, which testifies to his superb memory – he must have seen the specimen collected by Banks and Solander on the *Endeavour* voyage with this name attached in the herbarium at 32 Soho Square. As would become evident only later, both specimens belong to the genus *Josephinia*, with Brown's specimen representing *J. imperatricis*. Étienne – Pierre Ventenat had validated the scientific name *Josephinia imperatricis*

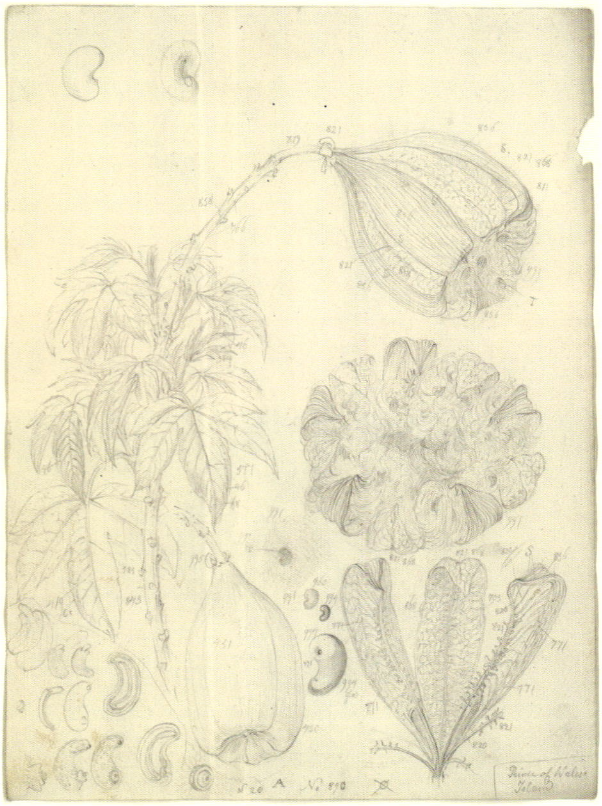

FIG. 10.10. Bush tucker, *Cochlospermum gillivraei*. Pencil drawing by Ferdinand Bauer, 1802. Paper, pencil. – Vienna, Naturhistorisches Museum, Archiv.

based on a specimen cultivated in the conservatory at Malmaison near Paris. The country house of Malmaison with its garden and conservatory belonged to Joséphine de Beauharnais, née Marie Josèphe Rose Tascher de La Pagerie, the first wife of Napoleon and soon Empress Consort of the French – hence the epithet 'imperatricis'. Seeds of this plant had been collected by Hamelin, captain of *Le Naturaliste*, at an unknown locality in Australia and arrived quickly in Paris. Again a French botanist had eclipsed not only the works of the naturalist on the *Investigator* but also of the naturalists on the *Endeavour*, because Banks and Solander had also collected the plant. Ferdinand, like Parkinson before him, prepared a pencil drawing of the *Josephinia imperatricis*; the latter was published posthumously by Endlicher (see Chapter 18). In a sense Ferdinand's work had also been eclipsed. Pierre-Joseph Redouté (see Chapter 16) had prepared the first finished watercolour of *Josephinia imperatricis* based on a specimen cultivated in Malmaison, which was published in 1804 in Ventenat's *Jardin de la Malmaison*. In this case, as in so many others, no finished watercolour is known to have been prepared by Ferdinand.

Torres Strait lies between the northernmost extremity of Australia and the Malesian Island of New Guinea. About one hundred and fifty kilometres wide and very shallow, even today hazardous to navigate because of the maze of reefs, it posed a special challenge for Flinders and his crew in the *Investigator*, since the numerous islands, called the Prince of Wales Islands, scattered to the north of the Australian mainland were not yet properly surveyed. Brown visited on 2 November 1802 the island e in Flinders's log, named by the latter in the *Voyage* Goods Island, to commemorate the gardener on board. Brown noted in his diary: 'The hills are extremely barren & stoney . . . No trees of considerable size or fit for any purpose but as fuel . . . Green ant also very common. No very recent signs of inhabitants'. In the afternoon of that day Ferdinand, Flinders and Good followed, and on that very day Ferdinand started to prepare a pencil drawing of a plant that one of the party must have collected on that day – the bush tucker, *Cochlospermum gillivraei*, another strikingly large-flowered new species, which had to wait for sixty years until it got its scientific name. In this case the colour-coded pencil drawing (Fig. 10.10.) survives, annotated by Ferdinand 'Prince of Wales Island', which may be identical to Flinders's island e, because no record exists that the Prince of Wales Island proper was visited by the *Investigator*. The following day the ship entered the Gulf of Carpentaria and the survey of the northern coast began.

ALONG THE NORTHERN COAST OF AUSTRALIA

To westerners this was a largely unknown coastline with virgin lands behind. On his way to western New Guinea, William Janszoon in the *Duyfken* had made a landfall near the modern town of Weipa as early as 26 February 1606; extremely few western ships followed and the whole coastline remained unchartered until the *Investigator* arrived. Five months of intense work under tropical conditions lay ahead, with finding sufficient quantities of potable water and the deplorable condition of the ship constant worries for all on board. In the diaries Ferdinand is mentioned only occasionally, the main source of his activities again being his pencil drawings.

Dry and stoic as always, Brown noted in late November 1802: 'The Carpenter who were employd caulking the ships sides accidentally discoverd one of the timbers to be quite rotten. This led to a more particular examination the result of which was that the greater part of the timbers seen were rotten that the greater part also of the plank examind was rotton [sic] except about an inch of the surface which was sound & the Master in conjunction with the Carpenter being orderd to give in a report of the state of the Ship pronouncd it unsafe to keep the sea in her longer than six months even supposing her to continue in good weather & to meet with no accident. They were also of opinion that taking the ground in unfavourable circumstances would be fatal to her'. Others would panic, but Flinders did not and continued to fulfil his mission. At anchor between Bentinck and Sweers Islands, it was again Brown who wrote in his diary: '[at] the Island where we waterd a Bustard was shot perhaps not essentially different from Otis tarda, it weighd 12 ½ Lb. & the flesh was well tasted that of the legs was much whiter than the breast &c.'. This is confirmed by Good stating, 'This day [25 November 1802] Mr Bawer Shot a large Bustard'. Definitely this was a specimen of the Australian bustard, *Ardeotis australis*, a ground bird of grassland and woodland across northern Australia and southern New Guinea. Judging from the weight it is more likely that Ferdinand had shot a male rather than a female specimen. The Australian bustard is the largest flying bird in Australia, with the males attaining up to 1.2 metres with up to a 2.3-metre wingspan and significantly heavier than the females. Ferdinand produced a pencil drawing (Fig. 10.1. p. 246) annotated 'Gulf of Carpentaria' showing this common beast standing erect in a very natural pose. Why Ferdinand refrained from colour coding the yellow- to cream-coloured legs as well as the head and eye in this case, remains unknown.

The 4 December 1802 ended with an unpleasant experience; in this case Good offered the most detailed account: 'I took the opportunity of the Boats and landed [at Bountiful Island] & collected a few things. Found the Cycas circinalis in great perfection & the fruit being both pleasant to the taste and sight I eat some as also Mr Brown & Bawer. On coming on board Mr Bawer and I were taken with a violent reaching with sickness which continued with short intervals the greater part of the night it had an unpleasant effect with Mr Brown – but several other people in the Ship eat a little & most did so were affected in like manner'. Ever painstakingly recording the manifold 'productions of nature', Ferdinand produced pencil drawings of the cycad, now believed to belong to *Cycas angulata*. These are now lost, but two lithographs of the female strobus based on Ferdinand's record were published posthumously by Friedrich Anton Wilhelm Miquel (see Chapter 18). Back in London Ferdinand produced two finished watercolours, which had to wait even longer for publication, one of them as late as 1999.

An aboriginal burying site discovered on Bickerton Island offered some change to the dull routine on board, with Brown noting in his diary on 4 January 1803: 'we found trees wch had been hollowd in the centre deprivd of their bark [and] stuck up in threes perpendicularly. Their height was about 14 feet above the surface. All of one set had fallen & were much decayd, the other set[s] were all standing . . . It was painted externally with transvers[e] bands & stripes of ochre red. Each of the trees containd a human skeleton. That in the painted tree was also painted generally red but some of the bones with si[n]gular white stripes towards one of the extremities. A few skulls wer[e] also lying at the bottom of the trees . . . Mr Bauer who had walkd a different way also met with three hollow trees each containing the whole or at least the greater part of a human skeleton'. Petroglyphs were recorded by Flinders on 14 January 1803: 'I went in a boat to this last island with the botanical gentlemen [possibly including Ferdinand], intending to take bearings from the uppermost cliffs . . . This was called Chasm Island . . . In the steep sides of the chasms were deep holes or caverns, undermining the cliffs; upon the walls of which I found rude drawings, made with charcoal and something like red paint upon the white ground of the rock. These drawings presented porpoises, turtle, kanguroos, and a human hand; and Mr. Westall, who went afterwards to see them, found the presentation of a kanguroo, with a file of thirty-two people following after it. The third person of the band was twice the height of the others, and held in his hand something resembling the whaddie, or wooden sword of the natives of Port Jackson; and was probably intended to represent a chief. They could not, as with us, indicate superiority by clothing or ornament, since they wear none of any kind'. Westall's watercolour known under the denomination 'Chasm Island: Native Cave Painting' and now also in the National Library of Australia, represents a very early specimen of European interest in aboriginal art.

Whereas most of the islands chartered by Flinders and his crew in the Gulf of Carpentaria had been low islands, one later named by Flinders in his *Voyage* Morgans Island, was rising at one point to around eighty metres above sea level. While Good noted for 20 January 1803, 'went ashore with Captain & party & collected some new plants – found fresh water & saw the print of the feet of Natives very recent on the Sand', Brown is more specific, stating 'new species of Corypha'. Indeed the party had found a new palm, subsequently named by Brown *Livistona humilis*, a most common tree up to seven metres tall, but restricted in distribution to what is now Northern Territory. Flinders had the more practical aspects of this palm in mind when he wrote in the *Voyage* 'the cabbage is too small to be an interesting article of food to a ship's company; and the young leaves, drawn into slips and dried, the seamen made handsome light hats, excellent for warm weather'. In his travel account Flinders had already made use of this palm's scientific name published by Brown in advance of the *Voyage* (see Chapter 11). Such a spectacular tree found on a remote island off the northern coast of the Australian mainland made Ferdinand prepare pencil drawings, now lost because of the Oberhöflein fire. However, in this case three watercolours were painted by Ferdinand in London, which survive. They are masterworks of botanical illustration, showing precisely those details that the palm specialist needs to see and examine (see Chapter 11), testifying not only to the excellent eye of Ferdinand but also to his profound understanding of the scientist's requirements.

Brown's diary entry for 27 January 1803 gives an impression of what field work was like on the northern coast of Australia: 'Went on shore intending merely to examine a small part of the shore but destitute of a compass & incautiously losing sight of the beach & there being no sun I lost my way & after wandering for about an hour & half found myself on a beach where every object appeard new, neither ship or high land being visible . . . The darkness of the night & my ignorance of this being a narrow neck of land & my uncertainty even of its being main not a little confusd me & the sailor who accompanied me being quite to[o] much fatigued that I was under the necessity of carrying both tin boxes & paper books, I judgd it most advisable to remain where we were . . . Accordingly we laid down under a tree & both of us being tird we fell asleep. Before however lying down I heard a gun from the ship which in a great measure determind the direction we should have to take in the morning'.

Shortly after that, on 4 February 1803, another meeting with Aborigines took place in Caledon Bay, with Good writing into his diary: 'one of them suddenly snatched a Musket from Mr Browns servant & run off with it the others all run off at the same time. Mr Bawer fired his Musket at the man who Stole the Musket but it did not take effect & we soon lost sight of them'. Clearly there were different views held by the aborigines and the westerners on what is meant by property. On the following day, 5 February 1803, Brown penned down a list of 'native words for plants and parts of the body', a feast for the linguists studying the numerous languages spoken along the coast of the Gulf of Carpentaria. Less an intellectual, Good noted for the 7 February 1803: 'In morning the Natives came to the Tents bringing with them some fruits, honey & some little roots probably of a species of Grass which eat very pleasant two of them were taken & confined. one was soon liberated & given to understand it was expected he would return & bring with him the Stolen hatchet which would procure liberty to the other . . . In the afternoon Captain went on Shore & took the Native into the Boat & Sailed to the North of the Creek near the Natives. when Several of them came to the Beach & made Signs for our people to land. and to induce them presenting a young woman, called Bungery & gave him to understand he might have her if he would land'.

The spectrum of notes taken on 14 February 1803 on Gove Peninsula in what is now Northern Territory is typical for the scientific assistants hired by Banks, who indeed formed a team of highly competent naturalists. Good noted, 'In the morning landed with Mr Brown & party & collected some plants Captain employed Surveying the harbour. he as [sic] also Mr Bawer & Westall who were with him found some Quartz Crystals they were found in Caverns in Ironstone & Granite'. These quartz crystals, which could have been in the hands of Ferdinand, seem to have ended up in the Natural History Museum, (Fig. 10.11.). Brown also records the geology of the area, the scientific names of two vascular plants, plus three birds – Phasianus (possibly the coucal, *Centropus phasianinus*), white pigeons (presumably pied imperial-pigean, *Ducula bicolor*) and Psittacus splendens, the latter certainly a specimen of the northern rosella, *Platycercus venustus venustus*. The latter is a parrot found from the Gulf of Carpentaria through Arnhem Land to the Kimberleys. Needless to say, it belonged to another species new to science.

Meeting Malay fishermen in the English Company Islands off Gove Peninsula brought interesting information, although – according to Brown's diary – they denied 'having any of their celebrated Poison, wch they call 'Ippoo, on board'. As a matter of fact the latex of *Antiaris toxicaria* contains intensely toxic cardenolides,

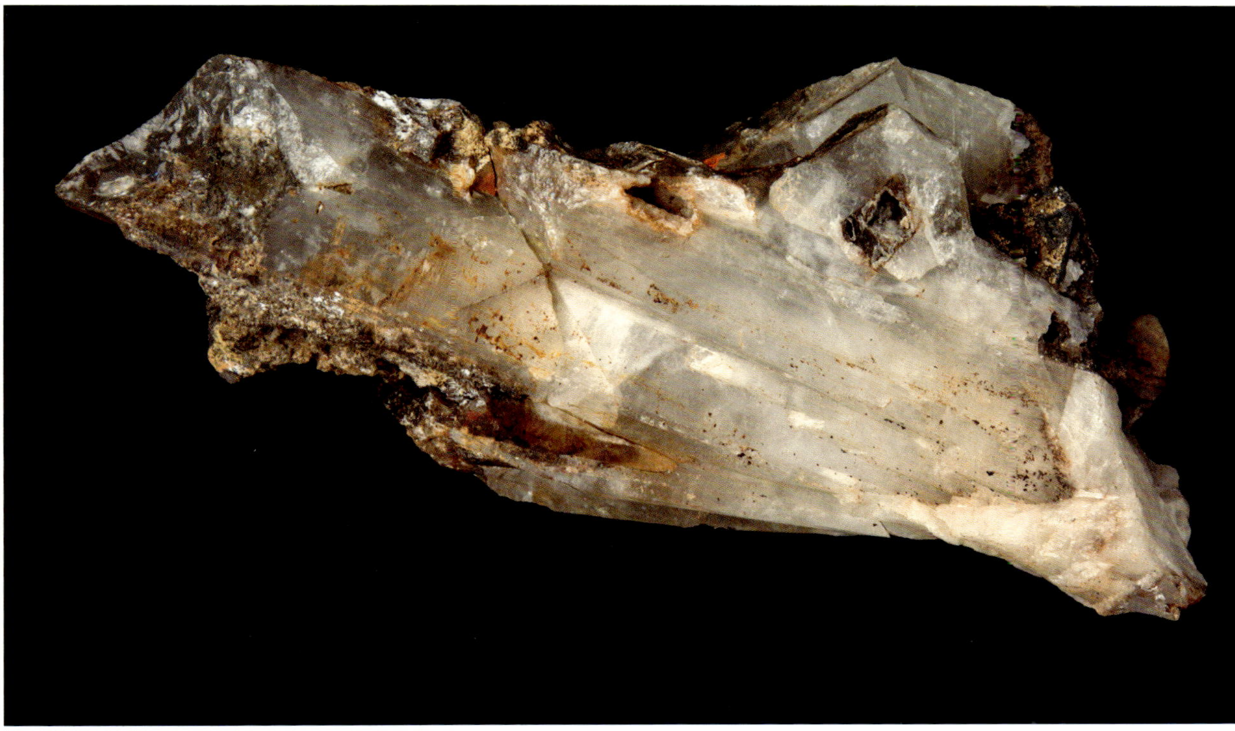

FIG. 10.11. Quartz crystal collected by Ferdinand Bauer, Matthew Flinders and William Westall, 1803. Australia, Gove Peninsula. – London, Natural History Museum, Earth Sciences. Call no. 75819.

used in arrow poison, generally called upas. Ferdinand prepared a pencil drawing of this upas tree, but we do not know where exactly the specimen had been collected. When this image was published in the *Voyage*, Brown gave the shores of the Company's Islands as provenance, which indeed is a plausible locality.

By now everyone on board must have been deeply worried about the *Investigator*. Brown, always cool-headed, noted on 5 March 1803: 'it was intended from the very crazy state of the vessel to leave the coast'. This effectively meant ending the survey of the northern coast of Australia at Arnhem Bay. However, they were still in totally uncharted and dangerous waters, with Good writing on 19 March 1803 in his diary, 'About 8 A. M. we were surprised to see the bottom as we had for several days had from 100 to 200 fathoms Soundings on Sounding we found 10 fathoms . . . hot sultry weather with light airs and Calms'. Another reality becomes evident in Brown's diary entry a week later: 'in consequence of incipient symptoms of scurvey [sic] in many of the crew & our Peas [provisions] being exhausted it was determined to put into Timor'. Captain, ship, crew and its company found themselves in a difficult situation.

FROM TIMOR TO SYDNEY COVE

As a consequence the leaky *Investigator* headed for Kupang (now Indonesia), the main harbour in the western part of the island of Timor, then belonging to the Batavian Republic. Although the British had unsuccessfully attempted to overrun the town six years before, causing heavy damage, Flinders, his crew and the scientific assistants were well received – maybe in an act of pity. Kupang had also been the final destination of William Bligh after the mutiny on the *Bounty* in 1797, having navigated an overcrowded seven-metre open launch on an epic forty-one-day voyage from Tofua in the Tonga Islands over a distance of 6710 kilometres.

On 30 March 1803, 'in sight of Timor' Brown wrote letters to Banks and Greville, which contain no reference to Ferdinand. A few days later the latter did the same. Although dated 'Timor 8 April 1803', the day of the departure of the *Investigator*, Ferdinand must have written his report to Franz at Kew earlier describing his past experiences (in German): 'We had high expectations of the Gulf of Carpentaria but found that the whole of the side facing east is very low-lying country with very shallow water for a long distance so that our ship could not come close . . . At the end of the Gulf towards the Strait you can, for the most part, only see the land from the mast of the ship. Turning to the west we came to many islands. The coast also begins to rise more steeply ... Amongst the islands we put down anchor several times in order to load water... During this time the ship was again examined but it was found that the decay of the ship's timber has not increased much since the last examination, whereupon Cap. Flinders decided to go to Timor for another cruise, he would like to finish the north and west coast of New Holland before going to Port Jackson [Sydney]. We will know this with more certainty when we reach Timor'. On a more personal note Ferdinand added, 'I have, during all this time, been in good health and spirits although one does suffer a good deal from the heat in this latitude. The ship's company was also fortunate in this regard and had few sick men but now, just recently, many show evidence of scurvy . . . I have, since we left Port Jackson, made sketches of 500 species of plants but only 90 of animals, mostly birds. I have not completed anything and will not be able to do either . . . I am very much looking forward to finding letters from you and hearing some news when we reach Sydney'. There are two postscripts to Ferdinand's letter, one of them referring to the Dutch Opperhoofden and resident at Timor, Johannes Giesler[26] 'who is a German' and the second containing the precognition 'I will leave it [writing a letter to Banks] till we reach Port Jackson and I can see what is going to happen with the expedition once the ship Investigator had been condemned'.

On 1 April 1803 Flinders accompanied by Brown waited on the Governor, who permitted botanizing, offered a European guide and, more importantly, agreed to have supplies delivered to the ship – as worded by Brown in his diary: 'oranges, limes, Shaddocks [*Citrus maxima*], cocoanuts, cucumbers & water melons . . . Carbaw [Carabao] & pigs & sheep'. He commented on the dinner given on that day by Giesler: 'our dinner was not splendid but to us who had been living on salt beef & putrid water, was excellent'. The stay on Timor lasting a single week brought the only encounter with the Malesian flora to Ferdinand, Brown and Good. Since a considerable part of their collections were common, widespread and well-known plants, they received little attention. Only two of Ferdinand's pencil drawings from Timor were later published – both posthumously in London (see Chapter 18). Even before the publication of the two Timor plants, an image of one of them was – hard to believe but true – included in one of the most famous flower pieces ever – named 'Jacquins Denkmal' and kept in Vienna in the Österreichische Galerie Belvedere (see Chapter 18).

When the *Investigator* left Timor on 8 April 1803, she headed in a southerly direction back to the point where the circumnavigation of Australia had started – King George III Bay. No one anticipated that they had taken a form of tropical dysentery on board, probably with fresh supplies from Kupang. This caused a catastrophe similar to that on board the *Endeavour*, which had taken in stores at Batavia when sailing to the Cape of Good Hope in 1771. In Brown's diary, now with several gaps, we find a first note added to the day of departure – 'Symptoms of Dysenter[y] am[o]n[g] the crew'. On 20 April 1803 Good confirmed this by writing 'with much Thunder & lightning & very dull moist atmosphere many of the Crew Complaining of disorders & dysenterie attacked several which seemed to increase'. On 15 May 1803 the *Investigato*r rounded the south-western tip of Australia. Two days later in the Recherche Archipelago Good had to confide to his diary, 'past Bay No I keeping to the South of all the Islands... departed this life Charles Douglas Boatswain of a Dysentery with which he had laboured since the middle of Aprile – Self and several of the Crew labouring under the Same disorder'. The subsequent journey along the south coast of Australia, recording only in the *Voyage*, turned out to be much more difficult than it had been seventeen months before. According to Flinders in the *Voyage* high surf made for a difficult landing on Middle Island, where Douglas was buried and William Hillier died. A hole dug in a gully yielded undrinkable saline water, only a few geese were found and the surgeon had another fourteen on the sick list. Unimpressed by the tragedy unfolding itself, Ferdinand continued to prepare pencil drawings – one of them annotated '18 May 1803 Salt Island', showed *Alyogyne hakeifolia*, an erect shrub, up to three metres high with large flowers variously blue, purple or creamy yellow, with Ferdinand recording a specimen with violet petals. The denomination 'Salt Island', certainly a synonym for Middle Island, may have been created by Ferdinand in reference to the saline water found on this very island. It was near the coast of this island that Flinders decided that it was necessary to lighten the ship by jettisoning two wrought-iron anchors. They have been found, recovered by divers and are now in the National Museum of Australia in Aton, Canberra and in the South Australian Maritime Museum in Adelaide. Flinders had to give up all hopes of continuing his charting of the islands to the north-west of Tasmania. Sailing day and night, he pressed back to Sydney Cove, with more deaths en route, among them John Draper, the quartermaster. Upon arrival on 9 June 1803, the first thing to be done was to place the sick from the *Investigator* in the care of 'the principal surgeon of the colony'.

For the following days the *Voyage* is the sole source of information. Four men are reported on 10 June 1803 'too much exhausted' to be moved from the ship and 'died in a few days . . . The first of them was Peter Good . . . a zealous, worthy man, who was regretted by all'. The *Sydney Gazette* reported 'On Monday last [13 June 1803], Mr Good, Botanist . . . was brought on shore for interment. A number of Officers attended in procession to the place of burial, where after the funeral ceremonies were performed, a party of marines fired three vollies over the grave'. Ferdinand may have attended; he was fortunate enough not to be afflicted by the tropical dysentery that had caused such heavy losses, but he too must have been tired of the long voyage. An inspection of the *Investigator* revealed a deplorable state of preservation, with the commission reporting to the Governor that 'we think it altogether unnecessary to make any further examination; being unanimously of opinion that she is not worth repairing in any country, and that it is impossible in this country to put her in a state for going to sea'. As a consequence the *Investigator* was condemned.

On 25 June 1803 Flinders wrote a long letter to his wife Ann, which was not finished and posted until more than a year later. However, the text written in New South Wales, possibly at the Hawkesbury settlement, contains material of considerable interest since he frankly characterises his associates. In this draft, now in the State Library of New South Wales in Sydney, Flinders is even franker. Concerning Brown, Flinders is formal in the final version, 'Mr. Brown is recovered from ill health and lameness', but in the draft he is more open 'we are not altogether cordial, but our mutual anxiety to forward far the complete success of the voyage is a bond of union; he is a man of abilities and knowledge, but wants feeling kindness'. Flinders calls Ferdinand 'your [Ann's] favourite' and continues by stating that he 'is still polite and gentle', this being the only comment on his personality known to exist. In the draft Flinders added 'and is so to a considerable depth, but I fear there is a dreadful disposition at the bottom'. This note in Flinders's draft, which was subsequently, however, not used by him in the final version, remains a total enigma. Ferdinand definitely was not flawless, but had good reason to feel dissatisfied because of the more relaxed life his three elder brothers were able to lead during their whole lives – Johann in Karlsberg (Karlovec; see Chapter 12), Joseph in Vienna (see Chapter 12) and Franz in Kew. All three were in permanent positions, Franz being rather nicely paid, while he, Ferdinand, the youngest of them, had been engaged just for the *Investigator* voyage with its harsh travelling conditions and rather uncertain long-term prospects after its conclusion. In addition, his two eldest brothers had found their jobs at home in a world familiar to them paid by the Reigning Princes of Liechtenstein, whereas he had to live among, after all, foreigners on the other end of the globe. All this may well have embittered Ferdinand. However, what Flinders really wished to hint at, remains obscure. It seems, at least on the surface, that Ferdinand had no enemies among the crew and the scientific assistants and that he was well liked. Flinders added that Ferdinand and Westall 'have always remained on good terms with me'.

With his crew further decimated at Sydney Cove and his vessel decommissioned, Flinders found himself in a difficult situation as to how to finish his task to survey the remaining west coast of Australia. In the end he decided to sail back in person to England accepting Governor King's offer of the *Porpoise* for transport – in order to organize a ship for the completion of the expedition. What were the options for the four remaining scientific assistants? They could either join Flinders, which is what eventually Allen and Westall decided to do, or remain for the time being in Australia, which is what Ferdinand and Brown applied for. They wrote to Flinders, 'we . . . think it our duty to state to you that in our opinion we would be much more advantageously employ'd [in the colony] than by going Home now, until your return in the prosecution of the voyage, or 'till we received instructions from the Lords commissioners of the Admiralty relative to the future proceedings'. On 17 July 1803 the permission was granted by Flinders in writing, adding 'I will leave eighteen months provisions for yourselves and servants', while on the same day the *Sydney Gazette* reported that Flinders had had mooring chains laid in Sydney Cove for the purpose of securing the *Investigator* 'as a Hulk'. Although still employed by the Admiralty, Ferdinand and Brown and their servants were now on their own. From a footnote in the records it appears that Gouvernor King made a house available for the two scientific assistants remaining behind, but no further details are available.

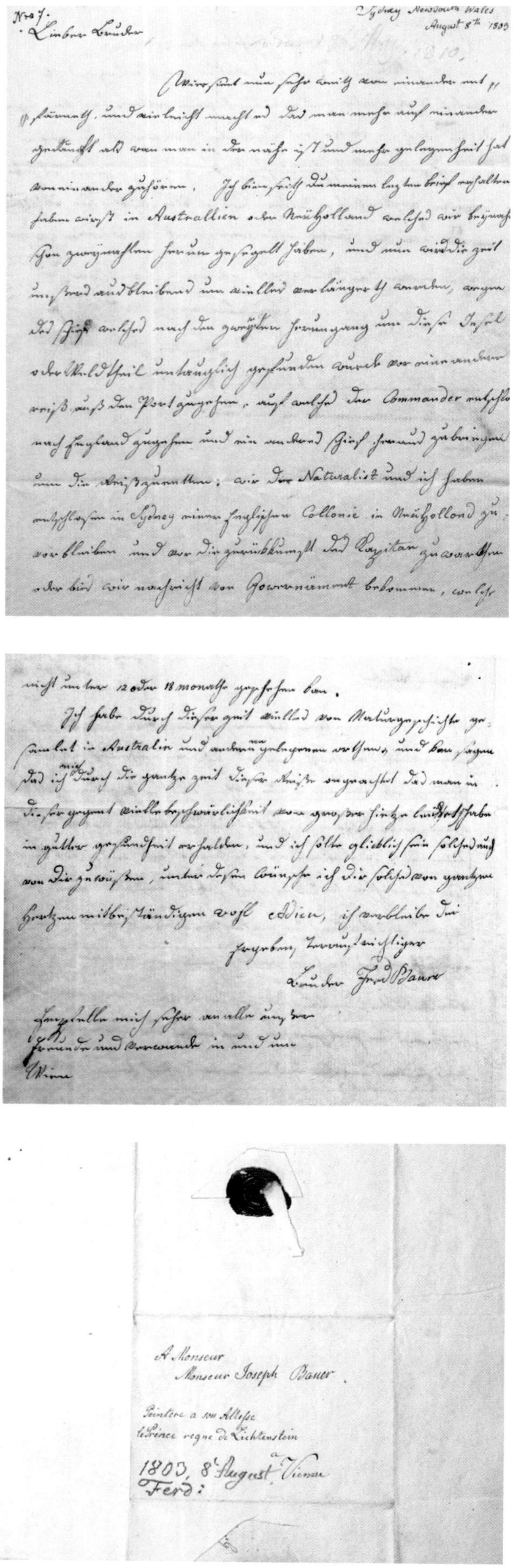

No 7.
Liebe Brüder
Sydney New South Wales
August 8th 1803

A Monsieur
Monsieur Joseph Bauer
Peintre a son Altesse
le Prince regne de Lichtenstein

1803, 8t August Vienne
Ferd:

Brown must have felt it time for another report to be sent to Banks. In his letter of 6 August 1803 he summarizes the achievements of the expedition, but in our context only the two sentences referring to Ferdinand are relevant: 'Mr. Bauer, who writes to you by the Porpoise, will give you an account of what he has done. I can only say that he has been indefatigable, and that considering his minute accuracy, the number of drawings he has made is astonishing'. Two days before the *Porpoise* sailed for England with Flinders as a passenger Ferdinand sat down to write two letters, the first to Banks, obsequiously beginning 'I am sorry and beg you will excuse that I have so long omitted to give you an account of my proceeding in the time of the Investigator Voyage to which you have been bleased to recommend me'. The following lines summarize Ferdinand's understanding of his role in the expedition and are therefore quoted in full, and in the original spelling: 'In any opinion I resolved that in such an expedition it will be the best by every opportunitis to preserve as many subjects of Natural history in sketches as shall be in my power to execute; for fear to loss some which might be new or rare before they are ascertained; in this manner I have prosid & now have collected from New Holland in sketches of plants above one Thousand and of Animals two Hunderd, from this can be select the new or best which may be at any time finished & I am sorry that I have not any finished Drawings to send by this opportunity for in the time of the voyage I was most occupied with the sketches until a little befor our arreival in Part Jackson & after Capt Flinders determination of going to England, things have unsettled that little mor could have been done'. This is convincing reasoning – in an age when the permanent preservation of many natural history objects such as fishes or birds was, at best, unsafe it was the image that counted. In stressing the 'fear of loss' Ferdinand is extremely modern in approach. Only a year later, on Norfolk Island, Ferdinand would record and collect a few species extinct by now. In his letter to Banks he continues, 'This unexpected determination of Cap Flinders; requested much consideration, to take such part which maight be to your department advantageous . . . our stay in New South Wales would add much to the collections and if not new objects could be procurt I would be able to finish some from them wath I have

FIG. 10.12. Letter by Ferdinand Bauer to his brother Joseph, Sydney, 8 August 1803. Paper, pen. – London, Linnean Society, Archives, MS 634, No. 7.

already made, and must bey done in England', ending on a servile tone, 'therefor our stay in New South Wales would be mor to the interest for our engagemend & has occasiond my resolution to remain here, in expectation I hope it will be approved of'.

The second letter written on 8 August 1803 (Fig. 10.12.) is the only extant proof of Ferdinand's contact with his brother Joseph in Vienna, calling him very appropriately 'Peintere [sic] a son Altesse le Prince regne de Liechtenstein' (see Chapter 12). It is much shorter and remarkable in other respects. There is some personal warmth not found in his few other letters, starting (in German), 'We are now far away from each other and perhaps that makes one think more about each other than when one is close and has more opportunity to hear from each other'. More interesting for etymology is the following sentence 'Since you received my last letter I have been in Australia or New Holland . . .', reading in the original 'Ich bin seith du meinen lezten brief erhalten haben wirst in Australlien oder NeüHolland. . . '.

The English word 'Australia' in its modern circumscription had already been used in print in George Shaw's *Zoology of New Holland*, published in London in 1794, containing the famous sentence 'The vast Island or rather Continent of Australia, Australasia, or New Holland . . . seems to abound in scenes of particular wildness and sterility'. Although it is generally assumed that Flinders popularized the name Australia, this is not so – in his *Voyage*, published as late as 1814, persuaded by Banks, he used the term 'Terra Australis', adding in a footnote 'Had I permitted myself any innovation on the original term, it would have been to convert it to Australia; as being more agreeable to the ear, and an assimilation to the names of the other great portions of the earth'. Nothing proves that the *Voyage* was instrumental in the general adoption of the name Australia, it rather came in use gradually over the next decade. Ferdinand must have heard the word 'Australia' on board the *Investigator* and transmitted the new term to Joseph – with no effect. The letter survived the wreck of the *Porpoise*, when the living plant material previously under the care of Good was lost. The letter was transported back to Sydney Cove, was taken on board the *Cumberland* and was detained together with Flinders in Mauritius. It arrived in Vienna on 13 May 1810, almost seven years after it had been written. In short, Ferdinand was definitely the first to ever use the German toponym 'Australlien'. Even more significant is the fact that Ferdinand's manuscript 'Australlien' predates Flinders's manuscript name 'Australia'. Detained in Mauritius Flinders told Banks in a letter dated 23 August 1804, re-dated 4 November 1804, 'I send you . . . a copy of my general chart of New Holland . . . The propriety of the name Australia or Terra Australis, which I have applied to the whole body of what has generally been called New Holland must be submitted to the approbation of the Admiralty and the learned in geography'. Two days later, on 25 August 1804, Flinders told his brother Samuel briefly and concisely 'I call the whole island Australia or Terra Australis' – this appellation was, to be precise, one year after Ferdinand had done so. Needless to say Ferdinand's letters are a feast for the linguist – he uses German phraseology in his English and English phraseology in his German, plus a spelling determined by Bavarian-Austrian phonetics in his English prose, for example 'Hunderd' (Hundert in German) instead of 'hundred'.

PORT JACKSON AND THE HINTERLAND

What followed for Ferdinand was a stay of thirteen months in Sydney and its surroundings – first together with Brown, who left on 28 November 1803 for a voyage to Tasmania, Port Dalrymple and Port Phillip, and then alone until his departure for Norfolk Island on 21 August 1804.[27] The original sources for this period are more meagre than those for the circumnavigation: Flinders had left Australia for England but got imprisoned on the island of Mauritius, then under French control; Good was dead; Brown's diary is incomplete; and Ferdinand kept no notes on his movements. Ferdinand the travelling illustrator transmuted into Ferdinand the sedentary illustrator who botanized in the surroundings of Sydney Cove for the following months. Again his pencil drawings are the most important guideline of his actions, and that annotated 'from the South of Port Jackson, Aug. 15. 1803', measuring 74.5 cm in length, and 51.8 cm in width (Fig. 10.13.), became later the basis for his most famous watercolour, of which reproductions are found to today on all continents.

In this case Ferdinand pre-empted the press, the *Sydney Gazette* writing on 21 August 1803, 'An Animal whose species was never before found in the Colony, is in His Excellency's possession. When taken it had two Pups, one of them died a few days since. This creature

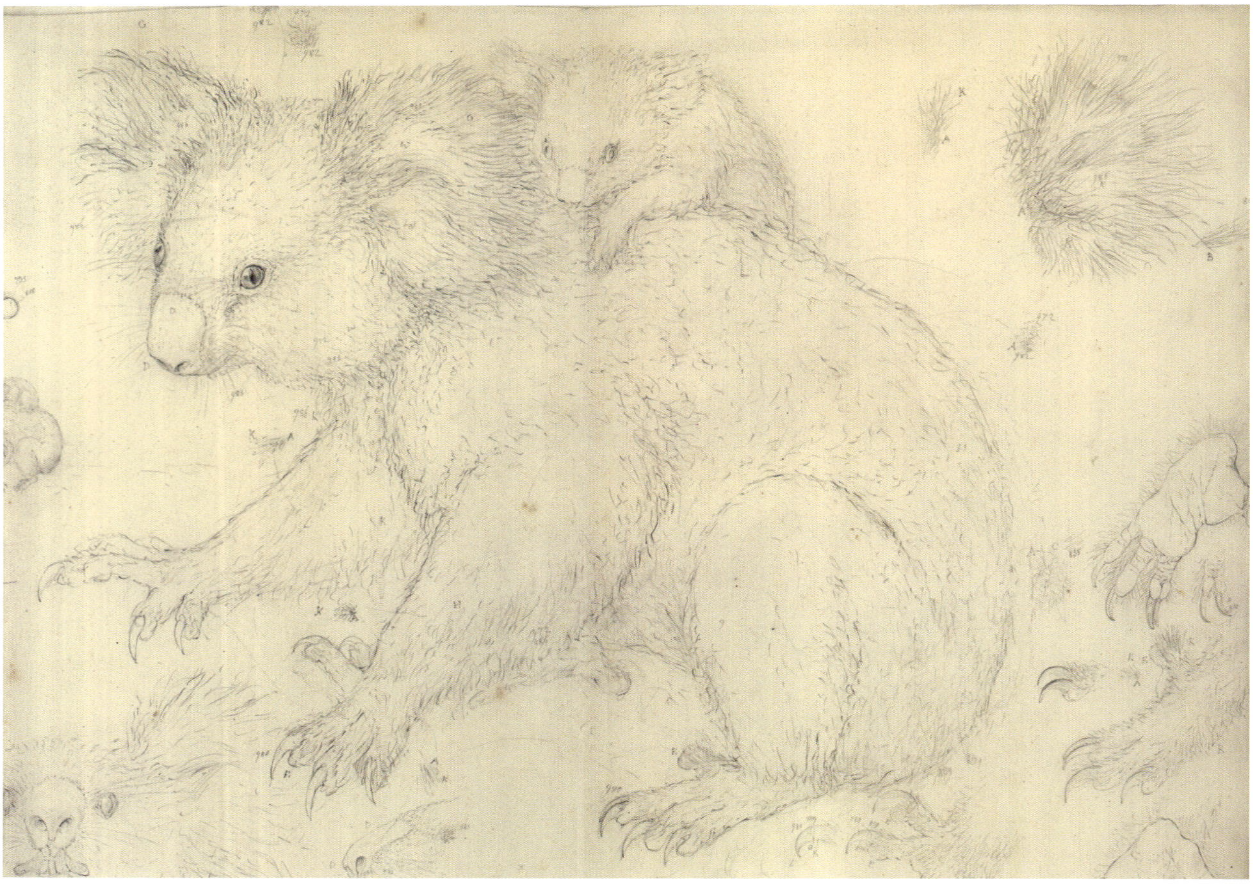

FIG. 10.13. Koala, *Phascolarctos cinereus*. Pencil drawing by Ferdinand Bauer, 1803. Paper, pencil. – Vienna, Naturhistorisches Museum, Archiv.

is somewhat larger than the Waumbut [Wombat], and although it might at first appearance be thought much to resemble it, nevertheless differs from that animal. The fore and hind legs are about of an equal length, having five sharp talons at each of the extremities, with which it must have climbed the highest trees with much facility. The fur that covers it is so soft and fine, and of a mixed grey colour; the ears are short and open; the graveness of the visage, which differs little in colour from the back, would seem to indicate a more than ordinary portion of animal sagacity; and the teeth resemble those of a rabbit. The surviving pup generally clings to the back of the mother, or is caressed with a serenity that appears peculiarly characteristic . . . and its food consists solely of gum [*Eucalyptus*] leaves, in the choice of which it is exceedingly nice'. This is regarded to be the first description of an animal appearing in print that had been seen earlier in the bush: the koala, *Phascolarctos cinereus*. The text probably came from the pen of William Paterson, then Lieutenant-Governor of New South Wales and a Fellow of the Royal Society. Considering the arrival of the First Fleet in 1788 (see Chapter 6), this is a rather late start for such a spectacular animal. But as has been rightly said 'curled high in the tall eucalyptus trees, eating tender leaves, sleeping largely by day and moving at night', the koala remained unknown, out of sight to European settlers, inconspicuous and simply overlooked.[28]

In September 1803 Brown reported to Banks about a 'new and remarkable species of Didelphis [that] has been lately brought in from the southward of Botany Bay. It is called by the natives coloo or coola', the name 'Koolah' first appearing in print in the *Sydney Gazette* of 9 October 1803. Brown added 'The Governor . . . sends a drawing made by Mr. [John William] Lewin. Mr. Bauer cannot on so short a notice finish the more accurate one he has taken. The necessity of sending my description, which is very imperfect, as the animal will not submit to be closely inspected, and I have had no opportunity to dissect one, is in a great measure superseded by Mr. Truman having purchas'd a pair, which from their present healthy appearance, will probably reach England alive, or if not, will be preserv'd for anatomical examination'. Years later and in Vienna – not in London, as generally assumed –

Ferdinand was to produce a total of five watercolours of this animal, today iconic for Australia (see Chapter 11). The first shows a koala with a young clinging to its back, the second one a koala sitting and another crawling, the third a koala sitting in the axil and another clinging to a branch in such a life-like posture that we can assume that Ferdinand had observed the two individuals in their native habitat. The fourth watercolour documents the head of the animal in two views.

Although Ferdinand's pencil drawing was not the first graphical record of the koala, it was definitely the first scientific figure of this animal, highly superior to the representations of Lewin, of which three variants are known - in the British Museum, the Natural History Museum and the State Library of New South Wales in Sydney. By contrast, Ferdinand's pencil drawing is one of the many treasures of the Naturhistorisches Museum. Recently a proposal has even been made to select Ferdinand's drawing as the type of the name *Phascolarctos cinereus.*[29] A specimen of the koala must have been later dissected, and Ferdinand recorded a front and a side view of the skull as well as several teeth, the paws and feet, which later became the basis for his fifth watercolour documenting the koala. The provenance of this specimen and its current whereabouts remain unclear. The possibility cannot be excluded that the skeleton was later sent by Paterson to Home in London (see Chapter 15) for further study. In his letter of 10 March 1804 the former told Banks, 'Mr Lewin made a very bad drawing of it which the Governor informed me he has sent to you . . . Since that time many [koalas] have been caught brought in and became very domestic . . . I had a male and a female for several months . . . but an unaccountable accident happened to him & he died soon after. The skeleton of the latter I have sent to Mr Home, and both their skins are in a box with some specimens directed to you'. In parallel with Ferdinand preparing the graphical record, possibly on the basis of this very skeleton, Brown penned a description of the male specimen, which remained unpublished. As its provenance Brown noted 'the foothills of the mountains near Hut Hill', which may also have been the provenance of the skeleton documented by Ferdinand.

A spiny anteater or echidna belonging to the Tachyglossidae, a small group of egg-laying mammals covered with coarse hair and spines occurring in Australia and New Guinea, attracted the attention of Ferdinand soon after. He annotated his colour-coded pencil drawing now in the Naturhistorisches Museum 'Stinktier Sydney Oct 13 1803', but did not base a finished watercolour on it. He seems to have documented a specimen exuding the stinking pheromone typically produced by the sexually receptive female, which is what Ferdinand's denomination would suggest. The drawing has never been properly studied by a monotreme specialist and its precise identity remains unknown.

Much less is known about Ferdinand's colour-coded pencil drawing documenting the common ringtail possum, *Pseudocheirus peregrinus*, another marsupial, about 30–35 cm long when fully grown, excluding the tail. It is restricted in distribution to the eastern and south-western coasts of Australia, as well as Tasmania; Ferdinand's pencil drawing is annotated by him 'South Coast' which excludes Port Jackson as a provenance. Being a nocturnal animal well adapted to arboreal life and rarely descending to the ground, it is remarkable that Ferdinand was able to produce a pencil drawing of it. Very much a naturalist, he documents the prehensile tail on which the animal is shown hanging from a branch, a lactating female with its young, and a possum pair copulating..

For Ferdinand and Brown, botany remained the first priority for the following months. Extremely few watercolours in the Natural History Museum, carry locality information, all referring to August–September 1803 and all representing species of the orchid genus *Diuris*. The representation of *D. punctata* is labelled in an unknown hand 'found near Sydney. Port Jackson. New South Wales. Augst. 27. 1803. Ferd. Bauer', that of *D. sulphurea* has 'found near Sydney. Port Jackson. New South Wales. Sepr 1803 Ferd. Bauer', and *D. aurea* is annotated 'found near Sydney Port Jackson New South Wales. Sepr. 25. 1803. Ferd. Bauer'. Some of the evidence on the botanical excursions by Ferdinand, partly undertaken in conjunction with Brown during this period, is clearly contradictory, but there are also straightforward examples. One of them is Brown's diary entry, 'George's River . . . Doryanthes excelsa'. This entry refers to the gymea lily, *Doryanthes excelsa*, a spectacular plant sending up a flower spike up to six metres high. The record is corroborated by a specimen collected 'Near Georges River', almost certainly by Brown and Ferdinand, now in the Natural History Museum, and by Ferdinand's colour-coded pencil-drawings, now lost. On their basis he must have produced two watercolours, now also lost. However, we know of their existence because Ferdinand published them in his *Illustrationes* (see Chapter 11). The use of the scientific name given above in Brown's diary is a surprise, because it had not been published before the departure of the *Investigator*. However, Brown may have read the name in an unpublished manuscript by Correa and even seen the respective specimens before leaving England. Joseph Correa de Serra, a Portuguese expat, Fellow of the Royal Society and one of Banks's cronies, had described the plant on the basis of: (1) a herbarium specimen sent by well-known navigator George Bass, immortalized in the denomination Bass Strait between the Australian mainland and Tasmania; (2) a specimen preserved in spirit sent by John Hunter, King's predecessor as Governor of New South Wales; and (3) a single flower from a plant cultivated in the Royal Garden at Kew 'from a portion of a stem without roots'.[30] Correa had published his paper accompanied by two copper engravings in the *Transactions of the Linnean Society* in May 1802 when Ferdinand, Brown and Good had just arrived at Sydney for the first time. Plant taxonomy was highly competitive not only for plants from the Levant.

Another clear-cut case is the entry 'towards South Head, Drosera pumila' in Brown's diary for 3 November 1803. This is another name then not validated and later used for a honeydew from tropical America, whereas the Australian plant to which Brown referred was described by Swiss botanist Augustin-Pyramus de Candolle as *Drosera pygmaea*. Its basis is a specimen sent by George Caley, a plant collector engaged and paid for by Banks who collaborated with Brown and Bauer in the field. Ferdinand must have prepared a pencil drawing of the tiny, carnivorous plant with rosettes of only 2 cm in diameter, which is native along the south-eastern coast of the Australian mainland, in Tasmania and on the western coast of New Zealand. Later it formed the basis of another of his finished watercolours. Sundews were nothing new to Ferdinand – three central European species had already been illustrated by the Bauer brothers in the *Codex Liechtenstein*. The large veined sun orchid, *Thelymitra venosa*, is another, unusually well-documented case. The watercolour bears the annotation 'found on the banks of a lake in the way going from Sydney to Botany Bay . . . Novr 12. 1803 F. Bauer', a somewhat obscure locality south of Sydney Cove.

With Brown leaving Sydney Cove on 27 November 1803 on board the *Lady Nelson*, the record of Ferdinand's further activities did not change significantly. On 8 December 1803 he recorded in Sydney a fruiting specimen of a shrub he had already portrayed the year before in flowering state at the same locality: *Notelaea ovata*, belonging to the Olive family. Ferdinand had to be economic with his paper – for the pencil drawing of the fruiting specimen in the Naturhistorisches Museum he used the verso of the drawing of the flowering specimen. A herbarium specimen of *Notelaea ovata* in early flowering stage collected by Brown and annotated by him is kept in the Natural History Museum, and testifies to the relationship between the material and pictorial record kept in the two institutions. On Christmas Eve 1803, a specimen of nardoo, *Marsilea drummondii*, a common and widespread aquatic fern, was recorded by Ferdinand in a colour-coded pencil drawing annotated by him 'Richmount' (Richmond), a locality in the hinterland of Sydney. This plant can form mats on the water surface and cover the ground in carpets when the floods recede. The Naturhistorisches Museum keeps many more similar proofs of Ferdinand's continuing work in recording plant diversity while Brown was in Tasmania.

One further document for this period is Ferdinand's letter written on 7 March 1804 from 'Farm Cove Sydney' to Franz at Kew reproaching, him for his long silence and telling him (in German) 'A long time has passed and there must have been opportunities to write to me but I have not received one letter from you'. Ferdinand seems to be somewhat tired of Australia, continuing 'now I have come to the point of wishing very much that he [Flinders] would soon come out with another ship and end this voyage'. Reminiscing about what had been achieved, he seems not to be unhappy: 'I have kept good health since our transportation [the return to Sydney Cove] and have increased my collection by several hundred [specimens] during this time. But now the plants are beginning to be rarer and one has to go further to find something new and anyway I think New Holland is not as rich in plants as I thought. I also went several times some distance inland and to the Blue Mountains as far as my feet would let me but did not manage to go a great distance. As far as I can see and hear no one has got very far. After our arrival in Sydney when we left the ship, Mr. Brown and I rented a house in Farm Cove a mile from Sydney where I live alone and undisturbed'. Wisely Ferdinand asked his brother to transmit his greetings to his friends as well as past commissioners – among them we find Hawkins and Lambert, as well as some as yet unidentified names, possibly associated with the Imperial Representation in London or the community of German-speaking expats.

It is the postscript that contains a reminder, with Ferdinand writing 'In the letters which Capt. Flinders brought you I asked you to send me half a ream of cartridge and half a ream of large brown wrapping paper when he came [sic] out again to conclude our voyage. I beg you not to forget it'. In short, Ferdinand seems to have run out of paper at Sydney Cove. This echoes Brown's earlier complaint to Banks, to whom he had written on 6 August 1803 'My paper is almost exhausted. What remains will, I am afraid, be scarcely sufficient for all my purposes while here. For a supply of this . . . I take the liberty – certainly an unwarrantable one – of applying to you, fearing my agents, or, indeed, anyone I could request to procure them, might neglect them altogether'. Needless to say, different paper qualities were needed. We are ignorant as to what came out of the two requests; in any case Ferdinand was faced with the same problem he had encountered on his tour to the Levant with Sibthorp.

By now Ferdinand had worked for more than three months completely alone. As in his previous travels with Sibthorp, he definitely needed neither surveillance nor advice from Brown. Only an in-depth analysis of the pencil drawings in the Naturhistorisches Museum will reveal what kind of graphic work Ferdinand undertook during these months. Herbarium material from Farm Cove, the shallow bay in Sydney harbour and very near today's Royal Botanic Garden Sydney, gathered on 23 March 1804 seems to indicate that he continued to collect. One of the more unconventional plants documented by Ferdinand was a weed from tropical America – the Brazilian button flower, *Centratherum punctatum*. The watercolour cannot be associated with a specimen collected by Brown in February 1805 at the Hawkesbury River, as it has been done in the literature, since at that time Ferdinand was still on Norfolk Island. This record proves two facts: the extremely early invasion of the Sydney area by weeds from overseas and the caution necessary in correlating Ferdinand's pictorial evidence with the material evidence from herbarium specimens and attached labels.

The hypothesis has been put forward that Ferdinand may have used his spare time in Australia in 1804–1805 or on his return voyage to paint sixteen finished watercolours on the basis of his colour-coded pencil drawings. These watercolours are kept today in the Natural History Museum, and form the so-called B series, documenting – with the exception of a watercolour of the Australian pitcher plant – only orchids native to Australia. Such an early origin cannot be excluded, but hard evidence is lacking. The B series is not on paper with datable watermarks and there is no documentary evidence, for example in Ferdinand's letters or in Brown's diary. Only an in-depth analysis of the pigments used in the B series, which could be of inferior quality, may help to solve this puzzle. In any event these are quite different from those used in the main, or A, series. This would be an indication of the work being of a different background. The paper of the B series seems to be also of an inferior quality and is often discoloured; however, it will be difficult to prove that this paper, almost certainly manufactured in Britain, could only have been painted on in Australia and not in England. Complexities arise when a more thorough analysis is undertaken to clarify which of the two variants had been deposited in the Admiralty and which remained in the hands of Brown (see Chapter 12).

On 27 or 28 March 1804 Ferdinand left Sydney Cove, the *Sydney Gazette* reporting under the headline 'Expedition to Newcastle', 'On Tuesday His Majesty's Armed Tender Lady Nelson, with Lieutenant [Charles] Menzies . . . and Mr Bawer, Natural History Painter on board, got under weigh.' Nothing is known of why Ferdinand went north, the purpose of the voyage being the resettlement of Coal Harbour – in the words of Governor King 'to Separate the Worst of the Irish [convicts] sent here [to Australia] for Sedition from the others, as well as the great public advantage that Settlement will be of'.[31] Arriving on 30 March 1804, Menzies named the settlement Kingstown, but over the years King's own choice, Newcastle, prevailed. On 19 April 1804 Menzies penned a letter to his superior writing, 'Mr. Bauer will present you with a sketch of this delightful Spot, which I have taken the liberty of naming after your excellency'. This view is reported to have been lost, but a copy, admittedly of inferior quality, survives in the State Library of New South Wales. It is easy to imagine that Ferdinand wished for a change from natural history illustration and seems to have found it in landscapes, echoing his earlier work done for Hawkins in producing landscapes of the Levant. Judging from a specimen collected in Farm Cove on 29 April 1804, Ferdinand was probably back at home by then. Later collections from 'Manangle' (Menangle) or 'White's Farm' near the

Parramatta Road testify to Ferdinand's continued activity in the hinterland of Sydney Cove. In his second letter to Banks written on the *Albion* (see below) Ferdinand was to report 'I am very glad that I remain here in this fine spring season in which many plants have been in flower which not have been seen before, and I found in places wher I have been many times befor this two years after another we have been in same season at Port Jackson. I also made various excursions in to the Country, so far as the Blue Mountings and the Cowpasturs to Mount Hunter, and the Rivers; Grose, Nepean, Hokesbury, Georges River . . . all this places have add and much enlarged our collection'. In this report Ferdinand offers Banks a new, and indeed convincing, reason why he did not prepare finished watercolours – because of the time needed for them 'I would loss many which might be New', and he continues 'I flatter myself when the Investigator Vojage should come to a happy conclution the pense which I have taken to give satisfaction in that pranch of science which was allotted to me will meet with approvation'. The erratic flowering observed by Ferdinand in the hinterland of Sydney Cove is another interesting observation.

Ferdinand, focused as ever on the manifold manifestations of life, also recorded a few fungi. A colour-coded pencil drawing documenting *Camarophyllus lilacinus* and kept in the Naturhistorisches Museum is particularly revealing: it is annotated by Ferdinand 'Silky. Sydney Jun 24 1804' and became the basis of a watercolour prepared by him, now in the Natural History Museum,[32] (see Chapter 11). However, the Naturhistorisches Museum surprisingly also keeps a watercolour based on the pencil drawing – not by Ferdinand, but by Nicolaus Zehner, a lithographer who worked a few decades after Ferdinand's death in Vienna for Heinrich Wilhelm Schott[33] (see Chapter 12). It differs markedly in colour – Ferdinand's watercolour shows a distinctly pink mushroom, Zehner, whose name appears on Ferdinand's pencil in the anonymous note 'Agaricus violaceus in Zehner icon.', a dark violet mushroom. How can this enigma be explained? Possibly the answer is simple: not fully understanding Ferdinand's colour code and misinterpreting the latter's pencil drawing Zehner may have thought it represented the edible, striking dark violet and, to the uninitiated eye at first glance, somewhat similar *Cortinarius violaceus* (*Agaricus violaceus*), a species common in Europe. As a consequence Zehner painted his watercolour in a strikingly dark violet tint. This image is clearly based on Ferdinand's pencil drawing then in Vienna, the former annotated by Zehner in pencil as 'Agaricus violaceus Zehn. Icon. Pauer 567 b' (see also Chapter 18).

When departing for Norfolk Island, Ferdinand left all his materials behind at Sydney Cove. Brown, back from his voyage on 24 August 1804 and missing Ferdinand by just three days, must have inspected the latter's collection, noting in his diary entry of 15 September 1804 'Spent the day with Caley, who showd me his Orchidae [orchids] . . . & several other plants. . . I believe Mr Bauer has a few more species than him'.

NORFOLK ISLAND

Norfolk Island is a true oceanic island, almost exactly midway between New Zealand and New Caledonia, each about six hundred and twenty-five kilometres away.[34] It lies at a distance of 1412 kilometres directly east of the modern town of Evans Head on the Australian mainland. Norfolk Island, today one of Australia's External Territories, consists of Norfolk Island proper, occupying thirty-four square kilometres with the town of Kingston, and Phillip Island, with only four square kilometres situated about six kilometres to the south of the capital. The island has a climate that is both tropical and oceanic and originally supported a rich flora and fauna. When discovered by Cook on his second circumnavigation of the globe in 1774, it was no longer inhabited by Polynesians and was still covered by forests. In 1788 Norfolk Island had become just one month after Port Jackson the first European settlement in this part of the world – with seven freemen and fifteen convicts starting to clear the land and to cultivate New Zealand flax, *Phormium tenax*, and other crops. Two years later a larger party of convicts had been dispatched to Norfolk Island, establishing the first penal settlement.

Sixteen years later and together with a cargo of whale oil and 13,000 seal skins, no doubt still stinking, Ferdinand left Sydney Cove on the *Albion* on 21 August 1804 and arrived on Norfolk Island at the beginning of September 1804. The ship's destination was England; this made Ferdinand write a letter to Banks (see above) on 27 August 1804, giving his motivation for the tour: 'Having heard that the produce of Norfolk Island are so different from this of New Holland, I tuk the opportunity to go with Capt. Bunker of the Ship Albion, the convoyer of this letter who is in his passage to England will leave me there, which will give me a time of two month to stay at the Island, when the Investigators Ship will be finished, and come to remove the settlers with whom I shall return to Port Jackson'. Who could have informed Ferdinand about the Norfolk flora being so different? There are several possibilities, one of them being Governor Philip Gidley King, the very first commandant of the penal settlement, commemorated in the toponyms Phillip Island and Kingston. Again the envisaged plan differed from what would turn out to happen – Ferdinand, almost certainly accompanied by his servant, stayed on Norfolk Island three times longer than intended. This offered him ample opportunity to collect plant and animal specimens and to record them in pencil drawings. The results made him the most important early naturalist on the island, although not the very first. In fact this was Ferdinand's high time as a diligent observer and collector, working independently from Brown or anyone else. Although he did not focus on the plants restricted in distribution to Norfolk Island, he collected and recorded very many of them – out of 44 species and subspecies known to be endemic today he recorded no less than 37. Of these, *Rapanea ralstoniae*, a small tree of the Myrsinaceae family, was described as new to science as late as 1990. Seen against a background of an island flora totalling 523 species of vascular plants, these figures prove better than many words that Ferdinand possessed an exquisite eye – as an illustrator and also as a naturalist. There is neither a diary, nor a report by Ferdinand nor an allusion to the work done on Norfolk Island. Again only the pencil drawings and the specimens testify to his activities.

When approaching the island, Ferdinand must have been staggered – from a distance it appeared dark green, covered with a very tall tree reminiscent of the Norway spruce that he must have known from the alley leading from Feldsberg to Eisgrub or from the woods on the Schneeberg. Coming nearer he will have realized that these were indeed conifers, though belonging to a different genus. Maybe Ferdinand even recalled having already examined such a tree at

FIG. 10.14. Landscape on Norfolk Island. Pencil drawing by Ferdinand Bauer, 1804–1805. Paper, pencil. – London, Natural History Museum, Library and Archives, Landscape Drawings from Norfolk Island, no. 4.

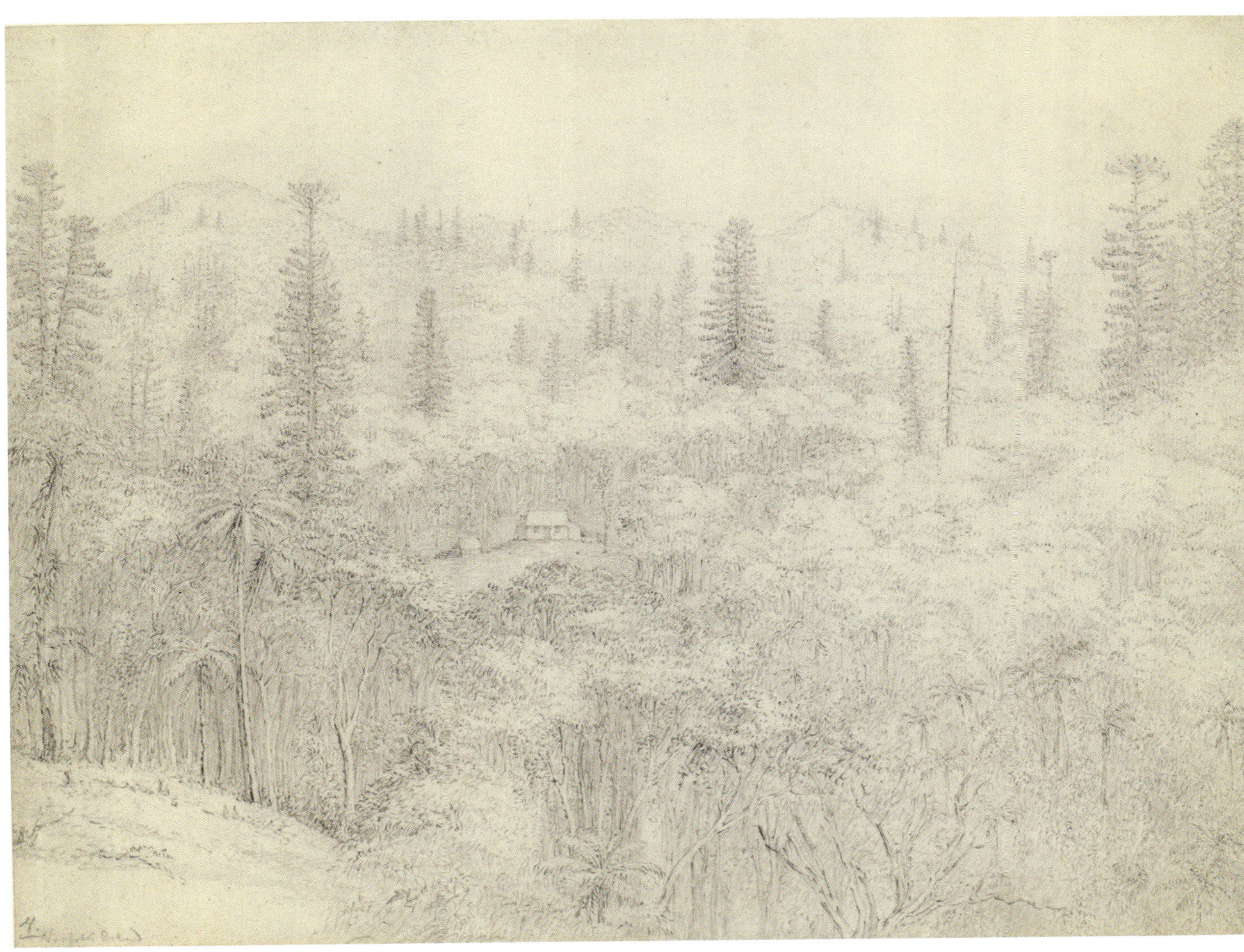

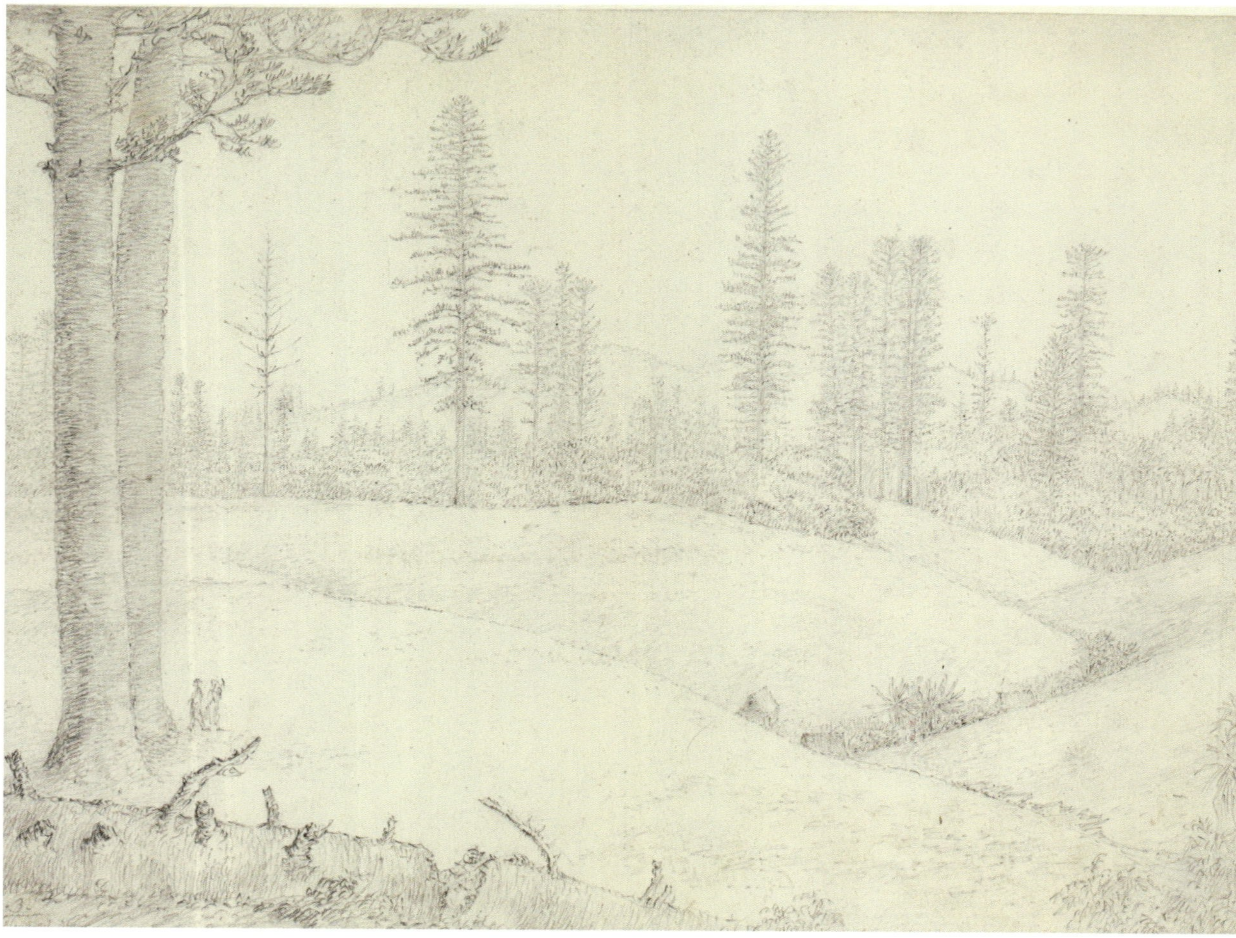

FIG. 10.15. **Forest clearing on Norfolk Island. Pencil drawing by Ferdinand Bauer, 1804–1805. Paper, pencil. – London, Natural History Museum, Library and Archives, Landscape Drawings from Norfolk Island, no. 3.**

FIG. 10.16. ***Cyathea brownii*. Herbarium specimen collected by Ferdinand Bauer on Norfolk Island, 1804. – London, Natural History Museum, Life Sciences, Herbarium.**

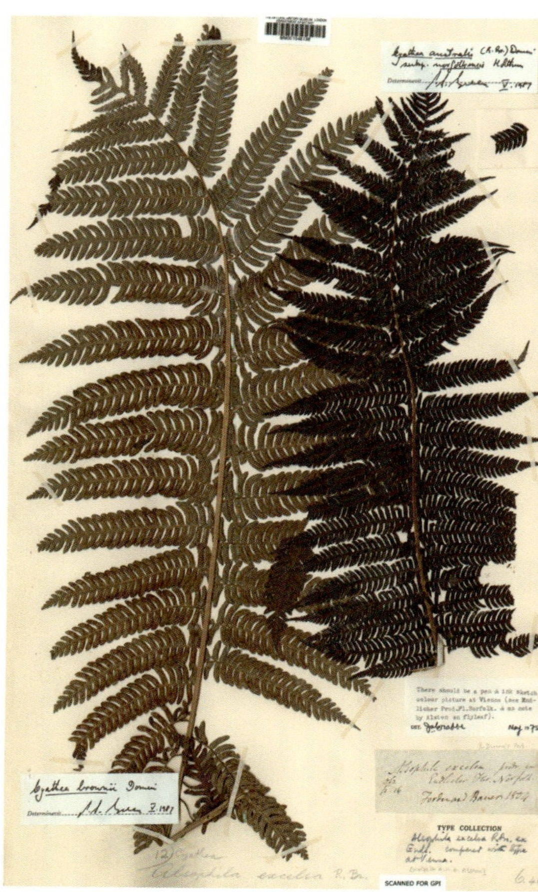

the Royal Garden at Kew – the Norfolk pine, *Araucaria heterophylla*, of which he probably had documented a twig for Lambert before his departure with the *Investigator* (see Chapter 9). It is likely that after landing Ferdinand prepared pencil drawings showing all the details, in particular of the cones, but these must have been lost in the Oberhöflein fire. However, four pencil drawings documenting the landscape of Norfolk Island survive in the Natural History Museum.[35] One of them shows the scenery dominated by Norfolk pines and tree ferns (Fig. 10.14.). The hut and the small building to the lower left indicate that this was not untouched forest; another drawing reveals a forest clearing with a cut-down tree trunk at bottom left, tall Norfolk pines in the background and – very Bauerian – two figures to indicate size (Fig. 10.15.). Among the tree ferns depicted in the pencil landscape the Norfolk tree fern, *Cyathea brownii*, another endemic of the island reported to reach no less than twenty-four metres, could have been one of them. More detailed is a pen-and-ink drawing by Ferdinand in large format kept in the Naturhistorisches Museum: it shows three gigantic specimens of the Norfolk tree fern with a root climber, probably *Freycinetia baueri* (see below), attached to it. Two detailed figures have been added to indicate proportions, maybe much later (see Chapter 18). A frond of the Norfolk tree fern collected by Ferdinand ended up in the Natural History Museum (Fig. 10.16.). Like the Norfolk pine, the Norfolk tree fern is cultivated on all five continents; however, if climatic conditions make cultivation in a conservatory necessary, both are rather impractical – because of the gigantic size they reach over time.

Two Norfolk Island endemics did not survive the impact of humans and have become extinct; both were collected and recorded in pencil drawings by Ferdinand, and both have created an exceedingly rich literature record. The more famous case is the Phillip Island glory pea, *Streblorhiza speciosa*, a liana originally endemic to the tiny Phillip Island; the pencil drawing is annotated 'Phillip [sic] Island near Norfolk Isl. Oct 20 1804' by Ferdinand. Although it had been stated in writing that the famous two specimens in the Naturhistorisches Museum belong to material cultivated in the botanical garden of Vienna University,[36] there is no evidence for this assumption. Instead they must have been gathered by Ferdinand on Phillip Island, as given in the annotation (Fig. 10.17.); a fragment of one of these specimens is in the herbarium of the Royal Botanic Gardens, Kew. Only very recently a small glass container with a few seeds of the Phillip Island glory pea has been located in the herbarium of Vienna University,[37] Ferdinand's colour-coded pencil drawing (Fig. 10.18.) provides additional evidence. Since there was limited interest in plant life of Norfolk Island and its offshore islets, no watercolour was prepared by Ferdinand back in London. The Phillip Island glory pea was definitely in cultivation in England, but has disappeared. An attempt to raise a plant from seeds collected by Ferdinand and conserved in the herbarium of Naturhistorisches Museum is reported to have failed in the 1980s.

FIG. 10.17. *Streblorhiza speciosa*. Herbarium specimen collected by Ferdinand Bauer on Phillip Island, 1804–1805. – Vienna, Naturhistorisches Museum, Botanische Abteilung, Herbarium.

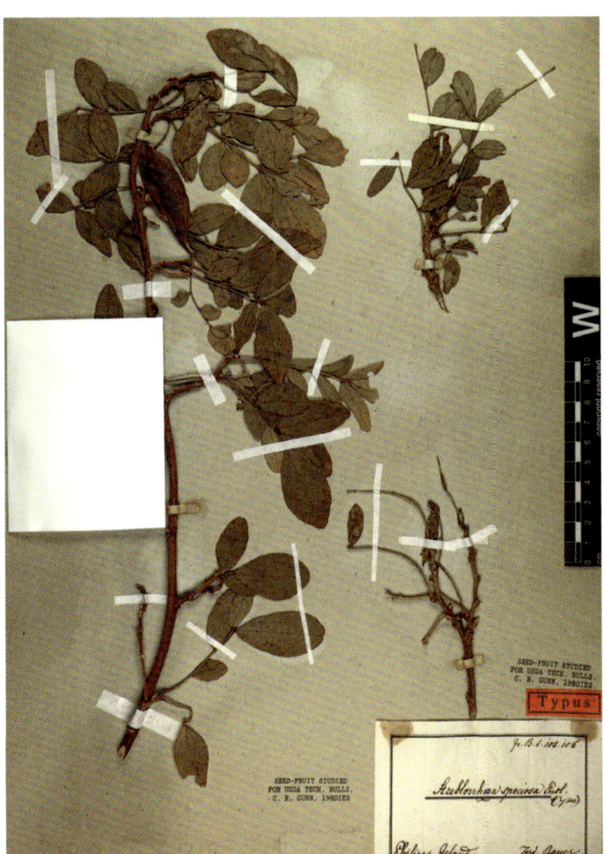

FIG. 10.18. *Streblorhiza speciosa*. Pencil drawing by Ferdinand Bauer, 1804–1805. Paper, pencil. – Vienna, Naturhistorisches Museum, Archiv.

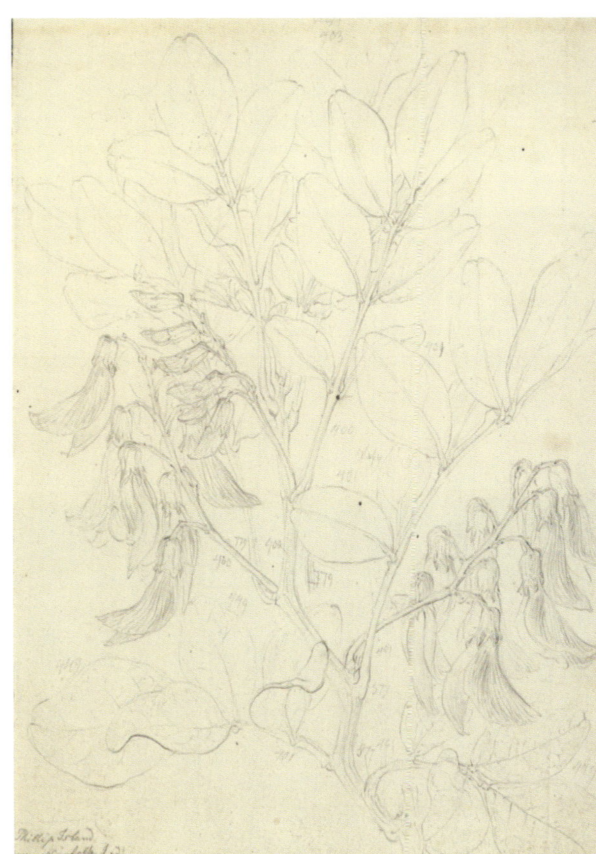

The same applies to *Solanum bauerianum*, restricted in distribution to Norfolk and Lord Howe Islands and apparently extinct. In this case the tree was 'revived': on the basis of Ferdinand's pencil drawings and making use of the reconstructed 'Australian' colour code, a watercolour was produced in c.2006 and published, showing this plant with its flowers and fruits for the first time in colour.[38] Before only herbarium specimens that had long since lost their natural colours had been known and two unconvincing watercolours. One had been prepared for Paterson by John Doody, a convict artist, who was on Norfolk Island from 1791 to 1793; the second by Adam Forster,[39] a botanical illustrator based in Sydney, showing a fruiting specimen dated June 1925 and annotated 'Solanum Bauerianum'. The former is in the State Library of New South Wales, the latter in the National Library of Australia.[40] Other plant species collected by Ferdinand and later described as new to science on the basis of his specimens and drawings fared somewhat better. The Norfolk Island abutilon, *Abutilon julianae*, although diligently searched for on Norfolk Island was thought lost, until it was rediscovered recently on Phillip Island, following the extermination of rabbits. The Phillip Island hibiscus, *Hibiscus insularis*, an endemic of Phillip Island, collected and documented by Ferdinand, is reported to have consisted in 1963 of just eight living specimens, but has survived and is now in cultivation, on Norfolk Island for example. By contrast, other endemics have done much better and are today even frequent, among them *Wickstroemia australis*, again recorded by Ferdinand in a pencil drawing (Fig. 10.19.). On Norfolk and Phillip Islands Ferdinand did not find untouched virgin plant life, but a vegetation cover already affected by feral animals, in particular pigs. When Endlicher (see Chapter 18) annotated the specimen of the Phillip Island hibiscus collected by Ferdinand on Phillip Island and now in the Naturhistorisches Museum, he made use of the name 'Pig Island'; how he arrived at this name remains a mystery, since Endlicher is not known to have ever been in personal contact with Ferdinand.

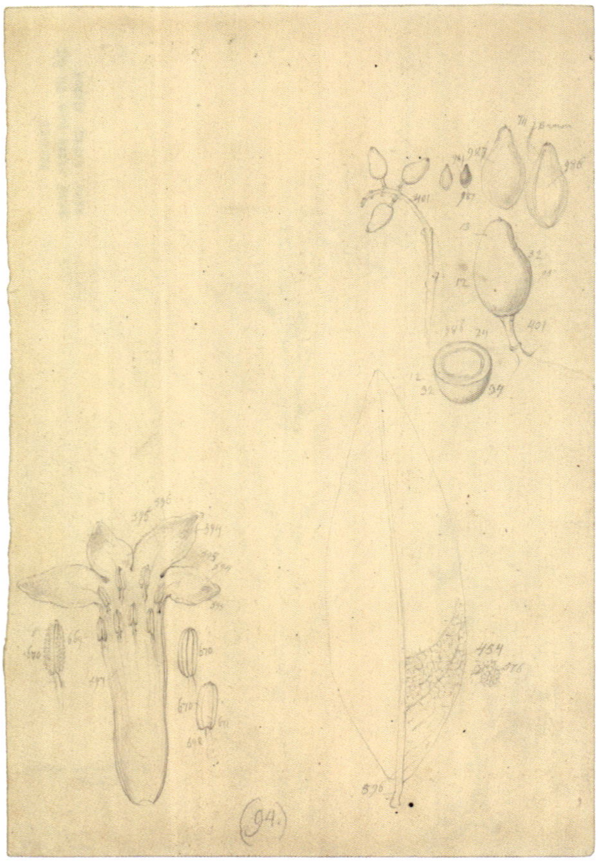

FIG. 10.19. ***Wickstroemia australis.*** **Pencil drawing by Ferdinand Bauer, 1804–1805. Paper, pencil. - Vienna, Naturhistorisches Museum, Archiv.**

Throughout life Ferdinand, like his elder brother Franz, continued to cultivate a special interest in orchids. One of them, the epiphytic *Oberonia titania*, long thought to be an endemic of Norfolk Island, but recently found also on the Australian mainland, was recorded in a colour-coded pencil drawing by Ferdinand annotated by him as occurring on 'Norfolk Island Feb[ruary 1805]'. It is most unconventional since three petals, albeit depicted as detached, are represented in watercolour only – in shades of coral red, one of them with a number. Clearly it is impossible to find out if this detail, covering approximately four square centimetres, was painted on Norfolk Island or added later, which would not be a surprise.

More widespread plants, such as the orange-petaled passion flower, *Passiflora aurantia*, first collected by the Forsters in New Caledonia and published as a species new to science by George Forster just two years before Franz had visited the former's father in Halle, were also included in the spectrum of specimens collected on Norfolk Island and recorded by Ferdinand.[41] His specimen is in the Naturhistorisches Museum; several years after collecting it Ferdinand made use of the pertinent pencil drawing as basis for a finished watercolour now in the Biblioteka Jagiellońska in Cracow. This was published by John Lindley, albeit with the incorrect name *Murucuja baueri* (see Chapter 12). *Dysoxylum bijugum*, a relatively common tree on Norfolk Island, has been recorded at least on three dates – 15 October 1804, 24 October 1804 and 14 January 1805, probably documenting different stages of flower and fruit development. Ferdinand even found time to prepare a pencil drawing of a flowering and fruiting specimen of the coffee bush, *Coffea arabica*, 'from the garden at Norfolk Island', according to his annotation. This may well have been the garden then supervised by commandant John Piper.

Today Ferdinand is remembered in the epithets of several plant species restricted in their distribution to Norfolk Island that he had collected – among them are *Coprosma baueri* and *Melodinus baueri*, while there is less evidence in the case of the famous Norfolk Island palm, *Rhopalostylis baueri*. The latter is endemic to Norfolk Island and the Kermadec Islands[42] (one of the outlying islands of New Zealand), attaining a maximum height of around fifteen metres with leaves up to 3.6 metres long. This palm had been described in 1853, twenty-seven years after Ferdinand's death, on the basis of a specimen cultivated in the Royal Botanic Garden at Kew originating from Norfolk Island, though the precise collection data are not given. Considering the longevity of this palm and the close contacts between Ferdinand and Banks, the unofficial superintendent of the Royal Garden at Kew at the time of his return from the *Investigator* voyage, it is difficult to exclude a connection. Today this fast-growing ornamental palm is in cultivation worldwide and is the most popular plant commemorating Ferdinand's name. *Freycinetia baueri*, an only slightly less spectacular, bat-pollinated root-climber belonging to the screwpine family, is known to occur both in Norfolk Island and in the moist subtropical to temperate forests of New Zealand.[43]

Although possibly overwhelmed by the diversity of plant life, Ferdinand also observed the rich fauna of Norfolk Island, birds being high on his agenda. One of them was the Pacific black duck, *Anas superciliosa*, a dabbling duck found in many regions of what is now Indonesia, New Guinea, Australia, New Zealand and many islands in the southwest of the Pacific, reaching Polynesia in the east. It is a sociable duck found in a variety of habitats, and it feeds while upending itself. Ferdinand shot at least one specimen; its state of preservation proves that he knew how to prepare it (Fig. 10.20.). Much more spectacular is another bird that he shot: the Norfolk kaka, *Nestor productus*, a large

FIG. 10.20. Pacific black duck, *Anas superciliosa*, collected by Ferdinand Bauer on Norfolk Island, 1804–1805. – Vienna, Naturhistorisches Museum, 1. Zoologische Abteilung, Vögelsammlung, Inv. No. 47.336, 1826.V.97.

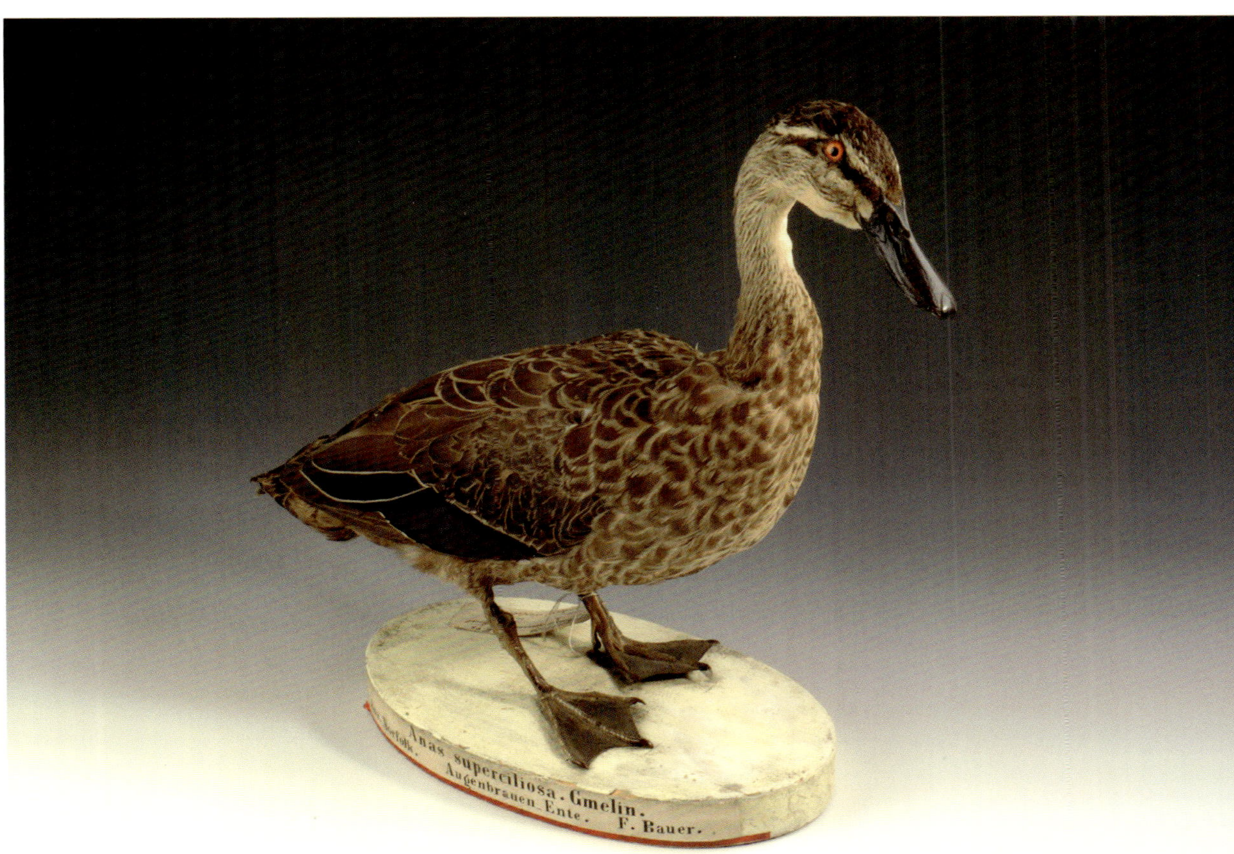

parrot formerly inhabiting Norfolk and Phillip Island, but long extinct. It is believed that the Polynesians inhabiting the island before the arrival of the first European already hunted the bird, and later with the arrival of the first settlers the population must have been further reduced in size. The bird seems to have become extinct during the time of the so-called second penal colony in the second quarter of the nineteenth century. Ferdinand was lucky not to be late: he managed to collect a specimen of the Norfolk kaka and prepare a very accurate pencil drawing of it (Fig. 10.21.). However, he was preceded by others: the first to observe the bird were the Forsters on board the *Resolution* in October 1774, the earliest illustration was prepared in 1789 for John Hunter and is kept today in the National Library of Australia in Canberra.[44] The scientific name of the bird was validated in 1836 by John Gould (see Chapter 6), without reference to Ferdinand's pencil and specimen, then already in Vienna.

During the circumnavigation of Australia Ferdinand had numerous occasions to study underwater life, but he seems to have concentrated on fishes and the occasional crab. With possibly more free time available on Norfolk Island, he also produced pencil drawings of further invertebrate species, among them several holothurians, which are still in need of more careful study. Some of the specimens must have been dissected by Ferdinand. One holothurian is annotated by him, 'issueing yellow Milk when irritated', with the colour of the milk given as 'orangeable [sic] 606 50 286'. More surprising is a pencil drawing showing several specimens of the walking sea hare, *Aplysia juliana*, a cosmopolitan marine gastropod mollusc occurring in warm seas. It is annotated 'Norfolk Island Dec 21 1804 see of Estners Mineral Tables for all this'. This reference is to *Versuch einer Mineralogie für Anfänger und Liebhaber*, an extremely rare introductory text-book on mineralogy by Franz Joseph Anton Estner, a theologian and mineralogist living in Vienna. Based on the extensive mineral collections of Rudolf Count Wrbna von Freudenthal (Rudolf Jan Bruntalský z Vrbna), subsequently a Knight of the Order of the Golden Fleece and one of the principal figures at the imperial court in the struggle against Napoleon, this work had been printed in Vienna and was almost completed before the *Investigator* left Spithead. In our context only the four coloured plates published in the first volume in 1794 are relevant. Each shows 56 hues of the rainbow with reference to mineralogy – t. 1, nos 1–24 stand for grey and black, nos 25–56 for blue and red, t. 2, nos 1–48 for red, nos 49–56 for yellow, t. 3, nos 1–32 for yellow, nos 33–56 for green, t. 4, nos 1–24 for green, nos 25–36 for brown. The pencil drawing documenting the walking sea hares is colour coded in a special way – it carries on the one hand Ferdinand's conventional three-digit numbers and on the other hand additional annotations unequivocally referring to Estner's work, such as T III 7 standing for table 3, fig. 7 of the *Versuch*. A colour-coded pencil drawing documenting a specimen of the banded moray eel, *Gymnothorax rueppellii*, annotated 'Norfolk Island Jan 12 1804' even carries a decoder – a strip of 14 squares, all in different hues with pencil notes superimposed, such as T I 32 ('milky white') or T IV 41 ('dark chestnut brown'), undoubtedly all again referring to the plates in Estner's *Versuch*. Needless to say, this strip could also have been added later by Ferdinand, but since the pencil drawing itself is also annotated with the references to Estner's four colour charts, Ferdinand must have had the latter with him on Norfolk Island. Nothing is known about why Ferdinand added this new approach to his well-established method of colour coding in the case of the two pencil drawings from Norfolk Island, but a hypothesis may be put forward: glistening surfaces in underwater organisms such as the walking sea hare and the banded moray eel, a fish found in coral reefs, may be more easily recorded with the help of a mineralogical colour code (referring as it does also to objects with somewhat shiny or glossy surfaces) than with a conventional colour code used for the hues of flowering plants. This special challenge is corroborated by Ferdinand's note 'glasig' (glassy) on the pencil drawing of the banded moray eel, a term rarely applicable to a part of a flowering plant. Why Ferdinand did not test Estner's tables before remains unknown.

It would be quite wrong to picture Ferdinand alone on a desolate island like Daniel Defoe's legendary Robinson Crusoe on the island in the Caribbean Sea. He was almost certainly accompanied by his servant, and there was the commandant Piper, the men of the 73rd Regiment, convicts and settlers. Ferdinand could have even met on Norfolk Island engraver Barnard (later Bernhard) Walford, who was born, hard to believe but true, in Vienna.[45] He was a Jew and had come to England as an adult five years before Ferdinand. Convicted on 28 October 1789 for stealing a basket of laundry in Petticoat Lane, he was sentenced to seven years transportation, arriving with the third fleet in New South Wales.

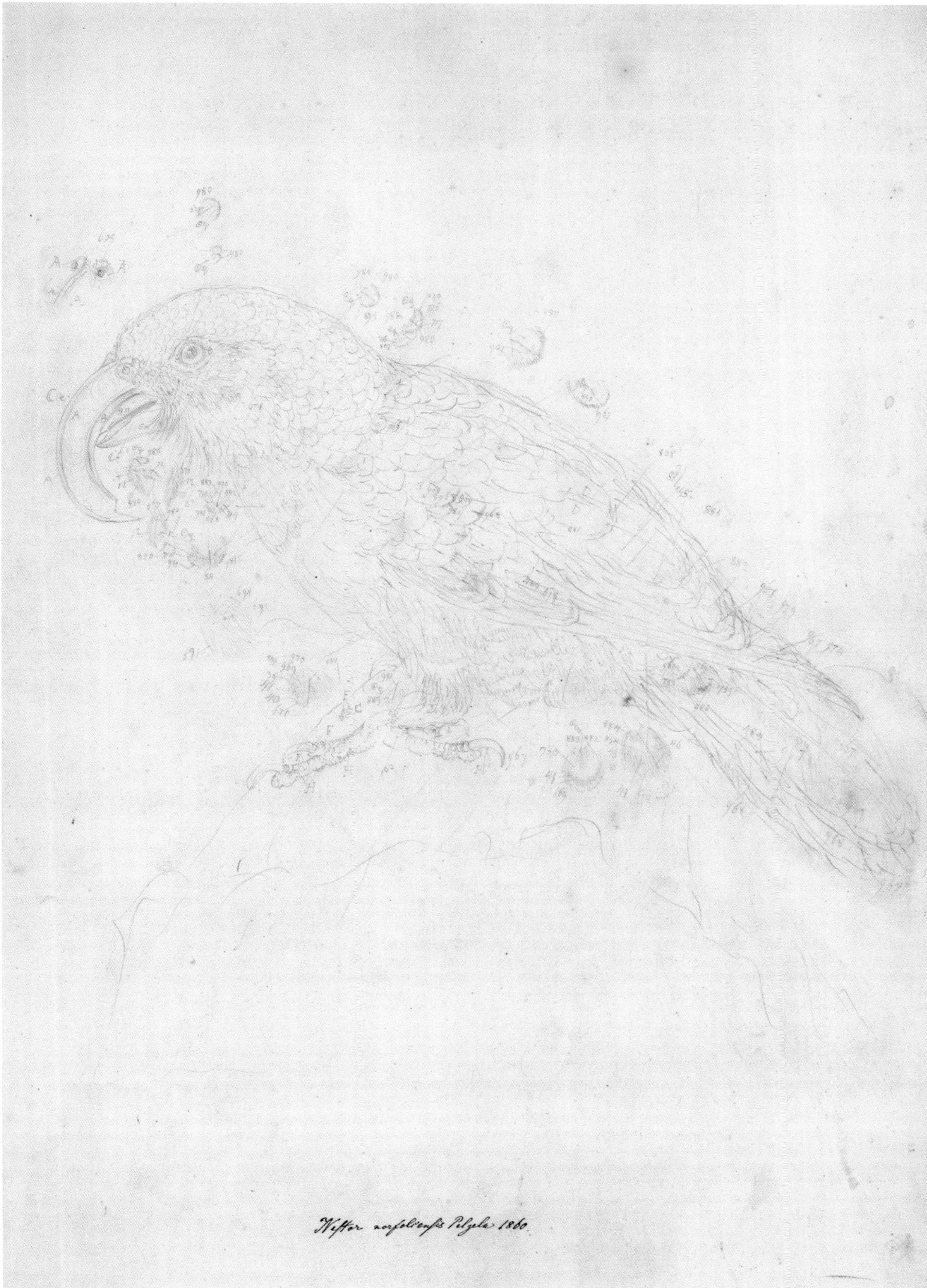

FIG. 10.21. Norfolk kaka, *Nestor productus*. Pencil drawing by Ferdinand Bauer, 1804–1805. Paper, pencil. – Vienna, Naturhistorisches Museum, Archiv.

After serving his sentence he was allowed to settle on Norfolk Island and was granted forty acres (c.16,000 square metres) of land to be farmed. Living there as an emancipist from February 1796 until February 1808 with his Irish wife Jane Malley and five children, he could very well have encountered the German-speaking naturalist who had lived for a few years in Vienna.

Ferdinand left Norfolk Island in the reactivated *Investigator* with its cut-down front deck on 1 March 1805, accompanied by prisoners to be moved to the Australian mainland and a military detachment arriving in Sydney Cove on 11 March 1805. His luggage may not have been extensive, but it contained the rarest plant and animal specimens ever collected by him.

FROM PORT JACKSON TO LIVERPOOL

Suspected piracy and politics lay behind an official proclamation that had been published in the *Sydney Gazette* on the day before Ferdinand's return. The statement concludes, 'And it is further requested, that the Persons who are or may be in charge of these Vessels do not in any manner degrade the Spanish Flag, but hoist it in its proper place, until it is known whether War did exist at the above period [26 September – 2 October 1804]'. The background of this is complex. The brig *Harrington* had been active along the coast of the Spanish Viceroyalty Peru, and her commander had seized two Spanish vessels, claiming to believe that Spain and the United Kingdom were at war. Suspicious erasures in the log of the *Harrington*, which had reached Sydney on 4 March 1805, were brought to the attention of Governor King, who was uncertain if indeed a state of war had existed or, alternatively, the taking of the two Spanish vessels was rather an act of piracy. Fearing political repercussions, King had the proclamation published. When further inquiries brought serious evidence for piracy he thought it necessary to report the incident to the Secretary of State for War and the Colonies, then John Pratt, 1st Marquess Camden, a Knight of the Garter and very high up in hierarchy. Since there was no vessel in Sydney Cove preparing to sail for England, the Governor's attention fell on the repaired *Investigator*. First an opinion on the capacity of the ship to sustain a voyage to England was sought by William Kent, who reported back that she was 'fit and Safe', although one of the inspectors, who had brought her to Norfolk Island and back, thought otherwise. On 18 March 1805 Kent received King's commission to carry to London the dispatches relating to the *Harrington* affair.

On 16 March 1805 Brown, maybe accompanied by Ferdinand, had been to Toongabee, probably hunting orchids, with one of them, *Chilioglottis reflexa*, marked 'Sydney – Parramatta area March 1805'. As was so often the case before, Ferdinand must have prepared a pencil drawing that later was to form the basis of a finished watercolour. A few days later King informed Brown that he and Ferdinand were to return to England by the *Investigator*. Both knew the ship from the circumnavigation with Flinders and in Ferdinand's case – as an already cut-down vessel – from his voyage back from Norfolk Island. We have written evidence that Brown was very unhappy about the proposal, as he believed the ship had scant promise of safe transport, and Ferdinand will have felt likewise. The exchange of letters between King and Brown unveil a crescendo of arguments, but in the end, unsurprisingly, the Governor prevailed. Brown's main argument was 'the extreme wetness of the ship' and the few opportunities he would have during the return trip to inspect the collections. As was evident from Brown's letter to King of 23 March 1805, paper was another problem: 'I take leave also to add that the paper w'ch I brought from England for the purpose of preserving specimens is nearly expended; and as in all probability I shall during the passage be obliged, for the safety of the collection, to remove a considerable part of it into fresh paper, I find it necessary to apply to your Excellency for at least one ream of cartridge paper from the store'. King's reply of 9 May 1805 was formal: 'as I am averse to interpose any authority in the direction of your collection, otherwise than in facilitating its safety, it will remain with you and Mr. Bauer to decide on the propriety of yourselves and collection going with the Investigator or not; and it is necessary I should inform you that it is out of my power to say when another opportunity within my authority may again occur'. Two days later Brown felt obliged to persevere in declaring, 'I beg leave to repeat my decided opinion of the absolute unfitness of the Investigator's after-hold for any part of my collection, except minerals and such other subjects of natural history as are but little liable to suffer in so humid an atmosphere'. In the end a compromise seems to have been reached as evident from the Governor's order to Captain Kent: 'you will receive and victual them [Brown and Ferdinand] and a servant each, the magazine of the

Investigator and a cabin being allotted for the reception of the most perishable part of the collection, and a part of the after-hold for the less perishable part, together with a cabin for each of those gentlemen; and as those objects have been procured at a great expanse, and their being carried safe a very desirable circumstance to the public, you will cause those gentlemen to have every facility and assistance to inspect the different parts of their collection during the voyage'.

While preparations for the departure were under way, Ferdinand recorded probably his last bird observed and shot in Australia – a cockatoo, possibly the sulphur-crested cockatoo, *Cacatua galerita galerita*, depicted in a colour-coded pencil drawing in large format, annotated 'Sydney [M]ay 16th 1805' and kept in the Naturhistorisches Museum. However, this determination still needs to be confirmed and the sex of the specimen depicted ascertained by a specialist of the group. Surprisingly, the large bill has been painted in colour, but as in the orchid from Norfolk Island, it is impossible to find out when this was done. The bill's ebony-like tint raises some doubts on the correctness of the preliminary identification. Ferdinand must have become familiar with this bird and its distinctive, very loud and raucous call during his wanderings in the woody hinterland of Sydney.

Meanwhile, a letter from Flinders written as a prisoner at Mauritius on 8 August 1804 to the Governor of New South Wales had ended the uncertainties that long had prevailed about his fate. Flinders informed King how he had been branded a spy after his arrival on the island without official documents confirming his mission on the *Cumberland*. In order to avoid a similar problem, King prepared an open letter dated 30 April 1805 for captain, crew and the two remaining scientific assistants on the *Investigator*, explaining 'As the Chance and Hazard attending the unfortunate Warfare in which Great Britain is engaged may occasion His Britannic Majesty's Vessel Investigator to fall into your power, it is a duty I owe to the cause of science to represent to you the Circumstances in which Mr. Robert Brown, naturalist, and Mr. Ferdinand Bauer, painter of natural history, are placed . . . ' and only then referring to the ship carrying papers relative to an 'injury committed contrary to the Law of Nations'. This was a most sensible move: after the interval caused by the Peace of Amiens hostilities between Britain and France had started again. In the end this document, surprisingly given to Brown instead of Captain Kent, was not needed.

Otherwise we have no detailed information on the preparation of the return voyage. Ferdinand must have taken with him a drawing desk made by a convict, either Laurence Butler or Thomas Williams, using Australian timbers.[46] It is a unique object, since very few pieces of furniture made locally have survived the early decades of the settlement in New South Wales. Although first described as a travelling desk, this item, 51.6 centimetres long, 23.5 centimetres high, 25.5. centimetres wide, with two brass lifting handles at each end and kept today in the Economic Botany Collection of the Royal Botanic Gardens, Kew, is clearly a drawing desk. Comprising a large drawer for storage of paper and five compartments for drawing utensils along the upper edge, it is similar to the slightly older drawing desk conserved in the Stiftung Weimar Museum (see Chapter 6), but by contrast Ferdinand's acquisition could be folded. It is highly likely that it was purpose-made for him, using eucalyptus wood (probably red mahogany, *Eucalyptus resinifera*) with a veneer of casuarina (possibly swamp oak, *Casuarina glauca*, or river oak, *C. cunninghamiana*). Inset from the margins, the top is strung with a pale wood, possibly Port Jackson pine, *Callitris rhomboidea*. Upon his return to England Ferdinand presented this special souvenir of his prolonged stay in Australia to a person he seems to have been especially fond of – his brother Franz. It is the only personal object owned by the Bauer brothers that has survived. Ferdinand's luggage must have been considerable, since it comprised in addition to the masses of pencil drawings all the birds shot, the platypus and the herbarium specimens from Norfolk Island. From a letter by Banks to William Marsden, Secretary to the Admiralty, dated 19 October 1805 we learn the details on the scientific cargo on the *Investigator* – eleven boxes 'of Drawings of all sorts' all attributable to Ferdinand, two packages of 'Birds & Beasts Preserved', twelve packages of herbarium specimens, one box of insects, three boxes of minerals. No mention is made of the living material on board of which we hear about only later – seeds and, rather surprisingly, a male wombat, which survived the passage and was subsequently kept for a few years in London before it was dissected. Only the conservatory brought to Sydney on the *Porpoise* and later removed to shore was to remain in Australia, with the plants left in the care of Caley. After the compromise had been reached between Brown and King, unsatisfactory as it may have seemed to the former, the refitted *Investigator* left Sydney Cove on 23 May 1805. Three days later the

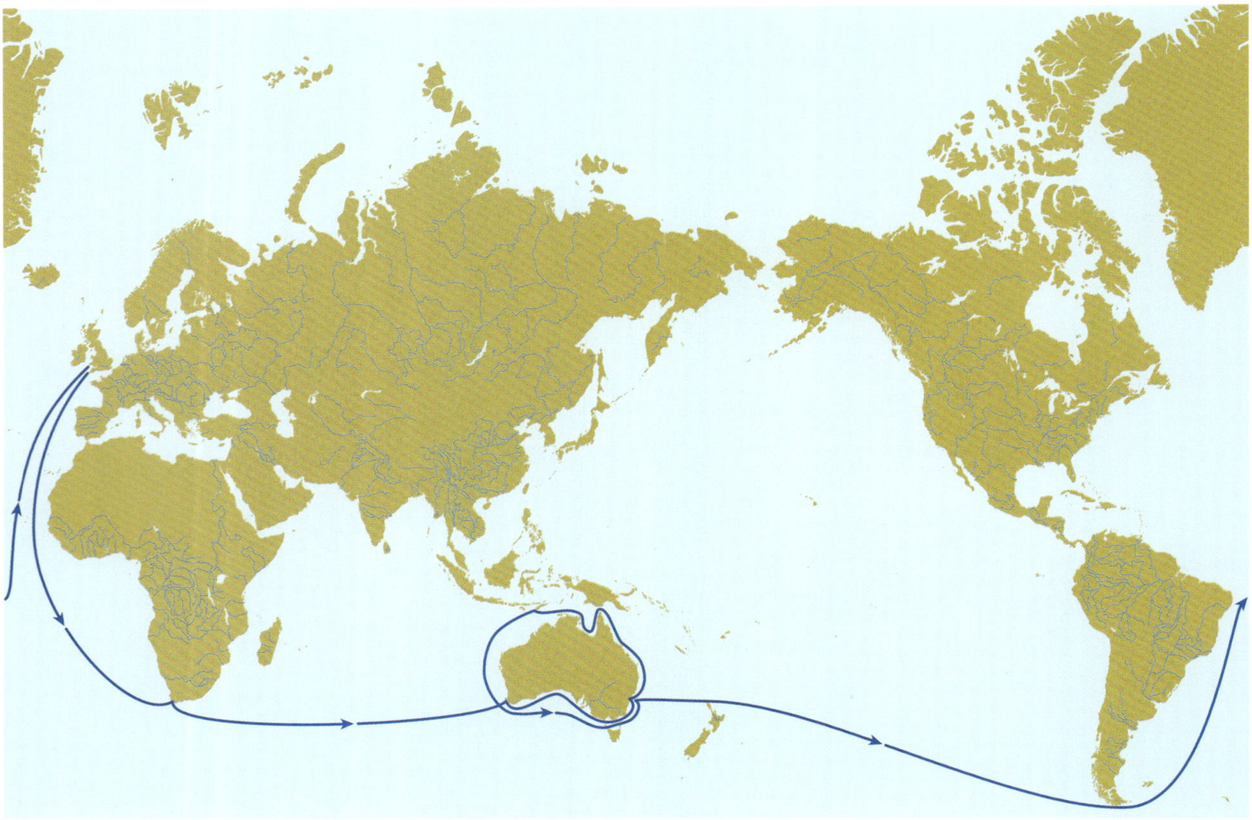

MAP 5. Route of the *Investigator* circumnavigating the world 1801–1805.

Sydney Gazette reported, 'The whole collection was sent on board the Investigator in her present reduced state, with every apparent reason to expect it will safely arrive in England, where it will be very acceptable to the Amateurs of Natural History'. Considering the tensions between Brown and the Governor there seem to have been no farewells, and so the most extensive scientific survey of Australia undertaken up to that moment ended quietly. Maybe Ferdinand and Brown were not entirely unhappy to leave Sydney Cove.

What followed is dealt with often as a postlude and not analyzed in detail, although it was the most critical and most dangerous part of the whole expedition. There is no evidence for diary notes kept by Brown, though he later copied the latitudes and longitudes from *Investigator*'s log. His comments need to be cited in full: 'By the above dates it appears this extensive Voyage was performed in four months & sixteen days in the crazy low cut down Investigator, perhaps the most deplorable Ship in all the World; and if the season of the Year be considered, it being the depth of Winter when Cape Horn was rounded, that the Main Topmast was sprung & a Reef of eight feet taken in it; the Top-gallant Mast upon Deck & the Topsails double reefd the whole Passage, every Seaman will admit it to be a great undertaking'. It is difficult to imagine that Ferdinand found the necessary concentration and peace of mind to produce finished watercolours under these circumstances, as has been suggested in the literature.

It was a non-stop voyage – Brown noted having passed Diego Ramírez, a group of islands west-south-west of Cape Horn, Staten Island and Sea Lion Island, off the coast of the East Falkland Islands. He had seen Ilha da Trinidade in the South Atlantic on his way to the Cape of Good Hope and now he had occasion to see it again. Indeed this was an outstanding piece of seamanship accomplished by Captain Kent and his crew, worthy of the highest praise. What the realities must have been like transpires from a letter sent by Kent to Marsden: 'Although the sufferings of the men have been considerable in rounding Cape Horn in the dead of winter in this low cut-down ship, whose single deck was almost constantly under water, yet we have had the good fortune not to lose any of them'.

The arrival at Liverpool on 13 October 1805 had not been planned. It was rather that the *Investigator* was forced to seek shelter because of a storm – according to Kent with 'but three days water, the men without clothing, and the masts, sails, and rigging in a shatter'd condition'. A friend of the Liverpool banker and naturalist William Roscoe recalled many years later 'the extraordinary appearance of this wonderful old ship, her sides being covered with barnacles and seaweed, and her sails, masts, and rigging presenting the usual signs of a vessel that had been abandoned'. Invited on board 'a sight presented still more astonishing – plants we had never before beheld, black swans, and other curious birds and animals surrounded us on every side'.

Ferdinand and Brown were finally back in England, exhausted perhaps, but safe and sound. By now Ferdinand was forty-five years old. He had twenty-one more years to live, a considerable part of them dedicated to work on the graphical record of the voyage just completed.

NOTE

The literature on the voyage of the *Investigator* is vast, and only the main sources are listed in the endnotes. A valuable summary has recently been published in the volume edited by Wege el al. (2005), and considerable original material as well as general background information is provided by http://flinders.rmg.co.uk.

ENDNOTES

1 Unless otherwise stated, this chapter is based on Mabberley (1985), Norst (1989), Lack (1997a), Watts, Pomfrett & Mabberley (1997), Mabberley & Moore (1999), Vallance, Moore & Groves (2001), Mabberley (2002), Wege (2005). The scientific names follow Mabberley & Moore (1999).
2 Chambers (2000)
3 Rigby (2005)
4 Péron (1998)
5 Norst (1990)
6 Estensen (2002)
7 Findlay (1998)
8 Brown (1844)
9 Wheeler & Moore (1994)
10 London, Natural History Museum, Library and Archives, B 97.
11 Pignatti, Riedl-Dorn & Mabberley (2000a, 2000b)
12 e.g. Hoeppe (1999)
13 Riedl-Dorn (1999)
14 Edwards (1981)
15 Lack (1998 b)
16 Ker (1818)
17 Riedl-Dorn (2001)
18 Riedl-Dorn (2014)
19 D. Stuart (pers. comm., 2014)
20 e.g. Stearn (1960)
21 Lack (2003)
22 Ibáñez (1992)
23 Moyal (2001)
24 Home (1800, 1802a, 1802b)
25 Schifter (1992)
26 http://www.topix.com/forum/world/east-timor/T12S0MPMCF2623EHI (accessed 12 July 2014)
27 Wright (1988)
28 Moyal (2008)
29 http://www.ouw.edu.au/morgan/koala.htm (accessed July 2014)
30 Correa de Serrea (1802)
31 Flowers (1967)
32 May & Pascoe (1996)
33 Nissen (1951)
34 Green (1994)
35 Moore (1998)
36 Schrire (2008)
37 H. Rainer (pers. com. 25 August 2014)
38 Mabberley, Pignatti & Riedl-Dorn (2007)
39 Mattingley (2010)
40 http://nla.gov.au/nla.pic-an6155162 (accessed 16 July 2014)
41 Satterthwait (1982)
42 Lange et al. (2005)
43 Lord (1991)
44 Olsen (2010)
45 Norst (1988)
46 George (2006)

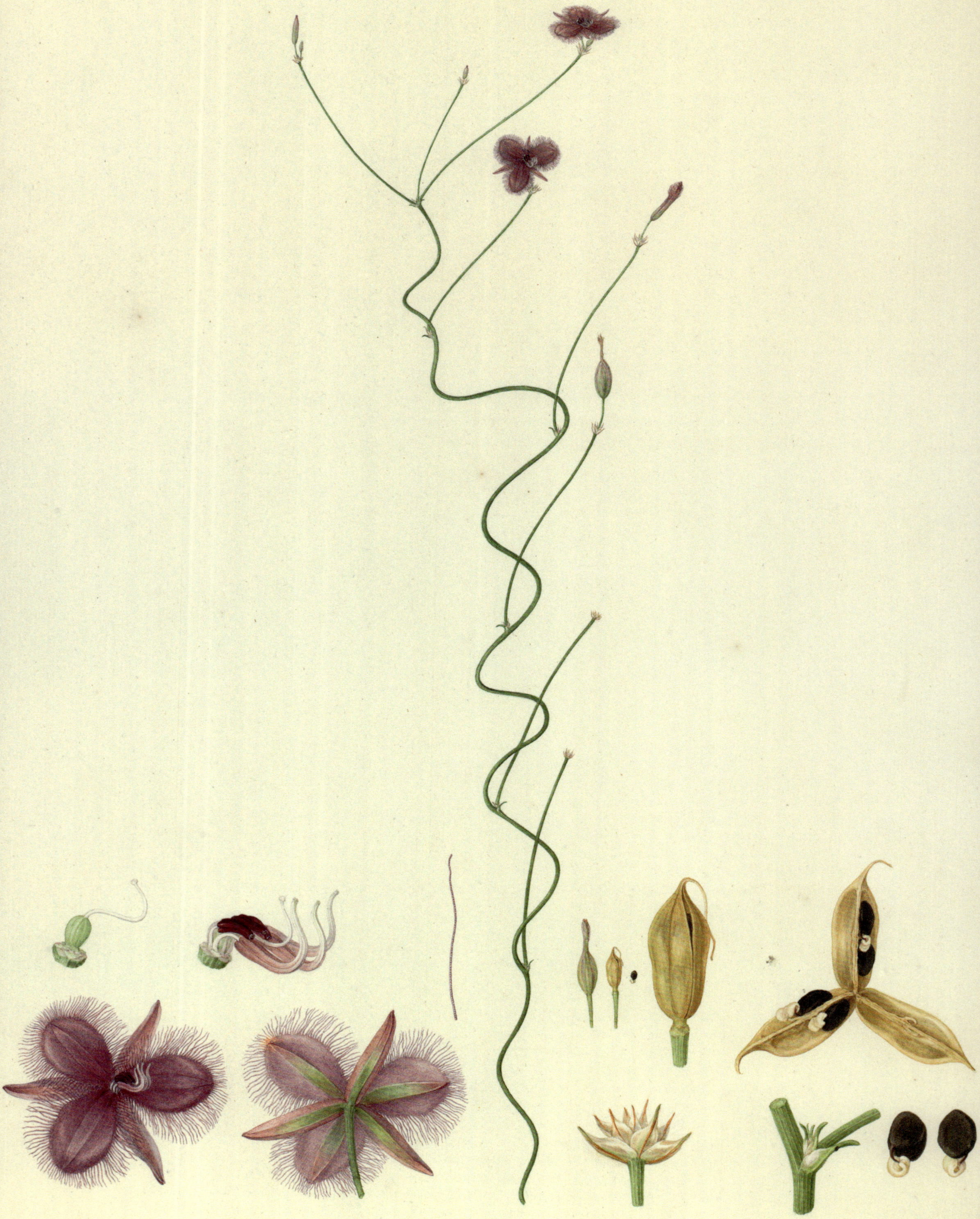

CHAPTER 11

Ferdinand Bauer in the circle of Sir Joseph Banks in London[1]

MIRED IN BUREAUCRACY

UPON THEIR ARRIVAL in Liverpool one of the first things to be done was the writing of letters – those by Brown to Banks and by Kent to Marsden are extant; it is possible that Ferdinand also reported his return to his brothers Franz in Kew and Joseph in Vienna; however, if they did exist, these documents are untraceable.

In his typical understated style Brown started by informing his patron of 'a tedious and uncomfortable voyage of nearly five months' before dealing with the first priority – the transfer of the collections and drawings to safe storage. Here is his reasoning: 'As to the collection embark'd in the Investigator, it was . . . provided for as well as the reduc'd state of the vessel and her crazy condition would admit of, and since it had been on board I have receiv'd from Cap'n Kent every assistance in his power to enable me to preserve it. While within the tropics the plants were carefully examin'd and those that most requir'd it were chang'd into dry paper, but such has been the wet state of the ship that they must again be suffering, and that I fear considerably. I am therefore most anxious to have them removed on shore, and I earnestly beg they may not again be put on board the Investigator for the purpose of being brought round to Portsmouth or the river [Thames, i.e. to London], as she is not only a crazy but absolutely a defenceless ship'. Indeed this was the state of the *Investigator*, and no doubt Brown's proposal to transport the 'cases containing the collection, as also that of Mr. Bauer' as well as the personal luggage by land to London was a very sensible one. Banks, however, was at Revesby Abbey, and there were a few days of delay until he could act.

FIG. 11.1. *Thysanotus volubilis.* Watercolour by Ferdinand Bauer, signed, 1806–1810. – London, Natural History Museum, Library and Archives, Australian Botanical Drawings, pl. 212.

Meanwhile, Ferdinand and Brown had met a man in Liverpool of whom we have only the initials – P.M.T. More than half a century after the arrival of the *Investigator* in the harbour this man reminisced in *The Athenaeum* that he had been on board 'while lost in wonder two gentlemen came up from the cabin, who the officer we had been previously speaking to informed us were Messrs. Brown and Bauer . . . Having been then recently spending more than two years in Germany, I immediately addressed Mr. Bauer in his native language, to his intense delight, he not having heard it spoken for (I think he said) upwards of twelve years. This was at once a 'Bündnip' [Bündnis, i.e. bond], and we were together some part of every day while he and Mr. Brown remained in Liverpool'. Ferdinand's calculation was another understatement – he had left Vienna in March 1786 and probably had not heard his native language spoken for a total of almost twenty years, not counting his meetings with Franz at Kew and his friends in London mentioned in the letters from Australia. Several dinners at Allerton Hall, Roscoe's country house near Liverpool, followed with Ferdinand, Brown and P.M.T. as guests, the latter writing 'I never passed *such* [his italics] an evening before. And so it went on for three days'.

Finally, on 19 October 1805, Banks sat down and wrote an important letter to Marsden that was decisive for both Brown and Ferdinand. The first part deals with the transport of the materials to London, with Banks suggesting that this should be handled by Brown and Ferdinand. The second part contains one of those Banksian proposals that reflected his position as President of the Royal Society, London club member and adviser to the royal couple on all sorts of matters. He wrote 'for the people [Ferdinand and Brown] I will request to look at the Precedents of the draughtsmen of Capt Cook in his Second & third voyage: they were, as I remember, retained for some time after their arrival in the service of the Public to Finish the Sketches they had made on the voyage; & they delivered in the Pictures with which the Admiralty is now furnished: if this is agreed to, Brown may be employed in making a compleat assortment of the natural history he has collected, & Bauer in Finishing the most interesting of his Sketches; the whole done in the time they are thus employed to be at the disposal of the office [the Admiralty]'. Clearly this letter referred to the famous Memorandum signed by Ferdinand and Brown in the presence of Banks in April 1801. As was so often the case with Banks, this was a grandiose scheme, in particular concerning 'the compleat assortment of the natural history' and it meant an engagement of several years for both Ferdinand and Brown. Brown definitely received payment from the Admiralty until the end of 1810 and a similar arrangement seems to have been found for Ferdinand, although its termination is less clear. As a matter of fact Ferdinand's first biographer, the well-informed Jan Lhotsky (see Chapter 20), stated that the Ferdinand's employment continued up until 1813.[2] This means that Ferdinand was employed by a succession of First Lords of the Admiralty (see Dramatis personae).

Meanwhile Kent had been to London, presumably to deliver Governor King's documents to Whitehall on the *Harrington* affair, and was instructed to bring the *Investigator* to Plymouth. Thus the collections, drawings and personal luggage had to be removed from the vessel. Unloaded goods, however, became a concern to Customs, and Kent, when advising the Admiralty of the arrival of the collections, drawings and luggage, had refrained from securing the necessary dispensation from the Board of Customs. Marsden seems to have been inactive in this matter, Banks was out of town, and therefore Brown had to write an urgent letter from Liverpool on 29 October 1805 to Dryander, Banks's librarian at 32 Soho Square. From this letter it appears that the material was already at the Customs House in Liverpool, 'where we have been obliged to deposit it', with Brown asking to arrange for its release and speedy transport to London, if possible by wagon as this was regarded as safer though more expensive than by canal. In the end, the materials were dispatched on 1 November 1805 by wagon to the Customs House in London, as substantiated in a note by Brown. Four days later Ferdinand and Brown arrived in London, and the collections, drawings and personal items were finally released.

Back after having spent fifty months on the remotest shores, Ferdinand and Brown did what everybody would have done after such a long absence – report to their superiors, the Admiralty. To their considerable surprise, they had to wait for more than two hours and in the end penned the following letter to Marsden: 'Robert Brown, Botanist & Ferdinand Bauer Painter of Natural History . . . have the honor of acquainting Mr Marsden, for the information of their Lordships, of their arrival in London . . . They are anxious also to receive their Lordships instructions both with respect to themselves . . . An instruction from Mr Marsden at what time they can have the honor of receiving their Lordships Instructions may be addressd to them care of Mr Dryander at the Rt Honble Sir Joseph Banks Soho Square'. Their not being received by Marsden was not an act of impoliteness; it occurred for a totally different reason, to which Dryander hinted in his letter of 7 November 1805 to Banks, writing 'he [Brown] desired him [Marsden] to direct his answer to me [Dryander], which I have not yet heard of, but don't wonder at it, as the good news yesterday would too much occupy them at the Admiralty, to think of these things'.

The good news was the crushing victory of the British Navy over the combined French and Spanish Navies at Trafalgar, which determined British superiority over the oceans for the next century. Champagne must have been served at the Admiralty, and the arrival of the two strange naturalists coming back from a prolonged stay down-under, which as far as the elite were concerned was at best the dumping-ground for convicts, was of no relevance at that moment. In addition, the survivors of the *Investigator* expedition had finally returned from the Cape, Canton and Port Jackson – but without the captain, and now the cut-down vessel had also arrived. All this could hardly excite much notice at the Admiralty.

FERDINAND IN 32 SOHO SQUARE

For the following years there is a substantial problem already well known to the reader – the almost complete lack of sources for the years 1806 to 1814, when Ferdinand focused on producing the finished botanical watercolours based on his colour-coded pencil drawings prepared in Australia and started the publication of some of them. Today, these include his most popular works, globally acclaimed, but we have only a London address for Ferdinand's last year in the British capital – '10 Russell Street, Bloomsbury', probably 10 Little Russell Street.[3] Furthermore, not a single letter written by him during this period has been traced, although he may have written one – in retrospect Endlicher (see Chapter 18) was to note, although incorrectly, '[Ferdinand] has never ever written anything [hat überhaupt nie etwas geschrieben]'.[4]

Ferdinand's social circle may have been small. One important member was his brother Franz at Kew. Upon Ferdinand's return aged forty-seven years old, Franz attached an engraved metal plate to the drawing desk from Sydney Cove – with the text reading, 'This desk was manufactured by a convict at Sydney in New South Wales intirely [sic] of the native woods of that country and brought from thence and presented to me by my brother Ferdinand Bauer 1805'.[5] Another member of his circle was Ferdinand's former travel companion Brown as well as possibly his former commissioners Hawkins and Lambert. It was under the frequently grey London sky that Ferdinand recreated the often-incredible brilliance and brightness of the Australian flora he had just encountered.

The more general scene is described in Banks's letter of January 1806 to Marsden, stating that the 'whole of the collections of Mr Brown are in my House [32 Soho Square], I can therefore be answerable for their safety'. The material was probably kept as a separate collection in the herbarium. The following lines written by Banks describe an incident concerning Ferdinand: 'Mr Bauer has however during my illness removed his Scetches & has them at present in his own Custody: I must therefore request their Lordships to give particular Orders to Mr Bauer to return these Scetches to me, that they may be kept in my House, not only for the sake of secure Custody, but because it is absolutely necessary that Mr Brown, who does all his business in my Library, should have before him the Scetch of every Plant during the time it is under his Examination, lest some misunderstanding of the Structure of those minute parts, on which systematic arrangement depends, should have taken place either on his part or on that of Mr Bauer, as not infrequently happens to the most expert naturalist, in such case Mr Brown & Mr Bauer must consult together till they agree in the same opinion, which can never fail to happen with Persons well versd in the Science of Botany, this is the only mode by which it is possible to prevent an occasional variation between the Drawings & the Description, which [never] fails to throw a great degree of discredit on all the parties concerned'. Although we are unaware of the effect this letter had at the Admiralty, it is likely that Ferdinand was asked by Marsden to return his pencil drawings to 32 Soho Square, which he must have done. More relevant in our context is the fact that Banks described the close cooperation between Ferdinand and Brown by stating '[they] must consult together till they agree in the same opinion'. In fact this is a most adequate description of this collaboration, which brought together two completely different personalities united in their work on the Australian flora. Notably Banks stressed that 'some misunderstanding' could have taken place on the part of Brown or of Ferdinand, underlining in a sense that he regarded them on a par in their observational skills. This collaboration is not mentioned in the watercolours, which were to be signed by Ferdinand. It is also passed over in silence in the introduction to Brown's main publication on the botany of the *Investigator* voyage, the famous *Prodromus Florae Novae Hollandiae*, a work which remained a torso, with only pages i–viii and 143–590 plus addenda published. Only Ferdinand's herbarium specimens made available to Brown are acknowledged in the *Praemonenda* to this work.

The division of labour between Ferdinand and Brown was clear right from the beginning of their work in London – Ferdinand produced the watercolours, and Brown produced the text consisting basically of scientific names, plant descriptions and notes on distribution. This work was interlinked and had the following elements as its basis: (1) the herbarium specimens, (2) the colour-coded pencil drawings of plants, (3) the field descriptions of plants and (4) the miscellaneous notes in the diary. Items 3 and 4 were exclusively the result of Brown's field work, item 2 exclusively the outcome of Ferdinand's activities, and item 1 was the result of work done by several people on board (see Chapter 10). There must have been instances when Ferdinand had depicted a detail not easily observable in the herbarium specimen and not mentioned in Brown's field description; in other cases Ferdinand had not depicted a detail visible in the herbarium specimen or a specimen cultivated in the Royal Garden at Kew that was now found to be relevant. These complexities led to watercolours that included elements first observed in the field in Australia and elements first observed in London or Kew. Several colour-coded pencil drawings now in the Naturhistorisches Museum document Australian plants cultivated at Kew; they are annotated 'Kew', often together with an Australian locality such as 'King Georg [sic] III Bay', and dated 1807, 1808 or 1809, corroborating this complex approach.

FERDINAND'S PLANT ILLUSTRATIONS FROM THE VOYAGE IN THE *INVESTIGATOR*

In September 1806 the readers of the *Annals of Botany* were informed in an anonymous note, 'We are happy to find that Mr. Brown and Mr. Ferdinand Bauer are sedulously employed in arranging their important materials for a work which cannot fail to prove a lasting monument to both their indefatigable zeal and the talents by which they were so eminently distinguished, the former as one of the most philosophical and accurate botanists of the day, the latter as an artist whose performances (like those of his brother) unite, with a truth hitherto unseen in botanical paintings, all the neatness, grace, and effect, so much admired in the works of a [Abraham] Mignon and [Jan] Van Huysum: It is scarcely necessary to add, that the work alluded to is the result of their botanical researches on the expedition of Captain Flinders to New Holland . . . The mode in which their publication will be offered to the public is unknown to us; we suppose however, that it will be preceded by a prodromus by Mr. Brown, which will convey a better idea of the variety and singularity of his materials than we are able to give in this place'.[6] The explicit use of the term 'prodromus' may imply a reference to another *Prodromus* then in the making – that stipulated in Sibthorp's testament (see Chapter 9), written by Smith with the first instalment published in late 1806, the very year in which the note above appeared. Comparing Ferdinand with two of the greatest names in Dutch still-life flower painting shows the high appreciation Ferdinand's work had already achieved at that moment.

Unlike Sibthorp, Banks had no friend to whom he regularly reported on the progress of Ferdinand's work. As a consequence we do not know whether Ferdinand followed the Linnaean order, as he had in his work for Sibthorp, or followed a different sequence. The only evidence for the start of the work seems to have been a botanical watercolour sent by Banks in early 1806 to the Admiralty as a specimen of Ferdinand's art. After all, he was on their payroll and his watercolours would end up in their 'Deposit' as specified in the Memorandum of understanding. In the same month Brown told Smith 'Mr. Bauer whose abilities & industry you are well acquainted with has made about 1800 drawings all of which are accompanied by minute dissections. In what manner these are to be given to the public, if indeed we should ever have it in our power to publish them, it is at present impossible to say'. As will be seen later in this chapter, this was a very realistic analysis of the situation.

The most likely place for Ferdinand's drawing table to be placed was the engraver's room in 32 Soho Square. He acquired water-colourist paper of the best quality, unfortunately only occasionally exhibiting watermarks that would enable a more precise chronology of the work. He must also have bought at a London artist's colourman a broad spectrum of pigments, which have not yet been submitted to an in-depth chemical analysis. Otherwise Ferdinand will have felt like he had done in the years 1788–1794 in Oxford. The main difference was that Ferdinand produced, at an early stage, a manuscript list of all his pencil drawings following the Linnaean sequence, numbering each individual entry, and furthermore added the respective number, as a rule in parenthesis, in pencil to each of his pencil drawings; these could be called the London numbers. The advance over his earlier work done for Sibthorp, in particular the even more sophisticated colour code, has already been discussed (see Chapter 10), but a few additional notes seem appropriate.

The first step in the process was the selection of those pencil drawings that were to form the basis of the repeats in watercolour. Nothing is known about the persons involved and the criteria applied, but is seems sensible to assume that Brown and possibly also Ferdinand were involved. A cross section of the plants recorded was selected, without a particular emphasis on a group of plants or a geographical area visited, although Madeira, the Cape Region, Timor and Norfolk Island were excluded (see Chapter 10). Judging from the manuscript list of pencil drawings written up by Ferdinand (see Chapter 10), only a relatively small portion of the pencil drawings was chosen for further work. The second step must have been to decide on the sequence to be followed. On this we have no details, but for this part of the chapter the assumption is made that the same order was followed as in Oxford. There are several reasons to make this plausible: the Linnaean sequence in Ferdinand's manuscript list, the practicality of the Linnaean system for a first orientation of plant diversity, and the fact that Brown could hardly have developed all his taxonomic concepts as early as 1806 for all the plants encountered in Australia. This means that a plant such as the purple-flowered *Thysanotus volubilis* (Fig. 11.1.) from the southern coast belonging to the Linnaean Triandria would have been among the very first to be documented as a watercolour. These decisions taken, the work on the individual plant portraits could start, and Ferdinand had to follow another step-by-step approach – more than two hundred times.

Firstly, he had to extract from the masses of pencil drawings those that referred to the species to be documented, since often more than one sheet of paper, mostly in rather small format, had been used in the field for a single taxon. Locality information that was sometimes, though not consistently, noted down in the field will have been of some help, but as a rule no genuine collection numbers in the modern sense were given on the pencil drawings with the London numbers offering additional guidance. Secondly, Ferdinand seems to have compared his pencil drawings with the herbarium specimens kept in 32 Soho Square and with the respective living specimens then already in cultivation in the Royal Garden at Kew. At first glance this may seem surprising, but the seed material gathered by Good had reached England on two different vessels before the *Investigator* had returned. On 6 June 1802 the *Speedy* had left Port Jackson, bringing the seeds from the southern coast of Australia to London; in July 1804 the *Calcutta* was back in England with the seeds gathered on the eastern and northern coast of Australia. In the same way as it had been done in the Physic Garden at Oxford, a plantation sacred to Australia was raised at Kew and Ferdinand must have had access to these plants (see above). As a result Ferdinand's finished watercolours were in some cases mosaics consisting of elements based on different specimens: (1) living plants recorded in the wild as pencil drawings, (2) herbarium specimens collected in the wild and (3) plants raised from seeds collected in the wild and now cultivated at Kew. This complex approach led at least in one case to an error: Ferdinand combined a male flowering shoot of *Dendrocnide excelsa* in the upper part of one of his watercolours with floral structures of *D. moroides* in the lower part. Needless to say, it is impossible to distinguish in a watercolour elements based on a pencil drawing documenting a plant cultivated at Kew from elements based on a pencil drawing of a plant encountered in the wild in Australia – unless both pencil drawings are found in the Naturhistorisches Museum. The watershield, *Brasenia schreberi*, is an example of a plant species recorded by Ferdinand in the wild in a pencil drawing that became the basis of one of his watercolours. Seeds of this very species collected in the wild were successfully raised in the Royal Garden in Kew, and the resulting plants were documented by Franz in a watercolour. The two images, one by Ferdinand, the other by Franz, are now kept in the Natural History Museum.

Having decided what to document in watercolour, Ferdinand must thirdly have decided how to arrange the plant illustration and the often numerous dissections on the sheet of paper. The rest must have been routine for the experienced watercolourist that he was. Significantly for Ferdinand, he signed all finished watercolours in the right-hand lower corner in his neat 'Ferd. Bauer del.', but refrained from giving a date, a mode he occasionally took up in his final years, and a locality. In January 1806 Banks sent a specimen of Ferdinand's activity to the Admiralty choosing a watercolour of the Western Australian pitcher plant, *Cephalotus follicularis* (Fig. 11.3.). In his covering letter, Banks indicated that Ferdinand had prepared 1541 pencil drawings from Australia, eighty-nine from the Cape, forty pencil drawings of animals from Norfolk Island, and two hundred and sixty-three pencil drawings of animals from Australia. The work proceeded at 32 Soho Square where Brown, then deep in the preparation of his *Prodromus Florae Novae Hollandiae* and working on the same material,

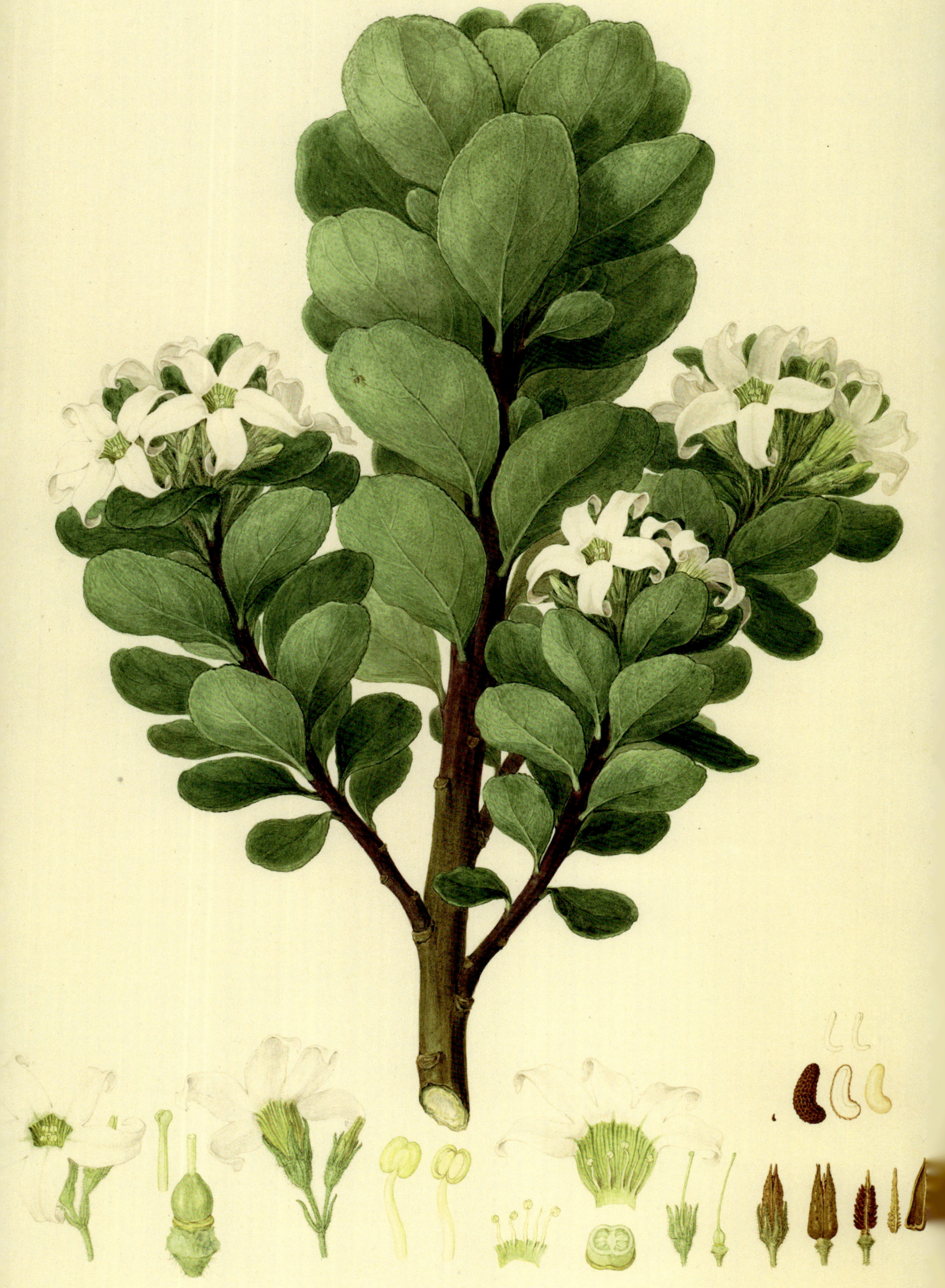

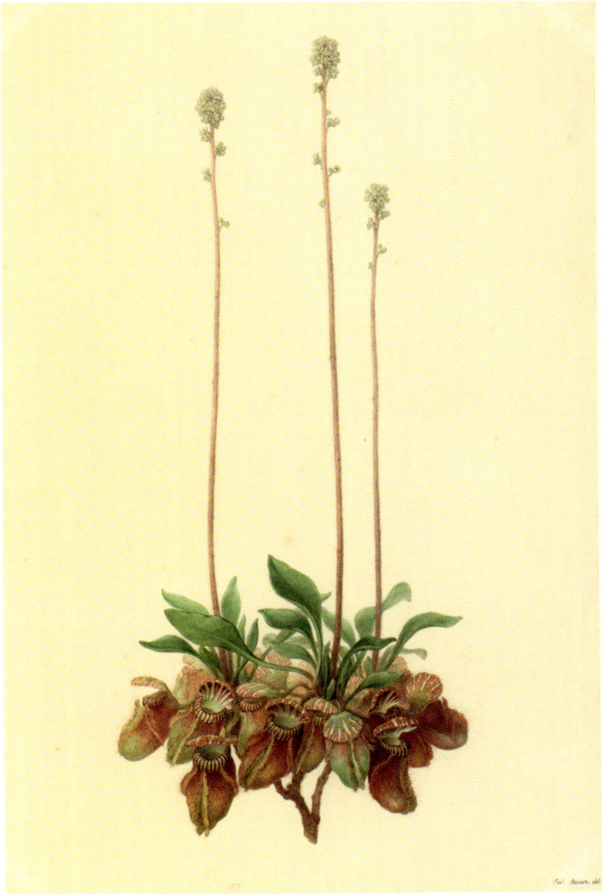

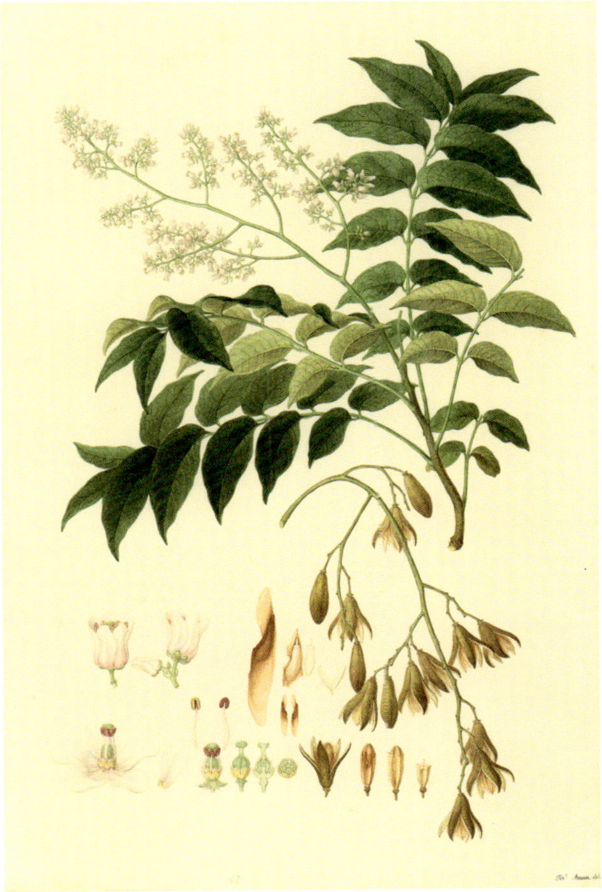

FIG. 11.3. *(above)* Australian pitcher plant, *Cephalotus follicularis*. Watercolour by Ferdinand Bauer, signed, 1806–1810. – London, Natural History Museum, Library and Archives, Australian Botanical Drawings, pl. 41. [A].

FIG. 11.4. *(above right)* Crow's ash, *Flindersia australis*. Watercolour by Ferdinand Bauer, signed, 1806–1810. – London, Natural History Museum, Library and Archives, Australian Botanical Drawings, pl. 23.

FIG. 11.2. *(opposite)* Sticky tailflower, *Anthocercis viscosa* subsp. *viscosa*. Watercolour by Ferdinand Bauer. signed, 1806–1810. – London, Natural History Museum, Library and Archives, Australian Botanical Drawings, pl. 109.

could act as a point of reference. Since they worked in close vicinity, though probably on different tables, communication must have been by word of mouth, of which nothing was ever put on paper.

In this atmosphere of concentrated work, the watercolour of the sticky tailflower, *Anthocercis viscosa* subsp. *viscosa* (Fig. 11.2.), was prepared, with Ferdinand possibly reflecting his first landings in King George III Sound, only a few days before he got critically dehydrated in the scorching summer sun near the south-western tip of Australia. Different associations may have been on Ferdinand's mind when preparing the watercolour depicting crow's ash, *Flindersia australis* (Fig. 11.4.), from the coast of what is now Queensland. By then Brown had probably already coined the generic name *Flindersia* later to be validated in his *Prodromus* and Ferdinand possibly knew about it – but Flinders was still detained by the French Governor of Mauritius (see Chapter 10). The flowering branch of *Eucalyptus pruinosa* (Fig. 11.5.) from the Gulf of Carpentaria offered a special challenge: how to document the distinctive whitish waxy surface of the leaves? Ferdinand buried the grey-greenish pigments under a fine layer of opaque white, which makes this watercolour a most perfect example of accurate documentation. The

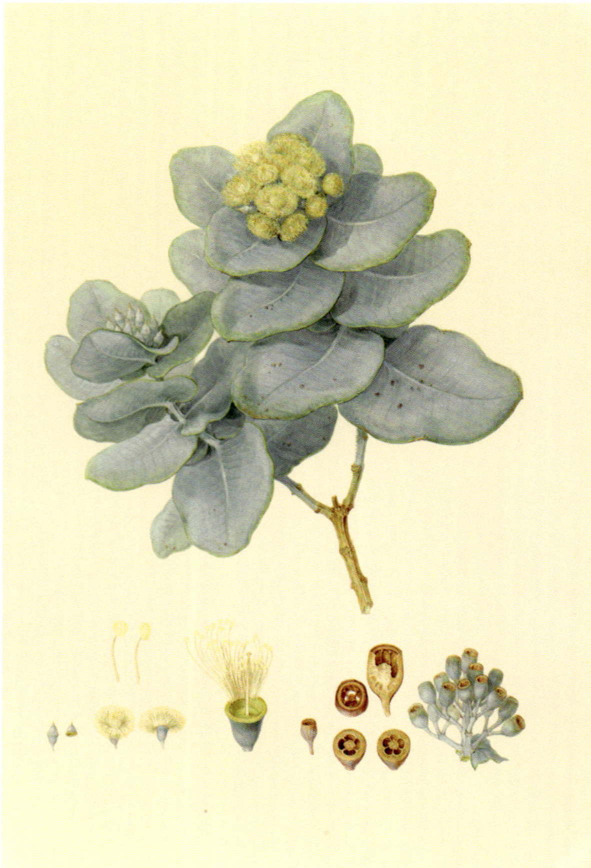

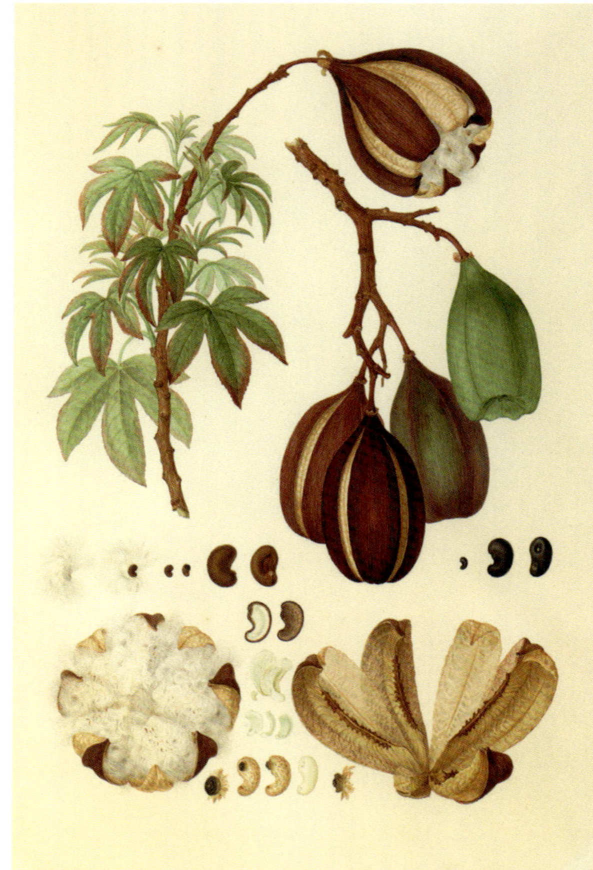

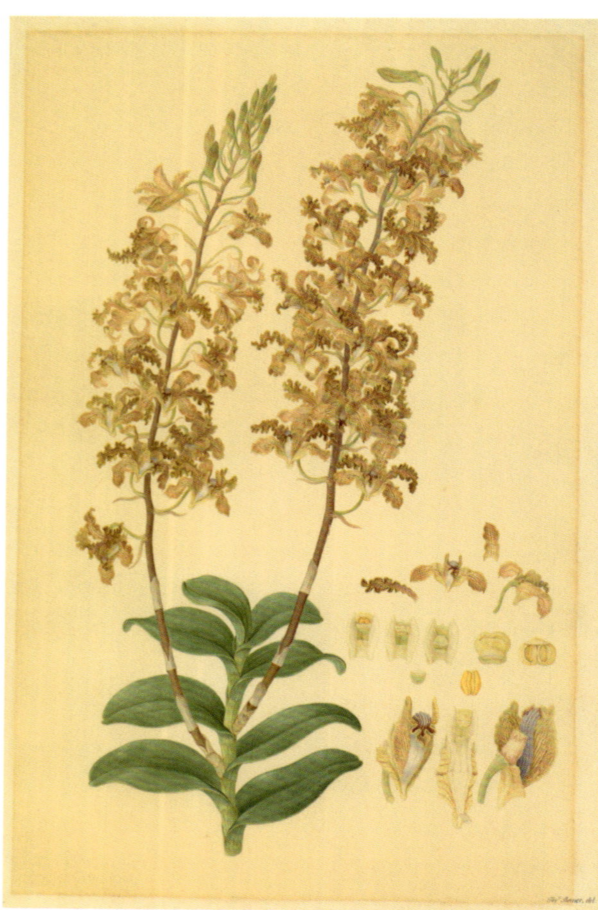

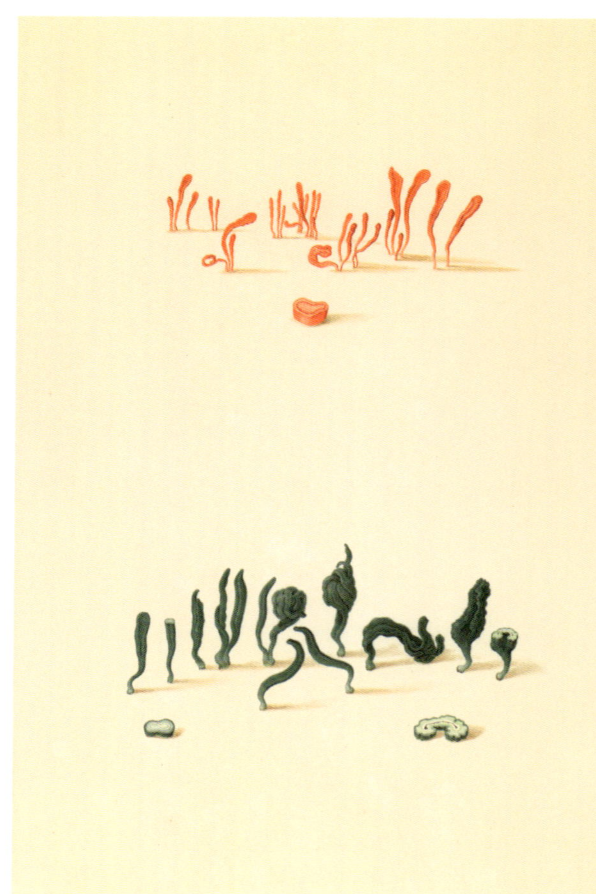

hairy seeds of the bush tucker, *Cochlospermum gillivraei* (Fig. 11.6.), offered another challenge: how to record the white fluffy seed hairs on a white sheet of paper? Ferdinand did not need to rely on painted shadows as other inferior hands would do, but added a tiny quantity of grey to the representation of a few seed hairs, which gave the necessary substance and shadow to the object itself. This is another masterpiece of precise observation, which is evidence of his superb workmanship. For orchid lovers, Ferdinand's representation of *Dendrobium discolor* (Fig. 11.7.), probably recorded in Keppel Bay on 10 August 1802, is another triumph, even greater when compared to the fascinating, so far unpublished pencil drawing in the Naturhistorisches Museum, with its very many numbers documenting the different stripes, patches and dots in brown and yellow on the tepals. Ferdinand also produced a few watercolours documenting fungi encountered by him in Australia. One of them is *Clavulinopsis miniata* (Fig. 11.8.), of which no hint is found in the diaries of Brown and Good, nor in the *Voyage*. Since no specimen has been found in the Natural History Museum, and the Naturhistorisches Museum to correlate with this watercolour, it is the only record of this fungus from the voyage of the *Investigator*.

The scientific value of Ferdinand's watercolours is outstanding, in particular for those cases in which no pertinent specimen or pencil drawing has survived. A good example is *Thysanotus volubilis*, a name based on a specimen collected by Ferdinand, as explicitly mentioned by Brown in the *Prodromus*. No such collection exists in the Natural History Museum and the pertinent specimen as well as Ferdinand's pencil drawing deposited in the Naturhistorisches Museum were destroyed in the Oberhöflein fire in 1945.

Ferdinand's botanical watercolours referring to the voyage in the *Investigator* have been compared to those commissioned by Sibthorp and found to be of superior quality. Here is the reasoning of one commentator: 'The plates for the *Flora Graeca*, while admirably portraying the plant's habit of growth, show in general only such enlarged dissections of the flower as are necessary to allocate it within the Linnean [sic] system of classification based essentially on the number of floral parts and only the external appearance of the seed or fruit. The illustrations made on the Flinders voyage not only give details of such parts of the flower as the calyx, corolla and stamens much more enlarged, but show the internal structure of the ovary and the seeds themselves. At this period few . . . paid such attention to those microscopical details, which are now considered to be of fundamental importance . . . Bauer also came to appreciate the intimate structure and to reveal this in enlarged drawings of unsurpassable accuracy. For that alone these Bauer drawings would constitute a landmark in the history of botanical illustration'.[7] In this context 'superior quality' clearly means competitive with new concepts in science, which required the documentation of the details that were until then left unrecorded – not quite what is usually meant by the term 'quality'. It has been argued that the advance over Ferdinand's botanical work done for Sibthorp is due to the influence of Brown. There may be some truth in it, but this is hardly the complete story – Franz had penetrated deep into the anatomy of orchid fruits and seeds as early as 1799, before Brown had even been invited to become scientific assistant on the *Investigator*, and Franz almost certainly had shared his observations with his younger brother before the voyage started in 1801. So it may be said that Brown could also have been strongly influenced by Ferdinand in his attention to internal details of ovaries and seeds. Praise for Ferdinand's botanical work is universal, with William T. Stearn writing, 'for . . . their meticulous accuracy and their wealth of exquisite detail . . . the illustrations of Australian plants are the most beautiful and informative that have ever been made'.[8] Similar words have been used by Peter Whitehead, calling Ferdinand an 'almost giant figure',[9] while Margaret Jones, a contemporary plant illustrator, noted 'it goes beyond technique'.[10] The latter remark

FIG. 11.5. ***(top left) Eucalyptus pruinosa.*** **Watercolour by Ferdinand Bauer, signed, 1806–1810. – London, Natural History Museum, Library and Archives, Australian Botanical Drawings, pl. 56.**

FIG. 11.6. ***(topright)*** **Bush tucker,** ***Cochlospermum gillivraei.*** **Watercolour by Ferdinand Bauer, signed, 1806–1810. – London, Natural History Museum, Library and Archives, Australian Botanical Drawings, pl. 4.**

FIG. 11.7. ***(bottom left) Dendrobium discolor.*** **Watercolour by Ferdinand Bauer, signed, 1806–1810. – London, Natural History Museum, Library and Archives, Australian Botanical Drawings, pl. 162.**

FIG. 11.8. ***(bottom right) Clavulinopsis miniata.*** **Watercolour by Ferdinand Bauer, signed, 1806–1810. – London, Natural History Museum, Library and Archives, Australian Botanical Drawings, pl. 236.**

echoes Banks who had told Marsden in January 1806 that Ferdinand's work 'is beyond, what I confess I thought possible to achieve'.

However, the unsurpassable accuracy had its price: the speed of production of an individual watercolour was slower in 32 Soho Square compared with the work done at Cowley House. From December 1787 until early 1794 Ferdinand had produced nine hundred and sixty-six botanical illustrations, two hundred and ninety-three zoological illustrations and seven frontispieces. From October 1805 until August 1814 Ferdinand produced two hundred and three botanical illustrations, ten engraved copper plates, and coloured about two hundred copper engravings for his *Illustrationes* (see below).

FERDINAND'S ANIMAL ILLUSTRATIONS FROM THE VOYAGE IN THE *INVESTIGATOR*

It was believed by some that all watercolours based on the pencil drawings from the *Investigator* voyage were prepared by Ferdinand while in England. However, there is evidence to the contrary, which is corroborated by Lhotsky. When Ferdinand sat down in Hietzing near Vienna to write his testament[11] (see Chapter 17) he stipulated that 'the already finished drawings from Greece and Australia are to go to my brother Franz'; there is little doubt that by finished drawings Ferdinand meant watercolours. While the inventory of Ferdinand's belongings drawn up after his death is not very specific, though listing 'two books bound in red morocco with animal and plant drawings' as well sixteen 'Greek drawings', the report by Nikolaus Joseph Host (see Chapter 17) to Francis I, Emperor of Austria, on Ferdinand's estate is precise. It stated, 'There are 49 complete plant drawings, which are to go as legacy to Franz Bauer in London'. The second report sent to Franz I, almost certainly by Caspar Maria Count Sternberg (Kašpar Maria hrabě ze Šternberga), first president of the Patriotic Museum for Bohemia in Prague and the founder of modern paleobotany, notes: 'One portfolio of about 40 watercolours of plant illustrations, mostly orchids, with very excellent dissections. The same with quadrupeds, birds and fishes executed in a masterly way, all from New Holland. Both portfolios contain [together c.] 90 illustrations, which would form an exquisite ornament for the already brilliant library of His Majesty the Emperor'. The most plausible interpretation is that Ferdinand had referred in his testament to sixteen finished landscape drawings from his trip to the Ottoman Empire and a total of around ninety finished botanical and zoological watercolours referring to the *Investigator* voyage (while the botanist Host had only referred to the botanical watercolours, mostly orchids). It seems likely that the latter, i.e. the plant and animal watercolours mentioned by Sternberg, were prepared during Ferdinand's years in Vienna and not in London.

So what happened to the '49 complete plant drawings' mentioned by Host? They definitely did not turn up at the auction of Franz's belongings. Instead Franz must have passed them directly on to Brown. Upon the latter's death precisely forty-nine plant and fungus illustrations by Ferdinand, which had never been to the Admiralty, were acquired by the British Museum – three representations of the Western Australian pitcher plant, seven of fungi and thirty-nine of orchids. In the end these forty-nine botanical watercolours in the possession of Brown were united, with the two hundred and three botanical watercolours coming from the Admiralty and deposited in the British Museum.

But what happened to the 'quadrupeds, birds and fishes'? They did not enter the library of Emperor Francis I being finished drawings covered by the clause in Ferdinand's will and therefore went to the latter's brother at Kew. They were not listed among the belongings of Franz to be auctioned in London in 1841. Instead they must have passed from Franz before that year directly - or indirectly via Brown – to the Admiralty, from where they were passed to the British Museum in 1843.[12] In short, the production of watercolours based on pencil drawings prepared in Australia started in London and was almost certainly continued during Ferdinand's final years in Vienna (see below and Chapter 12).

No information exists on the chronology of Ferdinand's work on the animal watercolours done in Vienna – for example whether or not he started with the mammals. The five watercolours documenting the koala, *Phascolarctos cinereus*, were particularly exciting for the scientific community, although by that time the animal was no more a novelty (see Chapter 10). One of Ferdinand's images depicts a female koala with her young clinging

FIG. 11.9. *(opposite)* Koala, *Phascolarctos cinereus*. Watercolour by Ferdinand Bauer, signed, after 1814. – London, Natural History Museum, Library and Archives, Australian Zoological Drawings, pl. 9.

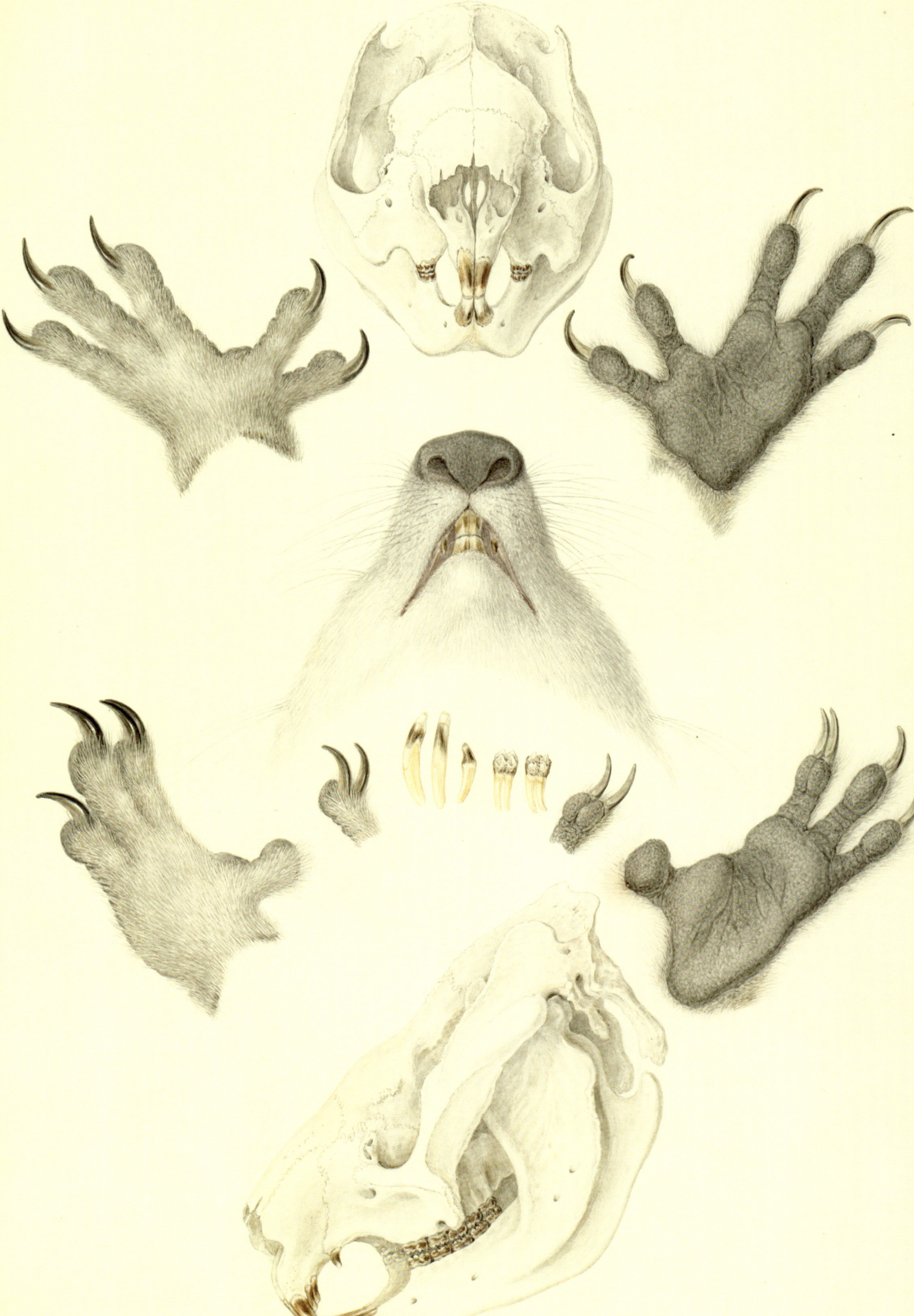

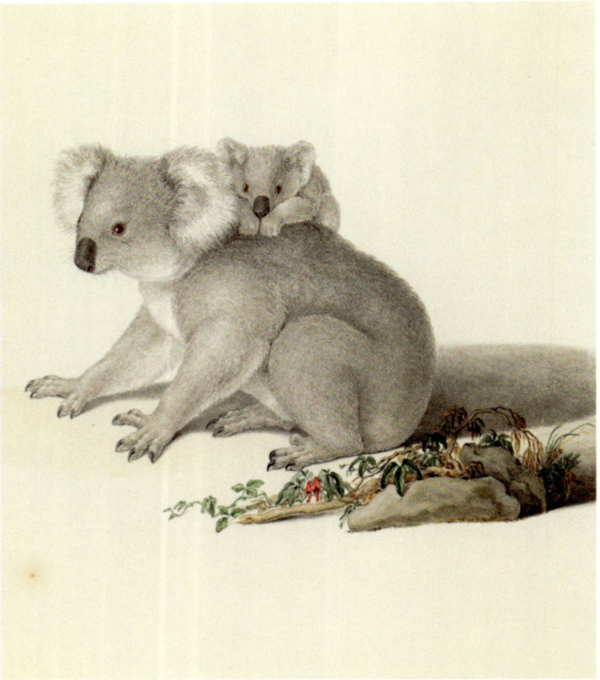

FIG. 11.10. Koala, *Phascolarctos cinereus*. Watercolour by Ferdinand Bauer, signed, after 1814. – London, Natural History Museum, Library and Archives, Australian Zoological Drawings, pl. 6.

FIG. 11.11. Platypus, *Ornithorhynchus anatinus*. Watercolour by Ferdinand Bauer, signed, after 1814. – London, Natural History Museum, Library and Archives, Australian Zoological Drawings, pl. 14.

its back (Fig. 11.10.); another represents the skull of the animal in front and side views above and below, in the centre the head from below together with five extracted teeth, plus the palms of the four extremities (Fig. 11.10.). From his earlier zoological work for Sibthorp, Ferdinand knew the importance of an accurate rendering of the eyes. As a consequence he took infinite pains to precisely document the details of the often relatively small eyes in his pencil drawings (see Chapter 10). Now he put a minute shiny spot on the eye of his mammals and birds to increase vivacity. The watercolour of the platypus (Fig. 11.11.) merits attention because of Ferdinand's strict adherence to a model seen decades ago, almost certainly in the Imperial Library in Vienna – the *Recueil d'Oyseaux* (see Chapter 2), today *Codex Miniatus 53* of the Österreichische Nationalbibliothek. For Ferdinand's earlier examples illustrating this approach see Chapter 6. Now he combined the aquatic platypus with the aquatic fern *Marsilea*, almost certainly the common nardoo, *M. drummondii*, restricted in distribution to Australia (see Chapter 10). This is another plant that had to wait for decades until it received its scientific name – in this case by Alexander Braun, then Professor of Botany at Berlin University and Director of the Royal Botanic Garden in Schöneberg near Berlin, which had been visited by Franz in 1788. The fur of both the koala and the platypus, rather difficult to render in a lifelike way, are depicted in utmost refinement (see Chapter 12).

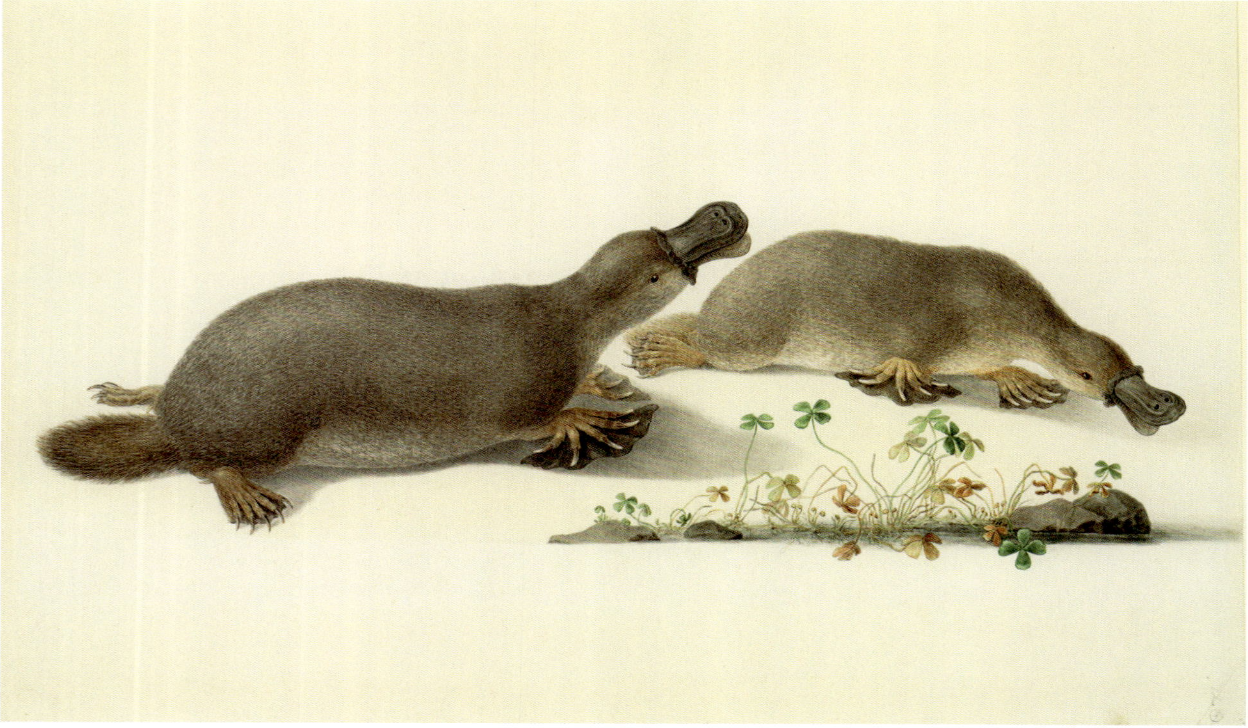

The two rainbow lorikeets, *Trichoglossus haematodus moluccanus* (Fig. 11.12.), one facing left, the other right, are another masterwork of accurate documentation. It is the 'jizz' of both birds and the exserted tongue of one of them that show Ferdinand as the accomplished observer. He did not even forget the hairy protuberances on the tongue's tip. The northern rosella, *Platycereus venustus venustus* (Fig. 11.13.), a parrot from the Gulf of Carpentaria, may have made Ferdinand think of the tropical heat and humid atmosphere during the

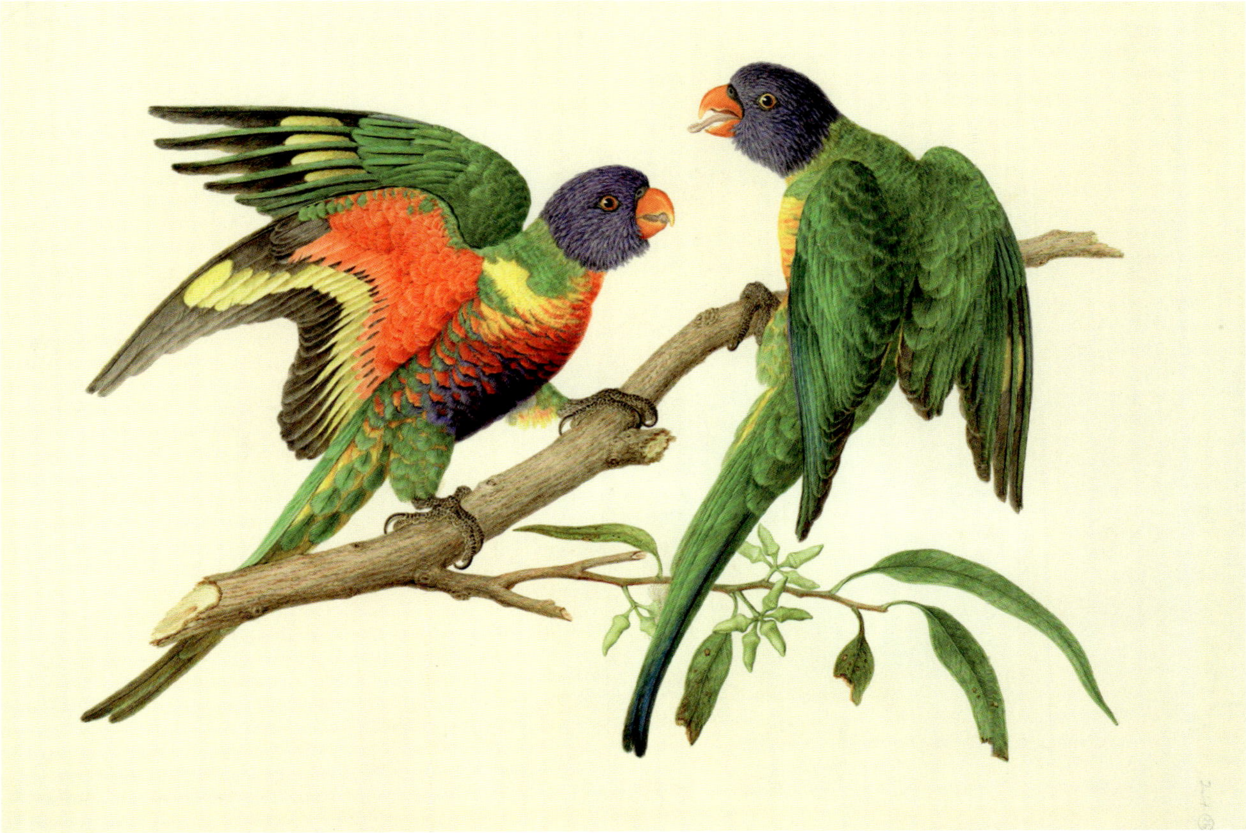

FIG. 11.12. Rainbow lorikeet, *Trichoglossus haematodus moluccanus*. Watercolour by Ferdinand Bauer, signed, after 1814. – London, Natural History Museum, Library and Archives, Australian Zoological Drawings, pl. 26.

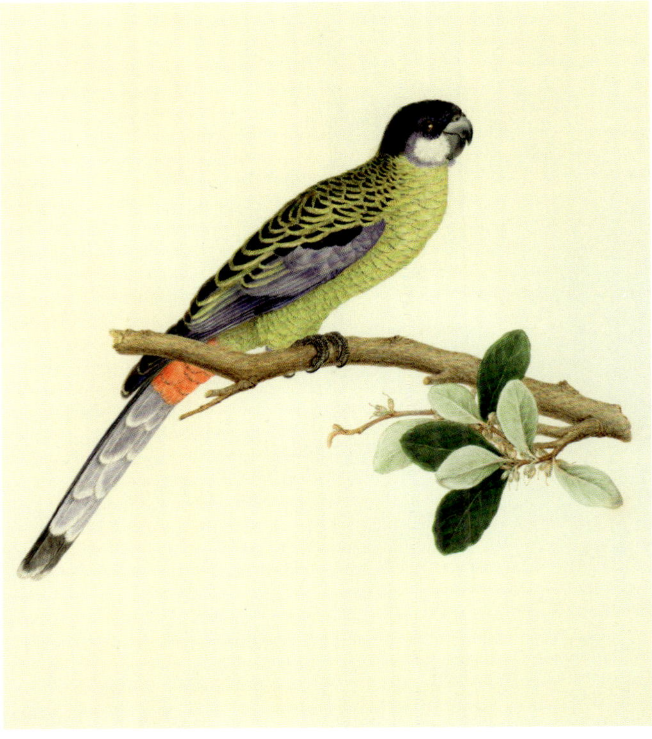

FIG. 11.13. Northern rosella, *Platycereus venustus venustus*. Watercolour by Ferdinand Bauer, signed, after 1814. – London Natural History Museum, Library and Archives, Australian Zoological Drawings, pl. 21.

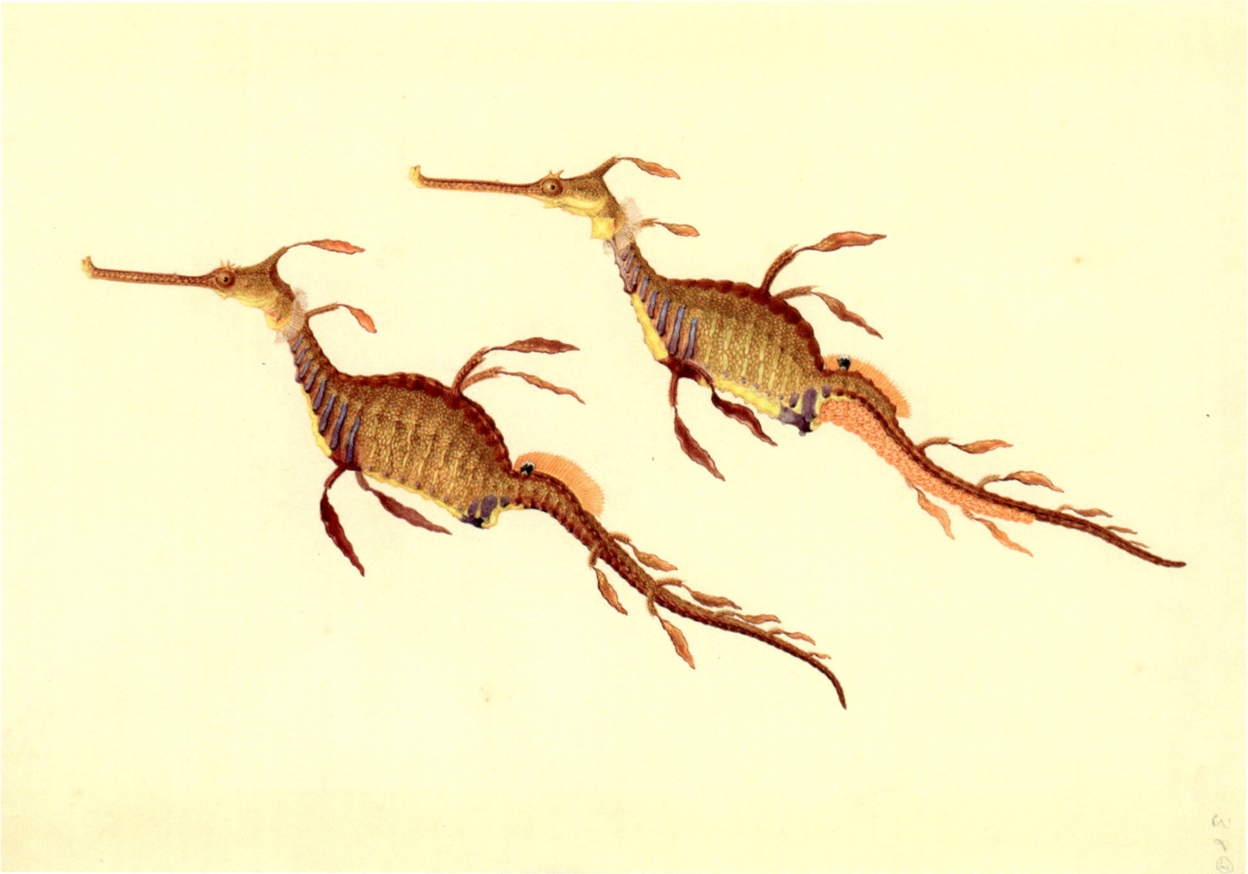

seemingly endless and dull trigonometric survey of the northern coast of Australia. Compared to what might have been done during the extended stay of the *Investigator* in Australian waters, very few fishes were recorded by Ferdinand, who may have suffered from hydrophobia as Brown did. No doubt the most spectacular of all was the weedy seadragon, *Phyllopteryx taeniolatus* (Fig. 11.14.), which became one of the most popular watercolours prepared by Ferdinand. Sex in fishes is not easy to discern, but in this case a female is shown to the left and a male carrying the brood to the right, both turned to left and drawn from the side. The butterfly cod, *Pterois* spec. (Fig. 11.15.), caught near the Strong Tide Passage cannot be determined to species level. The pectoral (ray) counts refer to one species, the dorsal and anal counts to another, while the pictorial shape refers to a third species. In this case Ferdinand must have been either confused or misled in his pencil drawing in the Naturhistorisches Museum, which has only very recently been published.[13] The famous male blue swimmer crab, *Portunus pelagicus*, has been recorded by Ferdinand in three watercolours – in a dorsal view (Fig. 11.16.), in a ventral view and in a detail depicting the mouthparts only.

FIG. 11.14. Weedy seadragon, *Phyllopteryx taeniolatus*. Watercolour by Ferdinand Bauer, signed, after 1814. – London, Natural History Museum, Library and Archives, Australian Zoological Drawings, pl. 48.

FIG. 11.15. *(opposite top)* Butterfly cod, *Pterois* sp. Watercolour by Ferdinand Bauer, signed, after 1814. – London, Natural History Museum, Library and Archives, Australian Zoological Drawings, pl. 40.

FIG. 11. 16. *(opposite bottom)* Male blue swimmer crab, *Portunus pelagicus*. Watercolour by Ferdinand Bauer, signed, after 1814. – London, Natural History Museum, Library and Archives, Australian Zoological Drawings, pl. 50.

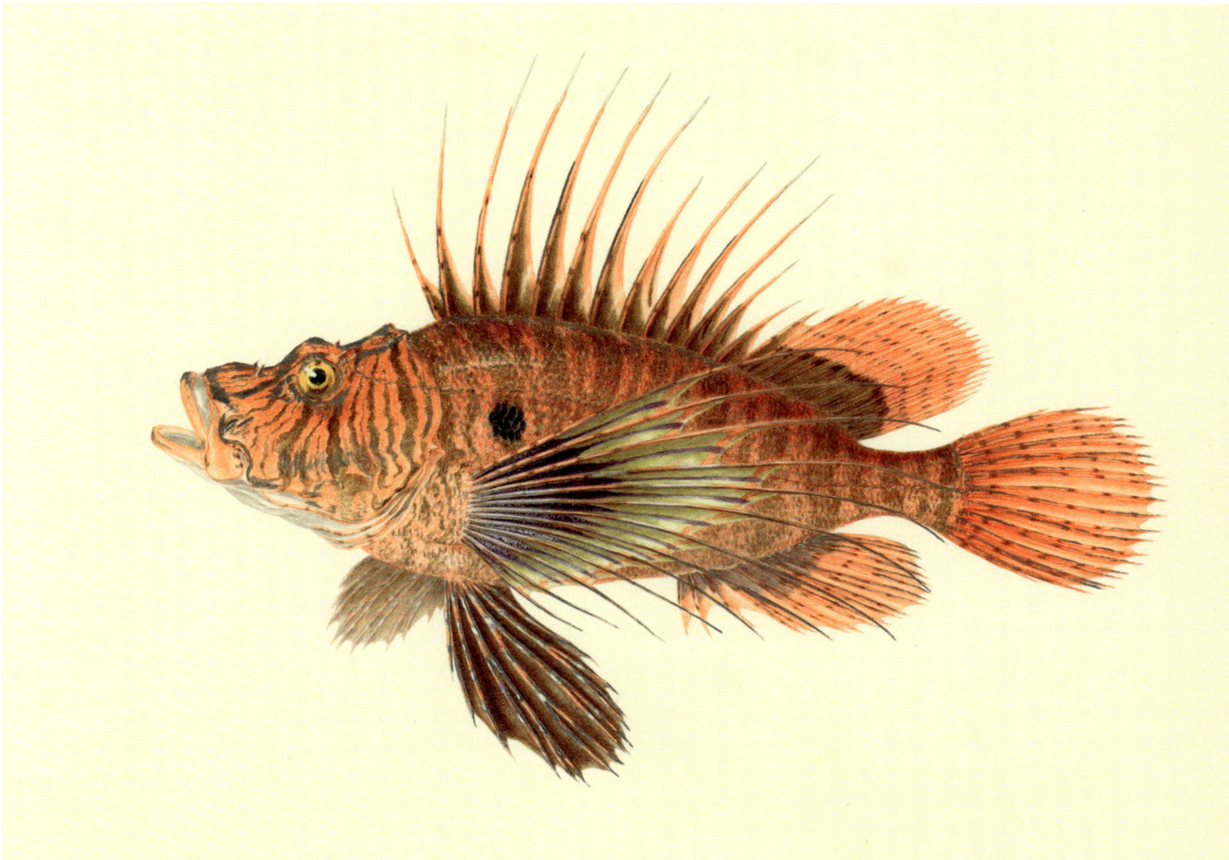

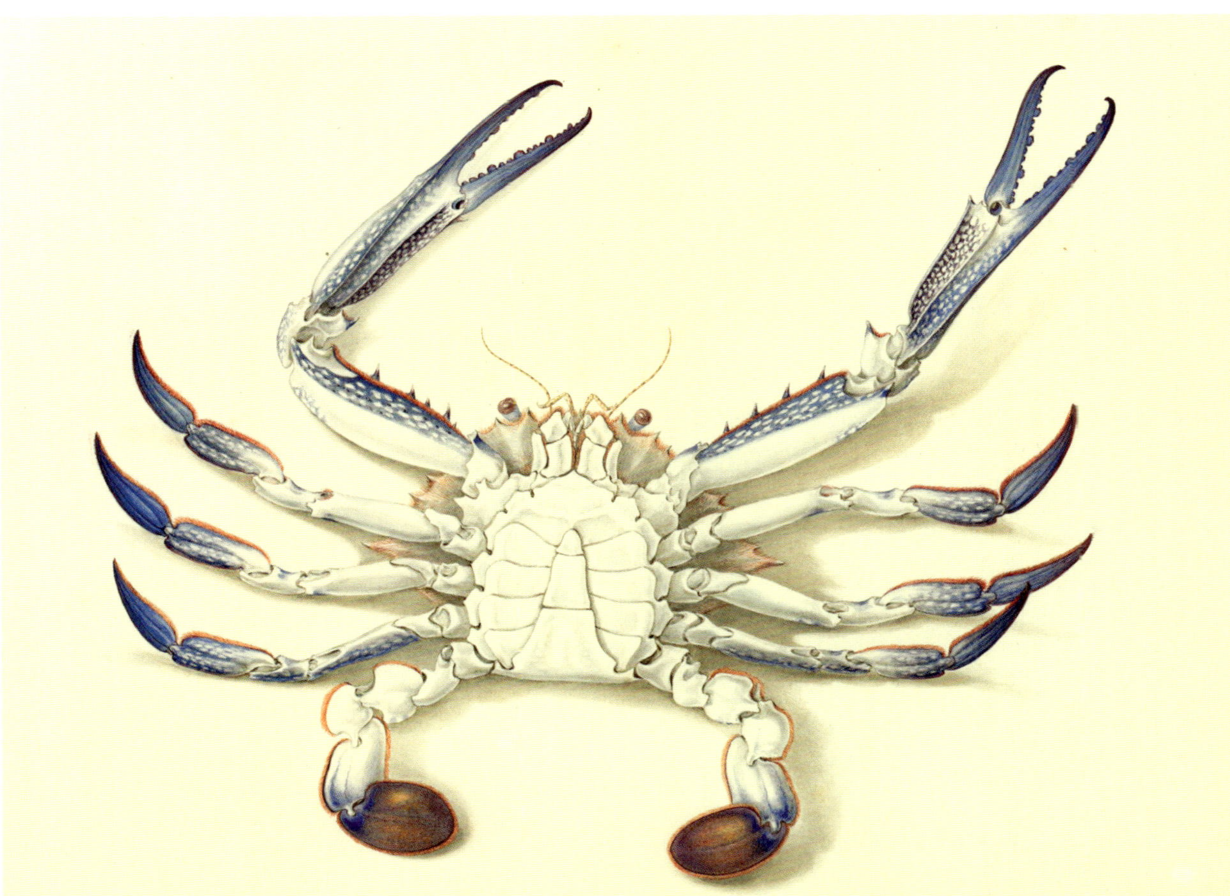

The final destination of Ferdinand's botanical watercolours was clear from the beginning – not Banks's Library at 32 Soho Square or the British Museum in Montagu House as one might suspect, but the Admiralty, who had paid Ferdinand's salary. The details of the delivery to the Admiralty, in particular the chronology, remain obscure, however. Clearly the recipient of this material had priorities different from research on plant diversity, which meant that Ferdinand's watercolours, the result of at least eleven years of his life, were shelved and remained outside the sphere of the botanical community for three decades. In effect this meant non-publication. In an incompletely published manuscript kept in the archives of the Royal Botanic Gardens Kew, botanist Allan Cunningham complained in a footnote, 'It is a known fact that for some time after the drawings were [sent] to the Admiralty, they were laid on the table of the drawing room of the then first Lord [Melville, i.e. Robert Dundas, 2nd Viscount Melville, Keeper of the Signet of Scotland, First Lord of the Admiralty from 1812 to 1827 and a Knight of the Thistle]'. Cunningham continues that there 'they were exhibited as Portfolios as most Souvenirs etc. now usually are – Thus exposed to inquiry in the hands of a person, utterly ignorant of Botanical Science, & therefore totally incompetent to appreciate their real worth'.

In the minutes of the Trustees of the British Museum for 8 April 1843 (twenty-three years after the death of Banks, and seventeen years after the death of Ferdinand), Cunningham's account is confirmed and record made of 'three volumes of very exquisite drawings of plants [and animals] of New South Wales, made many years ago by Mr Bauer which were preserved at the Admiralty'. Soon afterwards the three volumes were given to the British Museum and must have been divided: the two hundred and three botanical illustrations were placed under the custody of the Keeper of Botany (Brown), the forty-nine zoological illustrations under the custody of the Keeper of Zoology (Asa Gray). Upon Brown's death, Ferdinand's illustrations of Australian plants purchased by the former from Franz were added to the botanical series previously in the Admiralty (see above). In a chronicle of the Natural History Departments of the British Museum these figures – two hundred and three botanical and forty-nine zoological drawings donated by the Admiralty – are corroborated.[14]

The deposition of this precious material at the Admiralty for such a long time was a deplorable decision for the botanical and zoological community, but not felt as long as Ferdinand remained in London with his pencil drawings, because on the basis of these he could produce repeats at any time. This is the context for a few pencil outline drawings prepared by Ferdinand to act as exemplars for a small number of copper engravings, one of them published by Smith showing the blue pincushion, *Brunonia australis* (Fig. 11.17.). It is worth stressing that in this very special case a watercolour based on the same pencil drawing prepared in the field had been made or was to be made for the Admiralty.

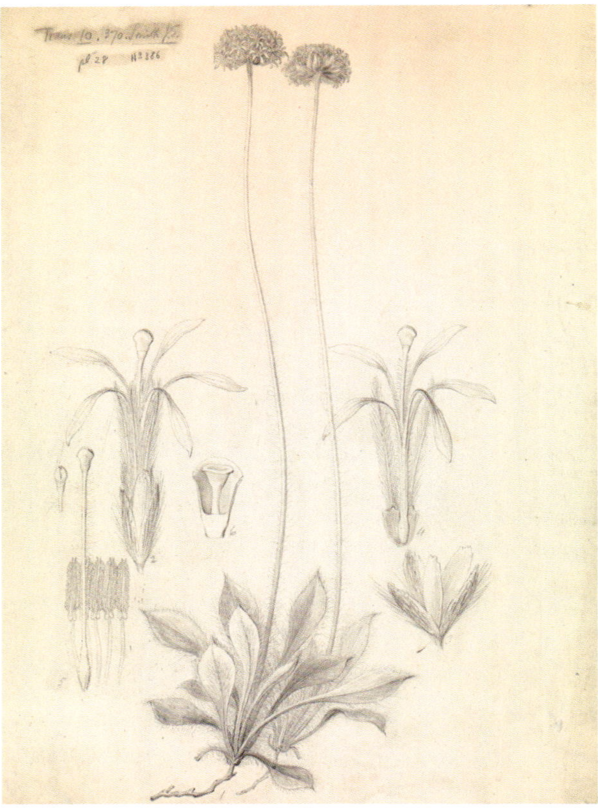

FIG. 11.17. Blue pincushion, *Brunonia australis*. Pencil drawing by Ferdinand Bauer, c.1809. – London, Linnean Society, Archives.

Another extremely narrow channel of publication for Ferdinand's pencil drawings and watercolours was the Atlas to accompany Flinders's *Voyage*, the travelogue published in 1814, which came out on the day following the death of its author. The botanical appendix included ten copper engravings prepared by several hands based on eleven of Ferdinand's pencil drawings and watercolours, among them the crow's ash, *Flindersia australis*, the Western Australian pitcher plant, *Cephalotus follicularis*, and the orchid *Corymbas fimbriatus*. Ferdinand is likely to have been dissatisfied by this tiny number of copper engravings and by the fact that all his meticulous attention to the hues of the rainbow had been disregarded, since the ten plates were published uncoloured. However, he may have received an advance warning of what the appendix would comprise.

Praise for Ferdinand's watercolours documenting the Australian flora and fauna is universal – often eclipsing, partly or completely, his earlier and later work. Words such as 'the Leonardo of natural history painting'[15] or 'superb examples of botanical illustration, the work of one of its greatest masters at the height of his powers'[16] refer to this material, now one of the great treasures of the Natural History Museum, and regularly shown to prominent visitors. There are very good reasons to do so since 'these [watercolour] drawings please both eye and intellect by conveying so much information of scientific value in an aesthetically enjoyable form'. Indeed this is the very cause for their outstanding success. However, the complex origin of the watercolours, in particular their relationship to the colour-coded pencil drawings in the Naturhistorisches Museum, are often either forgotten or not understood. Again we are faced with an amalgam of contributing factors: Ferdinand's superb eye; his exceptional capacity for accurate observation and analysis; his precise recording in colour-coded pencil drawings prepared in the field; his capacity for concentrated work on the drawing table in the studio; his feeling for an aesthetic arrangement of the dissections; and his profound knowledge of plant colours and their reproduction on paper. All this resulted in botanical and zoological illustrations of outstanding quality, rightly regarded as among the best ever made.

In a sense Ferdinand's watercolours of Australian plants have become iconic for this continent: enlargements were used as murals in Sydney Airport, where they were seen by millions of passengers for several years.

ILLUSTRATIONES FLORAE NOVAE HOLLANDIAE

In 1792 Hawkins had discussed with Sibthorp Ferdinand's idea of publishing the birds of the Levant, a plan that was immediately rejected by Ferdinand's master and therefore had come to nothing (see Chapter 6). Supported by Banks, Franz had successfully started the publication of the *Delineations*, but this project had been stopped by the realities of the market after merely three instalments (see Chapter 8). By contrast, Ferdinand's nascent project of publishing a work on foxgloves seems to have aborted at an even earlier stage, with a single plate engraved, dated '1 Oct 1799' and possibly used for printing (see Chapter 12). The Memorandum signed by Ferdinand in 1801 allowed him under the clauses of this regulation (see Chapter 10), to publish his drawings at his own risk, and he decided to do so. This is how the first publication that had Ferdinand's name on the title page started – the famous *Illustrationes Florae Novae Hollandiae*, a work in large format very much in the Banksian style. It consisted of (1) a title page, (2) a dedicatory page, (3) a page with a brief introduction, (4) a page listing the abbreviations used in the plates, for example 'I 2' standing for stigma, and (5) the copper engravings, but, significantly, no text. As with many other botanical works of the time the *Illustrationes* exist in a coloured and an uncoloured version.

Ferdinand may have been induced into undertaking this venture because of the start of the *Flora Graeca*, which began to appear in late 1806, ten years after the death of his former master, and which had proceeded very slowly. Seeing the first instalments, each comprising fifty coloured copper engravings and text, no doubt in the library of Banks, who was one of the subscribers, may well have been an extremely frustrating experience: of course Ferdinand could not have expected his name on the title page, but it was also suppressed on all the copper plates – in the old style as practised by Jacquin. Furthermore, this procedure was to be continued in all later instalments. A feeling of dissatisfaction and mutilated pride may have played a role when Ferdinand started with his new project, which must have been in c.1811. This was not an auspicious period in time for the production of a botanical work destined for the inner circle of the elite. Europe was at war, and the domestic economy was suffering from these debilitating general circumstances. Two more factors served to act strongly against Ferdinand: firstly he was not the beneficiary of a foundation, effectively Sibthorp's estate in Oxfordshire with its regular proceeds, and secondly he was a perfectionist. The latter fact made his brother Franz report to Jacquin the son in Vienna on 19 August 1814 [in English], 'my brother begun a most excellent work on his own account, from the Sketches an[d] Herbarium which he made and collected during his Voyage and which are now his own property . . . This work is the size of the Hortus Schönb [runnensis, i.e. *Plantarum rariorum horti caesarei schoenbrunnensis descriptiones et icones*, the last great work of Jacquin with five hundred

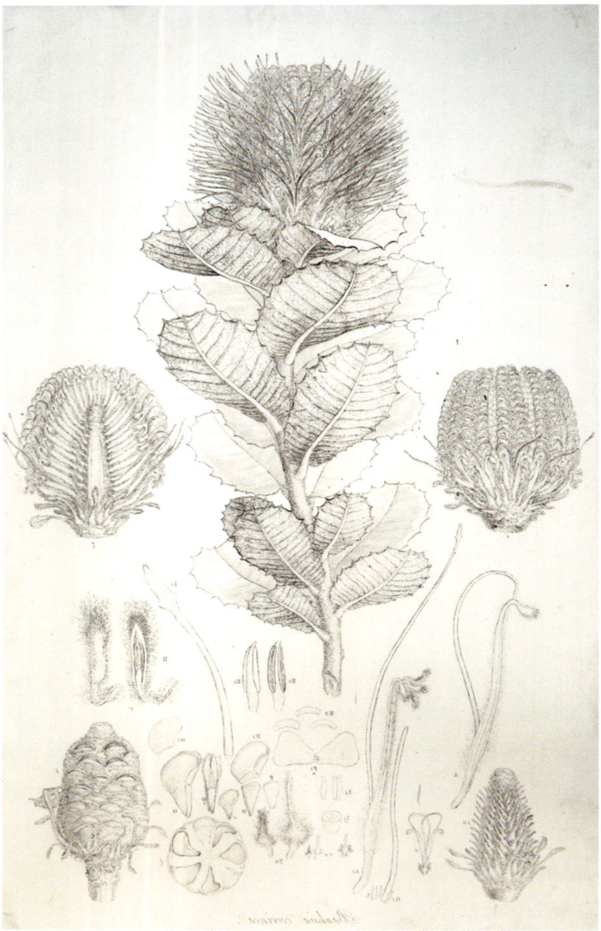

FIG. 11.18. Scarlet banksia, *Banksia coccinea*. Copper plate engraved by Ferdinand Bauer based on a watercolour by his own hand, c.1813, chromium plated, c.1980. – London, Natural History Museum, General Library.

coloured copper engravings in folio format]; drawn, engraved and colord in a highly finished style, by himself; it appears in numbers of five plates, at one Guinea and half pr number the colord and at five Shillings pr number the plain – there are already two numbers published'.[17] Everything indicates that Ferdinand did not rely on assistance by graphical workers, but engraved the copper plates himself (see Fig. 11.18.), had them printed and coloured a proportion of the print run. This meant a huge workload for Ferdinand: assuming the production of twenty coloured copies of the two instalments published in London he had to colour two hundred copper engravings, and this he did to utmost perfection.

The *Illustrationes* were clearly intended as a companion to Brown's *Prodromus Florae Novae Hollandiae*, since each plate contains a page reference to the later work, incomplete as it was and destined to remain. Contrary to what one might expect from such a work, Ferdinand did not follow Brown's sequence of genera: this resulted in another, much smaller cross section of the Australian flora, albeit limited to the portion dealt with in the *Prodromus*. All this resulted, for example, in the two orchids represented in the *Illustrationes* appearing on plate 2 and 8, while the scarlet banksia, *Banksia coccinea* (Fig. 11.19), found its place on plate 3 and *Grevillea banksii*, which belongs to the same family, on plate 9. The plates of the first two instalments are only partly numbered, which indicates that there was limited planning in the project, obviously open-ended right from the beginning. Considering the realities of the production process and Ferdinand's perfectionism, only a small print run was produced and an even smaller number of coloured sets. A list of subscribers in Ferdinand's hand for the first two instalments exists in Ferdinand's own copy of the *Illustrationes*, which ended up in the Library and Archives of the Natural History Museum. Uncoloured copies were to go to Banks, Brown, Lambert, Aiton the son, and Home, but also to figures associated with Ferdinand's tour to the Ottoman Empire – Hawkins; Humphrey Sibthorp the son, John's half-brother; John Montague Cholmeley, the husband of John's half-sister Sarah; John's successor in the chair, Professor George Williams of Oxford University. Coloured copies were to go to Banks, Brown, Menzies, and fewer than ten copies to prominent booksellers in London. In short, the *Illustrationes* were intended for the happy few and were totally dependent on subscriptions.

What is more important for the history of botanical illustration is Ferdinand's steady advance in documenting details as evidenced by his portraits of the blue pincushion, *Brunonia australis*, of which he had produced three versions – the first a watercolour for the Admiralty, the second a pencil outline drawings for Smith, belatedly published as copper engraving in the *Transactions of the Linnean Society*, and the third an engraved plate for his *Illustrationes* (Fig. 11.20.). The first has enlargements of the flowers and of flower parts, as well as details of the fruits and seeds, including the embryo. The second is arranged in a different way and comprised several additions, among them representations of (1) the

FIG. 11.19. Scarlet banksia, *Banksia coccinea*. Coloured copper engraving by Ferdinand Bauer based on a watercolour by his own hand, 1813. Illustrationes Florae Novae Hollandiae, London, 1813, t. 3. – London, Natural History Museum, Library and Archives.

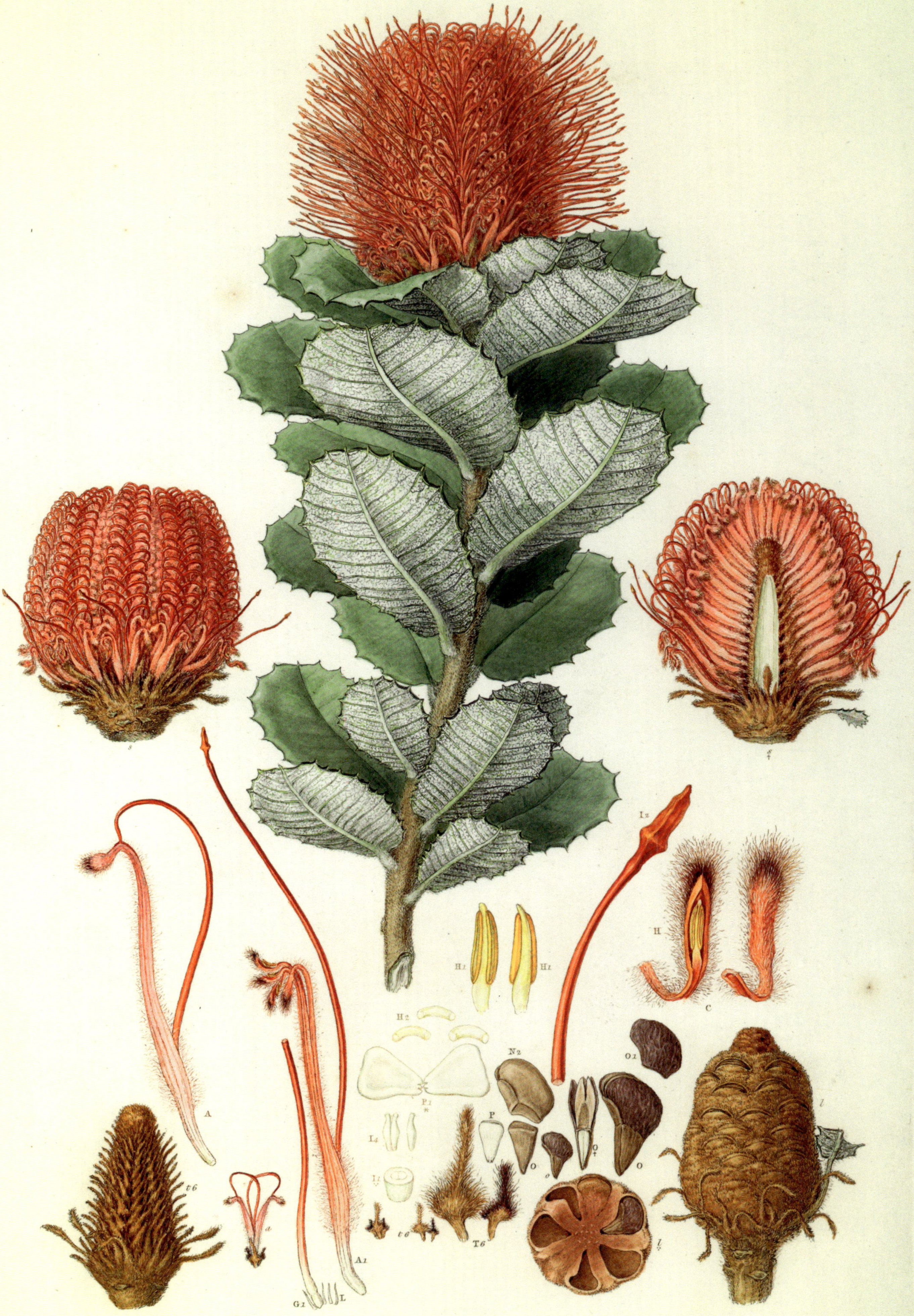

Banksia coccinea.

Brown prod. fl. nov. holl. p. 394. n. 17.

Ferd. Bauer.

Original in Bauer's drawings 139

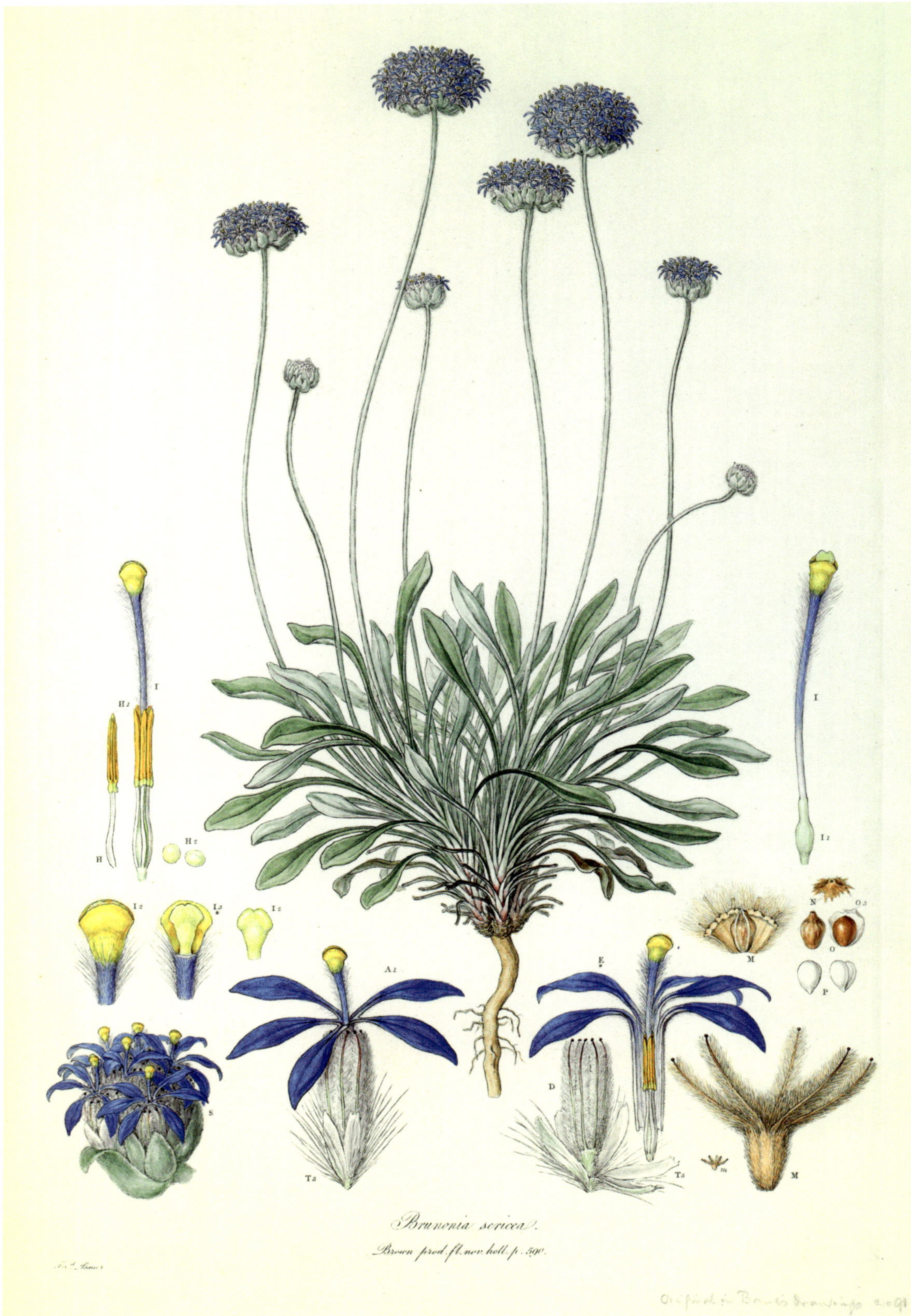

FIG. 11.20. Blue pincushion, *Brunonia australis*. Coloured copper engraving by Ferdinand Bauer based on a watercolour by his own hand, 1813–1814. Illustrationes Florae Novae Hollandiae, London, 1813–1814, t. 10. – London, Natural History Museum, Library and Archives.

style surrounded by the stamens, (2) the style without the stamens, (3) the tip of the style enlarged and (4) the tip of an anther enlarged. The third variant is the printing form for plate 10 of the *Illustrationes*: it is arranged in a third way and shows additionally two enlarged pollen grains marked 'H 2', another abbreviation explained in the front of the book. To consider pollen in plant illustrations was novel and has to be seen in the context of the palynological studies at that time undertaken by Franz (see Chapter 14) and later by Brown.

In November 1813 the first instalment of Ferdinand's *Illustrationes* was appropriately reviewed by the *Monthly Magazine* stating, 'No letter-press will accompany the plates but with the first number is given a general table, to which the letters, figures and marks on each table refer. Thus, by always preserving the same characters to denote similar parts, one general table of explanation serves the whole . . . The drawings, we are informed in the Preface, have been, for the most part, made from the living plants in their native soil . . . and some few [drawings] will necessarily be taken from dried specimens, preserved in the herbaria of this botanist [Ferdinand], or that of Mr Brown'. Who wrote the Latin preface of the *Illustrationes*? Twelve years after Ferdinand's death Endlicher in Vienna offered the answer – it was Brown, an extremely erudite man and competent Latinist, although of markedly inferior linguistic competence than Endlicher. The latter was fluent not only in his native German and Hungarian, but also in several other languages, and published a substantial textbook – on Chinese grammar – a few years later (see Chapter 18).

FERDINAND'S ENIGMATIC OIL PAINTING

On 22 June 1999 an oil painting on panel came up for auction at Phillips, New Bond Street, London that created a great surprise among the specialists and fetched a very considerable price.[18] Measuring 34.3 cm × 25 cm it was a flower piece showing three different species of passion flowers, a tropical butterfly and three carefully painted water drops on the leaves of the passion flowers, in the lower left-hand corner annotated 'Ferdinand Bauer 16 May 1812'. So far it is the only oil painting signed by Ferdinand known to have survived and ended up in the Art Gallery of South Australia in Adelaide (Fig. 11.21.). It had originally been exhibited in the Royal Academy in London in 1813, the year before Ferdinand returned to Vienna. The signature on the painting has not yet been properly analyzed, but it seems not to have been forged.

Conventional flower pieces show a plethora of unrelated plant and animal species often together with a memento mori, such as a fly (see Chapter 3). However, in this case we observe a strictly botanical approach, since all plant elements represent passion flowers: the stem-winged passionflower, *Passiflora alata*, the blue passionflower, *P. caerulea*, and *P. holosericea*. All three are native in the neotropics, a region Ferdinand had never visited, but he could easily have seen all of them cultivated in conservatories in England or earlier in and around Vienna. It is the butterfly that is the sensation. Ferdinand was not the type of 'botanical draughtsman', using Admiralty speak, to associate the passionflowers with a gratuitous butterfly, but instead he chose a butterfly whose larvae feed on passionflowers. In fact he painted a member of the *Heliconii* sitting on a flower of *P. holosericea*. This is Ferdinand the accurate observer, although it cannot be said that the oil painting is a

FIG. 11.21. Passion Flowers. Flower piece by Ferdinand Bauer. Oil on panel, signed, 1812. – Adelaide, Art Gallery of South Australia.

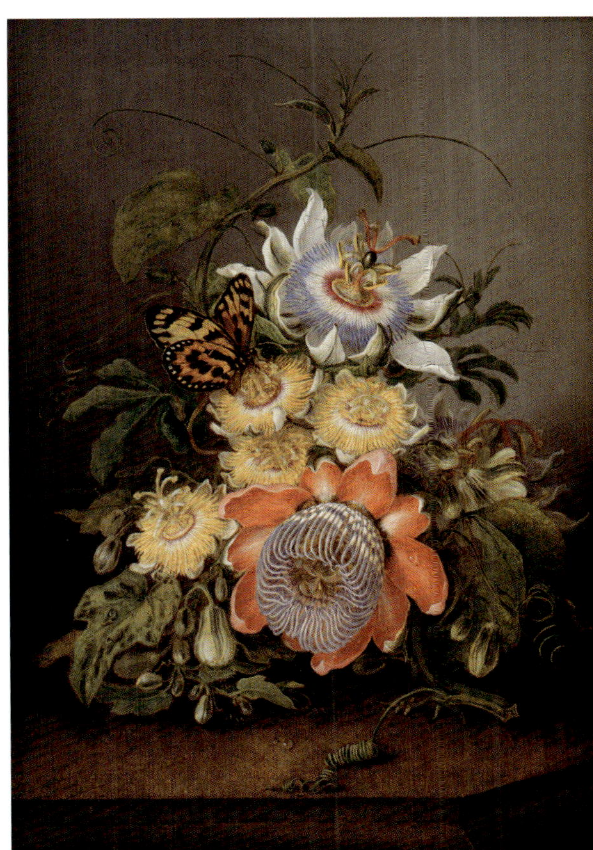

masterpiece when compared to the works of the artists to which his watercolours had been compared in the *Annals of Botany*. He clearly had limited experience in oil painting, and some of the details, for example the diameter of the individual stigmatic branches, are incorrect. However, in any event the technique chosen does not result in the accuracy and precision that can be achieved in watercolour.

The context of Ferdinand's Passionflowers in oil is unclear. It is a fact that the Bauer brothers had already portrayed passionflowers for Boccius, Ferdinand had recorded *P. aurantia* on Norfolk Island (see Chapter 10), and he must have had ample opportunities to see other passionflower species in the Royal Garden at Kew. But why on earth a painting in oil? It has been stated that the 'Royal Academy [in London] only accepted works in watercolour [sic, recte oil], and therefore it is likely that Bauer executed 'Passionflowers' in oil hoping to be recognised by this highly esteemed organisation', but this may not be the full story. Considering the extreme paucity of facts currently available on this oil painting, it is difficult to make more than a guess about a commissioner, possibly from the circle of connoisseurs in the Horticultural Society. Joseph Sabine, long-serving Inspector General of Assessed Taxes, Fellow of the Royal Society, Fellow of the Linnean Society and Honorary Secretary of the Horticultural Society of London, comes to mind. He was a passionflower enthusiast and in 1822 published in the *Transactions of the Horticultural Society* the first report of a successful cross between two passionflower species, one of the parents being the blue passionflower (see Chapter 12). The rare dark violet hybrid was documented by Ferdinand in his very last datable watercolour painted on paper watermarked '1825' (see Chapter 12). The hypothesis is tentatively put forward that the oil painting of passionflowers stands in some form of relationship to Sabine. There is little doubt that he would have had the financial means to pay Ferdinand a decent honorarium.

The fact remains that little is known about Ferdinand's flower pieces, among them the pair painted for Canwick Hall (see Chapter 6) and the one including the touraco bird (see Chapter 9). Together with the Passionflowers three more are listed as having been exhibited at the Royal Academy of Arts in 1813.[19] They are entitled 'Group of Flowers', 'Group of Flowers from New Holland' and 'Group of Flowers from Cape of Good Hope', but their whereabouts are currently unknown. Two more flower pieces by Ferdinand appeared in the auction catalogue of Franz's belongings under the heading 'Pictures'; they are also untraceable. Six more flower pieces are explicitly mentioned as property of the Naturhistorisches Museum in a standard bibliography for botanical illustration,[20] of these four, all pencil drawings and not colour coded, are kept in the archives of this institution.[21] Needless to say, further flower pieces by Ferdinand painted in London or in Vienna may exist in private collections.

It seems helpful to pause for a moment and reflect on the more general political circumstances under which the Passionflowers were painted. Twelve days after the painting had been signed and dated by Ferdinand in London, Mikhail Illiarionnovich Kutuzov, the Russian commander at the infamous Battle of Austerlitz [Slavkov u Brna] (see Chapter 12), signed in Bucharest another treaty between Russia and the Ottoman Empire. Under this agreement the eastern half of the Principality of Moldavia between the Prut and Dniester Rivers was ceded by the Ottoman Empire and Russia obtained trading rights on the Danube. The reason behind the conclusion of the treaty and its rapid ratification by Alexander I, Emperor and Autocrator of all Russias, on 5 July 1812 in St Petersburg was clear. Russian intelligence had learned that the *grande armée* was to cross the Neman (Nemunas) River the next day and enter Russia; therefore peace with the archenemy in the south was absolutely necessary. The *grande armée* of 1812 under the command of Napoleon was not a French army, but a European army with troops from all the different states on the continent except the Kingdoms of Portugal, Sweden and the Ottoman Empire. Of the estimated 690,000 soldiers, until then the largest army ever, only a small remnant of c.93,000 returned to Prussia alive in the early winter months of 1812. Although Moscow went down in flames, this was a brilliant victory due to the scorched earth principle advocated by Michail Bogdanovich Barclay de Tolly, the Russian Minister of War, and put into effect by the commander-in-chief Kutuzov. Whereas all this led to huge casualties and enormous economic problems on the continent, life in Great Britain, where Ferdinand and Franz dedicated themselves to their work of documenting plant diversity, was only marginally affected. After all, peace was the prerequisite for botanical illustration.

In the following year, when Ferdinand published the first instalment of the *Illustrationes* in London, the Napoleonic wars culminated in the bloodiest battle of all. In October 1813 a coalition army consisting of Russian, Austrian, Prussian and Swedish troops under the Austrian commander-in-chief Karl Philipp Prince Schwarzenberg (Karel Filip kníže ze Schwarzenbergu) decisively defeated a coalition army led by Napoleon near Leipzig in the Kingdom of Saxony, with an estimated 80,000 to 110,000 soldiers killed. Alexander I, Francis I and Friedrich Wilhelm III, King of Prussia, had been present on the battlefield and a good deal of diplomacy had been necessary on the part of Schwarzenberg to keep them in harmony. On 31 March 1814 Russian and Prussian troops headed by Alexander I and Friedrich Wilhelm III entered Paris. Two days later, the brother of the beheaded King Louis XVI was declared King of France. He named himself Louis XVIII and returned from his exile in England. After a suicide attempt Napoleon abdicated for himself and his son on 13 April and was escorted into exile on the island of Elba. Finally, on 30 May 1814 the First Peace of Paris was signed, ending the Napoleonic Wars, which had devastated the European continent and resulted in a collapse of the established political system. The structure of a new one had to be designed at a congress to meet in Vienna in September 1814. These monumental facets of news must have reached Ferdinand in London via the press or his contacts at the Austrian Embassy, then headed by the newly appointed envoy extraordinary Maximilian Count Merveldt. All this may have made him realize that a window of opportunity was now open for him to leave England without risking imprisonment or his life – Napoleon had finally been defeated and a lasting peace agreement had been found.

LEAVING ENGLAND

A single letter explains Ferdinand's proximate reasons for leaving England after a stay that had lasted, not counting the fifty months on the *Investigator* voyage, for no less than twenty-two years. Written in August 1814 (see above) Franz told Jacquin the son in Vienna, 'though the work [*Illustrationes*] is highly approved off, the sell [sic] of it here is so inconsiderable, that my brother could not clear the fiveth [sic] part of his expenses; he therefore determined to leave England and packet up his whole collection of Drawings, consisting of more 2000 of plants, several hundrets of animals, Birds, Fishes, Snakes etc etc also his [valuable herbarium and skins]'. The term 'therefore' sounds very much like frustration, in particular when considering the fact that Brown had been engaged by Banks as his private secretary in 1810, whereas Ferdinand had not been given a position, let alone a permanent one. All this may have contributed towards his decision to leave England, with Lhotsky (see Chapter 20) writing that Ferdinand was 'excessively and unduly disheartened by this failure [of the *Illustrationes*]; so much so, that, fearing he should never be able to do any thing else'.

This sounds like a personal crisis in Ferdinand's life, but there may have also been other ulterior reasons, on which some speculation may be permitted. Judging from the few letters extant, Ferdinand seems to have never envisaged remaining permanently in England, as expressed by him in a letter to Hawkins in 1797 (see Chapter 9). There was also no special reason for him to remain there – the *Flora Graeca* project had passed into other hands, the engagement with the Admiralty had come to a conclusion, and Banks was either unwilling or unable to organize a prolongation of the arrangement with the Admiralty that he had initiated. In addition, Ferdinand must have been very unhappy that several hundred more watercolours could be very easily prepared by him on the basis of his pencil drawings from Australia, Madeira, the Cape, Timor and Norfolk Island, but there was no more interest in them. In short, Banks with all his wealth was seemingly not prepared to engage a second Bauer, either temporarily or on a permanent basis. Additionally, Banks was clearly not a sponsor in the conventional sense – no doubt he would have possessed the means to support the *Illustrationes*, dedicated to him like so many other works, but these funds he did not make available to Ferdinand. In this the latter was not alone: Brown's *Prodromus florae Novae Hollandiae*, another work dedicated to Banks, also did not find the necessary financial support by his dedicatee for continuation and because of the extremely meagre sales of the first instalment remained a torso (see above). All this led to the well-known fact, that many plants first collected during the *Investigator* voyage received their scientific names on the basis of specimens collected by other botanists, often several decades later.

Any major decision of long-term consequences such as changing the country of residence has financial, organizational and emotional aspects. We know very little about Ferdinand's financial situation, apart from two facts: firstly he had received 300 guineas per year during his engagement by the Admiralty, from 1801 to c.1810; and secondly he died as a man in comfortable circumstances in 1826. At the moment of his death he owned a house in Hietzing near Vienna (see Chapter 12), where he kept his library, collections and drawings; there were also 1070 Austrian guilders in cash, but the real surprise is that he had no less than 7610 Austrian guilders in the Bank of England. It is somewhat difficult to believe that Ferdinand would have gained this amount of money during his final years in Vienna – even after taking into account the revenue from his work for Cattley and the sale of his passionflowers on paper to the Horticultural Society of London (see Chapter 12). No precise calculation concerning exchange rates has as yet been made, but it seems possible, at least at first glance, that Ferdinand had possessed an additional source of income during his final years in England.

One of them was painting animal watercolours for the lucrative London art market. As subjects Ferdinand chose exotic birds and, in at least two cases, large format. On 19 July 1978 seven such watercolours came up for auction at Christie, Manson & Woods Ltd. in London[22] and some reached hammer prices more than double the estimated values. Judging from the very brief description in the auction catalogue Ferdinand had documented Australian birds, among them the rainbow lorikeet, *Trichoglossus haematodus moluccanus*, and the sulphur-crested cockatoo, *Cacatua galerita*, which he had recorded during the voyage on the *Investigator*, as well as birds from other continents. The latter were probably observed by him in captivity in London. Here we find e.g. the African grey parrot, *Psittacus erithacus*, a companion parrot prized for its ability to mimic human speech, and the green-winged macow, *Ara chloroptera*, wide-spread in the forests and woodlands of northern and central South America. The Alexandrine parakeet, *Psittacula eupatria*, one of the oldest captive parrot species in Eurasia, naturally docile, peaceful and playful, is another bird depicted by Ferdinand. The watermarks given in the auction catalogue for three of these bird watercolours indicate that they were painted after the return to London in 1805.

In addition, Ferdinand could have made money by producing flower pieces, taken up engagements like the work done for the Duke of Marlborough (see Chapter 9), and he could have also given lessons in flower painting to the elite like his brother had done or gained money in a field outside his professional activity as a natural history illustrator. We simply do not know. In any case the assumption is permitted that Ferdinand, aged fifty-four in 1814, deeply frustrated, a bachelor with some reserves and no private obligations, could have had simply enough of England and, tired of London, left for Vienna. Moving his belongings would not have caused insurmountable problems, in particular since he had already ample opportunity to collect experience during his travels on the cramped *Investigator*. Ferdinand's luggage must have been extensive, with Lhotsky mentioning '14 large cases': these included botanical and zoological collections from the *Investigat*or voyage, his colour-coded pencil drawings from this expedition, to which he was entitled according to the Memorandum, the finished views from his tour to the Ottoman Empire, the ten engraved copper plates for the first two instalments of the *Illustrationes* and even his waddy from the southern coast of Australia.

How the botanical and zoological collections were divided in London and what sort of criteria were used remains an enigma, apart from the fact that Ferdinand rightly regarded all his collections from Norfolk Island as his property and took them with him to Vienna. When after Ferdinand's death his botanical collections were accessioned by the Vereinigte k. k. Naturalien-Cabinete – United Imperial Royal Natural History Cabinets in Vienna (see Chapter 17) – Leopold Trattinick, the Keeper of the Botanical Cabinet, noted 'Pflanzen angeblich 3339' (plants supposedly 3339), largely from the voyage on the *Investigator*. On the same occasion Joseph Natterer, the Keeper of the Zoological Cabinet, recorded '114 birds from Australia, Norfolk Island and Cayenne' as well as four mammals and invertebrates not listed. Of these specimens, none appear to have been labelled by Ferdinand; all annotations are from a later date and by different hands, for the botanical specimens notably Endlicher's (see Chapter 18). All this makes the Naturhistorisches Museum the second recipient of botanical and, surprisingly, the first for zoological specimens from the voyage in the *Investigator*. The following figures may illustrate this rather general statement: the record by Natterer for one hundred and fourteen birds (see above) stands in contrast to the eighty-one bird skins extant in the Naturhistorisches Museum in 1993,[23] while only twenty-five[24] and, more recently in 2005, twelve,[25] have been recorded for the Natural History Museum.

Nothing is known of Bauer's feelings when leaving England, but we should not automatically exclude this aspect. In short, the ultimate reasons for Ferdinand's departure, for which we have no date, were probably complex. Why Vienna? Again there may have been purely emotional reasons, such as the desire of the tired expat to come back home and rest, which does not necessarily mean that one suffers from home-sickness: to speak one's own language and no longer being immediately spotted as a foreigner because of imperfect pronunciation; to see again the countryside where one has been born and grown up, the city where one has passed part of one's younger years; to enjoy the markedly warmer and drier climate; for Ferdinand more specifically to meet his brothers Johann and Joseph, whom he had not seen for twenty-eight years; or simply to try a new start with the money that he had put aside. In addition, Ferdinand could have had heard in London that Francis I and two of his brothers, the Archdukes Anton Victor and Johann, employed natural history illustrators (see Chapter 12) and that they would be potential customers for his watercolours.

There may have been also a financial expectation behind Ferdinand's move – the great Congress, which was about to convene in Vienna in September 1814. Among the first to arrive in the imperial city in that month were Cardinal Secretary of State Ercole Consalvi representing Pope Pius VII, Karl Robert Count Nesselrode and Robert Stewart, Viscount Castlereagh, the ministers of foreign affairs representing Alexander I and George III. The top stratum of the European aristocracy was expected, two Emperors, several kings and many diplomats would come together for several months, among them possibly one or two who might be induced to purchase flower pieces or coloured copies of the *Illustrationes*. Maybe Ferdinand was also thinking of Johann I Joseph, 10th Reigning Prince of Liechtenstein, who was employing Johann and Joseph (see Chapter 12); or Franz Joseph, 8th Reigning Prince of Dietrichstein, whose mother had employed Franz and whose sister Therese had married Merveldt, then based in London. As will be shown in the following chapter in this attempt Ferdinand had but limited success. Or is it just a coincidence that Franz announced the return of his brother to Vienna in August 1814, just a month before the great Congress began?

ENDNOTES

1 Unless otherwise stated, this chapter is based on Mabberley (1985), Wheeler & Moore (1994), Mabberley & Moore (1999), Mabberley (1999a), Vallance, Moore & Groves (2001), Mabberley (2002) and Lack (2003).
2 Lhotsky (1843)
3 J. Compton (pers. com., 2014)
4 Endlicher (1836)
5 George (2006)
6 Anon. (1806)
7 Blunt (1976)
8 Stearn (1960b)
9 Whitehead (1988)
10 M. Jones cited in Watts, Pomfrett & Mabberley (1997)
11 Lack (2003)
12 Sawyer (1971)
13 Riedl-Dorn (2014)
14 Anon. (1904)
15 Smith (1988)
16 Stearn (1976)
17 London, University College London, Library, Ms Add 254, Lord Odo Russell Collection. – Franz Bauer to Joseph Franz Jacquin, 19 August 1814 (unpublished letter)
18 Anon. (1999)
19 Graves (1905)
20 Nissen (1951)
21 C. Riedl-Dorn (pers. comm., 2014)
22 Anon. (1978)
23 Schifter (1992)
24 Edwards (1976)
25 Moore (2005)

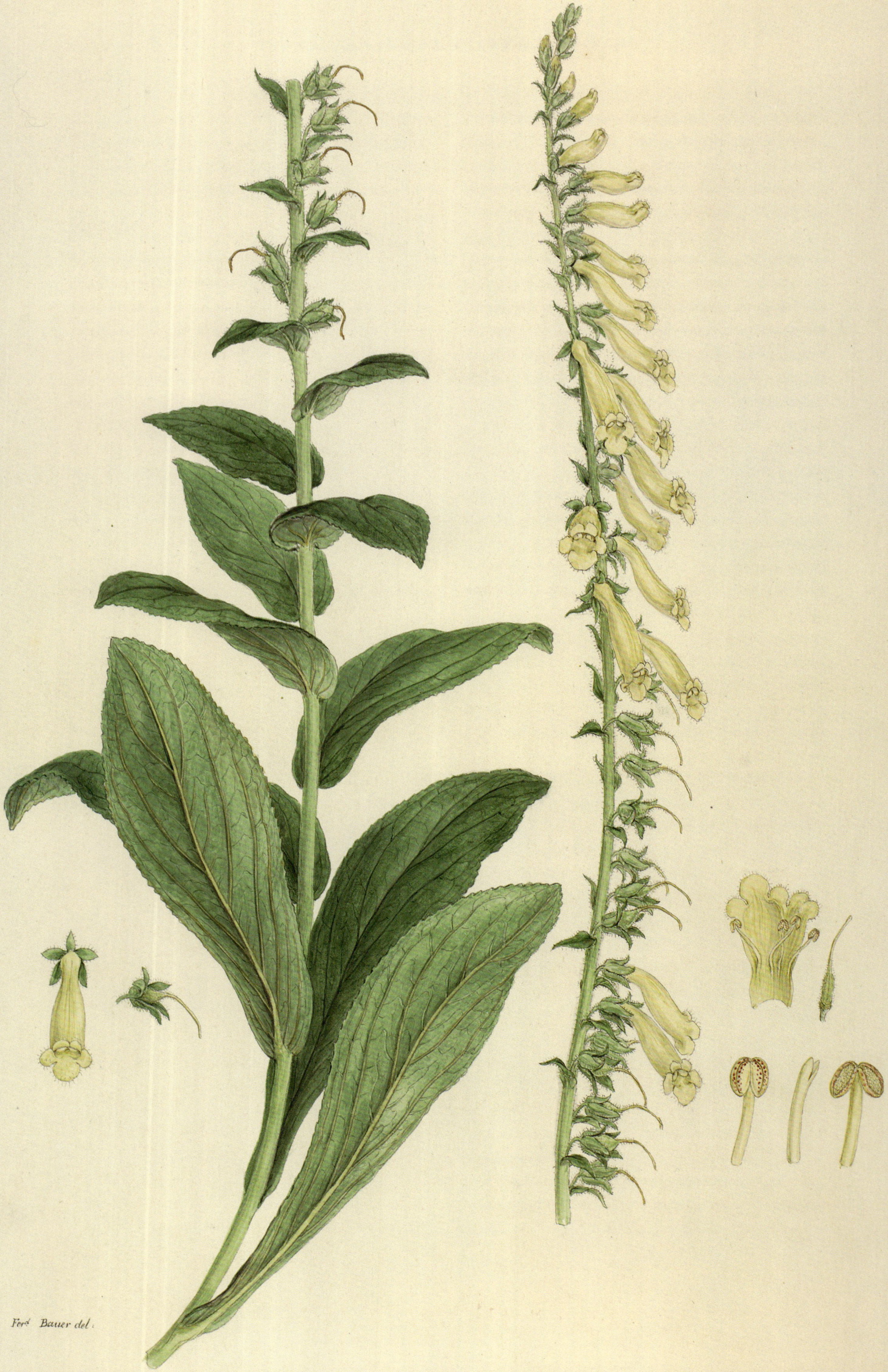

Ferd. Bauer del.

CHAPTER 12

Joseph and Ferdinand Bauer back in Vienna[1]

JOSEPH RETURNS TO VIENNA

IN EARLY 1788 JOSEPH'S stay in Rome (see Chapter 4) had come to an end and he returned to Vienna. During his seven years of absence from the 'Kaiserliche Reichshaupt- und Residenzstadt' the Hereditary States had undergone a period of drastic reforms following the principles of enlightened rule. These were all directed by Emperor Joseph II, but basically they were reforms within the framework of the *ancien régime* (see Chapter 3). In 1787 another war between the Russian and the Ottoman Empire had broken out, with the Hereditary States and the Kingdom of Hungary joining the Russian side. The Emperor led his troops in what is now Serbia and Romania, but due to epidemics the military success of the campaign was meagre, and he died on 20 February 1790 before the conclusion of a peace. Among the officers that were then quickly promoted to colonel was a young man who was to play a central role in the second part of Joseph's life – Johann Joseph, Prince of Liechtenstein, the second son of Franz Joseph I (see Chapter 2); in 1781 the latter had sent Joseph to Rome.

From 1 May 1788 Joseph entered the service of the head of the Liechtenstein family, thereby following the model of his elder brother Johann, about whom we know so little (see below), and in a sense also that of his father Lucas (see Chapter 2). However, Joseph was not engaged as an artist or artisan as one might expect but as a valet with a salary of 250 guilders per year.[2] For the following thirteen years he appears in the records of the 'Majorats-Kassa' as 'Kammerdiener Bauer[or Pauer]', which clearly means that he belonged to what later would be called 'downstairs'. Although a permanent position and associated with an impressive livery, about which we have no details, Joseph was definitely placed on the fringe of the Princely Court. At that time the latter comprised nine key persons: the 'Garde Captain' (Commander of the Guard), 'Second Captain' (Deputy Commander of the Guard), 'Bibliothecaire' (Librarian), 'Physicus' (Physician-in-ordinary), 'Galerie Inspector' (Director of the Gallery), 'Adjunct' (Assistant to the Director of the Gallery), 'Hofkaplan' (Chaplain), 'Garten Controlleur' (Controller of the Gardens) and 'Schloßinspector' (Castellan). They formed the inner circle around the Reigning Prince. In May 1788 this was no longer Franz Joseph I, who had died in 1781 in Metz in the Kingdom of France, but his eldest son Aloys I Joseph (see Chapter 2). Aged twenty-two, he had taken over and two years later married Karoline Countess von Manderscheid-Blankenheim, nine years his junior. Both had their portraits made by well-established painters – Aloys I Joseph by Peter Eduard Ströhling (Fig. 12.2.) and Karoline by Élisabeth-Louise Vigée-Lebrun, both portraits are now in the Fürstliche Sammlungen in Vienna.[3] She had produced, among others, famous portraits of Marie Antoinette, Queen of France, and of the young Heinrich Prince Lubomirski as Genius of Fame, the latter now in the Gemäldegalerie in Berlin.[4]

FIG. 12.1. *Digitalis purpurea* × *D. lutea*. Watercolour by Ferdinand Bauer, c.1820. – London, Royal Horticultural Society, Lindley Library.

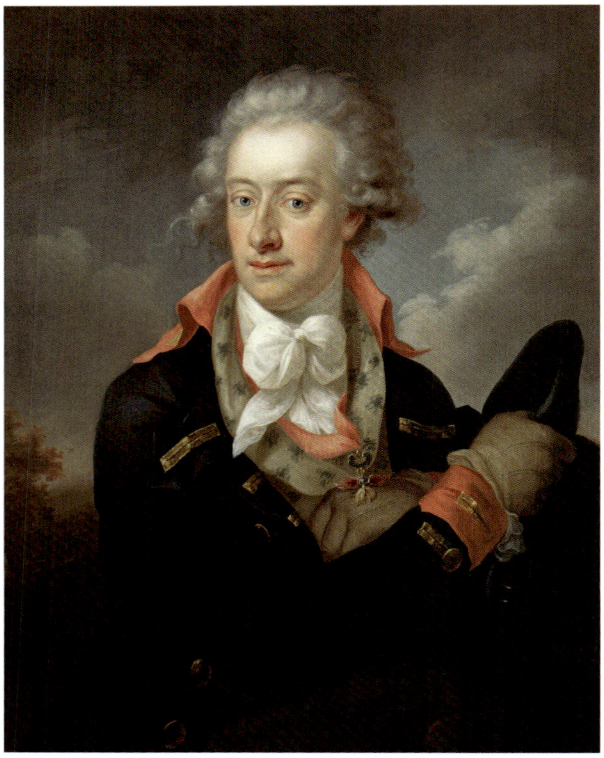

FIG. 12.2. **Aloys I Joseph, 9th Reigning Prince of Liechtenstein. Oil on canvas painted by Peter Eduard Ströhling, 1795. – Vienna, Fürstliche Sammlungen.**

Immensely rich, Aloys I Joseph led the privileged life of a member of the aristocracy, dedicating himself to the development of his estates, his collections of art, his library and to music. On 23 January 1782 Mozart informed his father Leopold in Salzburg that the young prince contemplated forming his own private orchestra and that music pieces were to be composed for him. Although nothing came of this idea, another project became reality – the foundation of a private theatre in Feldsberg with the erection of a building that still exists adjacent to the castle. A dedicated bibliophile, Aloys I Joseph had a large wing added to his city residence in Vienna's Herrengasse as a receptacle of his magnificent library. He was also prepared to create a foundation for the hospital run by the Brothers of Mercy in Feldsberg in order to acquire from Boccius that perfect florilegium, very appropriately known as *Codex Liechtenstein* (see Chapter 2).

As a valet Joseph had occasion to listen to many conversations and may well have heard about the admission of his employer to the Knights of the Order of the Golden Fleece in 1790, with his name added under No. 836, following a tradition that had started with Johann Adam I, followed by Joseph Wenzel and Franz Joseph I. At that time Emperor Leopold II had already succeeded his elder brother Joseph on the throne and become the new Grand Master of the order.

Naturally Joseph would also have taken part in the traditional *villeggiatura* of his employer, as a rule spent on the Feldsberg-Eisgrub estate. This would have given Joseph ample opportunity to meet his mother Therese, who was to die in 1790 in Feldsberg (see Chapter 8), and to see with his own eyes all the embellishments of the estate commissioned by Aloys I Joseph. The Chinese Pavilion,[5] erected in 1795 in Eisgrub by well-known architect Joseph Hardtmuth, became the repository of silk wallpaper from the Petit Trianon in Versailles, the favourite country house of Marie Antoinette, who had already been beheaded. While the pavilion has long since been demolished, the obelisk to commemorate the Treaty of Campo Formio modelled on the Obelisco Flaminio in Rome still stands. The same is true of the Hansenburg (Janův Hrad), an artificial ruin typical of the fashion of the time, and the Turkish Tower, also called the minaret. Erected with no purpose other than acting as a *point de vue* for Aloys I Joseph's country house in Eisgrub, it reached sixty metres in height and was at the time the tallest minaret in the non-Muslim world, only four metres shorter than the six minarets of the Sultan Ahmed Mosque in Istanbul. At a cost of more than one million guilders constructed in the morass of the Thaya wetlands, this was one of Hardtmuth's masterpieces and a reference to the much smaller mosque erected by Chambers in the Royal Garden at Kew, long since demolished, and that built by Nicolas de Pigage in Schwetzingen, still standing.

In a note on his allowance in the form of wine,[6] no doubt from the extensive vineyards in Feldsberg belonging to the Reigning Prince, Joseph is first mentioned in 1797 as 'Kammer Mahler Bauer' – literally painter in the chamber (of the prince). Five years later, when he had reached his forty-sixth year, Joseph also received this job description in the list of the princely employees. Thereby Joseph had obtained the position held some forty years ago by his father Lucas. It is likely that he had to fulfil similar duties, such as painting and restoring altar pieces in the many churches of which Aloys I Joseph held the patronage; for the latter's grand-nephew their number is given as two hundred and thirty-seven. Joseph may also have undertaken similar work that his father had done in the several country houses and castles belonging to the Liechtenstein family. His employer seems to have been satisfied with Joseph's work: in 1794 the latter had received a donation of four hundred and fifty guilders, his emolument was increased significantly in 1804 and 1805, and upon the early death of Aloys I Joseph his court painter received a substantial annuity for life.

Back from Rome Joseph seems to have drastically reduced his activities as an artist. A single original painting in oil called 'Mars seated',[7] now kept in the Fürstliche Sammlungen in Vienna, is known from his hand. It is a work in neoclassical style following the example of his 'Deploration of Hector by Andromache' (see Chapter 4) and may date from this period of his life. A few aquatints by Joseph have been traced, probably all based on works of art made by others, all undated und largely unrecorded in the literature. Two of these prints are kept in the British Museum and have been described as 'Two studies of heads from Rubens's painting of St Ambroise standing in the doorway and denying emperor Theodosius the Great access to Milan Cathedral'[8] (Fig. 12.3.) and 'Portrait of a boy'[9] (Fig. 12.4.). Both are in small format – for example, the latter measures 108 × 79 mm – they exist signed 'Jos Bauer' and in the *avant la lettre* stage. The first of the two is indeed based on the painting by Rubens mentioned above, then already in the Imperial collections in Vienna and on show in the Belvedere, now in the Kunsthistorisches Museum.[10] Two further aquatints have been listed in the catalogue of Andreas Andresen's art collection in Leipzig[11] described as 'Old Armenian' (possibly also based on Rubens) and 'Soldier', respectively, of which copies recently have surfaced on the German art market.[12] Joseph's aquatint 'Deposition' shows Christ lowered from the cross with Joseph of Arimathea wearing a turban standing left, and the crying Virgin, Mary Magdalene and St John standing on the foot of the cross. It also is kept in the British Museum[13] and was based by Joseph on a painting by Christian Wilhelm Ernst Dietrich, now in the Staatliche Kunstsammlungen Dresden.[14] Copying also took the form of copying in oil. On 3 December 2011 Joseph's

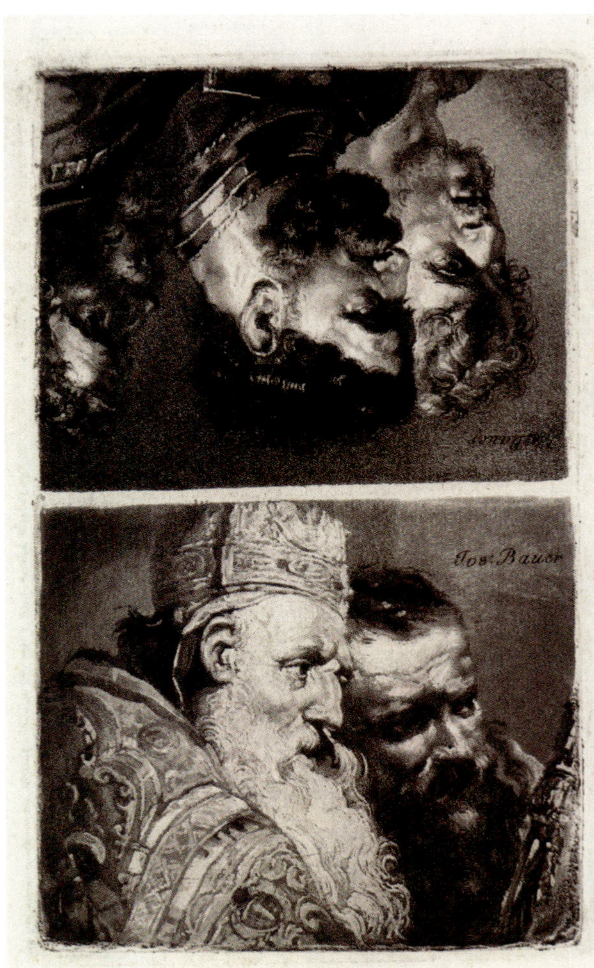

FIG. 12.3. Two studies of heads from Rubens's painting of St Ambrose standing in a doorway. Aquatint by Joseph Bauer, c.1795, lettered. – London, British Museum, Department of Prints and Drawings 1978,0121.276.

FIG. 12.4. Portrait of a boy. Aquatint by Joseph Bauer, c.1810, lettered. – London, British Museum, Department of Prints and Drawings 1978,0121.278.

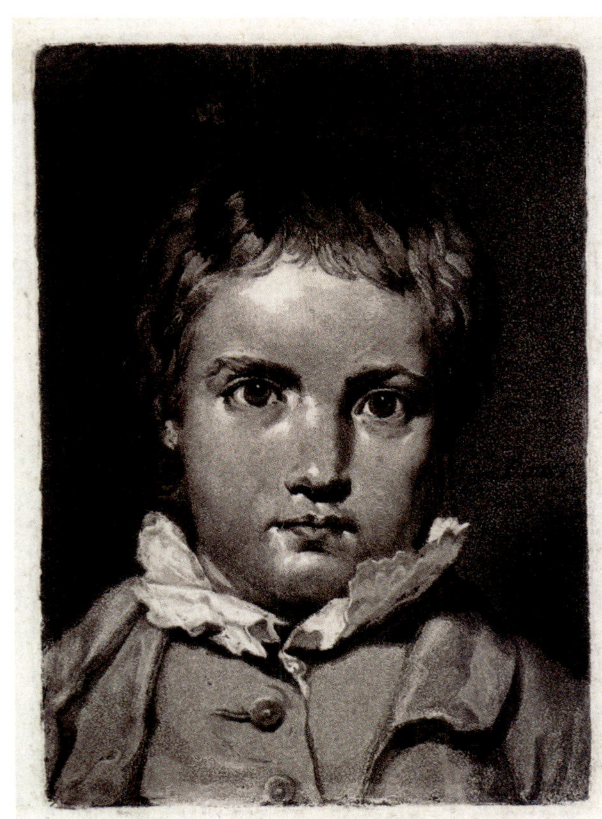

copy of a well-known painting came up for auction in Leipzig. It is based on 'The Dream of Saint Joseph' by Mengs (see Chapter 4), then in the Imperial collections in Vienna and now also conserved in the Kunsthistorisches Museum.[15] Even the dimensions are very similar – 114 × 86 cm in the original, 113 × 86 cm in Joseph's copy, which was signed by him and dated 1817, therefore originating from a later phase of his life.

At 45 years old Aloys I Joseph died on 24 March 1805 in Vienna and was laid to rest in the burial chapel of the Liechtenstein family in Wranau in Moravia, and we may hypothesize that Joseph attended the funeral. Since Aloys I Joseph had no legitimate offspring his younger brother succeeded him as Johann I Joseph, 10th Reigning Prince of Liechtenstein.[16] In 1792 the latter had married Marie Josepha Sophie Landgravine of Fürstenberg-Weitra, sixteen years his junior, and continued his military career reaching the rank of a Lieutenant General. Within eight months of the death of his elder brother Johann I Joseph was faced with the first very major challenge of his life, requiring first the talents of a military leader and, subsequently, the skills of a diplomat. On 2 December 1805 the Battle of Austerlitz (today Slavkov u Brna, Czech Republic), also known as the Battle of the Three Emperors, brought Napoleon one of his greatest victories crushing the Third Coalition. After nine hours the Russo-Austrian army led by Alexander I, Emperor and Autocrator of all Russias, and Francis I, Emperor of Austria, was decisively defeated. For Leo Nikolayevich Count Tolstoy the battle fought in Moravia was the main event of his novel *War and peace*, rightly mentioning among the principal military leaders Kutuzov, Franz von Weyrother and Johann I Joseph. Although Archduke Karl, a younger brother of Franz I, had been Chief Commander of the Austrian army, Johann I Joseph as the Commander of the Cavalry played a very prominent role in the battle. When selected as one of the two chief negotiators for the peace agreement with Napoleon, Johann I Joseph became a weighty political figure in the struggle of the Habsburg Empire for survival. The Peace of Pressburg (today Bratislava, Slovak Republic), co-signed by him on 26 December 1805 in the Palace of the Primate of Hungary, the Archbishop of Gran (today Esztergom, Hungary), in Pressburg, was a bitter pill for Francis I to swallow and the penultimate nail in the coffin of the Holy Roman Empire. The Tyrol and Vorarlberg, ancient Habsburg territories, had to be ceded to Bavaria, while Venice, Venetia, Istria and Dalmatia, acquired only in the Peace of Campo Formio, were incorporated into the Kingdom of Italy.

JOSEPH APPOINTED DIRECTOR OF THE LIECHTENSTEIN GALLERY

Only a few days later, on 6 or 8 January 1806, Johann Dallinger (also spelled Tallinger, Thallinger, or Dallinger von Dalling) died in Vienna.[17] Born on 13 August 1741 he had been appointed in 1771 director of the Liechtenstein Gallery by Joseph Wenzel (see Chapter 2) and remained in this position for life. To Joseph his death opened the one and only chance in his life to make one further step upwards in the hierarchy, similar to the moment in late 1789 when Franz received the generous offer by Banks to remain in England. Joseph was lucky enough to become Dallinger's successor in the job. This is clearly evident from a note dated 'Wienn ano 1806' and signed 'Joseph Bauer Gallerie Inspecteur' on the front page of the gallery's inventory book H, stating that Johann I Joseph had intrusted Joseph with the care of the 'hochfürstl: en Bilder Gallerie' (High Princely Picture Gallery).[18] Joseph's advancement went with a rise in his salary to 700 guilders per year, gradually augmented until it reached its maximum of 1068.44 guilders per year in 1816. By 1806 Joseph, aged fifty, had thus become a member of the inner circle of the princely court – 'upstairs' in nineteenth century terminology. Like his predecessor, he was to hold this position for life and must quickly have become a prominent figure in the circle of art collectors and art dealers in Vienna.

The new job meant the care and responsibility of one of the most splendid collections of works of art then in existence and in Vienna only eclipsed by that of the Emperor. This was not just a huge collection of paintings but also included a vast spectrum of other objects of art, such as sculptures, porcelain, weapons, furniture, tapestries, coins and medals. All this was not exclusively kept in the city residences of the Reigning Prince in Vienna and Prague, but scattered over his numerous country houses and castles in the Archduchy below the River Enns and elsewhere. It is difficult to find the appropriate words for this collection – it contained and still does contain a long series of exquisite paintings by Rubens, in particular the Decimus Mus cycle acquired in

1691 in Antwerp, a no less extensive series of portraits by Anthony van Dyck and of many equally outstanding masters. The top piece, however, had not yet been correctly attributed during Joseph's time in office: a portrait by Leonardo da Vinci, conventionally said to show Ginevra de' Benci,[19] now in the National Gallery of Art in Washington, where it is the only Leonardo on public view in the Americas. Needless to say, the gallery contained a long series of family portraits, such as Hyacinthe Rigaud's portraits of Wenzel,[20] and an extensive collection of views documenting the various possessions of the family, among them the two views of the summer palace in the Rossau district of Vienna painted by Bernardo Bellotto (see Chapter 3). Only two printed catalogues of the paintings and sculptures were available to Joseph – that by Vincenzo Fanti published in 1767 and the anonymous one printed in 1780, attributed recently to Dallinger (see Chapter 3).

Whereas Joseph had reached the final stage of his career with the new appointment, this was not the case for his employer, four years his junior. While Johann I Joseph's admission to the Knight of the Order of the Golden Fleece on 12 February 1806 followed established tradition, Napoleon broke with all traditions when on 12 July 1806 he created the Confederation of the Rhine. This was nothing but a confederation of small client states of the First French Empire with Napoleon as protector and acted as the ultimate nail in the coffin of the Holy Roman Empire. On 6 August 1806 Emperor Francis II, then on his traditional *villeggiattura* in Baden bei Wien, chose to declare the Holy Roman Empire defunct, but continued to rule as Francis I, Emperor of Austria. On the explicit request of Napoleon and without informing Johann I Joseph, the small Principality of Liechtenstein on the Upper Rhine had become one of the original sixteen states of the newly created confederation, which was after all a defensive alliance against Austria. In order to minimize any potential conflicts with the state where he was a rich landowner, Johann I Joseph on paper resigned as Reigning Prince and nominally made his youngest son Karl Johann, then three years old, his successor ad interim. When the allies opposing Napoleon dissolved the Confederation of the Rhine and a new political order for Europe had been found, Johann I Joseph resumed his title and reign. In doing so, surprisingly, he became for the first time in the long history of his family a sovereign in his own right, since the Holy Roman Empire had meanwhile ceased to exist. All this would have long-lasting consequences for our story, since the Principality had become a state independent from the Austrian, later Austro-Hungarian Empire, forming a custom union with Switzerland only after the end of the First World War (see Chapter 19).

On 18 December 1806 Johann I Joseph was promoted to 'Commanding General in Lower Austria and Commander of the Fortress and City of Vienna' and less than two years later also to General of the Infantry. The year 1809 brought to him two momentous battles – the Battle of Aspern-Essling on 21–22 May, which resulted in the first victory ever over Napoleon, followed by the Battle of Wagram on 5–6 July, a decisive defeat of the Austrian Army that destroyed the Fifth Coalition, the Austro-British alliance. In both battles Archduke Karl had been the supreme commander, but as in Austerlitz Johann I Joseph was a key figure. This fact is convincingly expressed in a famous painting by Johann Peter Krafft, entitled 'Archduke Karl and his staff at the battle of Aspern' now in the Fürstliche Sammlungen.[21] Staged as an icon of patriotic (that is, anti-Napoleonic) feelings, it shows the two military leaders riding side by side.

After the debacle at Wagram and the armistice concluded on 12 July 1809 in Znaim on the Thaya River, Johann I Joseph was commissioned for a second time to act as plenipotentiary for the conclusion of the peace agreement. This situation bordered on catastrophe – Napoleon requested the immediate resignation of Francis I and the dissolution of the Austrian Empire, to which Johann I Joseph is reported to have replied diplomatically that he possessed no authority to listen to such a proposal. In the end Francis I was forced to agree to a deeply humiliating peace, named the Peace of Schönbrunn, after the imperial summer residence then used by Napoleon as his headquarters. For Francis I, who had fled to Komorn (today Komárno, Slovakia and Komárom, Hungary) it meant losing several provinces or parts thereof inhabited by one-sixth of the population and having to pay an indemnity of 85 million francs in instalments. It seems that Francis I had expected better peace conditions, a rather unrealistic view considering Napoleon's position at the height of his power, blaming his plenipotentiary for the resulting problems. In order to avoid the first tranche of six million francs, Johann I Joseph offered his private gallery to Napoleon, but the proposal was rejected. Since the finances of the Austrian Empire and the Habsburg family were then at a low ebb, as was often the case, Johann I Joseph offered his vast private property as a guarantee for the astronomic sum requested as an indemnity by Napoleon, and this time the plan was accepted.

When Francis I was obliged in 1812 to send an Austrian auxiliary force to join the *grande armée* in order to support Napoleon in his campaign against Russia (see Chapter 11), Johann I Joseph was asked to become the commander. Wisely he refused the offer and instead proposed Karl Philipp Prince Schwarzenberg to do the job. This led to alienation with Francis I and Johann I Joseph's resignation, and it put an end to the latter's career in the service of the Emperor of Austria. At that moment he was a man of fifty-two years who had twenty-four more years to live. Like his elder brother before his early death, Johann I Joseph now dedicated himself to the development and embellishment of his estates and to the acquisition of several more of them. He managed to create no less than three postremogenitures for his sons Franz, Karl and Friedrich in the Duchy of Styria, the Archduchy of Austria below the River Enns and the Duchy of Carinthia, respectively. Another field of interest was his collection of art. It is not easy to conceive the financial momentum of Johann I Joseph, no doubt the richest man in the Austrian Empire, who spent gigantic sums of money on a wide spectrum of activities.

As director of the Liechtenstein Gallery, Joseph was very successful in attracting the attention of Johann I Joseph. This is best illustrated by a single figure – during the latter's reign the number of paintings rose from eight hundred and forty to 1630.[22] This almost coincides with Joseph's time in office, since he had got his job in the first year of Johann I Joseph's reign and predeceased him by less than five years.

Here we are faced, however, with a major problem: no one has ever studied in detail the accession and deaccession profile of the Princely Collections in the years 1806 to 1831. While Dallinger is considered to have published a catalogue of the gallery's holdings (see above), Joseph refrained from following his predecessor's example. Therefore it is currently impossible to present an enumeration or synopsis of all accessions. Instead, five selected examples out of the seven hundred and ninety paintings known for certain to have been acquired during the twenty-five years when Joseph was in office are given here to illustrate the kind of works that passed through his hands. For convenience these examples are arranged here in the sequence of their putative origin, not in the sequence of their accession.

Everybody would agree that 'A portrait of a man',[23] for good reason attributed to Raffaelo Santi, better known as Raphael, dated 1502–1504 and previously in a private collection in Bologna, is the item that would attract the highest price at auction today. Another Old Master acquired during this period was Pieter Brueghel the Younger's 'The Numbering at Bethlehem'.[24] This is a copy of the homonymous work of his father Pieter Bruegel the Elder, kept today in the Museés Royaux de Beaux-Arts de Belgique in Brussels. The narrative richness of detail and the realistic depiction of the numerous activities taking place in a village during winter may have been motives for purchasing this painting. Franz Xaver Petter's 'Still-life with fruits and parrot'[25] is an important work of the early Biedermeier period, though inferior to Knapp's 'Jacquins Denkmal' (see below). Petter shows a mixture of fruits, among them watermelons, grapes, cherries, pomegranates, apricots and a pineapple, two sulphur-crested cockatoos, *Cacatua galerita*, a parrot, a buffy-tufted marmoset, *Callithrix aurita*, and in the background dramatic alpine scenery. A portrait painted by Johann Baptist Lampi the Elder has an entirely different background. It depicts Johann I Joseph in the white uniform of an Austrian general wearing the decoration of the Golden Fleece and the Grand Cross of the order of Maria Theresa.[26] It is the military leader on which Lampi the Elder had focused his attention, showing a canon in the background. 'Prometheus bringing fire to mankind',[27] an oil painting measuring 221 × 156 cm, stands for a painting of the neoclassical movement. Its author Füger had, like Joseph, been in Rome and after his return to Vienna become, like Joseph, director of a gallery – the Imperial Gallery then exhibited in the Belvedere Palace (see Chapter 3).

It would be quite wrong to assume that the gallery of the Reigning Prince of Liechtenstein would simply have grown during Joseph's years in office without any deaccessions. Instead, influx and efflux continued until recently. A 'Boucquet of Flowers', an oil painting by Jan van Huysum, had been acquired in 1819 by Johann I Joseph, and after being sold in 1950 was purchased in 2002 by Hans Adam II, the 15th Reigning Prince of Liechtenstein, and as a consequence is now back again in the Fürstliche Sammlungen in Vienna.[28] Earlier, at the auction of the belongings of Wilhelm Graf von und zu Sickingen in Vienna, this flower piece had been acquired by a 'Bar: v: Paris' with a minor financial assistance from Johann I Joseph, who exchanged it with his portrait of 'Lady Spencer vom Raynolds Br: Bild

[Sir Josuah Reynolds Brustbild]'. It is not quite clear to which of Reynold's portraits of Lady Caroline Spencer, the eldest daughter of the 4th Duke of Marlborough, this brief description refers, but one aspect of the transfer is clear – it was Joseph who wrote the pertinent note, signed it 'Wien 1819 14 Apr: J: Bauer GallerieInspector' and glued it into the inventory book H. This inventory book is a mine of information not yet fully analyzed and published. What kind of information it contains becomes evident when looking at a single line written on the same page 'Herrngasse 730 Rubens Ein Kleiner Mädchen Kopf [Small head of a girl]'. This refers to one of Rubens's most famous paintings and, admittedly, one of the most touching portraits of a child: painted in c.1616 it shows his daughter Clara Serena Rubens, aged about five, in small format. Today it is among the top objects of the Fürstliche Sammlungen,[29] acquired at the latest in 1712 by Johann Adam I, the 3rd Reigning Prince of Liechtenstein. Of course 'Herrngasse' refers to the city residence of the Liechtenstein family in that street in Vienna, and '730' is the inventory number.

Joseph was certainly involved in all these accessions and deaccessions, but to what degree we do not know. It seems likely that important decisions were taken, as they are today, by the Reigning Prince, but proposals and all technicalities such as payment, transport, framing, repair work and hanging belonged to the circle of Joseph's obligations.

THE REARRANGEMENT AND OPENING OF THE LIECHTENSTEIN GALLERY

Johann I Joseph is known to have commissioned Joseph to create a gallery in the modern sense in the summer palace in the Rossau district, his *villa suburbana* outside the city walls of Vienna. Unfortunately, this topic has not yet attracted the attention it deserves from art historians, and as a consequence we know very little about the sequence of events – neither do we know precisely when the project started nor when it came to a conclusion. However, there is a prosaic fact: in 1812 the administration of the Reigning Prince argued for a tax reduction on the ground that the gallery, and also the garden adjacent to the summer palace, were open to the public for free.[30] This implies that the rearrangement had been completed before that date. Furthermore, we do not have the instructions issued by the Reigning Prince, which may not have been given in writing. However, it is difficult to imagine that Joseph was at liberty to decide on matters of principles such as the style of hanging without consultation with his employer. In addition, very little is known about the philosophy underlying the new arrangement, and we also seem to have no information on the logistics of moving a very considerable number of paintings, statues and furniture. Quite by contrast we know a great deal about the result – the wall plans for the paintings in very large format, now kept in the archives of the Fürstliche Sammlungen in Vienna and recently dated 1815.

Therefore a first attempt is made here to reconstruct the various steps Joseph had to take in order to create a proper gallery for his employer. Clearly this became the most important project in his life and it also had a long-lasting effect. The gallery opened in c.1810 and remained in place until the advent of the Second World War made the evacuation of the works of art to safe storage sites and later their transfer to the Principality unavoidable (see Chapter 19). It was only after the fall of the Berlin wall that the collections were brought back from Vaduz to Vienna and placed in the old premises: The summer palace opened to the public again on 29 March 2004.

At the start of the undertaking, probably soon after having taken over the job from Dallinger, Joseph must have consulted with Johann I Joseph and made an agreement with him over the following basic assumptions for his work: Firstly, a high percentage of the paintings should be exhibited, which automatically meant using both floors of the palace and Russian hanging, also known as Petersburg hanging – not the particular hanging that is now the golden standard. Secondly, at least one room was to be created a *Gesamtkunstwerk*, that is a combination of paintings, sculptures and furniture, following the models Joseph had seen in Rome in the city residences and country houses of the nobility. Thirdly, Joseph must have been aware that he would not be permitted to strip the family's city residences in Herrengasse and Bankgasse of all their paintings, nor should the several country houses and castles lose their ornaments, since they would all be continuously or discontinuously used and lived in by members of the family and their guests. A good example of the last point was a series of four paintings in oil by Friedrich Oelenhainz firmly installed in the country house at Eisgrub. These were portraits of four of the children of Franz Joseph I – Leopoldine Adelgunde,

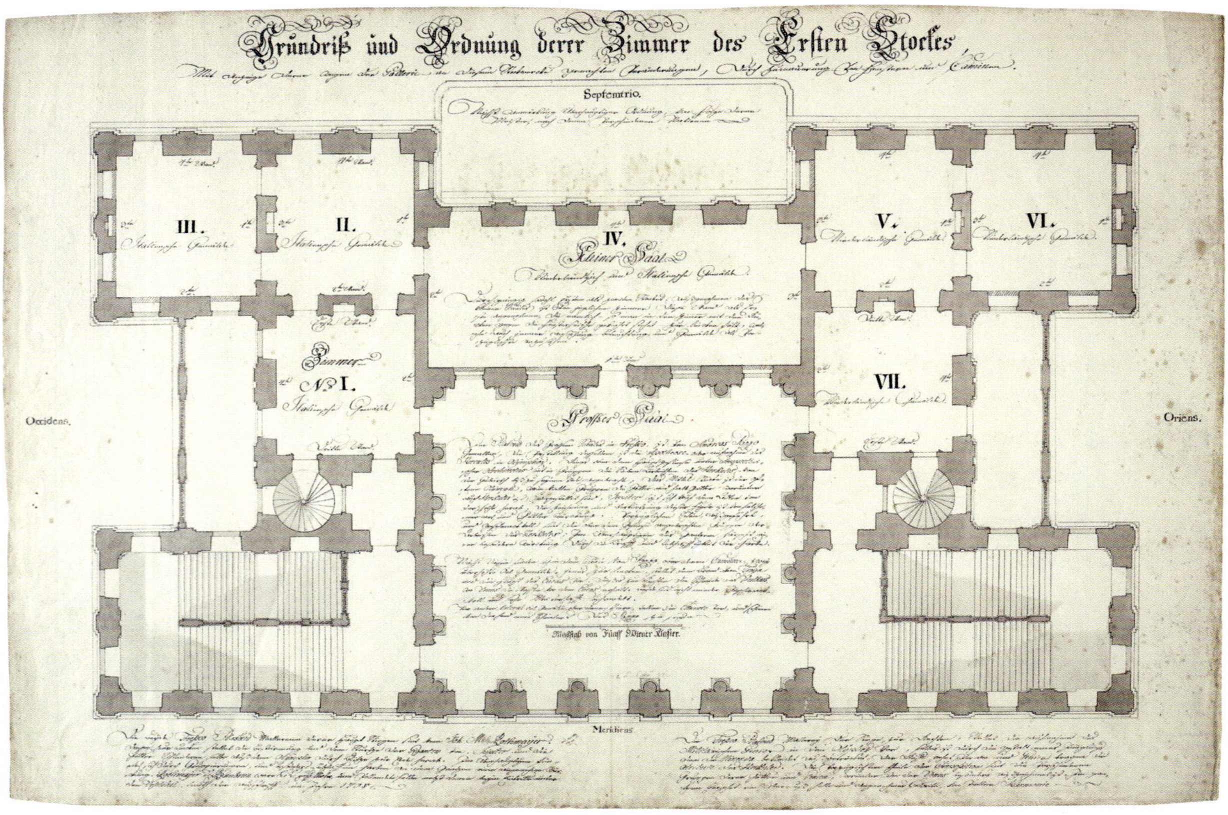
Septemtrio.
III.
II.
IV.
V.
VI.
Zimmer
No I.
VII.
Occidens.
Oriens.
Meridiens

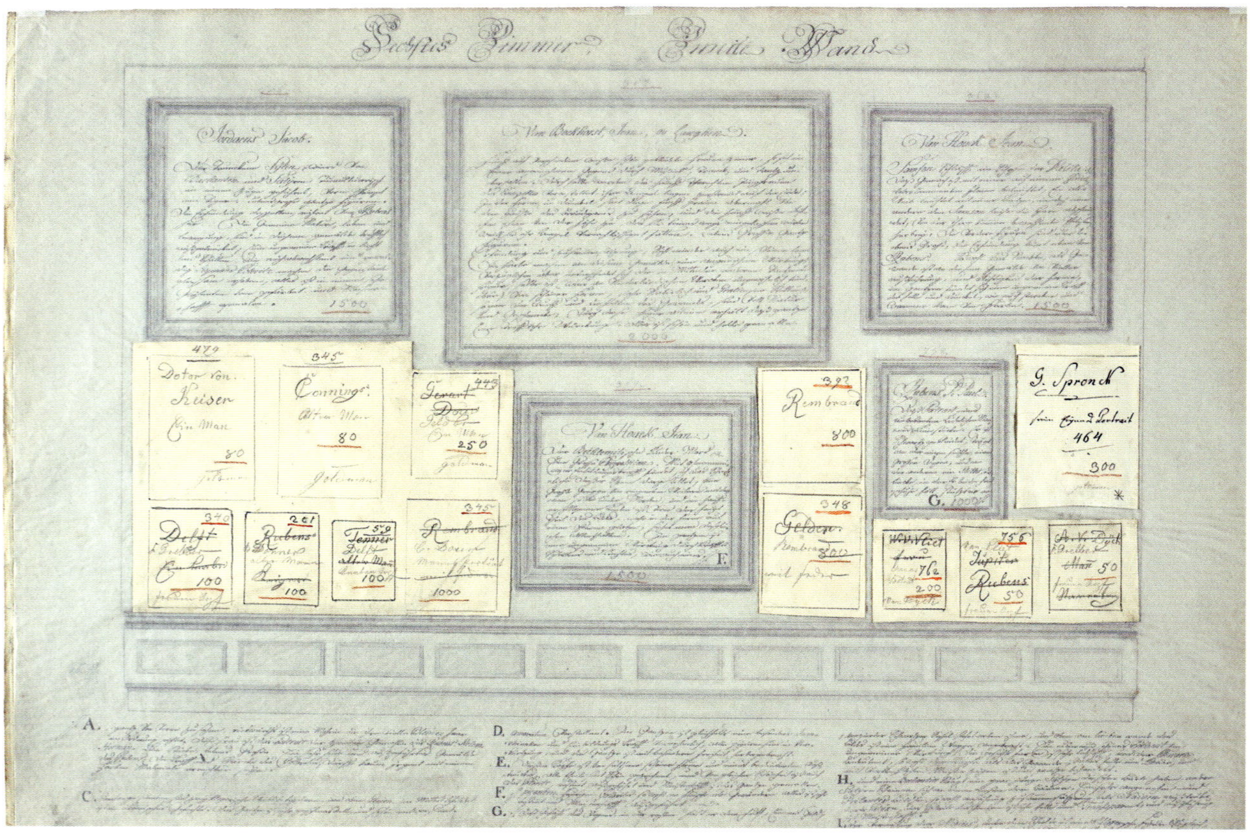
G. Spronck
Rembrant
Rubens
A.
C.
D.
E.
F.
G.
H.

later Landgravine of Hesse - Rotenburg, Aloys Joseph, Johann Joseph and Marie Josepha Hermenegilde, later a Princess Esterházy and the dedicatee of no less than six masses by Haydn. For the time being all four paintings were to remain in situ; they were evacuated only during the Second World War (see Chapter 19) and are now in the Fürstliche Sammlungen in Vienna.[31]

The first step on Joseph's agenda must have been to take accurate measurements of the twelve large rooms and the so-called small hall, later called the Great Gallery, on the first floor of the summer palace (Fig. 12.5.) – in particular the size of the walls. Similar measurements must have been taken for the rooms on the second floor. In addition Joseph must have studied the geometry and symmetry of the building, the light conditions in the course of the day and the seasons, the position of the windows and doors – key considerations before the invention of artificial light. All this would have given him a feeling for the potential of the summer palace as an exhibition space and its capacity to host paintings, sculptures and furniture. The items to be exhibited must have been the subject of the next step. Joseph had taken over the manuscript catalogues from Dallinger, containing as a rule only the most fragmentary information on the identity of the paintings and their current whereabouts. A long and tedious treasure hunt in dozens of buildings must then have started – in Vienna, and the several other country houses and castles in Moravia and the Archduchy of Austria below the River Enns, among them Eisgrub and Feldsberg as well as in Bohemia and the part of Silesia belonging to the Hereditary States. When in the end Joseph had located a particular painting he had to take measurements, note the subject depicted, and try an interpretation of the image and an attribution of the artist.

Intellectually the third step was the most challenging work: to select what should remain and what was to go to the summer palace, where in the end a total of maybe 1000 paintings were exhibited on both floors by Joseph. To decide how to form groups, according to topics (thematic hanging), following putative chronology (chronological hanging) or hanging according to artists or schools of artists; how to arrange the items in the different halls of the summer palace so that they would 'fit together' and at the same time create both a logical and a pleasing impression for the visitor; and how to combine paintings with sculpture and furniture. Synthesizing all these considerations and drawing wall plans for the paintings must have been the final step, followed by annotations and corrections, clearly recognizable today with cut-out pieces of paper affixed to the plans. All this material is still in need of an in-depth analysis.

This was just the planning phase, to be followed by the implementation of the project, which also comprised several steps. Firstly, a few windows and doors in the summer palace had to be bricked in order to increase the hanging space. Secondly, many walls in the summer palace had to be prepared to receive the paintings (that is, painted over, with Joseph choosing a green colour). Thirdly, very precise drawings had to be made for the manufacturers of the pedestals and the respective tables for the sculptures and their work controlled. Fourthly, the transport of all the objects had to be organized and executed: not an easy job considering the carriages of the time, the realities of the roads, the considerable distances to be covered, the sometimes very large format of the paintings and the delicacy and sensibility of very many items. Fifthly, the state of preservation of the paintings had to be controlled upon arrival and the necessary restoration work organized. Finally, the actual hanging and the display of the sculptures had to be organized and undertaken – requiring manpower for the sometimes large canvases and a considerable time budget allowed in order to cope with last minute changes and adjustments.

Joseph's very precise wall plans (Fig. 12.6.) seem to mark the end of the project. They contain detailed descriptions and interpretations of the paintings in calligraphy, which have been found to be exceedingly difficult to decipher. One of them documents the wall opposite the windows in the Great Gallery (Fig. 12.7.) of the palace and is particularly noteworthy. In addition to showing the presentation of four paintings belonging to Rubens' Decimus Mus series, Joseph's pen-and-ink drawing contains the precise image of several sculptures, pedestals and tables. The two sculptures standing next to the door merit attention: to the left we see a representation of Massimiliano Soldani-Benzi's bronze statue of the

FIG. 12.5. *(top)* Ground plan for the first floor, Summer Palace in the Rossau. Pen-and-ink drawing by Joseph Bauer. c.1815. – Vienna, Fürstliche Sammlungen, Hängepläne für Palais in der Rossau.

FIG. 12.6. *(bottom)* Sixth Room, second wall, Summer Palace in the Rossau. Pen-and-ink drawing by Joseph Bauer, c.1815. – Vienna, Fürstliche Sammlungen, Hängepläne für Palais in der Rossau.

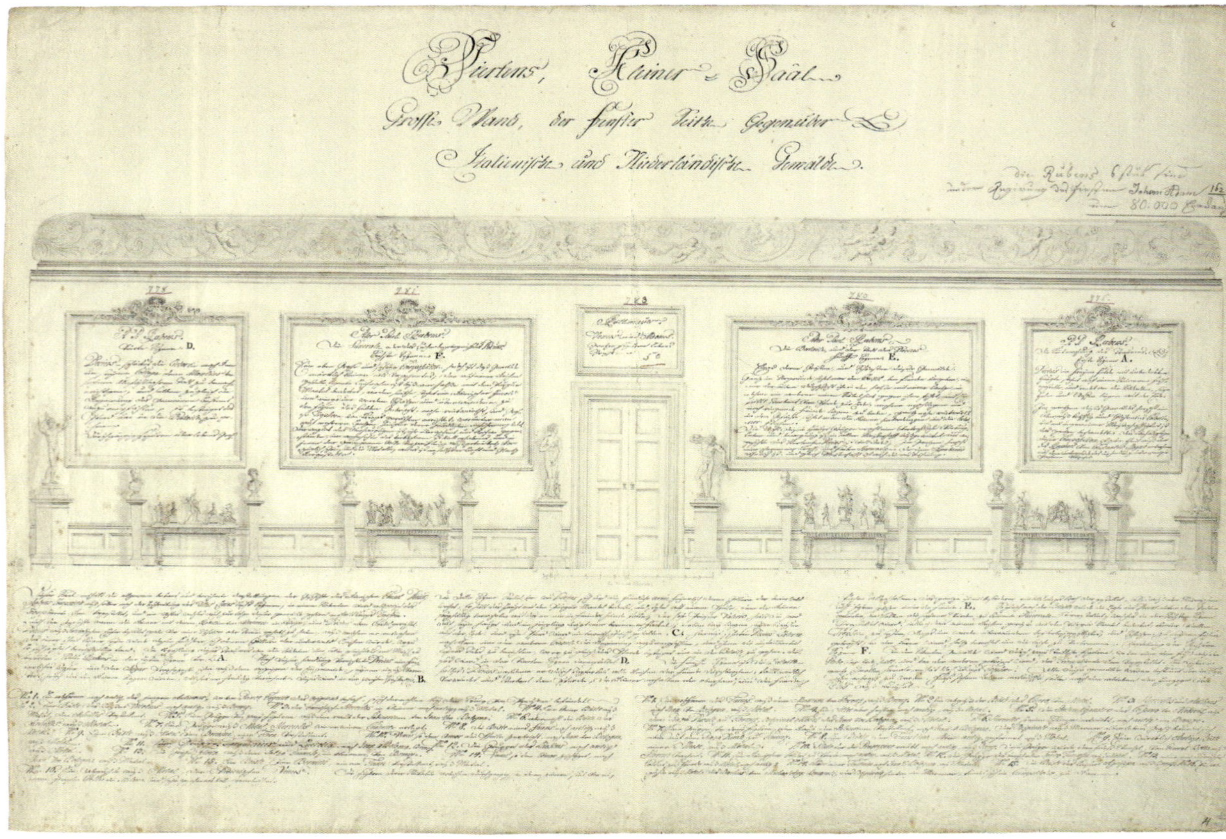

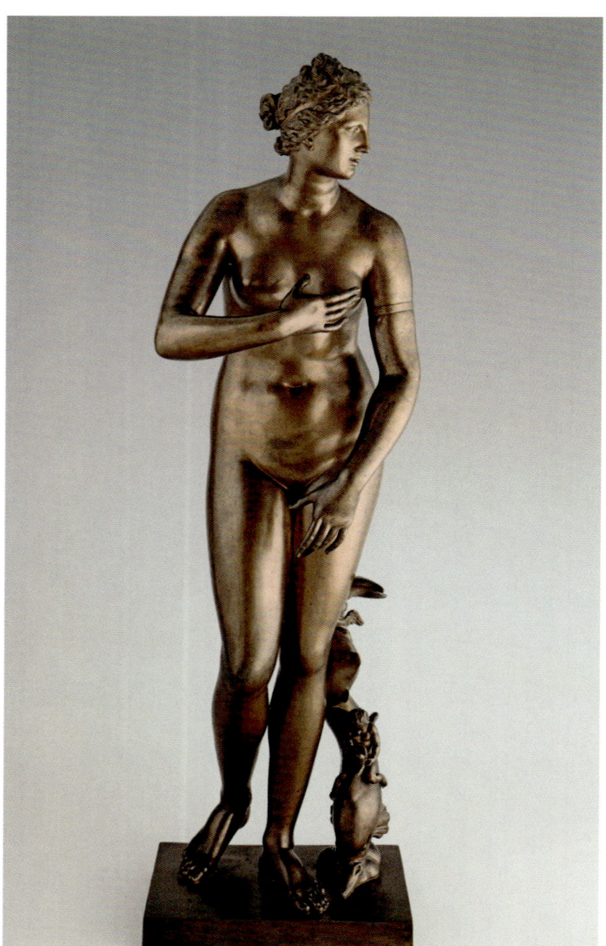

FIG. 12.7. Great Gallery, large wall opposite the windows, Summer Palace in the Rossau. Pen-and-ink drawing by Joseph Bauer, c.1815. – Vienna, Fürstliche Sammlungen, Hängepläne für Palais in der Rossau.

FIG. 12.8. Venus Medici. Bronze sculpture by M. Soldani-Benzi, c.1700. – Vienna, Fürstliche Sammlungen, SK 537.

Venus Medici (Fig. 12.8.), to the right a representation of the bronze statue of a Dancing Faun by the same artist. Both patinated casts had been acquired by Johann Adam I in 1702. Several items of the collection had already been moved several times, a good example being Rubens's large 'Ascension of Mary' measuring 501 × 351 cm[32] – it was acquired in Brussels in 1643, transferred to the parish church in Feldsberg (see Chapter 2), brought in 1764 to the city residence of the Liechtenstein family in Vienna's Bankgasse and now relocated to the summer palace in the Rossau district, where Joseph placed it in Hall 7, the Great Gallery.

Nothing seems to be on record about a formal inauguration of the gallery, but the transfer seems to have come to an end in about 1810. In travel reports the gallery was criticized as being too dark, not in a proper order and poorly labelled. Joseph cannot be blamed for the light conditions in the building; but he was definitely responsible for the spatial arrangement and the inscriptions on the paintings, never an easy job to be done. Nothing is on record about subsequent improvements. The art collections under Joseph's care are likely to have become a neverending story in his life – there was not only the constant influx and efflux of items during Joseph's time in office; he must also have been faced with the realities of pieces of art steadily moving from one place to another, as requested by the changing predilections of Johann I Joseph and his family. A good example to this point is Schloss Liechtenstein, a country house erected by Joseph Georg Kornhäusel and Joseph Franz Engel in 1821–1822 very near to Liechtenstein Castle in the vicinity of Mödling just south of Vienna, a country house which soon later became the favourite spring residence of Johann I Joseph. It is inconceivable that this building remained without paintings, necessitating, as we may hypothesize, frequent journeys by Joseph to this new destination.

In 1902, seventy-one years after the death of Joseph, Raimund Freiherr von Stillfried und Rathenitz, a painter specialized in the documentation of interiors, recorded the Great Gallery (Fig. 12.9.). Almost certainly he had been commissioned by Johann II, the 12th Reigning Prince of Liechtenstein, a grandson of Johann I Joseph. It is hard to believe but true that the room with the sequence of Rubens's paintings and the various sculptures presented itself totally unchanged, precisely as recorded in the drawing by Joseph almost ninety years earlier. Johannes Kräftner, currently director of the Princely Collections and in charge of the reopening of the summer palace in 2004 after the return of the works of art from Vaduz to Vienna, told the present author that he had tested alternatives to Joseph's hanging of the paintings in the Great Gallery. In the end he reverted to precisely the arrangement adopted by Joseph, because the room cried out for symmetry and because of the internal logic in the sequence of paintings. In short, though hardly an artist of fame in his own right, Joseph had no doubt done an excellent job in the arranging and hanging of the Liechtenstein Gallery.

FIG. 12.9. Rubens-Hall in the Liechtenstein Gallery. Watercolour by R. v. Stillfried, 1902. – Vienna, Fürstliche Sammlungen.

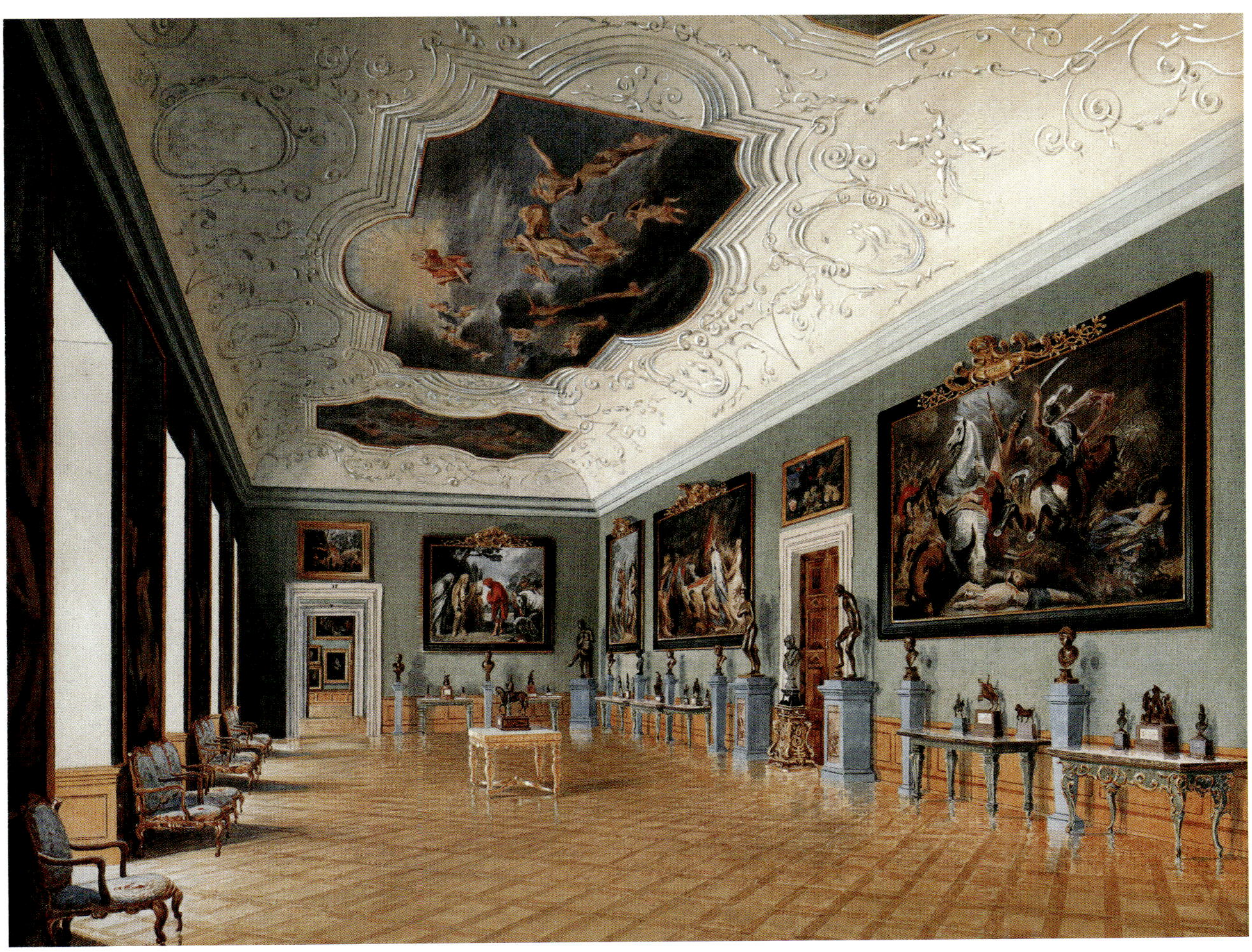

FERDINAND RETURNS TO VIENNA

Nothing is known of when Ferdinand left London, when he arrived in Vienna and which route he chose, though it is safe to assume that he was back in the imperial city in autumn 1814. The political geography of Europe had changed fundamentally since he set out with Sibthorp for the Levant in March 1786 (see Chapter 5). The Holy Roman Empire was no more, and Francis I ruled as Emperor of Austria, which meant a totally different political concept and in fact was a hereditary title referring only to the Hereditary Lands, excluding the Kingdom of Hungary, where he continued to reign as I. Ferenc. The Hereditary Lands had changed their contours only insignificantly, while the ecclesiastical territories had totally disappeared from the political map of Europe, with the exception of the Papal States, soon to be formally re-established thanks to the efforts of Cardinal Consalvi. However, the final borders in Europe were still to be settled at the Congress of Vienna, and Ferdinand must have crossed central Europe while the political future of the different political entities was still in limbo.

We know somewhat more about Ferdinand's last twelve years than we do about his second stay in London from late 1805 until mid-1814. The evidence consists of three herbarium specimens, a single, rather long letter, his testament and the inventory of belongings taken upon his death, though the graphical oeuvre of his final years again provides the most significant information. We also have an address for the years starting with 1818, since in this year Ferdinand bought himself the house Hietzing 155.[33] Before that he may have lived with his brother Joseph in the Rossau quarter of Vienna, but this is hypothetical.

At that time Hietzing was still a small village west of the city centre, then belonging to the abbey of Klosterneuburg, comprising in 1822 merely four hundred and fifty-six houses.[34] Situated in the immediate vicinity of Schönbrunn Palace and its extensive parks and gardens, the village had profited enormously over the centuries from the traditional *villeggiatura* of the imperial family with Francis I regularly spending the months of June and July in Schönbrunn.[35] Becoming gradually a fashionable summer residence for people of superior social position, Hietzing nevertheless still comprised many peasant houses, and Ferdinand bought one of them. Hietzing 155 does not exist anymore, but from the 'Franziceischer Kataster', the catastral map commissioned by Francis I, and various documents kept in the Wiener Stadt- und Landesarchiv,[36] we know precisely where it was situated and what it looked like. Standing on a narrowly rectangular piece of land measuring approximately seven hundred and five square metres,[37] Ferdinand's house was an L-shaped, single-floor building with the short front lining the street, then called Schmidgasse, and included a back garden – described as 'Obstgarten im Dorfried' (orchard within the limits of the village) in the documents. Trauttmannsdorffgasse 56 is the modern address, situated within walking distance to the Hietzing entry of the Schönbrunn garden, precisely where the Dutch garden was located (see Chapter 3). This echoes very much Franz's habitation at Kew near the entrance to the Royal Garden from The Green, though this was a rented flat, whereas Ferdinand had now become a man of property in Hietzing.

Nothing could be discovered about the purchase price in 1818, a good moment after the hype caused by the Congress of Vienna had come to an end. By contrast we know that Ferdinand's property was sold on 1 October 1826 by his legal heirs (see Chapter 17) for 4300 guilders,[38] a very considerable sum. The prestige and value attributed to this house and garden becomes evident by its new owner – Franz Joseph Baron Bretfeld zu Cronenburg, later Bretfeld – Chlumčanský. An Imperial Royal Chamberlain, 'Staatskanzleirat' (Councilor at the State Chancellery), Professor of Law at Vienna University and its Vice-chancellor in 1822, he came from a prominent family with his uncle Vaclav Leopold Chlumčanský z Přestvlk a Chlumčan the Archbishop of Prague and Primate of Bohemia since 1814. It was Bretfeld zu Cronenburg, an owner of estates in Bohemia including Weselinczko castle, and a well-known collector of art, books and, notably, coins, who had Ferdinand's house torn down and a fine neoclassical villa[39] erected before 1832, today a listed building in private hands with a deep-reaching garden behind.[40]

It is most likely that Ferdinand's first contact in Vienna in 1814 had been his brother Joseph (see above), but we have no evidence for this assumption. We may hypothesize that Joseph showed Ferdinand round the Liechtenstein Gallery to which he had dedicated so much time and effort, but again we have no evidence for this. Jacquin, aged eighty-seven in 1814 and advanced in rank from Edler to Freiherr (Baron), was still active and contributing to the fourth fascicles of his *Stapeliarum in hortis vindobonensibus cultarum descriptiones* (Descriptions of Carrion Flowers cultivated in Viennese gardens). Three years before Ferdinand's return he had his portrait painted by Füger, now in private hands.[41] As planned (see Chapter 7), Jacquin the son had become his father's successor in the Chair and as Director of the Botanical Garden of Vienna University. Since Ferdinand's departure in March 1786 the Jacquin family had grown – Jacquin the son had married Marie Barbara Freiin von Natorp, one of Mozart's pupils (see Chapter 3). Their only child Isabella had married Carl Franz Anton Ritter von Schreibers, whom Sibthorp had met on his way to Istanbul in 1794 (see Chapter 9). Appointed director of the United Imperial Royal Natural History Cabinets in 1806,[42] the former was a key figure in the natural sciences in Vienna. Leopold Trattinick, one of the curators at the Imperial Royal Natural History Cabinets and the man responsible for the botanical collections, may have been another early contact of Ferdinand in Vienna. We may image the repatriate walking through the centre of the 'Kaiserliche Haupt- und Residenzstadt' reminiscing his first stay in Vienna. He could have been to St Stephen's Cathedral, where Cardinal Migazzi had been succeeded by the Jesuit Sigmund Anton Count Hohenwart from Carniolia as Archbishop of Vienna. At a short distance he could have paid a visit to St Michel's Church, where on 10 December 1791 the requiem for Mozart had been performed and the Introitus and the Kyrie of his own requiem (K. 626) had been sung for the first time. Ferdinand might have been surprised to find Zauner's monumental equestrian statue of Emperor Joseph II (see Chapter 4) newly erected on the square in front of the Imperial Library. There Gottfried Baron van Swieten had been succeeded as prefect by Joseph Maximilian Count Ossolinski (Jósef Maksymilian Ossoliński), Count of Tenczyn, an honorary member of the Königliche Gesellschaft der Wissenschaften in Göttingen.[43] The latter, a historian of Polish descent, writer, Maecenas and friend of Beethoven, was to die in Vienna on the same day as Ferdinand. In the nearby church of the Augustinians, where the Archduchess Marie Louise had been married off by proxy to Napoleon, the latter in August 1814 already exiled to the island of Elba, Ferdinand would have been able to see another newly erected monument – that by Canova for Archduchess Maria Christina (see Chapter 4). The windows above the main entrance to this church were belonging to the rooms where the collections of the United Imperial Royal Natural History Cabinets were preserved at the time.

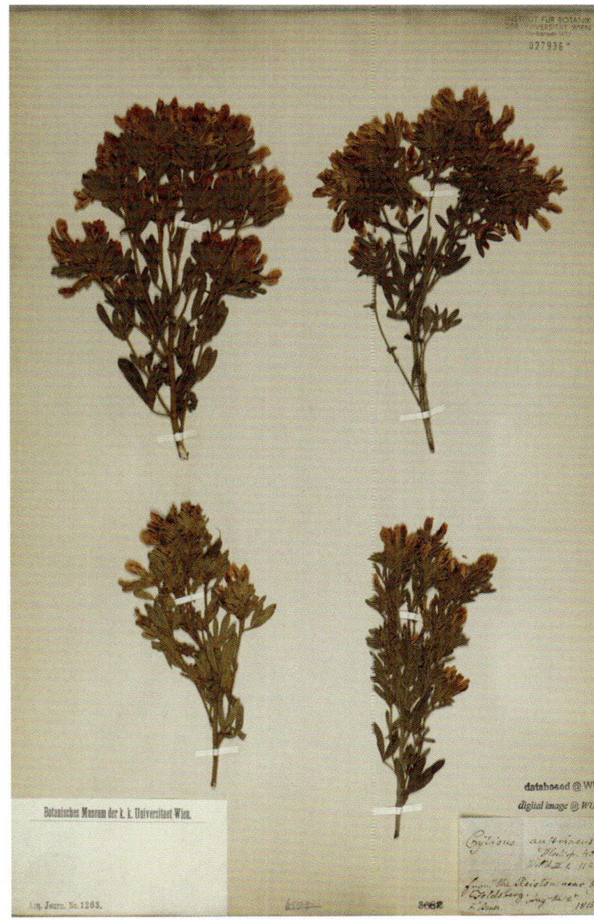

FIG. 12.10. ***Chamaecytisus austriacus*****. Herbarium specimen collected by Ferdinand Bauer near Feldsberg, 1815. – Vienna, Universität Wien, Institut für Botanik, Herbarium.**

The three herbarium specimens collected by Ferdinand, two with original labels, testify to his renewed interest in the very diverse flora of the surroundings of Vienna, quite different from what he had seen for so many years around Oxford and London. In 1815 Ferdinand, perhaps accompanied by Joseph, made an excursion to his native Feldsberg, where he collected *Chamaecytisus austriacus* (Fig. 12.10.), one of the many eastern elements growing in this area dominated by a Pannonian climate. Rather amusingly, Ferdinand's specimen kept today in the herbarium of Vienna University carries a label in English reading 'From the Reisten near by Feldsberg August the 2nd 1815'. Raisten is a hill south of Feldsberg with a fine view looking down on the town where Ferdinand had been born and had grown up (see Chapter 2). He must have seen the nearby vineyards, the village of Eisgrub, the riverine forests along the Thaya River and the extensive agricultural lands beyond. No doubt Ferdinand was staggered to see in the distance the gigantic minaret or Turkish tower, which may have made him remember the pagoda in the Royal Garden at Kew, where he had been so often with his brother Franz. For Ferdinand this may have been a moment of thoughtfulness: he had seen more than almost all of his contemporaries, but now the wanderer was back home and the time had come to settle down. His mother Therese had died twenty-five years earlier while he had been in Oxford and was buried in Feldsberg in the old cemetery, which he possibly visited on that occasion, situated near the Franciscan Monastery then already dissolved. Boccius, whose hospital had been flooded by the wounded after the battle of Austerlitz during his last months, had died nine years earlier, when Ferdinand had returned to London. Now he lay buried in the cemetery belonging to the Convent of the Brothers of Mercy, which one could see with the naked eye from the Raisten. Thanks to Boccius the enormous labours of Ferdinand's youth and that of Joseph and Franz had entered the library of the Reigning Prince of Liechtenstein in Herrengasse in Vienna. But how much more time would be allocated to him? What would come of the *Illustrationes* and the masses of new plants recorded in pencil drawings in that remote country for which he had first used the name 'Australlien'? What would happen to the botanical and zoological pencil drawings he had taken with him to Vienna? Would he have the energy to keep going in this endless project? Would there be anyone prepared to publish his extensive pictorial record of the Australian flora and fauna?

Considering the relatively easy and frequent communication from Vienna via Wilfersdorf, another Liechtenstein estate in Weinviertel, to Feldsberg, Ferdinand could also have visited his native town again, to see the latest improvements and embellishments of the estate. He would have found that the Raisten had become a long-term building site. Johann I Joseph had commissioned Hardtmuth to have a Colonnade (Kolonnada na Reistně) constructed there, echoing Maria Theresa, who had commissioned Hetzendorf von Hohenburg (see Chapter 3) to have the Gloriette erected as the last addition to the Schönbrunn Gardens. Hardtmuth followed models in Paris and London, for example the frontage of Carlton House, which Ferdinand must have remembered from his years in this city. While the Gloriette had been dedicated to the 'just war', the colonnade is a memorial to four members of the Liechtenstein family, of which two were important for Ferdinand's brothers Johann and Joseph. The inscription on the northern front, then probably not yet affixed, was to read 'Der Sohn dem Vater, der Bruder den Brüdern' (The son to the father, the brother to the brothers). By this Johann I Joseph had referred to his father Franz Joseph I and to his two brothers who had died before him – Philipp Joseph and Aloys I Joseph. Befitting his ego, Johann I Joseph had his own statue added to the three statues of his father and the two brothers placed in niches of the colonnade on Raisten.

Ferdinand could have wished to see again the ponds of his youth, now embellished with an aqueduct following models in Rome commissioned by Aloys I Joseph, or to gaze at the Rendez-vous, also called the Temple of Diana, an oversized hunting lodge in the woods near Feldsberg. The latter has the form of a Triumphal Arch echoing the Arch of Titus in Rome, was planned and built under the orders of Johann I Joseph by Hardtmuth, though finished by his successor in the job Kornhäusel. In a sense both were Joseph's colleagues in the service of Johann I Joseph. No less than 11.4 metres high with a colossal relief of thirty figures, it has about one-quarter of the size of the famous Arc de triomphe in Paris, then also under construction. This fine neoclassical building, close to the road from Feldsberg to Lundenburg, was finished in 1813. It still exists and contains a large and magnificently decorated reception room located on the top floor of the arch, offering a majestic view above the crowns of the extensive surrounding oak forest. In the same year Kornhäusel, rightly called the Viennese Palladio, had finished another spectacular monument commissioned by Johann I Joseph. Near the town of Mödling on the top of a hill, a neoclassical Greek temple was built to commemorate the famous Battle of Aspern-Essling. It is called Husarentempel, because Johann I Joseph had five hussars who had died in action buried in this building. Ferdinand is likely to have been to this spot, since he collected a specimen of *Trinia glauca* annotated 'Aichkögelberg [Eichkogel] near Mödling May the 21st 1816', not very far from the temple. As a matter of fact this area south of Vienna was and is the traditional hunting ground for botanists based in Vienna, with possibly Clusius in the seventeenth century being one of the first. While this umbel could also have been found by Ferdinand in the southern parts of England, this is definitely not possible for the submediterranean *Himantoglossum adriaticum*, a very tall terrestrial orchid with twenty-five to forty flowers, each with a trilobate lip, the central lobe filiform, twisted and up to sixty millimetres long. Ferdinand collected this specimen in 1815 on Leopoldsberg, a famous mountain just a few kilometres north of the centre of Vienna, overlooking the Danube and the city, and forming the north-easternmost corner of the Alps.

By now Ferdinand had become a genuine botanist, though definitely not of Jacquin's calibre. The two extant herbarium labels from the former's hand have page references to two standard-works, both in Ferdinand's library – the fourth edition of Linnaeus's *Species plantarum* by Willdenow, whom Franz had met in Berlin in 1788, and the *Synopsis plantarum in Austria provinciisque adiacentibus sponte crescentibus* by Host, physician-in-ordinary to Francis I (see Chapters 10 and 17).

CONTINUING THE *ILLUSTRATIONES FLORAE NOVAE HOLLANDIAE*

Ferdinand's only extant letter from his years in Vienna was written on 26 November 1816 and directed to Brown in London.[44] It starts in a submissive way, 'With infinit pleasure I received Your letter which convinced me although after a long silence which I am happy to fergive, that I have been in an error which made me think that you have quit[e] forgot me and that I must now beg you will fergive me', and ends almost in the expression of an inferiority complex, 'I am sorry that I giveing you greattill [great deal] to much trouble but I trust of the goodnis in Your friendship that you will fergive me and remain with sincerly wishing you health and happiness and am Your ever affectionet friend Ferd. Bauer'. The letter deals exclusively with the continuation of the *Illustrationes*, and more particularly the third instalment comprising the plates 11–15. Ferdinand tells Brown that at that time he is able to send him only four coloured sets (a total of twenty plates) to go to Banks, Lambert, John Walker and Brown himself. In addition Ferdinand reports that ten plain sets are also on their way to London (a further fifty plates), with one set each to go to Banks, Menzies, Franz and Brown, the rest 'to them which like to receive it [the *Illustrationes*] farther [further]'. Three sets on expensive 'Swiss paper' had been printed and are also sent to Brown. This proves that the third instalment of the *Illustrationes* was produced in Vienna with Ferdinand acting as publisher and Ferdinand, Brown and the bookseller White (see below) in London as distributors. Indirect evidence for the publication of the third fascicle after 7 March 1816 is found in an unpublished letter by Jacquin the son to Mariano Lagasca y Segura in Madrid from that date, stating (in French), 'the famous painter Ferdinand Bauer . . . is at the moment in his homeland . . . starts to publish his drawings of the new genera described by Brown, there are already two instalments published'.[45]

John Hunneman, a London bookseller and agent who had been in Vienna, arranged the transport of the seventeen copies of the third instalment from Vienna to London. Ferdinand also tells Brown that otherwise he had settled an arrangement with the bookseller White in London, who had received '6 copies of the wholl [sic] 3 No and one Copie [sic] of the first No Colourd, for which I have from them a Book of 16 shilling which I have not paid for'. All this underlines the fact that the *Illustrationes* was a privately published series of copper engravings with a title page and the extremely brief introduction by Brown (see Chapter 11). A fourth instalment of the work was planned but did not materialize.

Of the five plates in the Vienna instalment, three are dedicated to the gymea lily, *Doryanthes excelsa*, while the last one, t. 15, shows *Stylidium calcaratum*, *S. pygmaeum* and *Levenhookia pusilla*, all three from King George Sound in what is now Western Australia (Fig. 12.11.). As he had in the two London instalments of the *Illustrationes* Ferdinand added references to Brown's *Prodromus*, stressing the fact that they were intended as an illustrated complement to the latter work, which was also given up because of poor sales. What came on the market was a mixture of completely coloured sets of fifteen plates, uncoloured sets of fifteen plates and all intermediates as well as fragments of the work. The copy in the Universitätsbibliothek Leipzig illustrates the latter point: plates 3 and 9 are coloured only with the reds and greens, table 4 only with the greens, and all other plates remained uncoloured.[46]

However, Ferdinand at least managed to sell two coloured copies of the three instalments – to Francis I and to Johann I Joseph. Both copies survive in the

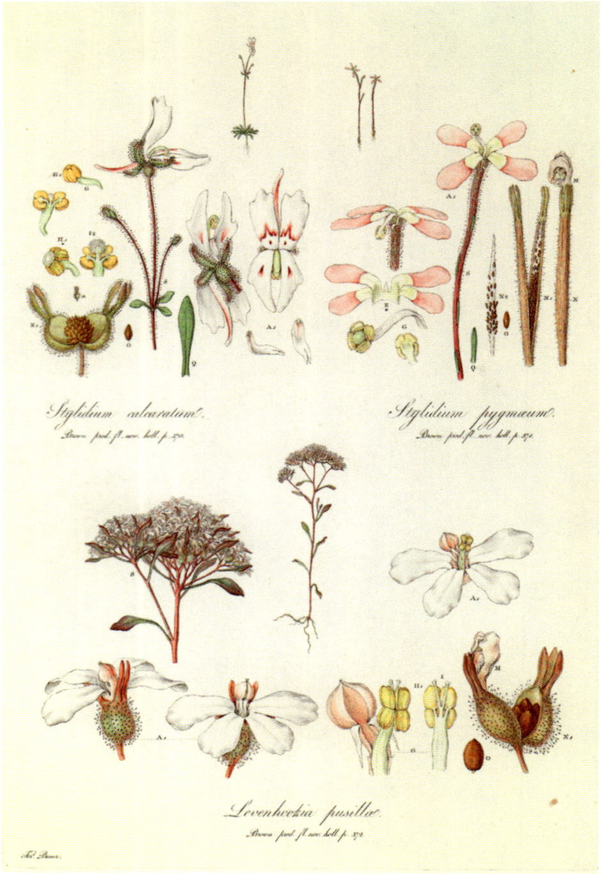

FIG. 12.11. *Stylidium calcaratum* (top left), *S. pygmaeum* (top right), *Levenhookia pusilla* (bottom). Coloured copper engraving by Ferdinand Bauer based on watercolours by his own hand. *Illustrationes Florae Novae Hollandiae*, t. 15. Vienna, 1816. – London, Natural History Museum, Library and Archives.

Österreichische Nationalbibliothek and the Fürstliche Sammlungen, respectively, and both have very recently been shown in major exhibitions in Vienna.[47] For Ferdinand the former Russian Ambassador in Vienna, Andrey Kirillovitch Prince Razumovsky, chief Russian negotiator at the Congress of Vienna, one of Beethoven's benefactors and later permanently living in the imperial town, would have been a potential purchaser for the *Illustrationes*, but we have no evidence that he bought a copy. By contrast, his brother Alexei Kirillovitch Count Razumovsky, minister for education in St Petersburg and another honorary member of the Königliche Gesellschaft der Wissenschaften in Göttingen, subscribed to the immensely expensive *Flora Graeca*.His copy is now one of the many treasures of the Botaniceski Institut im. Komarova of the Russian Academy of Sciences in St Petersburg.[48]

Ferdinand may have been disappointed about the failure of the *Illustrationes*, but nevertheless for the rest of his life the graphical record of the *Investigator* voyage remained on his agenda. For this we have only a single report by two eye witnesses. When Carl Friedrich Philipp Martius (see Chapter 18) and Johann Baptist Spix spent a few weeks in February 1817 in Vienna, they met Ferdinand. In their travelogue *Reise in Brasilien*[49] they recorded that the latter was busy with 'the depiction of the strange forms of plants and animals from those regions [i.e. Australia]'. The explicit note on the illustrations of animals is noteworthy, since no watercolours of Australian animals had been deposited by Ferdinand at the Admiralty prior to his departure for Vienna, which means that the latter must have been prepared in Vienna.

Whereas we know from Ferdinand's letter to Brown that the third fascicle of the *Illustrationes* was completed before 29 November 1816, we have absolutely no chronology for his other two jobs: (1) to continue with his botanic watercolours and start with the mycological and zoological watercolours referring to the voyage in the *Investigator*; and (2) to produce pen-and-ink tracings of his colour-coded pencil drawings. For the first project we have reliable figures: forty-nine botanical and mycological watercolours and forty-nine zoological watercolours (see Chapter 11). The seven representations of fungi are unique items (Fig. 11.8., for example) as are the twenty-two images of orchids and all the images of animals. By contrast sixteen illustrations of orchids and three illustrations of the Western Australian pitcher plant, *Cephalotus follicularis*, are repeats or variants of watercolours deposited at the Admiralty before Ferdinand left for Vienna. It has been rightly observed that the respective pairs have been painted on paper of different quality. However, when analyzing the two 'series', labelled A and B, there is no clear-cut correlation: sometimes a watercolour on inferior quality paper had been deposited in the Admiralty, more rarely one on the superior quality paper had been deposited there. Maybe Ferdinand had indeed prepared a short series of orchid watercolours on inferior paper in Australia (see Chapter 11) and deposited several of these at the Admiralty and later, in Vienna, produced variants on better quality paper. In other cases he may have produced variants on better quality paper in England and retained the inferior copy until his death. In the end all the finished orchid watercolours in the possession of Ferdinand, irrespective of where they had been painted, unique items or variants/repeats, arrived via Franz in the hands of Brown who retained them. Only after the latter's death were they combined with the Admiralty series then already at the British Museum, which led to parallels in sixteen orchids (see Chapter 11). The situation with the Western Australian pitcher plant is obscured by the fact that one watercolour had been sent by Banks earlier to the Admiralty (see Chapter 11), while Ferdinand may have prepared repeats in Vienna, which then found their way back to London after his death.

The orchids, with the complex architecture of their flowers (for example *Lyperanthus suaveolens*), were no doubt a minor challenge compared with the fur of the mammals (Fig. 11.9., 11.11.) and the strikingly coloured feathers of the birds (Fig. 11.12., 11.13.) from Australia. For the latter, worked up by Ferdinand after his return to Vienna, he had a reference point nearby – Albrecht Dürer's 'Young Hare' and his 'Wing of the European Roller',[50] both masterpieces of observational art, priceless, today omnipresent and for some the very icons of the great change in nature observation in the Renaissance. Ferdinand must have heard of the two watercolours, dated 1502 and 1512, respectively, which had counted for centuries among the great art treasures of Vienna. First recorded in the Imperial Treasury, they were kept since 1796 in the collection of Albert Prince von Sachsen, Duke of Teschen, one of the sons-in-law of Maria Theresa, who had amassed a spectacular collection of prints and drawings in his city residence in Vienna. Today this material forms the cornerstone of the Albertina, still located in his palace just opposite the city residence formerly owned by the Dietrichstein family (see Chapter 7). Albrecht also possessed a copy of Lambert's *The genus Pinus* and may therefore at least have seen the name Ferdinand Bauer printed (see Chapter 9).

THE PEN-AND-INK TRACINGS

Considering the failure of his *Illustrationes* on the one hand and the masses of unused colour-coded pencil drawings from the *Investigator* voyage on the other hand, Ferdinand seems to have looked out for alternatives to the long-established method of engraving copper plates and using them as printing forms. He may have heard from his brother in Kew about a new form of printing on stone, called lithography, which Franz used as a true pioneer when publishing his book *Strelitzia depicta* as early as 1818 (see Chapter 8). In the same year the first textbook on the new printing method *Vollständiges Lehrbuch der Steindruckerey* came out in Munich. Lithography is based on the principle that oil and water do not mix; the printing form is a smooth level lithographic limestone plate, with the line drawing (1) either directly made on the stone or (2) transferred with the help of an intermediate covered with graphite or red chalk on the lower surface and a pen on the stone. Tracing paper is often used as the intermediate carrying the image. Lithography has several advantages over copper engraving, in particular the superior speed, and a print run that is basically unlimited and consequently much less expensive.

Based on his colour-coded pencil drawings made in the field, Ferdinand produced hundreds of tracings in pen and ink, bringing drawings from different sheets of paper together and rearranging the dissections. In addition, he often added to his tracings written information copied from his pencil drawings, the latter always in pencil. In one group of the tracings Ferdinand copied the numbers referring to his colour chart in pencil in his hand; in another group these were not added. Basically two types of support were used: (1) thin, smooth, semi-transparent tracing paper and (2) thin non-transparent paper. It is somewhat difficult to imagine that this gigantic work was done by Ferdinand alone, but no convincing evidence has been found that he made use of an assistant, although this cannot be excluded. Even in cases which have been studied in detail, for example the illustration of the sticky tailflower, *Anthocercis viscosa* subsp. *viscosa*, from one of the first landings in Australia,[51] it is not completely clear why two pen-and-ink tracings with different arrangements of the dissections were prepared: one colour-coded and annotated with a number in pencil by Ferdinand (Fig.12.12.); the other without a reference to the colour chart and without the number. Since the format is slightly different, as is the arrangement of the dissections, one variant may have been selected for publication. However, the fact remains that Ferdinand never saw a single of his pen-and-ink repeats ever published (see Chapter 18). It must be remembered that all those pen-and-ink tracings, with a few exceptions in the Naturhistorisches Museum, still wait for a thorough analysis, not an easy task considering the masses of material to be compared and analysed. Special attention should be given to the different numbering systems:

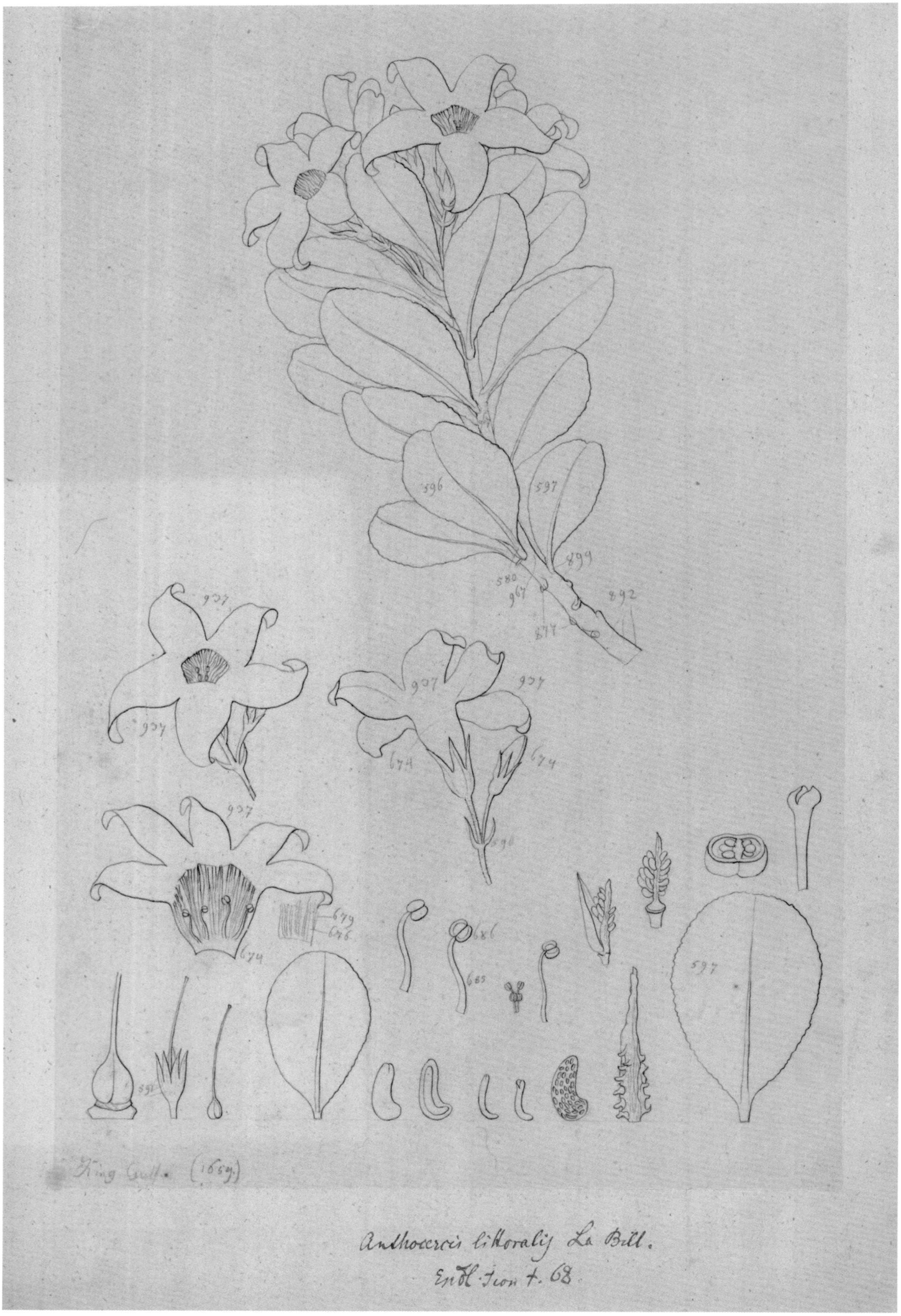

FIG. 12.12. *Anthocercis viscosa* subsp. *viscosa*. Tracing in ink with colour-code numbers in pencil by Ferdinand Bauer, c.1820 (?). – Vienna, Naturhistorisches Museum, Archive.

Ferdinand used at least two – one consisting of numbers, and a second one consisting of numbers in parentheses, one of them referring to Ferdinand's manuscript list prepared after his return to London (see Chapter 11). Later, upon accession in the Imperial Royal Natural History Cabinets, each and every drawing received an inventory number, such as 'Herb. Palat. Vindob. Ferd. Bauer Icon. Nr: 1536 Nova Hollandia' for the colour-coded pencil drawing of a species of *Correa* and 'Herb. Palat. Vindob. Ferd. Bauer Icon. Nr: 1536 – a Nova Hollandia' for the pen-and-ink tracing of it. It is still to be elucidated whether such 'a' numbers were consistently used for the repeats; 'b', 'c', and 'd' numbers also exist, their meaning is still to be clarified. As so often is the case in botanical illustration, the written annotation, apart from the clearly attributable numbers referring to the colour code by Ferdinand and his locality information such as 'XV South coast' for the *Correa* species mentioned above, cries out for the graphologist. There are different hands, one of them probably Endlicher's (see Chapter 18), who published more pen-and-ink repeats by Ferdinand than anybody else, albeit after the latter's death. Another hand remains so far unidentifiable and was possibly written by an amanuensis. Furthermore, it seems that pen-and-ink tracings have not been prepared for all pencil drawings made in the field.

The question may be asked why Ferdinand did not find a publisher for his material. Printing lithographs of unnamed and undescribed plant species from the other end of the globe proved totally unattractive to every entrepreneur in Vienna. So the question remains of why Ferdinand was unable to find an author to name and describe his very accurate pictorial record and his herbarium material. Two reasons become immediately evident: a mind with the qualification of Brown was needed, and neither Jacquin the son, forty-eight years old when Ferdinand returned to Vienna, nor Trattinick possessed Brown's standing. Secondly, soon after the arrival of Ferdinand the botanical world in Vienna was focused on Brazil and its riches, which flooded the conservatories in Schönbrunn and Vienna thanks to material sent back from the great Austrian Expedition to the Portuguese colony on the verge of its independence. Endlicher, born in 1804 in Preßburg (then Poszony, now Bratislava), had the brains, but when Ferdinand died Endlicher was a student of theology and only twenty-one years old. This is how it came about that another chance was missed – to publish the pictorial record of the voyage in the *Investigator* as lithographs in Vienna. Ferdinand could not have been happy about this state of affairs. In a sense he was in the same situation as Franz Schubert, one of his compatriots in Vienna. His iconic Symphony No. 9 in C major, D. 944, known as the Great, was premiered in Leipzig eleven years after Schubert's death, his no less famous Symphony No. 8 in B minor, the Unfinished, D. 759, had to wait for thirty-six years until it experienced its first performance ever, in Vienna.

FOXGLOVES

In their youth the Bauer brothers had painted four species of foxglove for Boccius – two common and widespread species, the red foxglove, *Digitalis purpurea*, and the yellow foxglove, *D. lutea* – as well as two species from the Iberian Peninsula, which they could have observed only as cultivated specimens – the sunset foxglove, *D. obscura*, and the small-flowered *D. parviflora*.[52] Of these, the first three species were already known to Linnaeus, while the fourth had first been described by Jacquin in 1770 on the basis of cultivated material in the Botanical Garden of Vienna University. All four illustrations are included in the first volume of the *Codex Liechtenstein*, which was completed in 1776 at the latest; at that time Ferdinand, the youngest of the three contributing brothers, had been only sixteen years old (see Chapter 2). When touring the Levant with Sibthorp, Ferdinand had observed on the Bithynian Olympus (Ulu Dağ) the rusty foxglove, *D. ferruginea*, on Mount Athos a foxglove new to science, *D. viridiflora*, as well as a subspecies new to science, *D. lanata* subsp. *leucophaea* (see Chapter 5). In addition, seed material of the second foxglove from Mount Athos had been collected, which upon his return to England in 1787 Sibthorp had given to William Aiton, who raised plants from them in the Royal Garden at Kew and very likely kept them for years in cultivation there. Probably on the basis of these Ferdinand produced a watercolour of *D. lanata* subsp. *leucophaea*, now in the archives of the Royal Horticultural Society in London. Since it is not annotated, we are at a loss to say when it had been prepared.

During the interval between the end of his engagement at Oxford and the beginning of his work for the Admiralty, Ferdinand must have produced at least one watercolour of the woolly foxglove, *D. lanata*, the most important supplier of digitoxins, because he engraved a copper plate with an image of this plant, adding the conventional 'Ferd. Bauer del et sculp.' as well as the unusual note 'Pub. as the Act direct Oct 1 1799'. It remains unknown if prints were taken from this very plate in 1799, but they were definitely included as an unnumbered plate in the *Digitalium monographia* published by John Lindley in 1821[53] (see below). Surprisingly, the short note on the copper plate was not erased in all copies taken, which may indicate that a stock of pre-existing prints was used for the publication of the book. Since the woolly foxglove is not native in England, Ferdinand must have based his work on a cultivated specimen. The Royal Botanic Garden at Kew is a plausible source of the material, since the species is listed in the second edition of the *Hortus Kewensis*, with the relevant volume, however, published as late as 1812. More than a century later Sacheverell Sitwell compared in poetic prose the illustrations of *The genus Pinus* with those of *Digitalium monographia* and wrote 'Ferdinand [is] in just the same trauma of calm ecstasy before the foxglove' when preparing the watercolours and continues, 'Each and every spire is drawn by him as though he were drawing Magdalen Tower or the Giralda of Sevilla. The marvellous stippling and mottling of each flower is rendered with patient accuracy, each mullion is drawn; and the spire of bells, when finished, could be an architect's despair but the pride and delight of the campanologist'.[54]

However, the two illustrations of *D. lanata* and *D. lanata* subsp. *leucophaea*, respectively, did not remain the only watercolours of Ferdinand representing this genus: seventeen more followed, among them again the purple foxglove and the yellow foxglove known to Ferdinand from Feldsberg and *D. viridiflora* from Mount Athos. However, the chronology of the *Digitalium monographia* remains totally obscure. Of the illustrations included, the watercolours documenting the *D. ferruginea* and *D. lanata* subsp. *leucophaea* are particularly interesting: they are not based on the pencil drawings from the tour to the Levant when he accompanied Sibthorp as one might suspect, but on completely different specimens, probably also originating from the Royal Garden at Kew, where both species were listed as in cultivation in 1812. This is no surprise: Sibthorp with his enormous ego and possessive attitude had not permitted Ferdinand to take with him the pencil drawings of plants and animals used for the watercolours when his work was completed, fearing that the images could be reused. The provenance of the remaining foxgloves documented by Ferdinand is often more difficult to establish; we find abbreviations such as 'v. v. c. hort. Kew.', standing for 'I saw it as a living specimen cultivated in the [Royal] Garden at Kew'. There is a single reference to the botanic garden of Cambridge University, from where Ferdinand may have received a specimen or visited the place, and to a garden of one Mr Bell, possibly John Bell, the land steward for the Duke of Northumberland at Syon House, across the Thames from Kew, who had received foxglove seeds from Siberia. In another case Lindley's text is more explicit, for *D. thapsi* it is stated that the plants had been raised in the Royal Garden at Kew from seeds obtained from a herbarium specimen conserved in Lambert's herbarium, which had originally been collected by Abbé Antonio Cavanilles on the Iberian Peninsula. All this indicates that Ferdinand's foxgloves had been cultivated in England – with a single exception. The hybrid *D. purpurea* × *D. lutea* (Fig. 12.1. p. 322) is recorded as having been cultivated in the botanical garden at Heidelberg. This is not impossible. Having settled down in Vienna in late 1814 Ferdinand twice visited his brother Franz at Kew – in 1819 according to an obituary[55] and a second time in 1824 according to a manuscript note on one of Ferdinand's watercolours in monochrome in the Natural History Museum (see below). No details are available, but it is plausible to assume that on one of these occasions Ferdinand stopped to visit the tiny botanical garden of Heidelberg University. At that time it was situated on the left bank of the Neckar River near the monastery of the Dominicans and very near to the centre of the small town.[56] At the most plausible time for Ferdinand's visit, in 1819 or 1824, the territory to which Heidelberg belonged was ruled over by Ludwig I, Grand Duke of Baden.

Ferdinand's foxglove watercolours are all kept today in the archives of the Royal Horticultural Society; six of them carry the signature of William Cattley on the verso. It is plausible to assume that the complete material, or at least the six watercolours, were in the possession of Cattley, an international merchant based in London engaged in importation of grain from Russia. Ferdinand could have been commissioned by him to prepare the watercolours or, alternatively, sold them to him, which would help to explain Ferdinand's relaxed financial situation in his later years. Franz had also contributed to the *Digitalium monographia*: he had prepared a watercolour showing the flower of *D. purpurea* dissected resulting in a plate with a total of thirty-five details, while Lindley contributed eight watercolours of several more species, among them *Isoplexis sceptrum* from Madeira and *I. canariensis* from the Canary Islands. Lindley's watercolours are also kept today in the archives of the Royal Horticultural Society; that by Franz is currently untraceable.

Following the preparation of the watercolours the engraving work started, with William Hooker, Lindley and Ferdinand being involved. Lindley engraved eight copper plates, Hooker a single one, and Ferdinand no less than fourteen, all based on watercolours by himself. It is not known who was responsible for the engraving of Ferdinand's remaining watercolours; and it is also unclear when and where Ferdinand had prepared his engravings, in England in the years 1794 to 1801 or 1805 to 1813, or in Vienna between 1814 and 1820. Lindley's preface is not precise in this respect: he only relates that Ferdinand had prepared 'many years ago' watercolours and engravings after his return from the Ottoman Empire, which makes the first period indicated above more likely than the later periods. Maybe Ferdinand had nurtured the idea of a fully illustrated work on foxgloves, but was forced to discontinue work due to lack of funds. However, at least the watercolour of the foxglove hybrid from Heidelberg must have been prepared significantly later, since Ferdinand is not known to have left England before 1814.

As so often, a Maecenas, in this case Cattley, was required to provide funds for publication – (1)for engraving the remaining copper plates, and (2) for printing and colouring the impressions, and (3) for someone to write the text, which in the end was done by Lindley, and (4) have it printed, which was done by Richard & Arthur Taylor in London, the printers of the *Flora Graeca*. Cattley may also have bought the remaining watercolours by Ferdinand and the copper plates engraved by him. The title page, often the reason for heated arguments among collaborators, is carefully worded and gives Lindley merely as editor ('cura J. Lindley') while the complete title reads 'Digitalium monographia: sistens Historiam naturalem generis, tabulis omnium specierum hactenus cognitarum illustratam, ut plurimum confectis ad icones Ferdinandi Bauer' (Monograph of foxgloves: being a Natural History of the genus illustrated with tables of all species discovered so far, with many of them made from the images of Ferdinand Bauer). The work is dedicated to Cattley, and the publication in large format was completed in London in 1821. Ferdinand must have been satisfied with the end result, particularly by seeing his name for a second time on a title page. Indeed he had come a long way from his boyhood in Feldsberg. It is possible that Ferdinand collected his copy in London, when he visited his brother Franz. What is maybe more important for the history of science is the fact that this work fulfils all modern criteria and requirements for a taxonomic monograph in dealing with all entities of a systematic group in a comprehensive way. In short the publication was the result of the combined efforts of three highly qualified people – Lindley, an ambitious and dedicated young botanist (see below), Ferdinand, a highly respected plant illustrator, and Cattley, Lindley's benefactor and the sponsor of the project.

On Ferdinand's last visit to London three years after publication, he prepared watercolours in monochrome of the parasitic *Hydnora africana* (Fig.12.13.) and dated them '1824'. They were published more than two decades later together with several watercolours, largely in monochrome, by his brother Franz, who had documented the parasitic *Rafflesia arnoldii*. In a sense this was their last joint project, with Brown as author of the erudite text (see Chapter 13).

FIG. 12.13. ***Hydnora africana*****. Monochrome watercolour by Ferdinand Bauer, 1824. Paper, watercolour. – London, Natural History Museum, Library and Archives, Unfinished Drawings of Plants, vol. 2, no. 9 t.**

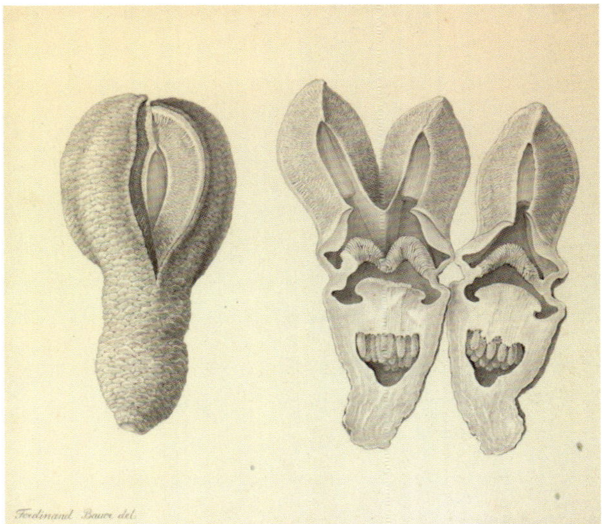

PASSION FLOWERS, FERDINAND'S SWAN-SONG

Like the foxgloves the Bauer brothers had portrayed in their youth several species of passion flower for Boccius, among them the blue passion flower (see Chapter 11). Ferdinand had no possibility of seeing passion flowers on his tour to the Levant, but there must have been ample opportunities later to see them cultivated in conservatories in Oxford, Kew or elsewhere in England. The only chance to observe and record passion flowers in the wild presented itself on the voyage of the *Investigator* and on Norfolk Island (see Chapter 10).

Late in life Ferdinand seems to have been commissioned by the Horticultural Society of London or one of its members, maybe Sabine (see Chapter 11), to produce a painted documentation of the passion flowers of the world.[57] The results are two volumes of exquisite watercolours. The larger of these (volume A) measures 59.5 × 48.5 centimetres and the smaller (volume B), 33.5 × 26.0 centimetres. Both are portfolios bound in half leather, volume A containing forty, and volume B forty-one, watercolours. Volume A has 'Bauer's Passion Flowers', volume B 'Bauer's Dissections of Passion Flowers' and 'Horticultural Society of London' in gold lettering impressed on the spine. Volume A has a cut-out from a printed auction catalogue of Messrs Sotheby & Wilkinson in London from 1859, reading 'Bauer (Ferdinand) . . . Forty exquisite Drawings in colour of this admirable Botanical Artist . . . These were executed expressly for the Society, and are unique. Their execution is a model of botanical drawing in water colours'. Added are two pencil notes, 'from the Library of the Horticultural Society' and '£ 24. 0. 0', probably the price given in the auction catalogue. In the two volumes only the plates A: f. 1–38 and B: f. 1–36 have been signed in black ink mostly 'Fer.d Bauer del.' and 'Drawn by Ferd. Bauer'. The complete series is very homogenous; all drawings are very true to nature, of the highest quality, with the dissections showing minute details in utmost perfection. Nothing indicates that the seven unsigned drawings might have been made by an illustrator other than Ferdinand. Of the eighty-one drawings, only three show dated watermarks: 'B. E. & S. 1825'. Since Ferdinand died on 19 March 1826 (see below), these watercolours must have been among his last works.

Ferdinand illustrated passion flower species from the New World, among them popular species such as the blue passion flower, *Passiflora caerulea* (Fig. 12.14.), which he had previously depicted in his oil painting (see Chapter 11), and the red passion flower, *P. racemosa* (Fig. 12.15.), both of which he could only have seen in cultivation. However, there are also species from the Old World, such as the orange-petaled passion flower, *P. aurantia*, on which Ferdinand noted that it is 'the plant I found on Norfolk Island' (see Chapter 10). For A: f. 12 showing the latter species we even have a *terminus ante quem*: Lindley, then Assistant Secretary of the Horticultural Society and later the man to have several orchid watercolours by Franz printed (see Chapter 13 and 14), published this illustration in his *Collectanea Botanica*, of which the sixth instalment came out in London in 1821. This image remained the only watercolour of Ferdinand's 'Passion Flowers' to have been printed while he was still alive; three more were published by the present author one hundred and seventy-eight years later.

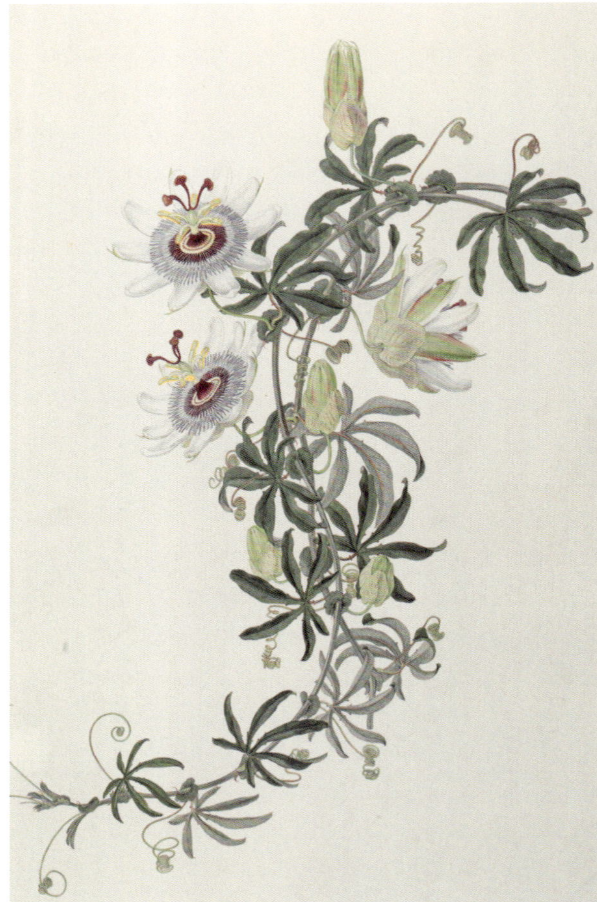

FIG. 12.14. Blue passion flower, *Passiflora caerulea*. Watercolour by Ferdinand Bauer, signed, c.1825. Berlin, Stiftung Preußischer Kulturbesitz, Staatsbibliothek zu Berlin, Handschriftenabteilung, Libri Picturati A 102, volume A: f. 28 (deposited in Cracow, Biblioteka Jagiellońska, Berlinka).

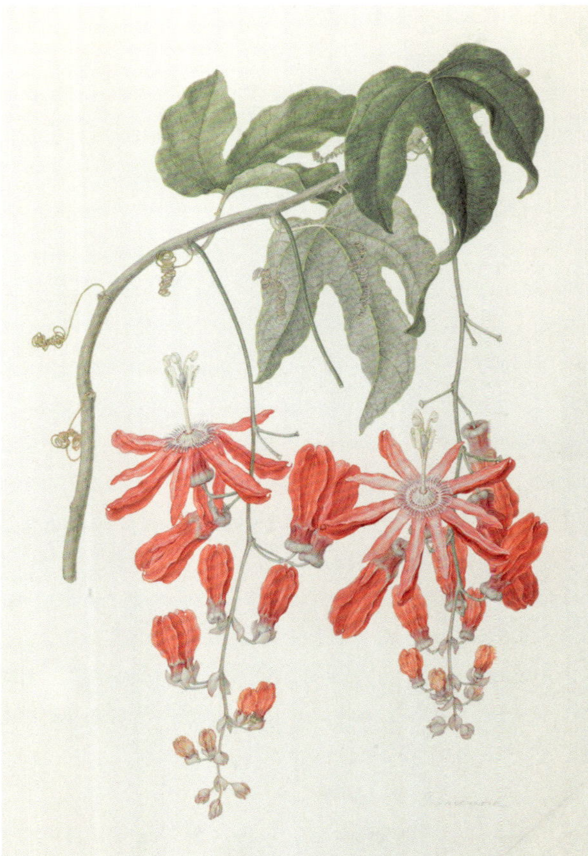

FIG. 12.15. **Red passion flower, *Passiflora racemosa*. Watercolour by Ferdinand Bauer, signed, c.1825. Berlin, Stiftung Preußischer Kulturbesitz, Staatsbibliothek zu Berlin, Handschriftenabteilung, Libri Picturati A 102, volume A: f. 37 (deposited in Cracow, Biblioteka Jagiellońska, Berlinka).**

For a few more watercolours, similar notes have been added by Ferdinand, for example A: f. 6 'from a plant which flowered in the Imperial Garden at Schönbrunn', A: f. 7 '[watercolour] made in England ... [at] Mr. Lee [of Vineyard Nursery, in Hammersmith]' and A: f. 8 'from Mr. [W. T.] Aiton of the Kings Garden at Kew'. The former provenance is notable since this company had previously supplied plants to Banks, W. T. Aiton and Joséphine, Empress of the French.[58] For other watercolours the provenance is more difficult to elucidate and only the background of the illustration showing *P. amethystina*, a neotropical passion flower with amethyst petals, is analysed as an example in full.

In 1817 Archduchess Leopoldine, whose elder sister Marie Louise had been married off to Napoleon, faced the same procedure with a different husband, at the same church and with the same proxy, her uncle Archduke Karl. She was married to Pedro, the eldest son of João VI, King of Portugal, Algarve and Brazil. Because of the Napoleonic wars the Bragança family had fled from Lisbon to Rio de Janeiro, which became the destination of the bride's voyage. She was accompanied by a group of naturalists, sent out by her father Francis I to study the vast Portuguese colony and to collect, among other things, plants and animals for the Imperial gardens and the menagerie in Schönbrunn as well as for the United Imperial Royal Natural History Cabinets in Vienna. Among the gardeners we find Heinrich Wilhelm Schott (see Chapter 18), the son of Heinrich Schott, Jacquin the son's head-gardener in the botanical garden of Vienna University. The scientists included Johann Christian Mikan, director of the botanic garden and professor of botany at Prague University, whose father had hosted Jacquin the son and Franz in 1788. Among the rich collections brought back by Mikan to Vienna in 1818, there were seeds of a passion flower which had not yet received a scientific name, stated to have been collected near Tocaja. These passion flowers were raised in Vienna, possibly in the conservatories at Schönbrunn, and the flowering plant was documented by Ferdinand in the watercolour forming A: f. 19. Another watercolour, no longer extant, seems to have been used as the basis for an unnumbered coloured lithograph included in Mikan's *Delectus florae et faunae brasiliensis*, where the species had its scientific name validated. Although this particular plate is attributed to someone called Stoll, this may not be the name of the illustrator. The similarity with A: f. 19 is so striking that a single authorship is plausible – the drawing must have been made by Ferdinand or at least under his guidance.

The watercolour A: f. 38 (Fig. 12.16.) is one of those made on the 'B. E. & S. 1825' paper. It shows a passion flower with petals in a most unconventional colour – bright dark violet. It is annotated in pencil 'Passiflora caerulea-racemosa' and represents the progeny of the first successful cross within the genus, with the blue passion flower as the male and the red passion flower as the female parent. This hybrid is now called *Passiflora × violacea* and was first made in spring or early summer 1819 by Thomas Milne in the Fulham Nursery of Whitley, Brames and Milne in Fulham, Middlesex with the hybrid progeny raised in December 1819. It flowered, and Sabine immediately recorded the occurrence of the spectacular plant in print,[59] naming the plant *Passiflora caerulea-racemosa*. All this was reported to the Horticultural Society on 7 November 1820 and was subsequently published in the *Transactions of the Horticultural Society*.

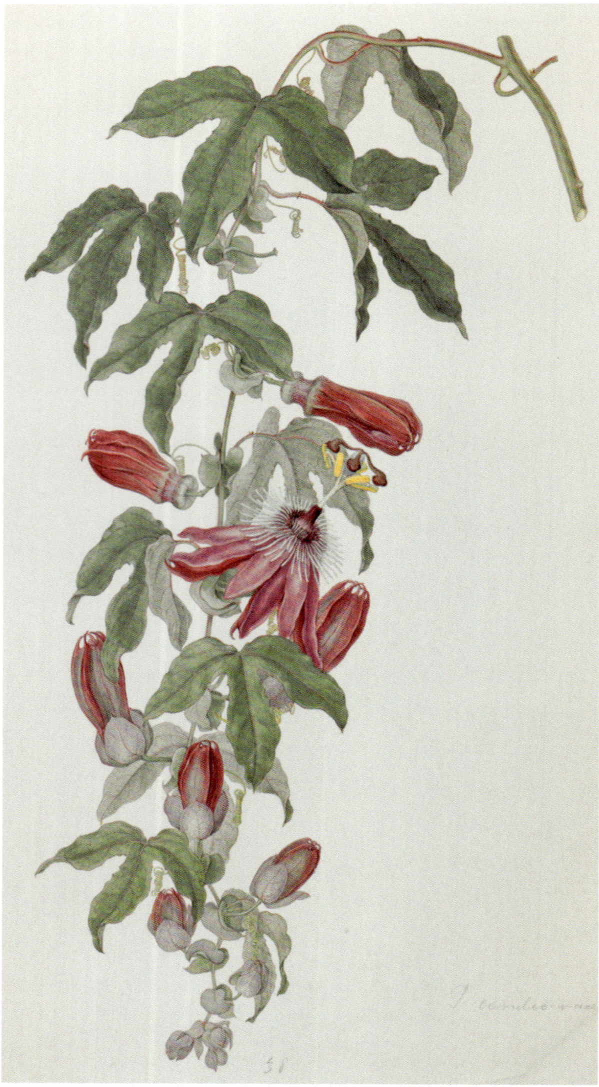

FIG. 12.16. *Passiflora* × *violacea*. Watercolour by Ferdinand Bauer, signed, 1825–1826. Berlin, Stiftung Preußischer Kulturbesitz, Staatsbibliothek zu Berlin, Handschriftenabteilung, Libri Pictiurati A 102, volume A: f. 38 (deposited in Cracow, Biblioteka Jagiellońska, Berlinka)

Due to the detailed record it is known that not all the flowers which had been crossed produced fruits, that Milne was only able to raise seven hybrid plants and that of these a single plant flowered, was exhibited at the Horticultural Society and was drawn by Lindley. This drawing was used as the basis for a coloured illustration included in the published account.

How on earth could a second watercolour of this extremely rare hybrid have been made by Ferdinand Bauer then living in Vienna? Two interpretations offer themselves: he could have received cuttings of the hybrid plant, which would not be very surprising since *Passiflora* × *violacea* was quickly distributed among passion flower enthusiasts. Alternatively, Ferdinand could have documented Milne's hybrid while in England in 1824, in any case after the specimen had been exhibited in the Horticultural Society. Maybe he had even seen the plant in the Fulham Nursery, a thirty minute walk from the Vineyard Nursery in Hammersmith (see above). Since Ferdinand is not known to have been to England in 1825 or 1826 and because of the dated watermarks he must have prepared the finished watercolour during his final months in Vienna. Preliminary versions of the watercolour A: f. 38 showing the hybrid and the female parent ended up among the unfinished drawings by Franz in the Natural History Museum. Not being colour coded, but partly coloured, they may well have been the result of studio work, possibly prepared in Franz's studio at Kew. Clearly this was a deviation from Ferdinand's standard working scheme, but easily attributable to the fact that by then he had become a resident plant illustrator like his brother had always been.

LAST CONTACTS

Ferdinand's illustration of *Passiflora amethystina* was not his only contribution to the *Delectus florae et faunae brasiliensis*, the first of the natural history books to appear documenting the findings from the imperial expedition to Brazil. The work in large format was dedicated to Clemens Wenzel Lothar Prince Metternich, Minister of State, who had been the key figure at the Congress of Vienna, remained in office for three decades and was also nominally the person responsible for the expedition. Rather unsurprisingly it was in this book that the name *Metternichia principis* was validated, preceded by *Stifftia chrysantha* dedicated to Andreas Baron Stifft, the long-serving physician-in-ordinary of Francis I. The lithograph illustrating this shrubby composite with strikingly orange flowers (Fig. 12.17.) was based on an illustration by Ferdinand, now lost. Like his image of *Badusa corymbifera* done for Lambert (see Chapter 9), it may also have been painted in monochrome since it had a herbarium specimen as the subject. The same applies to the illustration of *Vellozia candida*, a showy petaloid monocot native in Brazil. To the plate showing *Conchocarpus macrophyllus* Ferdinand added only the illustration of the dissected flower parts which were included in the lithograph.

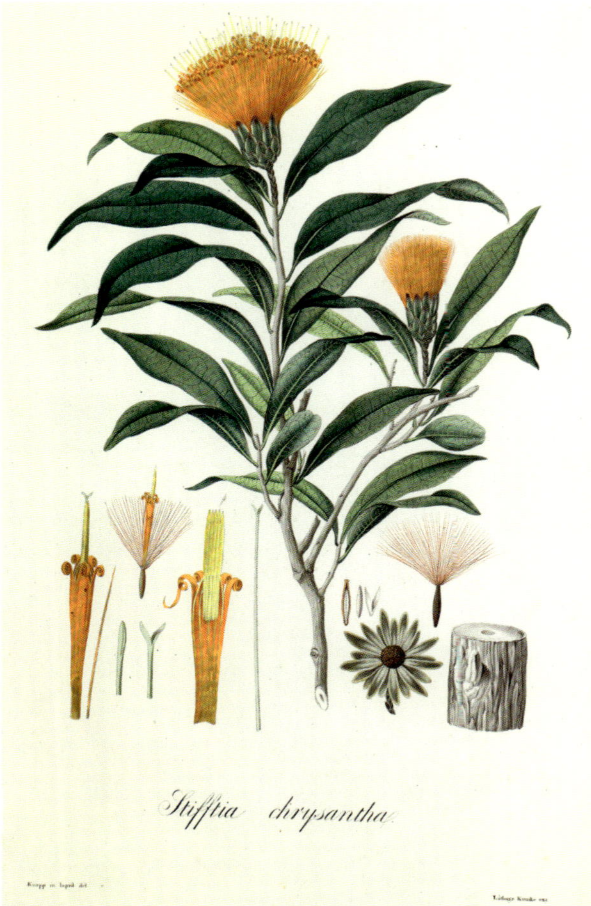

FIG. 12.17. *Stifftia chrysantha*. **Coloured lithograph based on watercolour by Ferdinand Bauer. J. C. Mikan, Delectus florae et faunae brasiliensis, t. [1]. Vienna, 1820. – London, Natural History Museum, Library and Archives.**

The introduction to the *Delectus florae et faunae brasiliensis* contains four names which are directly or indirectly relevant for our story: Jacquin the son, who possessed one of the complete copies of the *Illustrationes Florae Novae Hollandiae*[60] and was possibly in contact with Ferdinand during his final years and remained in correspondence with his brother Franz in Kew (see Chapter 16); Heinrich Schott, whose son Heinrich Wilhelm was to receive Ferdinand's library as a legacy (see Chapter 17) and to play a role, albeit marginal, in publishing several pencil illustrations by Ferdinand after the latter's death (see Chapter 18); Trattinick, who became responsible for Ferdinand's botanical collections and drawings after their purchase by Francis I (see Chapter 17); and finally Johann Knapp, given in Mikan's text as 'rerum naturalium pictor c. r. aulico Knapp' (Imperial Royal natural history court painter Knapp), who had acted as the lithographer for the illustration of *Stifftia chrysantha*.

The rather pompous title was most appropriate – Johann Knapp[61] was the natural history illustrator of Archduke Anton Victor, a younger brother of Francis I. Following his uncle Archduke Maximilian Franz (see Chapter 7) Anton Victor had been chosen the last Prince Elector of Cologne, but did not take up this office because his territory was at that time partly occupied by French, Prussian and Hessian troops. Subsequently Anton Victor succeeded his elder brother Archduke Karl (see above), who had married, as Grand Master of the Teutonic Order in the Hereditary Lands, a position for life, which focused the former on Vienna. Later, from 1816 to 1818, Anton Victor acted as his eldest brother's viceroy for the Lombardo-Venetian Kingdom in Milan. Like his youngest brother Cardinal Archduke Rudolf, Archbishop of Olmütz (now Olomouc, Czech Republic), Anton Victor was one of Beethoven's dedicatees. The two younger brothers, Anton Victor and Rudolf, spent their summers together with their two elder brothers, Francis and Karl, in Baden bei Wien, where many others, among them Ossolinski, had their summer residences. In parallel with his oldest brother Francis I, Emperor of Austria, Anton Victor possessed a genuine interest in botany and horticulture, initiated the foundation of the Imperial Royal Horticultural Society in Vienna and became its first protector. Similar interests, although with more emphasis on fruit-growing and viticulture, were cultivated by one of Anton Victor's younger brothers, Archduke Johann. Based in Styria, he was the great modernizer of the Duchy and became Knapp's second commissioner. In the same way that Banks had engaged Franz in 1789 for life (see Chapter 7), Archduke Anton Victor had recruited Johann Knapp in 1804 for life. The latter even enjoyed a special benefit: an official residence in a building opposite the winter riding school belonging to the Schönbrunn complex.

Knapp produced a rich oeuvre of watercolours documenting an extremely broad spectrum of plants and animals, many of them probably cultivated in the garden and park of Schönbrunn or kept in the menagerie there. Today largely conserved in the Museum für Kunst und Gewerbe in Hamburg, they are pleasing to the eye, of excellent quality but lack the many analytical details and dissections so characteristic of the work of Ferdinand and Franz. Not correlated with specimens, the value of Knapp's watercolours for science is therefore somewhat smaller. However, Ferdinand and Knapp, who lived in Hietzing in walking distance from each other, must have found considerable common ground. They probably had many occasions to meet in the botanic garden and conservatories at Schönbrunn or in the garden and park, for example at the fête de fleurs given annually on 25 May when the imperial court moved out of the Hofburg to Schönbrunn Palace.[62]

Though mainly a watercolourist like Ferdinand, Knapp's masterpiece is a large painting in oil now in the Österreichische Galerie Belvedere,[63] signed 'Johann Knapp fecit MDCCCXXII Schoenbrunn'. Since upon completion it was quickly acquired by the Imperial Royal Gallery in Vienna and soon later exhibited, Knapp's fame is based largely on this painting and not upon his watercolours done for the two Archdukes, which remained their private property. Measuring 218 × 164 centimetres, Knapp's painting in oil is an opulent and extravagant flower piece known as 'Jacquins Denkmal' (Jacquin's monument) and is rightly regarded as the icon of Vienna Biedermeier flower painting standing on a par with the greatest French and Dutch flower pieces. The complex, not yet fully analysed iconographical programme of this painting mirrors on the one hand plant and animal diversity on all five continents with representatives of all 24 Linnaean classes plus a few animals and minerals added. On the other hand, many of Jacquin's colleagues, correspondents and associates are referred to in Knapp's painting – either by plants described by them or by plants with generic names, which are eponyms based on their names. Among them are Franz and Ferdinand, referred to in the generic name *Bauera*. Of the very many different plants depicted by Knapp in this flower piece, one is particularly relevant in our context – the small-flowered rhynchoglossum, *Rhynchoglossum obliquum*, a widespread palaeotropical gesneriad ranging from China in the north to Indonesia in the south, and India in the west to the Philippines in the east. When the oil painting was completed and publicly exhibited in the Aula of Vienna University, this plant had not yet received a scientific name. How then can this surprising fact be explained? Ferdinand must have lent his pencil drawing or the pen-and-ink tracing based on it, both now on the Naturhistorisches Museum, to his colleague Knapp, who produced a copy catalogued as No. 1079 in Hamburg. Later a repeat of this was integrated into his masterpiece. The small-flowered rhynchglossum had been documented by Ferdinand in April 1803 in Timor, and Brown had noted on his label referring to the respective herbarium specimen in the Natural History Museum, 'Fig. Pict D Bauer Loxotis obliqua R B . . . in rupibus cumbrosis prope oppidum Coopang Insulae Timor April 1-7 1803 frequens'.

In Hamburg at least one further proof for contacts between Ferdinand and Knapp is conserved – a watercolour by Knapp, very accurately copied either from the coloured copper engraving published by Ferdinand in his *Illustrationes Florae Novae Hollandiae* showing the scarlet banksia, *Banksia coccinea*, from Western Australia or, alternatively, on Ferdinand's colour-coded pencil drawing now lost. Catalogued as No. 1076 in Hamburg, this illustration is considerably simplified – a single flower twig, a single capitulum and two flower details are depicted, while Knapp omitted all the other details of flower and pollen illustrated by Ferdinand. Later Knapp integrated an image of the scarlet banksia into 'Jacquins Denkmal', no doubt the most spectacular monument erected to the memory of Jacquin and the apotheosis of flower painting in Vienna. Ferdinand must have been aware of Knapp's painting and asked Brown to use the generic name *Antonia* for the gesneriad from Timor, almost certainly intended as a dedication to Archduke Anton Victor. This is stated by the Brown in the entry to the genus *Aikinia* in Wallich's *Plantae asiaticae rariores* (see Chapter 18), noting in retrospect 'it been intended to introduce a figure of the Timor plant under the generic name in a wreath, ornamenting a bust or portrait of the late Baron Jacquin at Vienna'. This proposal came to nothing, since *Antonia* had already been used for a plant from Brazil. The brief entry '*Knappia* Ferd. Bauer mpt. [manuscript]' in the second edition of Steudel's *Nomenclator Botanicus*, almost certainly intended to honour his colleague Knapp, is a further hint to contacts between Ferdinand and Knapp; however, this generic name was never validated and seems to be the only albeit abortive and obscure excursion of Ferdinand into botanical nomenclature.

Knapp's masterpiece seems also to contain a zoological reference to Ferdinand – an illustration of a sulphur-crested cockatoo, *Cacatua galerita galerita*, documented by Ferdinand in a colour-coded pencil drawing annotated by him 'Sydney [M]ay 1805' (see Chapter 10). No watercolour by Ferdinand based on this drawing is known, but Knapp produced a watercolour catalogued as No. 1110 in Hamburg. It may have based on a living specimen in the aviary of Schönbrunn palace, on Ferdinand's permanently preserved bird specimen or his pencil drawing. In the latter case Ferdinand would have been obliged to disclose his Australian colour code to his colleague in Vienna.

During his last eleven years in the 'Kaiserliche Haupt- und Residenzstadt' Ferdinand must also have had ample opportunity to meet plant illustrator Mathias Schmutzer[64] and his successor Johann Jebmeyer,[65] both born in Vienna. They had an even more prominent commissioner than Knapp – Francis I, the eldest brother of the Archdukes Anton Victor and Johann. As a consequence Schmutzer had the grandiose title of 'k. k. Hof-Botanik-Mahler und Hofzeichner' (Imperial Royal Court Botanical Painter and Court Drawing Master), and Jebmeyer 'Hofpflanzenmaler' (Court Plant Painter). Schmutzer, who died in 1824, and Jebmeyer concentrated on plants cultivated in the strictly private gardens of the Emperor: the garden adjacent to the Hofburg, now forming the Burggarten, the conservatory on top of one of the wings of the Hofburg famous for its collection of succulents (no longer extant), the reserve garden in the suburb Ungarische Vorstadt (no longer extant) and the garden dedicated to the flora of the Hereditary States situated under the care of Host. This garden is situated on a piece of land bordering the Belvedere garden and the botanical garden of Vienna University and now forms part of the latter. Jebmeyer continued the tradition of a pictorial documentation of the Imperial Gardens during the reign of Emperor Ferdinand I, the eldest son of Francis I and his successor to the throne. An extensive series of finished watercolours by Schmutzer und Jebmeyer is kept in the Österreichische Nationalbibliothek in Vienna together with a catalogue of Jebmeyer's drawings written by Jacquin the son. By contrast, Schmutzer's more fully annotated preliminary drawings ended up in the Staatsbibliothek in Berlin and the Biblioteka Jagiellońska in Cracow. Both hands lack the ultimate refinement of Ferdinand, nor do they reach the accuracy of his colours and the precision in his dissections, which Schmutzer recorded much more rarely than Jebmeyer. However, no trace of a personal contact has yet been found between the two plant illustrators working for the two emperors and Ferdinand. If Ferdinand had met his two colleagues he may well have been again deeply frustrated. Inferior hands had found permanent employment in Vienna while he, Ferdinand, had been obliged to work for so many years abroad and had never attracted the attention of a crowned head of state.

FAREWELL

Ferdinand's last twelve months started on 17 March 1825 when he had already completed his sixty-fifth year.[66] He lived in his house in Hietzing with its large back garden and it was granted to him to be able to continue to work until late in his life. Endless hours must have been spent on the watercolours documenting the plants, fungi and animals from Australia (see above), with the passion flowers becoming Ferdinand's last works (see above).

Precisely during this period of time two other artists also based in the suburbs of Vienna were in a similar way deeply involved in producing their final works: one of them was Beethoven, who was to outlive Ferdinand by twelve months and to die in the Schwarzspanierhaus (now Schwarzspanierstraße 15), formerly a monastery, in Alservorstadt. The other was Schubert, who was to outlive Ferdinand by twenty months and to end his days in Lumpertgasse (now Kettenbrückengasse 6) in Wieden, then also not yet part of Vienna. Beethoven was composing his last string quartets, No. 15 in A minor op. 132 and No. 13 in B major op. 130, both commissioned by Nicolai Borissowitch Prince Golitzin in St Petersburg, although a considerable part of the work was actually done in Baden bei Wien, where the composer had stayed from May to October 1825.[67] Schubert was making the last corrections to another famous string quartet at that time, his No. 14 in D minor 'Der Tod und das Mädchen' (Death and the Maiden), D. 810, which experienced its first private performance just six weeks before Ferdinand died. Needless to say, all three string quartets were published posthumously.

On 12 March 1826 Ferdinand must have felt the time had come to sign his testament (see Chapter 16), which he did in a trembling hand. It is a wise and carefully worded document, starting and ending with expressions of deep religiosity (see Chapter 2) and regulating his estate 'in order to avoid a dispute among my relatives'. Legacies were to go to the two physicians who had treated Ferdinand in his terminal illness, to his warden, to his servant couple, to the poor of the parish of Hietzing and a small sum to the local school fund. Ferdinand arranged for a funeral first class and six services to be held in his memory in the parish church to the Nativity of Mary in Hietzing. Having received the last ointment he died on 17 March 1826 and was buried two days later in the new cemetery at Hietzing (now Maxinggasse 15) very near to his house. The church record gives gangrene as the cause for Ferdinand's death, a brief mortuary note indicate 'dropsy and gout',[68] the former a rather imprecise diagnosis.

Ferdinand's grave no longer exists. Years later the burying ground of the village of Hietzing became a cemetery for Vienna prominenti and space was needed. Among the graphical artists resting there is a single one of Ferdinand's caliber – Gustav Klimt, the most prominent member of the Vienna Secession movement, who had his last studio in this suburb. In Hietzing, today an elegant residential district to the west of the city centre, only the interior of the parish church, previously the favorite church of Maria Theresa next to Schönbrunn, retains the atmosphere Ferdinand must have enjoyed during his final eight years back home in Vienna. No doubt the column with the statue of Maria Immaculata in front of the parish church must have reminded Ferdinand of the similar, though more impressive and older, column on the main square (now náměstí Svobody) in Feldsberg. Otherwise only some private gardens in Hietzing, which have remained unchanged since the Biedermeier period, like the one in Trauttmannsdorffgasse 54 bordering on Ferdinand's orchard in Trauttmannsdorffgasse 56, are reminiscent of what it was like when he lived there.

For Joseph the last year of his life began on 14 November 1830. Seven months before he had finished his seventy-fourth year and lived in Rossau, the suburb to the east of Alservorstadt, in house No. 140 (now Pramergasse 19), almost certainly belonging to his employer Johann I Joseph and located in the immediate vicinity of the latter's summer palace. Joseph was attended by Dorothea Dallinger, the daughter of his predecessor as director of the princely gallery and sister of his successor in the job Johann Baptist Michael Dallinger, often called Johann Dallinger the younger, a much more productive artist than Joseph had been. Joseph died on 14 November 1831 in his lodgings in Rossau 140; judging from information in the papers concerning Joseph's estate it is very likely that he died of abdominal cancer. This is confirmed by a very brief note which appeared in the daily *Wiener Zeitung* of 18 November 1831, though giving Joseph's last address as Rossau 130.[69] No details are available on Joseph's burial, which probably took place in one of the several cemeteries for Catholics in what is now known as Alsergrund, the ninth district of Vienna, all long ago discontinued and built over. The inventory of Joseph's estate lists several bank obligations and shows that he had been an artist – 'several plaster casts, six copper plates, patterns in gypsum, i.e. heads, feet, hands, . . . an easel, four portfolios with insignificant copper engravings, drawing tables, paints, shells, brushes' are listed. How the estates of Ferdinand and Joseph were divided, will be explained later (see Chapter 17). Here it is sufficient to say that the surviving brothers were the main beneficiaries in both cases, one of them Johann, who appears in the papers concerning the estate of Joseph as 'Rentmeister' (administrator of properties rented out), no doubt in the service of the Reigning Prince of Liechtenstein.

Johann's last twelve months started on 14 March 1833. He lived at that time in Feldsberg, the town still surrounded by its medieval wall, and in the same house near the church dedicated to the Assumption of the Virgin Mary where his mother Therese had died forty-three years ago and which may also have belonged to the Reigning Prince of Liechtenstein. Johann may have lived in retirement in Feldsberg, but for this we have no evidence. If this were so, he would have had the chance to observe the continued embellishment of the Feldsberg–Eisgrub estate with further buildings commissioned by Johann I Joseph. As his final additions he had erected the Temple of the three Graces (Chrám Tří grácií) and the fine Border Chateau (Hraniční Zámeček), both based on plans by Engel, the latter palace with references to the summer residences planned by the famous architect Andrea Palladio in the hinterland of Venice. Both buildings survive and belong to the masterpieces of the neoclassical style in Central Europe.

Johann must have been in a reduced state of health prior to 1833, signing some of the papers regulating the estate of his brother Joseph now in the Wiener Stadt- und Landesarchiv no longer with his name but with three crosses instead.[70] On 27 June 1833 he had his seventy-ninth birthday, but there were probably no celebrations. Death came on 14 March 1834, the church record noting infirmity of old age as the cause of his death and that he had not received 'the last ointment because of neglect of his wardress'. It is this document now in the Moravský zemský archiv in Brno[71] that reveals Johann's former occupation – he had been princely administrator of the mineral springs in Karlsberg (Karlovec, now a part of Bruntál), an estate in the Duchy of Jägerndorf in the Austrian part of Silesia. It belonged to the Reigning Prince of Liechtenstein following the redistribution of properties in the Kingdom of Bohemia after the battle of Bíla Horá (see Chapter 2). The church where Johann's requiem must have been held still bears the inscriptions 'Carnoviae Dux' (Duke of Jägerndorf) on its majestic front, a title Hans Adam II continues to hold today. Johann must have been buried in the old cemetery in Feldsberg, where his mother and possibly also his father had found their final resting place before. This burying site has long since also been given up and used as building ground.

Two important figures in our story were to die in the two years following the death of Johann: Francis I, the monarch who had bought the collections and drawings from Ferdinand's estate (see Chapter 16), died on 2 March 1835 in Vienna and was buried in accordance with family tradition in the Imperial Crypt beneath the Capuchin Church and Monastery in Vienna. Johann I Joseph, Johann's and Joseph's long-term employer, died on 20 April 1836 in the imperial city and was buried, again in accordance with family tradition, in Wranau, and more particularly in a new crypt he had arranged to be erected and where he was the very first to find his final resting place. Larger than life-size statues exist for both of them in Vienna: for Francis I in the outer courtyard of the Hofburg, for Johann I Joseph in the Heeresgeschichtliches Museum. While the first monument carries the political motto 'Amorem meum populis meis' (My love to my peoples), Johann I Joseph as a man is probably best characterized by the inscriptions on the arch leading to his summer palace in Vienna. When entering the *cour d'honneur* it reads 'Der Kunst, den Künstlern' (To the arts, to the artists), when leaving 'Der Natur und ihren Bewunderern' (To nature and its admirers).

Franz, in Kew, survived all three brothers and the employer of his two elder brothers. It must have been the notaries in Vienna and Feldsberg, respectively, who informed him about the death of Ferdinand, Joseph and Johann, and the money he would inherit. It is now time to return to Franz's life and achievements.

ENDNOTES

1 Information on the holdings of the Fürstliche Sammlungen is based on Kräftner (2004) and on www.liechtensteincollections.at (accessed 5 x 2014)
2 Vienna, Fürstliche Sammlungen, Hauptbuch zugleich Majorat Hauptkaßa Rechnung 1788–1831
3 Inv. No. GE 1535, 1787
4 Inv. No. 74.4.
5 Zatloukal, Krejčiřík & Zatloukal (2012)
6 Lack (2003)
7 Inv. No. GE 1360
8 Reg. No. 1978,0121.276
9 Reg. No. 1978,0121.277
10 Inv. No. GG 524
11 Anon. (1873)
12 Galerie Franz Schoen, Bonn; offered for sale via ebay in 2014.
13 Reg. No. 1978,0121.280
14 Inv. No. 93/02.
15 Inv. No. GG 124
16 Criste (1905)
17 Wacha (1999)
18 MS 2016
19 Inv. No. 1967.6.1.a
20 Inv. No. GE 1496
21 Inv. No. GE 1873
22 G. Wilhelm (1998)
23 Inv. No. GE 36
24 Inv. No. GE 720.
25 Inv. No. GE 1334.
26 Inv. No. GE 465.
27 Inv. No. GE 1362
28 Inv. Nr. GE 540
29 Inv. No. GE 105
30 Kräftner (2008)
31 Inv. No. GE 1756, 1757, 1758, 1760
32 Inv. No. GE 80
33 Vienna, Wiener Stadt-und Landesarchiv, Grundbücher, Grundbuch Klosterneuburg 20 / 99 b: f. 421
34 www.oeaw.ac.at/vid/research/historslexikon.shtml (accessed 5 x 2014)
35 Pfundheller (1881)
36 Vienna, Wiener Stadt- und Landesarchiv, Franziszeischer Kataster 1817 – ca. 1831; B1 – Parzellenprotokoll, Hietzing; P1 – Pläne, Hietzing
37 Consisting of Bauparzelle 162 (45,08 QKl) and Bauparzelle 227 (150,92 QKl): pers. comm. C. Sonnlechner, 2014.
38 Vienna, Stadt- und Landesarchiv, Grundbücher, Grundbuch Klosterneuburg 21, Art. 1392, 1393
39 Weissenbacher (1998)
40 Berger (2004)
41 Tietze (1919)
42 Riedl-Dorn (1998)
43 Krahnke (2001)
44 Mabberley & Moore (1999)
45 Madrid, Real Jardín Botánico, Archivo, Unpublished letter by J. F. von Jacquin to La Gasca, 7 iii 1816
46 Gr. Fol. 1144.
47 Lack (2008a), Kräftner (2008)
48 Tchernaja, pers. comm. (2014)
49 Spix & Martius (1823)
50 Koreny (1985)
51 Lack (1998b)
52 Lack (2000a)
53 Lindley (1821)
54 Sitwell (1956)
55 Lhotsky (1843)
56 Leins (1994)
57 Lack (1999b)
58 J. Compton (pers. com., 7 x 2014)
59 Sabine (1822)
60 Anon. (1840b)
61 Cuveland (2006)
62 Frodl & Frodl – Schneemann (2010)
63 Inv. No. 3631
64 Lack (2006b)
65 Birke (2005)
66 Lack (2003)
67 Willander (1980)
68 Anon. (1826)
69 Anon. (1831)
70 Vienna, Wiener Stadt- und Landesarchiv, Mag. Zivilgericht, Abh., Fasz. 2, 7425 / 31, Verlassenschaftsakt Joseph Bauer.
71 Brno, Moravský zemský archiv, Matriky, Valtice, Sign. 65 / 3533: f. 133.

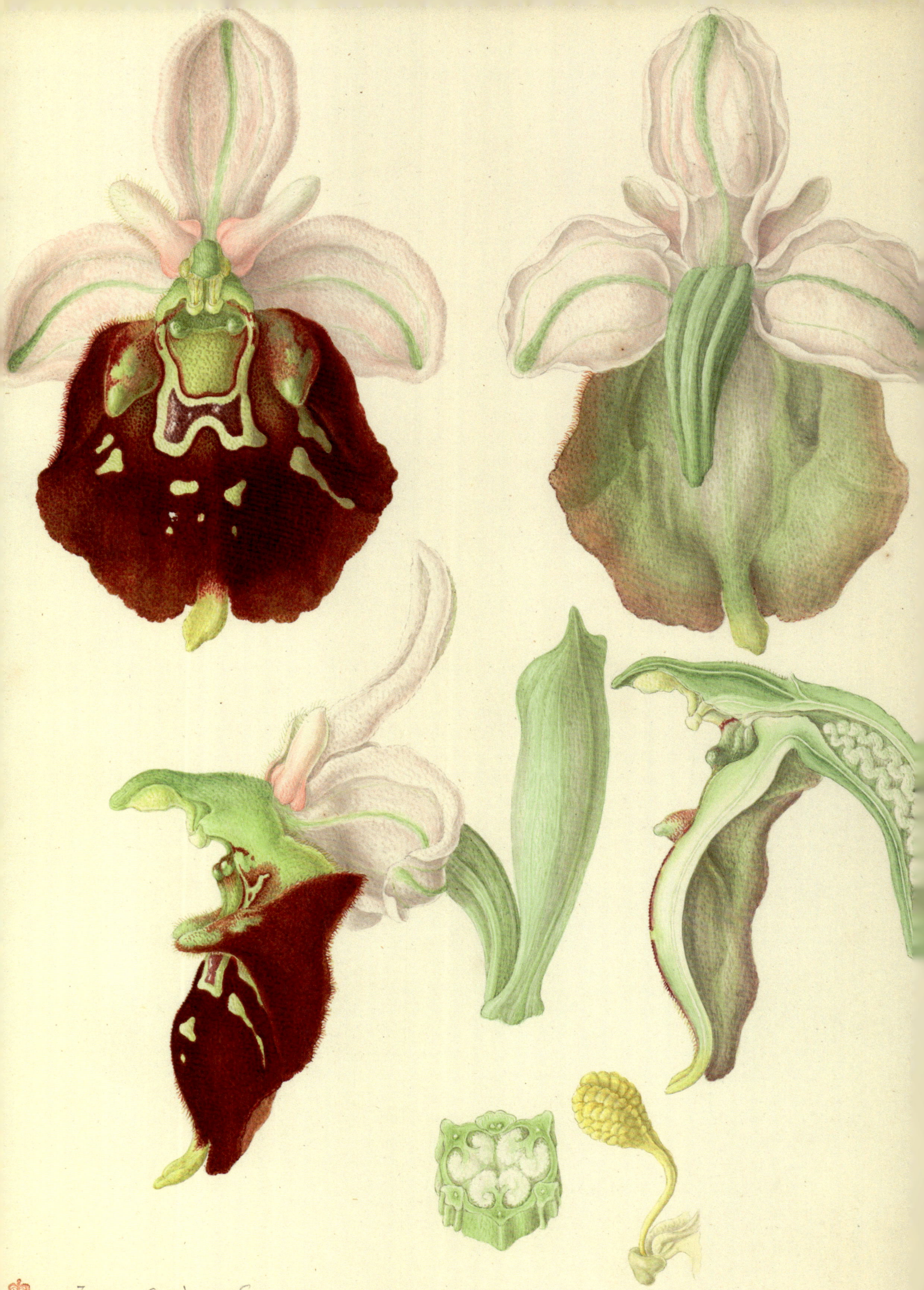

From — Swainson Esq
1807

Ophrys arachnites

CHAPTER 13

Franz Bauer examines red snow and orchids

THE ADMIRALTY, THE ROYAL SOCIETY AND THE ARCTIC

With the combined French and Spanish navies annihilated in the Battle of Trafalgar in 1805 and the status quo on a global scale confirmed by the Congress of Vienna in 1815, Britain remained the only colonial power of any importance and Russia the principal land-bound power in Eurasia. With British maritime supremacy established and the Royal Navy enjoying unchallenged sea power, a period of relative peace in Europe and the world at large began. In short, *Pax Britannica* reigned with Britain as the global hegemon. The key figure for naval affairs in London was the First Lord of the Admiralty, from 1804 to 1805, 1812 to 1827 and 1828 to 1830 this was Robert Dundas, 2nd Viscount Melville (see Chapter 11). However, with the peace agreements reached in Vienna he was faced with a major problem; drastic cuts in the navy budget forced him to a constant struggle against the parsimony imposed by Parliament and Treasury.

Elected Fellow of the Royal Society on 15 May 1817, Dundas received a letter a few months later from Banks, his President. It contained a proposal that efforts should be made 'to endeavour to correct and amend the very defective geography of the Arctic Regions more especially on the side of America. To attempt the Circumnavigation of old Greenland, if an island, as there is reason to suppose. To prove the existence or non-existence of Baffin's Bay; and to endeavour to ascertain the practicability of a Passage from the Atlantic to the Pacific Ocean, along the Northern Coast of America. These are objects which may be considered as peculiarly interesting to Great Britain not only from their proximity and the national advantages which they involve but also for the marked attention they called forth and the Discoveries made in consequence thereof in the very earliest periods of our foreign navigation. . .'[1] This again was a Banksian scheme on a grandiose scale, with the focus on a persisting dream in the Royal Navy – to find the north-west passage, the alternative to the Straits of Magellan on the southernmost tip of America, which always caused an enormous detour when sailing from England to the Pacific. Wisely Banks had hinted at 'national advantages' and there was also a reference to precedence – the voyage of Constantine John Phipps to Sjuøyane (Seven islands) north of Svalbard (Spitsbergen) in 1773. It is difficult to imagine that such a formal proposal from the President of the Royal Society to the First Lord of the Admiralty had not been agreed beforehand. In any case the Admiralty was enthusiastic about the plan since it meant a new commission for the Royal Navy, and a Parliamentary Act was passed 'to reward with any ship reaching a longitude 100° West (about half way across . . . Canada)'.[2]

Very quickly a decision was reached and on 8 January 1818 Banks was able to inform the Council of the Royal Society that not one but two expeditions

FIG. 13.1. Late spider orchid, *Ophrys holosericea*. Flowers. Watercolour by Franz Bauer, 1807. Paper, watercolour. – London, Natural History Museum, Library and Archives, British Orchids 1798–1817, no. 8.

were to be sent out by the Admiralty to sail up each side of Greenland. Clearly the prime consideration was geographical exploration and the second magnetic observation, but some attention was paid to biology and the Inuits' languages and culture. In our context only the expedition along the western coast of Greenland is relevant, since it brought back specimens to be studied by Franz in Kew. The latter communicated the results of his researches in the form of two letters: the first was directed to William Thomas Brande, a professor of chemistry at the Apothecaries' Society in London and Fellow of the Royal Society, and appeared in print in July 1819 in the *Quarterly Journal of Literature, Science and the Arts;*[3] the second, directed to Banks, was read before the Royal Society on 11 May 1820, just five weeks before the death of Franz's great benefactor, and published in the *Philosophical Transactions*[4] for that year. At the same time these two letters were of personal importance to Franz: they constitute the first texts from his pen ever published. Their topic was novel and may have been a reason for Sir Everard Home, with whom Franz had already been collaborating for a few years (see Chapter 15), to nominate him for election to fellowship of the Royal Society on 29 June 1820, just a few days after the death of Banks. Among Franz's sponsors we find Brown and the Milan-born gynaecologist Augustus Bozzi Granville (see Chapter 15), with both of whom he had also previously collaborated.[5] Franz was balloted and elected on 8 February 1821, which brought him to the inner circle of natural scientists in England, in a sense the High Table of the scientific establishment. No doubt this was an outstanding achievement for the country lad from Feldsberg whose mother was almost certainly analphabetic, who had lost his father at the age of four, but who had steadily moved upwards in the system by dedicated, strictly focused work. At the moment of his election Franz had reached the age of sixty-two; twenty more years were to be given to him.

RED SNOW

In the spring of 1818 two specially strengthened whalers were commissioned to sail along the western coast of Greenland – the *Isabella* of three hundred and eighty-two tons and Captain John Ross as commander, and the *Alexander* of two hundred and fifty tons and Lieutenant William Edward Parry in charge.

For 17 August 1818 Ross's travel account *A voyage of discovery . . . for the purpose of exploring Baffin's Bay* published in London in 1819 records a puzzling observation: 'We have discovered that the snow, on the face of the cliffs, presented an appearance both novel and interesting, being apparently stained, or covered, by some substance, which gave it a deep crimson colour. Many conjectures were afloat concerning the cause of this appearance; it was at once determined it could not be the dung of birds, for thousands of these, of various descriptions, were seen repeatedly sitting on the ice, and on the snow, but without producing any such effects . . . I sent a boat . . . to bring off some of the snow . . . The boat returned . . . with a quantity of the snow, together with specimens of the vegetation, and of the rocks. . . the snow was immediately examined by a microscope, magnifying 110 times, and the substance appeared to consist of particles like a very minute round seed, which were exactly of the same size, and of a deep red colour: on some of the particles a small dark speck was also seen. It was the general opinion of the officers who examined it by the microscopes, that it must be vegetable, and this opinion seemed to gain strength, by the nature of the places where it was found; these were the sides of the hills, about six hundred feet high, on the tops of which was seen vegetation of yellowish green and reddish brown colours. The extent of these cliffs was about eight miles; behind them at a considerable distance high mountains were seen, but the snow which covered these was not coloured; during the calm I took the view of this remarkable land'. This spectacular view showing a chain of mountains covered with crimson snow was published as the frontispiece of Ross's report of this expedition (Fig. 13.2.). According to a note in Franz's first letter (see above) an officer on the *Alexander* had written in his diary, 'It is worthy to remark, that this colouring matter, be it what it may, *does not penetrate more than an inch or two* [Franz's italics] beneath the surface of the snow'.

For the day of this observation Ross gave his position as 'Lat. 75° 74' N, Long. 67° 15' W', which was quite wrong. He was not in the middle of Baffin Bay, but instead either off the western coast of Greenland or off the eastern coast of Ellesmere Island, in both cases on the northern end of Baffin's Bay. This can be proved by the published view, which shows a chain of mountains denominated 'Crimson Cliffs'. In the evening of 17 August 1818 Ross had some of the red snow dissolved and bottled 'when the water had the appearance of muddy port-wine; in a

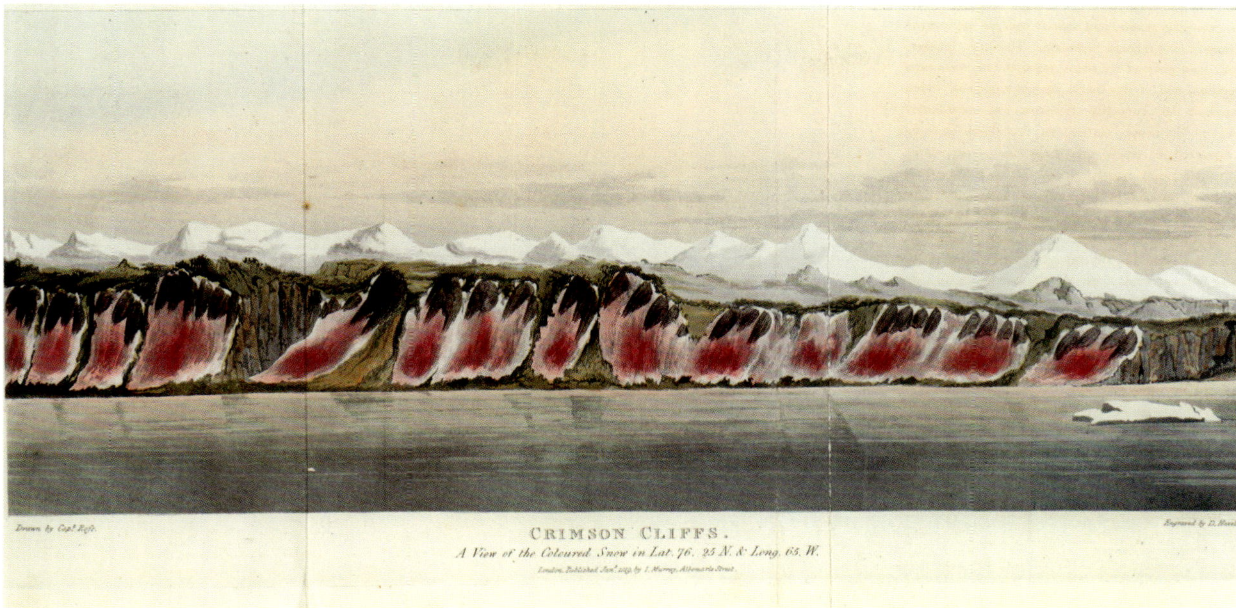

FIG. 13.2. Crimson Cliffs. Coloured lithograph. Frontispiece of J. Ross, A voyage of discovery, London, 1819. – London, Natural History Museum, Library and Archives.

few hours it deposited a sediment, which was examined by the microscope; some of it was bruised, and found to be composed wholly of red matter; when applied to paper it produced a colour nearest to Indian red. It was preserved in three states, viz., dissolved and bottled, the sediment bottled, and the sediment dried . . . It [the red snow] cannot be a maritime production, as in several parts we saw it at least six miles from the sea, but always on the face or near the foot of mountains'.

Public interest in the specimens of red snow brought back to London was considerable, with *The Times* writing on 4 December 1818: 'Captain Sir John Ross has brought from Baffin's Bay a quantity of red snow, or rather snow-water, which has been submitted to chymical analysis in this country, in order to the discovery of the nature of its colouring matter. Our credulity is put to an extreme test upon this occasion, but we cannot learn that there is any reason to doubt the facts as stated. Sir John Ross did not see any red snow fall; but he saw large tracts overspread with it. The colour of the fields of snow was not uniform; but, on the contrary, there were patches or streaks more or less red, and of various depths of tint. The liquor, or dissolved snow, is of so dark a red as to resemble red port wine . . . It is suggested that the colour is derived from the soil on which the snow falls: in this case, no red snow can have been seen on the ice'. Three days later, *The Times* published another explanation, reporting 'Some doubt has been expressed as to the red snow observed by Sir John Ross and his associates in the newly discovered arctic region; but when it is known that the iron which was also found there, lying on the surface, in heaps, and in considerable quantities, was all meteoric, the doubt will cease, and the fact will admit an easy solution. Sir John Ross brought home small specimens of this iron, which has been subjected by Mr. Professor Brande to the usual tests, and it is found to be precisely of the kind of the meteoric stones that occasionally fall in more southern latitudes. It is impregnated with nickel, which is never found in earth iron. That, therefore, which loads the atmosphere with the fluid which composes this meteoric iron, serves to colour the snow . . .'. Even a satirical print appeared in London entitled 'Landing the treasure or results of the Polar Expedition!!!' which is now kept in the British Museum[6] showing a frostbitten crew bearing specimens to the British Museum, among them a box labelled 'Red Snow for B. M.'

As a matter of fact the botanical materials were not transferred to the British Museum but to 32 Soho Square for further study and analysis by Brown, among them the specimens of red snow. The latter's note on this striking finding in his 'List of plants collected by the officers, & c., in Capt. Ross's voyage on the coasts of Baffin's Bay'[7] is laconic. Under the heading 'Cryptogamia' Brown simply wrote 'Algarum genus?? Confervis simplicissimis et Tremellae cruentae (Eng. Bot. 1800) quoammodo affine?? Minute globules, the colouring matter of the Red Snow, of which extensive patches were seen in lat. 76° 25' N, and long. 65° W'. In our context only the four question marks expressing considerable doubt are relevant: Brown was very uncertain about his interpretation of the minute globules as algae, which reminded him of *Pamella cruenta* and *Conferva* spec., both unicellular green algae. As we now know, Brown was right, but he refrained from giving a name to the strange red organism and from publishing a proper description.[8]

In his first letter, dated 24 April 1819 (see above), Franz reports having received on 28 February 1819, before the publication of Brown's List, 'a quarter bottle full of the melted snow' from the Crimson Cliffs to 'decide whether the colouring matter was an animal or vegetable substance'. In Dolland's compound microscope (see Chapter 17) Franz was able to see 'extremely minute organized spherical corpuscules, or globules, of different sizes, perfectly colourless and transparent' in the clear water. When shaking the bottle the sediment was disturbed, the whole content changing to a light crimson colour, and Franz observed 'some hundreds of similar globules, of various sizes, most of them opaque, and of a fine dark red colour, all of which soon sank to the bottom, but the transparent and colourless globules remained floating on the surface of the water'. This observation is difficult to interpret: nobody in the early nineteenth century was able to work under strictly sterile conditions, a notion not yet known of, and the sample studied by Franz may therefore have contained (1) more than one species and/or (2) organisms of a single species in different physiological or developmental stages, for example aplanospores and colourless forms. Franz compared what he saw with 'either the pollen of some plants, or the minute fungi of *uredo*'. Seen against the background of his biography this is understandable: he had studied pollen before (see Chapter 14) and was familiar with fungal spores, in particular of cereal pests (see Chapter 8), but he had never seen unicellular algae. Unsurprisingly, Franz came to the conclusion that he was observing fungal spores, writing 'the substance which gives the red appearance to the snow is a new species of *uredo*, and for which I think the most proper scientific name is *nivalis*'. Although this is undoubtedly a borderline case for the *International Code of Nomenclature of Plants, Fungi and Algae*, there is general agreement that

FIG. 13.3. Red snow alga, *Chlamydomonas nivalis*, and fungal spores. Colour print based on watercolour by Franz Bauer, 1819. Quarterly Journal of Science, Literature and Arts 7: t. 6. – London, Natural History Museum, Library and Archives.

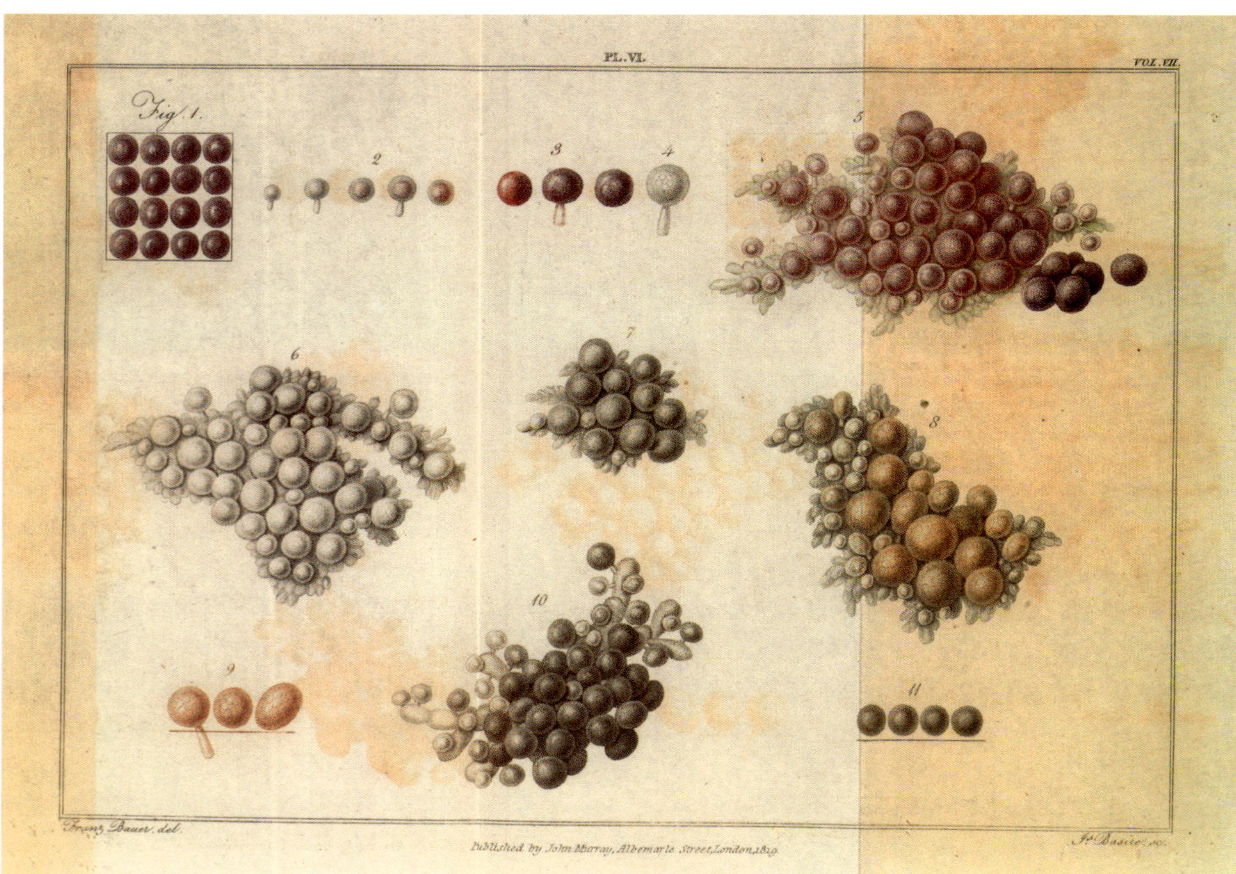

Franz has published the name *Uredo nivalis* according to the rules and attached it to the red globules he had seen in his microscope. The pertinent type specimen is kept today in the Natural History Museum. In order to ascertain the nature of his globules, Franz dried a quantity of them, put them 'upon some hot iron', sniffed the fumes and from experience came to the conclusion that the globules were 'vegetable matter'. No doubt this was a very crude test, but Franz saw it as a confirmation for his earlier interpretation.

As always Franz produced a painted record of the new organism together with spores of two fungal species he knew already – probably belonging to the stinking smut, *Tilletia tritici*, and to the stem rust, *Puccinia graminis*. This watercolour, now in the Niedersächsische Staats- und Universitätsbibliothek in Göttingen,[9] shows (1) the red cells of what Franz called *Uredo nivalis*, (2) black spores, probably teliospores of stinking smut, and (3) brownish spores, probably uredospores of stem rust. No doubt Franz kept specimens of the two parasitic fungi in his personal collection. This admittedly somewhat doubtful attribution is based on the legend referring to the colour print appended to Franz's first letter published in the *Quarterly Journal*. There smut and red rust are explicitly mentioned as subjects for comparison. Otherwise this print seems to be an example for the metamorphosis of a watercolour into a slightly different plate (for another example see Chapter 18) – the colours are rendered incorrectly in the print (Fig. 13.3.), details are added seeming to represent what Franz calls 'jelly-like white substances, with many of the full-grown red fungi adhering to it' and 'small flakes of the jelly-like spam'. In addition, Franz measured the diameter of the red cells as well as the fungal spores and had them correctly represented in the colour print.

The sequence of Franz's further experiments and manipulations is not entirely clear from his two publications (see above), but he observed 'newly-produced colourless fungi' which, when left to dry, exhibited – just like the red fungi - a 'brownish-grey colour', which is plausible. The 'red fungi, whilst fresh' turned out to 'form a pigment of the most brilliant vermillion, or red lead colour' which when rubbed on the skin of the head or face remained unchanged 'until washed off with soap and water', which is another correct observation. It would be no surprise if Franz, then a man of sixty-two years of age, had made the test on his own body. By April 1819 he did not seem to have observed a colour change of the new organism from red to green, which would have made him reconsider his interpretation. As we now know the presence of the red pigment astaxanthin protects the aplanospores, the immobile stages of the organism, against intensive radiation. Under less adverse ecological conditions astaxanthin is metabolized and the unicellular organism turns greenish and begins to reproduce. As a matter of fact Franz's *Uredo nivalis* is the green algae now called *Chlamydomonas nivalis*. To put it succinctly, Brown had been right and Franz, explicitly arguing against Brown's interpretation, had been wrong. Nevertheless the snow alga (*C. nivalis*) causing red snow – because of its crimson colour more appropriately called watermelon snow – carries the specific epithet coined by Franz in a combination validated decades later.

This cold-loving organism, which thrives in freezing water, became Franz's pet subject in the winter of 1819–1820. The experiments described by him in his second letter, dated 7 May 1820 (see above), are best understood against the background of his time. Firstly Franz had no apparatus available to produce and permanently maintain a given low temperature; instead he exposed his algal cultures to the oscillating temperatures at Kew from November 1819 to May 1820. As evidence he appended meteorological tables based on Aiton the son's measurements for December and January. Secondly Franz's algal cultures certainly contained the snow alga, but possibly also other algae, fungi and bacteria for the reasons given above. This weakness in experimental approach becomes evident in his note that his algal cultures had 'formed a kind of whitish crust at the top of the red sediment', which indicates an infection. Thirdly Franz did not control the purity of the water used in his experiments in the form of snow, which he repeatedly added to his algal cultures.

A single observation stands out: towards the end of January 1820 Franz observed 'a kind of crust, consisting of a light yellowish green substance, which on examination in the field of the microscope, proved to consist entirely of the same sized and shaped globules as the original red fungi. I immediately decanted the water and red sediment into another glass; when, on examining the empty glass, I found the appearance of the green substance was not only confined to the top of the glass, above the water mark, but I found many very minute clusters scattered, not only over the sides, but also on the bottom of the glass . . . I examined the glass

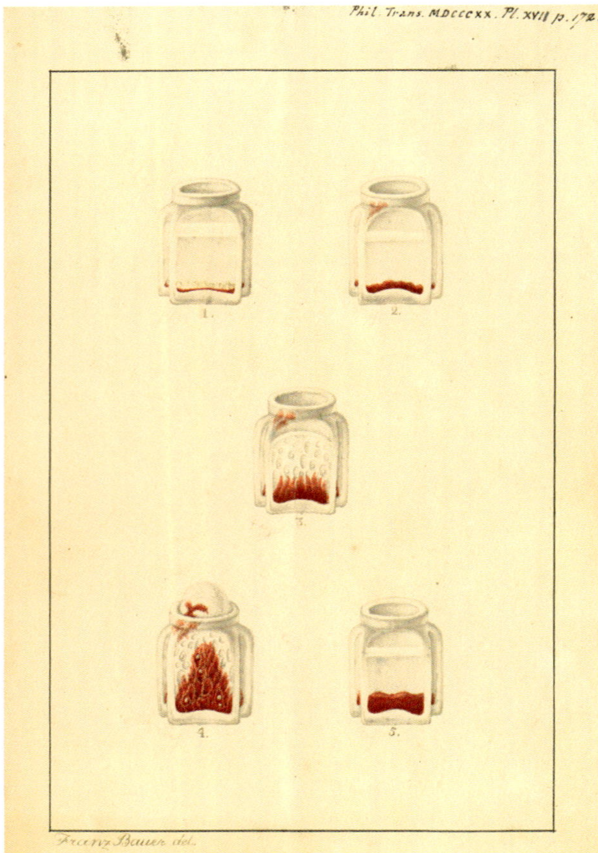

FIG. 13.4. Red snow alga, *Chlamydomonas nivalis*, cultivated in bottle. Watercolour by Franz Bauer, 1820. Paper, watercolour.– London, Royal Society, Archives, Image No. RS 8697.

this morning [7 May 1820], and find that a considerable increase of the green fungi had taken place since the 31st of January . . . and the whole glass is now of a greenish tinge'. Unsurprisingly Franz was unable to understand the conspicuous colour change, which either documents the transition of the red aplanospores to the green morph of the snow alga or the consequence of an infection of the original algal culture with a different green alga, possibly air-born. Again, Franz had made an interesting observation, but the time was not yet ripe to understand the switch in the physiological state of a unicellular alga and/or the infection.

Furthermore, Franz had observed an increase of the red sediment in his algal cultures over the winter months (Fig. 13.4.). This made him measure at intervals the diameter of the individual cells in his microscope, making use of a micrometer. The results were astonishing: the diameter had increased significantly by the middle of January 1820 compared with the start of the experiments, and so had the percentage of larger cells in his algal culture. This observation may be explained as a result of photosynthesis in the snow alga over the winter months. Of course Franz could not know that the intensively red cells contained chlorophyll hidden by astaxanthin and the solid cell wall. With the arrival of the warm season in 1820 Franz's excursion into cryobiology came to an end. However, for specialists of snow algae – organisms occurring globally, being particularly striking in the Arctic, Antarctic and all regions at least temporarily covered with snow – Franz is known as a pioneer in the study of these tiny organisms with exotic ecological requirements.

MELVILLE ISLAND PLANTS

The voyage of the *Isabella* and the *Alexander* cried out for further Arctic expeditions. One of the reasons was the suspicion that what Ross had regarded as the Crocker Mountains, blocking his route in a westerly direction, had only been the effects of superior mirage, a complex type of Fata Morgana caused by temperature inversion occurring in Arctic and Antarctic regions. As a matter of fact it would be shown in the following year that Ross had indeed been deceived and that it was actually possible to sail through what he had called the Crocker Mountains. The dedicatee John Wilson Croker, First Secretary of the Admiralty, was not amused by the fact and made sure that Ross's further naval career progression was delayed.

The first of the follow-up projects to pursue exploration of the Arctic regions west of Greenland was lead in 1819 by Parry as commander of the expedition with Edward Sabine as captain of the *Hecla*, a bomb vessel of three hundred and seventy-five tons. She was accompanied by the *Griper*, a slow gun-brig of one hundred and eighty-one tons and rather unfit for the voyage, under the command of Lieutenant Matthew Liddon. The two vessels headed straight to Lancaster Sound, passing Ross's farthest west point and reached 110° West, which entitled the team to the £5000 award offered by Parliament.

With the Arctic winter approaching, the anchors were dropped early. Parry's *Journal of a Voyage for the discovery of the north-west passage*, published in London in 1821, the main source for this subchapter, reports for 1 September 1819: 'I had great reason to be satisfied with our having anchored the ships, as the wind shortly after blew a hard gale from the northward. In the evening I sent Captain [Edward] Sabine . . . on shore to examine the country, and to collect specimens of its natural productions; . . . several flocks of ducks were seen, and some glaucous gulls and tern; the dung and foottracks of the deer and musk-ox were also observed in many places . . . This island . . . I honoured with the name of Melville Island, after Viscount Melville, the First Lord of the Admiralty'. No one could envisage at that moment that they had stepped on a vast, totally unknown island with an area of 42,129 square kilometres, the thirty-third largest island in the world. Then, as now, it is completely uninhabited, the region dominated by a harsh, arctic climate. Captain and crew had to prepare themselves for spending the winter in what became known as the Winter Harbour on the southern coast of Melville Island. Soon they were frozen in for a period that eventually lasted for ten months, a long period of total darkness followed, but with the late arrival of the Arctic spring captain and crew were able to observe the reappearance of plant life. For 3 June 1820 Parry's *Journal* reported, 'we have passed, during our last march, a good deal of rich soil, consisting principally of decaying moss and other vegetable substances mixed with sand; and the sourel and saxifrage, (*Saxifraga oppositifolia*), were more abundant than before'. Summer was approaching on 20 June 1820, with 'the poppy (*Papaver Nudicaule* [sic]) and some other plants . . . afforded us with a degree of enjoyment that made us for a while forget the rigor of this severe climate'. With good luck, the two vessels were able to float out of Winter Harbour on 1 August 1820, just a few weeks before the harbour was starting to freeze up again.

Before departure, the ephemeral flora of Melville Island was studied; in addition Parry and four members of the crew collected herbarium specimens, which survived the crossing of the Atlantic and ended up on the desk of Brown. Today they are mostly kept in the Natural History Museum, with a single specimen of *Parrya arctica* known to be conserved in the Gray Herbarium at Harvard University.[10] As for the Ross expedition, Brown must have been asked to write another botanical appendix for the collections made by Parry and his crew which, however, were much more substantial.

In February 1821, Parry, who was elected a Fellow of the Royal Society in that year, sent a reminder to Brown telling him that the Natural History Appendix was ready to be sent to the printers and added, 'If you think I can expedite the engraver by any means in my power, pray, let me know, as I begin to fear they will detain us'.[11] As was so often the case in such matters there were further delays for a variety of reasons, among which was Brown's health, which resulted in the belated publication of the Natural History Appendix independently from Parry's official account *Journal*, which came out in London in 1821. Two years later the botanical part entitled *Chloris Melvilliana. A list of plants collected in Melville Island . . .in the year 1820* appeared in London as a preprint, with the text by Brown and the copper engravings based on watercolours by Franz. The latter are now conserved in the Natural History Museum. Finally in 1824 *Chloris Melvilliana* reappeared in London with minor corrections as 'Appendix IX. Botany' to Parry's *Supplement to the Appendix of Captain Parry's Voyage for the discovery of the north-west passage.*

Four species had been selected by Brown to be illustrated by Franz, among them two crucifers – *Parrya arctica* and *Eutrema edwardsii*. The generic name of the first referred to the commander of the expedition, the epithet of the second to John Edwards, the surgeon on

FIG. 13.5. ***Parrya arctica*. Canada, Melville Island. Six herbarium specimens collected by W. E. Parry, J. C. Ross, A. Fisher, J. Edwards and E. Sabine respectively. – London, Natural History Museum, Life Sciences, Herbarium.**

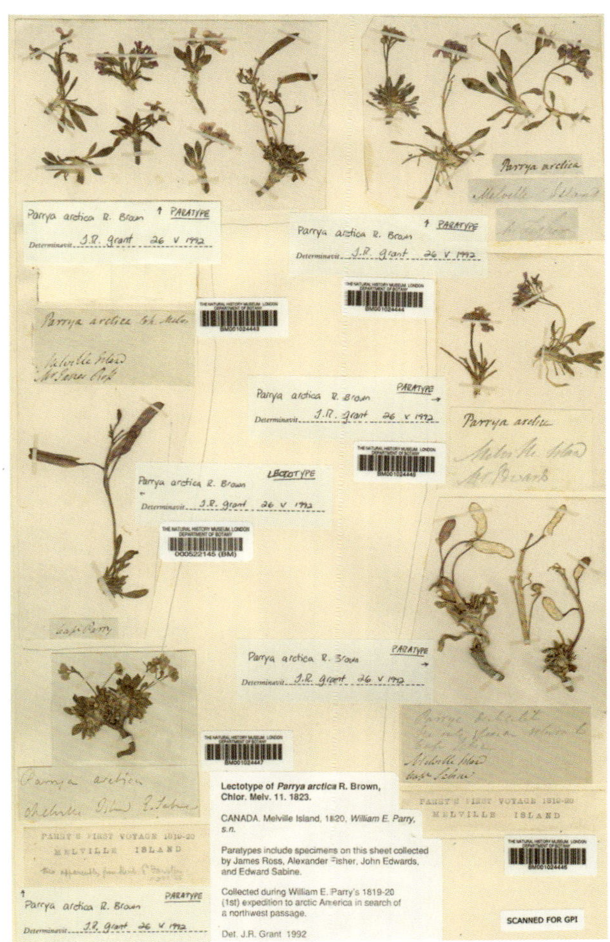

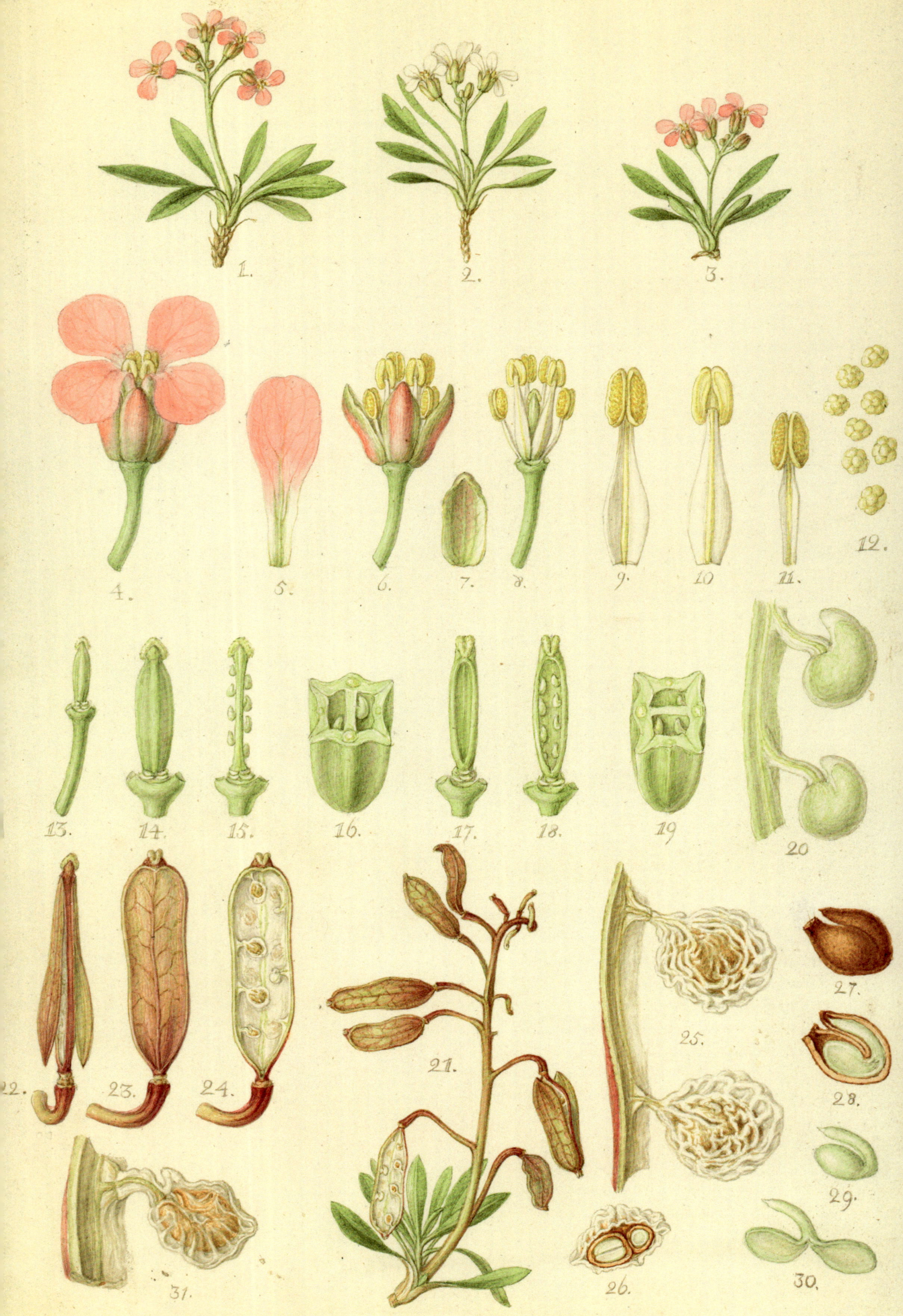
1.
2.
3.
4.
5.
6.
7.
8.
9.
10
11.
12.
13.
14.
15.
16.
17.
18.
19
20
22.
23.
24.
21.
25.
27.
28.
29.
30.
31.
26.

the *Hecla*. A total of six specimens of *Parrya arctica* were collected on Melville Island by Parry and four members of his team (Edwards, Sabine, Alexander Fisher, the assistant surgeon, and John Clarke Ross, one of the midshipmen); four specimens of *Eutrema edwardsii* were gathered on the same island by the Captain and Edwards. Brown had good reasons to select the two species to be depicted – they were new to science and also belonged to genera new to science. Both taxa have stood the test of time, and Brown's circumscription continues to be accepted by the scientific community. As we currently know, both species are widespread – *Parrya arctica* occurs in what is now Arctic Canada, *Eutrema edwardsii* from Alaska to Greenland. Franz and Brown worked with loose specimens, those belonging to *Parrya arctica* were later mostly mounted on a single herbarium sheet now kept in the Natural History Museum (Fig. 13.5.). Franz produced a watercolour of *Parrya arctica* (Fig. 13.6.), a masterpiece of accurate observation showing precisely all the characters relevant for the botanist, together with a draft text for the legend (Fig. 13.7.), and he did the same for three more species. By contrast Brown wrote the complete text of the *Chloris Melvilliana*.

For both Franz and Brown this was strenuous work with rather small specimens in all states of maturity, which made Franz write to Brown on 4 March 1821 in a letter kept among the *Chloris Melvilliana* papers in the Library and Archives of the Natural History Museum: 'I avail myself of Sir Everard Home's [see Chapter 15] kind offer, to call tonight at twelve o' clock, to convey to You another drawing, for the occupation of the Engraver. I regret very much I cannot send You more to night, but by Tuesday next will send You another, and as soon as I have a drawing finished will send it to You immediately. I find the investigation of those little beings infinitley [sic] more tedious that [sic] at first sight I thought them to be'. Meeting Brown at midnight in London to deliver a watercolour is a sign of outstanding dedication to this project, the more so as Franz was no longer a young man. To prepare the four watercolours was also a challenge

FIG. 13.6. ***(opposite) Parrya arctica*****. Watercolour by Franz Bauer, signed, 1821. Paper, watercolour. – London, Natural History Museum, Library and Archives, Chloris Melvilliana (loose).**

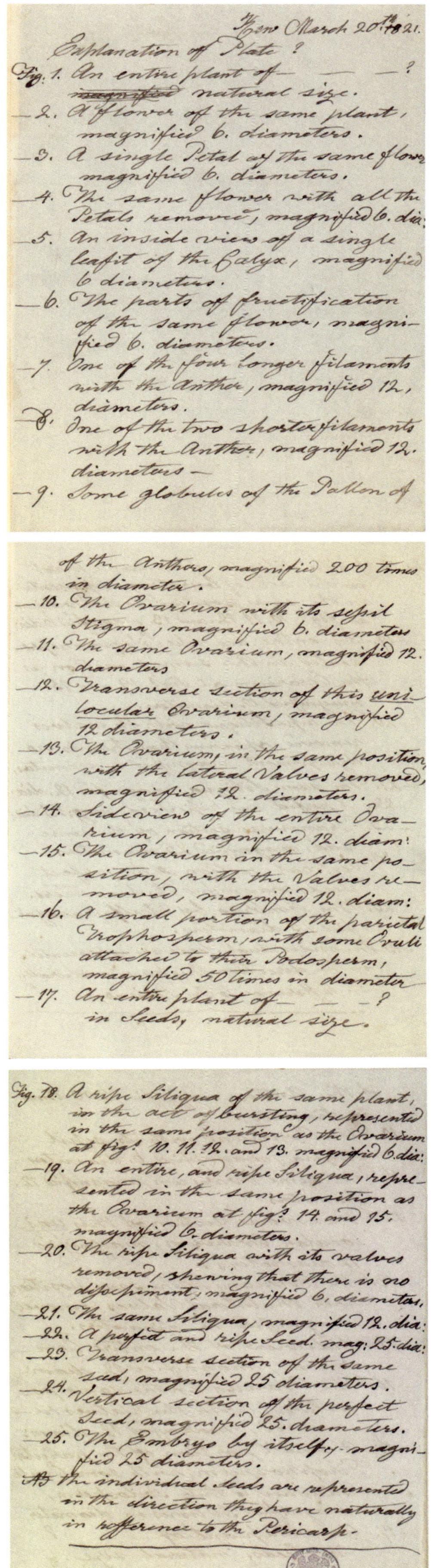
Kew March 20th 1821.
Explanation of Plate ?
Fig. 1. An entire plant of — — ?
~~magnified~~ natural size.
— 2. A flower of the same plant, magnified 6. diameters.
— 3. A single Petal of the same flower magnified 6. diameters.
— 4. The same flower with all the Petals removed, magnified 6. dia:
— 5. An inside view of a single leafit of the Calyx, magnified 6 diameters.
— 6. The parts of fructification of the same flower, magnified 6. diameters.
— 7. One of the four longer filaments with the Anther, magnified 12. diameters.
— 8. One of the two shorter filaments with the Anther, magnified 12. diameters —
— 9. Some globules of the Pollen of the Anthers, magnified 200 times in diameter.
— 10. The Ovarium with its sessil Stigma, magnified 6. diameters
— 11. The same Ovarium, magnified 12. diameters
— 12. Transverse section of this unilocular Ovarium, magnified 12 diameters.
— 13. The Ovarium, in the same position with the lateral Valves removed, magnified 12. diameters.
— 14. Sideview of the entire Ovarium, magnified 12. diam:
— 15. The Ovarium in the same position, with the Valves removed, magnified 12. diam:
— 16. A small portion of the parietal Trophosperm, with some Ovuli attached to their Podosperm, magnified 50 times in diameter
— 17. An entire plant of — — ? in Seeds, natural size.
Fig. 18. A ripe Siliqua of the same plant, in the act of bursting, represented in the same position as the Ovarium at fig.s 10. 11. 12. and 13. magnified 6. dia:
— 19. An entire, and ripe Siliqua, represented in the same position as the Ovarium at fig.s 14. and 15. magnified 6. diameters.
— 20. The ripe Siliqua with its valves removed, shewing that there is no dissepiment, magnified 6. diameters,
— 21. The same Siliqua, magnified 12. dia:
— 22. A perfect and ripe Seed. mag: 25. dia:
— 23. Transverse section of the same seed, magnified 25 diameters.
— 24. Vertical section of the perfect Seed, magnified 25. diameters.
— 25. The Embryo by itself, magnified 25 diameters.
NB The individual seeds are represented in the direction they have naturally in reference to the Pericarp.

FIG. 13.7. ***(right)*** **Legend to illustration of** ***Eutrema edwardsii*****. Manuscript by Franz Bauer, 1821. Paper, pen and ink. – London, Natural History Museum, Library and Archives, Chloris Melvilliana (loose).**

for Franz in another respect. He worked with small, miserable herbarium specimens collected by amateurs and may have been obliged to put them first into warm water to revive their three-dimensionality and then dry them again, as is the standard practice among botanists. However, he managed to record in a perfect manner many details, like the replum in the ripe fruits and the insertion of the seeds on it. Furthermore, there was the challenge of petal colour; for this Franz must have possessed field notes from the collectors, otherwise he would have been unable to devise the colours, which at least in the pink variant of *Parrya arctica* is on the dark end of the variation found in nature, but the tint is very correct.

Brown had made use of the occasion and included in the *Chloris Melvilliana* the description of a few other species from the Arctic. Among these was a specimen of *Parrya nudicaulis* collected in 1778 by David Nelson, the gardener from Kew who had accompanied Captain Cook's third expedition, which had been unsuccessfully searching for the north-west passage from the Pacific Ocean. The tiny plant was found on a small island off the coast of Alaska – Sledge Island, thousands of kilometres to the west of Melville Island. Needless to say, all these additions were a drain on the time allotted to the project and reasons for further delays, but the end result is a classic, though rare, publication on the flora of Arctic North America. A few of the species treated, among them *Geum rossii,* likewise illustrated by Franz, also occur in the Kamchatka Peninsula on the other side of the Bering Strait.

THE STUPENDOUS FLOWER OF SUMATRA

When Sir Thomas Stamford Raffles, just knighted and elected Fellow of the Royal Society, was appointed Governor-General of Bencooden (now Kota Benkulu in Indonesia) and set sail to take the new post, he was accompanied by his wife Sophia, née Hull, and his personal physician Joseph Arnold.[12] The latter was to assist Lady Sophia Raffles when she delivered her first child at high sea. On 19 March 1818 the party arrived in the small settlement on the west coast of Sumatra, a colonial backwater whose main export was pepper, with Fort Marlborough, erected by the East India Company, its only fortification. On a tour beyond the Manna River south-east of Bencooden, Raffles, his wife and Arnold encountered a most extraordinary plant at Pulau Lebar, with Arnold recording the sequence of events on 9 July 1818 on board the *Lady Raffles* in his penultimate letter: 'But here I rejoice to tell you I happened with what I consider as the greatest prodigy of the vegetable world; I had ventured some way from the party, picking specimens of plants, when one of the Malay servants came running to me with wonderment in his eyes, and said, Mari tuan ada bunga tĕrlalu bĕsar, bagus, indah, mari [Come with me, Sir, come, a flower very large, beautiful, wonderful]. I immediately went with the man about an hundred yards in the Jungle, and he pointed to a flower growing close to the ground under the bushes, which was truly astonishing. My first impulse was to cut it up and carry it to our hut, I therefore seized the Malay's prang [parang] (a sort of instrument like a woodman's chopping hook) & began to chop away. I then found that it sprang from a small root (perhaps as large as two fingers or a little more) which ran horizontally I therefore soon detached it & removed it to our hut. – To tell you the truth, had I been alone, and had there been no witnesses, I should I think have been fearful of mentioning the dimensions of the flower, so much does it exceed every flower I have ever seen or heard of, but I had Sir Stamford & Lady Raffles with me, and a Mr. [Edward] Presgrave a respectable man resident at Manna, who tho' equally astonished with myself, yet are able to testify as to the truth. – The whole flower was of a very thick substance, the petals & nectary being in but few places less than a quarter of an inch [0.62 centimetres] thick, and in some places three quarters of an inch [1.86 centimetres], the substance of it was very succulent; when I first saw it a swarm of flies were hovering over the mouth of the nectary, and apparently laying their eggs in the substance of it, it had precisely the smell of tainted beef. The calyx consisted of several roundish dark brown concave leaves, which seemed to be indefinite in number, & were unequal in size. There were five petals attached to the nectary, which were thick and covered with protuberances of a yellowish white, varying in size, the interstices being of a brick-red colour; the nectarium was cyathiform, becoming narrower towards the top the centre of the nectarium gave rise to a large pistil which I can hardly describe, but is was something of this shape [two figures in the text]; at ye top of which were about twenty processes somewhat curved and sharp at the end, resembling cows horns; there were as many smaller, very short processes; a little more than half way down a brown cord, about the size of common whip cord, but quite smooth, surrounded what perhaps is the germen, & a little below it was another cord somewhat moniliform. – Now for the dimensions, which are the most astonishing part of the flower. It measured a full yard [91 centimetres] across; the petals (which were sub-rotund being twelve inches [30.5 centimetres] from the base to the apex, and it being about a foot [30.5 centimetres] from the insertion of the one petal to the opposite one. Sir Stamford, Lady Raffles & myself taking immediate measures to be accurate in this respect, by pinning four large sheets of paper together, and cutting them the precise size of the flower. The Nectarium in the opinion of all of us would hold twelve pints [11.4 litres], and weight of this prodigy we calculated to be fifteen pounds [6.75 kilograms]. – I have said nothing about the stamina, in fact I am not certain of the part I ought to call stamina; if the moniliform cord surrounding the base of the pistil were sessile anthers, it must be a polyandrous plant, but I am uncertain what the large germen contained, perhaps there might be concealed anthers within it as is observd in some Asclepiadeae. – But you will say, why did not you ascertain this on the spot? – In truth, my dear Sir, now I wish I had done so, but Sir Stamford suggested the Idea of preserving the flower in spirits, so that I hoped to be able to examine it more at leisure on my return to Manna, we therefore packed it up in the moist grass, and sent off two men carrying the box containing it, hung on a bambo[o], and we hoped from the directions given with it to find it in good state when we returned from Pasummah.– Alas! when I returned to Manna I found the flower still in the box, the petals dark brown, wasted, full of Maggots like a rotten mushroom, the only part that seemed to retain its form was the pistil. I therefore put this in spirits and sent it to Fort Marlborough , together with two large buds of the same flower, which I found attached to the same root, each of which is about as large as two fists. The buds are now in spirits, and I propose to send them to Sir Jos Banks, to whom I promised every thing remarkable. There were no leaves or branches to this plant, so that it is probable that the stems bearing leaves issue forth at a different period of the year . . . I think the drawings & description with every thing attached to this flower ought not be shoved aside among Sir Joseph's collection. I should be glad to see it properly described, named, delineated, and published in the Linnean transactions . . . You may of course suppose I was not Idle in making delineations and memorandums on this plant, indeed Sir Stamford Lady R. and myself each employed ourselves in taking different views of it; I have now nearly finished a coloured drawing of it, as large as I could procure drawing paper to admit of, but it is still considerably under the natural size. I propose also to make another drawing of the Pistil removed from the Nectarium, both of which I shall send to Sir Jos. Banks. – But, my dear Sir, surely Sir Jos. will not shove them aside on one of his shelves, and there let them forgotten. I intend to propose to him, or rather request of him, to beg one of his friends to dissect carefully the buds, when they may discover something which from haste or inattention, or fatigue escaped me . . . Most probably Sir Jos. would leave it to Mr. Browne [sic] (than whom certainly he could not leave it to a more proper man) . . .' Arnold had indeed made very careful arrangements for his outstanding trophy, collected on 19 or 20 May 1818, and he had done so just in time. On 26 July 1818, only seven days after having written this letter, he succumbed to a tropical fever in Padang, a harbour north-west of Bencooden.

Whereas Arnold's letter, from which the long quotation was taken, reached its addressee Dawson Turner, another Fellow of the Royal Society (see Chapter 16), some two years later, the poorly preserved specimens of the spectacular plant from Sumatra were soon transported by the *Lady Raffles* to Portsmouth, arriving on 12 July 1819 and then forwarded to 32 Soho Square. Banks passed them on to Brown, exactly as proposed by Arnold. In the same manner as the treatment of the fern genus *Woodsia* (see Chapter 8) and the Melville Island plants, Brown must have asked Franz to produce the graphical record of the material, while he would produce the textual record. Whereas the cabbage-like buds seem to have been in a tolerable state of preservation, this was not the case for the opened-up flower, which had started to decay even before Arnold saw it again at Manna. As before in the Melville Islands plants, Franz's experience in reconstructing badly preserved, incomplete specimens, in this case the largest flower in the plant kingdom, was requested. Since only pickled material was available in London, colour posed a problem. Only in the representation of the reconstructed flower at anthesis could Franz make use of Arnold's meagre notes on colour, but he left the colouration incomplete in order to indicate his fragmentary knowledge of the living plant

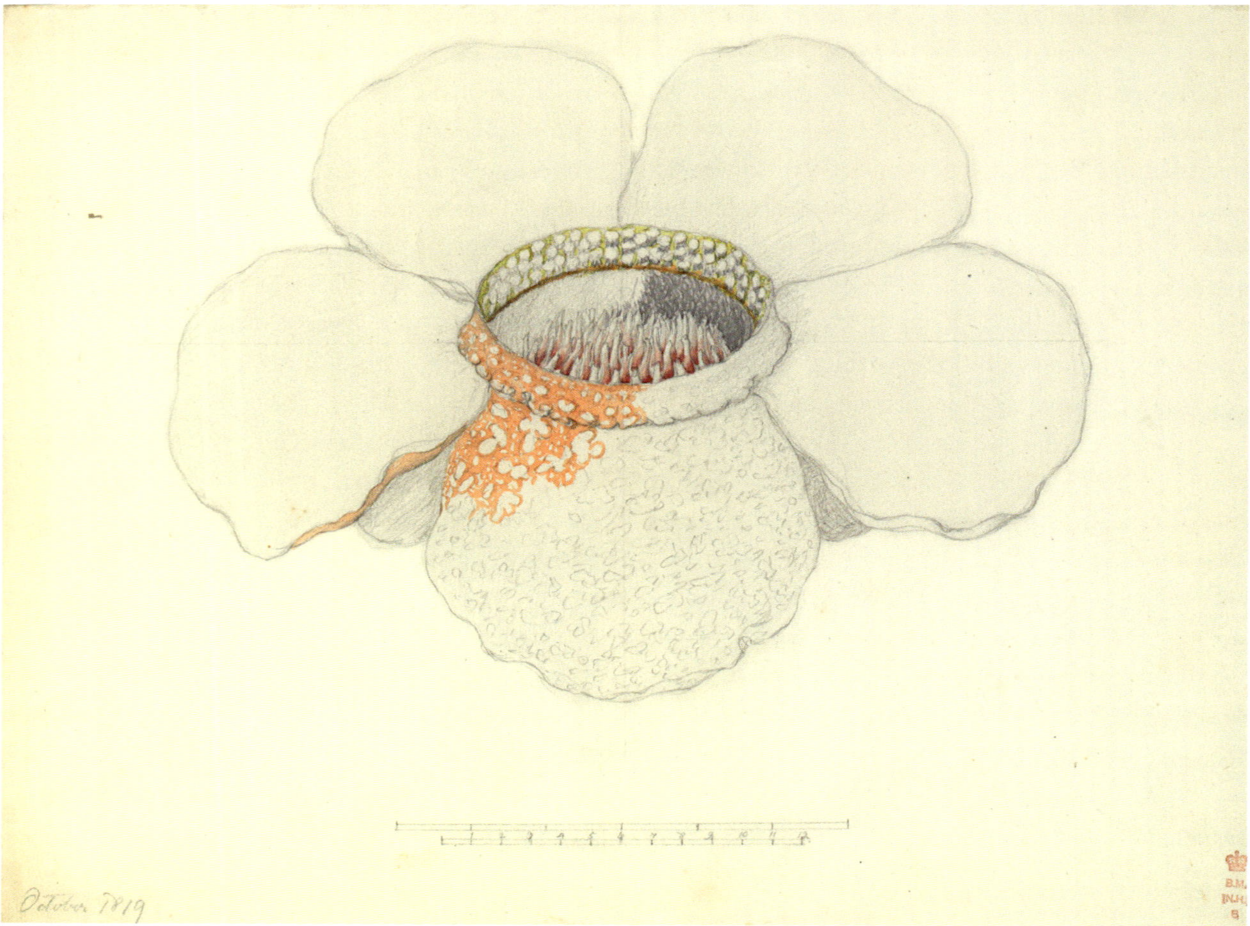

FIG. 13.8. *(above) Rafflesia arnoldii.* Watercolour by Franz Bauer, 1821. Paper, pencil, watercolour. – London, Natural History Museum, Library and Archives, Unfinished Drawings of Plants, vol. 2, no. 68.

FIG. 13.9. *(right) Rafflesia arnoldii.* Watercolour by Franz Bauer, 1821. Paper, watercolour. – London, Natural History Museum, Library and Archives, Unfinished Drawings of Plants , vol. 2, no. 67.

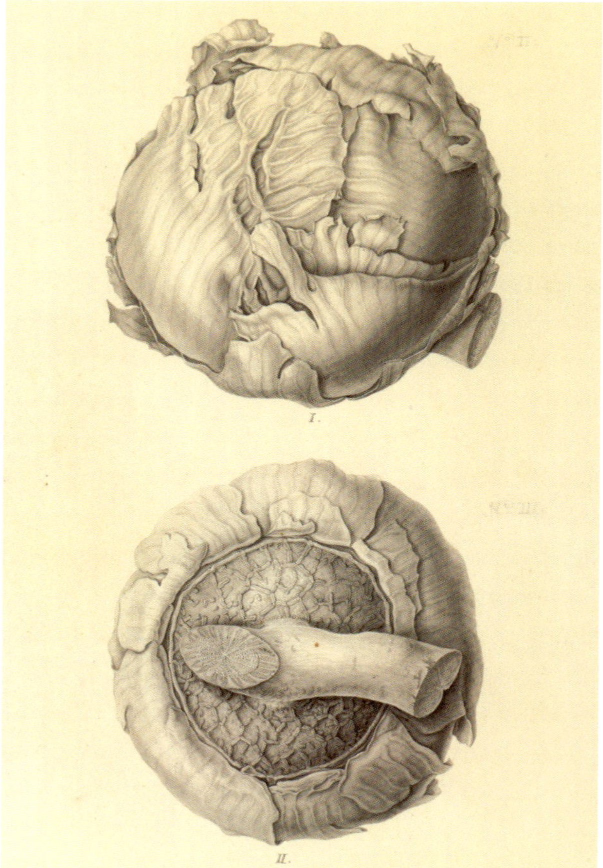

(Fig. 13.8.); by contrast Franz recorded the huge buds (Fig. 13.9.) in monochrome. On the basis of Franz's drawings seven copper engravings were produced, six documenting the two buds, one the flower at anthesis. These were the first-ever-produced prints of *Rafflesia arnoldii*, the largest and arguably the most famous *Rafflesia* species, which is now known to occur not only in Sumatra but also in Borneo.[13]

Arnold's plant became a sensation despite there being several unknown facts about it, with Banks writing in 1819, 'by far the most extrordinary [sic] vegetable production I have seen . . . Brown is now busily engaged in Examining ever Part, Neither he nor I can refer it to any Known Family of Plants or ascertain to our Satisfaction whether or not it is a Parasite as Brown suspects it to be'. However, the latter presented the results of his studies at a meeting of the Linnean Society in London on 30 June 1820, just eleven days after Banks's death, with 'Additional observations' at another, on 21 November 1820. Arnold's specimens and Franz's drawings were reported to have been deposited in 32 Soho Square in August 1820, though Franz is said to have been still piecing together fragments of the specimen later. Brown's paper entitled 'An account of a new genus of plants, named *Rafflesia*' appeared in April 1821 in London as an independently paginated preprint[14] and at the earliest in late May 1821 also in the *Transactions of the Linnean Society.*[15] Both versions contained the seven copper engravings based on Franz's watercolours. Twenty-eight copies of the preprint had been produced and were issued personally by Brown. Franz, Ferdinand and several botanists on the continent received complimentary copies, and so did the widowed Maria Feodorovna (see Chapter 3), the mother of Alexander I, Emperor and Autocrat of all the Russias. Via the Imperial Russian Embassy to the Court of St James she had explicitly asked for a printed description of the spectacular plant from Sumatra. For his paper Brown later received a letter of thanks from her secretary in Pavlovsk together with a ring of gratitude. The contents of Brown's contribution was reprinted, summarized and translated, which resulted in Franz's illustrations becoming known to an unusually large number of readers.

Brown had recognized the odd plant as new to science, named it *Rafflesia arnoldii* and given a detailed description based on Arnold's specimens from Sumatra, which survive in the Natural History Museum. By contrast, the watercolours of the spectacular flower prepared by Arnold, Raffles and Lady Raffles in Sumatra mentioned in Arnold's letter (see above) remain untraceable. However, Brown's knowledge of the sensational plant lacking chlorophyll and rightly regarded by him to be a root parasite was still incomplete, since he had been able to describe and Franz able to depict only the male flower, the two buds also having turned out to be male. Among specimens of *Rafflesia arnoldii* sent by Raffles in August 1821 from Sumatra to Brown in London there were also female flowers, which became the subject of a further communication by the latter, read at the meeting of the Linnean Society on 17 June 1834. Again Franz had been asked to prepare the pictorial documentation, which resulted in a few more monochrome watercolours prepared between 1825 and 1831. Several letters seem to have been exchanged between Brown in London and Franz in Kew, of which two survive in the Library and Archives of the Natural History Museum. One of them, dated 11 July 1821, has 'No, 14' and is from Franz to Brown, asking the latter to deposit a drawing at the porter, adding 'I will send my Waterman for it; perhaps on Wednesday next'. As in Brown's first paper, the watercolours by Franz were also used as exemplars for a total of five copper engravings, prepared by James Basire the son, the engraver for the Society of Antiquaries, but these were not produced until 1834.

Brown was strongly interested in elucidating the position of the anomalous genus *Rafflesia* within the system of flowering plants and had devoted a substantial number of pages to this topic in his first paper. In the second contribution[16] he included a full treatment of the genus *Hydnora*, another root parasite lacking chlorophyll, represented as we now know by a small number of species in sub-Saharan Africa and the Arabian Peninsula. The only species known at the time, *H. africana*, had been documented by Ferdinand during his ultimate visit to England (see Chapter 12). Little is known about the provenance of the specimens studied, but Ferdinand's monochrome watercolours survive in the Natural History Museum. On the basis of these watercolours, four copper engravings were prepared by Basire, all with the note 'Ferdinand Bauer del. 1824'. They formed additional illustrations to Brown's second paper on *Rafflesia*.

It remains totally unknown why Brown waited for more than two decades to have his contribution on the female flower of *Rafflesia arnoldii* and on the flower of *Hydnora africana* published. In the meantime he had travelled extensively and had demonstrated the engravings illustrating the two unconventional flowers at several scientific congresses. One of them was the Versammlung Deutscher Naturforscher und Ärzte, which met in September 1835 in Bonn, then belonging to the Kingdom of Prussia; another was the Terza Riunione degli scienzati italiani, which came together in September 1841 in Florence, the capital of the Grand Duchy of Tuscany. Martius in Munich (see Chapter 18) was also shown the engravings, and when on 5 September 1843 Brown was received by Friedrich Wilhelm IV, King of Prussia, in Pillnitz on the River Elbe in the Kingdom of Saxony, the former also demonstrated the printed illustrations. In 1845 in Paris Alexander Baron Humboldt was presented with Brown's two publications, which had finally appeared in the *Transactions of the Linnean*

Society. Another recipient of the two publications was Archduke Leopold, as Grand Duke of Tuscany Leopoldo II, a grandson of the short-reigning Emperor Leopold II, when he received Brown in Florence in 1847. For Franz and Ferdinand all this brought only posthumous publicity for their work on the two root parasites.

Brown's complete text accompanied by the copper engravings appeared under the title 'On the female flower and fruit of *Rafflesia arnoldi* [sic] and on *Hydnora africana*' as an independently paginated preprint in London, of which Brown distributed copies in September 1844 at the latest. When the respective volume of the *Transactions of the Linnean Society* finally appeared in November 1844, Ferdinand had already been dead for 18 years and Franz for almost four years.

The flowers of *Rafflesia arnoldii* and *Hydnora africana* produce an intensive carrion-like stink in order to attract pollinators, and the same applies to the inflorescences of the elephant yam, *Amorphophallus paeoniifolius*, a widespread aroid occurring from Madagascar in the south-west to Oceania in the east and China in the north. Franz produced two undated watercolours[17] of this tuberous plant, the former previously owned by John Lindley (see below) and now forming part of the Cingalese Drawings kept in the Lindley Library in London. Another watercolour by Franz dated 1805 and belonging to the same collection[18] depicts a species of the aroid genus *Typhonium*, which also produces an inflorescence with a revolting smell. So far no specialist for this group has determined the identity of the plant represented on this image.

TROPICAL MILKWEEDS

The Memorandum of Agreement signed by Brown and Ferdinand on 29 April 1801 clarified their relationship during the voyage in the *Investigator* with Brown being Ferdinand's superior (see Chapter 10). The situation was totally different with Franz, who had been recruited by Banks on a permanent basis in early 1790. At that moment Brown was sixteen years old and still had to wait for almost eleven years before being invited by Banks on 12 December 1800 to become the naturalist on the *Investigator*. It was only upon the death of Dryander in 1810 that Brown succeeded him as Banks's librarian, with a salary of £200 per year. In short Brown had become Franz's colleague, though with a significantly smaller stipend that was subsequently not raised by Banks. Clearly Brown was the junior member of staff. On the other hand, as librarian at 32 Soho Square Brown was nearer to Banks, while Franz was based in Kew. In addition, Brown managed to have his written findings, for example his *Prodromus florae Novae Hollandiae*, published much quicker than Franz's painted findings on the plants cultivated in the Royal Garden in Kew. This is easy to explain – printing text is incomparably cheaper than printing images, irrespective of the technique used. With science so strongly focused on the printed word (see Chapter 14), it is no surprise that Brown was elected Fellow of the Royal Society on 12 December 1811 aged 38, while Franz was elected only nine years later, on 8 February 1821, aged 62. Although their backgrounds differed considerably, their work was in a sense complementary – Brown contributing the written, Franz the painted record of the various manifestations of life.

The relationship between Brown and Franz was straightforward in all those fields where Brown invited his colleague to provide the graphical record for his work, as in Brown's paper on *Woodsia* (see Chapter 8), his *Chloris Melvilliana* (see above) or his two papers on *Rafflesia arnoldii* (see above). In all these cases Brown acknowledged Franz's work in print, for example by writing in the introduction to the *Chloris Melvilliana* 'the delay till now has deprived Botanists of the excellent figures so admirably illustrating the structure of the plants selected for engraving, and for which it is hardly necessary to add that I am indebted to the friendship of Mr. [Franz] Bauer'.

The situation is more complex in all those cases where Franz had prepared a painted record independently from Brown and, more importantly, before Brown became interested in a particular topic. There are several examples that document Franz's early and total independence from Brown, such as his painted record of the flower of tropical milkweed, *Asclepias curassavica*. This is an evergreen perennial subshrub native in tropical America, known already to Linnaeus and cultivated widely in the eighteenth century. In 1805, definitely before Brown's return to London on 5 November of that year, Franz had produced two watercolours documenting in minute detail the flower of tropical milkweed from different angles. Franz also included several flower dissections showing among other features the pollinia, the coherent masses of pollen grains that are the product of single locule, to be transferred during pollination as a single unit. In addition he had documented a longitudinal and transverse section of the ovary, etc. (Fig. 13.10.). Typically for Franz, the two watercolours remained unpublished,

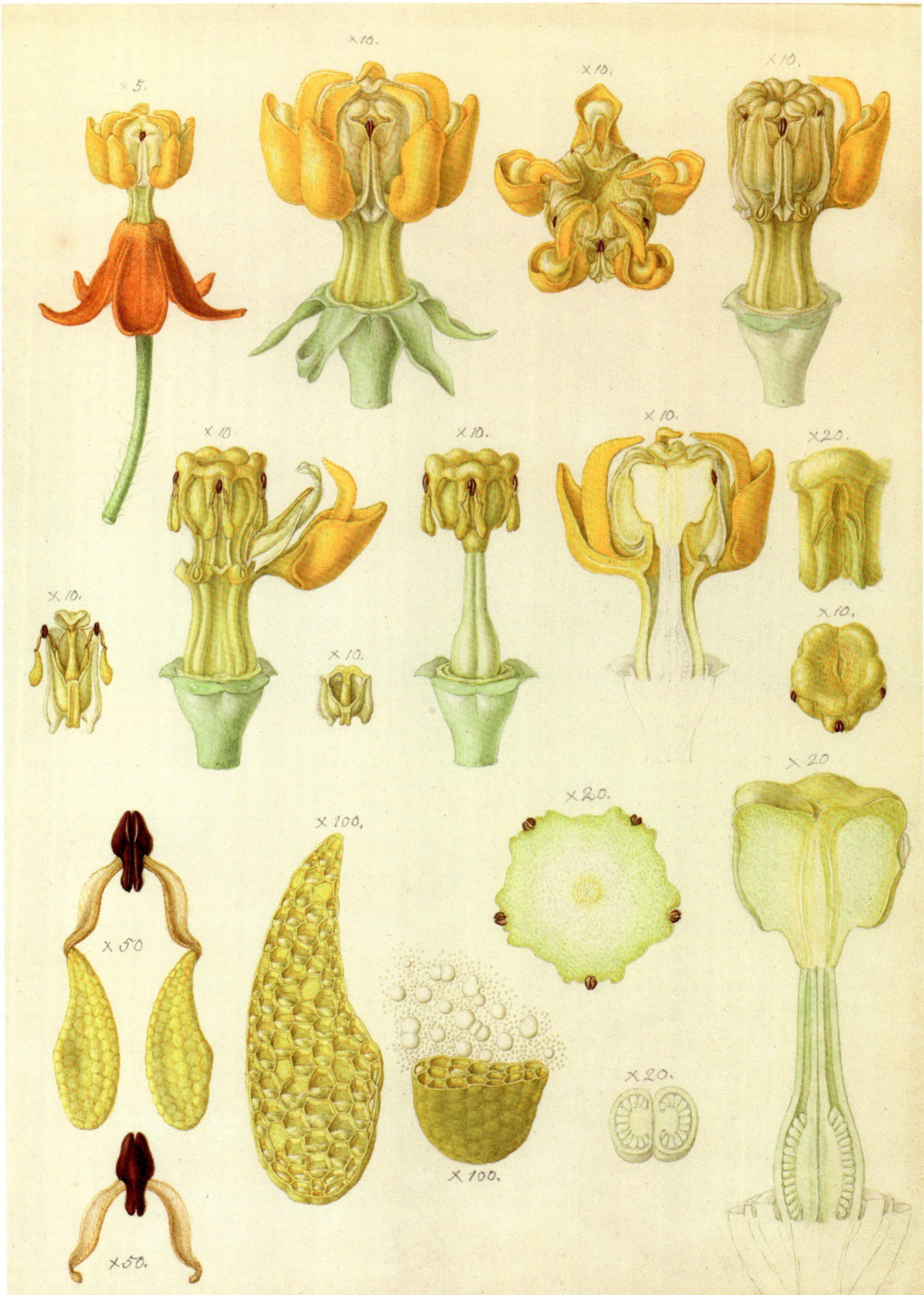

FIG. 13.10. Tropical milkweed, *Asclepias curassavica*. Flowers. Watercolour by Franz Bauer, 1805. Paper, watercolour. – London, Natural History Museum, Library and Archives, Unfinished Drawings of Plants, vol. 2, no. 12.

since for a plant illustrator of his calibre and a personality of his profile it was highly likely that publication simply was not that relevant (see Chapter 14).

When in the early summer of 1831 Brown set about examining germination of pollen in tropical milkweed and undertook pollination experiments, Franz seems to have shown him the two watercolours that had been shelved for 26 years. In his preprint 'Observations on the organs and mode of fecundation in Asclepideae and Orchideae', which appeared in the same year, and was published in 1833 with several additions and a somewhat differently worded title in the *Transactions of the Linnean Society*,[19] Brown referred to Franz's earlier painted record. He wrote, 'Among these [Franz's] drawings, exceeding perhaps in beauty and in the completeness of their details all other productions with which I am acquainted even of this incomparable artist [Franz], an extensive series, exhibiting the gradual development of the parts of the flower in *Asclepias curassavica*, were the most important'. In this case there was neither overlap nor conflict – Brown had his contribution accompanied by copper engravings based on the work of another illustrator showing other species of *Asclepias*. In addition, Brown's focus was on pollen germination, not documented by Franz in tropical milkweed in 1805.

However, in other cases the overlap between Franz and Brown was considerable, for example in Franz's observation of cell nuclei in the orchid *Phaius tankervilleae* (see Chapter 14). In such cases the impression prevails that Brown knew of Franz's unpublished illustrations, simply repeated the latter's observations and published the results without acknowledgement. Though Franz may not have been very happy about Brown's paper dealing with tropical milkweed, there was nothing wrong in the world of scientific ethics as we understand it today. However, to the non-scientist a somewhat embarrassing feeling prevails.

A LOVE AFFAIR WITH ORCHIDS

Franz's work on red snow, Arctic plants, and the stupendous flower of Sumatra were clearly circumscribed projects with his watercolours intended for publication. By contrast, orchids had become his life-long love affair, which had started in Feldsberg and ended more than sixty years later at Kew. Why this was so, is difficult to say. Most botanists are more attracted to certain plant families than to others and this also applies to plant illustrators, in particular to those who were, like Franz (see Chapter 7), free to choose their subjects of study. It seems that Franz had, by contrast, little inclination to record representatives of two major families of flowering plants – composites and grasses – whereas he was extremely fond of orchids, which, as a matter of fact, has made his painted record of the Royal Garden at Kew in a sense biased.

In their youth the Bauer brothers had become familiar with terrestrial orchids when working for Boccius. The first volumes of the *Codex Liechtenstein* (see Chapter 2) contain accurate representations of a considerable number of these, all native and occurring in different habitats in the Archduchy below the River Enns and adjacent Moravia. Like many before and again after them the Bauer brothers may have been attracted by the special aura of these plants – they flower at irregular intervals and in markedly variable numbers; they possess intricate flowers with their sexual parts strangely coalesced, most of them with a single stamen united to a three-carpelate ovary. In addition, the subterranean structures of terrestrial orchids native in Europe are as a rule reminiscent of testicles, a similarity well known since antiquity.

In the *Codex Liechtenstein* species of the genera *Orchis* and *Ophrys* are particularly well represented, among them widespread and locally common plants such as the military orchid, *Orchis militaris*, the green-winged orchid, *Orchis morio*, the lady orchid, *Orchis purpurea*, and the fly orchid, *Ophrys insectifera*, with the latter's name referring to the fact that at first glance parts of its flower resemble a fly. In addition we find several more native orchids of the area, such as the pyramidal orchid, *Anacamptis pyramidalis*, and the non-photosynthetic bird's nest orchid, *Neottia nidus-avis*, an epiparasite living on hyphae, which are in contact with trees from which this strange orchid receives its nutrients. Franz may have seen his first tropical orchids in Vienna, in particular those introduced from the Caribbean Islands and the northern fringe of South America by Jacquin, later cultivated in the conservatories in Schönbrunn and in the Botanical Garden of Vienna University (see Chapter 3). In addition the Bauer brothers must have seen the drawings of orchids prepared by Jacquin on his voyage and later published as copper engravings in the *Selectarum stirpium americanarum historia* or copied in the second, luxury edition of this work, to which Franz had contributed a few painted title pages (see Chapter 3). In Feldsberg Franz and his two brothers had already seen and copied some of the copper engravings documenting orchids published in Jacquin's *Florae Austriacae icones*, and in Vienna he may have been able to study Scheidl's watercolours, too, which had served as exemplars. Franz seems to have been so impressed by the quality of Jacquin's and Scheidl's orchid representations that in Kew and many years later he produced a further set of copies, which are kept today in the Niedersächsische Staats- und Universitätsbibliothek in Göttingen.[20] Franz's interest in orchids is testified to by another fact: he specifically collected watercolours prepared by other hands documenting species belonging to this plant family. The most remarkable collection previously owned by Franz is also in Göttingen: a set of twenty-two watercolours labelled 'Plants found near Port Jakson [sic, i.e. Sydney] Orchideous New Holland by Captain Paterson'. Although unsigned, these watercolours may have been produced for Paterson, who had stayed in Australia from 1791 to 1810 and acted as Lieutenant Governor of New South Wales in 1794–1795 and 1800–1808. They are probably the oldest orchid illustrations in colour prepared on the spot on the fifth continent.

Focused on the precise documentation of orchids since his Feldsberg period and with extensive experience in this extremely narrow field, Franz settled down in Kew, where we witness an explosion of his activity in recording the diversity of this plant family.[21] Critics may argue that a much higher percentage of Franz's orchid watercolours should have been included as illustrations in this book to give a more balanced impression of his oeuvre. However, it was the intention of the present author to offer a cross section of the works by the Bauer brothers, of which orchids form only one, albeit important element. In short, the accurate documentation of orchids remained Franz's favourite occupation – he painted more orchid species than species of any other plant family, and as a consequence his name enjoys the highest esteem among orchidologists, with Merle Reinikka dedicating a whole chapter of her *History of the orchid*[22] to Franz, for example.

The fragmentary annotations on Franz's pictorial record of orchid diversity are a serious impediment in

FIG. 13.11. ***(left)*** **Marsh helleborine,** ***Epipactis palustris*****. Flowers. Watercolour by Franz Bauer, 1799, Paper, watercolour. – London, Natural History Museum, Library and Archives, British Orchids 1798–1817, no. 36.**

FIG. 13.12. ***(right)*** **Marsh helleborine,** ***Epipactis palustris*****. Fruits and seeds. Watercolour by Franz Bauer, 1799. Paper, watercolour. – London, Natural History Museum, Library and Archives, British Orchids 1798–1817, no. 30.**

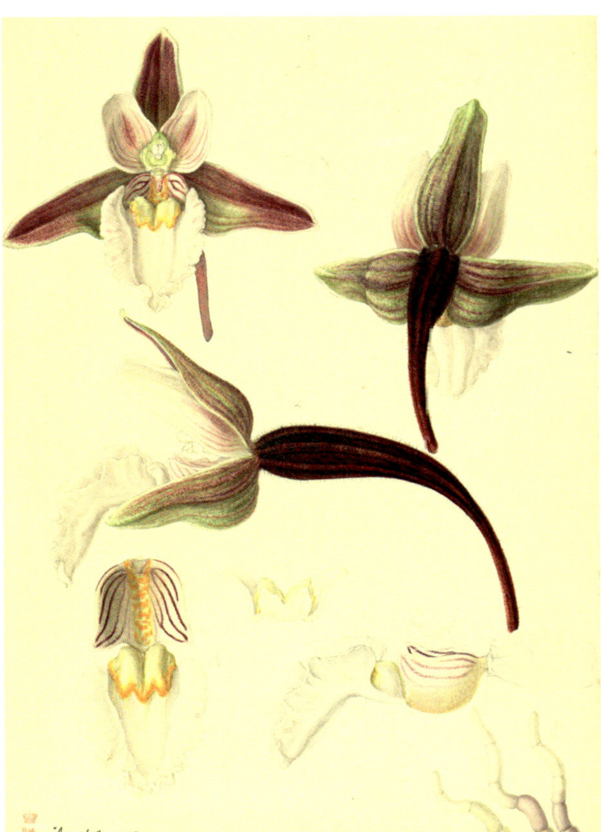

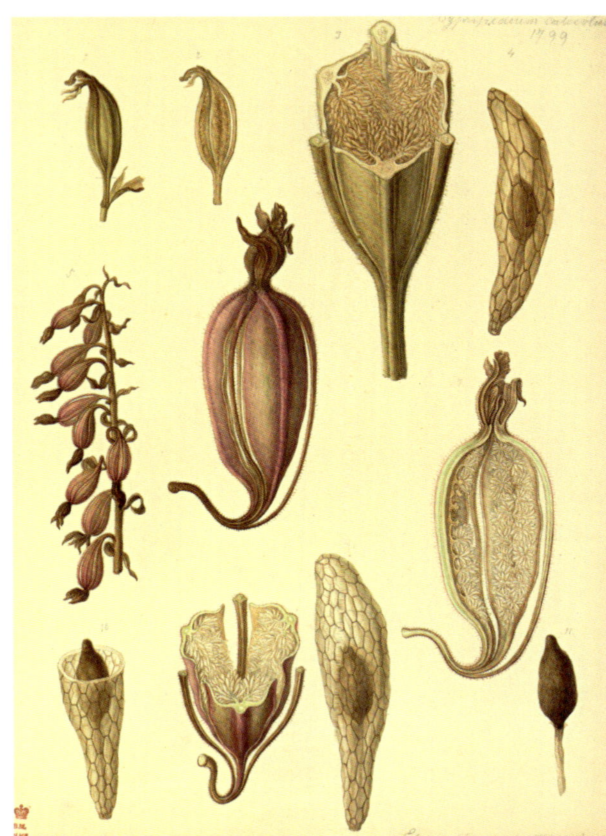

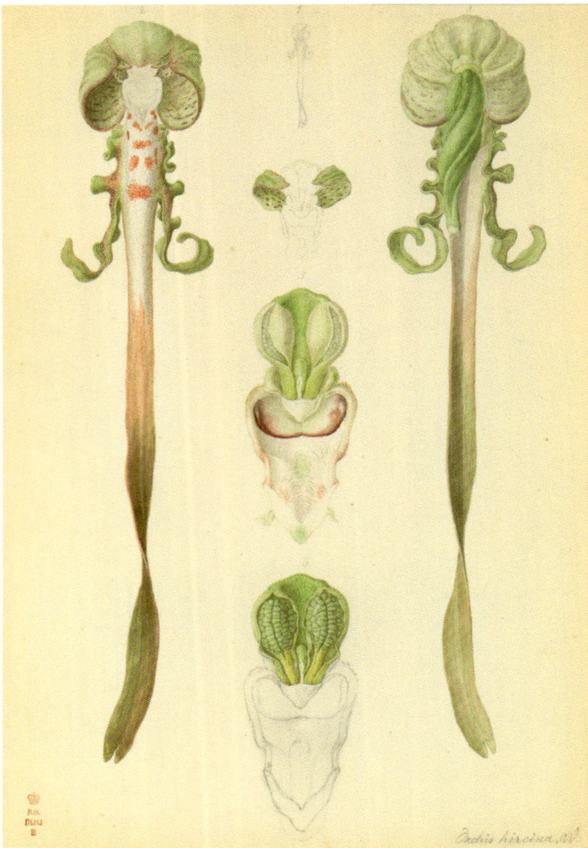

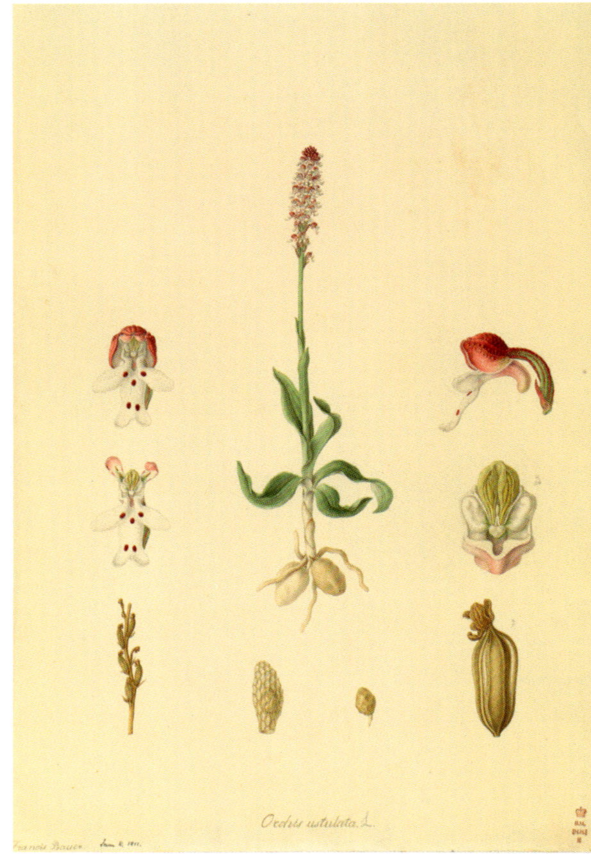

FIG. 13.13. ***(left)* Lizard orchid, *Himantoglossum hircinum*. Flowers. Watercolour by Franz Bauer, undated. Paper, watercolour. – London, Natural History Museum, Library and Archives, British Orchids 1798–1817, no. 82.**

FIG. 13.14. ***(right)* Burnt orchid, *Neotinea ustulata*. Watercolour by Franz Bauer, undated. Paper, watercolour. – London, Natural History Museum, Library and Archives, British Orchids, no. 67.**

elucidating the precise chronology of this work. However, the impression prevails that very soon after his settling down at Kew Franz began to record orchids native in Britain, all terrestrial, some of which he had previously documented for Boccius. We find representations of the marsh helleborine, *Epipactis palustris* (Figs. 13.11., 13.12.), a common plant in England, and the very tall and spectacular lizard orchid, *Himantoglossum hircinum* (Fig. 13.13.), which is of greatest rarity in Britain. Both the late spider orchid, *Ophrys holosericea* (Fig. 13.1. p. 352), and the rather small burnt orchid, *Neotinea ustulata* (Fig. 13.14.), possess a wide distributional range all over Europe, while in England they occur only in the southern part of the country. As with Ferdinand when documenting the flowering specimen of the sawfly orchid, *Ophrys tenthredinifera*, in Cyprus in spring 1787 (see Chapter 5), Franz in Kew could not imagine that the flower of the late spider orchid induces male hymenoptera to attempt copulation with the lip. Taking the extensive changes in land use and water regime which occurred over the last two centuries into consideration, it is very unfortunate that Franz refrained so much from giving adequate provenances and dates. All five watercolours are outstanding masterpieces meriting the highest praise. In the representation of the somewhat rancid-smelling flower of the lizard orchid the botanist will appreciate the attention given to the two lateral elements of the long, slightly twisted mid-lobe of the lip and the delicate colouring on the inner surface of the tepals. In the late spider orchid the detached pollinium, and in the burnt orchid the ripe seed, are equally noteworthy for the botanist. Considering the fact that orchid specimens lose their colours quickly when dried and become uniformly brownish, flat herbarium specimens, the high opinion of Franz's recording capacity among orchidologists is immediately understandable.

Terrestrial orchids from temperate zones, in particular those from the former British possessions in North America, form a second group which was documented by Franz. One of them was the ram's head lady slipper, *Cypripedium arietinum*, which Franz had cultivated in his own garden (see Chapter 8). Franz had dug up a specimen in order to pay special attention to the accurate documentation of the massive underground parts, which he first must have put into water to become fully observable. In the same way he treated a specimen of the lady slipper, *Cypripedium calceolus*, widespread in Eurasia, but very rare in Britain. Significantly one of Franz's watercolours is annotated 'From Yorkshire', then as today the only known natural habitat of this iconic orchid in England.

THE CHANGE IN THE HEATING SYSTEM AND ITS CONSEQUENCES

The change of the heating system in conservatories from hot air in channels to hot water in cast-iron tubes enabled a much easier regulation of the humidity and thereby greatly amplified the chances of successfully cultivating orchids from tropical and subtropical countries, both terrestrial and epiphytic. What previously had been the slaughterhouse and cemetery became paradise, or at least a good place to live, for orchids imported from warm climates. The consequence was an unprecedented influx of numerous orchids, of which very many first arrived in England, and notably in the Royal Garden at Kew and the newly created garden of the Horticultural Society in Turnham Green (see below). Both institutions became important sources of orchids for Franz. A considerable number of these were new to science, had not yet received a proper name and Franz was the very first to study and paint them in watercolour. However, Franz also recorded orchids that had been known already for several decades, among them in 1824 *Bonatea speciosa* (Fig. 13.15), a terrestrial orchid from the Cape of Good Hope which had received its oldest binomial from Linnaeus the son.

The vanilla orchid, *Vanilla planifolia* (Fig. 13.16.), is a climber native in the Atlantic part of Mexico, then belonging to the Viceroyalty of New Spain, reigned over by Carlos IV, King of Spain. Bringing this orchid in a conservatory to flower was a sensation first achieved in

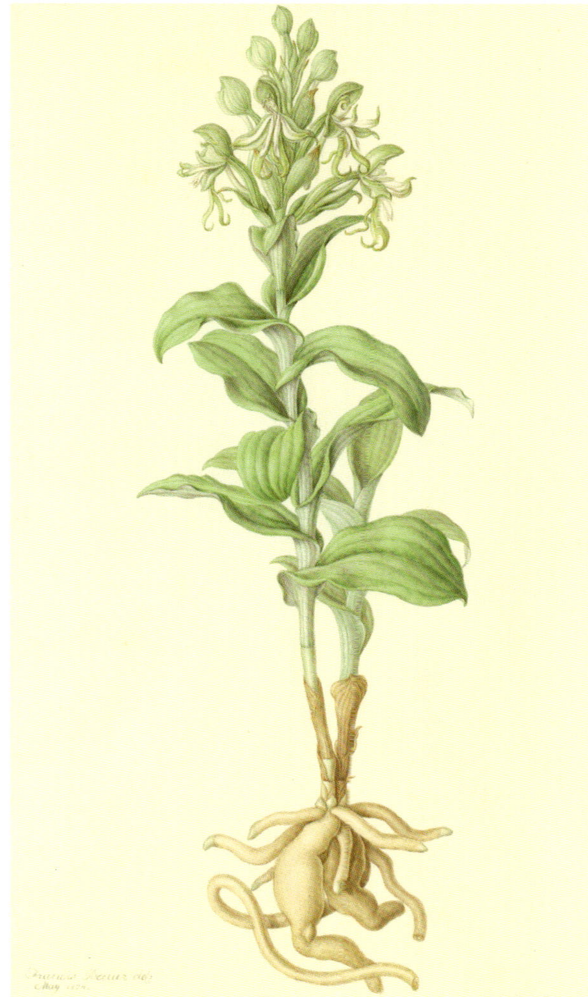

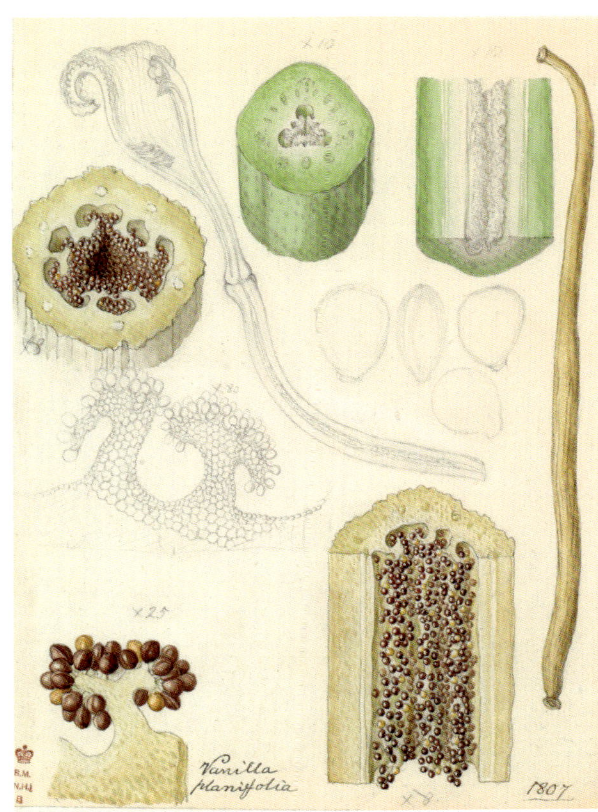

FIG. 13.15. ***(top)*** ***Bonatea speciosa*****. Watercolour by Franz Bauer, undated. Paper, watercolour. – London, Natural History Museum, Library and Archives, Finished Drawings of Plants, pl. 127.**

FIG. 13. 16. ***(bottom)*** **Vanilla orchid, *Vanilla planifolia*. Watercolour by Franz Bauer, Paper, watercolour. – London, Natural History Museum, Library and Archives, Unfinished Drawings of Plants, vol. 4, nos. 87 and 88.**

June 1807 by an anonymous gardener of Charles Francis Greville (see Chapter 10).[23] The latter was the second son of Francis Greville, 1st Earl of Warwick, a nephew of Sir William Hamilton, formerly British Ambassador to Naples (see Chapter 5). Elected a member of parliament, he had previously served as a Vice Chamberlain of the Royal Household. More relevant for science, Greville, one of Banks's cronies, was a Fellow of the Royal Society, like Banks a founding member of the Horticultural Society of London, and had developed a taste for growing rarities in his garden at Paddington Green, London. Naturally the conservatory was stocked with plants from many sources, with the vanilla orchid coming from George Spencer Churchill, the Marquess of Blandford, later 5th Duke of Marlborough, who was then developing an enormous and extravagant garden with several conservatories at White-Knights near Reading. Like a modern-day cameraman, Franz must have been called to Paddington Green to produce a watercolour of the vanilla orchid in flower, which he annotated 'R. H. Charles Greville July 18th 1807'; today it is kept in the Library and Archives of the Natural History Museum. During the period this drawing remained unpublished until it appeared in 1834 as a coloured lithograph in the *Illustrations of orchidaceous plants* (see below). Two other images of this plant were quickly published – one based on a watercolour by William Hooker (see Chapter 20) appeared in Richard A. Salisbury's *Paradisus Londinensis*, London, 1807–1808, the second appeared in Henry C. Andrews's *Botanist's repository* in March 1808. The latter two illustrations represent only the flowering plant, while Franz had added to his image a ripe fruit, including longitudinal and cross sections of it showing both ripe and unripe seeds on the placenta. Only after a complex and lengthy fermentation process, not yet understood at the time, does the ripe fruit of the vanilla orchid produce the delicate flavour and fine aroma that made it a plant of some economic importance.

When today a novelty or rarity blossoms for the first time in a botanical garden, the curator of the living collection first asks a photographer to document the event. This was the job Franz had to fulfil over decades in the Royal Garden in Kew, and we may assume that as a rule the respective plant was brought to his studio. Only exceptionally and in the case of extremely delicate plants he may have been asked by Aiton the son, the head gardener, to produce a preliminary drawing in situ within the conservatory. Maybe this was the case in 1827, when one more epiphyte from tropical Southern America, which had not yet received a scientific name, opened its spectacular dotted flowers with a wonderful fragrance for the first time. It was only two years later that this robust orchid with pendent, short-lived inflorescences received the name *Stanhopea insignis* in a publication by William Jackson Hooker of Glasgow University (see Chapter 16). It appeared in Curtis's *Botanical Magazine*, of which he was then the editor.

Since the Portuguese colony of Brazil had been for a very long time almost completely sealed off to non-Portuguese people,[24] many orchids from this vast country were introduced into cultivation only after independence, when access to travellers started to become much easier. One of the orchids from Brazil was *Christensonella subulata* (Fig. 13.17.), a rather widespread epiphytic or rupicolous orchid occurring from Santa Catarina to southern Bahia with reddish to brownish flowers and a strongly fruity, melon-like fragrance.[25] Lindley (see below) made use of a specimen cultivated in the Loddiges Nursery when he validated

FIG. 13.17. *Christensonella subulata*. Watercolour by Franz Bauer, 1818. Paper, watercolour. – London, Natural History Museum, Library and Archives, Unfinished Drawings of Plants, vol. 4, no. 41.

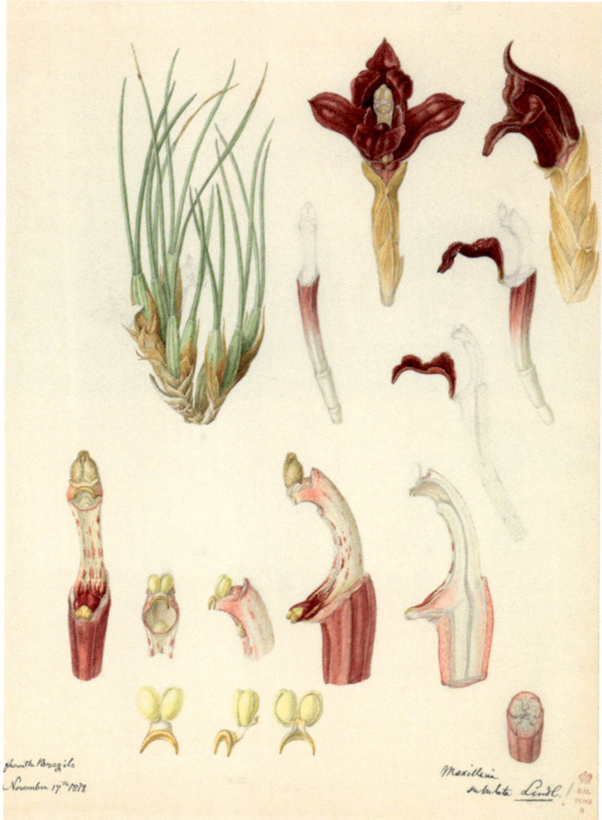

the name *Maxillaria subulata* in 1832. Interestingly, Franz produced a watercolour of this plant well before that, possibly from one cultivated in the Royal Garden in Kew, which he annotated 'from the Brazils 17 November 1818', to which Lindley added later 'Maxillaria subulata Lindl.', as he would have done on a herbarium specimen. Franz drew the flowers on the leafy plant only in pencil to indicate their position, while he documented the individual flowers enlarged, from two angles as well as dissected. A specimen of *Christensonella subulata* which strongly resembles the upper left-hand figure in this watercolour is kept in Lindley's orchid herbarium in the Royal Botanic Gardens Kew. *Oncidium sphegiferum* (Fig. 13.18., centre and below) is an orchid with an even smaller area of distribution in Atlantic Brazil. In this case Franz explicitly noted that he painted a specimen cultivated in the garden of the Horticultural Society on 25 May 1834, nine years before Lindley described this orchid species as new to science, again on the basis of a plant from the Loddiges Nursery. Franz's last orchid watercolour showed an undetermined specimen of the genus *Catasetum*. It is dated 8 September 1838, when he had reached the age of 80. It was to remain his last botanical illustration; in his final years Franz discontinued this kind of work (see Chapter 16).

The geographical coverage of Franz's work on orchids is indeed global, with a few originating in Australia, among them the terrestrial red rustyhood, *Oligochaetochilus rufus*,[26] its common name referring to the reddish-brown flowers with translucent white areas. Interestingly, Ferdinand had also produced watercolours of three species belonging to this and two closely related orchid genera native in the fifth continent, with two of these prepared for the Admiralty and now in the Library and Archives of the Natural History Museum.[27] The third one depicting the cobra greenhood, *Pterostylis grandiflorum*, a widespread and locally common plant occurring from what is now Queensland to Tasmania, is currently untraceable, but has been published as a coloured copper engraving in Ferdinand's *Illustrationes Florae Novae Hollandiae* (see Chapter 11).

Franz produced watercolours of two more terrestrial orchids from Australia, which were later published in Curtis's *Botanical Magazine*.[28] One of them possessed an unconventional provenance: From the sod of a pot containing the Australian pitcher plant, *Cephalotus follicularis*, sent from King George III Sound to Kew sprang up a specimen of the western onion orchid,

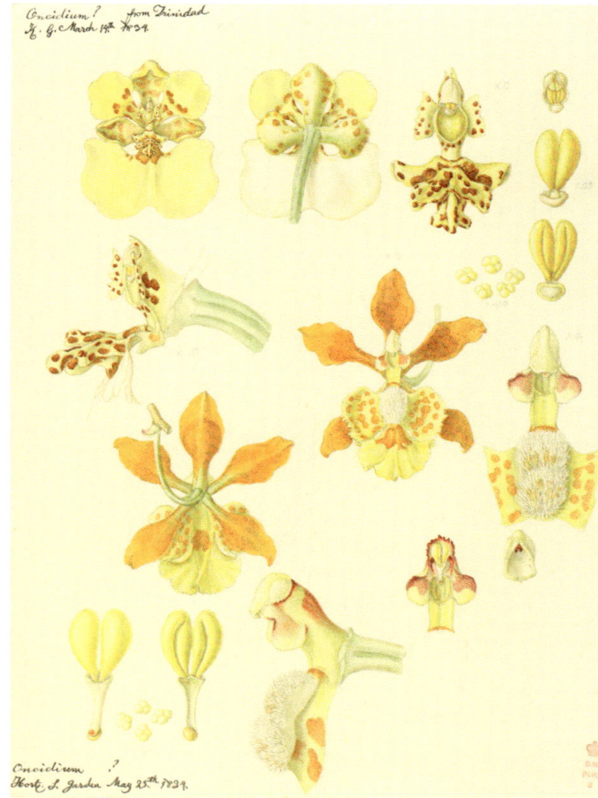

FIG. 13.18. *Oncidium ampliatum* (above) and *O. sphegiferum* (centre and below). Watercolour by Franz Bauer, 1834. Paper, watercolour. – London, Natural History Museum, Library and Archives, British Orchids, no. 98.

Microtis media. This inconspicuous plant with numerous small pale green to yellowish flowers had been first collected on the voyage of the *Investigator* and subsequently described by Brown in his *Prodromus Florae Novae Hollandiae*. The collector of the specimen of the Australian pitcher plant was Philip Parker King, a man born on Norfolk Island, where his father Phillip Gidley King was Lieutenant-Colonel of the convict settlement visited by Ferdinand twelve years later (see Chapter 10). Very occasionally Franz also recorded herbarium specimens of orchids, among them an undetermined specimen of the genus *Diuris* from Australia, deposited in Lambert's herbarium. Just like Ferdinand, who had previously depicted specimens from this collection (see Chapter 9), Franz used monochrome wash as the colours had been long lost. For additional information on a terrestrial orchid from New Zealand annotated 'K.[ing's] G.[arden] October 8 1828' (from the Royal Garden at Kew) documented by Franz, see Chapter 14.

The continuity in Franz's work on orchids, and in particular on their flowers, is best illustrated by his

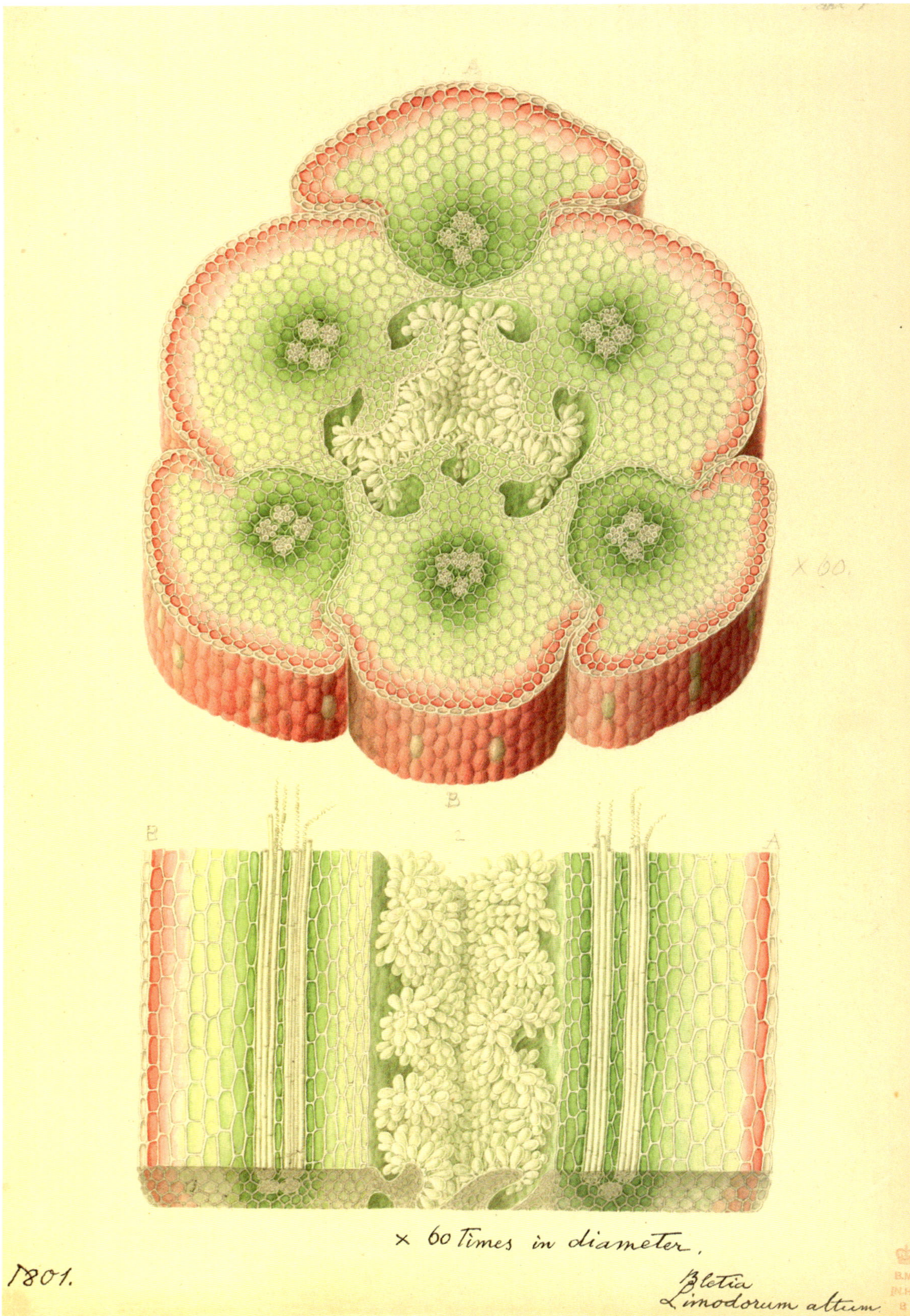

FIG. 13.19. *Bletia purpurea.* Longitudinal and traverse section of the ovary. Watercolour by Franz Bauer, 1801. Paper, watercolour. – London, Natural History Museum, Library and Archives, Unfinished Drawings of Plants, vol. 3, no. 41.

watercolours of the lady slipper, *Cypripedium reginae*. He first recorded a specimen he had seen in 1788 in the garden of Anton, Prince of Saxony, in Dresden (see Chapter 7), and later specimens cultivated in the garden of Lady Dorothea Banks in Spring Grove in 1798 and 1802. The anatomy of orchid fruits had also been for a long time on Franz's agenda. By 1801 he had documented a cross section of the gynoecium of pine pink, *Bletia purpurea* (Fig. 13.19.), showing the developing ovules on the Y-shaped placentas with even the minute vessels being accurately represented. In order to demonstrate the nature of the waxy pollinia of this orchid, Franz removed the set of four pollinia from the single anther of this orchid species, immersed them in water and studied the result in his microscope. Subsequently he recorded the elastic tissue uniting the individual pollinia as a web of material containing thousands of pollen tetrads (see Chapter 14). Both watercolours formed the exemplars of the lithographs published in the *Illustrations of orchidaceous plants*.

ILLUSTRATIONS OF ORCHIDACEOUS PLANTS

There were several ways for John Lindley to come into contact with Franz, almost 41 years his senior. One of them was the former's election to the Royal Society on 17 January 1828, seven years after Franz. Another occasion was Lindley's appointment on 25 February 1822, then a young man of 23 years, as Assistant Secretary of the Garden of the Horticultural Society,[29] of which Franz was a fellow. In addition both were fellows of the Linnean Society. A third chance existed even earlier, when in October 1819 Lindley became assistant librarian to Banks, working at 32 Soho Square and reporting to Brown, an appointment which was soon to come to an end with Banks's death. As superintendent of the plant collection cultivated in the garden of the Horticultural Society in Turnham Green, Chiswick Lindley worked and lived very near to Franz in Kew, just a short ride on horseback crossing the Thames at Kew Bridge. Therefore communication was probably mostly oral and only occasionally by letter. As a consequence almost nothing is known about their personal relationship, but it is clear that Franz and Lindley were united in their love of orchids.

When Lindley wrote his first ever description of an orchid to be published in the first instalment of his *Collectanea botanica* (see Chapter 12) in 1821, he added, full of enthusiasm: 'If we were requested to select the most interesting from the multitude of vegetable tribes, we should, on the whole, perhaps be willing to give preference to the natural order of Orchideae. Whether we consider general elegance of individuals, durability of blossoms, splendid colours, delicious perfume, or extraordinary structure, it would be difficult to select any order superior to Orchideae in these respects, and few even equal to them. To the cultivator, who esteems plants for their beauty only; to the botanist, who, rejecting outward attractions, is chiefly captivated by anomalous, or intricate organization; or the amateur, who wisely prefers to have these requisites combined; to all, or any of them, we conceive that accurate figures of foreign Orchideous plants cannot fail to be particularly acceptable'.[30] Even more important for the close contact between the two orchid enthusiasts were two facts – Lindley was the son of a gardener and therefore had a social background similar to that of Franz, but dissimilar to that of Banks, Brown, Home and many others; furthermore, Lindley was a qualified and dedicated botanical illustrator like Franz and had also started early to draw plants accurately, with his first watercolour published in the *Botanical register* in 1819[31] when he was a man of twenty years.

We therefore can well imagine Lindley's amazement when Franz showed him his extensive collection of watercolours documenting orchids prepared in Kew since at least 1791, plus those made for Paterson (see above). This would have been the amazement of the connoisseur and Lindley proposed what anyone would: to instigate the publication of at least a selection of Franz's works on orchids. When all this took place remains unknown. However, in November to December 1830 – definitely after the appointment of Lindley as Professor of Botany at the newly founded University College, London – the first instalment of a joint publication appeared in London. The title page is a good specimen of tact: 'Illustrations of orchidaceous plants; by Francis Bauer, Esq. F. R. S.; L. S. & H. S. with notes and prefactory remarks by John Lindley, Ph. D. F. R. S.'. This is a balanced wording indicating mutual respect of two personalities and their qualifications. To the considerable irritation of the bibliographer, the wrappers of the four instalments carry another title: 'The genera and species of orchidaceous plants by John Lindley, F. R. S.; L. S.; & H. S. illustrated by drawings on stone from the sketches of Francis Bauer, Esq. F. R. S.; L. S. & H. S.'. Admittedly, this variant is more conventional, but it transmits a different, and indeed incorrect, interpretation of this collaborative project – Franz did not illustrate Lindley's text, but Lindley added text to Franz's pre-existing illustrations, several of them prepared years before Lindley was born. Maybe this imbalance led to the more appropriate wording on the final title page given above.

In addition to this slight anomaly, the *Illustrations of orchidaceous plants* possesses a peculiar structure: the preface and the prefatory notes are followed by two chapters called 'Fructification' and 'Genera', each containing legends and lithographs, with separate numbering for the plates in each chapter. Each of the four instalments contained legends and lithographs from both chapters. The division between the two chapters is not entirely satisfactory, but there is a tendency to document the dissected flower and fruit parts in 'Fructification', while intact flowers and dissected flower parts are to be found under 'Genera'. The number of dissections shown on a plate is often high; for example, twelve different anatomical details are presented on t. 4. of 'Fructification' for the flower of *Stenorrhynchos speciosum*, a widespread terrestrial orchid from tropical America with bright red tepals first described by Jacquin in his *Icones plantarum rariorum*. In the lithographs Franz stresses the complex anatomy of the orchid flower but also deals with ripe orchid capsules and, interestingly, also with the minute seeds they contain, with their net-like testa. In the documentation of the latter, for example in the bee orchid, *Ophrys apifera*, Franz was far ahead of his time.

The division of work between Lindley and Franz was complex: preface and prefatory notes are by Lindley, all lithographs are based on watercolours or pencil drawings by Franz, some of them, as explicitly stated, redrawn by Lindley before being transferred on the lithographic stone by Maxim Gauci. It has been generally assumed that the legends are all by Lindley, and this may indeed be true, with Franz having a corrigendum added, printed on a separate sheet. Collaboration between Franz and Lindley continued for several years. One of the results was Lindley dedicating a species new to science to his senior partner – *Oncidium baueri* – validating the name in his *Genera and species of orchidaceous plants* in 1833. The plant is a massive pseudobulb epiphyte with erect to pendulous multi-flowered inflorescences, up to 2.5 metres long, occurring widely in tropical America. This spectacular orchid is the most lasting monument of the friendship between the father of orchidology and Franz, near neighbours and both men of abundant productive energy. With the publication of the fourth fascicle of the *Illustrations of orchidaceous plants* in late 1838, this collaboration seems to have come to its end. A small collection of orchid watercolours by Franz, which ended up in the possession of Lindley and is now among the Cingalese Drawings in the Lindley Library in London, bears additional testament to the close contacts between the two specialists. Here we find, for example, a representation of the butterfly-pollinated jewel orchid, *Ludisia discolor*,[32] a native of southeast Asia, and of the green-winged orchid, *Orchis morio*,[33] widespread in Europe and southwest Asia.

The preface of the *Illustrations of orchidaceous plants* dated December 1837 merits being read since it explains why only such a small fraction of Franz's work on orchids has ever been published and why lithography had been chosen as a printing technique. Lindley wrote: 'To make the whole of such splendid specimens of art public by means of engravings capable of expressing the beauty of the originals, would be an object worthy of a liberal Monarch and an enlightened Government; to a private individual [Lindley] the large expense must necessarily prove an insuperable impediment. To publish a selection of such as throw the greatest light upon structure is all that so humble an individual as myself could venture to undertake, and even this the little encouragement given to such undertakings has obliged us to extend through a period of some years. It has been found indispensable to represent Mr. Bauer's drawings by means of lithography instead of engraving on copper'. The subtext is clear: in contrast to their father George III, neither George IV nor William IV possessed a genuine interest in botany or horticulture, and no foundation comparable to that created by Sibthorp was available for the publication of the orchid watercolours prepared by Franz. Lindley knew the facts: for the completion of the *Flora Graeca* he was to receive in the end a total of £350, a large sum then. In addition, much larger funds from the foundation based on the profits from Sibthorp's estate in Oxfordshire were necessary to cover the cost of printing the text and plates, plus the extremely tedious colouring by hand of the latter.[34] In short no government funds were available to support the *Illustrations of orchidaceous plants*, quite in contrast to the situation in Napoleonic France. There considerable sums were spent by the government to sponsor the publications of Pierre-Joseph Redouté (see Chapter 16).

Predominantly *Illustrations of orchidaceous plants* deals with epiphytic orchids from the tropics and subtropics. Among them we find Franz's early favourite *Phaius tankervilleae* from China, which he had integrated into his flower piece now in the Fitzwilliam Museum (see Chapter 8), and the vanilla orchid, but also a small number of terrestrial orchids native in Britain, such as the bee orchid.

The prefatory notes discussed the divided opinions on the manner in which pollination takes place in orchids: Lindley stated that Franz 'adduces a great many facts to show that it does not take place, as in other plants, by actual contact between the pollen and the stigmatic secreting surface, but that it is effected by absorption of fecundating matter by the stigmatic gland from the pollen though the caudicula'. The opposite, and as we now know, correct analysis was offered by Brown and others: after the pollen adheres to the stigmatic surface, it germinates and the pollen 'tubes descend the stigmatic canal, and mix themselves among the ovules over the surface of the placenta'. Lindley had initially followed Franz's views, but changed them for the interpretation by Brown. As in the case of the parasitic stem rust, *Puccinia graminis*, a very accurate pictorial record does not automatically lead to a correct interpretation of the facts (see Chapter 8).

Lindley meticulously noted the year in which Franz had produced a particular watercolour in the legends and often additionally on the lithographs: for example, t. 3 of the chapter 'Fructification' showing the heath spotted orchid, *Dactylorhiza maculata*, is accompanied by a legend stating 'from a drawing of Mr. Bauer in May 1792'. This insistence on Franz's dates and historical chronology is reflected in a recommendation included in the preface: 'Should the reader take an interest in such investigations [of the chronology of discoveries of orchid structure and anatomy] I refer him to the dates of Mr. Bauer's drawings, and by comparing them with the dates of other publications he can judge for himself to what amount of credit this most admirable and original observer [Franz] is entitled'. This very sentence with precisely the same wording had also previously appeared in Lindley's earlier *An introduction to the natural system of botany*, published in London in 1830.[35] An anonymous reviewer of this book suggested that the wording casts doubt on Brown's integrity by implying that the latter's understanding of orchid morphology and anatomy were the result of Franz's watercolours, a fact that was neither mentioned nor acknowledged by Brown. As so often, there were different views on the concept of priority. This caused an acrimonious feud between Brown and Lindley, lasting until the former's death in 1858. Brown had asked Lindley for an apology, which was not given, with Lindley claiming that he had not intended to offend Brown.

In theory Franz was a third party in this conflict, but his position becomes evident in his letter to Lindley dated 7 February 1832, when he wrote 'I wish you could contrive that Brown should not see the [orchid] drawings before they are published'. The background for the opposing views on priority as exemplified by Lindley and Brown is dealt with in the next chapter.

ENDNOTES

1 Carter (1988)
2 Hall (1984)
3 Franz Bauer (1819)
4 Franz Bauer (1820)
5 Brown (1813), Granville (1820)
6 PD. 1859.0316.142
7 Brown (1819)
8 For a more complete account see Werner (2007)
9 gr. 2° Cod. Ms. Hist. Nat. 94, VI, f. 1.
10 Grant (1994)
11 Mabberley (1985)
12 Bastin (1973)
13 Nais (2001)
14 Mabberley (1999b)
15 Brown (1821)
16 Brown (1844)
17 Inv. No 37096-1001, 37097-1001.
18 Inv. No 67219-1001.
19 Brown (1833)
20 Lack (2003)
21 Stearn (1993), Stewart & Stearn (1993)
22 Reinikka (1995)
23 Ecott (2004)
24 Lack (2014)
25 Koehler, Singer & Amaral (2012)
26 Jones (2006)
27 Mabberley & Moore (1999)
28 Hooker, W. J. (1835)
29 Stearn (1999)
30 Lindley (1821 – 1826)
31 Allford (1999)
32 Inv. No. 67226-1001
33 Inv. No. 67224-1001
34 Lack (1999c)
35 Cribb (1999)

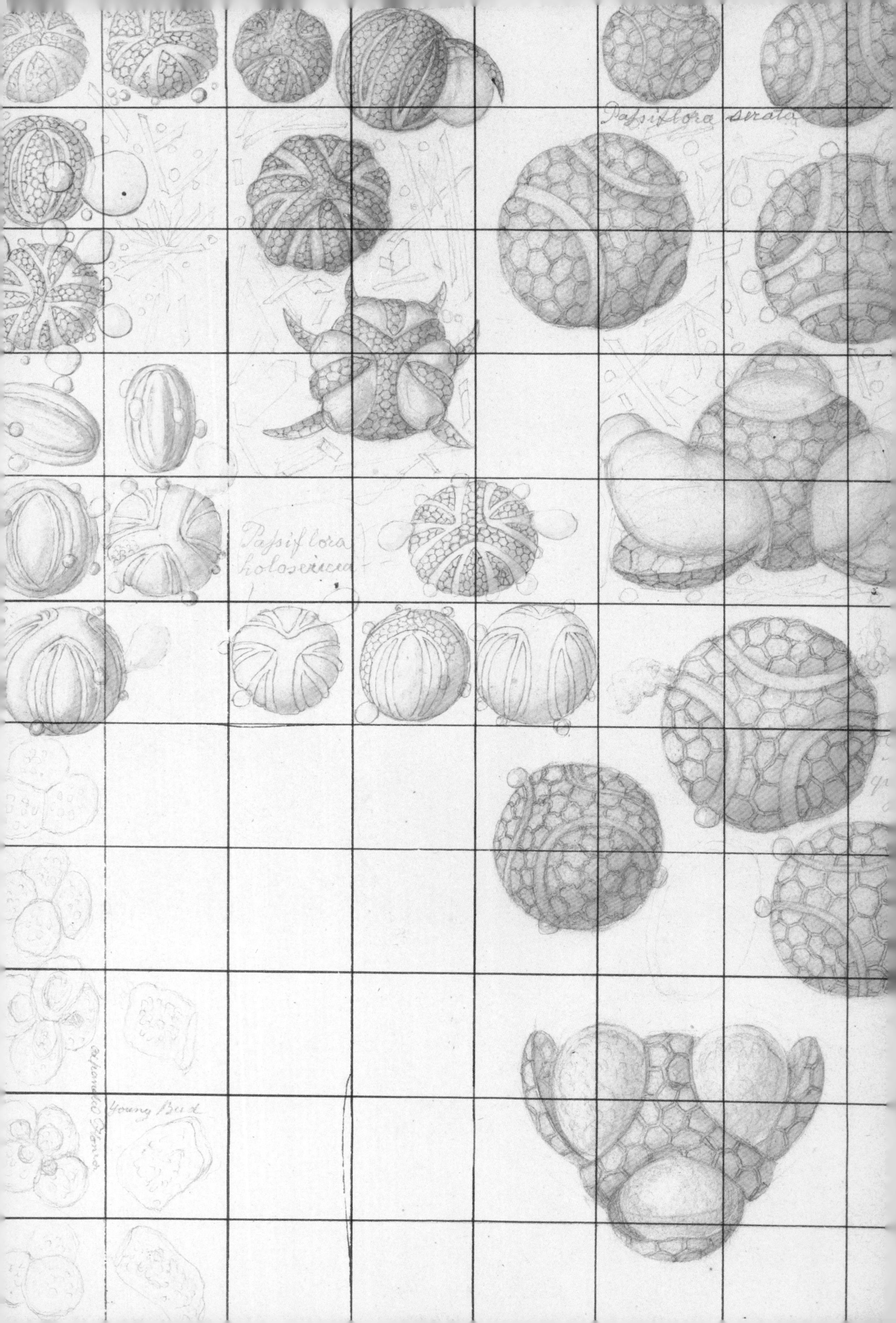
Passiflora serata
Passiflora holosericea
Expanded Flower
Young Bud

CHAPTER 14

Franz Bauer observes the cell nucleus and pollen diversity[1]

PRIORITY

THE *OXFORD ENGLISH DICTIONARY* offers several explanations for the term 'priority', the most relevant in our context being 'the fact or condition of being earlier in time; the fact of preceding something or someone else'. In science priority has a much narrower meaning – the credit given to the individual or group of individuals who first made a discovery or proposed a theory and who made the discovery or theory known by publishing the finding as a text, either in print or electronically. Fame and honours consequently go to the first person or group of people to publish; they are the legitimate winners, not the first person who made a particular observation or finding. This concept is often called the priority rule, nicely summarized in such a phrase as 'publish or perish', because there are no second prizes in science. To use a metaphor, the early bird that catches the worm, i.e. the first person who made the discovery, is quite irrelevant in the world of science. The laurels exclusively go to the early bird that caught the worm and told the world in writing that it had done so. In short, it is always the published text that counts, not the first observation made. Along similar lines, it is not the first image made that counts – not even the first image published – a proper text is necessary, not just a printed legend. Clearly there is a marked difference between what may be called 'historical priority' (as exemplified by Lindley's position, which had stressed Franz's priority on the basis of unpublished, but dated, watercolours) versus priority in science (as exemplified by Brown's view on properly published evidence; see Chapter 13).

Priority disputes in science can easily be decided in fields such as the scientific naming of plants, fungi and algae, where a set of universally accepted rules (the *International Code of Nomenclature for Plants, Fungi and Algae*) exists. Much more difficult are priority disputes to be resolved when the priority of an understanding or a discovery comes into question. Historians of science often disdain retroactive priority disputes, since the opposing parties often lack the complete understanding for the nature and relevance of the scientific change involved and for the context in which the discovery was made. In addition, discovery is rarely achieved by a single outstanding individual in isolation, although such cases do exist. It is against this general background that Franz's observations of the cell nucleus and of pollen cells are to be seen.

FIG. 14.1. Pollen of various species of passion flowers, *Passiflora* spec. Watercolour by Franz Bauer, undated. Paper, watercolour. – London, Natural History Museum, Library and Archives, Drawings of Francis Bauer of Epidermis, Hairs, Pollen Grains, Monstrosities, no. 41.

GREENISH YELLOW SPECKS

It has rightly been said that the discovery of the cell nucleus is only marginally less important than the discovery of the cell itself. The reasons for this are straightforward: without the understanding of the structure and function of the nucleus, no coherent explanation of the mechanisms that transmit hereditary characters is possible. Generally this discovery is attributed to Antoni van Leeuwenhoek,[2] a draper, city official in Delft in the Dutch Republic and since 1680 a Fellow of the Royal Society. Familiar with glass processing, he had developed a method to produce minute, high-quality lenses, which became the key component of his powerful microscopes. With their help Leeuwenhoek managed to discover, for example, 'infusoria' ('protists' in modern zoological terminology), bacteria and the vacuole in cells, and sent letters describing his observations to the Royal Society. In London these were translated from Dutch into English and Latin and quickly published, before they appeared in the original Dutch version.

On 3 March 1682 Leeuwenhoek had written to Robert Hooke, another Fellow of the Royal Society, a letter reporting (in Dutch): 'Thus I came to observe the blood of a cod and of a salmon, which I also found to consist of hardly anything but oval bodies and however closely I tried to observe these I could not make out of what parts these oval particles consisted for it seemed to me that some of them enclosed in a small space a little, round body or globule, and at some distance from this body there was round the globule a clear ring and round that ring again a slowly shadowing contour, forming the circumference of a globule and, as well as I could, represented in fig. 5'. This text needs an exegesis. By 'oval bodies' and 'oval particles' Leeuwenhoek means erythrocytes, which in fish indeed contain a cell nucleus, called by him a 'globule'. The 'clear ring . . . and a slowly shadowing contour' are best interpreted as phenomena due to deflections of light. Surprisingly, Leeuwenhoek noted a nucleus only in some erythrocytes, whereas he should have observed nuclei in all; further doubts are raised by his subsequent note that he had seen '3, 4, 5, 6, nay as many as 8 globules, much smaller than the first globule' in the erythrocytes, which is even more difficult to understand. However, his drawing in red chalk, Fig. 5, preserved with the letter in the archives of the Royal Society in London, is unequivocal: it shows three erythrocytes, each with a central body, which we would unhesitatingly call a nucleus. In short, Leeuwenhoek's convincing drawing, together with his less convincing description and their publication in Hooke's *Philosophical Collections* in 1682, marks the discovery of the nucleus in animal cells.

One hundred and twenty years passed until Franz studied a subject that he and others had already depicted several times before – the flower of *Phaius tankervilleae* (see Chapter 8). This striking orchid from shaded forests in tropical and subtropical Asia and Oceania was by then already in cultivation in several conservatories in England,[3] among them those in the Royal Garden at Kew and in Spring Grove, Banks's nearby summer residence, and thus easily accessible to him. Franz, then 44 years old, focused his attention on the stigma, the receptive tip of the carpel in the flower. It is this frequently sticky area that receives the pollen at pollination and is where the pollen germinates and starts to grow towards the ovaries, often passing through the so-called stigmatic canal. For his preparations Franz seems to have produced thin slices of the carpel's tip and studied them in his microscope, about which we know in this case nothing apart from the fact that it must have been manufactured in or before 1802. He was lucky: in cells detached from the stigmatic surface he could clearly discern what later were called 'granular, more opaque greenish yellow specks' – cell nuclei. Most importantly Franz produced a very accurate pencil drawing of the detached cells (Fig. 14.2.) and dated it '1802'; it is preserved in the Library and Archives of the Natural History Museum.

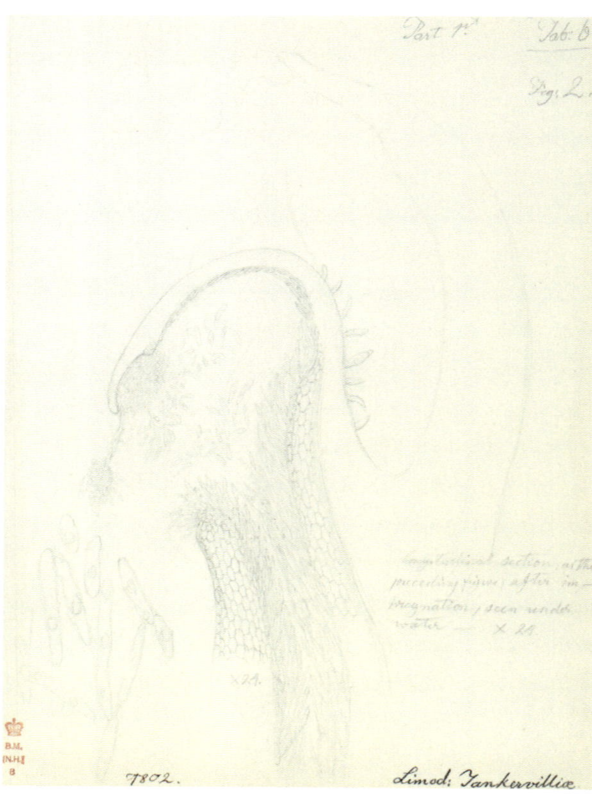

FIG. 14.2. ***Phaius tankervilleae*. Stigma and stigmatic surface. Pencil drawing by Franz Bauer, 1802. Paper, pencil. – London, Natural History Museum, Library and Archives, Unpublished Drawings of Plants, vol. 3, no. 53.**

At that time Ferdinand and Brown, who was later to play an important role in this context, were sailing in the *Investigator* along the southern coast of Australia.

Franz's drawing shows eight cells, of which six clearly have a single nucleus, one has two nuclei, and one has three nuclei; most significantly, there is no cell drawn without a nucleus. This was the first observation of the nucleus in plant cells. In 1802 Franz seems to have been unable to realize the momentousness of his discovery. He could have looked for similar globular structures in the cells on the stigmatic surface of other orchids; he could have searched for similar spherical bodies in the cells of other tissues such as the epidermis in orchids and in other plant families; he could have looked out for such specks in pollen cells in orchids or in other plant families. But Franz did none of this and instead shelved his pencil drawing. Additionally, and in contrast to Leeuwenhoek, Franz, never a man of letters, neither wrote a report to be published nor considered sending his pencil drawing to someone else to describe and have the text published. This is how he missed the unique chance of becoming as famous as Leeuwenhoek.

In the race for priority the winner on this subject was Robert Brown. On 1 and 15 November 1831, almost three decades after Franz had first drawn the nuclei in the cells of *Phaius tankervilleae*, Brown's paper 'Observations on the organs and mode of fecundation in Asclepiadeae and Orchideae' was read at a meeting of the Linnean Society in London. The text had already appeared on 30 October 1831 as a privately distributed preprint[4] and was published in March 1833 in the *Transactions of the Linnean Society*, soon reprinted in other journals and translated into German and French. Although the title contains the term 'fecundation', which is a totally different matter, Brown gives – embedded in the text – a report on the discovery of the cell nucleus. Because of Brown's reference to Franz, it is necessary to quote from Brown's paper[5] extensively: 'In each cell of the epidermis of a great part of this family [the orchids], especially of those with membranaceous leaves, a single circular areola, generally somewhat more opake [sic] than the membrane of the cell, is observable. This areola, which is more or less distinctly granular, is slightly convex, and although it seems to be on the surface is in reality covered by the outer lamina of the cell. There is no regularity as to its place in the cell . . . As only one areola belongs to each cell . . . This areola, or nucleus of the cell as perhaps it might be termed, is not confined to the epidermis, being found not only in the pubescence of the surface . . . but in many cases in the parenchyma or internal cells of the tissue . . . The nucleus is manifest also in the tissue of the stigma . . . I may here remark, that I am acquainted with one case of apparent exception to the nucleus being solitary in each utriculus or cell, namely in *Bletia Tankervilliae* [i.e. *Phaius tankervilleae*]. In the utriculi of the stigma of this plant I have generally, though not always, found a second aureola . . . Mr. Bauer has represented the tissue of the stigma in this species of *Bletia*, both before and as he believes after impregnation and in the latter case the utriculi are marked with from one to three areolae of similar appearance'. This last sentence indicates that Brown must have seen Franz's drawing of 1802, which is not surprising: both were fellows of the Royal Society, they had collaborated on *Rafflesia arnoldii* and were fond of research with the microscope (see Chapter 13). In addition, Brown seems to have repeated Franz's observation (see Chapter 13) using the very same plant – *Phaius tankervilleae*. Brown continues in noting 'The nucleus may even be supposed to exist in the pollen of this family [the orchids] . . . This nucleus of the cell is not confined to Orchideae, but is equally manifest in many other Monocotyledonous families; and I have even found it, hitherto however in very few cases, in the epidermis of Dicotyledonous plants, though in this primary division it may perhaps be said to exist in the early stages of the development of pollen'. In order to correlate this text with modern terminology, it may be helpful to add that Brown used 'areola' as a synonym for his new term 'nucleus', 'utriculus' for 'cell', and that his 'lamina' and 'membrane' stand for what we now call cell wall.

Brown seems to have been convinced that the nucleus, which he had searched for in different tissues and different plant families, possessed an important function in the cell, but he did not envisage in 1831 that it would occur in every eukaryotic cell and that it possessed an absolutely indispensable function for each and every one of them. In fact, Brown was certainly not yet convinced about the ubiquity of the nucleus, stating that in dicots it existed in pollen. For reasons unknown Brown refrained from publishing a single drawing of a cell showing a nucleus; this applies both for his preprint and the paper as it appeared in the *Transactions of the Linnean Society*.

Before the publication of Brown's preprint, the first instalment of Franz's *Illustrations of orchidaceous plants* had appeared in November to December 1830 in London (see Chapter 13). Plate 6, a lithograph by Gauci, contains five figures all based on illustrations redrawn by Lindley on the basis of Franz's pencil drawings of 1802. One of them, No. 3

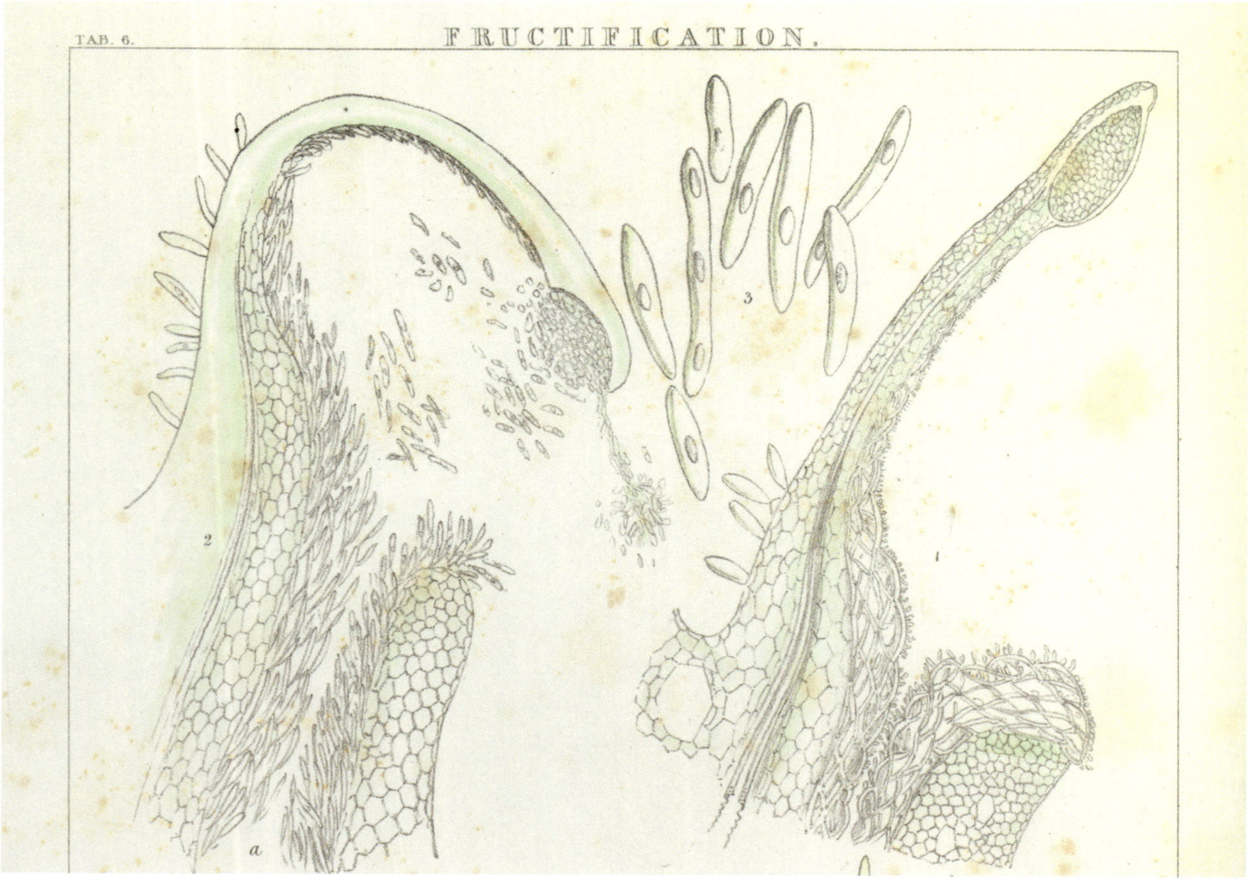

FIG. 14.3. ***Phaius tankervilleae*. Stigma and stigmatic surface. Lithograph by M. Gauci based on a redrawn illustration by J. Lindley which is founded on pencil drawings by Franz Bauer. – Illustrations of orchidaceous plants, London, 1830, t. 6, Fig. 1-3. – London, Natural History Museum, Library and Archives.**

(Fig. 14.3.), is the first published illustration showing plant cells with their nuclei. The respective legend by Lindley reads: 'Some of the same bodies magnified 200 times. They are perfectly transparent, and appeared to be cellules of an oblong to fusiform figure, with one, two or three granular, more opaque greenish yellow specks, looking like young seeds of an Orchis in the midst of loose reticulate testa'. This is a perfect description, with 'bodies' and 'specks' standing for what Brown was to call 'nuclei'. At the same time the legend offers a most convincing comparison of the cell with its nucleus on the one hand and the testa and the embryo in orchids on the other hand. For this chapter only one more figure, No. 4, is relevant, also showing nuclei with Lindley's legend reading, 'Longitudinal section of a portion of the surface of the upper part of the stigmatic canal . . . magnifies 100 times. This is seen under water'. The illustration again shows Franz the precise observer – the cells lining the stigmatic canal are depicted with their nuclei.

This is how it came that Franz lost his priority to Brown. The term 'nucleus', coined by Brown and observed by him in several types of tissue and several plant families, is now inseparably connected with his name and not with the name of Franz. The latter had failed to grasp the relevance of his discovery in 1802 and did not publish what he had observed. Furthermore Lindley's legends of 1830, clearly published before Brown's paper, do not compare favourably with the latter's text. Furthermore they did not contain a new term (see above). In short, Franz was the loser and Brown the winner in this race for a new discovery.

POLLEN DIVERSITY

In palynology (which literally means 'the study of dust', but refers generally today and in our context to the study of pollen), science was the loser not Franz, and there was no winner. The well-known palynologist Roger P. Wodehouse dedicated in his textbook *Pollen grains*[6] no less than eight pages exclusively to the achievements of Franz and wrote, 'He recognized the extent to which [plant] family likenesses are borne by pollen grains and other points which were not clear until half a century later. I believe that if [Franz]

Bauer's work could have been published during his life time, the science of pollen morphology would now [in 1935] be considerably in advance of what it is today'.

Over a period of more than three decades, from August 1802, when the *Investigator* with his brother Ferdinand on board was sailing along the eastern coast of Australia, until March 1836, Franz documented pollen grains of numerous plant species often – though not always – annotated with the scientific name and the date. This material was later bound into a volume labelled 'Drawings by Francis Bauer, F. R. S. Epidermis, Hairs, Pollen Grains, Monstrosities' and is now kept in the Library and Archives of the Natural History Museum. Franz illustrated pollen of more than one hundred and seventy-five species in one hundred and twenty genera and fifty-seven families, belonging to gymnosperms or angiosperms; typically for him there is no verbal annotation. Often squared paper is used as a scale of dimension for the pollen grains (Fig. 14.4.). Wodehouse rightly observed: 'His [Franz's] method of working appears to have been to sketch the grains as circumstances offered opportunity, probably in connection with his illustrations of the plants themselves drawn for other purposes, for there is no order in the arrangement. It was apparently his intention to redraw these with all their details at a later date. This was, however, fulfilled in only a few cases'. One of these was the watercolour of a pollen grain of the tiger lily, *Lilium lancifolium* (Fig. 14.5.), an ornamental lily native in China, Japan and Korea. Franz documents the pollen grains with their net-like surface exuding copious pollenkitt, the adhesive material making pollen grains sticky, although in the case of the lower pollen grain his presentation of pollenkitt is somewhat stylized.

In his pollen studies Franz, the sedentary plant illustrator, preceded Ferdinand, the circumnavigating plant and animal illustrator. However, considering the close collaboration between the two brothers it is likely that back in London in 1805 Ferdinand soon caught up. In his illustration of the showy dryandra, *Banksia formosa*, engraved by Sowerby and published in 1810 by Brown in his paper 'On the natural order of plants called Proteaceae' in the *Transactions of the Linnean Society*, we find a representation of two elliptic pollen grains. In the introduction to this monograph of the Protea family, Brown had written: 'The figure of the pollen has been attended by a few theoretical, but hardly any practical botanists; yet I am inclined to think, not only for its

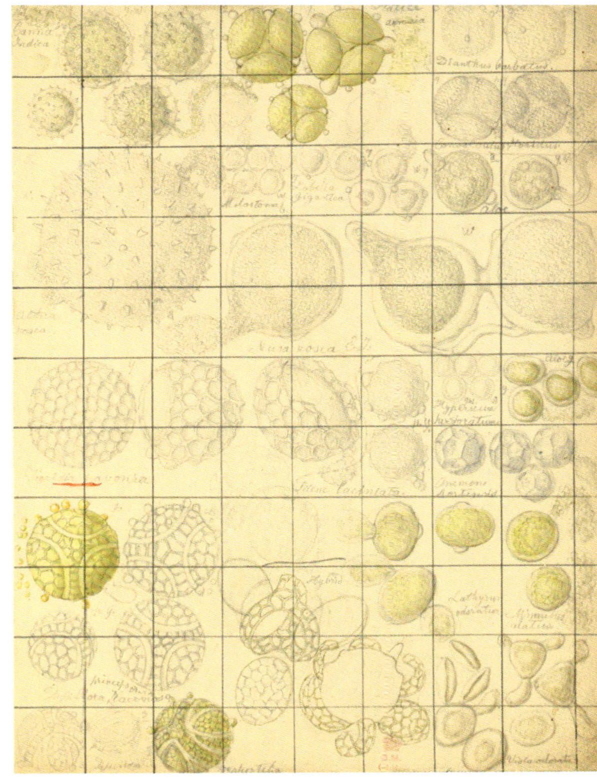

FIG. 14.4. Pollen of various angiosperm species. Watercolour by Franz Bauer, 1800–1833. Paper, watercolour. – London, Natural History Museum, Library and Archives, Drawings of Francis Bauer of Epidermis, Hairs, Pollen Grains, Monstrosities, no. 57.

consideration in this family [i.e. Proteaceae], but in many others, that it [the pollen] may be consulted with advantage in fixing our notions of the limits of genera'. Wodehouse stated that it is quite plain that 'in this case the idea of the importance of the pollen character came from [Franz] Bauer rather than from Brown'. This assertion has no factual basis; however, it is difficult to imagine that Brown, back from the voyage in the *Investigator*, did not learn about Franz's earlier pollen studies, which had begun at the latest at a time when Brown and Ferdinand were still in Australian waters.

In contrast to what has been said by Wodehouse, Franz also prepared representations of pollen diversity in a considerate and well-structured arrangement; one of them, for example, is dedicated to six species of passion flowers,[7] among them *Passiflora holosericea*, which Ferdinand had depicted in his flower piece in oil (see Chapter 11). In a sense this can be regarded as the beginning of comparative palynology. Another presentation is suitable for being printed as a plate in a modern textbook showing pollen grains of six species,[8] among them *Bauera rubioides*, the bigleaf hydrangea,

Hydrangea macrophylla, and the sensitive plant, *Mimosa pudica*, in two different magnifications. Franz's interest in pollen of the pea family, for example species of *Acacia* and *Mimosa*, is commented on by Wodehouse, who observed: 'The 16-celled grains of the acacias in their characteristic geometrical arrangement are shown again and again, but along with them are the exceedingly rare forms of *A.[cacia] ciliata* with 12 cells and of *A. marginata* with 10. Besides these he [Franz] showed the 16-celled *Acacia* grains in most of their irregular arrangements. One wonders, in looking over his sketches, if he understood that the normal 16-celled grain possessed its peculiar-looking arrangement because it is made up of four tetrahedral tetrads. It is likely that he did, for his drawings show that he knew that the grains of *Mimosa*, which is closely related to *Acacia*, occur only in fours – always in tetrahedral tetrads'. As a matter of fact, Franz possessed a marked interest in pollen tetrads, documenting normal and abnormal forms[9] in the non-photosynthetic bird's nest orchid, *Neottia nidus-avis*.

Several finished watercolours by Franz, in particular of orchids (see Chapter 13), also include representations of pollen cells or pollen tetrads. These echo Ferdinand's images of selected species collected on the voyage in the *Investigator* and published in his *Illustrationes Florae Novae Hollandiae*, which also include representations of pollen cells (see Chapter 11). Similarily, only a tiny fraction of Franz's finished watercolours of plant species were published while he lived, in particular in his *Illustrations of orchidaceous plants* and Brown's earlier *Chloris Melvilleana* (see Chapter 13), while his separate collection of pollen drawings remained unpublished (see above).

Another exception is a coloured lithograph published in Curtis's *Botanical Magazine* documenting a flowering specimen of tutukiwi, *Pterostylis banksii*,[10] a terrestrial orchid endemic in New Zealand and first collected by Banks and Solander on the voyage in the *Endeavour*. Franz had recorded a plant cultivated in the Royal Garden at Kew. As a very attentive observer and recorder he had found pollen tetrads in almost all species of orchids studied until that moment, but in the tutukiwi there were single pollen cells. In the coloured copper engraving six are recorded in water, and one is exsiccated, which may show that Franz knew how to extract pollen samples from herbarium specimens. This is Franz's report, found in a note to Allan Cunningham, and published by William Jackson Hooker: 'I have now, on the 2d of May [1832] examined the Pollen Grains [of *Pterostylis banksii*] with Ploessel's grand microscope, and, to my great surprise, have found a total deviation from those of all the hundreds of specimens of Orchideous plants I have yet investigated. These grains [i.e. of the hundreds of specimens], in their ordinary form, consist of three or four-celled corpuscles [tetrads] . . . This [the occurrence of single pollen cells in *P. banksii*] I consider an important circumstance, and could not be detected by Botanists possessed only of glasses of more moderate power'. Interestingly, the copper engraving based on Franz's watercolour has been selected the type of the name *Pterostylis banksii*, since no herbarium specimen has been made from the plant cultivated in the Royal Garden in Kew.[11]

In 1801 Franz had observed and recorded a totally different phenomenon in the sword-leaved helleborine, *Cephalanthera longifolia*, a common and widespread orchid, which he may have remembered from his years in Vienna and its surroundings. In this plant, single pollen grains are entangled by viscin threads holding them together. It is uncertain if Franz understood what he saw and recorded. In any case Lindley's legend published together with Gauci's lithograph based on Lindley's re-drawn illustration, the latter having Franz's watercolour as exemplar, is nonsensical since he wrote 'a group of pollen grains, bursting at the time of fecundation'.[12] Instead swollen viscin threads and pollen grains are represented, which is exactly what one would expect – Franz had put his preparations in a drop of water under his microscope. The crystal-clear photographs of modern scanning electron microscopy showing thin, cable-like viscin threads were as yet in the remote future and not even dreamed of.

The pollen representations of Franz's *Illustrations of orchidaceous plants* (see Chapter 13) had a prominent admirer – Charles Darwin. Like Linnaeus, Darwin needs no introduction. However, it may be helpful to pause for a moment and consider Darwin's interest in orchids in the broader context. In his iconic *On the origin of species*, published in London in November 1859, he had only briefly mentioned evolutionary interactions between insects and orchids, for example by stating, 'Many of our [British] orchidaceous plants absolutely require the visits of moths to remove their pollen-masses and thus fertilize them'. It was only logical for him to explore this topic later in more detail. Field studies and practical investigations developed gradually into experiments and led to a demonstration of the mechanisms of natural selection, as exemplified in the coevolution of insects and orchids. At the same time the evolution of extreme specialization was analyzed. In the long run Darwin's studies open up new areas of pollination research and, more importantly perhaps, reproductive biology. In our context a single aspect matters: like every decent scientist, Darwin condensed and digested information from an extremely broad spectrum of sources: (1) his own observations in the field, in gardens and conservatories; (2) reports gathered from correspondents, field naturalists, middle class ladies and country gentry; and (3) pre-existing records in the vast literature on pollination, insects and orchids. When in 1862, twenty-two years after the death of Franz, Darwin published in London his results in the book *On the various contrivances by which British and foreign orchids are fertilised by insects, and on the good effects of intercrossing*, he referred in footnotes twice to Franz's *Illustrations*. Darwin confirmed that the 'grains [in the sword-leaved helleborine] are tied together by a few weak elastic threads' and referred to the existence of waxy pollen masses cohering by means of elastic threads in the pine pink, *Bletia purpurea*,[13] a neotropical orchid Franz had documented sixty-one years before. Nothing indicated that Darwin had seen Franz's watercolours, then already in the British Museum in Montagu House; instead he explicitly refers to the coloured lithographs in the 'magnificent Illustrations of Orchidaceous Plants'.

Neither before or after Franz has anyone studied in utmost detail the spectacular flower of the bird-of-paradise, *Strelitzia reginae*, from which he had published his exceedingly rare *Strelitzia depicta* (see Chapter 8). Therefore it is no surprise that he also documented the pollen grains, including the cellular chains weaving them together in a manner similar to the viscin threads, although structurally totally different. Again Franz was far ahead of his time; the phenomenon was properly described and published only in 1891, fifty-one years after Franz's death, by Eduard Palla,[14] then Privatdozent at Graz University.

POLLEN STERILITY, POLLEN GERMINATION AND SULPHURIC ACID

A mule is the offspring from a male donkey, *Equus asinus*, and a female horse, *Equus caballus*, a hinney is the offspring from the reciprocal cross – that is, from a male horse and a female donkey. Donkeys and horses belong to separate species with different numbers of chromosomes; horses have 2n = 64, donkeys 2n = 62. Mules and hinnies, both males and females, have normal sexual appetite and a chromosome complement of 2n = 63, but they are sterile. This means that mules and hinnies are unable to produce viable gametes – egg cells and sperm cells – and therefore cannot reproduce. Such sterile hybrids are also known in the plant kingdom and are easily detected by their abnormal pollen grains.

Ferdinand documented the blue passion flower, *Passiflora caerulea*, the red passion flower, *P. racemosa*, and the hybrid (*Passiflora × violacea*) in watercolours of outstanding accuracy (see Chapter 12). Both species have a chromosome complement of 2n = 18, but belong to two different subgenera, the blue passion flower to *Passiflora* subgen. *Passiflora*, the red passion flower to *Passiflora* subgen. *Calopathanthus*.[15] As in mules and hinnies, the hybrid is sterile and does not produce viable pollen. It has been explained that Ferdinand's preliminary drawings were almost certainly made in England in 1824, while the finished version must have been prepared during his final months in Vienna (see Chapter 12). At Kew, Franz also studied pollen of passion flowers (Fig. 14.1. p. 376). Most interestingly his collection of pollen representations contains also an image of a pollen grain annotated 'Passiflora princeps or racemosa', i.e. *Passiflora racemosa*, and of a pollen grain annotated 'hybrid'. The former is a sound pollen grain, the latter is disorganized and aborted, a typical example of the 'bad pollen' of a hybrid. Franz was unable to understand the significance of his finding and, additionally, refrained from publishing what he had seen – another missed chance for fame. At the same time the two drawings prove the close collaboration that Franz and Ferdinand maintained throughout their lives.

Pollen grains are strongly dehydrated before they are released from the anther, which facilitates their transport from one plant to another. Once the grains have landed on the stigma of a receptive carpel, or a cone in gymnosperms, the pollen takes up water and starts to germinate. At that moment at the latest the pollen grain in angiosperms consists of three cells, one of them elongates, forming the pollen tube, which grows towards the ovule, while the other two, the sperm cells, are localized within the first cell. Fecundation takes place when these two cells fuse: one with the egg cell to form the embryo, the second with the central cell to form the endosperm. While Franz was unaware of the famous double fecundation in angiosperms, which was discovered and understood only

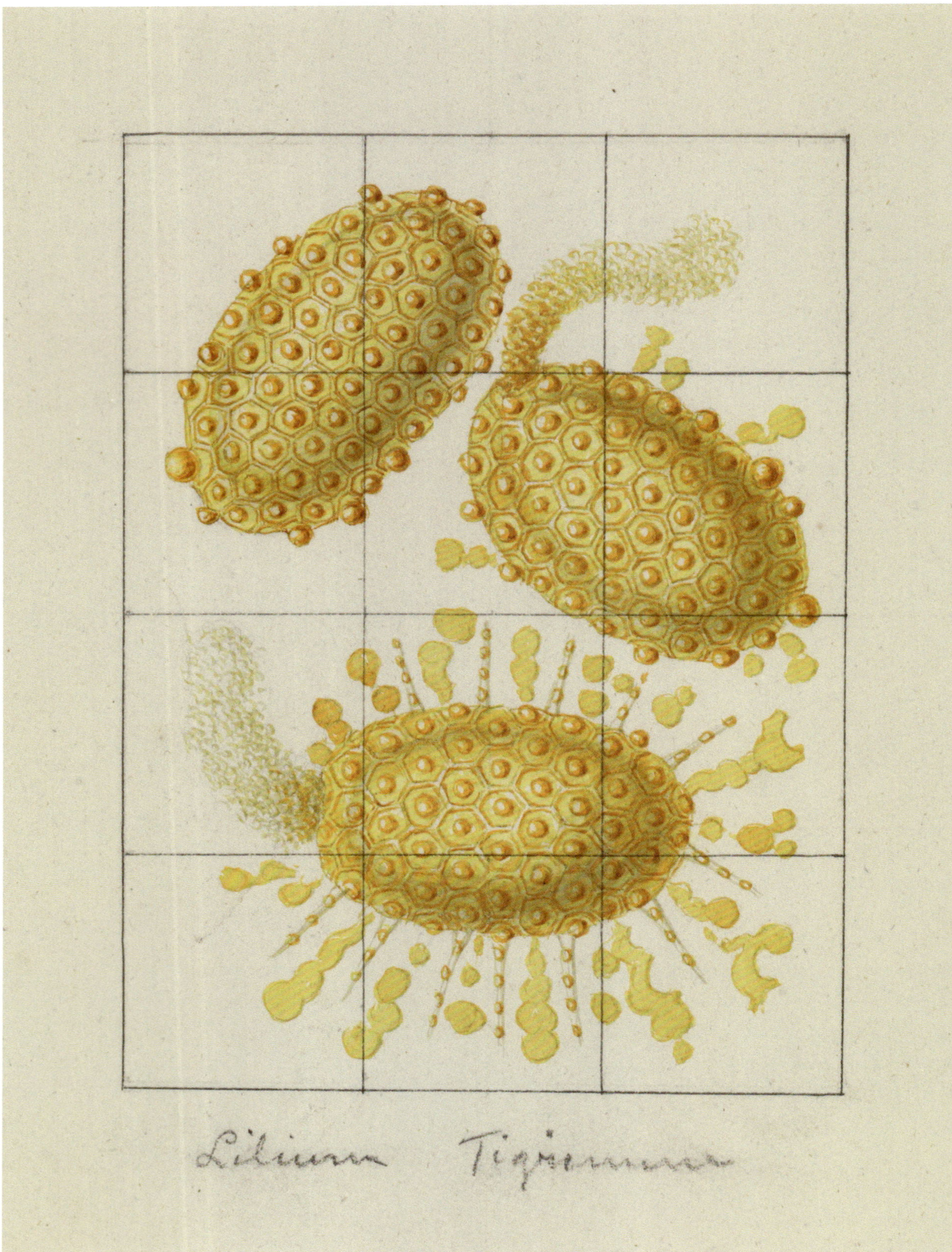

FIG. 14.5. *Lilium lancifolium*. Pollen grains exuding copious pollenkitt. Watercolour by Franz Bauer, undated. Paper, watercolour. – London, Natural History Museum, Library and Archives, Drawings by Francis Bauer of Epidermis, Hairs, Pollen Grains, Monstrosities, no. 59.

decades later, he documented several germinating pollen grains. One of them is annotated 'Kaempferia Roscaeana' (*Kaempferia roscoeana*) but remained undated (Fig. 14.6.). It is unclear if Franz had observed the phenomenon *in vivo*, on the stigma of the plant, and had the germinating pollen subsequently removed to be studied under his microscope, or *in vitro*, by having stimulated the pollen grains to germinate, for example in water or in a sugar solution. Since no date is given, it remains open to speculation whether or not Franz was preceded in the first observation of a germinating pollen grain by Giovanni Battista Amici.[16] The latter, an extremely versatile microscopist and naturalist, lived at that time in Modena, then reigned over by Archduke Franz, as Duke of Modena and Reggio Franz IV, a cousin of Francis I, Emperor of Austria. In 1822 Amici had observed the germination of pollen in pigweed, *Portulaca oleracea*, and published a report in his paper 'Osservazioni microscopiche sopra varie piante', which appeared in Modena in the *Memorie di Matematica e di scienze fisiche della Società Italiana delle Scienze*. In any case the winner in this case was Amici, since Franz once more did not publish what he had seen and documented in his drawing.

Acetolysis, the heating and boiling of pollen in a mixture of sulphuric acid and anhydrous acetic acid, is a standard technique in palynology and paleobotany. Its purpose is to destroy all non-sporopollenin substances completely, so that the acetolyzed pollen grains are cleared and thereby provide excellent topographical information, since the outer diagnostic cell wall of pollen grains consist exclusively of sporopollenin. The method is invariably associated with Swedish palynologist Gunnar Erdman, first proposed by him in his textbook *An introduction to pollen analysis* published in 1943 in Uppsala, which became instrumental for the development of this branch of botany. It came therefore as a great surprise to find next to a

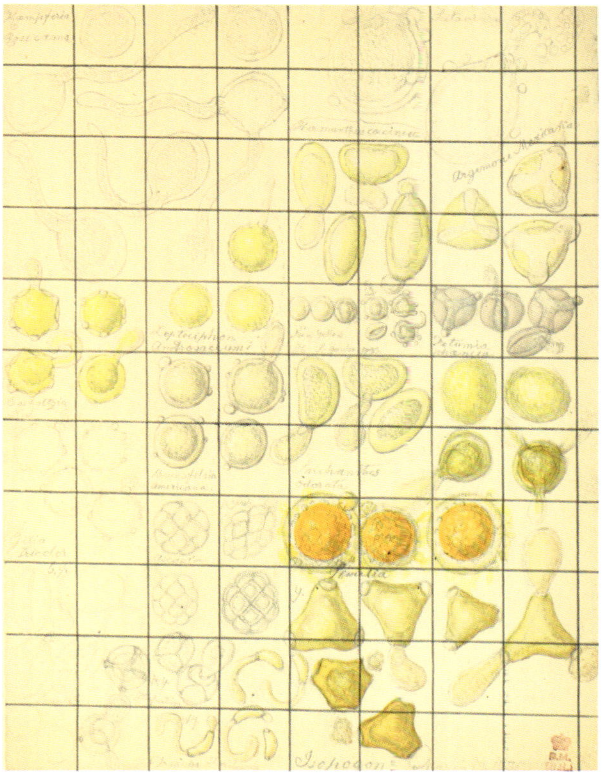

FIG. 14.6. Pollen of various angiosperm species. Upper row: *Kaempferia roscoeana*. Germinating pollen cells. Watercolour by Franz Bauer, undated. Paper, watercolour. – London, Natural History Museum, Library and Archives, Drawings by Francis Bauer of Epidermis, Hairs, Pollen Grains, Monstrosities, no. 75.

drawing of a pollen grain of an undetermined pelargonium species a brief note in Franz's microscopic hand, reading 'June 1833 in sulphuric acid'.[17] Therefore we have to imagine Franz, then aged 75, using this aggressive acid to clear his pollen samples before making a drawing. The little note is the only indication that he preceded Erdtman in this preparatory method by no less than one hundred and ten years. Again Franz had refrained from publishing his discovery.

ENDNOTES

1 Unless otherwise stated, this chapter is based on Baker (1951) and Harris (1999).
2 www.lensonleeuwenhoek.net (accessed 21 x 2014)
3 Mabberley (2011)
4 Mabberley (1985)
5 Brown (1831, 1833)
6 Wodehouse (1935)
7 London, Natural History Museum, Library and Archives, Drawings by Francis Bauer. Epidermis, Hairs, Pollen Grains, Monstrosities, f. 62 r.
8 London, Natural History Museum, Library and Archives, Drawings by Francis Bauer. Epidermis, Hairs, Pollen Grains, Monstrosities, f. 53 r.
9 London, Natural History Museum, Library and Archives, Drawings by Francis Bauer, Epidermis, Hairs, Pollen Grains, Monstrosities, f. 102 r.
10 Hooker (1832)
11 Jones, Clements & Molloy (2002)
12 Bauer (1830–1838), Genera, t. 19, 20
13 Bauer (1830–1838), Fructification, t. 8
14 Palla (1891)
15 De Melo, Cervi & Guerra (2001)
16 www.gbamici.sns.it (accessed 22 x 2014)
17 London, Natural History Museum, Library and Archives, Drawings by Francis Bauer, Epidermis, Hairs, Pollen Grains, Monstrosities, f. 63 r.

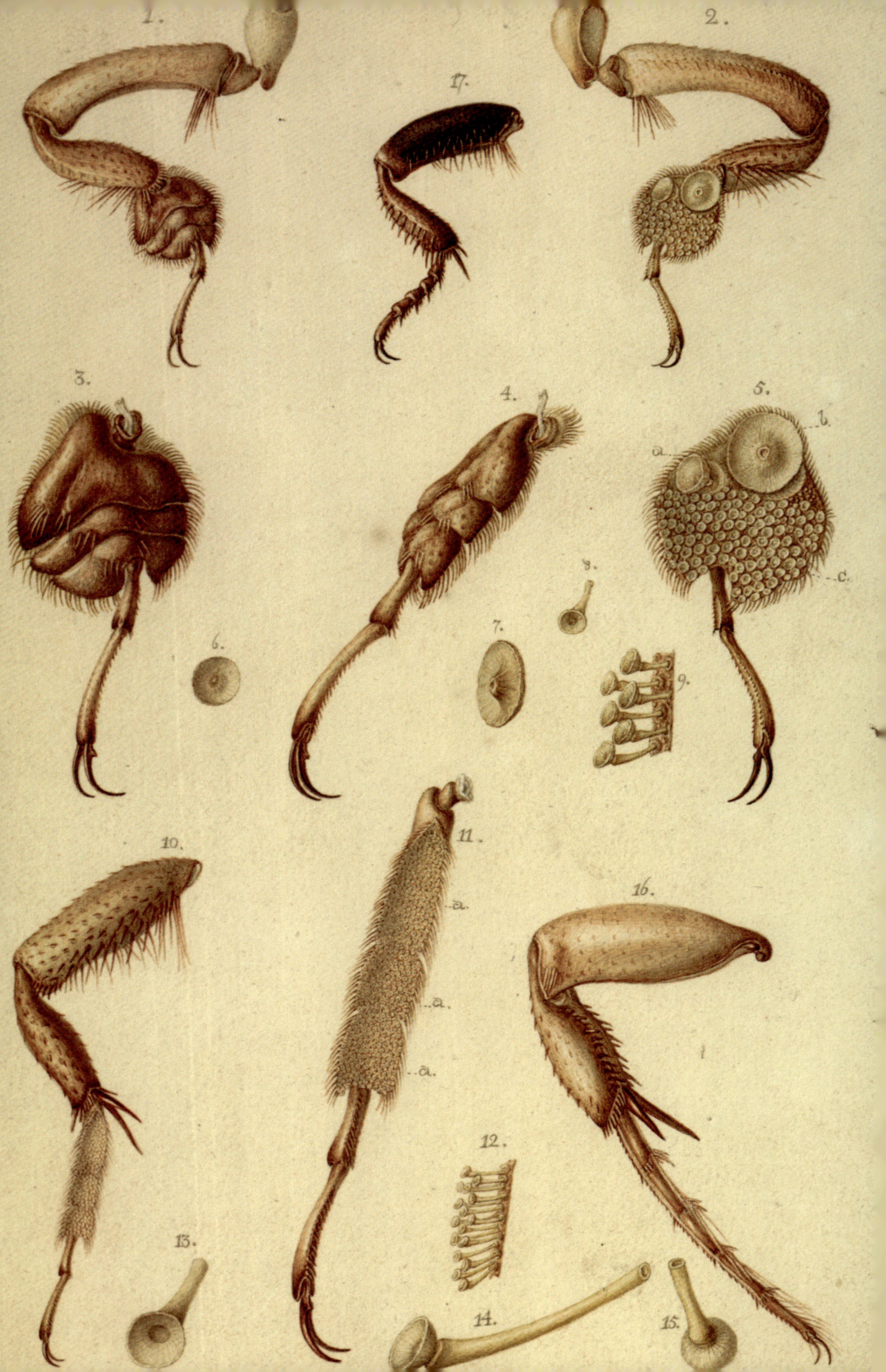
1.
2.
17.
3.
4.
5.
a
b
c
6.
7.
8.
9.
10.
11.
a.
a.
a.
16.
12.
13.
14.
15.

CHAPTER 15

Franz Bauer in the circle of Sir Everard Home in London and Kew

SIR EVERARD HOME,
BT, VPRS, FSA, FRCS, FRCP, FLS

ACADEMIES OF SCIENCE are headed by elected presidents with elected or appointed vice presidents acting as their deputies. The Royal Society is no exception to this general rule, subscribing to the concept of appointed vice presidents.[1] During his forty-two year tenure of the presidency, Banks appointed several vice presidents, among them in 1814 his personal medical attendant Sir Everard Home, 1st Baronet, who had been elected Fellow of the Royal Society in 1787. The latter remained in this appointed position until 1830, acting as deputy for Banks and his successors in the chair – Wollaston (see Chapter 10), Sir Humphry Davy, 1st Baronet, and Davies Gilbert, né Giddy. Starting in c.1815, Home became Franz's most important commissioner and brought him into contact with an entire range of topics that were totally new to him – human anatomy, embryology, physiology and pathology plus the comparative anatomy of animals: vertebrates as well as invertebrates. With Franz being elected Fellow of the Royal Society in 1821, all this took the form of collaboration between two fellows, with Home the author of the text and Franz the person responsible for the illustrations. Furthermore, Home (who had a lease of the Manor House in Ham, Surrey) and Franz (based in Kew) were near neighbours, so communication may have been at least partly by word of mouth; a few letters by Franz, not yet submitted to analysis, are kept in the General Library of the Natural History Museum. Franz could demonstrate his experience in microscopic work to one of his vice presidents who was two years his senior and a key figure in the medical world of London, an aspect not to be forgotten by a man 56 years old when the collaboration started. Testing the power of his expensive microscopes (see Chapter 17) on non-botanical specimens may have been another motive for Franz, who indirectly moved back to the medical world, which he knew from his association with the hospital of the Brothers of Mercy in his early years in Feldsberg (see Chapter 2).

Since the results of the joint studies undertaken with Home appeared quickly in a long list of publications,[2] in the *Philosophical Transactions of the Royal Society*, in Home's *Lectures on comparative anatomy* and his *Supplement to the foregoing lectures on comparative anatomy* (see below), it seems advisable to provide some background information and sketch briefly the life and work of Franz's commissioner. There are two further reasons to do this: firstly, Home seems to have suggested the establishment of a botanical museum at Kew and the appointment of Franz as its librarian at a salary of £200 per year[3] to George IV, although the whole plan came to nothing (see Epilogue); secondly, the total number of copper engravings based on Franz's drawings made for Home in the years 1815 to 1830 eclipses the total number of copper engravings based on Franz's botanical and mycological drawings which appeared up to 1830.

FIG. 15.1. Great diving beetle, *Dytiscus marginalis* (♂). Front, middle and hinder legs. Watercolour by Franz Bauer, signed, 1816. Board, watercolour. – London, Royal Society, Archives, Image No. RS 8635

Everard Home was born on 6 May 1756 in Hull, the fourth son of the surgeon Robert Boyce Home of Greenlaw Castle, Berwickshire and his wife Mary née Hutchinson.[4] As a boy he was sent as one of the King's scholars to Westminster School in London and came to know his future brother-in-law while the latter was still courting Everard's oldest sister Anne.[5] His name was John Hunter, the celebrated anatomist and surgeon,[6] Fellow of the Royal Society and the only figure in our story whose remains were later transferred to Westminster Abbey, the most prestigious burying place in Britain and pantheon of its elite. Home reported in an autobiographical text to have won a scholarship for Trinity College, Cambridge where he could have met John Hawkins (see Chapter 5), but surprisingly he resigned and became a resident pupil of his famous brother-in-law, later based on Leicester Square in London. On completing his apprenticeship Home was admitted to membership of the Company of Surgeons, subsequently the Royal College of Surgeons in London, in 1778 and entered the naval medical service. First posted to the naval hospital in Plymouth and later as staff surgeon to Jamaica, he returned to London where he became Hunter's assistant and, with him as one of his sponsors, was elected Fellow of the Royal Society. Even before that Home had started to publish in the *Philosophical Transactions*, his first paper appearing in 1785, on a marine animal thrown up by a hurricane on the coast of Barbados; many more were to follow, among them his paper on the anatomy of the koala[7] (see Chapter 10). Having obtained a permanent position as lecturer in anatomy at St George's Hospital in London, where his brother-in-law was surgeon, Home married and moved out of Hunter's house. Hunter hired as his new amanuensis, laboratory assistant and illustrator William Clift from Cornwall, who would also take care of the very considerable collection of specimens and preparations which had accumulated at Leicester Square. These comprised not only the conventional contents of an anatomical and pathological cabinet of a well-established surgeon but also a plethora of other objects, mainly zoological, among them numerous items donated by Banks – when the latter divided his non-botanical collections, half of them went to the British Museum and the other half to Hunter.

When Hunter died in 1793, Home became one of his co-executors and effectively the scientific heir of the great surgeon. It was Home's duty to fulfill two tasks of his brother-in-law: to open his body to perform the autopsy and to take care of his scientific collections. The latter was a much more tedious matter, but in the end Parliament voted on 13 June 1799 to afford £15,000 for the purchase with the custody offered and accepted by the Company of Surgeons. The money went to Hunter's widow, a London socialite and minor poetess, best known perhaps for the texts of eight songs to which Haydn wrote the music (Hob. XXVI a: 25–32) during his first stay in London in 1794–1795. During the years of uncertainty for the Hunterian collection Clift had continued to work at Leicester Square preparing and arranging specimens and taking copies of some of his former master's manuscripts. As Hunter's scientific heir it was easy for Home to become his apostle as well: one of his first jobs was seeing his brother-in-law's *Treatise on the blood, inflammation and gun-shot wounds*[8] through the press. Today this is considered a landmark medical text, because Hunter described in it the discovery of the erythrocyte sedimentation rate (ESR) and gave a very correct interpretation of its diagnostic value.[9] This common laboratory test is based on the increased sedimentation of red blood cells observed in vitro in many diseases, in particular those with an inflammatory component, and continues to be used in everyday clinical routine. Although Hunter did not provide a detailed timescale for the standard measurement of the ESR, he was able to correlate an increase in ESR with various conditions, such as local inflammation, 'inflammatory fevers' and 'symptomatic fevers'.

Home's rise in the medical hierarchy continued. In 1801 he was appointed to the Court of Assistants (later the Council) of the College of Surgeons and in 1807 received the prestigious Copley Medal of the Royal Society 'for his various Papers on Anatomy and Physiology printed in the Philosophical Transactions'. In the following year Home was appointed sergeant surgeon to George III, shortly before the advisors of the unfortunate monarch were obliged to declare a regency in the name of his eldest son, the future George IV. On 31 May 1810 Home was called to operate on a head wound of Ernest Augustus, Duke of Cumberland (see Chapter 17), removing a bone fragment that was pressing on the exposed brain. While the exact background of the attack causing the wound was never clarified – with the suspicion that Ernest August had killed his servant Joseph Sellis in the fight, although the inquest declared that the latter had committed suicide[10] – the successful operation endeared Home to the Royal Family. This led to an 'augmentation of honour' to Home's arms dated 23 November 1812 'in the Name and on the behalf of His Majesty' signed by George, Prince Regent.

Elected Master of the Royal College of Surgeons in 1813, its museum, now the Hunterian Museum, was opened on 18 May 1813; six weeks later Home was made a baronet, accurately echoing Banks's earlier rise to the same rank. In that year the Prince Regent visited the newly opened museum, and Home, together with Matthew Baillie, arranged for the institution of a Hunterian Oration, to be delivered at the Royal College of Surgeons on the birthday of Hunter. Needless to say, Home was the first orator. The Hunterian Museum quickly attracted several prominent visitors, among them the Duchess Ekaterina Pavlovna, a sister of Alexander I, who later was to accompany her brother to the Congress in Vienna and was to die as Queen of Württemberg. Karl Joseph von Hamel (Josif Hristianovic Gamel), a member of the Imperial Academy of Sciences in St Petersburg, was another visitor, later a collector of the correspondence of Franz (see Chapter 16). By that time Home was a busy man with a large practice and in demand as a lecturer, giving the Croonian Lecture at the Royal Society no less than twenty times, a record in its long history.

Banks and Home were now old friends and boon companions, an association developed against their shared interest in natural history, a field to which the latter had contributed considerably. Indeed a steady stream of papers from Home's pen had appeared in the *Philosophical Transactions*, largely illustrated with copper engravings based on Clift's drawings. It was Banks who recommended Franz to Home and the latter's first commission had a special subject – the feet of the bluebottle fly, *Calliphora vomitoria*, a dipterous insect similar to the common housefly, *Musca domestica*, already accurately depicted by Franz on the title page of the copy of *Selectarum stirpium americanarum historia* painted for Jacquin in Vienna (see Chapter 3). The result of this collaboration was so very satisfying for both Home and Franz that they continued along this line for years. The division of work between Clift and Franz was more or less straightforward, with the former engaged mainly with macroscopic work, the latter mainly with microscopic work; only the toes of the gecko, *Gekko* spec., (see below) were, surprisingly, documented by both. Occasionally Home noted the superiority of Franz's work in comparison to that of Clift.

STRUCTURE AND FUNCTION, OR THE AGGRESSIVE BEHAVIOUR OF THE MALE GREAT DIVING BEETLE

Organisms possess structures which help them to easily perform a certain function or many functions within the structures: for example, the fins of a fish help the fish to propel itself through the water. Home was interested to find out 'the mechanisms by which the feet are fitted for supporting the weight of the animal against gravity'[11] and focused his attention on the gecko and on insects. Franz used his microscope to study the feet of the blue bottle fly, the fever fly, *Dilophus febrilis*, and a species of sawfly, *Cimbex* spec., while Home observed the function of the tarsi in the living animals and described their structure. For example he noted that on the feet of the bluebottle fly 'the suckers are two in number . . . they are connected to the last joint of the toe, immediately under the root of the claw, and have a narrow infundibular neck attached to the toe, which has the power of motion in every direction'.

The great diving beetle, *Dytiscus marginalis*, a voracious predator native in Europe and the northern part of Asia, attracted the special attention of Home because of the pronounced sexual dimorphism characteristic of this species. On the first pair of legs the males possess a small number of large suckers (Fig. 15.1.), and on the second pair of legs many distinctly smaller suckers, whereas the females totally lack suckers. Franz documented the legs showing these structures accurately and from both sides as well as considerably enlarged. Very correctly Home observed that the suckers 'are used to retain the female in the embrace of the male'. Nothing indicates that Home had observed underwater copulation in the great diving beetle, but his interpretation is correct. As we now know, the males of this species use violence and harassment to force the females to copulation: with his suction cup structures a male grabs a female, attaching himself to her dorsal surface with her beneath him, which enables him, and only him, to get access to oxygen at the surface of the water. To get the female to submit, the male shakes her violently and keeps her submerged. Unable to get air, the female submits to the male's advances in order to avoid drowning. Once the male is attached, copulation can occur, but males continue to guard females for several hours, probably to see that the eggs fertilized with their sperm are actually laid. For good reason this form of aggressive mating behaviour is called sexual coercion, a widespread phenomenon among animals, which helps the males to acquire mates and at the same time offers the females a chance to test the stamina of their partners.

Following the philosophy of 'structure and function', then fashionable, Home also compared the structure of the hind flipper of the walrus, *Odobenus rosmarus*, a social marine mammal living in the Arctic and sub-Arctic seas, to that of the foot of a fly and that of the human hand.[12] Franz provided only images of the left hind flipper with 'the palms upon the ground in the expanded and contracted state'; they were combined in a single plate, while the remaining illustrations of Home's contribution have been prepared by other illustrators.

THE PHYSICAL BASIS OF HUMAN SEX AND REPRODUCTION

Even for a prominent surgeon like Home sex and reproduction were still largely veiled in mystery. Among several other aspects Home was fascinated by the erectile structures of the human body – the penis and the clitoris – and had them studied in detail. This took the form of collaboration with Franz and resulted in the publication of Home's paper 'Observations on the Human Uretha, showing its internal structure, as it appeared in the Microscope of F. Bauer, Esq.'[13] which appeared in 1820, illustrated by six copper engravings based on watercolours by Franz. From the text it appears that it was Franz who made the observations based on Home's instructions, wrote the protocol and produced the graphical record, while Home wrote the text of the paper and saw the manuscript and the watercolours through the process of publication. In the introduction Home acknowledged Franz's share in the work by writing 'For the knowledge I have now acquired, which it [sic] is the object of this paper to communicate to the Society, we [Home] are indebted to Mr. [Franz] Bauer, who has given the most patient and laborious attention to the examination of these parts, magnified in different ways, and has faithfully represented them as they appear to his view'. In this rather short paper Home mentioned the name of Franz no less than four times, for example by stating 'It was found by Mr. Bauer . . . that the cellular structures of the corpora cavenosa [erectile tissues of the penis, excluding the corpus spongiosum; see below] is made up of an infinite number of thin membraneous plates, exceedingly elastic, so connected to form a kind of trellis work . . .'.

In order to enable Franz to observe these structures, rather crude methods had been used; one of these was injecting the arteries of the penis with quicksilver. Later the preparation was left to dry and then placed into turpentine. When a section of the rigid, turgescent penis was made, the quicksilver ran out and Franz made his watercolour documenting the trellis work which had become visible. Neither the watercolour nor the preparation survived – the former did not end up in the archives of the Royal Society in London, where many of Franz's watercolours published in the *Philosophical Transactions* are conserved today, and is currently untraceable; the latter, initially deposited in the Hunterian Museum, was destroyed when during the Second World War the blitz hit the building of the Royal College of Surgeons. Unfortunately Home's text

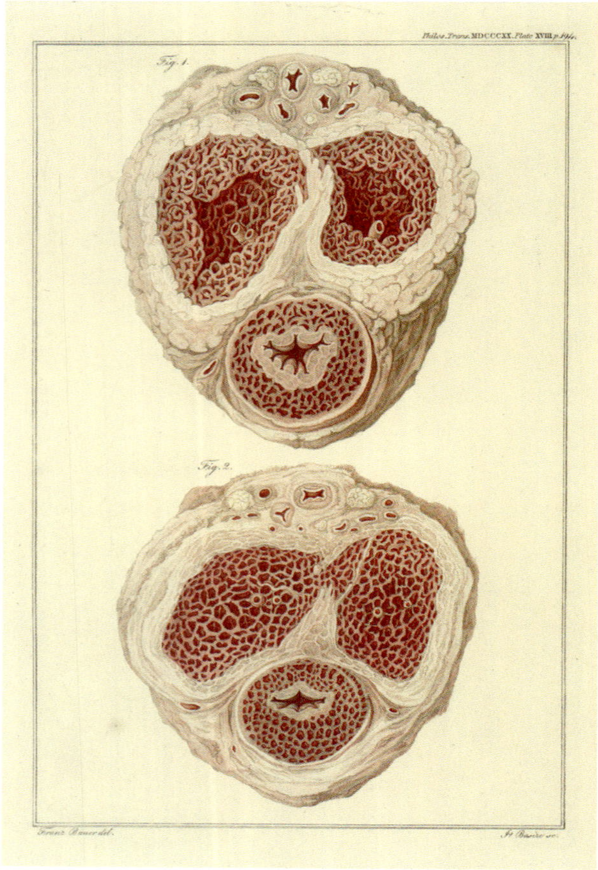

FIG. 15.2. *Homo sapiens* (♂). Penis in cross section, anatomical preparation. Colour print by P. Basire based on watercolour by Franz Bauer, 1820. – Göttingen, Niedersächsische Staats- und Universitätsbibliothek, Handschriftenabteilung, gr. 2° MS Natur. Hist. 94, XVIII, f. 53.

often offers insufficient detail on the way the various preparations were treated. In the present case, for example, the coagulated blood was still in the interstices since in the moment of death the penis was 'found filled naturally, by the blood that had been retained there' (Fig. 15.2.). Home described penile erection correctly: 'When the nerves that supply the corpora cavenosa and corpus spongiosum [the erectile tissue surrounding the uretha and forming the head of the penis] (and they [the nerves] are very numerous) are not affected by the state of mind of the individual, the blood is not carried into this trellis work, but returns by the veins, as in other parts of the body; but when the mind receives such impressions as excite these nerves, the small arteries have their action so much increased, as not only to pour blood into the interstices, and the open space which they inclose[sic], but to distend them, overcoming the elastic power that, under ordinary circumstances, keeps them collapsed'. Needless to say Home possessed not the slightest idea of the role played by nitric oxide in this process. No comparable description is given on the erection of the clitoris, which Franz had also recorded. The respective print appeared in the *Supplement to the foregoing lectures on comparative anatomy* (see below) with a meagre legend appended, mentioning the 'terminal nerves of the papillae', which in the absence of adequate fixing and staining procedures Franz, however, could not have seen.

A good example of the division of work between the surgeon and the illustrator is a case described in Home's paper 'On the passage of the ovum from the ovarium to the uterus in women'.[14] A servant maid, aged 21, is reported to have had sexual intercourse on 7 January 1817 and died eight days later, and on post-mortem was found to be pregnant. Assisted by Clift, Home opened the cavity of the uterus and found a blastocyst (which Home called 'ovum') 'concealed among the long fibres of the coagulated lymph . . . and was brought to view by separating them with the point of a needle'. Since the blastocyst was 'so very small as to admit of dispute whether it was one or not, I [Home] carried it immediately to Kew to Mr. [Franz] Bauer, who after examining it, said it looked like the egg of an insect' – which is indeed a very pertinent comparison. After some further discussion by Home a lengthy text – taken verbatim from Franz's protocol and marked with quotation marks – reads: 'On closely examining the subject under the microscope, I [Franz] found it consisted of membrane, which, considering the minuteness of the subject, is of considerable thickness and consistence, very little transparent, quite smooth, and milky white, forming a kind of bag or pounch of an irregular oval shape not quite 19/200 parts of an inch in length, and at its middle about 9/200 parts of an inch broad; on one side it has as elevated ridge or large fold . . . along the whole length . . . but has no appearance of being in torn, the edges of the membrane being smoothly rolled upwards, which gives it much the shape of a little shell of the genus *Voluta*'. Exactly this stage of the blastocyst was illustrated by Franz (Fig. 15.3.); he added two enlarged dorsal and lateral views, plus the result of his manipulations with the needle. This was a sensational finding which created astonishment among Home's fellow embryologists,[15] for example Alfred Armand Louis Marie Velpeau, then surgeon in various hospitals in Paris and later a professor at Paris University, member of the Académie de Médicine and the Académie des Sciences.

FIG. 15.3. *Homo sapiens*. Blastocyst, second week. Pen-and-ink drawing by Franz Bauer, signed, c.1817. Paper, pen and ink. – London, Royal Society, Archives, Image No. RS 8651.

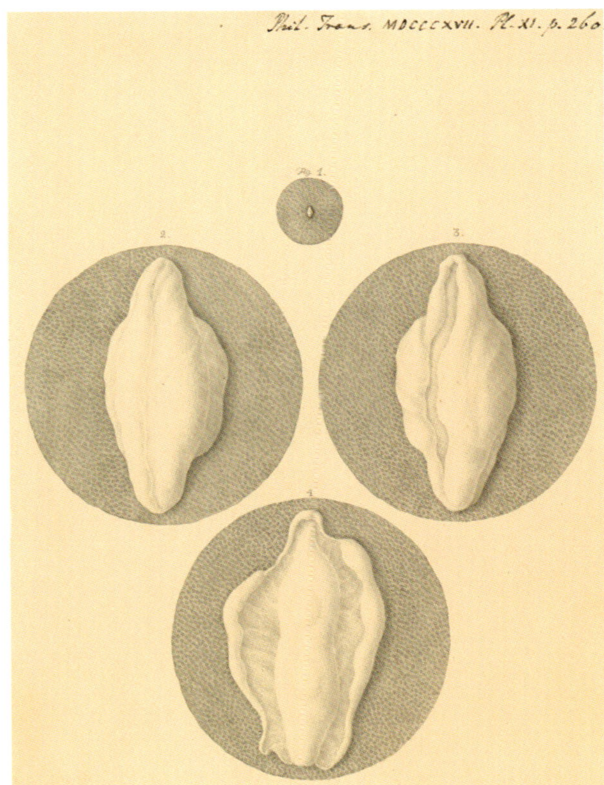

With the aid of a fine camel hair pencil Franz laid the structure open and subsequently 'attempted . . . to extract the corpuscules, but on piercing, with the point of a very small needle, the upper extremity, a thick slimy matter, like honey, came out . . . so that I [Franz] could not further proceed; and fearful of spoiling the whole, I gave up the attempt, and left the subject on the glass to dry'. Clearly Franz was not the father of embryology, but he was critically observing and describing what he had seen in the microscope or in the words of Home, 'had it [the blastocyst] not been brought under the eye of Mr. [Franz] Bauer, the only person, I may say, in this or any other country, who could correctly apply to it the power of the microscope, as to display its organization; and so to delineate what he saw as to convey distinct notions that it was the first rudiment of a child'. At the same time Home's text shows the basic dilemma: whereas Franz's drawings and description are totally clear, although very primitive by modern standards of human embryology, Home was totally wrong in several of his interpretations: what he regarded as the ovum was in fact a blastocyst at the beginning of the second week after fertilisation and Home's statement that menstruation is 'totally unconnected with the formation of the ovum, the process of leaving the ovarium, or its impregnation' turns out to be very wrong. However, in his ignorance of the cyclic nature of ovulation and menstruation as well as their interrelationship Home was not alone. It was to take several more decades until this hormone-driven cyclic process was completely understood. The egg in mammals, first found in a dog, was only discovered ten years later in 1827 by Karl Ernst von Baer, then professor of zoology and anatomy at Königsberg University (now Kaliningrad, Russia). He was to report his famous finding in the Latin letter *De ovi mammalium et hominis genesi* printed in Leipzig, which was directed to the Imperial Academy of Sciences in St Petersburg. In addition to the blastocyst, Franz had also documented in a separate watercolour its precise position – on the posterior wall of the uterus, as is the standard for implantation.

FIG. 15.4. *Homo sapiens* (♀). Ectopic pregnancy, fourth month. Uterus with fallopian tubes and ovaries. Anatomical preparation. Watercolour by Franz Bauer, signed, 1819. Board, watercolours. – London, Royal Society, Archives, Image No. RS 8685.

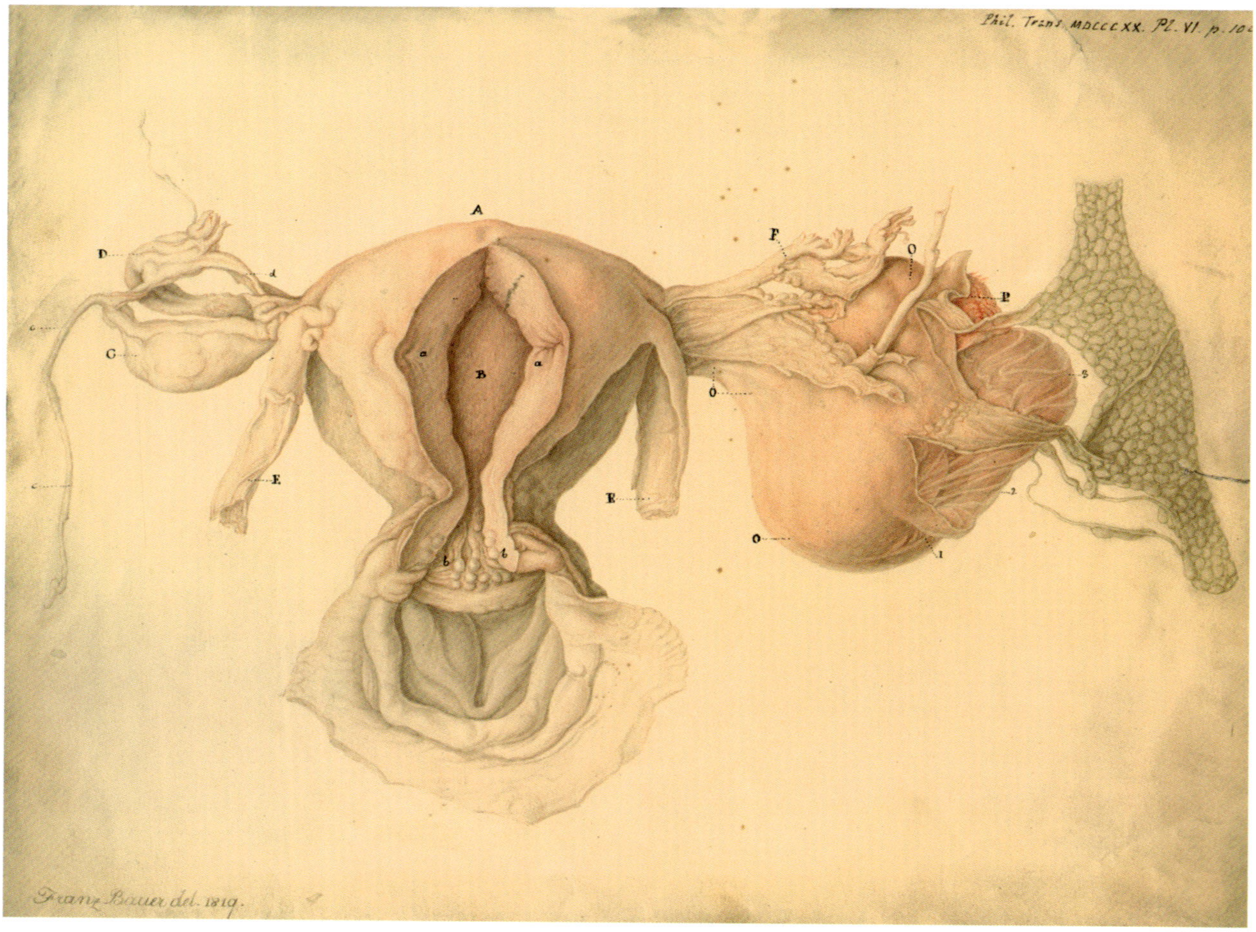

Two years later, in June 1819, Granville (see Chapter 13) wrote a letter to Home on a case of ectopic gravity,[16] which was subsequently published in the *Philosophical Transactions*. Franz must have been asked to produce two watercolours to document the complication: both survive in the archives of the Royal Society. A lady aged 39 years, the mother of seven children, had menstruated regularly until December 1817, when she again became pregnant. 'After having experienced various and severe sufferings' she died on 9 June 1818 and after post-mortem a foetus was found in one of the ovaries. Granville reported, 'I proceeded to detach the parts from the pelvis, and brought them to that state of demonstration, which Mr. [Franz] Bauer has so beautifully represented in the drawing' (Fig. 15.4.). It was on Franz's explicit suggestion that the dissection of the pathologically enlarged ovary was further continued by Granville to show amnion, chorion and placenta, which were again documented by Franz (Fig. 15.5.).

FIG. 15.5. *Homo sapiens* (♀). Ectopic pregnancy. Foetus in ovary. Anatomical preparation. Watercolour by Franz Bauer, signed, 1819. Board, watercolours. – London, Royal Society, Archives, Image No. RS 8686.

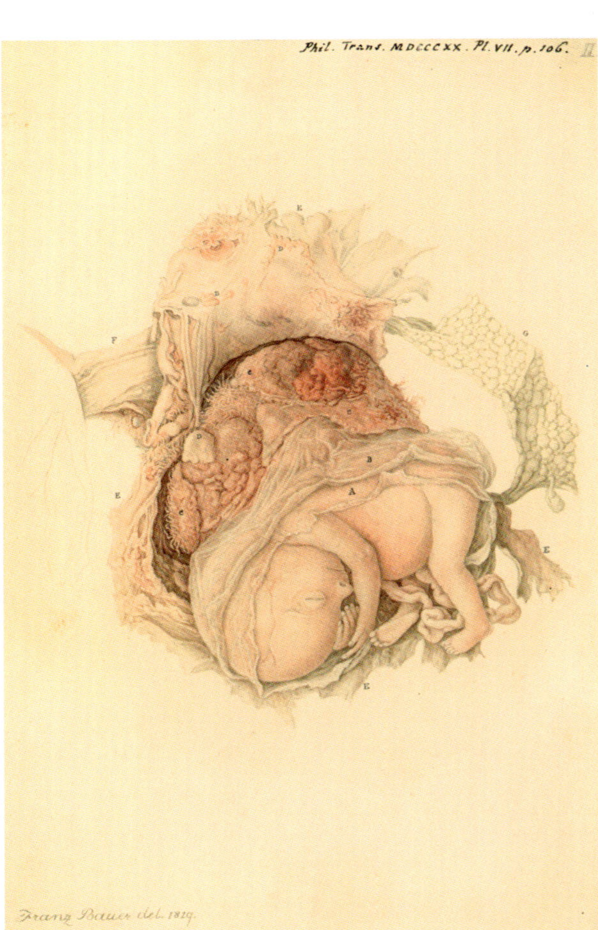

BLOOD, ANEURISMS AND TUMOURS

Blood coagulation is the process by which blood changes from a liquid to a gel, which as a rule leads to the cessation of blood loss from a damaged vessel. This was the topic of the Croonian Lecture delivered by Home to the Royal Society on 20 November 1817, later published in the *Philosophical Transactions*[17] with a subsequent addition.[18] To the latter Franz contributed several watercolours, which were used as exemplars for copper engravings in order to illustrate Home's text. In the first contribution Home praised Franz, who clearly had again done a substantial portion of the observations, with the text offering an accurate analysis of the work done by a microscopist 'His [Franz's] whole life . . . has been employed in investigations of a similar nature in plants, observing first the natural appearances, and then magnifying them in different degrees, and comparing, with the nicest discrimination, what was exhibited by one magnifying power, with what was shown by that immediately above it, and, where they did not exactly correspond, employing the whole energies of his mind, with a patient labour, almost beyond what is natural, in ascertaining the cause of the deception which must in one of them have taken place'. By modern standards the experiment conducted by Home was naïve: he wounded the skin of one of his arms with the point of a lancet, collected a drop of blood in a watch glass and placed it in the field of Franz's microscope. Here is an extract from Home's text: 'The first thing that happened, was the formation of a film upon the surface, that part beginning to coagulate sooner than the rest. In about five minutes, something was seen to be disengaged in different parts of the coagulum'. Franz documented the blood drop from Home's arm in the moment of coagulating (Fig. 15.6.); from the accompanying legend it becomes evident what that 'something' was believed to be – 'carbonic acid gas' (carbon dioxide). This is another correct observation, since venous blood had been extracted. However, other, clearly erroneous observations have to be understood against the background of the time. In the case of other observations, for example on pus,[19] both Home and Franz seem to have been quite at a loss as to what they saw and what Franz recorded in his watercolour and Home described in his text.

An aneurysm is a localized dilatation in the wall of a blood vessel, which results in a blood-filled balloon-

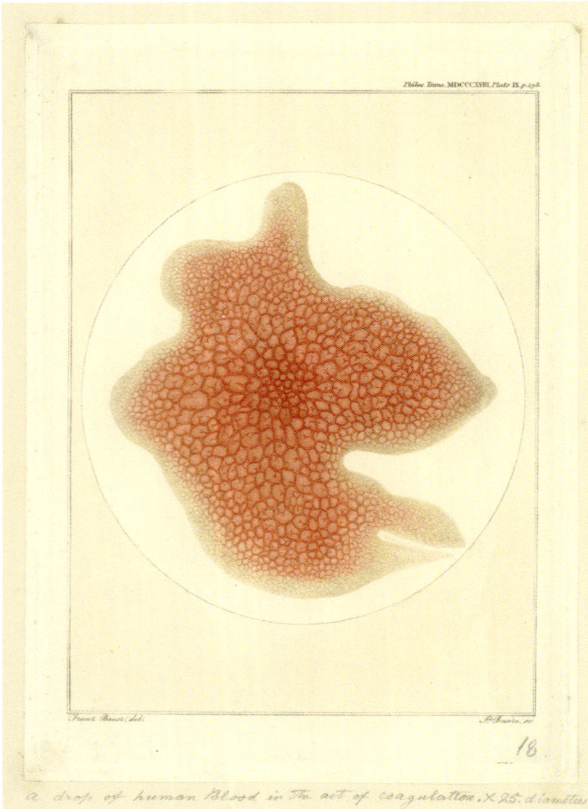

FIG. 15.6. ***Homo sapiens*. Blood agglutinating. Colour print by P. Basire based on watercolour by Franz Bauer, 1818. – Göttingen, Niedersächsische Staats- und Universitätsbibliothek, Handschriftenabteilung, gr. 2° MS Natur. Hist. 94, XVIII, f. 18.**

like bulge. The condition can occur in any such vessel, common examples being the circle of Willis in the brain, the aorta and the renal and popliteal arteries. With aneurysms increasing in size, the risk of rupture increases. As a rule, a ruptured aneurysm leads to excessive internal bleeding and decreased blood volume, often resulting in immediate death. The popliteal artery is located behind the knee. When affected by an aneurysm the prognosis is absolutely disastrous, leading either to amputation because of blood clots building up to the point where the blood flow down the leg is shut completely or to death when the condition is left unattended. Popliteal aneurysms are rare, mainly atherosclerotic in origin and tend to occur in older men with significant co-morbidity; nevertheless they have been a topic for surgeons since antiquity.[20] One approach was the proximal ligation of the aneurysm, i.e. a ligation of the diseased artery nearer to the heart, which leads to collateral circulation in the leg thereby making amputation unnecessary. Hunter was not the first to carry out proximal ligation, but he did so quite remotely from the aneurysm, possibly provoked by his desire to deal with normal, rather than diseased, vessels. First carried out in December 1785, Hunter performed four more procedures with three of the five operations successful, which indicates the risks involved.

The first patient with a popliteal aneurysm to be operated on by Hunter at St George's Hospital in London survived for fifteen months and died of an unrelated cause.[21] At the autopsy the femoral artery with the aneurysmal sac was removed and put into Hunter's collection. Home, not Hunter, published this very case quickly in the *London Journal of Medicine*[22] with a supplement the following year,[23] the latter including a copper engraving showing the preparation, which still survives in the Hunterian Museum. Home must have asked Franz to produce a new illustration of this famous preparation, and a copper engraving based on his watercolour was included in the former's paper 'On the coagulation by heat of the fluid blood in an aneurysmal tumour'.[24] The two copper engravings differ not only in the orientation of the preparation but also in the fact that for reasons unknown Franz did not represent the ligatures Hunter had placed on the artery. This is difficult to understand, since the ligatures are the key steps of the operation. Maybe Home failed to explain to Franz the purpose of the operation and only expected an illustration ready for being published, or maybe the ligatures were no longer in place. On the right side of the copper engraving (Fig. 15.7.) published in the *Philosophical Transactions* the dilatation of the popliteal artery into the 'aneurysmal sac' is easily recognizable. Home noted in his text the slow progress that had been made in the absorption of its contents in the months after the operation until the death of the patient.

The more sinister realities of vascular surgery at the time become evident from the case of an aneurysm of the external iliac artery described by Home in that very paper. A heavy drinker, 43 years old, was operated on by Home on 16 September 1825 in Chelsea Hospital. Home made a ligation distal to the aneurysmal sac, which quite naturally did not stop increasing its size after the intervention. By any standard the next step in the operation was hair-raising: Home placed a needle into the aneurysmal sac and heated it, resulting in a wax-like coagulum on the needle. This caused intense pain, which was treated with laudanum, a tincture containing opium. It is hard to believe that this cruel and, as we now know, completely senseless treatment was repeated, but

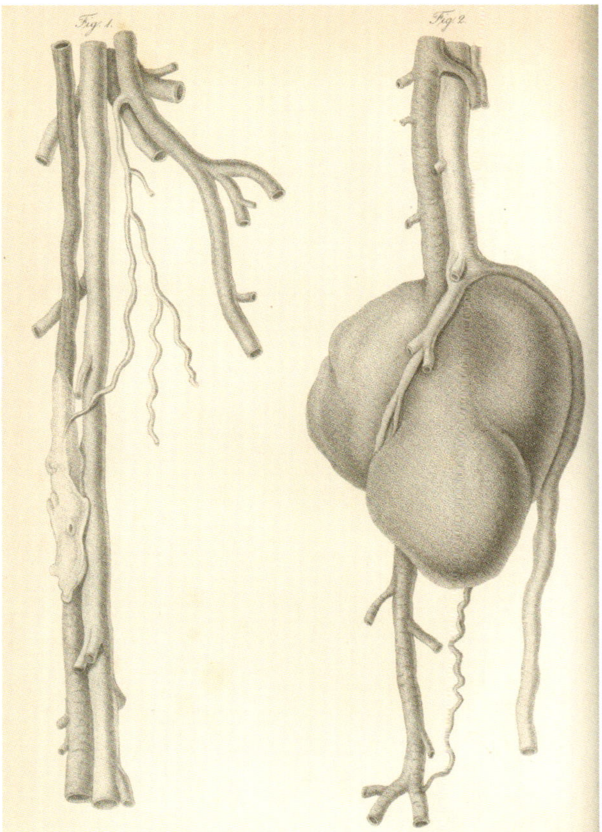

FIG. 15.7. ***Homo sapiens*** **(♂). Aneurysma of the femoral artery. Anatomical preparation from Hunter's Museum. Copper engraving by P. Basire based on Franz Bauer, 1826. Philosophical Transactions of the Royal Society vol. 116, t. 6. – London, Natural History Museum, Library and Archives.**

on the fiftieth day the foot became cold to the touch, the leg oedematous, the skin lost its sensibility and on the ninetieth day the patient was dead. It was Franz's job to document the preparations taken upon post-mortem, showing the aorta and the external iliac artery with the attached aneurysmal sac. Although the treatment had ended in the loss of the patient, Home proudly announced in his paper in the *Philosophical Transactions*, 'I trust I have fully proved that the heated acu puncturium [sic; i.e. the needle] can coagulate the fluid blood contained in the aneurysmal sac; and in doing so, is not productive of any local or constitutional symptoms that should prevent its being employed for that purpose'.

For any woman breast cancer, an ancient disease visible even to the untrained eye, is an absolutely devastating diagnosis. Depending on the aggressiveness of the malady, today a broad spectrum of procedures and treatments is available – to combat the tumour, to inhibit or slow down the development of metastases and, most importantly, to reduce pain in the final phases of the disease. In the early nineteenth century only two options existed: mastectomy, including the removal of infected lymph nodes, in an era before effective anaesthetics and antiseptic surgery a complete nightmare with very high mortality, or local treatment. In a footnote to his Croonian Lecture held on 4 November 1819, entitled 'A farther investigation of the component parts of the blood',[25] Home related the case of a woman of unknown age who suffered from a tumour in her breast. When first diagnosed it was the size of a pea, in six months had increased to the size of an orange and in fourteen more days had doubled its size. Home's harsh treatment – 'arsenic was applied once in twenty-four hours' – had to be stopped because of intense pain after two applications. Ten days later the tumour was 'black and shrunk, but not dead to any depth'. Further treatment with 'white arsenic and sulphur' resulted in even more excessive pain, but 'half of the tumour came away'. A specimen of this tissue was sent to Franz, who documented the overall structure of the colourless piece – without fixation, staining or sectioning. Histology was in its infancy and so was the understanding of breast cancer. The same applies to a tumour of the prostate gland also dealt with by Home in this Croonian Lecture. Franz found 'the texture ... soft and spongy; it was made up almost wholly of red globules free from colour, very few other those of lymph, and some of a transparent jelly'. Needless to say, prostate cancer, first described in 1536 by Venetian physician Niccolò Massa and two years later illustrated by well-known anatomist Andreas Vesalius, professor at Padua University, remained a total mystery to Home and Franz. Neither of the two papers contributed in any way to an increase in knowledge or understanding.

In 1830, just two years before Home's death, his book *A short tract on tumours and the peculiarities that are met with in the structure of those that become cancerous* appeared in London. It contains four illustrations all based on watercolours by Franz. Two of these had previously been published in the *Philosophical Transactions*. Two more were new, each documenting cancerous outgrowths of the skull and are currently untraceable. They witness Franz breaking his own taboo – documenting a human face. However, in this case his attention was focused by necessity on the malformation, not on the face itself and the facial expression of the girl who was suffering almost certainly from an osteogenic sarcoma.

ANIMALS ON THE DISSECTING TABLE

Like his more famous brother-in-law, who had published in the *Philosophical Transactions* on such diverse animals as the common cuckoo, *Cuculus canorus*, the electric eel, *Electophorus electricus*, and the honey bee, *Apis mellifera*, Home exhibited a strong interest in comparative anatomy. This term stands for the study of differences and similarities in the structure of different species, a field of science today closely related with evolutionary biology and phylogeny, both still undreamed-of concepts at the time. Unsurprisingly a broad spectrum of animal creatures ended up on the dissection table, where Home produced countless preparations. Of these several were documented by Clift, Franz and others in watercolours later published as copper engravings illustrating Home's various contributions. Home deposited some of his preparations in the Hunterian Museum, but owing to the partial destruction of this collection during the Second World War only a fraction survived.

The comparative aspects are clearly evident and appropriately visualized in the two copper engravings illustrating the Croonian Lecture held by Home on 20 November 1823 and entitled 'On the internal structure of the Human Brain, when examined in the microscope, as compares to that of fishes, insects, and worms'.[26] Here a piece of the human brain is compared with the nervous systems of a tench, *Tinca tinca*, a garden snail, *Cornu aspersum*, a bumblebee, *Bombus* spec., a caterpillar of the silkworm, *Bombyx mori*, an unspecified lobster and an unspecified earthworm, possibly the common earthworm, *Lumbricus terrestris*. The pertinent copper engraving demonstrates two facts. Firstly, it shows the similarities between the two insects, the annelid and the lobster, all four possessing a ventral cord. This on the one hand became evident, just as did the totally different nervous systems of the fish and the snail on the other hand. Secondly, while Franz accurately records all nervous systems from his preparations of the six animals, he is quite lost with a slice of the human brain 'immersed in distilled water'. He recorded 'globules, in straight lines, which pass uninterruptedly across the cortical substance into the medullary', and drew largely fictitious structures which he hardly could have seen in the soft tissue.

Home studied eyes, as revealed in his Croonian Lecture of 15 November 1821 entitled 'On the anatomical structure of the eye; illustrated by microscopical drawings by F. Bauer'.[27] This time Franz's participation is even mentioned in the title. Home chose a similar approach and compared the human eye with that of a bullock, *Bos taurus*, and of a greylag goose, *Anser anser* (Fig. 15.8.). A front and side view of the head of the goose are given by Franz, a front and a side view of the eye as well as a ventral and traverse section of the eye. More remarkable is his recording of the vitreous humour and the lens, because very careful sectioning and manipulation of the eye are mandatory to achieve such a perfect preparation. The hypothesis is put forward that at least in this case the anatomical work may have been done by Franz in Kew, because transporting a vitreous humour and the lens in a carriage from London would have easily ruined the preparation. Molluscs were not forgotten with Home comparing in his Croonian Lecture delivered on 16 November 1826 the reproduction of the marine 'common oyster' (an unspecified representative of the Ostreideae) with that of the freshwater mussel, *Margaritifera margaritifera*.[28] The year before the formation of pearls in the freshwater mussel, with their concentric layers of nacre, had been the subject of another communication

FIG. 15.8. Greylag goose, *Anser anser*. Watercolour by Franz Bauer, signed, 1821. Board, watercolour. – London, Royal Society, Archives, Image No. RS 8714.

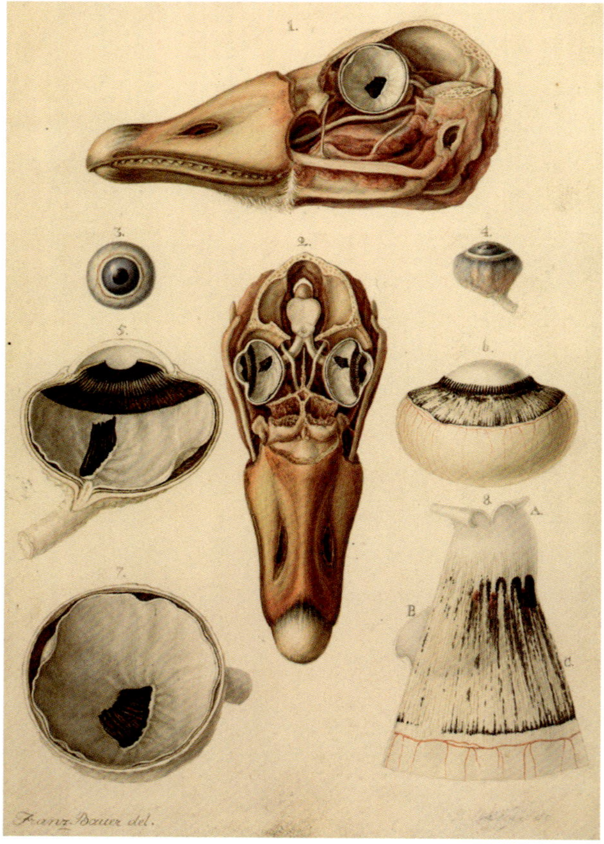

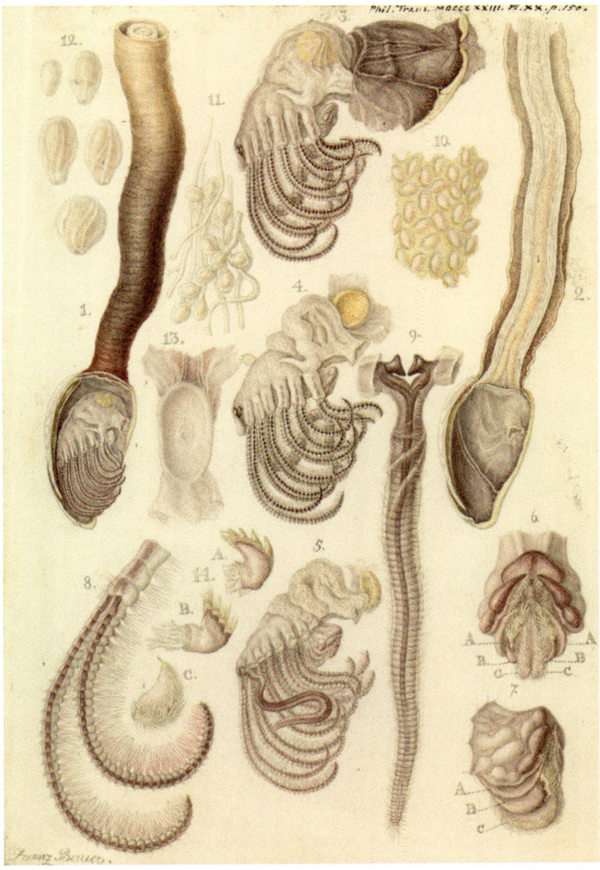

FIG. 15.9. Pelagic gooseneck barnacle, *Lepas anatifera*. Watercolour by Franz Bauer, signed, 1823. – London, Royal Society, Archives, Image No. RS 9907.

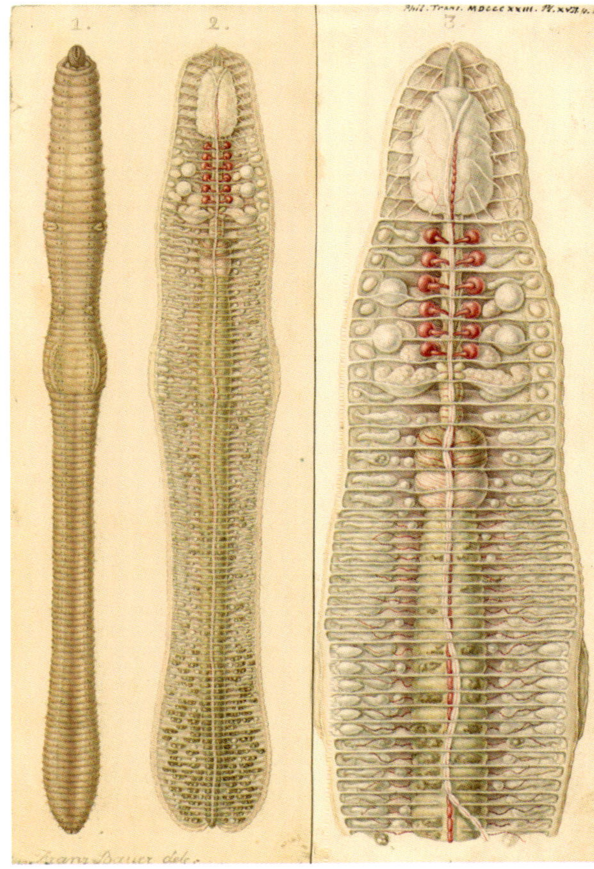

FIG. 15.10. Earthworm, *Lumbricus* spec.. Watercolour by Franz Bauer, signed, 1823. Board, watercolour. – London, Royal Society, Archives, Image No. RS 9904.

by Home;[29] both papers were illustrated with copper engravings based on watercolours by Franz.

Home's paper on hermaphrodite animals[30] also follows a comparative approach, focusing the attention of the reader on the gonads and pertinent sexual structures of an unspecified lamprey (any jawless fish of the order Petromyzontiformes), at a conger eel, *Conger conger*, a common eel (possibly *Anguilla anguilla*), a pelagic gooseneck barnacle, *Lepas anatifera* (Fig. 15.9.), and at an unspecified earthworm (Fig. 15.10). While lampreys are ectoparasitic fishes that suck blood from other fishes, both eels are bony fishes that feed on prey. By contrast, the barnacle belongs to the crustaceans, and the earthworm to the annelids. Franz's part was the pictorial record and Home composed the text, which may well have meant that Home actually described and commented on the watercolours prepared by his associate. Similarities and differences in the reproductive structures between the five species were striking; however, when reading Home's interpretations we have to consider that he was living in a period when hermaphroditism, protandry and protogyny were still incompletely understood phenomena, let alone their proximate and ultimate causes. We may well imagine that both Home and Franz were surprised about the deeper biological reasons for copulating earthworms, which belong to the group of simultaneous hermaphrodites, and they may also have been stunned by the gigantic penis in the sessile hermaphroditic barnacles, several times the size of the adult animal.

In other contributions to the *Philosophical Transactions* authored by Home, the comparative aspects are more concealed. A good example is Home's last paper published in this journal,[31] which is dedicated to the stomach of the giraffe, *Giraffa camelopardalis*, a well-known even-toed ungulate native to Africa. In the text, surprisingly dedicated to George IV, who was to die in the year of its publication, comparisons are made with the stomach of cattle and sheep, but only a single copper engraving based on a watercolour by Franz, now lost, exhibits the differences between the inner lining of the second cavity in the giraffe, cattle and sheep.

Clearly Home profited from the fact that voyagers brought to London from the remotest corners of the world exotic animal specimens, some of which were acquired by Home for further study. Lake Texcoco, a natural formation at an altitude of 2236 metres in the *alto plano* of Mexico, where the Aztecs had built the city of Tenochtitlan, was a slightly unconventional provenance. In June 1822 William Bullock, an English traveller and naturalist, had collected in this lake specimens of a curious salamander, already known to Alexander Baron Humboldt, who had visited central Mexico in 1803 and stayed for nine months in what is now Mexico City. Although sometimes called the Mexican walking fish, this creature, now known by its Nāhuatl name axolotl, *Ambystoma mexicanum*, is not a fish at all, but an amphibian. Home was able to distinguish male and female specimens[32] and Franz documented them accurately. However, Home's interpretation of the axolotl as a 'perfect animal' is utterly wrong. Franz had correctly depicted the gills in the sexually mature amphibian, but Home seems not to have paid due attention to them. Today the Mexican salamander is the textbook case for a phenomenon called neoteny, meaning in this case that an amphibian reaches sexual maturity without undergoing or completing metamorphosis. As we now know, it is the lack of the hormone that stimulates the thyroid in the axolotl which results in the failure of metamorphosis. In fact axolotls are larvae, as rightly observed by Mauro Rusconi in his *Amours des salamandres aquatiques,*[33] published in Milan in 1821, a fact explicitly and wrongly refuted by Home in his paper published in the *Philosophical Transactions.*

Specimens of the common purple snail, *Janthina janthina*, a marine gastropod mollusc, came from an expedition to the 'Congo',[34] and nests of the Java swallow or edible-nest swiflet, *Aerodramus fuciphagus*, were brought from Java by Raffles[35] (see Chapter 13). From Australia specimens of a platypus and a short-beaked echidna, *Tachyglossus aculeatus*, both depicted by Ferdinand during the voyage in the *Investigator* (see Chapter 11), arrived in London. They were dissected with the preparations, in particular their ovaries, depicted by Franz; the text was written by Home.[36] The latter is correct in describing the penis of the platypus as 'bifid, throwing the semen into both oviducts at the same time', though his additional note that this occurs 'through several points like a watering pot, as to scatter it all over the cavity [i.e. the vagina]', belongs to fiction, and the same is true about his idea that the ova (ovaries) in the platypus are 'aerated' by the vagina. Home also searched for teats and found none stating briefly 'the mother not giving suck'. When in 1831 further female specimens became available for dissection, two were passed on to Franz,[37] who reported on 3 October 1831 in a letter to Home that he 'could not find the slightest trace of anything like mammillary glands, internally, nor nipples externally'. Whereas the second observation was correct, Franz was wrong in his statement on the absence of mammillary glands. Why Franz was misled in this case remains unknown; maybe the two female specimens studied by him had not been impregnated. As we now know, milk is released through pores in the skin and pools in grooves on the abdomen allowing the blind and hairless young to lap it up. Home's brief paper on the nonexistence of nipples in *Ornithorhychus anatinus* based on a total of three specimens was read to the Royal Society on 24 November1831; it remained unpublished and was his last contribution.

FIG. 15.11 Chicken, *Gallus gallus domesticus*. Changes during incubation. Watercolour by Franz Bauer, signed, 1822. Board, watercolour. – London, Royal Society, Archives, Image No. RS 8747.

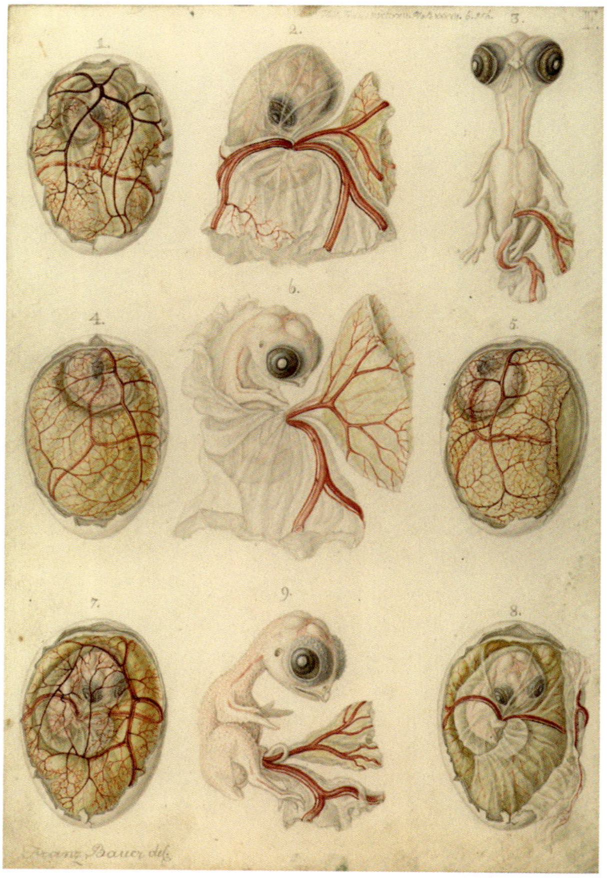

Home did not exclude standard topics of biology from his papers published in the *Philosophical Transactions*, such as the changes the egg of common fowl, *Gallus gallus domesticus* undergoes during incubation[38] (Fig. 15.11.) or the metamorphosis of the frog.[39] There was nothing new in the latter two communications, but the preparation of the pertinent watercolours must have kept Franz busy for a few weeks. Finding an unpublished watercolour by Franz kept in the Senate House Library[40] in London and documenting *Campanularia volubilis* (Fig. 15.12.), a widespread pelagic hydrozoan already known to Linnaeus, was a considerable surprise to the present author. Nothing is known as to where and by whom the specimen depicted had been caught, but it is very likely that Home was the commissioner. This watercolour has been found among a convolute containing among others one further watercolour by Franz documenting the eclipse of the sun, which occurred on 7 September 1820 in London. A copper engraving based on it was published in Home's paper 'On the black rete mucosum of the negro, being a defence against the scorching effect of the sun'.[41]

FIG. 15.12. *Campanularia volubilis*. Watercolour by Franz Bauer. Paper, watercolour, undated. – London, Senate House Library, MS 1015, f. 10.

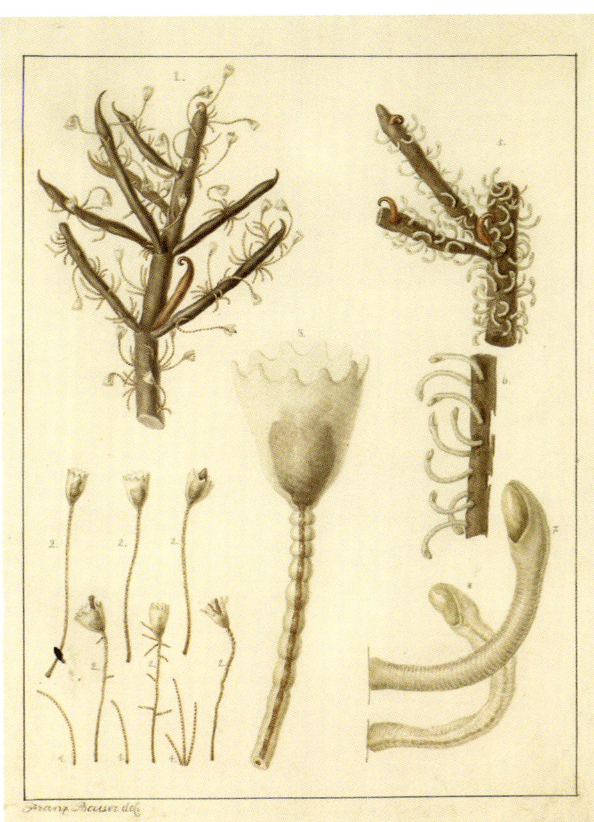

LECTURES ON COMPARATIVE ANATOMY

The Trustees of the Hunterian Collection were a prominent committee, having among their several members the Chancellor of the Exchequer, the Speaker of the House of Commons and the President of the College of Physicians.[42] With William Eden, 1st Baron Auckland, previously the President of the Board of Trade, in the chair the trustees met on 1 May 1813, just a few days before the inauguration of the Hunterian Museum was to take place (see above). They asked Home to publish his lectures which he had held previously on a number of occasions, for example at the meetings of the Royal Society. This is the background for Home's *Lectures on comparative anatomy*, of which the first two volumes appeared in the following year comprising two courses, each with twelve lectures. Of the two volumes, both published in London, the first contained the text and the second the copper engravings with the legends. In this work Franz had no share, the copper engravings being based almost exclusively on drawings by Clift with a few by another illustrator, Howslip. During spring 1822 Home had given further lectures in the theatre of the Royal College of Surgeons and a second time he was asked for publication,[43] again by the Trustees, this time chaired by Alleyne FitzHerbert, 1st Baron St Helens, a Member of the House of Lords, retired diplomat and a Lord of the Royal Bedchamber. The President of the Royal Society, one of its future presidents and Home himself were also among the committee members. The result was the speedy publication of the volumes three and four of the *Lectures on comparative anatomy* in 1823, comprising the contents of fifteen more lectures, with the third volume containing text and the fourth copper engravings and legends, which both also appeared in London. Many of the copper engravings based on Franz's watercolours, which had been already published in the *Philosophical Transactions*, were republished in the fourth volume, some with deletions and rearrangements, while the rest of the copper engravings were largely based on Clift's works. Only three watercolours by Franz – two depicting colonies of barnacles, one showing copulating earthworms (Fig. 15.13.) – were printed here for the first time as copper engravings.

Five years later Home, then seventy-two years old, published a *Supplement to the forgoing lectures on comparative anatomy*, which contains ten chapters and

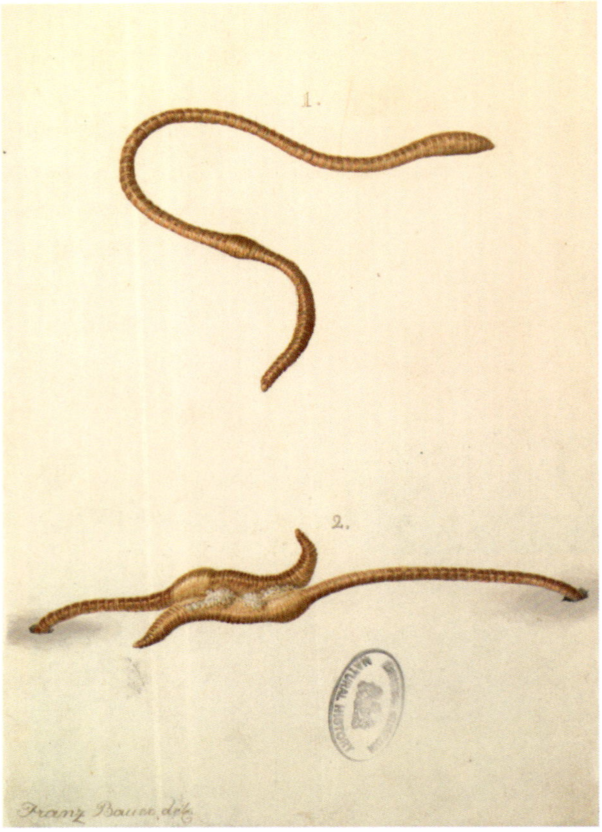

FIG. 15.13. Earthworms, *Lumbricus* spec., copulating. Watercolour by Franz Bauer. – London, Natural History Museum, Library and Archives, Series of original watercolour drawings by W. Clift, Franz Bauer, Mrs. Marsh, and others, no. 145.

FIG. 15.14. *(top)* Urinary stones. Copper engraving based on drawing by Franz Bauer. – E. Home, *Supplement to the lectures on comparative anatomy* 2, London, 1828, t. 6. – London, Natural History Museum, Library and Archives.

FIG. 15.15. *(bottom) Homo sapiens* (♀). Clitoris. Watercolour by Franz Bauer. – London, Natural History Museum, Library and Archives. Series of original watercolour drawings by W. Clift, Franz Bauer, Mrs. Marsh, and others, no. 112

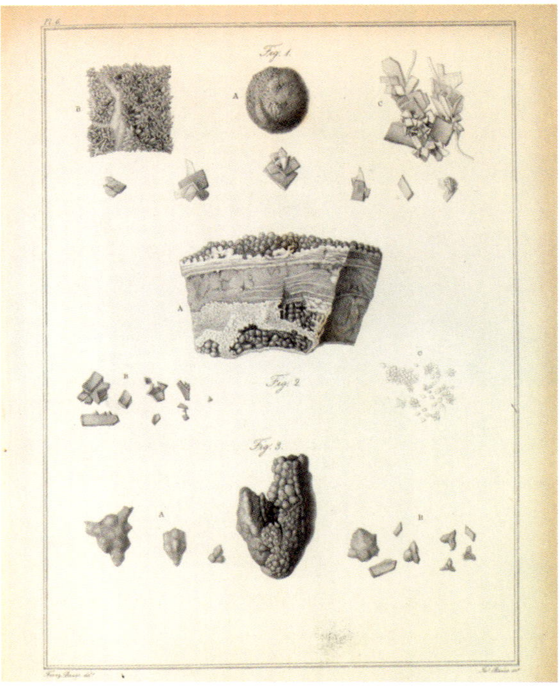

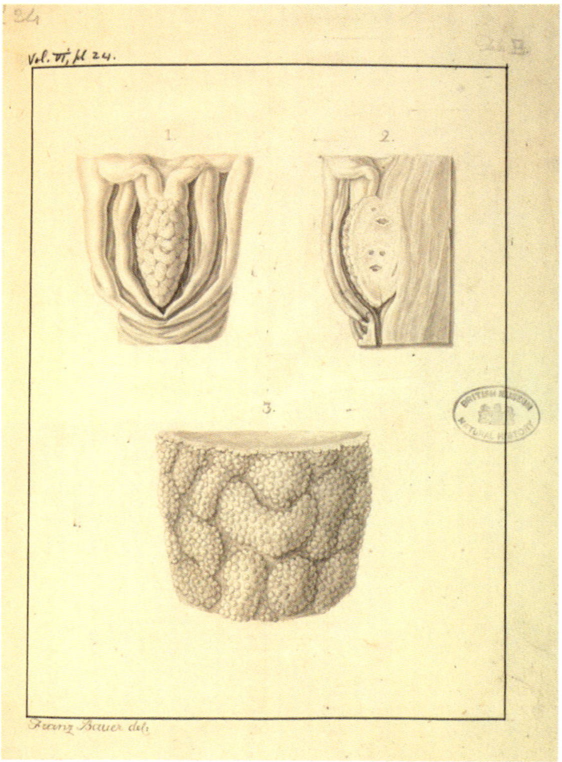

followed the pre-existing model, with the first volume containing the text and the second the copper engravings, lithographs and legends. This work also appeared in London. As before, many of the prints based on Franz's watercolours were republished, but also eighteen new ones printed for the first time, largely as lithographs. The latter were a curious mixture typical of Home: crystals of uric acid, kidney stones extracted during operations by the great surgeon (Fig. 15.14.), adrenal glands, and a clitoris (see above; Fig. 15.15.). A considerably larger number of the new plates were dedicated to animal anatomy. Franz recorded, surprisingly, in this sequence: (1) the stomach of the alpaca, *Vicugna pacos*; (2) the internal structure of the lung of an unspecified turtle; (3) the roe of several fishes (among them the shad, *Alosa* spec., and the sprat, *Sprattus* spec.); (4) the ovary of the ocean sunfish, *Mola mola*; (5) dissections of the female and male of the largescale four-eyed fish, *Anableps anableps*; and (6) foetuses of the rock hyrax, *Procavia capensis*. Of these, the largescale four-eyed fishes are particularly interesting since Franz correctly documented the penis, an organ consisting of anal rays, greatly elongated and fused into a tube, acting as an intermittent organ to deliver sperm to the female – a rather unconventional structure in fishes. Due to its surface-dwelling lifestyle, the eyes of this species are specially adapted, with the upper half for vision in air and the lower half for vision in water, and the two halves divided by a band of tissue. These unique characters are very correctly documented by Franz; in a transverse section of the eye he also paid attention to the huge lens and the upper underwater retina and the lower air retina, typical for *Anableps anableps*. The provenance of the specimens is indicated by Home only for the more spectacular material – for example, the specimens of the largescale four-eyed fish had been sent by 'Dr. Austin, from Demerara', probably William Piercy Austin, later the inaugural bishop in Georgetown, now the capital of the Republic of Guyana.

The texts of the *Lectures on comparative anatomy* and the *Supplement* are partly novel, partly based on Home's papers previously published in the *Philosophical Transactions*; Franz's text on nematodes forming his Croonian Lecture[44] (see Chapter 9) is also dealt with. Only in volume three did Home explicitly refer from his text to the illustrations based on Franz's watercolours in footnotes, but not in the first volume of the *Supplement*, where pertinent background information is insufficient and at best hidden in the legends.

Compared to Home's accounts published in the *Philosophical Transactions*, Franz's name appears more rarely in the *Lectures on comparative anatomy* and the *Supplement*. However, there is a most interesting account in the first volume of the *Supplement*[45] on a frustration that Franz and Home had shared. When studying human seminal fluid Franz had realized that after emission it forms a jelly that dissolves spontaneously and forms 'a liquor which does not again coagulate'. Irrespective of whether it was taken from man, the epididymis of a quadruped or from the testes of a bird, the same change occurred - in the seminal fluid of rabbits it was just the same as in that of geese. In his microscope Franz could only discern globules, but did not see the 'animalculae' claimed to exist by Leeuwenhoek and others. Even when sperm was diluted with saliva, distilled water or exposed to light, there was no difference. Subsequently the observations were extended to fallow deer, *Dama dama*, in Richmond Park. Home reported that he had 'made arrangement with the park-keeper to send to Mr. [Franz] Bauer weekly, from the 3d September [1824?], the testicles of a buck, the vasa deferentia of which were secured by ligatures before they were removed from the body, and being carried by a man on horseback in his bosom, wrapped up in warm flannel, the contents were examined while the parts continued warm'. Again Franz looked out for the tadpole like appearances, but in vain.

As a consequence, Home and Franz decided to travel to Paris in 1825, which meant for Franz leaving England for the first time since his arrival in late 1788 (see Chapter 8). Their aim was to see instrument-maker Chevalier and plant illustrator Pierre Jean François Turpin, who was reported to have made drawings of the 'animalculae'. However, the two visitors from England were shown only 'dried specimens on glass', but no living sperm cells.

With the beginning of the rut in fallow deer in early October 1826, Franz continued his observations making use of Chevalier's microscope, but again no 'animalcules' could be discerned. It is interesting to note that Franz asked Home and Rodolphe Viverot, a visiting physician from Vienna, to confirm his negative evidence, thereby in a sense subscribing to the modern two- or even three-man-rule. Home published this negative evidence in the first volume of his *Supplement* stating, albeit wrongly, that 'the appearance of living animaculae' in seminal fluid is a microscopic deception. However, it is not clear why Franz was unable to see the spermatozoa.

HOME'S 'PLAGIARISMS'

It is unavoidable that in our context one must deal briefly with Home's 'plagiarisms' and his burning of Hunter's papers. This event led to very bad press for Home, strong feelings against him from the medical fraternity and the refusal of the Royal College of Surgeons to purchase a precious and unique element of Franz's estate in 1841 (see Chapter 17). The context is remarkable: while all copper engravings based on Franz's watercolours prepared for Home had been printed in black, in the *Philosophical Transactions*, the *Lectures on comparative anatomy* and the *Supplement*, a single series had been printed in colour. This luxury set had not been coloured by hand, but printed in colour following the example of the great masters of this technique based in Paris who had worked for the luminaries of the Premier Empire, like Josephine, Empress of the French, or later Alexander Baron Humboldt. This approach, printing à la *poupée*, is characterized by the application of several coloured inks directly on the copper plate's surface using cotton daubs called dollies, in French poupée. The main problem is the careful wiping out of the printing form without mixing the neighbouring colours. Who coloured the printing forms for this set of copper engravings documenting human and animal anatomy remains unknown. It would be a surprise if Franz had done so, because he is not known to have ever been involved in the printing process. However, it would be inappropriate to close one's eyes to the fact that the real colours must have been different – many anatomical preparations had been soaked for a considerable time in liquids, like the penis in turpentine, and must have long lost their natural colours.

In any case, after Franz's death this unique set was rejected by the Royal College of Surgeons, for which it may have been originally intended. Instead it went up for auction at Christie's, was acquired for Ernest Augustus, King of Hanover, donated by him to Göttingen University and deposited in the University Library[46] (see Chapter 17). It is from this notable set that two illustrations pertinent to this chapter have been taken, that is Fig. 15.2. and Fig. 15.6.

In order to understand Home's surprising actions in 1823 it is necessary to provide the background: on 18 May 1820 a special Council meeting of the Royal Society had been summoned by Banks, at which the President himself was not present. On that occasion Home announced 'that they [the members of Council] were called together by desire of the President who in consequence of the failure of his general health felt himself too much oppressed longer to perform the Duties of President, and had therefore with deepest regret requested Sir Everard to give the Council his resignation of that most honourable office which ha[d] constituted the pride and happiness of his life'. Home, always an ambitious man, no doubt displayed an ardent desire for the post, and indeed at the death of Banks on 19 June 1820 he was a strong candidate. However, Wollaston was elected, accepting the Presidency only until the Anniversary Meeting in November 1820, when Davy, a brilliant emeritus professor of the Royal Institution, began his seven years in the chair of the Royal Society. Home must have been deeply disappointed. However, he remained Vice President and was permitted to continue to monopolize the Croonian Lecture and its emolument until 1829, when this tradition was discontinued for twenty years. There was a single exception – in 1822 had Franz held the prestigious lecture with his text on nematodes published in the *Philosophical Transactions* (see Chapter 8).

In 1821 Home was elected for a second timer Master of the Royal College of Surgeons and later secured another prominent appointment – as surgeon to Chelsea Hospital. Rumour has it that his presence during the operation of a common encysted tumour on the scalp of King George IV in 1823 played an important role. In the same year Clift was elected a Fellow of the Royal Society, shortly before Home told him when driving out to Kew that he had burned Hunter's manuscripts that were then on loan to him at his private home in Ham. Although one may argue that these materials actually belonged to the Hunterian collection, which had become the property of the nation, this may not have been seen as such at the time. In any case Home's action was at best unwise and definitely incorrect. Furthermore, it was imprudent to inform Clift, who may have been jealous of Home's rise in fame after the death of Hunter, Clift's former master. In any event Clift blamed Home for having made use of Hunter's manuscripts without due acknowledgments in his publications. Furthermore Clift presented transcripts of Hunter's manuscripts at a hearing, proving that Home had indeed made use of them in his publications, for example in his papers published in the *Philosophical Transactions* and his *Lectures on comparative anatomy*.

Seen from a distance of almost two centuries, the scandal has much smaller dimensions now than were felt at the time.[47] Firstly, Hunter and Home had collaborated for years, with Home publishing, for example, the details of treatment of the popliteal aneurysm, known as Hunter's operation. Secondly, Home repeatedly referred to Hunter's achievements in his numerous contributions to the *Philosophical Transactions* and in the full title of his *Lectures on comparative anatomy; in which are explained the preparations of the Hunterian collection*. Although Home did not complete a catalogue of the Hunterian collection, a job requested from him by the Trustees, he effectively published a considerable part of the collections in the *Philosophical Transactions*. The date of the burning, which took place in the Manor House in Ham, is indicative – after having finished the proofs for the third and fourth volumes of the *Lectures on comparative anatomy*. Here is Home's involute reasoning for the destruction of Hunter's manuscripts in his letter of 19 February 1825 to the Trustees: 'The board had heard the severest censures against me [from

Clift] for having destroyed the unfinished manuscripts, which Mr. Hunter considered unfit to be laid before the public; but my unceasing labours to uphold his posthumous fame, by endeavouring to complete the investigations that remained imperfect at his death, most of which are now brought to a successful issue, are passed over in silence, although they amount in 30 years to 100 papers in the Philosophical Transactions, in which the engravings cost the Royal Society 4,000 £, being 320 in number; and every one of these papers forms materials for a catalogue raisonné of the Hunterian Collection as it is now arranged'. Furthermore, it should be noted that part of the material studied by Home and Franz had been definitely acquired after the death of Hunter. In short, Clift's claim of plagiarism was only partly founded.

What was Franz's position in this conflict? He simply had none, being the man to produce the watercolours for Home to be used as illustrations for his texts. These had been taken from the Hunterian Museum, like the preparation of the popliteal aneurysm, or from newly accessioned specimens, like the urinary stones extracted by Home or the largescale four-eyed fish (see above). The collaboration between Franz and Home continued unaltered after the 'clarifying letter' (see above) with Franz's last works done for the surgeon published in 1830, two years before the latter's death on 31 August 1832 in his official residence in Chelsea Hospital. Enviousness and distrust towards Home continued in the medical community for several years, with the *Lancet* publishing in 1836 an anonymous letter, writing about Home, 'Always partial to the bottle, he became at last a regular sot and never went to bed sober', with a further explosion in a footnote, 'Liar as well as thief – double, treble, quadruple liar, – as the further evidence will too truly prove'.

ENDNOTES

1 Hall (1984)
2 Lack (2004)
3 Desmond (2007)
4 Beasley (2000)
5 Livesley & Pentelow (1978)
6 Qvist (1981)
7 Home (1808)
8 Hunter (1794)
9 Madrenas, Potter & Cairns (2005)
10 Wardroper (2002)
11 Home (1816)
12 Home (1824b)
13 Home (1820b)
14 Home (1817a)
15 Schmuck (2009)
16 Granville (1820)
17 Home (1818a)
18 Home (1818b)
19 Home (1819a)
20 Galland (2007)
21 Schechter & Bergan (1986)
22 Home (1786)
23 Home (1787)
24 Home (1826a)
25 Home (1820a)
26 Home (1824a)
27 Home (1822a)
28 Home (1827)
29 Home (1826b)
30 Home (1823a)
31 Home (1830b)
32 Home (1824c)
33 Rusconi (1821)
34 Home (1817b)
35 Home (1817c)
36 Home (1819b)
37 Gruber (1991)
38 Home (1822b)
39 Home (1825)
40 GB 96 MS 1050
41 Home (1821)
42 Home (1814)
43 Home (1823b)
44 Franz Bauer (1823)
45 Home (1828)
46 Handschriftenabteilung, gr. 2° MS Natur. Hist. 94, XVIII, XIX; for details Lack (2003)
47 Oppenheimer (1946)

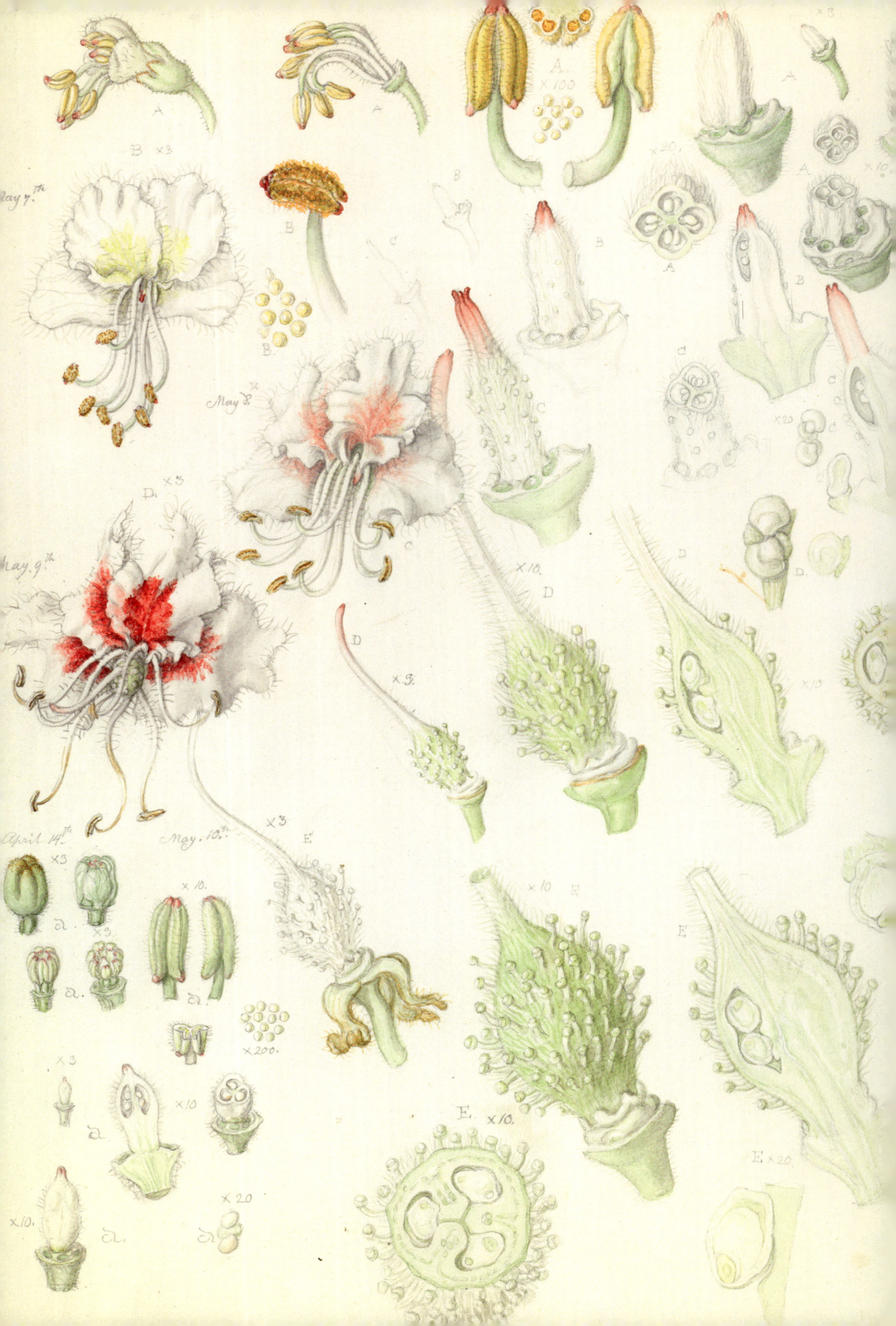
May 7th
May 8th
May 9th
April 14th
May 10th
A
B ×3
×100
×20
×10
×3
×200
E ×10
E ×20

CHAPTER 16

Franz Bauer, the Nestor of Kew

THE ROYAL GARDEN AT KEW IN DECLINE

THE DEATHS OF GEORGE III on 29 January 1820 at Windsor Castle and of Banks on 19 June 1820 in Spring Grove had marked the end of an epoch. Their successors, George IV on the throne and the chemist Wollaston in the chair of the Royal Society – the latter quickly followed by the chemist Davy (see Chapter 15) – had completely different interests, and as a consequence the traditional support for the Royal Garden at Kew dwindled. As was so appropriately worded in the standard chronicle of the institution, 'years of neglect' followed, with the absence of royal patronage.[1] Aiton the son's expertise was now conscripted to projects outside Kew, among them the gardens of Buckingham Palace in London, which earned him the title of Director General of His Majesty's Gardens, but resulted in less attention to his previous responsibility. As a consequence the direct accountability fell on four foremen – in charge of the four sections of the Royal Garden at Kew: the Botanic Garden, the Pleasure Grounds, the Kitchen Garden, and the Fruit and Forcing Department.

Comments by visitors to the Botanic Garden proper in Kew, then comprising merely nine acres[2] (3.6 hectares) compared to currently three hundred acres (one hundred and twenty-one hectares), were not always complimentary, with August Wilhelm Schultes, then professor of Natural History and director of the surgical school at Landshut University, noting in 1824: 'The Botanic Garden at Kew is surrounded by high walls . . . With regard either to its plan, or its nine or ten stoves, it will not bear a comparison with Malmaison [near Paris] as it had been, or with the garden of [Karl August] Grand Duke of Weimar, of [Nikolaus II, the 7th] Prince Esterhazy [Esterházy Miklós II] in Eisenstadt [then Kismarton in the Kingdom of Hungary], let alone with the botanical division of the Imperial Garden at Schönbrunn . . . A supplement to the Hortus Kewenis, under the inspection of Sir [sic] Robert Brown, will soon be published [which never happened]: some of the species which were formerly cultivated there [i.e. at Kew], are said to be lost'.[3] In addition, the herbaceous plants continued to be displayed in the Linnaean order and the labels identified only genera, while the species were tagged only with numbers, which related to names in a ledger. This outdated arrangement echoes that in the Botanical Garden of Vienna University half a century ago (see Chapter 3), but offered a special advantage – identification was difficult for thieves, always a topic in botanic gardens, though rarely addressed in the literature. Maybe Schultes's report was opinionated, in particular when considering the very considerable staff still working on this royal estate. However, the Royal Garden at Kew suffered additionally from the Treasury's habitual economies, with the number of Kew collectors on duty overseas reduced to one – Allan Cunningham, upon the death of Banks traveling in Australia. The

FIG. 16.1. Horse chestnut, *Aesculus hippocastanum*. Watercolour by Franz Bauer, signed, 1819. Paper, watercolour. – London, Natural History Museum, Library and Archives, Franz Bauer Miscellany (loose).

decline of the financial support which began under George IV and continued under William IV – from £1,800 per year in 1824 to £1,277 for the period from 1832 to 1836. In retrospect Frederick Scheer (see below) noted 'Kew descended somehow from the high station which it had held; it had, in fact, ceased to *lead* [Scheer's italics] in matters of Botany, and the idea of its receiving or disseminating any novelties, was gradually shelved amongst the obsolete reminiscences of by-gone days'.[4]

This does not mean that royal interest in Kew had come to a complete stop, but it shifted away from botany. George the Prince Regent had bought Hunter's House on Kew Green and the adjacent properties to the east of it and authorized repairs and improvements. From 1830–1 this house became the residence of George IV's brother Ernest Augustus (see Chapter 17) and the latter's wife Frederica Princess of Mecklenburg-Strelitz, later the King and Queen of Hanover. After the death of Franz it was Ernest Augustus who purchased a considerable proportion of Franz's estate (see Chapter 17) and contributed to his epitaph (see below).

Since a very considerable proportion of Franz's plant illustrations is undated, it is difficult to establish a clear-cut correlation with this general decline of the Royal Garden in Kew. The fact remains that after 1820 and, in particular after 1830, the number of plants documented by Franz and dated is clearly smaller than for the previous three decades. At the same time the number of orchids (see Chapter 13) and ferns (see below) portrayed by Franz increased. In addition we observe a marked shift towards an interest in the development of plants.[5] This is best exemplified by Franz's day-by-day record of the flower development of a tree he knew from his childhood in Feldsberg – the horse chestnut, *Aesculus hippocastanum* (Fig. 16.1.). From 6 to 9 May 1819 he documented the anthesis and followed the colour change of the nectar guides, from yellow through yellow-red to carmine. Franz hardly understood the significance of this: pollen- and nectar-providing flowers with yellow nectar guides are much more frequently visited by pollinators such as honey bees or bumblebees then those with a red nectar-guide. The development of the gynoecium of the horse chestnut from primordia to young fruit also attracted his attention as did the mature seed and the plumule of the huge embryo it contains (see p. 468). In a similar way Franz observed the flower and fruit development of the rose of Jericho, *Anastatica hierochuntica*, an annual crucifer widespread in the Near East. He followed the germination of seeds, among them those of the horse chestnut from November 1833 to February 1834, and of acorns in spring 1833. This interest is paralleled by Franz's attention to the development of the one-seeded cone of yew, *Taxus baccata* (Fig. 16.2.) with its red aril, six stages have been recorded between 20 February and 10 April 1835. No doubt living at Kew made these observations particularly easy. Fungi were not forgotten. Over three months, from December 1833 to February 1834, Franz documented the different developmental stages of the common truffle, *Tuber aestivum* (Fig. 16.3.), which he had found in Kew and may not have paid attention to before. Vertical and transverse sections were made and recorded, as well as spores.

Franz did not avoid curious specimens, among them an indeterminable plant, possibly an umbelllifer, which could not be induced to flower by the gardeners at Kew. Franz's watercolour is annotated in an unknown hand, 'This drawing of the Plant supposed to be the Silphium of the Ancients was taken from the original Plant presented to his Royal Highness [Augustus Frederick] The Duke of Sussex [a future President of the Royal Society, see below] to [sic] His Majesty [i.e. George IV]'s Garden at Kew'. Franz sent another drawing of this specimen (Fig. 16.4.) with the following note dated 21 June 1830 to Konig at the British Museum: 'till then You may dispose of this part of the drawing as you please; wether [sic] You will show this part to H: R: H: [His Royal Highness] or not, I leav [sic] entirely to your pleasure'. A considerable amount of speculation seems to have been involved here: the *Silphium* of the ancients, probably an umbelllifer native in what is now the eastern part of Libya, is believed to have been extinct since Roman times.

FIG. 16.2. *(top)* Yew, *Taxus baccata*. Watercolour by Franz Bauer, 1835. Paper, watercolour. – London, Natural History Museum, Library and Archives, Unfinished Drawings of Plants, vol. 6, no. 29.

FIG. 16.3. *(left)* Common truffle, *Tuber aestivum*. Watercolour by Franz Bauer, 1833–1838. Paper, watercolour. – London, Natural History Museum, Library and Archives, Unfinished Drawings of Plants, vol. 6, no. 96.

FIG. 16.4. *(right)* Undeterminable plant (cf. *Laserpitium* spec.?). Watercolour by Franz Bauer, 1830, with letter by Franz Bauer attached. Paper, watercolour. – London, Natural History Museum, Library and Archives, Unfinished Drawings of Plants, vol. 1, no. 159.

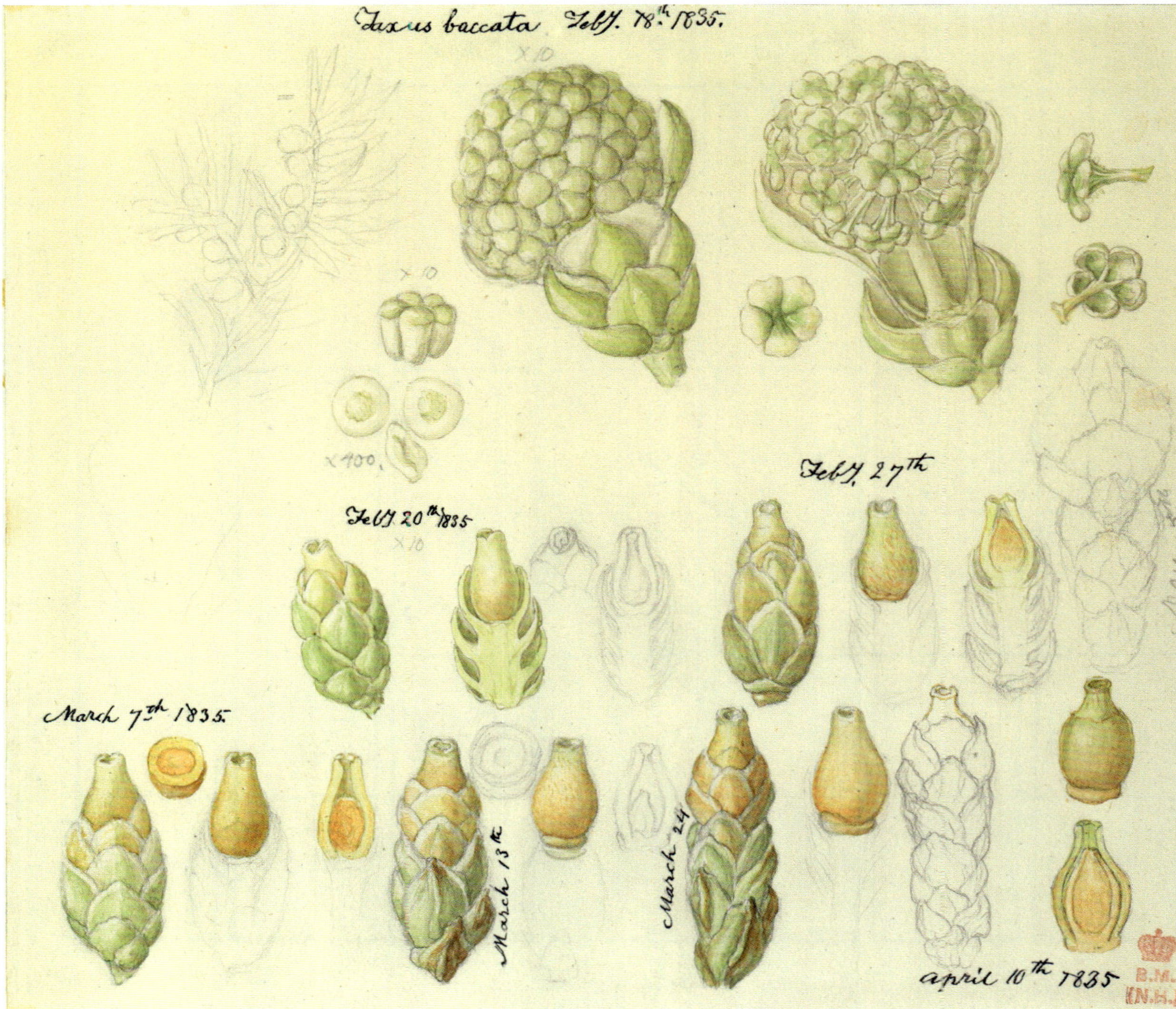
Taxus baccata Feby. 18th 1835.
x10
x10
x400.
Feby. 20th 1835
x10
Feby. 27th
March 7th 1835.
March 13th
March 24
May
april 10th 1835
B.M.
(N.H.)

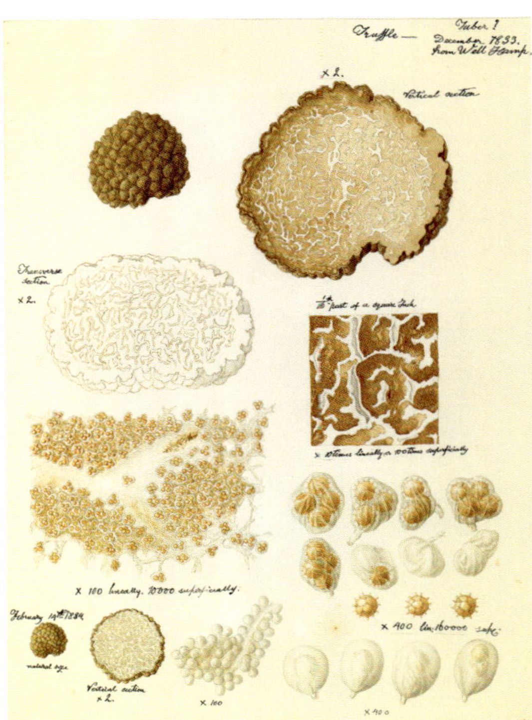
Truffle —
December, 1833.
x 2.
Vertical section
Transverse section
x 2.
x 100 linearly. 10000 superficially.
February 14th 1834.
natural size
Vertical section
x 2.
x 100
x 400
x 400 lin. 160000 superf.

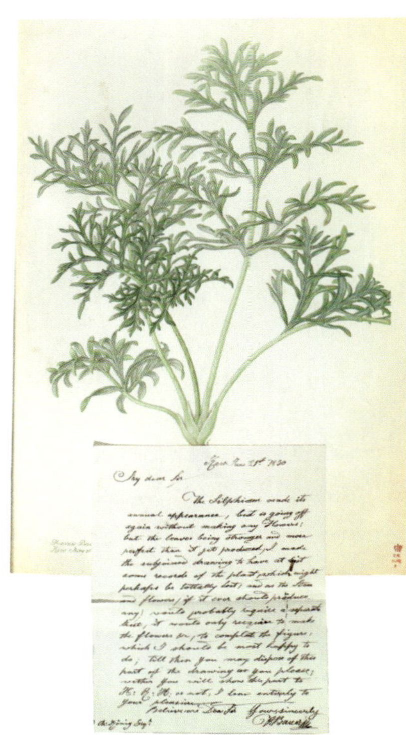

The crisis of the Royal Garden at Kew culminated during the final years of Franz's life and had a decisive influence on his testament (see Chapter 17). At that time, in 1837–1841, the plan to convert this royal estate into a kitchen and fruit garden for the Royal Household was a plan supported by the Lords of the Treasury with closure as an alternative.[6] The reasons for all these considerations had nothing to do with the garden itself but solely with the accession of Queen Victoria to the throne on 27 June 1837. This immediately raised the question of the future royal income in the form of the Civil List to be voted on by Parliament. As a consequence royal expenditure had to be examined. Since after the death of William IV the royal income from Hanover would cease and the Queen herself had no means, there was pressure to economize the royal budget. The Royal Household administered by the Board of the Green Cloth was a major drain of resources and therefore its expenditure was analysed in detail by the Treasure Board, with the Prime Minister as the chairman, and the Chancellor of the Exchequer and the First Commissioner of Woods & Works among its members. Not unexpectedly, several anomalies were revealed and a conventional solution found: the establishment of a commission consisting of a representative of the Board of the Green Cloth and a representative of Woods & Works, both Members of Parliament, who looked out – again not unexpectedly – for specialist advice. They turned to Lindley, then already Professor of Botany at University College London and Assistant Secretary of the Horticultural Society of London (see Chapter 13), who formed a working committee with himself as the chairman and two gardeners as members: John Wilson, gardener of Bernard Howard, the 12th Duke of Norfolk, and Joseph Paxton, gardener of William George Spencer Cavendish, the 6th Duke of Devonshire. However, it seems that the whole arrangement with Lindley and the two gardeners reporting to two noble lords, the latter both invested by the Order of the Garter, had been agreed upon beforehand.

In the end the committee's report arrived at the Treasury and was shelved for two years, for a variety of reasons, which again had nothing to do with the garden itself. The Prime Minister William Lamb, 2nd Viscount Melbourne, resigned, although he was soon back in office; the Lord Steward, the chairman of the Board of the Green Cloth, died, and was succeeded by William Hay, 18th Earl of Erroll, a Knight of the Thistle and Privy Councillor; the coronation of Queen Victoria furthermore preoccupied the administration. Finally, the decision was taken to move the Royal Garden at Kew – excluding the Royal Kitchen Gardens and the grounds of Kew Palace and the Queen's Cottage – from the Civil List and to transform this royal estate into a national institution under Woods & Works. All these arrangements took their time, since the government was divided on the issue. A campaign for a national botanic garden had started, with petitions to Parliament from University College London, the Linnean and the Horticultural societies and a further petition to the Treasury from the Richmond Literary and Scientific Institution, with the usual reverberations in the press. In Franz's last year there were further alarming items of news on the planned closure, which made him apply to the Crown (see Chapter 17). The matter was raised in the House of Commons on 4 May 1840, an extract of the shelved report was published, and questions were asked in the House of Lords followed by a semi-official denial of the plan to close the Royal Garden at Kew. Meanwhile the various departments involved were negotiating a solution which resulted in the formal transfer to Woods & Works as of 25 June 1840, the actual birthday of the Royal Botanic Gardens Kew in their modern sense. Meanwhile, Sir William Jackson Hooker, Regius professor of botany (see below), was waiting for the fate of the Royal Garden at Kew to be disclosed and to learn if his long-held hopes of the directorship would finally be fulfilled. As was so often the case, there were further delays, caused by the necessity to decide from what kind of funds the salary of the new director would be paid. In our context a single aspect is relevant – Hooker was to receive £300 per year, precisely the sum Banks had paid to Franz as an annuity for life. These figures, including the £200 per year in lieu of a house for Hooker, had of course first to be approved by the Treasury, which meant another delay of several months. Up to that moment confidentiality was maintained. It was only on Lady Day, 25 March 1841, that Aiton the son, aged 75, resigned and Hooker succeeded him as director. On 9 June 1838 Franz had reported on the developments at the Royal Garden in Kew to Jacquin the son, aged 75, in Vienna,[7] but both did not live long enough to learn about the final solution – they died months before the appointment of Hooker was made public.

THE ROYAL SOCIETY AND THE MUMMY CLOTH

Since Franz's election to the fellowship in 1821 (see Chapter 13), the Royal Society had undergone drastic changes.[8] Firstly, fellowship, previously divided almost equally between active scientists and wealthy amateurs, among them a copious number of the aristocracy, gradually shifted towards those actively practising science. Secondly, the leisurely atmosphere of intellectual life changed to a more strictly observed focus on research in the natural sciences, which evidently accompanied a diminution of the fellows' previously broad range of interest. Thirdly, the profile of the presidents changed. Davy was followed by Gilbert, who was succeeded by August Frederick, Duke of Sussex, the youngest brother of George IV and William IV. During the last years of Franz's life a representative of the upper crust of aristocracy, Spencer Joshua Alwyne Compton, 2nd Marquess of Northampton, was in the chair. Although this might give at first glance the impression that Banks's profile – the wealthy baronet coming from the landed gentry, Eton, Oxford and at home at Court, authoritarian and ready to express strong opinions – was perpetuated forever, this was not so. Instead, the reforming scientists gradually gained support and the President's power was truncated. Improvement of the standard of papers read to the meetings of the Royal Society was another topic of reform, which must have met with considerable resistance for a single reason – it implied that the individual judgement of a fellow was no longer enough to determine the suitability of a contribution. In short, the Royal Society was becoming more serious, its officers became more conscientious and the presidential autocracy was weakened.

Franz did not get involved and seems to have followed all these developments from a safe distance. What mattered to him were not the politics and dissentions within the Council of the Royal Society, but the contact and collaboration with a few individuals – among them William Brockedon, Home (see Chapter 15), Hooker, Lambert (see Chapter 8 and below) and James Thomson. The first of them, a man with many qualifications, was a painter who contributed regularly to the exhibitions of the Royal Academy of Arts and the British Institution. At the same time he was a writer interested in geography, which resulted in him becoming one of the co-founders of the Geographical Society, started as a London dinner club but soon becoming under the patronage of William IV the Royal Geographical Society of London. In 1831 Brockedon, an inventor, developed a form of pen called the 'oblique'. This may have brought him into contact with Franz and resulted in the latter's pencil-and-chalk portrait (Fig. 16.5.), now in the National Portrait Gallery in London.[9] It is dated 25 June 1834 and shows Franz as a man in his 76th year, seated, bald, with a domed forehead, gazing with greyish, wide open eyes at the artist.

Thomson was another contact. As an industrial chemist he became at an early date associated with Wollaston and Davy, being even prepared to inhale the latter's nitrous oxide, due to its euphoric effects now known as 'laughing gas'. However, it was Thomson's later career and reputation in calico printing which created his interest in ancient textiles. Via Home he contacted Franz, but we do not know when this happened. The result of their cooperation was a paper by Thomson[10] published in the *London and Edinburgh Philosophical Magazine* as late as 1834, when its author was 55, and Franz, the illustrator, was 76 years old. At that time Egyptology, and more particularly the mummies of

FIG. 16.5. Franz Bauer. Portrait by William Brockedon, 1834. Pencil and chalk on paper. – London, National Portrait Collection, Inv. No. 2515 (63).

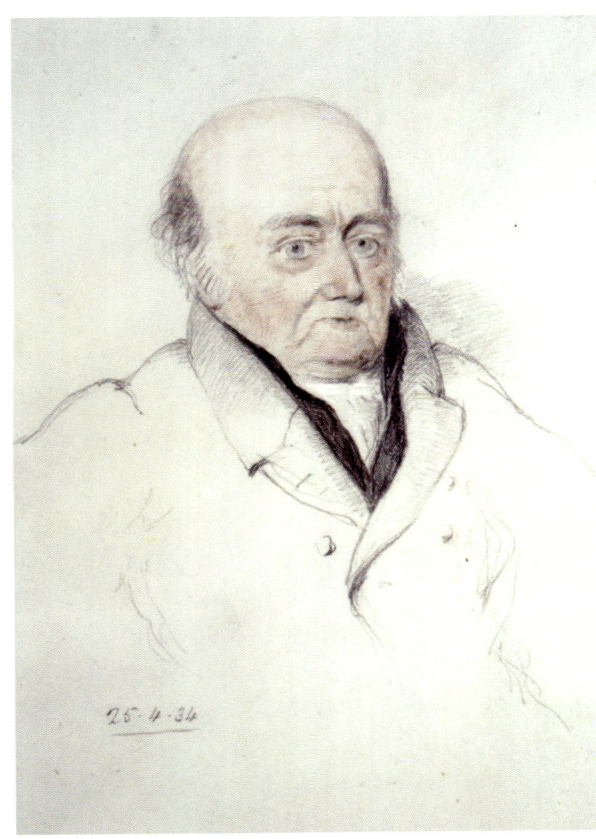

ancient Egypt, had become a fashionable topic in society, and the stripping of a mummy was a special event. This is the general background for the study of 'mummy cloth' undertaken by the two fellows of the Royal Society.

At that time the opinion prevailed that the material used for mummy binds was cotton. This claim was made by both Egyptologists and botanists, among the latter Forster the father in his tract *De Bysso Antiquorum*, published in London in 1776 just after the return from his circumnavigation of the globe in the *Resolution*. As we now know, this view was totally wrong,[11] and the first to prove this fact convincingly was Franz. Thomson had received specimens of mummy cloth from several sources, among them Giovanni Battista Belzoni, an adventurer and pioneer archaeologist born in Padova, then belonging to the Serenissima Repubblica di San Marco, who was to die in Ughotou in present-day Nigeria. Belzoni became famous for having removed the colossal bust of the pharaoh Ramesses II from Thebes in Egypt in 1816 and transported it to the British Museum,[12] where the 7.25 ton heavy subject became an iconic object of the Egyptian Department. Later Belzoni donated specimens of mummy cloth to Thomson, and other material was received from Henry Salt, then British Consul-General in Cairo and Belzoni's partner, as well as from others. At a time when archaeology was in its infancy, with pillaging the standard of the day and in the absence of radiocarbon dating, neither the age nor the precise provenance of the textile fragments acquired by Thomson are known.

How to differentiate between cotton and linen when the samples available are as small as a coin? Normal tests such as the sense of touch, differences in weight, smell of burned fibres, the degree of polish when the fibres were rubbed with a glass stopper, did not provide answers. Consequently, Thomson asked Franz to study the various specimens with his microscope, described as 'fibres of cotton and linen, both manufactured and in their raw state, as well as fibres of unravelled mummy cloth'. The results were clear cut. Franz told his colleague in a letter now lost but quoted by Thomson that every specimen of mummy cloth was linen and attached a drawing, now in the Library and Archives of the Natural History Museum. This shows the cotton fibres as twisted ribbons, while the fibres of linen and mummy cloth always consisted of straight, cylindrical tubes. More specimens made available to Franz from the British Museum, the Royal College of Surgeons in London, the Wernerian Museum in Glasgow and other institutions confirmed the observation. Franz's use of an 'improved achromatic microscope of Ploessl' from Vienna – Plössl's Grosses Mikroskop No. 1 (see below) – is clearly indicated in Thomson's paper, who stressed the marked differences: flattened ribbons, twisted round their own axes, in cotton, *Gossypium herbaceum*, transparent, cylindrical articulate tubes in flax, mainly *Linum usitatissimum*, weaved into linen. This is an unsurprising result to the botanist: cotton consists of dead seed hairs which have collapsed while linen consists of stiff fibres extracted from the skin of the stems in flax plants.

Two pencil drawings were prepared by Franz and printed as a lithograph, dated 11 February 1822 and 24 April 1834, respectively, which was appended to Thomson's paper. Two aspects merit attention: firstly, Franz depicts not only mummy cloths but also juvenile seed hairs of the cotton plant, which are stiff, hollow and untwisted tubes and could only be observed on young seeds or in the living plant, possibly cultivated in the Royal Garden at Kew or in herbarium specimens; secondly, Franz, possibly the author of the legends, offers some information, albeit vague, on the provenance of the specimens of mummy cloth, among them an *Ibis* mummy and the mummy of 'an ox's head', possible the specimen kept in the British Museum.[13] Needless to say, today we know much more about textiles in ancient Egypt, which were made from sheep's wool, goat hair, palm fibres and, in the great majority of cases, flax fibres, but not of cotton, since the latter did not come into general use before the first century AD. Thomson even claimed that Franz could 'ascertain whether cotton rages [rags] have been mixed with linen in any manufactured paper whatever'. This is also not surprising, since cotton and flax fibres retain their characters though the operation of spinning, weaving, bleaching, printing and dyeing as well as washing or reducing the rags to making paper pulp.

The Library and Archives of the Natural History Museum keeps a remarkable pencil and pen-and-ink drawing by Franz (Fig. 16.6.) referring to this work done in collaboration with Thomson. It shows dry flax fibres annotated in calligraphy 'Flax Magnified 400 times' with the following text added 'The undersigned certify that the above is a correct representation of a group of fibres of unravel'd thread of manufactured Flax as seen

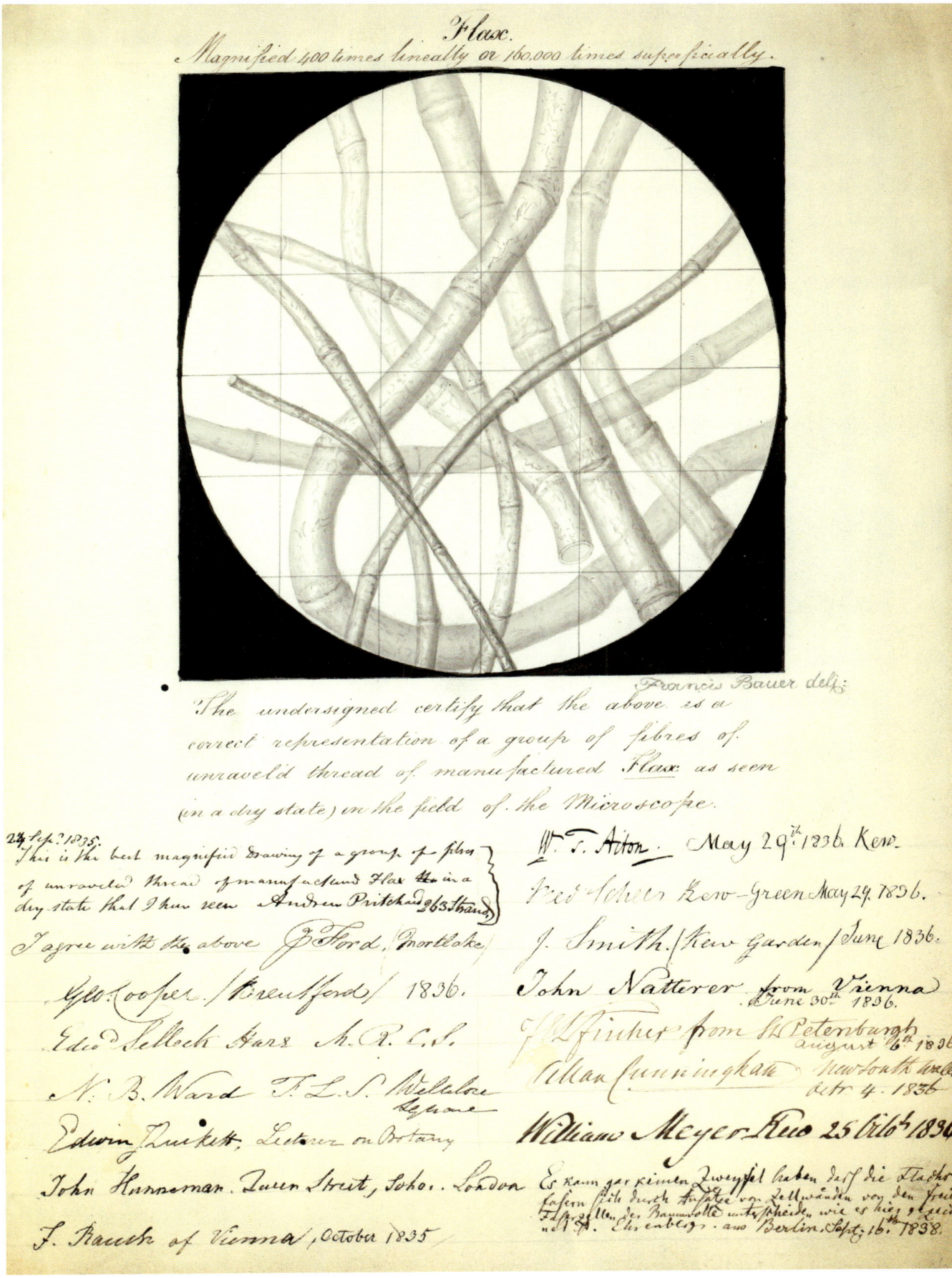

FIG. 16.6. Fibres of flax, *Linum usitatissimum.* Pencil drawing by Franz Bauer with several signatures, 1835–1836. Paper, pencil, ink. – London, Natural History Museum, Library and Archives, Franz Bauer Miscellany (loose).

(in the dry state) . . .'. Sixteen signatures follow, among them several from figures in our story like Aiton the son, Cunningham, William Meyer (see Chapter 17) and others; many are dated, some with brief notes added. Foreign visitors include Christian Gottfried Ehrenberg, then not yet the great professor of Berlin University and without a doubt one of the most influential biologists of his time, and Johann Natterer from Vienna, who had accompanied Archduchess Leopoldine in 1817 to Rio (see Chapter 12) and was now, more than nineteen years later, on his return voyage from Brazil to Vienna for a stay of several months in England.[14] In horticultural circles the most famous person who signed Franz's drawing was no doubt Nathaniel Bagshaw Ward, a member of the Royal College of Surgeons and popularizer of a case for growing and transporting plants, which became known as the Wardian case. This most practical invention, actually the direct forerunner of the modern terrarium, made George Loddiges, the owner of a nursery in Hackney, write in 1833 'where I used formerly to lose nineteen of twenty of the plants I imported during the voyage [from Australia], nineteen out of twenty is now the average of those that survive'.[15] Soon after having made this invention, which unleashed a revolution in the mobility of economically important plants, Ward must have been a guest of Franz in Kew and confirmed the correctness of his drawing of flax fibres.

Fibres are not hairs, but research on the former created in Franz an interest in the latter. A carefully annotated pencil drawing deposited in the Niedersächsische Staats- und Universitätsbibliothek Göttingen is inscribed by him 'Human Hair of various ages. July 1836' documenting hair surfaces from No. 1 'Hairs of an Infant Still born at the 8th month' to No. 16 'Hair from the Whiskers of a man of 65 y. old'.[16] In the context of our story rather than for science, No. 7–10 are particularly relevant: No. 7 has 'E[lizabeth] Baker 26 ys. of age', No. 8 'Same E[lizabeth]. B [aker]. 33 ½ years of age', No. 9 'F[ranz]. B:[auer] a Mon[th] 65 ys. of age', No. 10 The same F[ranz]. B[auer]: 76 Ys old'. No far-reaching hypothesis is attempted on the basis of these brief notes. However, a confirmation is provided for the statement in Franz's testament (see Chapter 17) that Elizabeth Baker, later adopted by him, had lived for 'upwards of twenty five years' in some form of association with him. She was definitely his junior by decades, first mentioned when she was 26 years old.

THE VISIT TO PARIS

Very little is known about the trip to the French capital, possibly brief, undertaken by Franz and Home in 1825 in order to observe living sperm cells (see Chapter 15). The latter reported in his *Supplement* that a microscope from (Vincent) Chevalier (see Chapter 17) was bought and that Franz received an introduction of how to use it from a very prominent man who was to leave his beloved Paris in the following year – Alexander Baron Humboldt, the well-known Prussian polymath. Very much a political animal, Humboldt was a member of the Royal Prussian Academy of Sciences in Berlin and a great promoter of the sciences, a man thinking and acting very much along the same lines as Banks had done with a similar network of global contacts. It would be no surprise if Humboldt, who had employed Turpin as his botanical illustrator for years, had brought the two visitors from London also in contact with Carl Sigismund Kunth from Leipzig, the man to complete the botanical part of Humboldt's famous *Voyage aux regions equinoxiales du Nouveau Continent*. This appeared as 'Partie 6. Botanique' in Paris in 1805–1834,[17] with the grasses in 1826 still on Kunth's agenda.

For someone with a mind like Franz's, Paris would have offered very many things to visit. We may hypothesize that he visited the Jardin des Plantes, the French equivalent to the Royal Gardens in Kew, though situated in the centre of the city and effectively a government institution not a royal estate. In the pre-revolutionary period the Jardin des Plantes had been called the Jardin du Roi, and Franz may have been interested to see the collection des vélins, of which he or Boccius is likely to have seen repeats in the Imperial Library in Vienna (see Chapter 2 and 3). This is of course only a hypothesis, just like a possible contact with Franz's homologue in Paris – Pierre-Joseph Redouté.

Franz would be staggered by the many parallels in their lives. Born on 10 July 1759 in the small town of Saint Hubert in the Ardennes, then belonging to the Holy Roman Empire and more specifically the Habsburg Low Countries, Redouté had moved to Paris in 1782[18] like Franz had moved to Vienna in c.1781. There the former became botanical illustrator, working for a long list of commissioners starting with L'Héritier de Brutelle who had invited him to come over to London (see Chapter 8). Back in Paris Redouté became associated with royalty: first with Archduchess Marie Antoinette, Queen of France, who gave him the title of 'peintre du Cabinet de la Reine' (painter of the cabinet of the Queen), followed by Joséphine, Empress of the French, Napoleon's first wife. She paid him no less than 18,000 Francs per year for his services of documenting the botanical novelties and rarities in her gardens in Malmaison and Navarre. Joséphine was followed by Archduchess Marie Louise, Napoleon's second wife, and, at the time of Franz's visit, by Marie-Amélie, the wife of Charles X, King of France. Because of very generous support by his protectors, among them Napoleon, an extremely high percentage of Redouté's plant portraits had been published in luxury publications that Franz could only dream of. These include Redouté's *Les Liliacées*, Étienne-Pierre Ventenat's *Jardin de la Malmaison* and Henri Louis Duhamel du Monceau's *Traité des arbres et arbustes* (the so-called *Nouveau Duhamel*) just to name the most important ones. Largely refraining from the use of the microscope and the dissection of flower and fruit parts, Redouté had endeared himself to the public by a somewhat softer, less scientific approach than that of the Bauer brothers, which made him extremely fashionable during the Premier Émpire. In 1826 he was to finish the first reissue of his most popular work, *Les Roses*, which was to see one further new edition while he lived. There are further parallels: Redouté had also painted a few flower pieces and also had a younger brother – Henri-Joseph, a natural history illustrator like himself. The latter was involved in the *Description de l'Égypte*,[19] a publication in twenty-three volumes in mammoth folio (1 × 0.81 metres) appearing in 1809–1829. This work offered a comprehensive scientific description of ancient and modern Egypt, including its natural history, a project going back to Napoleon's expedition to that country in 1798–1801. In short, had such a meeting taken place, both Franz and Pierre-Joseph Redouté, who was to predecease him by only five months, would have had occasion for a long and interesting conversation.

FERNS

By the year 1833 at the latest, definitely after the death of Home, two fellows of the Royal Society began to collaborate on another project. One of them was Franz in Kew, the other Hooker in Glasgow. How their communication worked, remains unknown. The final outcome, forty-one watercolours by Franz, was in the end sent to Glasgow, but they came back and are now conserved in the Library and Archives of the Natural History Museum. However, it seems necessary to digress for a moment and deal with the life of Hooker, who effectively became Franz's last commissioner – albeit only briefly, since Hooker was, as far as Franz was concerned, definitely less important than Banks and Home. In addition the work to be done for Hooker was much more monotonous and repetitive than that undertaken for Home.

Like Sibthorp, William Jackson Hooker belonged to the landed gentry.[20] He had inherited an estate at the age of eleven and lived under comfortable circumstances. Born in Norwich, Norfolk on 6 July 1785 as the second son of Joseph Hooker of Exeter, he showed an early interest in natural history. Sir James Edward Smith (see Chapter 7), was later also living in this town and focused the boy's attention on botany. While it was conventional to receive some training in drawing, young Hooker got excellent and in-depth tuition, which enabled him to produce very fine plant illustrations and later to control with great competence the work of other draughtsmen as well as that of graphical artists, echoing in this respect Jacquin in Vienna (see Chapter 3). As a gentleman of private means, Hooker travelled to Iceland in 1809 at the suggestion of Banks and became a member of Banks's select circle, with free use of the library and herbarium at 32 Soho Square. In 1814 Hooker spent many months in France, Switzerland and the northern parts of the Apennine Peninsula. Back in England he married a wealthy woman – Maria Dawson Turner, the eldest daughter of the banker Dawson Turner (see Chapter 13). The couple settled down at Halesworth in Suffolk, where the Hookers raised a family and gathered a herbarium. He took a share in the local brewery and moved into the brewery house with a large garden and conservatories. Mosses and ferns, fields of botany only very incompletely known at that time, became Hooker's speciality, and he started to publish a most impressive series of publications on these plants. With Smith, Lambert, Konig and his father-in-law among his supporters, Hooker was elected a Fellow of the Royal Society in 1812. In 1820 Hooker was appointed Professor of Botany at Glasgow University and Director of the Botanical Garden, retaining the partnership in Halesworth, with the salary from the university supplementing his income from the brewery. Full of energy and ambition, Hooker continued to publish extensively. What is notable in our context is the simple fact that he had most of his works fully illustrated and established a link with several botanical illustrators.

A close collaboration between Franz and Hooker, a man twenty-seven years younger than Franz followed, with the subject of their studies being ferns,[21] a group of plants which had not before been in the focus of Franz's attention. The third element in this collaboration was John Smith, one of Aiton the son's foremen at the Royal Garden at Kew, who noted in retrospect: 'In the year 1822 I found the collection of Ferns at Kew extremely poor, especially as regards Tropical species, very many of those introduced in previous years having been lost, and very few new ones added. Any person who remembers the hothouses in existence forty years ago [i.e. in 1826], will have but little difficulty in accounting for the falling off of the Fern collection. In those days hot-water pipes were unknown, and the houses were exclusively heated by means of brick flues, too often imperfectly constructed, and the excessively dry and ungenial [uncongenial] atmosphere thus induced was quite unsuited for the good cultivation or even the mere preservation of these moisture-loving plants . . . I cannot estimate the entire Kew collection of exotic Ferns at that period at more than forty species'.[22] In the same year (1822) the fern collection at the Royal Botanic Garden in Schöneberg near Berlin, visited by Franz in 1788 (see Chapter 7), comprised ninety-one species.[23] A few years later Smith added: 'In 1825 I arranged the tender ones in a group at the end of the then lean-to house, the space they occupied being 12 feet by 6 feet . . . New species were occasionally imported, and others raised from spores, the spores being obtained from collections of dried specimens, chiefly from the West Indies, Brazil, and Australia, also from a collection of Indian specimens presented to me in 1829 by the late Dr. [Nathaniel] Wallich'.[24]

Franz's watercolours have been signed by him, but the annotations, albeit often fragmentary, are in the hand of Hooker. As a rule a date is given, which indicates that the work on ferns was done in the years 1833 to 1836, and came to an end when Franz was 78 years old. Both living material and herbarium specimens have been documented by Franz, among the former ferns cultivated in the Royal Garden at Kew, but also a specimen of wall-rue, *Asplenium ruta-muraria*, from St Anne's Churchyard in Kew. In several cases Hooker indicated the provenance of the specimen, always in general terms such as 'South Sea Islands', 'The Cape' or 'Jamaica', with a few provenances outside the British sphere of influence, such as 'Brazil' and 'Mexico'. Very occasionally Hooker gives the name of the collector, among them David Don, Wallich, who both had been to the Indian subcontinent, and Ward. The latter had gathered a specimen of the widespread Tunbridge filmy fern, *Hymenophyllum tunbridgense*, in Ireland. The Norfolk Island tree fern, *Cyathea brownii* (Fig. 16.7.), collected and recorded previously by Ferdinand (see Chapter 10), was now documented by his brother, not on the basis of a collection by Ferdinand, but one by Allan Cunningham, who had been on this island after him. In this watercolour Franz showed, for example, (1) the lower surface of a small proportion of the frond, (2) the upper and lower surface of a lobe of the frond, (3) a complete sorus (a cluster of caspsules), and a longitudinal section of it, (4) a transverse section of the basal part of the sorus with the capsules removed, (5) spore capsules and (6) spores.

Hooker took a global approach in his *Genera filicum* (see below) and consequently also included *Woodsia caucasica*, a native of the Caucasus area, with the drawing annotated by him '[Ludolph Christian] Treviranus, May 1835'. The latter, a Professor of Botany and Director of the Botanical Garden of Bonn University, was clearly the source of the specimen, not its collector. In the text Hooker, who in this case totally relied on Franz's drawing, stated that he had not seen the specimen and copied the protologue of the name of this fern, which had appeared in St Petersburg.[25] When Franz's illustration appeared in print it was the first representation of *Woodsia caucasica*, and the same applied to several other plates. Hooker is full of praise for Franz's work, writing for example: 'The admirable analyses by Mr. Bauer, in the accompanying figure, will illustrate the nature of fructification much better than can be done by words'. Another case is *Loxsoma cunninghamii*, collected by Allan Cunningham at the Kerikeri River near the Bay of the Islands on the Northern Island of New Zealand. Several of Franz's figures arranged in a plate were first published by Cunningham,[26] who added that he had profited from them when drawing up the generic description. In fact it was the drawing which was described by Cunningham, not the specimen. Not long after, Hooker published in the *Genera filicum* Franz's complete watercolour with a few more figures added and could not refrain from noting 'The admirable analysis of Mr Bauer . . . will illustrate the nature of the fructification much better than can be done by words', a statement indicative of

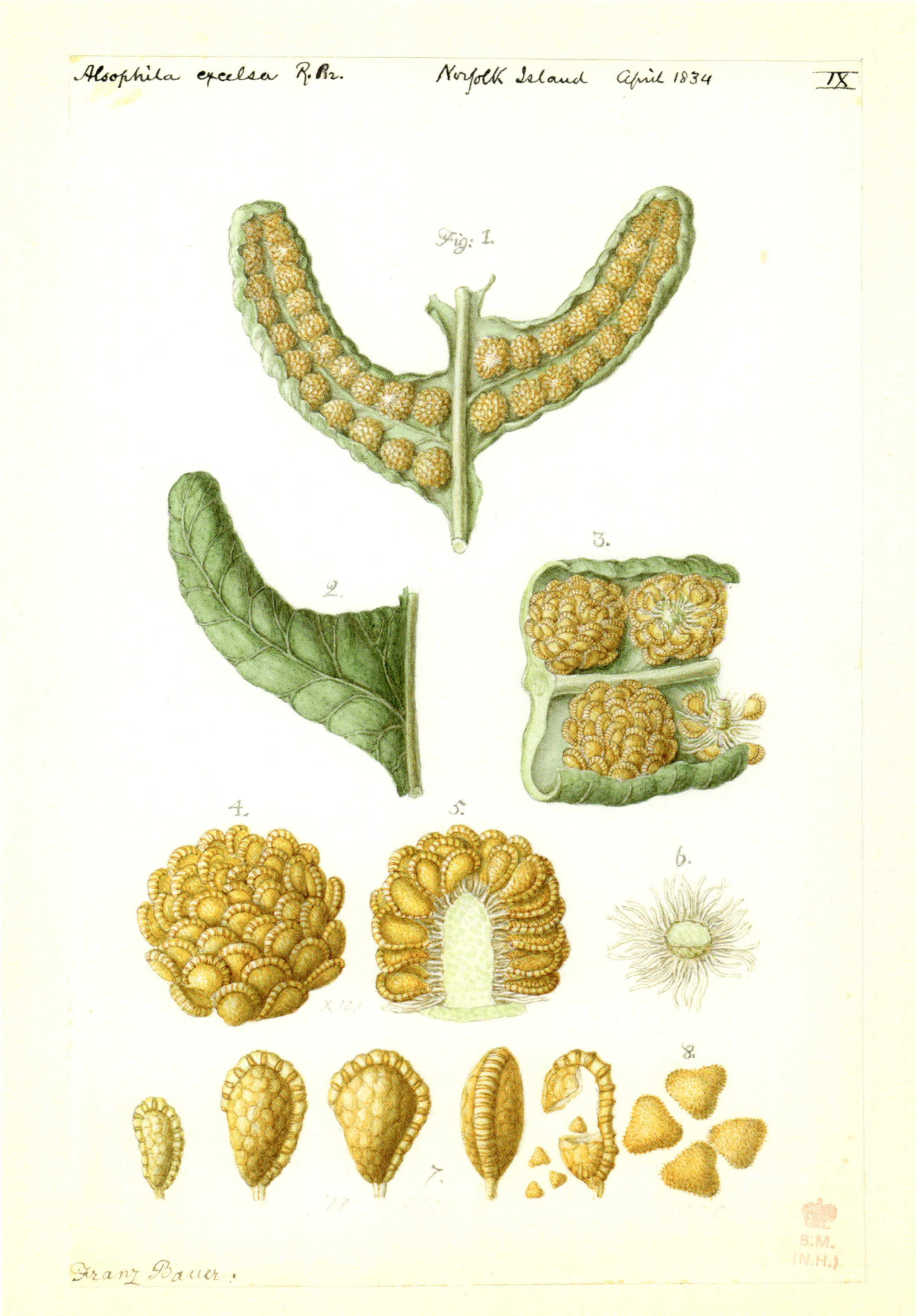

FIG. 16.7. *Cyathea brownii*. Watercolour by Franz Bauer, 1832. Paper, ink, watercolour. – London, Natural History Museum, Library and Archives. Original drawings by Francis Bauer for Hooker & Bauer's Genera Filicum, no. 9.

the inferiority of every botanical description versus an illustration of superior quality. However, in at least one case Hooker disagreed with a detail in the latter's watercolour, the shape of the annulus in the spore capsules of *Parkeria pteridoides*, which Hooker regarded as square but Franz had drawn as a rectangle. Hooker found a diplomatic solution: he wrote 'nor do I perceive a further trace of an annulus above or below that spot, as depicted by Mr Bauer, whose accuracy, however, in this, as in all his botanical analyses, cannot for a moment be called in question'. Upon publication, Hooker had a drawing which he thought to be correct added to Franz's illustration. It is impossible to find out who was right and who was not, since the factual basis of Franz's illustration has not been indicated, neither on the watercolour nor in Hooker's text.

Although documenting very different subjects, Franz's drawings for Hooker and those previously done for Home are similar in a particular way: right from the beginning they were intended to be published, which led to a smaller format of the image and an economic use of the space on the paper. Gone were the days of the documentation of nature in the Banksian style, the painted encyclopaedia of the diversity of plant life in large format kept in the library of a gentleman of private means – with neither necessity nor intention to publish.

From a pencil drawing on squared paper kept in the Niedersächsische Staats- und Universitätsbibliothek in Göttingen, it becomes evident that in the years 1834 and 1835 Franz also induced fern spores to germinate and recorded their growth in dated pencil drawings. In addition, he produced a watercolour, again on squared paper, documenting fern spores of twenty-seven species with a minimum of three spores per species in order to document intraspecific variation, which is now also in Göttingen and refers to the *Genera filicum*.

Hooker's intention was to give an illustrated overview of the worldwide diversity of ferns at generic level. Franz produced the watercolours for the plates 1–14, 16–40 and 50–51, and Walter Fitch prepared many more: 15, 41–49, 52–120. Printing took place in Glasgow. It is totally unclear why Hooker suppressed the information on the provenance of the specimens documented in the lithographs, which thereby reduced the scientific value of his opus. When Franz saw the title page of the first instalment of Hooker's *Genera filicum* he may have been satisfied. The title is nicely worded: 'Genera filicum or Illustrations of the genera of ferns; from the coloured drawings of Francis Bauer, Esq. Botanic Painter to Her Majesty; with descriptive letterpress, by William Jackson Hooker, K. H., LL. D., F. R. A., & L. S., &c. &c &c. and Regius Professor of Botany in the University of Glasgow'. However, it is not quite clear why Hooker with all his illustrious postnomials, among them the Knighthood of the Guelphic Order and his Fellowships of the Royal Asiatic and Linnean Societies, suppressed his and Franz's fellowship of the more prestigious Royal Society as well as Franz's fellowship of the Linnean Society. As a matter of fact, Franz's title of 'Painter to Her Majesty' (Queen Victoria) is not substantiated from the archival record, and may in fact be an informal continuation of 'Painter to His Majesty' (King George III) as given on the title page of the *Delineations* (see Chapter 8). In any case the *Genera filicum* contributed more to the rising fame of Hooker than of Franz. Publication began in 1838, but when the last lithographs based on Franz's watercolours, t. 51, appeared in 1840,[27] Franz had only a few more months to live. Ferns remained Hooker's main topic for life. His epitaph in St Anne's Church in Kew, a Wedgewood plaque, shows him in profile together with the fronds of several fern species.

TAKING PRECISE MEASUREMENTS WITH THE MICROSCOPE

The terms 'hard science' and 'soft science' are used colloquially to compare the individual fields of science on the basis of methodological rigour and legitimacy. Undoubtedly, biology belongs to the hard sciences, but it is definitely softer than physics or chemistry. Within the vast field of science summarized under the term biology, the description, analysis and interpretation of biodiversity stand at the soft end. However, over time this branch of study has gradually become somewhat harder, due to an increasing emphasis on accurate measurements and the statistical analysis of the resulting data. Franz contributed to this development by a precise description of how to take measurements of observations by the microscope. This took the form of a letter[28] dated November 1836 and sent to Andrew Pritchard, a naturalist and optician selling microscopes and micro-slide preparations based in London. Starting in 1830 the latter had got engaged in the publication of illustrated books on very small organisms. Franz's letter was published as an appendix to *Micrographia*, printed in London in 1837 and co-authored by Pritchard. Naturally the problem of taking precise measurements was an old one and many had contributed to it, among them Jacquin the son in Vienna in 1828.[29]

Franz's advice of how to arrive at accurate measurements with a microscope is crystal clear: 'The requisite apparatus for ascertaining with a microscope the correct dimensions and proportions of any minute object, are two glass micrometers, each of forty divisions to an inch, and crossed or squared over their whole surface . . . One of them ought to be pretty sharply engraven [sic], on a very thin and clear plate of glass: this micrometer is fitted into the eye-piece in the focus of the eye-lens; the other micrometer, which is used on the stage, ought to be very strongly engraven [sic], and its lines well blackened, that they might be distinctly seen when viewed through the micrometer in the eye-piece. When my microscope was thus armed and adjusted, I ascertained the acting magnifying powers of its objective lenses, which is effected by placing the stronger engraved micrometer on the stage, and viewing and carefully observing, through the micrometer in the eye-piece, how many divisions are occupied by one division, or fortieth part of an inch of the micrometer on the stage: I found that with the objective, No. 2 [probably of Plössl's Grosses Mikroskop No. 1, see Chapter 17 and below], it occupied precisely ten divisions each way, and consequently divided one linear inch into 400 parts . . . Having [this ascertained] . . . I now retain this adjustment of the microscope permanently, and have no more occasion to use any micrometer on the stage for future operations'. The following text describes Franz's process of documenting his microscopic preparations in a drawing: 'I trace on my drawing paper a number of squares, similar to those on the micrometer, of which each division is an inch (linear), which consequently is 400 times longer than a 400th linear division on the micrometer in the eye-piece . . . , I then place the minute object . . . on the stage, view it through the squared micrometer in the eye-piece, and move the object on the stage, till it comes into the proper position to be examined; this is, till one extremity of the object touches one line of the square on the micrometer of the eye-piece. I then trace on my ruled drawing-paper a correct outline of the object . . . '. Franz also knew the advantages of sliding tubes for the eye-piece. This is how he explained the calibration of a microscope: 'I must also notice, that in adjusting the microscope for ascertaining its magnifying power, great care must be taken that the lines of the square of the micrometer on the stage exactly join and encompass the lines on the squares on the micrometer in the eye-piece, which is not always the case, or with some objectives the image may perhaps be found to occupy *nine divisions and a fraction* [Franz's italics], which would occasion much calculation and trouble to produce a correct drawing; but this difficulty is easily obviated, when the microscope is constructed with sliding tubes, for by elongating or shortening the tubes, the instrument can be adjusted to the greatest exactness, and on that adjustment depends the correctness of the whole.

In an appended lithograph Franz shows on squared paper different fossil diatoms, spores of common bunt (see Chapter 8) and erythrocytes isolated from human blood. While the source of the spores and erythrocytes is not indicated, we are informed about the provenance of the mixed sample of diatoms – tripolite, then called tripoli, a diatom containing sediment from Karlsbad. When preparing this outline drawing Franz may have remembered his stay together with Jacquin the son in this famous Bohemian spa in May 1788, 48 years earlier (see Chapter 7). By modern standards the accuracy of Franz's measurements following the method described above is not overwhelming, but he was more precise than several of his contemporaries, among them such celebrities as Wollaston and Young. For the diameter of a human erythrocyte he gives '1-2500th part of an inch', which is approximately 0.010 millimetres or 10 micrometres, while in fact it is approximately 0.0075 millimetres or 7.5 micrometres. More remarkable than this deviation is another fact. In his letter to Pritchard Franz corrects his own measurements published in 1817 and gives a convincing reason: 'an improved achromatic microscope'. Four more objects are depicted in the lithograph: a pollen grain of giant granadilla, *Passiflora quadrangularis*, with the three ring-like apertures clearly indicated (although ignoring the reticulate, lophate ornamentation not yet visible even with his excellent microscope); a triporate pollen grain of a species of evening primrose, *Oenothera* spec.; and a ripe and a burst-open capsule of the fern *Aspidium trifoliatum*, with particular attention to the annulus and the form and size of the spores thrown out of the capsule.

What kind of improved achromatic microscope did Franz use? This was almost certainly Plössl's Großes Mikroskop No. 1,[30] which was described in detail in the auction catalogue of Franz's belongings (see Chapter 17) published four years after he had written the letter to Pritchard. Simon Plössl was a well-known optician in Vienna, supported by Jacquin the son and others, with this very same model of microscope on the market since 1828. It has been reported that until 1835 he delivered it to several prominent institutions,[31] such as the Imperial Academy of Sciences in St Petersburg, the Royal Academy of Sciences in Berlin, and to several universities. Baer (see Chapter 15) made use of it, as did two famous physiologists – Jan Evangelista (later Ritter von) Purkyně, then at Breslau University (now Wrocław, Poland) who was to describe the Purkyně fibers on the tip of the human heart, and Theodor Schwann, then at Leuven University, who was one of the fathers of the cell theory, a major advancement in the field of biology. Unsurprisingly a considerable number of Plössl's Großes Mikroskop No. 1. was purchased by customers in Vienna, among them Archduke Ferdinand, the eldest son of Francis I, Emperor of Austria, and since 1835 as Ferdinand I his successor on the throne; Archduke Karl, who had won the Battle of Aspern-Essling against Napoleon; Metternich and Kornhäusel, the architect engaged by Johann I Joseph, 10th Reigning Prince of Liechtenstein (see Chapter 12). Only two scientists based in the 'kaiserliche Haupt- und Residenzstadt' are mentioned here: Jacquin the son and Endlicher (see Chapter 18). The latter is reported to have given botanical lessons to Emperor Ferdinand I,[32] and it is likely that these contained demonstrations with the microscope. Jacquin the son, Franz's travel companion in the years 1788–1789 and later his correspondent, may well have had a hand in organizing the transport of the famous instrument from Vienna to Kew. Considering its high price and Franz's contacts it is very likely that the transport from Vienna to Kew was handled by the Imperial Embassy in London, then headed by Paul III Anton Prince Esterhazy [Esterhazy Pál Antal]. In fact in September 1829 Plössl's Großes Mikroskop No. 1 had won the first prize at the Achte Versammlung der Naturforscher und Ärzte (Eighth Meeting of the Naturalists and Physicians) which had met in Heidelberg,[33] then in the Grand Duchy of Baden, with Sternberg and Brown attending[34]. The latter was to purchase this very microscope from the estate of Franz in 1841 (see Chapter 17). Today complete early specimens of Plössl Großes Mikroskop No. 1 are exceedingly rare outside Vienna, one such dated 1840 is in the private collection of Timo Mappes[35] in Eggenstein-Leopoldshafen in Germany.

In an age before methods for fixing and embedding animal tissues were developed and before the sectioning of such preparations with a microtome was routinely done, close microscopic study remained difficult. Needless to say, taking accurate measurements was impossible, since micrometers could not be used. Franz was inventive: in his letter to Pritchard he reports to have had several 'silver plates (of about a square inch [6.45 centimetres squared], and almost the thickness of a shilling) constructed, and in the middle of them regular and accurate square holes or perforations cut out, from half an inch [12.7 millimetres] to a 1–10th of an inch [2.54 millimetres] linear'. Opaque preparations were examined in water and then one of the silver plates was placed upon the spot to be examined. Subsequently a contour was taken on the squared drawing paper following the principles outlined above. Franz developed the excellent idea to improve his 'silver plate micrometer, by dividing, with some fine silver wires, the orifices into smaller divisions'; however, he could not find anyone who could create such a drawing aid. Most relevant in the context of our story is a small note in Franz's text stating that he had used this approach to taking measurements for more than thirty years (since c.1805), when he had produced the extensive series of watercolours documenting cereal pests for Banks (see Chapter 8). In the preface to the *Micrographia* Pritchard stated that Franz had prepared more correct drawings of microscopic objects 'than any other person in England'. This may have been no exaggeration, although he definitely was not the first or last to publish on taking precise measurements.

The letter by Franz to Jacquin the son written in the year following the publication of the *Micrographia* reveals Pritchard's other field of interest – dealing in natural history objects. A curious mixture of subjects is sent by Franz to his former travel companion in Vienna, among them 'a very fine section of common Flint stone . . . sections of the human bone . . . Quartz with moving fluits [sic] . . . Beautiful Section of the Stem of *Doryanthes excelsa*', that is, the gymea lily documented by Ferdinand in Australia (see Chapters 10 and 12). Furthermore 'four copies of private papers on the Improvement in the Microscope mostly by [Cornelius] Varley' were announced to be sent to Jacquin the son. Today known primarily as a watercolourist and original member of the Society of Painters in Watercolours, Varley had before worked as a scientific instrument maker and published on this subject. All this demonstrates the breadth of Franz's interests, which one can of course also interpret as a lack of focus, now anathema in scientific research. However, for Franz, a man in his eighty-first year, there was no obligation on him to focus. Unfortunately the letters sent by Jacquin the son to Franz during this period could not be traced. They would possibly have contained information on the last developments in the world of science in Vienna. Francis I, Emperor of Austria, had been succeeded by his son Ferdinand (see Chapter 12), a figure with limited political power, but a man strongly interested in science, in particular botanical studies with the microscope (see above). Supported by Metternich, Ferdinand I was to act in 1847 as founder of the Imperial Royal Academy of Sciences with Endlicher (see Chapter 18), Jacquin the son's successor in the chair of botany and as Director of the Botanic Garden at Vienna University, a strong candidate for the first president.

Over the years a feeling of confidence and trustfulness seems to have developed between Franz and Jacquin the son, with Franz informing his friend in the letter of 9 June 1838 (written in English) about the state of his health 'I assure You that I have been extremely ill this 14 or 15 days, suffering most violent paroxisms every night (of fits) of five to six hours of spasmodic atacks [sic] of which I think each of them is the last of my existence, last night was one of my worst sufferings, You must therefore excuse any omission or neglect, though I have been near three days scrawling this letter together'. Although at that moment no doubt quite ill, Franz was to outlive his former travel companion, who was to die on 26 October 1839 in Vienna. The latter's son-in-law Schreibers had played an important role when Ferdinand's estate was purchased by Franz I, Emperor of Austria (see Chapter 17).

NIÉPCE'S EPOCH-MAKING INVENTION

On 26 February 1839, aged eighty, Franz sat down and wrote a long letter to the Editor of the London weekly *The Literary Gazette and Journal of Belles Lettres, Arts, Sciences & c.*, which appeared in print on 2 March 1839.[36] It referred to a visitor from Chalon-sur-Saône in eastern France whom Franz had met in Kew more than eleven years before and to the gifts he had received from him. The text resulted in comments in the press, in London and elsewhere, and made Franz a well-known figure among specialists for the early phase of photography.

In order to understand the context of Franz's letter it is necessary to set out briefly the sequence of events in the small town in Burgundy, which led to the invention of heliography, heliogravure and the physantype process. This will provide the necessary background to understand what subsequently happened in Kew, Windsor and London, where Franz played an important role. The materials[37] documenting this fascinating story are widely spread today – one key object, previously in the possession of Franz, is now in the Harry Ransom Center at the University of Texas in Austin, Texas; three more items owned by Franz are the property of the Royal Photographic Society in Bath, but permanently deposited in the National Media Museum in Bradford; the very few other specimens of the new techniques, such as those in the Musée Nicéphore Niépce in Chalon-sur-Saône, have no reference to Franz and are not dealt with here. The most significant part of the pertinent correspondence was acquired by Hamel (see Chapter 15) and is now in the St Petersburg Branch of the Archive of the Russian Academy of Sciences, perfectly edited and commented.[38] Needless to say, the literature on this topic is vast and reference is made here only to the core story and the part played by Franz, who was definitely the first to fully grasp the novelty of the invention.

What kind of invention was this and who was the inventor? Nicéphore Niépce (né Joseph Niépce, Nièpce or Niepce), the son of a wealthy lawyer and council to the King of France, was born in Chalon-sur-Saône in 1765 under the reign of Louis XV and thus almost seven years junior to Franz.[39] Having served in the revolutionary army and acted as an administrator in the district of Nice, Niépce and his family returned to their native town, and more particularly to the family's property in Saint-Loup-de-Varennes, to pursue scientific research. This led to the world's first internal combustion engine, which attracted very limited interest and failed commercially, and to the invention of heliography, the world's oldest photographic process. While the idea of capturing images was an old one, Niépce was the first to master successfully all the processes – the capturing of the image, the fixing of the image and, in heliogravure, the transfer of the image to another support. In addition he also chose the substrate on which the image would finally rest. Needless to say, extensive trial and error was necessary, with first tests taking place in c.1813. Niépce tried several substrates, such as lithographic stone, copper, silvered copper, silver and glass, but some of his most important artefacts are on pewter, an alloy composed primarily of tin with additions of various other metals such as copper and lead in order to affect its colour, hardness and workability.

All four items have this material as support, the pewter plates conventionally known as 'Le Cardinal d'Amboise',[40] 'Christ Carrying his Cross',[41] 'Un Clair de Lune' (also called 'Ruines d'une abbaye'),[42] and, the most famous of the four, the 'View from the Window at Le Gras' (also known as 'Paysage à Saint-Loup de Varennes').[43]

Detailed analyses with sophisticated technologies have demonstrated that Niépce made use of hand-finished and polished pewter plates of different composition, size and thickness.[44] Finding a substance which would change its physical and chemical properties in light irreversibly that was fit for the purpose was a greater challenge. Niépce's solution to the problem was a thin layer of bitumen, a black resinous material also called asphaltum, obtained from crude oil deposits and a by-product of petroleum refining.[45] Dissolved in oil of lavender, bitumen varnish was applied to pewter plates and heated or left to dry. When exposed to light the thin layer would slowly harden, with the bitumen polymerizing; when unexposed it would remain soft, and no polymerization would occur. In short, the varnish was light-sensitive.

Niépce experimented with copper engravings, which he had made transparent with oil. These he placed on his bitumen-coated pewter plates so that they were in very close contact and exposed them to the sun. For good reason this approach has been also called the contact method.[46] Paper that became transparent would permit the light to reach the light-sensitive layer and harden it. By contrast, paper covered with printing ink would hinder light reaching the light-sensitive layer, which would remain soft. After exposure to light Niépce removed the varnished copper engravings and had a latent image on the light-sensitive layer. It was relatively easy to develop it (make it visible). With lavender oil he washed away the soft parts, while the hardened parts would adhere to the pewter plate. The result was a permanent image on a substrate produced by the sun; therefore the end result was called heliography, or 'sun drawing'. This was a very major breakthrough. Without having used a drawing pen or an engraving utensil Niépce had managed to arrive at a light-stable picture. In short, he had induced the light to produce an image.

However, Niépce did not stop at this stage. He put the pewter plate with the developed image (that is, the hardened bitumen spots) first into an acid etch, which would attack only the exposed portions of the plate, and then into a bitumen stripping bath, which resulted in the hardened bitumen spots being washed away from the plate. In the end he had arrived at an etched pewter plate, which carried the image and could be used as a printing form without further manipulation. This process is called heliogravure. The 'Cardinal d'Ambroise' and 'Christ Carrying his Cross' are very early examples of such etched pewter plates, a copper engraving of the Cardinal varnished by Niépce before he started the process is in the Musée Nicéphore Niépce, prints of the Cardinal d'Amboise taken from the etched plates are also in this museum and several other collections, among them the Royal Photographical Society, deposited in the National Media Museum. The latter print was previously owned by Franz. A variant of the heliogravure is a process called physantype. This was applied when producing the image of the copper engraving 'Un Clair de Lune'. The latent image was developed by placing the coated pewter plate, which had been exposed to light, in vapour above an alcohol bath, which resulted in the dissolution of the soft parts of the varnish. In this case no attempt of etching is known to have been undertaken, the varnished copper engraving of 'Un Clair de Lune' used by Niépce to prepare the physantype is now kept in the Cinémathèque Française in Paris. Niépce had prepared these three images by making use of the contact method.

Surprisingly, Niépce was not even willing to leave it at that. He had the ingenious idea of placing a coated pewter plate into a camera obscura provided with a lens. No doubt after some experimentation with lenses of different power and a very long exposure time, Niépce arrived at the first heliograph ever made directly from nature (without the light-sensitive layer having been in contact with a pre-existing image like a copper engraving). For this he had chosen the view from the window of the first floor of his property Le Gras. When the latent image was developed with a mixture of lavender oil and kerosene, it showed the chicken run and the adjacent farm building. No further processing was undertaken. This icon of the history of photography was also in Franz's possession. Although another major breakthrough was achieved, Niépce's images were far from satisfactory – the 'View from the Window at Le Gras', a heliograph, and 'Un Clair de Lune', a physantype, could only be viewed in a landscape orientation and in an inclined position; the reflection of the polished pewter plate caused additional problems and the images were rather faint. By contrast, the two heliogravures could be viewed in a portrait orientation and the prints taken from them used like a conventional lithograph.

Niépce remained extremely secretive about his spectacular inventions and exchanged his views with few friends, refraining from publishing a single word. In retrospect the reasons are clear: he wanted either to have the newly discovered heliographic process patented or, alternatively, have it sold, to the state for example, which would put an end to his life-long financial troubles. When Nicéphore Niépce decided to visit his seriously ill and senile brother Claude, who was living in Hammersmith near Kew, he took with him specimens of his inventions – heliographs, heliogravures and physantypes – and tried to attract the attention of the community of the learned in London. Accompanied by his wife, Niépce took lodging at Kew Lane (now Kew Road) at the house of Mr and Mrs Cassel.[47] Thanks to the letters in St Petersburg and Franz's letter of 28 February 1839 to the *Literary Gazette*, the subsequent movements can be followed almost day by day.

Niépce's first contact at Kew was Aiton the son, who having been shown the heliographic material brought along from France, suggested that it should be brought to the attention of the Royal Court. Indeed, the specimens were sent to Windsor Castle and displayed in the office of the Chamberlain of the Royal Household, then James Graham, 3rd Duke of Montrose, in the week beginning 12 November 1827. Whether this heliographic material was ever inspected by George IV remains unknown; in any case, the items were returned to Kew six days later. When Niépce came to collect them from Aiton the son, the latter recommended contacting Franz, who had received a first letter from Niépce written on 19 November 1827. This message contains the essence of Niépce's discovery 'je m'occupe des recherches sur la manière de fixer l'image des objets par l'action de la lumière' (I occupy myself with research about the way to fix the image of subjects by the action of light). This letter resulted in a visit of Franz, who seems to have been greatly impressed with the specimens. Niépce wrote to his children in France about Franz (in French), 'this is a man of a certain age and a distinguished naturalist. My discovery seemed to him something very extraordinary and of very great interest.' Wisely Franz suggested to Niépce that he should write a brief report of his invention to be sent to Home (see Chapter 15), which was delivered to Franz on 22 November 1827. However, in the covering letter Niépce exposed the Achilles Heel of his project in saying (in French), '. . . it was difficult to explain myself in a satisfactory manner on certain details without compromising my secret.' Images produced by the heliographic process were sent to Franz on 30 November 1827. After the failure at Windsor Castle, two alternatives were followed: visits to the Royal Society of Arts and the Royal Society were planned. However, both initiatives turned out to be unsuccessful – at the former because of trivial reasons, at the latter because of the secret Niépce was unwilling to disclose. Niépce's contact, the polymath and long serving Foreign Secretary of the Royal Society Thomas Young, wrote to Home. 'Mr Niépce's invention appears to me to be very real and elegant and I have no knowledge of any similar method having before been employed . . . How far it may become practically useful thereafter is impossible for me to judge especially without knowing the whole of the secret.' Wollaston, like Young deeply involved in optics, was prepared to see Niépce, but nothing seems to have come of this.

In February, possibly after the death and funeral of Claude, who was interred in St Anne's Churchyard at Kew, Niépce and his wife returned to France, where the experiments continued. Before leaving Kew, possibly disappointed by his failure to attract attention, Niépce presented specimens of his heliographic works – one to Mr. Cassel, one to Home, both untraceable today, and four to Franz. In his methodical manner the latter annotated his presents on the back, as every owner of an art gallery would do. For example, 'The View from the window at le Gras' (Fig. 16.8.) has 'L'Heliographe obtenu spontanément par l'action de la lumière. 1827. Par Monsieur Niepce De Chalon sur Saône. 1827. Monsieur

FIG. 16.8. **Reverse side of view from a window at Niépce's estate at Le Gras, 1826–1827. with annotations by Franz Bauer. – Austin, Texas, University of Austin, Harry Ransom Center.**

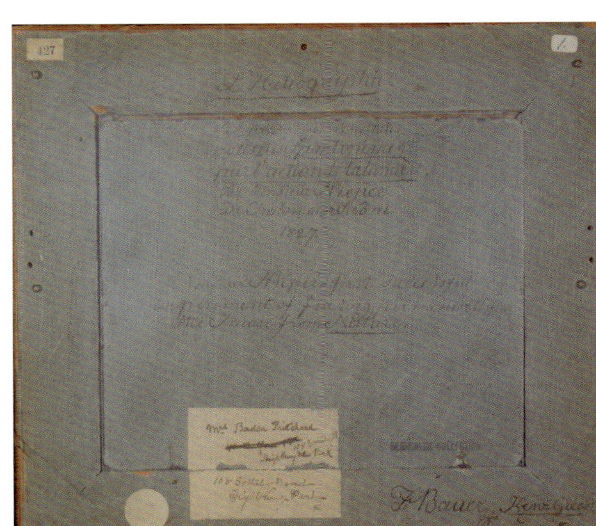

Niépce's first successful experiment of fixing permanently the Image from Nature. F. Bauer Kew Green'.

Back in France Niépce wrote at least three letters to Franz, all now in St Petersburg, but then the correspondence seems to have been discontinued. Without having published a single word about his revolutionary invention and without any appreciation of it, Niépce died in his native town on 5 July 1833. Only six years later his text 'Notice sur l'héliographie' appeared in the *Historique et description des procédés du daguerrographie* authored by Louis Daguerre, whose more successful, though quite different, method totally eclipsed the pioneering work of Niépce.

By annotating his productions of the heliographic process and by preserving them, Franz had done an extremely valuable service to the history of photography. His letter to the Editor of the *Literary Gazette* disapproving of Daguerre's claim for priority and stressing Niépce's pioneer role is similarly significant. In addition Franz's text offers a detailed description of the events in late 1827 in retrospect. Here is his summary: 'M. Niepce was soon introduced to some of the most efficient members of the Royal Society, to whom he presented his memoir and several very interesting specimens – the products of his discovery; but though he had several interviews with those gentlemen who had that memoir several weeks under consideration, because M. Niepce declined explaining the secret *then* [Franz's italics], the Memoir and all the specimens were returned to him, and the subject was never laid before the [Royal] Society'. Appended to Franz's letter published in the *Literary Gazette* is the English translation of a manuscript by Niépce entitled 'Héliographie' signed and dated 'Kew 8th December 1827'. The original version in French is now also in the Harry Ransom Center at the University of Texas in Austin. It is probable that Franz was the translator of Niépce's tract, but this is not certain.

The date 8 December 1827 is the key argument for Niépce's priority vs. Daguerre. An immediate effect of Franz's letter was to bring Niépce's images before the Royal Society. On 9 March 1839, a week after the publication in the *Literary Gazette*, Niépce's gifts to Franz were exhibited at a Royal Society soirée, given by the newly elected President, the Marquess of Northampton. Only a few days later the exhibition was repeated at a meeting of the Graphic Society in London, founded by Franz's portraitist Brockedon. However, because of his age Franz probably did not take part in the two events. Instead they were organized by Charles Wheatstone, a Fellow of the Royal Society, together with William Henry Fox Talbot, the latter deeply involved in another approach to photography. Both had come to see Franz at Kew and to collect the precious specimens he had kept – rightly recognizing them to be the first of their kind and certain that they were documents of a very major innovation for the arts and sciences.

FAREWELL

Franz's last twelve months started on 11 December 1839. He had already completed his eighty-first year, apparently for several years unable to physically leave his house, his body 'thus as it were chained down'.[48] Since 1837 he lived in Eglantine Cottage, 20 Waterloo Place on the north-eastern side of Kew Green, a building pulled down in 1939–1940 and replaced by the house that is now 70 Kew Green. For many years Elizabeth Baker had been Franz's housekeeper and had lived in his household (see Chapter 17). Among Franz's drawings of a private nature purchased by the British Museum in 1862 and now kept in the Library and Archives we find a delicate watercolour showing a wreath (Fig. 16.9.) of white roses and Siberian bugloss,

FIG. 16.9. Wreath of white roses and *Brunnera macrophylla*. Watercolour by Franz Bauer, undated. Paper, watercolour. – London, Natural History Museum, Library and Archives, Franz Bauer Miscellany (loose).

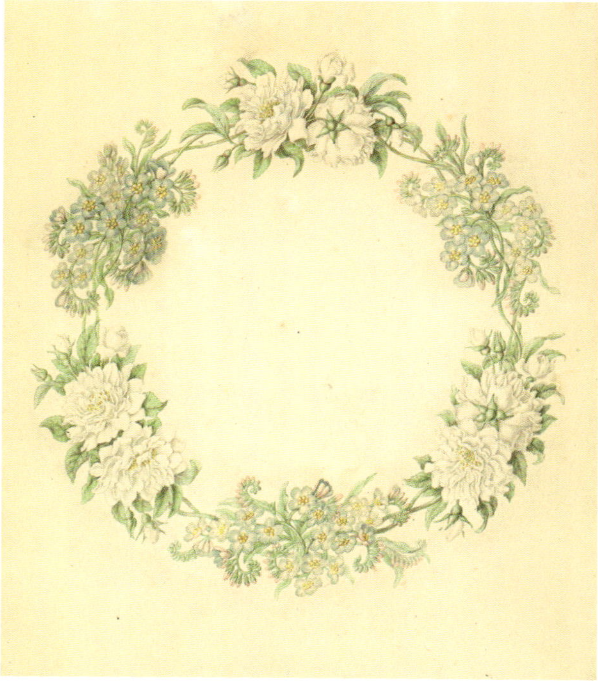

Brunnera macrophylla. The possibility exists that this image is in some way or other connected with Elizabeth, as it bears a marked resemblance to pencil sketches for a finger ring, which are also kept in the Natural History Museum among Franz Bauer's miscellaneous drawings. Needless to say the inscriptions for the bezel refer to botany – 'Pensée a moi' (pansy, *Viola tricolor* hybrids) and 'V. M. N.' (standing for Vergissmeinnicht or forget-me-not, *Myosotis* specp.). Along the same line is the note in calligraphy 'Je ne saurois [sic] l'oublier' (I cannot forget her/him/it) in Franz's hand on one of the sketches. Elizabeth's later life remains unknown apart from being mentioned as Miss B.[etsy] Baker Bauer in a letter by Meyer of 28 October 1845 sent to the solicitor H. D. Haverfield (see Chapter 17).

Like his younger brother in Vienna, Franz in Kew remained active until very late in life. On 14 December 1839 he dated his last manuscript, entitled 'On the ergot of rye', and had it sent to the Secretary of the Linnean Society, of which he had been a Fellow since 1804.[49] It was read at the meeting on 21 January 1840, with a summary published belatedly in the *Proceedings of the Linnean Society*,[50] though missing the key argument. The complete text with two appended lithographs, the first based on a drawing dated 1805, the second on a drawing dated 1839, appeared much earlier, in June 1840,[51] in the *Transactions of the Linnean Society*,[52] so Franz was probably still able to enjoy seeing his last contribution in print. The two texts are interesting for several reasons. Firstly they deal with a topic Franz was familiar with since his days as a boy in Feldsberg, where he must have seen fields of rye infected with ergot, *Claviceps purpurea*. Secondly because Franz made a final attempt to get published at least a fragment, albeit small, of his 'Diseases in corn' prepared for Banks, in particular his work on ergot forming t. 38 and t. 43–47 (see Chapter 8). However, when preparing the first plate Franz probably had to make use of his preliminary drawings, now in the Niedersächsische Staats- und Universitätsbibliothek in Göttingen, since Banks had deposited the final version in the British Museum long ago. Thirdly it is important because Franz's paper demonstrates the limited understanding of the life cycle of this parasitic fungus at the time.

Franz's paper consists of two parts, the first summarizing his earlier observations, clearly pointing to the main problem of studies in ergot: 'The difficulty of tracing the origin of this disease undoubtedly arises from the fact, that not the slightest indication of its existence is observable either in the infected ear or in the plant till the ergot actually emerges from the husk in which it is formed, when it is too late to determine its origin'. However, Franz is totally wrong in stating that ergot is 'a monstrosity, or transformation of the embryo [of the rye plant]'. As we now know, it is a compact mass of hardened fungal mycelium containing food reserves and numerous other substances forming an essential part of the life cycle of the fungus. This was first elucidated in 1851 by Louis René Tulasne, then based at the Muséum National d' Histoire Naturelle in Paris. As a consequence Franz did not follow up the purple-black ergots, which fall off the rye plant, hibernate and germinate in the following season by producing club-shaped spore bearing structures.[53] That the ascospores released from them would land on the stigma of the rye flowers, infecting the ovary, was outside Franz's mind and as a consequence he concentrated on the development of the ergot itself.

In October 1838 John Smith had observed 'several large drops of a brown-coloured fluid suspended from the spike of a species of *Elymus* [wild rye], in which were several full grown ergots' as well as 'the articulations of a minute filamentous fungus which is developed in the early stages of the flower and . . . impregnating the soil or the perfect seed, which on germination and subsequent development carry up some of the reproductive matter of the fungus . . . in the manner that Mr Francis Bauer has shown to take place in the propagation of the smut . . .'.[54] These observations, published first in more detail in the *Transactions of the Linnean Society*, were partly right and partly wrong. In the honeydew Smith had indeed found spores, to be more precise conidiospores, but these were only able to infect the flower of the rye plant, but not remain in the soil and infect the seeds as he thought. Less than a month later[55] Edwin John Queckett, then lecturer on botany at the London Hospital and today epitomized in the *Quekett* [sic] *Journal of Microscopy*, presented a paper at a meeting of the Linnean Society and had it published in the *Transactions of the Linnean Society*, as a matter of fact on the pages preceding Franz's paper.[56] Although giving several more details, Queckett effectively confirmed Smith's finding.

Franz, however, was not convinced by the observations made by Smith and Queckett and the interpretation offered by the latter. Franz's main arguments put forward in his very last paper were that firstly all grasses are infected by that 'minute filamentous fungus', but only very few produce ergots and that among the agricultural crops only rye is subject to this infection. Secondly Franz observed that in autumn 'all decaying plants are infected with such filamentous fungi and minute sporidia', quoting as an example a flower of India shot, *Canna indica*, brought by Smith on the same day as the infected *Elymus*. The inner wall of the anther and the pollen grains of the Indian shot were infected with filamentous fungi. Franz's latter observation was not incorrect in stressing the difference between the filamentous fungi and an element of the ergot's life-cycle, but, he was wrong in refuting the interpretation offered by Smith and Queckett. They had possibly indeed seen in the honeydew the asexually formed conidiospores of ergot that were able to infect rye flowers and thereby capable of spreading the ergot. However, the filamentous fungi on the whole plant could not have originated from these condiospores, but from another fungus. Without appropriate cultivation methods not yet available at the time, the identity of such fungal infections on the leaves of *Elymus* and other plants could not be determined.

An important source for Franz's final years and the situation in Kew at that time is provided by Frederick Scheer. He had been born as Friedrich Scheer in 1792 on the island of Rügen, when this region still belonged to the Holy Roman Empire and was ruled over by Gustav IV Adolf, King of Sweden and Imperial prince. It is very likely that Scheer grew up with German as his mother tongue, which may have eased his conversation with Franz, 34 years his senior. After years in Russia Scheer had taken up 'his residence in England as a City merchant, and for many years had lived on Kew Green, where his neat cottage and well-kept garden und greenhouse (the latter full of new plants imported by him) was well known to botanists'.[57] In Franz's final year Scheer published a booklet entitled *Kew and its gardens*. The main purpose of the book was to create public awareness of the plans of the Treasury dictated by parsimony (see above) and at the same time a nicely worded plea to Queen Victoria to maintain the Royal Garden in Kew: 'we humbly submit to Her Majesty's consideration, that it would be a wise and salutary policy to uphold what is cherished by so many of Her Majesty's subjects [the Royal Garden at Kew], what is so intimately interwoven with the private history of the Royal family during fully a century, and what is looked upon by the scientific men of all Europe . . . as of inestimable value'.

However, Scheer's text also contains a description of Franz's old age, albeit with a hagiographic touch: 'On the north eastern corner of the Green lives, still greatly lives, our venerable octogenarian friend Francis Bauer, Esq. F. R. S., the Nestor of Kew. Up to a very late date he continued his profound microscopic researches and exquisite drawings, but now he rests! Unwilling to risk even the possibility of leaving any work of his unfinished, he has determined to attempt no more. Seated near his Microscope, which long use has almost amalgamated with his very being, he spends his hours in re-examining the many miracles which his mind perceived, and which his pencil has so admirably perpetuated, and reviews in the monuments of his labor the history of his life. What surprise must be his, as he sees himself the author of such multitudes of magic performances! What sensations must press on him as they recall to his memory the, perchance, nearly obliterated circumstances and occurrences that brought them into existence! His was indeed a life of incessant activity, usefulness and greatness! The motives which stimulate common men never moved him! Vanity, selfishness, and illiberality, were foreign to his disposition, as friends and strangers in great numbers can testify; and that his virtuous toil has spared him self-reproach and remorse, those bitter companions of most men's earthly pilgrimage, his composed serenity – his cheerful resolve to await his time in peace – abundantly prove . . . Abler hands than ours will, no doubt, in due time lay before the world the life and the labours of this extraordinary and excellent gentleman'.

Scheer's somewhat enthusiastic prose does not mention the three problems Franz was faced with during his last year: the writing of a testament, his debts and the future of Elizabeth. How he solved them and what sort of long-term consequences his arrangements were to have, is dealt with in the following chapter.

At Franz's home in Kew an informal anatomico-botanical club is reported to have met regularly. We do not have a membership list, nor do we know when this tradition started and when it ended. In a footnote Scheer mentions that 'almost every Saturday, at Mr. Bauer's, many of the eminent man of the day, for purposes connected with Botany and other branches of Natural Philosophy' met 'for a friendly and social intercourse'. Home and Lambert (see Chapter 8) are explicitly mentioned in *Kew and its gardens*, and we may hypothesize that a few of those who signed the microscopic drawing of flax fibres (see above) also belonged to this informal club – Aiton the son, Pritchard, Scheer, Queckett, and John Smith.

Franz may also have found time for books. We do not have his reading list, but a collection of quotations taken from English and French literature in Franz's hand survives among his miscellaneous papers in the Niedersächsische Staats- und Universitätsbibliothek in Göttingen.[58] The few manuscript pages are undated and therefore cannot be correlated with any specific period of Franz's life. In his methodic way the authors are indicated – Jean Jacques Rousseau, Philip Stanhope, 4th Earl of Chesterfield known as Lord Stanhope, Henry Fielding, Henry Mackenzie and Jonathan Swift, largely authors form the Enlightenment period. Most quotations are taken from the original English editions and, in the case of Rousseau, from an English translation, but there is also a citation in French, which Franz must have understood. It is found impossible to deduce any more specific character traits of Franz from this selection of quotations – apart from a certain wit and slightly sarcastic undertone, as evident in quotations such as 'in France post-horses feel the whip in proportion to the favourable [sic] opinion the postilion has of his temporary master' (Rousseau) or 'It is the privilege of Englishmen to drink to the King's health, and to talk of his conduct' (MacKenzie). On the same line is 'To say the truth, he was of those compositions which nature makes up in too great hurry, and forgets to put any brains into their head' (Fielding).

Meanwhile a further temple had been constructed in the Royal Garden at Kew and Franz may have heard about the works. Destined as a 'Temple of Military Fame', a tetrastyle with doric colmns, it later became known as 'King William's Temple', as he had commissioned it. Designed by Sir Jeffrey Wyatville, this Pantheon contained cast-iron plates inside commemorating British victories from 1765 to 1815 and a parade of busts of George III, George IV, William IV, and of two military leaders in the struggle against Napoleon – Frederick, Duke of York and Albany, the second son of George III, and Arthur Wellesley, Duke of Wellington, the latter still alive. After Castlereagh had returned to London, Wellesley represented Great Britain at the Congress of Vienna, but left to lead the British troops in the battle of Waterloo, which ended in the final crushing defeat of Napoleon. The 'Temple of Military Fame' was almost completed when William IV died, aged seventy-one, ending what has been rightly termed the Georgian era, spanning the reign of the first five Hanoverian kings.

In late autumn 1840 Franz's end was approaching. He was looked after by Elizabeth and remained until his last days in contact with Robert Brown and William Meyer, two of his future co-executors (see Chapter 17), as well as Sir James Everard Home, 2nd Baronet, the son of Sir Everard Home. By then Franz may have forgotten the friction between Lindley and Brown over the priority of the *Illustrations of orchidaceous plants*, which as a matter of fact had never been a conflict between Franz and Brown. The latter had written already years ago, 'For in whatever way these [Brown's] investigations [on orchids] may affect your present theoretical views, your drawings must ever form a lasting Monument of your matchless talents which be assured no one has ever more highly valued than My Dear Sir Yours most sincerely R. Brown'.[59]

Franz may have heard in late November 1840 the good news that the *Flora Graeca* saga had finally come to an end. The twentieth and last fascicle comprising merely sixteen coloured copper engravings based on Ferdinand's watercolour were published, with the note 'quod exstat' (what remains) added.[60] For Franz this may have been a moment of reflection. The continuation of his *Delineations* had to be given up, *Strelitzia depicta* came out in a microscopic print run and *Illustrations of orchidaceous plants* comprised a meagre twenty plates. At least *Genera filicum*, with its rather poor prints in small format, seemed to become a success, just like the collaboration with Home, now long dead, who had managed to have so many of his illustrations published, and even republished. But what was this compared to the luxurious publication of nine hundred and sixty-six coloured copper engravings by his younger brother? On the other hand, did publication really matter? Was not the finished watercolour with all its elegance and lustre of colours superior to anything printed and coloured by inferior hands? What would happen to the immense labours of his very long life? Would his collection of watercolours ever be appreciated? And, if so, by whom?

In old age people often reflect on their early youth, and Franz may not have been an exception to this tendency. Fifty-two years had passed since he crossed over the River Thaya near Znaim and said farewell to the country where he had grown up, to his mother, to his two elder brothers and to Boccius (see Chapter 7). Maybe Franz was now recollecting the varied impressions of nature he had gathered in Feldsberg and its surroundings, the great diversity of plants and animals he had experienced early, and of which so much more he had been able to see and record with his pen and brush later in England. It is possible that Franz also recalled the principal buildings in Feldsberg, which must have impressed him, like the monumental parish church and the splendid Castle of the Liechtensteins, with its luxurious chapel, ballroom and extensive gallery containing several fine flower pieces. In the inner courtyard of Feldsberg Castle there is a sun dial, inscribed 'Tempora gaudendi sun[t] tempora certa dolendi', which Boccius may have translated to Franz as 'There are times of joy and certain times of suffering'. Now, in his eighty-third year of age, the difficult weeks had come. From letters exchanged between Brown, Meyer and Sir James Everard Home describing Franz's last days, it becomes evident that the latter was a man of deep religious feelings. Sir James Everard Home told Brown on 3 December 1840: 'He [Franz] complains rather of his prolonged existence in such a miserable state , but said that it was the will of God – he had lived in His fear all the days of his life, and His name be blessed and Praised'.[61]

It was in Kew in Eglantine House where Franz died on 11 December 1840. In the letter of Meyer to Brown written on the day of Franz's death 'a state of blessed calmness' is mentioned for the final hours, the symptoms being described as 'attacks of breath' and 'fever'.[62] They are too unspecific to enable a statement about his final illness and the reasons for his death. Six days later the funeral took place, of which no record could be traced.

On 16 December 1840 the remains of Franz were interred in St Anne's Churchyard in Kew,[63] which he must have passed thousands of times during his fifty years of residence in this village and in the immediate vicinity of his last domicile. Franz's grave is No. 62, located very near to that of Gainsborough (see Chapter 8), and persists, but his tombstone is no longer readable. Three years after Franz had settled down in Kew, William Aiton, with whom he had walked the melon ground in the Royal Garden and discovered a specimen of tree peony under the snow (see Chapter 8), was buried in the cemetery.[64] He was followed in 1810 by the painter Johann Zoffany (Zoffani, Zauffelij), whose well-known portrait of Emperor Francis Stephen of Lorraine and of the latter's daughter the Archduchess Maria Christina, both now in the Kunsthistorisches Museum,[65] Franz could have seen during his years in Vienna. Zoffany was followed in 1829 by George Engleheart (see Chapter 8) who had studied with Sir Joshua Reynolds and made copies of his paintings in miniature. Twenty years after Franz's funeral Sir William Jackson Hooker was to be interred in St Anne's Churchyard, and 71 years later his son Sir Joseph Dalton Hooker. The latter had followed his father as Director of the Royal Botanic Garden Kew and was elected in 1873 to the Presidency of the Royal Society. He was also a recipient of the Copley Medal, as well as of the order Pour le Merite for Sciences and Arts. Alexander Baron Humboldt, then long dead, had been the first chancellor of this most prestigious order of the Kingdom of Prussia. Naturally, for the Hookers pompous epitaphs were erected in St Anne's Church, but, surprisingly, Franz's epitaph is even larger.

Richard Westmacott III,[66] an associate of the Royal Academy of Arts and Fellow of the Royal Society, had been selected as a sculptor. He chose white marble and as Franz's attributes a microscope, a palette, brushes, wads of paper and a herbarium fascicle as well as leaves and a flower of the bird-of-paradise, *Strelitzia reginae*, plus leaves and infructescences of a species of Banksia. Meyer (see Chapter 17), had worded the very appropriate inscription 'In memory of Francis Bauer, Esq., F. R. S., F. L. S., & c., Botanical Painter to His Majesty George III and resident Draughtsman for fifty years to the Royal Botanic Garden at Kew, where he devoted himself to the advancement of natural science; under the munificent patronage of Sir Joseph Banks, Bart., the President of the Royal Society. In the delineation of plants he united the accuracy of a profound naturalist, with the skill of the accomplished artist, to a degree which has been only equalled by his brother Ferdinand. In microscopical drawing he was altogether unrivalled, and the science will be ever indebted for his elaborate illustrations of animal and vegetable structures, of which invaluable specimens are preserved in the British Museum and in the University of Gottingen. He was born at Feldsperg in Austria, on the 4th of October 1758 and accompanied his friend the Baron Joseph Jacquin to England in 1788. He settled at Kew in 1790, where he lived admired, loved and respected. He died on the 11th of December 1840, aged 82 years. The works of Francis Bauer are his best monument, Friendship inscribed this record of his honoured tomb "Lord, how manifold are Thy works, in wisdom hast Thou made them all" Psalms CIV v 25"; the draft version of this text is kept among the papers of the Bauer Unit in the Natural History Museum.[67] Franz's portrait was carved in half-relief by Westmacott, forming the central medallion, precisely as the former had envisaged as a boy in Feldsberg when painting the two-storey garden entrance with his own portrait and that of Ferdinand in medallions (see Chapter 2).

As usual the expenses were covered by subscription, with substantial contributions coming from Ernest Augustus and his younger brother Adolphus, Duke of Cambridge, a resident of Kew for a great part of the year; however, the largest sum was provided by Meyer, the son of Jeremiah Meyer[68] (see Chapter 3). Ten years after Franz, Adolphus was also buried in St Anne's Churchyard, before the latter's remains were transferred to the Royal Vault in St George's Chapel, Windsor Castle.

Today Franz's tomb is not a place of silence. At a distance of a few metres heavy transport on Kew Road passes by almost day and night and the former idyllic days of the village on the River Thames are long gone. The hundreds of thousands of visitors to the Royal Botanic Gardens, Kew passing by the small churchyard are neither aware of Franz nor of his grave, and the same holds true for most of the members of staff of this famous institution.

ENDNOTES

1 Desmond (2007)
2 Walker (1994)
3 Schultes (1825)
4 Scheer (1840)
5 Lack (2008b)
6 Meynell (1982)
7 Uppsala, Universitetsbiblioteket, Handskrifts- och musikenheten, Dörflers Brefsammlingar, Franz Bauer to Joseph Franz von Jacquin, 9 vi 1838 (unpublished letter)
8 Hall (1984)
9 Inv. No. 2515 (63)
10 Thomson (1834)
11 Vogelsang – Eastwood (2000)
12 EA 19
13 EA 6773
14 Schmutzer (2011)
15 Endersby (2008)
16 Lack (2003)
17 Lack (2009)
18 Lack (2015)
19 Dierkens (1993), Bruwier (1994)
20 Allen (1967)
21 Ware (2010)
22 Smith (1866)
23 Link (1822)
24 Smith (1875)
25 Meyer (1831)
26 Cunningham (1837)
27 Anonymous note in the copy of Hooker's *Genera filicum* kept in the Library and Archives of the Natural History Museum.
28 Franz Bauer (1837)
29 J. F. v. Jacquin (1828)
30 Bancher, Hölzl & Kotlan (1968)
31 Anon. (1835)
32 Riedl – Dorn (1989)
33 Steiner & Schulz (2005)
34 Anon. (1830)
35 www.musoptin.com/Ploessl_grosses_mikroskop.html (accessed 31 x 2014)
36 Franz Bauer (1839)
37 Catalogue: www.niepce.com/pages/catal.html (accessed 3 xi 2014)
38 Kravec (1949)
39 www.niepce.com/pages/page-bio.html (accessed 3 xi 2014)
40 Bradford, National Media Museum, Inv. No. 2003 – 5001/2/22332
41 Bradford, National Media Museum, Inv. No. 2003 – 5001/2/22661
42 Bradford, National Media Museum, Inv. No. 2003 – 5001/2/22662
43 Austin, University of Austin, Harry Ransom Center
44 Kaplan & Stulik (2013)
45 Stulik, Kaplan & Khanjian (2013)
46 Lack (2011a)
47 Smith (1983)
48 Anon. (1840b)
49 Anon. (1841)
50 Franz Bauer (1849)
51 Raphael (1970)
52 Franz Bauer (1840)
53 Tenberge (1999)
54 Smith (1840, 1849)
55 Queckett (1849)
56 Queckett (1840)
57 Seemann (1869)
58 4° Cod. Ms. Hist. Nat. 94: XI (unpublished manuscript)
59 Mabberley (1985)
60 Lack (1999c)
61 London, Natural History Museum, Library and Archives, Brown Correspondence 1: f. 39, Sir James Everard Home to Robert Brown, 3 December 1840 (unpublished letter).
62 London, Natural History Museum, Library and Archives, Brown Correspondence 1: f. 42, Meyer to Robert Brown, 11 December 1840 (unpublished latter)
63 Anon. (1840b)
64 www.findagrave.com (accessed 27 January 2015)
65 GG 6389, 1809
66 Stearn (1993)
67 Draft for Franz Bauer's epitaph in William Meyer's hand.
68 Simpson (1849)

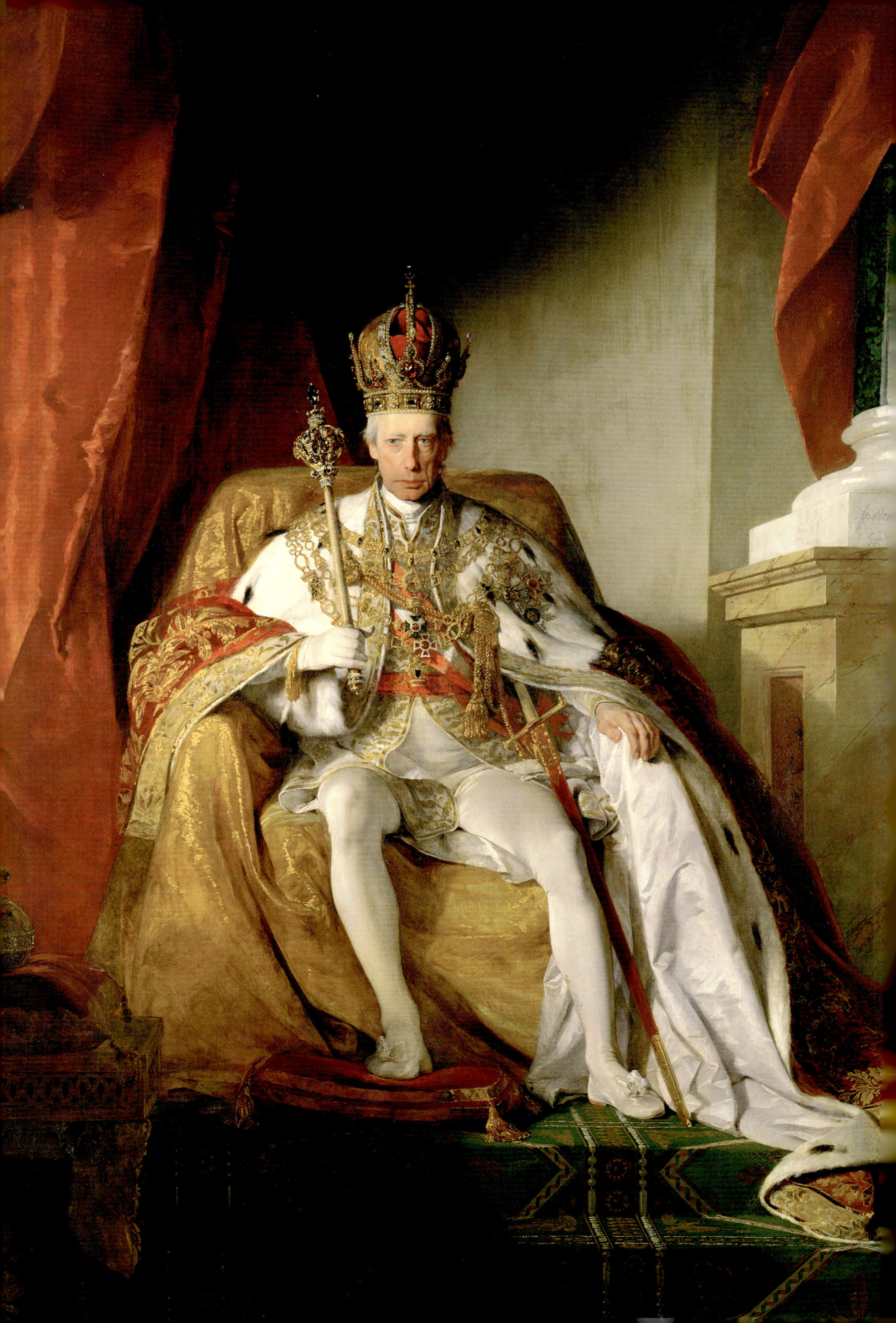

CHAPTER 17

Testaments and the dispersal of two estates[1]

FERDINAND'S TESTAMENT

THE *OXFORD ENGLISH DICTIONARY* defines a testament as 'a formal declaration, usually in writing, of a person's wishes as to the disposal of his property after his death'. Both Austrian and English laws of probate subscribe to this basic principle, although neither of them declares the testament to be an obligatory document. Ferdinand and Franz both expressed their wishes in testaments, while Joseph definitely refrained from doing so. Nothing is known of Johann's estate, nor indeed if he set out his last wishes in a written testament. In the case of Joseph, the Austrian legal regulations entered into force: since he had no legal issue and had not married, his estate (mainly savings in banks) was divided among the two brothers outliving him, Johann and Franz. It should also be noted that the stipulations laid down in a testament are legally binding unless the respective superior authority, in the case of Ferdinand the Abbey Court of Klosterneuburg, which possessed rights on the village of Hietzing, disagreed. This was, however, not the case. More importantly, Ferdinand, like his elder brother Joseph, had not married and had no children, which resulted in his unlimited freedom when drawing up his testament. The documents, all in German, relative to both brothers are kept today in the Wiener Stadt- und Landesarchiv in Vienna (see chapter 12).

FIG. 17.1. Franz I, Emperor of Austria. Portrait by Friedrich Amerling, signed, 1832 Oil on canvas. – Vienna, Kunsthistorisches Museum, Inv. No. GG 8618

Ferdinand chose a conventional approach. The second clause of his testament, signed and sealed in the presence of three witnesses just five days before his death (see Chapter 12), contains the heart of the document: he declares his three brothers – Johann in Feldsberg, Joseph in Vienna and Franz in Kew – as co-heirs with equal rights to his estate, explicitly specifying that invested money and cash had to be equally divided among them. Furthermore, Ferdinand stipulated that in the case that one of his brothers wished to acquire his house in Schmidgassse (now Trauttmannsdorffgasse; see Chapter 12) he would be obliged to pay out two-thirds of the estimated price to his other two brothers. The same principle was to be applied to 'my herbarium, birds, insects, shells and the respective drawings' as well as to the fifteen engraved copper plates for his *Illustrationes Florae Novae Hollandiae*, and furniture, clothes, etc. In addition, Ferdinand stipulated that his collections and drawings would have to be evaluated and – in the case that one of the surviving brothers wished to acquire this part of the estate – he would also be bound to pay out two-thirds of the estimated price to his other two brothers. Otherwise, all these materials were to be auctioned and the cash receipts divided. Ferdinand furthermore created in his testament two legacies: 'my already finished drawings from Greece and Australia' were to go to his brother Franz in Kew and 'the library with everything belonging to it' plus his microscope to go to Heinrich Wilhelm Schott (see Chapter 12). Effectively this meant that these objects were exempt from the regulations set out in clause two and did not become the property of the co-heirs.

Austrian law does not subscribe to the concept of an executor, a person appointed by the testator to execute his testament and carry it into effect, and consequently no executor is nominated in Ferdinand's testament. In agreement with Austrian tradition the curator of Ferdinand's estate, a person reporting to the Abbey Court in Klosterneuburg, prepared an inventory of all the items left behind and had it cosigned by Trattinick (see Chapter 12). This document, now also in the Wiener Stadt- und Landesarchiv, contains estimated prices for many items and testifies at the same time that Ferdinand had become a genuine collector of natural history objects – boxes of beetles and butterflies from Brazil and Australia are listed as well as three glass boxes with various other butterflies. Even the head and the skull of a chamois, *Rupicapra rupicapra*, in Vienna a conventional souvenir from excursions to the nearby Alps, turned up in the inventory. In addition to Ferdinand's herbarium a box with sixty-eight specimens of exotic fruits are listed, as are several maps, among them a plan of London and six maps of Central Europe, a region Ferdinand had crossed from London to Vienna and back at least five times in 1814, in 1819 and 1824 (see Chapter 12). A drawing table, drawing boards and pigments are also mentioned and specifically two drawing aides – the camera obscura plus curtain (see Chapter 2 and 20) and the camera clara (see Chapter 10 and 20). In addition, the contents of Ferdinand's library, which contained a broad spectrum of natural history books, is listed in a separate inventory. Here we find, among others, Brown's *Prodromus Florae Novae Hollandiae* (see Chapter 11) and *Chloris Melvilliana* (see Chapter 13), Flinders's *Voyage* (see Chapter 11), Lindley's *Digitalium monographia* and parts of his *Collectanea Botanica* as well as Mikan's *Delectus Florae et Faunae Brasiliae* (see Chapter 12). As one might expect, a complete copy of the *Illustrationes Florae Novae Hollandiae* (see Chapter 11 and 12) appears along with several papers published in the *Philosophical Transactions* and the *Transactions of the Linnean Society* authored by Brown, Home and Franz. Among the group of natural history books two items are outstanding – Basilius Besler's *Hortus Eystettensis*, admittedly one of the greatest flower books ever published and considered today as the great masterpiece of botanical illustration in the Baroque era,[2] represented in Ferdinand's library in the 1713 reissue; and Jacob Hoefnagel's *Archetypa studiaque*, a collection of forty-eight engravings of plants and animals, mainly insects, based on the works of his father Joris Hoefnagel, represented in the exceedingly rare first edition published in Frankfurt in 1592. However, a miscellany of other topics had also been covered by Ferdinand's library, among them playing chess and topography: for example, he owned an encyclopedia on churches and monasteries in the Archduchy of Austria in three volumes. We find books on costumes, and in particular art and art history; Estner's *Versuch einer Mineralogie* (see Chapter 10) appears in the inventory, as do guide books for London and Paris and several English, French, German and Italian dictionaries. Ferdinand possessed also a more unconventional object: it is described in the inventory as 'two books in large format in red morocco with painted birds', which must have passed to Schott the son with the library. Many years later this material was catalogued as 'a set of 335 well drawn original water-colour drawings of Birds of India and the Far East executed about the year 1800 by Ferdinand Bauer and others'. This brief description had appeared in the *Report of the Zoological Survey of India* and referred to two volumes now in the library of the Zoological Survey of India in Calcutta.[3] However, an analysis of these ornithological illustrations revealed that they had definitely not been painted by Ferdinand or Franz, who both had never visited India, but almost certainly by an anonymous Indian bird illustrator. How this material made its way from Vienna to Calcutta remains to be elucidated.

Judging from the inventory it seems that Ferdinand had died as a man of some property. In the Bank of England there were savings amounting to 7610 guilders (see Chapter 11), while the value of his belongings in Austria was estimated to be 6092 guilders. Of these, Ferdinand's house and garden were considered to have a value of 3000 guilders, a rather low estimate, since his co-heirs managed to sell this property for 4300 guilders in the very year of Ferdinand's death (see Chapter 12). Furthermore, Ferdinand had left behind 1070 guilders in cash, and his library was estimated to possess a value of seven hundred and twenty-two guilders, and his herbarium six hundred and forty guilders. Ferdinand's 'birds, animal skins and shells' were estimated at one hundred and forty guilders, the fifteen copper plates of the *Illustrationes Florae Novae Hollandiae* with the engravings based on them together with the microscope at one hundred and sixty guilders and the sixteen finished views from the Ottoman Empire merely at sixteen guilders. It is hard to believe but true that the masses of plant and animal drawings, mainly from the voyage in the *Investigator* including the tracings in pen and ink, but also several of the pencil drawings referring to the *Codex Liechtenstein*, scarcely attracted the attention of the valuators – sixty-four guilders was their combined estimate. Other colour-coded pencil drawings by Ferdinand included in this sum are still in need of an in-depth study.

Neither Johann nor Joseph nor Franz seem to have expressed an interest in acquiring Ferdinand's collections and drawings, and consequently the auction envisaged in the testament may have been seen by many as the only solution. However, in contrast to what has been repeatedly stated in the literature, there never was an auction. Instead the story took an unexpected turn. Ferdinand's death on 17 March 1826 had not remained unnoticed in the botanical circles in Vienna and resulted in a report by Host dated 5 May 1826 sent to Francis I, Emperor of Austria (Fig. 17.1.).

FRANCIS I, EMPEROR OF AUSTRIA, PURCHASES

At that time Host, a versatile botanist (see Chapter 12), was in charge of the botanic garden next to the Belvedere Palace in Vienna, which was dedicated exclusively to the cultivation of the plants native in the Hereditary Lands. With permission from the representative of the Abbey Court in Klosterneuburg, Host had inspected Ferdinand's collections and drawings and afterwards reported to Francis I that the vertebrate specimens had been carefully wrapped in paper, that the insects had remained undetermined, that the number of shells was insignificant. According to Host the Austrian herbarium was 'not so interesting', while few herbarium specimens from Australia had been determined. Host does not seem to have studied Ferdinand's drawings and mentioned only that according to Schott this material was kept in different containers.

A second, undated report on the collections and drawings was written almost certainly by Sternberg (see Chapter 12), who stated in the first paragraph that the collections from Ferdinand's tour to the Ottoman Empire with Sibthorp and from Ferdinand's voyage in the *Investigator* with Brown formed a significant part of it. Among others Sternberg mentions a total of three thousand plant specimens from the Cape of Good Hope, Australia and Timor, a European herbarium comprising twenty-six 'Hefte' (fascicles), a box of fruits, mainly of species belonging to the Protea family, and a box of two hundred specimens of Australian 'birds and animals'. The contents of two portfolios, the first containing watercolours of Australian plants, mainly orchids, the second described by Sternberg as containing watercolours of Australian animals, has already been dealt with elsewhere (see Chapter 11).

We may hypothesize that the two reports rang a bell in the mind of the addressee. Francis I may have remembered that he was one of the subscribers to *Flora Graeca*, of which at that moment five volumes and the first instalment of the sixth must have already arrived in his private library illustrated with a total of five hundred and fifty coloured copper engravings all based on watercolours by Ferdinand. Lambert's *The genus Pinus*, also kept in his private library, may have come likewise to his mind, illustrated with coloured copper engravings largely based on watercolours by Ferdinand. Maybe he even recollected the meeting with Jacquin or Jacquin the son when an exchange was arranged and the copy of the luxury edition of *Selectarum stirpium americanarum historia* with a superb title page was acquired for his private library (see Chapter 3). In any case Host's and Sternberg's reports made Francis I commission Schreibers, the director of the United Imperial Royal Natural History Cabinets, to write another report on Ferdinand's collections and drawings. This was done on the basis of an order dated 8 May 1826 from the Head Chamberlain of the Imperial Court in Vienna, Johann Rudolf Count Czernin von und zu Chudenitz, a Knight of the Golden Fleece and the man in charge of all the Imperial collections. Unfortunately Schreibers's report has not been traced; it was probably lost in the fire which ravaged the Augustinian Wing of the Hofburg in late 1848 (see Chapter 19), where Schreibers had his office. However, the report of the Head Chamberlain to Franz I, dated 17 August 1826, is based on it and probably follows the arguments put forward by Schreibers, and previously by Host and Sternberg.

The rather lengthy and somewhat verbose document, kept in the Haus-, Hof- and Staatsarchiv in Vienna, stresses that the zoological collections were of limited interest, with several specimens already in a poor state of preservation. By contrast, the botanical collection 'amounting to over 2500 species in several specimens, arranged systematically . . . in perfect . . . state of preservation' would be a most welcome acquisition to the United Imperial Royal Natural History Cabinets. The drawings are described as '2249 leaves in fol: 4to and 8°°, of which 1800 referring exclusively to the Australian plants are only pencil sketches and outlines with complete botanical accuracy and by the hand of the known master'. According to this third report submitted to Francis I, the co-heirs Johann, Joseph and Franz had asked for 1500 guilders for this part of Ferdinand's estate, which was significantly above the value estimated in the inventory. This price was considered appropriate in the Head Chamberlain's report for the botanical collections only and he advocated speedy action in order to ensure adequate preservation of the materials, adding, however, that no funds were available for the year 1826. The Head Chamberlain knew how to play the patriotic card and wrote, 'Very rarely the occasion occurs that plants from that part of the world come to Europe . . . Collections from that continent always pass through London and from there to Paris, to Berlin, and only what is deaccessioned there, usually arrives in Vienna. This collection is even more remarkable for the imperial cabinet . . . since it is rich, brought together by a native [a subject of the Emperor of Austria], to be acquired first hand and very cheap'.

When this report was signed in Vienna by Czernin's deputy, Francis I stayed in Baden bei Wien on the third part of his annual *villeggiatura*, from where he moved in late August to Persenbeug on the Danube River. It was in the nearby castle of Weinzierl on 3 September 1826 that a decision was taken which merits being quoted in full. Francis I had the following note on the proposal from the Head Chamberlain written: 'I grant permission that the complete natural history estate, including the collections of drawings, of the late plant illustrator Ferdinand Bauer is bought for 1500 guilders C. M., to be paid from my private purse'. This was a decent sum, although only about a third of the selling price of Ferdinand's house in Hietzing gained by his co-heirs in the same year (see above).

Three aspects of this brief note merit attention: firstly, it was written by Francis I in his own hand and signed by him, echoing his grandfather Emperor Franz Stephen, who had written in his own hand the instructions for Jacquin's expedition to the islands in the Caribbean Sea and signed them (see Chapter 3); secondly, in less than six months after Ferdinand's death a decision by the Crown had been reached; thirdly, Francis I was prepared to pay a price significantly above that estimated by Trattinick and the curator. Even when subtracting (1) the interment costs including the funeral music explicitly mentioned in the accounts,[4] (2) the infamous 'Mortuar', a concession tax to be paid by the coheirs, (3) the honorarium for the curator and (4) the inheritance tax, then five per cent of the estimated value of the estate, Johann, Joseph and Franz had acquired a substantial heritage from their youngest brother.

The new acquisition – Ferdinand's natural history collections and unfinished drawings – were not divided, as initially suggested by Host, into two parts – the collections going to the United Imperial Royal Natural History Cabinets and the drawings to the private library of Franz I. Instead, the complete materials entered the former institution and found its place in the Augustinian wing of the Hofburg, very near to the Imperial Court Library. There Moritz Count Dietrichstein-Proskau-Leslie, an important supporter of Beethoven like Archduke Rudolf, had succeeded Ossolinski as prefect. Shortly after its arrival, Ferdinand's material was divided between the zoological and the botanical cabinets; for the numbers of specimens recorded in the acquisition journal of the two cabinets see Chapter 12. While the total number of drawings accessed by the zoological cabinet is not known, Trattinick, the keeper of the botanical cabinet, recorded in his acquisition journal, 'botanical drawings allegedly 1876'.[5] The comparatively low number of botanical drawings given by Trattinick compared with the number in the report by the Head Chamberlain is puzzling. Maybe Trattinick referred only to the graphical record of the plants from the voyage in the *Investigator*, while the other botanical drawings, for example those referring to the work of the Bauer brothers undertaken for Boccius (see Chapter 2), were left unrecorded by him. As in standard herbarium practice, duplicate specimens of Ferdinand's botanical collection were later distributed by the United Imperial Royal Natural History Cabinets to other institutions; for example, a specimen of *Pteris kingiana* collected by Ferdinand on Norfolk Island is now in the herbarium of the Botanisches Museum in Berlin. By contrast, the specimen of *Alsophila excelsa* also collected by Ferdinand on this island and now in the Natural History Museum must have been given by Ferdinand to Brown, who filed it among the collections from the voyage in the *Investigator*. In a similar way a few pen-and-ink tracings colour-coded by Ferdinand were given by the United Imperial Royal Natural History Cabinets like duplicate specimens to other institutions, such as the Royal Botanic Gardens Kew,[6] while the respective pencil drawings remained in Vienna. A pen-and-ink tracing documenting the twining glycine, *Glycine clandestina*, a legume occurring widely in Australia, may act as an example.

In agreement with Ferdinand's testament, all his finished drawings (the plant and animal watercolours from the voyage on the *Investigator* prepared in Hietzing (see Chapter 12) and the sixteen views from the tour to the Ottoman Empire) were sent to Franz at Kew. He retained the views, which were auctioned in 1841 (see below), while the finished plant watercolours must have passed to Brown, who bequeathed them to the British Museum in 1858.[7] By contrast the finished animal watercolours were accessioned by the Admiralty and transferred in 1843 to the British Museum[8] (see Chapter 11).

FRANZ'S TESTAMENT

When Franz signed his testament, kept today in the Public Record Office in Kew, he was placed into totally different circumstances. Firstly he was faced with considerable debts; secondly he had 'lately', possibly in the year 1840, adopted his servant Elizabeth Baker; and thirdly he had to consider the complexities of the first and second codicil to the testament of his employer and great benefactor Banks, now already dead for twenty years. Franz was wise enough to ask for legal advice when formulating his testament, which he signed on 15 August 1840 together with the solicitor H. W. Haverfield and William Parker Baker, a relative of his adopted daughter Elizabeth Baker Bauer.

Banks's codicils, Franz's testament and a decision by the Crown caused a conflict between the Royal Botanic Garden Kew and the British Museum, later the Natural History Museum, with questions being asked in the House of Commons six decades after the death of Franz. Among the papers presented on this subject to the Departmental Commission on Botanical Work, which met in Westminster Palace Hotel in London on 29 November 1900, was even a letter by William Turner Thiselton-Dyer, the authoritarian third director of the Royal Botanic Garden Kew, bluntly claiming Franz's drawings of Kew plants at the British Museum as the property of the institution he headed. In this context and with little diplomacy, Thisleton-Dyer, a Fellow of the Royal Society and Knight Commander of the Order of St Michael and St George, called the Department of Botany of the British Museum 'a mere offshoot from Kew', no doubt causing acerbic reactions in the department. Therefore it seems appropriate to present the full background of this feud of such long duration between two major British institutions.

In the first codicil dated 21 January 1820 Banks had stated: 'And it is my will and I hereby declare that the said annuity is given to the said Frederic [recte Franz] Bauer upon condition that he continues to reside on Kew Green, and employ himself in making drawings of plants that flower in the collection at Kew in the same manner as he has hitherto done and the drawings which he shall so make be added to the collection now in his hands, and which revert to me or my representatives at the time of his death, as will appear by a paper in my possession, written and signed by the said Frederic [recte Franz] Bauer'. A few weeks later, on 7 March 1820, Banks had changed his mind and stipulated in his second codicil: 'I do hereby give and bequeath to his Majesty [then George IV] for the use of the establishment of the Royal Botanic Gardens of Kew all those drawings and sketches of plants that have grown in the said garden and have been made at my cost and charges by Mr. Frederic [recte Franz] Bauer, of Kew Green and are now deposited in his custody'. In a second clause of the second codicil Banks further stipulated that the drawings and sketches newly made by Franz had to be deposited 'in the hands of William Townsend Aiton, Esq. and his successors'.

Clearly there had been a marked shift in Banks's views: while all of Franz's illustrations documenting the plants cultivated in the Royal Garden at Kew, pre-existing as well as future, would end up according to the first codicil in the hands of Banks's legal heirs, according to the second codicil the pre-existing ones should go to the Crown for the use of the Royal Garden at Kew without permitting Franz to keep them for life, while the future ones should go directly to the head gardener at the Royal Garden at Kew. It appears that the second codicil had limited effect over the next two decades. Although Franz's 'drawings and sketches of plants that had been grown in the said garden' were now the property of the Crown, they remained with him. The drawings that were prepared after the second codicil had come into force may indeed have been delivered by Franz to Aiton the son. However, the only hint to this being so is an allegation made many years later by Sir Joseph Dalton Hooker, then second director of the Royal Botanic Gardens Kew, who stated in a letter to his son-in-law Thiselton-Dyer and presented by the latter at the Departmental Commission on Botanical Work, that these botanical illustrations intended to be delivered to the head-gardener were transferred by order of Brown to the British Museum.

For various reasons, in particular the uncertain future of the Royal Gardens at Kew during the reign of William IV and in the first years of Victoria (see Chapter 16), the first part of Franz's testaments deals exclusively with the drawings he had prepared and which should become property of the Crown for the Royal Garden at Kew. Here is Franz's line of argument (via his solicitor): 'First I [Franz] declare the eight Portfolios marked 'Kew Plants.' contain the whole of the Botanical Drawings finished and unfinished now in my possession and which in conformity with the first Codicil of the Will of the late Sir Joseph Banks Baronet ought to be delivered after my death to his representatives to be added by them to those other Botanical Drawings formerly made by me for Sir Joseph Banks and by him bequeathed to the Trustees of the British Museum . . . But by a second Codicil to his will Sir Joseph Banks bequeathed to His late Majesty King George the Fourth for the use of the Establishment of the Royal Botanic Garden at Kew All Drawings of the same description in my custody at the date of that second Codicil as well as all other drawings that might thereafter be made by me of Plants from time to time flowering in the Royal Gardens at Kew . . . As however the said Establishment at Kew is no longer on the same footing as it was when that Bequest was made namely the private Botanic Garden of the King . . . And more especially as the value of the Drawings which have been made by me for Sir Joseph Banks since my arrival in England upward of fifty years ago will be greatly increased if they be kept together in asmuch as they serve to explain and illustrate each other and being placed in a public Establishment of the Kingdom will be more accessible and useful for Botanists I may be allowed and I hereby express my most anxious wish that Her Majesty [Queen Victoria] would be gratiously pleased to order all the Drawings aforesaid to be placed with those which are already in the Botanical Department of the British Museum . . . And I direct my Executors hereinafter named to stand possessed of the said eight Portfolios marked 'Kew Plants' in the hope that the same may be deposited in the British Museum . . . '. The purpose of the first part of Franz's testament is clear; by applying to the Crown he wished Banks's two codicils to be overruled: neither the legal heirs of Banks nor the Crown for the Royal Garden at Kew should receive the eight portfolios marked 'Kew Plants' but instead should be received by the Crown for the British Museum. The success of this appeal will be subsequently described. However, it should be noted that the first part of Franz's testament contains an assertion which is incorrect: the first codicil to Banks's will does not contain an expectation or obligation of his legal heirs to pass the drawings and sketches of the Kew Plants prepared by Franz to the British Museum.

The second part of Franz's testament deals with his own property, that is 'all the other Portfolios marked alphabetically [see below], as well as all and every other articles and drawings and work of art books pictures and prints . . . furniture plate linen chin, and other effects' as well as 'cash stock funds and securities'. All these effects and funds were to go to his executors 'upon trust to sell such parts thereof as are in their nature saleable by public auction or private Contract'. Franz stipulated that the proceeds of the sales were 'in first place for payment of all my just debts and funeral and testamentary expanses' and in the second place to pay 'to Charlotte Ulise [Wise ?] my Housemaid One years wages and Five pounds for Mourning'. In case a residue remained, this was to go to 'my most faithfull Servant and truly esteemed and lately adopted Daughter Elizabeth Baker Bauer . . . who has for upwards of twenty five years been unremitting in her anxious endeavours and exertions to promote my comforts'. In his testament Franz nominated four co-executors and added a fifth in a codicil dated 8 October 1840: these were (1) 'William Meyer of Kew . . . Her Majesty's late Consul in Albania' who had been based from 1819 to 1835 in Preveza,[9] first belonging to the Ottoman Empire and since 1832 to the newly created Kingdom of Greece, (2) Robert Brown, (3) John Ford of Mortlake . . . Captain in the Army, (4) George Cooper Esquire . . . of Brentford, Surgeon and (5) 'Elizabeth Baker Bauer now residing with me'.

Like his younger brother Ferdinand in Hietzing, Franz had arranged for a carefully worded testament which was not open for interpretation.

FRANZ'S REQUEST

An interesting document headed 'Request !!', written by Elizabeth and signed by Franz on 24 October 1840, exists in the Niedersächsische Staats- und Universitätsbibliothek in Göttingen.[10] Firstly it is a plea to have at least some part of Franz's estate auctioned in

London and not at a country auction, where probably 'not a single bidder would appear' and, secondly, a recommendation for 'Mr. [William] Pamplin [the son] No. 9 Queen Street Soho Square', a dealer in natural history materials, to act as agent.

More relevant is a table of contents of Franz's alphabetically marked portfolios appended to the request. Written in Elizabeth's hand it is a mine of information and formed the basis for the printed auction catalogue, albeit in a complex way (see below). Lettered A to O Elizabeth described briefly the contents of the thirteen portfolios, almost exclusively works by Franz and, significantly, Ferdinand. Portfolio A and B, for example, contained the colour prints taken from the engraved copper plates prepared to illustrate many of Home's contributions to the *Philosophical Transactions of the Royal Society* (see Chapter 15 and 16), rightly regarded by her 'the only Copy of that work in existence'. By contrast Portfolio C comprises exclusively works by Ferdinand – the sixteen finished views from the tour to the Levant with Sibthorp, which Franz had received as a legacy from Ferdinand (see above), twenty-six pencil sketches of views and costumes in the Ottoman Empire (see Chapter 5), twenty pencil sketches of views on the continent, fourteen pencil sketches of views in England, and several pencil sketches of views on Norfolk Island (see Chapter 10). For the auction this portfolio was divided into lots 147–151.Elizabeth's notes on Portfolio D contain a gem of information: evidently referring to Ferdinand's colour-coded pencil drawings, she notes, 'it is a kind of short hand Drawing consisting of very correct Pencil sketches, with the natural colours of the several parts marked on the margin so that at any time finished Drawings might be made from them'. On the same line is her note on portfolio D likewise containing colour-coded pencil drawings '[a method] used by Ferdinand Bauer during his Voyages and Travels to save time'.

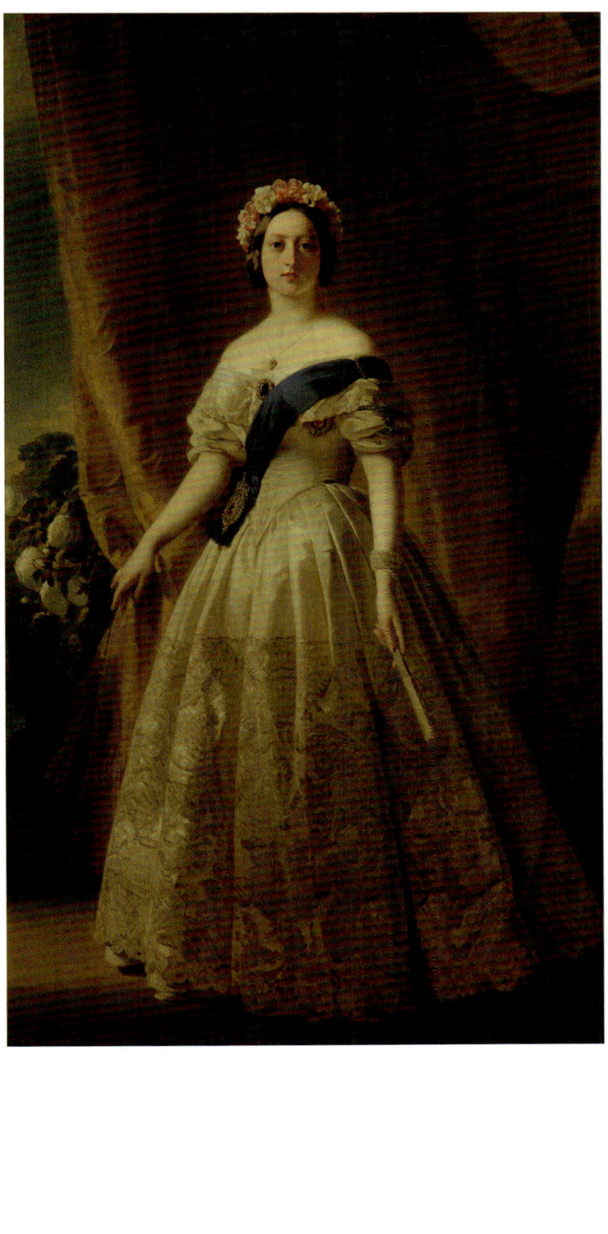

VICTORIA, QUEEN OF THE UNITED KINGDOM, DECIDES

Two weeks after the death of Franz, on 11 December 1840, his testament and codicil were proved in London and Meyer, Brown and Elizabeth Baker Bauer as co-executors were sworn in by the Worshipful John Daubeny, Doctor of Laws. Subsequently Franz's appeal to the Crown to overrule the second codicil of Banks's testament was accepted by Victoria, Queen of the United Kingdom of Great Britain and Ireland (Fig. 17.2.), with the actual transfer taking place on 19 May 1841 (see below). At that moment she was twenty-one years of age and expecting the birth of her second child Albert Edward, later as King Edward VII her successor on the throne. However, in the matter of the eight portfolios Queen Victoria was represented by her husband Albert Prince of Saxe-Coburg and Gotha, the Prince Consort, then twenty years old, who had married her just fifteen months before.

As a consequence the eight portfolios did not go to Aiton the son for the Royal Garden at Kew but instead to Brown for the British Museum. According to the latter's report to the Trustees of the British Museum dated 9 June 1841 the

FIG. 17.2. Victoria, Queen of the United Kingdom of Great Britain and Ireland. Portrait by Franz Xaver Winterhalter, signed, c.1845. Oil on canvas. – London, Royal Collections Trust, Inv. No. 403054.

transfer seems to have been more than a formality.[11] On 19 May 1841 Brown in his position of co-executor of Franz's will, took the eight portfolios (bound volumes) marked 'Kew Plants', to Buckingham Palace, Queen Victoria's new residence in London, where he received a receipt from George Edward Anson, the treasurer of Albert's household. The latter then handed back the eight portfolios and was given a receipt by Brown in his capacity as Keeper at the British Museum. This transfer is explicitly confirmed in a manuscript note added to one of the bound volumes, inscribed 'Unfinished Drawings of Plants Cultivated at The Royal Garden at Kew By Francis Bauer, F. R. S.', which is kept in the Library and Archives of the Natural History Museum. It reads 'These drawings and sketches were received after Mr Bauer's death, by command of H. R. H. Prince Albert, and in conformity with a request in Mr. Bauer's will'. Just as in Vienna within less than six months a final decision had been reached by the Crown, and Franz's Kew Plants had found a permanent place in a public institution of similar fame – the British Museum at Montague House, where Brown was the Keeper of the Department of Botany.

The eight volumes of 'Kew Plants' no longer exist in their original form. They have been disbound and Franz's illustrations rearranged. What were considered to have been finished illustrations have been conserved in acid free mounts and are stored in Solander boxes of large format and are kept in air-conditioned storage rooms (see Chapter 8). What were considered to have been unfinished illustrations remain in six large bound volumes. The grand total of this accession remains unclear. It has been reported that Brown counted 1532 drawings,[12] but whether this number includes preliminary material and unfinished works remains to be discovered. In any case no official number appears in the archival record.[13]

THE AUCTION

Considering Franz's debts, the co-executors were bound to have the saleable part of his estate auctioned, and they selected Messrs. Christie & Manson, 8 King Street, St James's Square, London as the company to deal with the matter. A detailed catalogue for the auction[14] was prepared and printed. In order to increase public awareness it contained again, for the last time, the incorrect statement that Franz had been 'Botanical Painter to his late Majesty George III'. The auction was scheduled for 24 November 1841; in good time, just six days before, a note appeared in the *Times* on the upcoming event, mentioning specifically the priceless materials from Niépce.

For obvious reasons the auctioneers were obliged to offer the saleable part of Franz's estate in palatable bits. This was a relatively easy job for his library and his collection of engravings, his microscopes and instruments, but more difficult for the very many drawings. Elizabeth's list of contents (see above) was no doubt useful in attributing the drawings to Franz, Ferdinand and Joseph, but several of the thirteen portfolios described by her were further divided and offered as individual groups of items. In the end a total of two hundred and nineteen lots came up for auction.

The Library and Archives of the Natural History Museum keeps an annotated copy of the auction catalogue, listing hammer prices and the names of the purchasers. Among them we find, as suggested by Elizabeth in her request, Pamplin the son, the dealer in natural history materials. Henry Bohn, antiquarian and publisher who had been involved in *The genus Pinus* (see Chapter 9) and would soon later organize a reissue of the *Flora Graeca*,[15] attended the auction, along with two co-executors, Meyer and Brown. The latter's assistant and subsequently his successor as Keeper of Botany John Joseph Bennett was present, as were Haverfield, Franz's solicitor, John Smith from the Royal Garden at Kew, and Sir Frederick Watson, who was to play a significant role in the further disposition of the materials he purchased.

Franz's library was very similar to that of Ferdinand; it also contained the second edition of the *Hortus Kewensis*, Brown's *Prodromus Florae Novae Hollandiae* and *Chloris Melvilliana*, Host's *Synopsis Plantarum* (see Chapter 12), Lindley's *Digitalium monographia* and the fourth edition of Linnaeus's *Species Plantarum* edited by Willdenow, as well as Ferdinand's *Illustrationes Florae Novae Hollandiae*. We also find a copy of Franz's *Delineations*, his *Illustrations of orchidaceous plants* and the first instalments of Hooker's *Genera Filicum*, all of them publications to which he had contributed the basis for all illustrations. Home's *Lectures on comparative anatomy* in six volumes referred to Franz's long association with this prominent surgeon and anatomist. Among Franz's zoological books we find Wilhelm Friedrich Freiherr von Gleichen-Russwurms's *Histoire de la mouche commune*, published in Nuremberg in 1766 and Ludwig Heinrich Bojanus's *Anatome testudinis Europaeae*, published in

Vilnius in 1819, the former standing probably in context with Franz's work for Home on the feet of the common house fly, *Musca domestica* (see Chapter 15). There are far fewer books on art in Franz's library compared with that of his young brother, but we find Giuseppe Canale's *One hundred and nineteen etchings after the original designs of Raffaelle, Parmigiano, Guido Reni and the other great masters*, printed in London in 1775. Its author, a son of Antonio Canal, was the first cousin of Bernardo Bellotto, called Canaletto, who had been to Vienna in 1759–1761 (see Chapters 3 and 20). In addition Franz possessed in his library no less than eighty-one prints showing views of a city he seems to have adored – Vienna (see Chapter 3).

Ferdinand's sixteen finished views from the Ottoman Empire (lot 151) achieved by far the highest hammer price – £25.4s, significantly above the estimate by Trattinick and the curator fifteen years before in Vienna; Watson bought this fine material, superior in quality to the set of views now in the Sherardian Library in Oxford. Based on watercolours prepared by anonymous artists in Calcutta documenting orchids from India and Nepal and collected by Francis Buchanan-Hamilton,[16] Franz had prepared a total of forty copies. These formed lot 138 and were acquired for the second highest price, £19.19s, by one Walker, who also bought Franz's drawing desk manufactured in Sydney Cove (see Chapter 11). Lot 194 achieved the third highest hammer price – £17.17s. Described in the auction catalogue as 'A Plössl Compound Microscope with Micrometer, and moveable stage, a series of six Achromatic object glasses, three Huyghenian and one Achromatic eye-pieces, prism for condensing the light, three Micrometer glasses, objects & c., in mahogany case' this was one of the best microscopes then on the market. Manufactured by Plössl, known for having greatly improved the achromatic microscope objective, the description applies exactly to what is usually called Plössl's Großes Mikroskop No.1[17] (see Chapter 16). A very considerable literature existed on this highly appreciated microscope, which no doubt was one of the reasons for the considerable price. Unsurprisingly Franz's colleague and co-executor Robert Brown, himself a microscopist of similar calibre, bought the instrument; its current whereabouts remain unknown.

Lot 117, a total of eighteen watercolours of Ericas by Franz, datable to his first years in England, was bought for £11; Watson had offered the highest bid. Bohn purchased Franz's set of the volumes of the *Philosophical Transactions of the Royal Society* beginning with the year of his election, 1821, to the year of his death, 1840, for £8. The sets of the *Transactions of the Linnean Society* and the *Transactions of the Horticultural Society*, learned societies to which Franz also belonged as a fellow, attracted much less attention. Ten watercolours of tulips reached a hammer price of £6 and were also bought by Walker. A relatively high price was attributed to a rare printed book – Daniel Wagner's *Pharmaceutisch-medicinische Botanik*, comprising two hundred and forty-nine coloured lithographs and published in Vienna in 1828–1829. Pamplin bought it for £5.7s. Two more compound microscopes, manufactured by Peter Dolland in London and Charles Chevalier in Paris, respectively, were also appreciated by bidders, but their hammer price was far below that of the Plössl. Twelve more microscopes and two portable telescopes attracted much less attention; they prove that Franz was fascinated, if not obsessed, by optical instruments, which had played a prominent role in his last letter to his old friend Jacquin the son in Vienna (see Chapter 16). As one might expect, lots 177–183, comprising various microscopical specimens of plants, animals and fossils prepared by Franz, as well as specimens of shells and madrepores, reached only low prices.

In our context the drawings by Franz forming lots 106–142 and 144–146 are more relevant. They form an amalgam of originals and copies, the latter based on illustrations published by Plumier, Sloane and Jacquin and several others, documenting mainly orchids, one of Franz's specialities (see Chapter 13). Among the original drawings we find lots described as 'fourteen Drawings in Colours of Insects and Plants made at Vienna' (see Chapter 3), 'twenty-two Mushrooms in pencil and colour', 'six Camellias in colour', 'twelve Roses in colour', 'forty-three microscopic drawings of Mummy cloth, Flax, &c. with MS. notes and autographs'. Lot 106 'a volume of drawings from a kaleidoscope' and lot 110 'Etruscian vases and ornaments' are among the heterodox works by Franz listed. At the auction a substantial number of pencil drawings by Ferdinand also changed hands, among them various views (see above), and a long list of pencil drawings of apples and pears, hyacinths, passion flowers, fox gloves and of birds – of the latter fifty from England and thirteen from Australia. Lot 143 had been wrongly described by the auctioneers as 'Ditto [i.e. copies] of New Holland Orchids with MS. Index', but in fact these are original watercolours They are not by Ferdinand as one might suspect but – with a single exception – by an anonymous illustrator and had been prepared for Paterson (see Chapter 13). The gifts from Niépce (see Chapter 16) to Franz formed lots 187–191, today iconic items largely kept in air-conditioned strong-rooms, were purchased for very little – £14.4s (see below).

In sum, the auction resulted in proceeds amounting to £444.8s.6d, somewhat less than what had been estimated when the co-executors were sworn in. This is evident from the note 'the Goods Chattels and Credits of the Testator do not amount in value to the Sum of Four Hundred andfifty Pounds' in the respective document. Considering the fact that the funeral expenses, the honorarium for the solicitor and the inheritance tax remain unknown, it is possible that Franz's estate was indeed negative. A letter by Elizabeth to Brown of 31 December 1840 kept in the Library and Archives of the Natural History Museum is indicative: she informs her co-executor that Haverfield had advised against removing Franz's estate from Kew to Ham, since this could raise the suspicion of Franz's creditors, it being uncertain that the proceeds of the sale of Franz's effects would cover his debts.

Quite naturally the auction resulted in the wide distribution of Franz's effects, of which many remain untraceable today (see Chapter 19). There are only two exceptions – the items purchased by Watson and by Brown.

ERNEST AUGUSTUS, KING OF HANOVER, PURCHASES

No less than 14 lots were acquired by Watson, 'the gentleman who manages the affairs of the King of Hanover in this country [England]' as described by Colonel Perceval in the House of Commons on 27 March 1840.[18] Ernest Augustus was the fifth son and eighth child of King George III and Queen Charlotte. When he was fifteen years old, he had been sent to Göttingen University and subsequently started a military career. In 1799, created Duke of Cumberland and Teviotdale, he entered politics, with a seat in the House of Lords and extreme Tory views, and became leader of the right wing of the party. Deeply involved in political and private scandal,[19] Ernest Augustus became thoroughly unpopular in England, with persistent allegations that he had murdered a valet and fathered a child with his sister Princess Sophia. Clearly the black sheep of the Royal family, he moved together with his twice-widowed wife Friederike, née Princess of Mecklenburg-Strelitz, later calling herself Frederica, to the continent, living mainly in Berlin. Back in London he was given a life-time residence in Kew, later called Hanover House (see Chapter 16). In those years Ernest Augustus, better known in England as the Duke of Cumberland, may well have met Franz, the Nestor of Kew, living on the same side of The Green.

Upon the death of his eldest brother William IV, King of the United Kingdom and King of Hanover, in 1837, their niece Princess Victoria (see Chapter 8) became Queen of the United Kingdom, while Ernest Augustus became King of Hanover. The reason for this is simple to explain: Salic Law, which barred women from succession, applied in Hanover, but not in the United Kingdom. When entering his new domain Ernest Augustus was sixty-six years old and the first ruler in over a century to live in Hanover. Immediately upon his arrival, the new King dissolved the parliament, which had convened under a liberal constitution, and on 1 November 1837 issued a patent declaring the constitution void. When all office holders (including the professors of Göttingen University) were required to renew their oaths of alliance to the King, seven professors – among them the Grimm brothers – refused and consequently lost their positions, with three of them even being expelled from the Kingdom of Hanover. All this created an outcry, both in central Europe and in Britain. Politically more significant was the protest of a number of towns, which refused to appoint parliamentary deputies. However, by 1840, a sufficient number of the latter had been appointed for the parliament in Hanover to be summoned by Ernest Augustus, and passed a budget.

It remains totally unclear why Watson purchased a significant portion of Franz's belongings for Ernest Augustus (Fig. 17.3.) and why the latter in January 1842 at the latest donated the new acquisition to his alma mater. Was the purchase an expression of goodwill towards his former neighbour in Kew? Was the donation an act of political reconciliation? Was this intended as a friendly gesture for Göttingen University? We simply do not know. The new acquisition was an amalgam consisting of drawings by Franz and Ferdinand, of prints based on the works of Franz and copious manuscript material by him. The whole material received the shelf mark gr 2° MS. Histor. Nat. 94, was carefully bound in twenty-one volumes with different auction lots reunited as well as separated, and decades later described in utmost detail in a printed catalogue.[20] The respective entry states

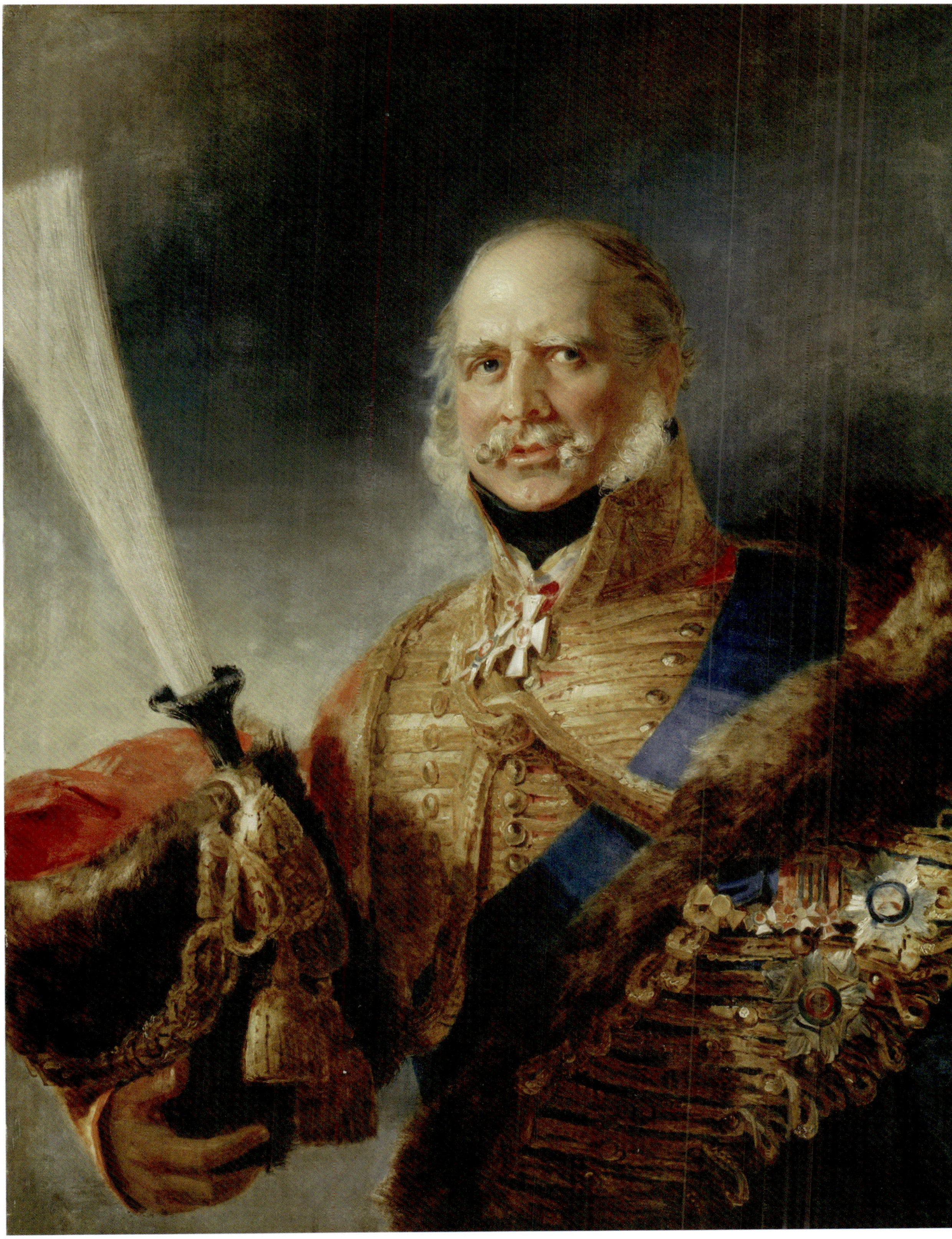

FIG. 17.3. Ernest Augustus, Duke of Cumberland, later King of Hanover. Portrait by George Dawe, c.1828. Oil on canvas. – London, National Portrait Gallery, Inv. No. 3309.

that the material was a gift from Ernest Augustus, King of Hanover, with the added note 'November 1841 (in remembrance of my stay there [i.e. in Göttingen or Kew ?]) and January 1842'. The acquisition journal of the Königliche Bibliothek in Göttingen is more precise, recording 13 December 1841 and 3 January 1842 for Ernest Augustus's donation.

Indeed this was a remarkable gift, containing a cross section of the works of Franz. It covers his years in association with Jacquin in Vienna – neither represented in the Naturhistorisches Museum nor the Natural History Museum – when Franz prepared very accurate watercolours of fruits, for example of peaches, *Prunus persica*, and grapes, *Vitis vinifera*. The Göttingen material also comprises ample manuscript material, excerpts from the literature and preliminary drawings on fungal diseases in cereals and the life cycle of wheat prepared for Banks (see Chapter 8 and 16) as well as on red snow (see Chapter 13). We also find an extensive pictorial documentation of the diversity of fern spores and their germination, several of them on squared paper (see Chapter 16), topics which are totally lacking in the Library and Archives of the Natural History Museum. The anatomy of the human eye and the diversity in human hair in relation to sex, age and position on the body (see Chapter 16) are additional topics covered by the miscellaneous drawings by Franz kept at Göttingen and absent in London.

The memory of the 'Göttinger Sieben', the seven dissenting professors dismissed from their chairs by Ernest Augustus continues to be very lively today, in particular in Göttingen and Hanover. The Niedersächsische Staats- und Universitätsbibliothek, where the King's gift is conserved today, has as its postal address Platz der Göttinger Sieben 1; near the building of the Landtag in Hanover a large monument for the seven professors has been erected, and there are currently plans for a further monument in Göttingen.

ROBERT BROWN'S ACQUISITIONS AND THEIR DISPERSAL

At the auction of Franz's saleable effects Brown had purchased nineteen lots, among them the Plössl compound microscope (see above). Of the other lots, those consisting of unique items such as drawings are particularly interesting. Franz's forty-three microscopic drawings of pollen, etc. with manuscript notes (see Chapter 14), Ferdinand's six pencil views of Norfolk Island (see Chapter 10) and his pencil drawings of passion flowers (see Chapter 12) and gymnosperms (see Chapter 9) were bequeathed by Brown to the British Museum and are now in the Library and Archives of the Natural History Museum while Ferdinand's flower piece, also purchased by Brown, is currently untraceable.

All of Niépce's gifts to Franz, forming lots 187–191, were also acquired by Brown. Unfortunately, only two lots, 187 and 191, have been carefully described in the catalogue, the former as 'An Original specimen of Héliographie (Christ bearing the Cross) by the original inventor of the art, Mons. Nicéphore Niepce, of Châlons sur Sâone, given to Mr. Bauer in the autumn of 1827'. This item passed from Brown via Bennett and R. W. Robinson to the Royal Photographical Society in Bath, who deposited this heliograph entitled 'Christ Carrying his Cross' in the National Media Museum in Bradford.[21] Lots 189 and 190 are both given in the auction catalogue as 'A Ditto Ditto', which is best interpreted as 'an original specimen of heliograph by Nièpce'. There is universal agreement that each of the two lots consisted of one heliograph by Nièpce, both took the same route as lot 187 from Brown via Bennett and Robinson to the Royal Photographical Society and were also deposited in the National Media Museum; they have been entitled 'Un Claire de lune'[22] or alternatively 'Ruines d'une abbeye' (see Chapter 16) and 'Le Cardinal d'Amboise',[23] respectively, but attribution of the two heliographs to lot 189 or 190 remains impossible.

The icon of the history of photography was auctioned as lot 188, given in the catalogue as 'A Ditto Ditto. A view', which is best understood as 'an original specimen of heliograph by Nièpce. A view'. No doubt this was the pewter plate measuring 16.2 × 20.2 centimetres and now called 'View from the window at Le Gras', or alternatively 'Paysage à Saint-Loup de Varennes' (see Chapter 16). It took a different route, passing from Brown to Bennett and from the latter to M. Baden Pritchard, who had it deposited in a trunk placed in a London store house. After an interval of almost half a century this precious item was found by a legal heir to Pritchard and given to Helmut and Alison Gernsheim, photo collectors and historians, who donated it to the Harry Ransom Center of the University of Texas in Austin, Texas where it remains.

More complex even is the fate of lot 191, which consisted of six elements, here numbered as No. 1–6, which are also top items for the early history of photography. In the annotated auction catalogue kept in the Library and Archives of the Natural History Museum the description of two of these, 'two autograph Notes of Instruction number 1 and 2, dated 25th and 27th January 1828' has been crossed out, which means that they (No. 2–3) were excluded from the auction. Of the remaining four elements, three could be traced. One of them (No. 1), described in the catalogue as 'The Autograph Memoir of M. Nicéphore Niepce on his own invention of Héliographie, dated Kew 8th Dec. 1827, and communicated by him to his friend Mr. Francis Bauer' (see Chapter 16) took the same route from Brown via the Gernsheims to the Harry Ransom Center. In the auction catalogue two further items (No. 4–5) are characterized as 'two impressions from a pewter plate engraved by the Heliographic process, presented by Mons. N. Niepce to Mr. Bauer the 22nd. November 1827'. These two elements are clearly two prints taken from the heliograph entitled 'Le Cardinal d'Amboise' (see above). Of these one passed from Brown via the Gernsheims to the Harry Ransom Center, the other from Brown via Pritchard to the Royal Photographical Society, which had this print deposited at the National Media Museum in Bradford.[24] 'Another impression of a landscape by the same process' (No. 6) remains an enigma; from this description in the auction catalogue it is evident that a print from a heliograph showing a landscape was meant, similar in technique to the print of 'Le Cardinal d'Amboise', but this item remains untraceable. However, it may have been the print mentioned in the literature as 'Paysage d'après une gravure de Claude le Lorrain'.[25]

The two testaments and the subsequent events in Vienna and London resulted in an important fact: the Naturhistorisches Museum and the Natural History Museum became the two most important depositories of the drawings and watercolours by Franz and Ferdinand, followed by the Sherardian Library at Oxford (see Chapter 9), the Fürstliche Sammlungen in Vienna (see Chapter 2) and the Niedersächsische Staats- und Universitätsbibliothek in Göttingen (see above). When, however, focusing on the botanical and zoological collections only, the Naturhistorisches Museum keeps by far the largest materials gathered, almost exclusively brought together in Australia by Ferdinand.

ENDNOTES

1 Unless otherwise stated, this chapter is based on Lack (2003).
2 Lack (2008a)
3 Coll. No. 596.5/024/I
4 Wien, Wiener Stadt- und Landesarchiv, Bestand Patrimoniale Verwaltung und Justiz A 13 Klosterneuburg über Hietzing, Verlassenschaftsakt Ferdinand Bauer, Erbsteuer Ausweis.
5 Riedl – Dorn (2014)
6 Verdcourt (1986)
7 Sawyer (1971)
8 Anon. (1904)
9 Prevelakis & Kalliataki Merticopoulou (1996)
10 gr. 2° MS Hist. Nat. 94, XVII.
11 Meynell (1983)
12 Stearn (1993)
13 A. Mendez (pers. comm. Sept. 2014)
14 Anon. (1841f)
15 Lack (1999c)
16 www.linnean-online.org/buchanan-hamilton.org (accessed 21 September 2014)
17 www.musoptin.com/Ploessl_grosses_mikroskop.htlm (accessed 21 September 2014)
18 Barrow (1840)
19 Wardroper (2002)
20 Anon. (1893)
21 Bradford, National Media Museum, Call No. 2003-5001/2/22661.
22 Bradford, National Media Museum, Call No. 2003-5001/2/22662.
23 Bradford, National Media Museum, Call No. 2003-5001/2/22332.
24 Bradford, National Media Museum, Call No. 2003-5001/2/22333.
25 www.niepce.com/pages/page-inv.html (accessed 23 September 2014)

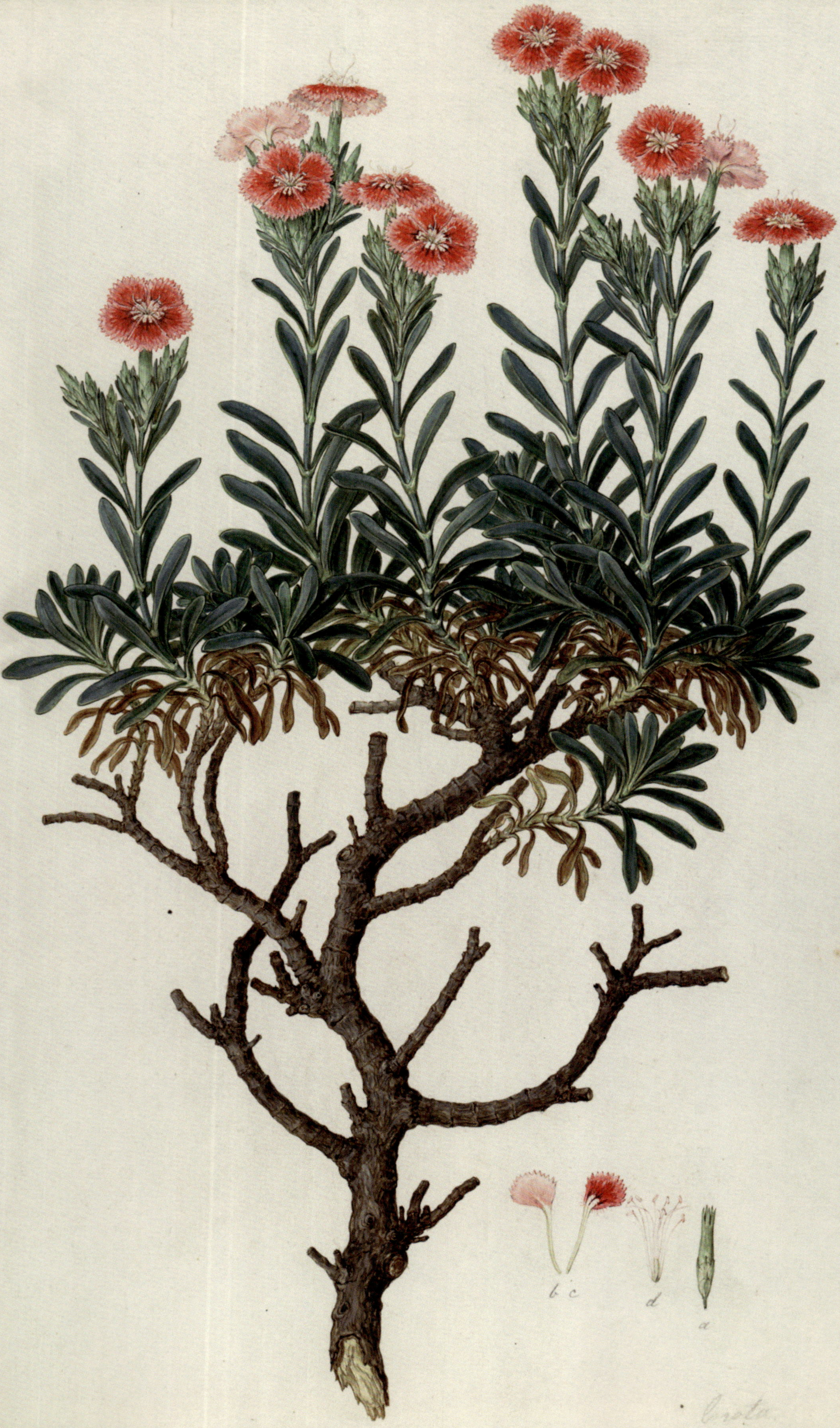
b c
d
a
Creta

CHAPTER 18

Posthumous publication[1]

A DOUBLE ANOMALY

FOLLOWING THE ARRANGEMENTS that were agreed on in 1826, a total of 2249 drawings by Ferdinand (including several by Franz and many from the earlier work of the three brothers for Boccius) were deposited in the United Imperial Royal Natural History Cabinets in Vienna, which meant that the material became available for consultation by the scientific community (see Chapter 17). In 1841 the eight portfolios labelled 'Kew Plants' by Franz arrived in the British Museum, comprising finished and unfinished watercolours (including a few by Ferdinand). Nothing indicates that the new acquisition, estimated today to comprise more than 2000 drawings, was ever counted[2] (see Chapter 17). This painted record of plant diversity on a global scale as reflected in the living collection of the Royal Garden at Kew covering the years 1790–c.1835 plus watercolours of other botanical and mycological objects now also became available for consultation by the scientific community. The same can be said of the donation by Ernest Augustus made at the latest in January 1842 to Göttingen University, which consisted mainly of Franz's materials on a wide range of topics (see Chapter 17). In the following year Ferdinand's finished watercolours of plant and animal life in Australia were transferred from the Admiralty to the British Museum, and in 1858 they were united with those bequeathed by Brown, who had acquired them from Franz (see Chapter 17). In short, some five thousand drawings of superb quality documenting plant and, to a limited extent, also animal and fungal life, almost all unpublished, were now open to study. Minor additions arrived in the British Museum later, among them a few drawings by Franz (e.g. those pertaining to the bezel) acquired from Elizabeth (see Chapter 16) and zoological watercolours prepared by Franz for Home, but purchased from Sir Richard Owens. The latter, a famous comparative anatomist and palaeontologist, who coined the term 'dinosauria', was the man behind the separation of the British Museum (Natural History), now the Natural History Museum from the British Museum.

One might have expected that the publication of the huge graphical documentation by Franz and Ferdinand would by now have been planned and effected; however, this has not happened. In the end only a small fraction of Ferdinand's drawings have appeared in print posthumously, while almost nothing of Franz's immense oeuvre has been published after the latter's death. This anomaly is difficult to explain. However, the period of the expensive and time-consuming copper-engraving process, restricted as it was to a short print run needing to be coloured by hand, was approaching its end. In addition, many works by the Bauer brothers had been produced in large format, which was another challenge and resulted in cost to cope with.

In producing hundreds and hundreds of line drawings, effectively pen-and-ink tracings based on his pencil drawings prepared on the voyage of the *Investigator*, Ferdinand had already done a considerable amount of preliminary work for the subsequent printing, making use of either copper engraving or lithography (see Chapter 12). By contrast Franz seems to have been unprepared to adapt his painted record of the Royal Garden at Kew in a way that would enable quick and cheap publication. With the exception of the copper engravings documenting the female flower of *Rafflesia arnoldii*, whose publication

FIG. 18.1. ***Dianthus fruticosus*** **subsp.** ***fruticosus.*** **Watercolour by Ferdinand Bauer, signed, c.1789. – Oxford, Sherardian Library, MS Sherard 241, f. 117.**

had been delayed (see Chapter 13), only two plates appearing after Franz's death contained figures based on his watercolours. A copper engraving documenting a specimen of native ivy, *Alchornea ilicifolia*, from Moreton Island in Australia cultivated in the Royal Garden in Kew was published in the *Transactions of the Linnean Society*[3] and four small figures in a lithograph forming part of Pritchard's *A history of Infusoria: living and fossil*, which appeared in print in London in 1842.

There is an additional anomaly: one might expect that the bulk of the publishing would have taken place in London, in the nineteenth century a central place for science and a city in which the exploration and economic development of the British overseas possessions always enjoined a high profile. However, despite the fact that all of Ferdinand's finished watercolours from the voyage in the *Investigator* were conserved there, publication did not take place in London, albeit with two exceptions, both plant illustrations from Timor (see below). By contrast, no fewer than one hundred and two pencil drawings by Ferdinand documenting plants from this voyage were published in Vienna, followed by nine in Munich and five in Utrecht.

LONDON

Why the interest in Ferdinand's finished watercolours from Australia was almost zero in London remains largely a mystery to this author. In the event Ferdinand's works, which are today so highly appreciated by scientists and the public alike (see Chapters 11, 12 and 20), were simply shelved in the British Museum, although they definitely were known to the inner circle of the cognoscenti in London, for example to people such as Brown, Bennett, Lindley (see below) and others. To those outside the museum world this may seem very odd. However, institutions like the British Museum keep vast collections of all kinds of objects from all corners of the world and a highly qualified and dedicated specialist for a clearly circumscribed group of items is necessary in order to study and publish research on a particular collection. It seems that in the decades following the death of Ferdinand no one on the staff of the British Museum possessed precisely that strict focus on the vascular flora of Australia which would have been necessary to fully appreciate Ferdinand's achievements and have them properly printed with the necessary scientific commentary. In a sense the same considerations apply to Franz's worldwide documentation of orchids (see Chapter 13), to mention just one example from his painted record of the living collections kept in the Royal Garden at Kew, since the *Illustrations of Orchidaceous plants* comprised only a tiny fraction of Franz's work on this large family (see Chapter 13).

On the basis of a pencil drawing and/or a pen-and-ink tracing of the small-flowered rhynchoglossum, *Rhynchoglossum obliquum*, both conserved now in the Naturhistorisches Museum, Ferdinand produced a single undated monochrome watercolour of this very plant inscribed 'Ferd Bauer del.', which survives in the Library and Archives of the Natural History Museum. This watercolour in grey washes served as the exemplar for a copper plate engraved by Franz,[4] from which a print run was taken. The plate was included in Bennett's *Plantae javanicae rariores*, a work published in instalments in London with the flora of Malesia as subject, including the island of Timor, where the respective specimen had been collected (see Chapter 10). By contrast, Ferdinand's illustration of *Epithema brunonis,* another gesneriad collected in Timor, could not be traced; on its basis, a coloured copper engraving appeared in the *Plantae Asiaticae Rariores,* a work on the flora of Asia in three volumes published in London by Nathaniel Wallich, superintendent of the Botanic Garden in Calcutta. While the illustration of the small-flowered rhynchoglossum appeared in 1838, that of the second gesneriad had been published six years earlier.

In London only the *Flora Graeca* project[5] continued to publish watercolours by Ferdinand. At the moment of his death in March 1826 the first five volumes with the coloured copper engravings t. 1–500 had appeared, in the following year the sixth volume with t. 501–600 was completed and published. After the death of Smith in 1828 the co-executors of Sibthorp's testament had recruited Brown to continue, but after the publication of the thirteenth fascicle including t. 601–650 the latter had given up and was superseded by Lindley (see Chapter 12). Intellectually most of the problems had already been solved by Smith; he had not only published the *Prodromus Florae Graecae*, but he had also labelled most of the herbarium specimens and annotated all of Ferdinand's watercolours with their scientific names. However, a considerable amount of work was still to be done – a few hundred plant species had to be described, their names and synonyms checked, the production of the copper engravings t. 651–966 and their colouring had to be supervised. Because of the decreasing income from Sibthorp's estate in Oxfordshire, the pace of the project slowed down somewhat, but it did come to an end in November 1840 when the twentieth instalment appeared, comprising t. 951–966 (see Chapter 16). This last fascicle contained two especially convincing coloured engravings – one showing the extremely widespread rough bindweed, *Smilax aspera*, a climber known for its sharp thorns, and the second the fern *Notholaena marantae*,[6] which Ferdinand had almost certainly recorded on Mount Athos as a young man in 1787. More than fourteen years after his death, the *Flora Graeca* saga had now reached its conclusion and the account was closed by Platt (see Chapters 9 and 19). Thanks to the foundation created by Sibthorp, who had shown such an unsympathetic attitude towards Ferdinand (see Chapters 5, 6), the *Flora Graeca* became the most lasting printed monument of Ferdinand's endeavours. He himself must only have seen the first fascicles in Banks' library at 32 Soho Square (see Chapter 11). The only copy in Vienna was kept in the private library of Francis I, to which he hardly had easy and direct access.

It remains a sad fact that nobody in England or elsewhere was prepared to follow the model of Sibthorp and create a foundation for the publication of the results of the voyage in the *Investigator* or at least for the publication of the graphical record of the botany and zoology of this expedition. This reluctance to publish the expedition's findings is even more surprising for three reasons: firstly, Australia formed a cornerstone of the British Empire and was not part of the western provinces of the Ottoman Empire; secondly, vast sums were otherwise spent on the exploration of the British Empire, thirdly, the first circumnavigation of Australia provided many more new and highly relevant scientific findings when compared to the number of discoveries made during the extended sailing tour by Sibthorp and Ferdinand in the Levant.

VIENNA

Heinrich Wilhelm Schott, or Schott the son, later the recipient of a legacy from Ferdinand (see Chapter 17), was born in 1794 in Brno in Moravia and started to work as a gardener in 1815 in the garden dedicated to the flora of the Austrian Hereditary States adjacent to that of the Belvedere Palace in Vienna, then supervised by Host[7] (see Chapter 17). Later Schreibers,[8] an imperial councillor and the son-in-law of Jacquin the son, recruited Schott the son to take part as gardener and assistant to Mikan (see Chapter 12) on the bridal voyage of Archduchess Leopoldine to Rio de Janeiro and the subsequent travels in the Portuguese colony of Brazil. Schott the son sailed from Trieste on 10 April 1817 and left Rio on 21 June 1821.[9] Back in Vienna he was appointed adjunct to the director of the Imperial Royal Court Gardens and promoted in 1828, two years after Ferdinand's death, to Imperial Royal Gardener. While still abroad, Schreibers had Schott's diaries published in Brno under the title *Tagebücher des k. k. Gärtners, Herrn H. Schott in Brasilien*. In Vienna Schott began to contribute to the knowledge of the immensely diverse flora of the neotropics by publishing the descriptions of species new to science, among them several begonias.[10]

Before 1832 Schott must have associated himself with a man with a rather different background - Stephan Ladislaus Endlicher (Endlicher István László), born in Pressburg and ten years his junior. He had studied theology at the seminary of the archdiocese of Vienna, had received minor orders but decided against a career in the Catholic Church and instead entered the Imperial Library in Vienna as a *scriptor* in 1828. A brilliant Latinist, Endlicher started to publish on a wide spectrum of topics, from theology to Latin poetry, and edited a manuscript on the early medieval history of Hungary. Concurrently with his job he studied natural history and East-Asian languages at Vienna University. In 1830 Endlicher produced his *Flora Posoniensis*, a Flora of the Pressburg area, printed in his native town when he was twenty-five years old.

Nothing is known about how it came that Schott and Endlicher began to collaborate on a work in large format entitled *Meletemata botanica*, which appeared in Vienna in 1832. However, there is a fateful coincidence and a simple fact: on Vienna's Josefsplatz the entrances to the Imperial Court Library and to the United Imperial Royal Natural History Cabinets, where Schott's and Ferdinand's collections were kept at that time, are less than twenty metres apart – Schott and Endlicher had to meet sooner or later. With a print run of fifty copies, the slim volume *Meletemata botanica* was and still is a very rare work intended to be purchased by the happy few; needless to say, there is a coloured copy in the private library of Franz I. This publication in large format contains a synopsis of the Balanophoraceae and Rafflesiaceae families, and the description and illustration of five species new to science – three from Brazil collected by Schott the son and two from Australia collected by Ferdinand (the Norfolk Island bastard oak, *Ungeria floribunda*, from Norfolk Island and *Helicteres cana* from the Gulf of Carpentaria). These were the first scientific names that were based upon Ferdinand's collections deposited in Vienna; several more were to follow. The two copper engravings illustrating the novelties from Australia look very Bauerian, but carry the printed note 'Zehner ex sicco del.'. The exemplar of the lithograph showing *Ungeria floribunda*, a pen-and-ink drawing in the Naturhistorisches Museum, is not colour coded and may indeed have been produced by Joseph Zehner. The latter very closely followed Ferdinand's style in depicting no less than twenty-six details plus a flower diagram, no doubt under the instructions of Schott and Endlicher.

Endlicher's next publication follows a basically similar line. It carries the title *Atakta botanica* and is reported to have started to appear in Vienna in January 1833, but remained a torso. Five fascicles appeared until 1835; however, the last one abruptly terminates in the middle of a word and the whole work lacks an index. The last lithograph carries the number 40, but five lithographs have never been published, and there is no text to accompany lithographs 27–40. As in the *Meletemata botanica*, a few species described as new to science have been based on specimens collected by Ferdinand on the voyage in the *Investigator*, but unlike the *Meletemata botanica* a number of lithographs have been based on pencil drawings and/or pen-and-ink tracings by Ferdinand. Among the latter we find representations of kiss me quick, *Portulaca pilosa*, a pantropical species known already to Linnaeus, of the beach myrtella, *Lithomyrtus obtusa*, a native of Australia and New Guinea, and of *L. retusa*, occurring only in Australia. All three were documented by Ferdinand on the northern coast of Australia. In validating in the *Atakta botanica* the generic name *Lippaya*, based on another plant from the Gulf of Carpentaria collected by Ferdinand, Endlicher could demonstrate his expertise of Hungarian ecclesiastical history. He dedicated the name to two brothers born in Pressburg: Georg Lippay Count Zombor (Lippay György Zombori gróf), archbishop of Gran, then belonging to the Ottoman Empire and therefore residing in Nagyszombat (today Trnava, Slovakia); and Johann Lippay Count Zombor (Lippay János Zomori gróf), a Jesuit, founder of the first garden all' italiana[11] in Pressburg and author of the extremely rare *Posoni kert*, published in 1664 in Nagyszombat. A few lithographs printed in the *Atakta botanica* are of particular interest, for example that showing *Dodonaea humilis*, a spreading shrub from the southern coast of Australia. This scientific name was not validated by being published together with a description of the plant, as is usually the case, but instead by being published on a printed illustration showing details, which directly go back to Ferdinand's meticulous documentation of the flower parts in the field. This echoes Jacquin's approach in his *Icones plantarum rariorum*, to which Franz and Ferdinand had contributed in the 1780s (see Chapter 3).

Having published a Flora of the Pressburg area, the polymath Endlicher, who would later produce an annotated catalogue of the Chinese, Manchu, Japanese and Korean books of the Imperial Library in Vienna, decided to write a Flora of an island he had never seen – Norfolk Island. Nothing is known about who suggested this project to him, but Schreibers mentioned in the preface, or Trattinick who had seen Ferdinand's estate as early as in 1826 (see Chapter 17), are likely candidates. Endlicher called his work *Prodromus Florae Insulae Norfolkicae* and had it published in Vienna, although the preface is marked 'Posonii Cal. Jan [Pressburg, January] 1833'. It is almost exclusively based on the comprehensive collection of herbarium specimens and accurate drawings by Ferdinand, in cases up to four per species, dealing also briefly with a few bryophytes and lichens collected by him. By that time Endlicher was still scriptor in the Imperial Library; it was only three years later that he succeeded Trattinick as keeper of the herbarium of the United Imperial Royal Natural History Cabinets.[12]

When studying Ferdinand's collections, pencil drawings and pen-and-ink tracings, Endlicher was in a very fortunate position. He had Ferdinand's complete material at hand; nobody before Ferdinand had done substantial collecting on the island, and nobody before Endlicher had studied the flora of Norfolk Island in a comprehensive manner. In addition he referred to a few specimens collected by Allan Cunningham, who had stayed on Norfolk Island from 10 May to 11 September 1830.[13]

Endlicher could be certain of a rich harvest of novelties, and indeed he described a long list of taxa new to science from this tiny spot in the Pacific Ocean, among them the Phillip Island glory, *Streblorhiza speciosa*, now extinct (see Chapter 11). Many of his new taxa are still accepted by the scientific community.[14] Endlicher's excellent understanding of plant diversity is mirrored by the fact that he correctly recognized all three orchids and the one lily that are endemic to the island as new to science, gave them scientific names and published their names together with accurate descriptions in his work. The *Prodromus Florae Insulae Norfolkicae*, a title echoing the *Prodromus Florae Graecae* (see Chapter 9), has Schreibers as dedicatee, by then also a councillor for the government of the Archduchy of Austria below the River Enns (see Chapter 17).

Although in the end Endlicher's *Prodromus* did not contain illustrations, its author had announced their production and publication in the preface. Sets of ten lithographs are conserved in at least four major botanical institutions – in the Natural History Museum in London, the Naturhistorisches Museum in Vienna, the Australian National Botanical Garden in Canberra and the Naturalis Biodiversity Center in Leiden.[15] All ten carry the printed note 'Ferd. Bauer del.' and are based on drawings that he prepared on Norfolk Island and that are now kept in the Naturhistorisches Museum, with the exception of those lost in the Oberhöflein fire (see Chapter 19). What argument is there to make such a claim for this connection? In the upper right-hand corner all ten lithographs carry a plate number. Although there are large gaps in the numerical sequence – the series starts with t. 40, followed by t. 54, t. 55, t. 74, etc. – this very sequence mirrors precisely the numbered sequence of species in Endlicher's text. Among the species represented we find the epiphytic soldier's crest orchid, *Oberonia titania*, the Phillip Island glory, the Norfolk pine and another Norfolk Island endemic, the climber *Capparis nobilis* of the Caper family. The present author was able to compare the sets of lithographs conserved at Kew, Leiden and Vienna and found that the degree and quality of colouring varies considerably. This seems to indicate that the plan for an illustrated supplement to the *Prodromus Florae Insulae Norfolkicae* was given up at an early stage and that a small number of uncoloured as well as incompletely coloured lithographs were distributed to a few selected institutions. In the text of this work and in that of the *Atakta botanica* Endlicher cites Ferdinand's graphical record by numbers, for example he listed four numbers for *Meryta latifolia*, and three numbers for *M. angustifolia*, both evergreen trees endemic to Norfolk Island. This information correlates perfectly with the numbers on seven pencil drawings by Ferdinand showing illustrations of the two species kept today in the Naturhistorisches Museum. However, for the time being the general philosophy of Ferdinand's numbering system on Norfolk Island as well as in Australia remains incompletely understood.

Endlicher's most important botanical work for the history of botany is, however, his *Genera Plantarum secundum Ordines naturales disposita*, which appeared in eighteen parts between 1836 and 1841 in Vienna, of which the last two parts formed the first supplement to the work. Later four more supplements were published, the last in 1850 shortly after Endlicher's early death.[16] At the time this was the most detailed and comprehensive survey of the vegetable kingdom on a global level, making its author stand on a par with giants like Augustin Pyramus de Candolle in Geneva before him, and George Bentham in London, Sir Joseph Dalton Hooker at Kew or Adolf Engler in Berlin after him. Following in the steps of Antoine-Laurent de Jussieu in Paris, Endlicher aimed at a natural arrangement of plant diversity, but refrained from providing any illustrations. In order to remedy this Endlicher started to publish an *Iconographia Generum Plantarum*, which again appeared in fascicles from 1837 to 1841 in Vienna, forming a companion to the main work, but otherwise consisting only of a title page, the legends with comments printed as introductory text, two registers and a total of one hundred and twenty-five uncoloured engravings. Eighty-nine of these have tracings based on Ferdinand's pencil drawings made on the voyage in the *Investigator* as exemplars; with the exception of two all were prepared by him in Australia. This makes the *Iconographia* a work biased in a funny way towards the flora of the fifth continent, and more particularly to the tiny Norfolk Island, for which Endlicher must have developed a special liking.

This is how it came about that a considerable proportion of Norfolk Island endemics made their way into a work dedicated to vascular plant diversity on a global level. Among these we find *Boehmeria australis* subsp. *australis*, popwod, *Myoporum obscurum*, and the fern *Tmesipteris norfolkensis*. The engraving published by Endlicher in his *Iconographia* showing Kurrajong (*Wickstroemia australis*), another endemic of Norfolk Island threatened by extinction, is of some interest. It is based on a pencil drawing by Ferdinand, but the dissections are arranged differently from the lithograph intended for publication in the projected supplement to the *Prodromus Florae Insulae Norfolkicae*, which came to nothing.

Several more engravings based on Ferdinand's pencil drawings and /or pen-and-ink tracings depict plants from the Australian continent and were also included in the *Iconographia Generum Plantarum*. Some of these carry names validated before the *Investigator* was back in England, like the Christmas bell, *Blandfordia nobilis*, with its showy, yellow and red flowers approximately four centimetres long, or the needlebush, *Hakea sericea*, widespread in Australia. Others have scientific names validated by Brown in his *Prodromus Florae Novae Hollandiae* Blackrod, *Buchnera urticifolia*, bloodroot, *Haemodorum spicatum*, a perennial grass-like herb almost certainly documented by Ferdinand in what is now Western Australia, and the yellow rush-lily, *Tricoryne elatior*, belong to this second group. Two more illustrations based on Ferdinand's colour-coded pencil drawing of this showy lily exist: a watercolour now in the

Natural History Museum and Ferdinand's coloured copper engraving which had appeared in the third instalment of his abortive *Icones Florae Novae Hollandiae*. Why two different illustrations both based on Ferdinand's pencil drawing or his pen-and-ink tracing appeared in Vienna at an interval of two decades remains a mystery, in particular since Endlicher's version is uncoloured and of notably inferior quality. Owing to the need for a different format used for the plates in the *Iconographia*, Ferdinand's pen-and-ink tracings had to be rearranged and redrawn, but by whom we do not know; these redrawn tracings do not carry Ferdinand's pencil numbers referring to his colour code, because no colouring of the lithographs was intended.

Endlicher's *Genera plantarum* refers to his *Iconographia* in the same way that Brown's *Prodromus Florae Novae Hollandiae* refers to Ferdinand's *Icones Plantarum Novae Hollandiae*. The sequence of plates in Endlicher's *Iconographia* and Ferdinand's *Icones* do not follow the text in Endlicher's *Genera plantarum* and Brown's *Prodromus*, respectively, although the legends on the plates individually refer to single entries in the two works. In other words, the individual plates of Endlicher's *Iconographia* and Ferdinand's *Icones* could also be easily rearranged to follow the sequence of the main works, that is, systematically by families as circumscribed by Endlicher and Brown, respectively. Endlicher even had two numbers printed on each plate of his *Iconographia* – the number in the upper right-hand corner of the plate is that under which the genus illustrated is described in the *Genera Plantarum*, the number in the lower right-hand corner is the plate number. Moreover, in the illustrations Endlicher follows Ferdinand precisely in the method of using letters for the names of flower and fruit parts (for example, d for chalice, h 4 for sterile anther). Endlicher calls his list of abbreviations 'Siglae Brunonianae' (Robert Brown's abbreviations), though they appeared for the first time in Ferdinand's *Icones Florae Novae Hollandiae* under the heading 'Tabularum explicatio' (Explanation of the plates). Certainly all this is not just a curious coincidence, but rather hints to the fact that Endlicher knew both Brown's *Prodromus* and Ferdinand's *Icones*. In Vienna he could have seen the latter exceedingly rare work in the private library of Francis I, but as a scriptor of the Imperial Library Endlicher would not have found it too difficult to consult this work – via his superior Moritz Count Dietrichstein-Proskau-Leslie. The alternative in Vienna was the copy kept in the library of Johann I Joseph, then in Herrengasse and in walking distance from the Imperial Library on Josefsplatz.

An example may help to illustrate the complexity of Endlicher's *Iconographia*: orchids are illustrated in this work on t. 5–8, 14–18, 29, 40–41, 51 and 99, which means that they have been placed in fascicles 1–5 and 9. By contrast, they are to be found in the main work under the generic numbers 1544, 1592, 1596, 1598, 1580, 1588, 1589, 1579, 1597, 1616, 1534 (*Bartholina burmanniana* from the Cape of Good Hope, see Chapter 10), 1594, 1594 (recte 1593) and 1369. Clearly the double numbering of the engravings and the concept of offering the user the chance to arrange them in two different ways mirror the mind of a librarian, the job in which Endlicher had started his career. When the last two fascicles of the *Iconographia* appeared, he had already reached his final position in academia: upon the death of Jacquin the son on 26 October 1839 Endlicher was appointed his successor as Professor of Botany and Director of the botanical garden of Vienna University. More honours were to come – Endlicher would receive, like Brown, the order Pour le Merite from Friedrich Wilhelm IV, King of Prussia.

On 3 June 1838 in Kew, Franz had received the first fascicles of the *Iconographia Generum Plantarum* from Vienna. Six days later he wrote his last letter to his former travel companion Jacquin the son, which is now conserved in the Universitetsbibliotek Uppsala. Full of joy he noted (in English) 'it is most gratifying to my feeling to se[e] so much justice done to my dear Brother's name, that [t]he udention [attention] he wout [would] have to expect in this country'.[17] Indeed, this is a most adequate comment on the general neglect of Ferdinand's work in England at that time. Others felt less satisfied than Franz. Six years after this letter had been sent to Vienna, John Lindley published a book review starting with critical notes on the botanical exploration of the South Sea combined with a latently xenophobic attack on the lack of publication of the results of the voyage in the *Investigator*: 'Although the English are the great discoverers in the annals of South-sea geography, and botanists have been attached to the more important of these perilous southern voyages, yet the English public has little for which to thank the scientific gentlemen. On the contrary, when their cost is compared with the result, it would have been better to have saved the money. In Cook's first voyage, large Botanical collections were made by Dr. Solander; but his materials are chiefly known from occasional allusions to them by writers who have access to the Banksian collection. With Flinders's voyage, a Botanist and a Botanical Draughtsman were associated, at the public charge; and both these officers discharged their duty of collecting materials most zealously. But if we inquire for the public Botanical advantage which has been derived from that expedition, either to ourselves or to the New South Wales colonists, we find it to consist of little more than – 1st, a Latin fragment [the *Prodromus Florae Novae Hollandiae*], dated 1810, the publication of which was so managed, that after a time it was withdrawn from circulation in this country, and is principally accessible through German reprints; 2d, some learned general remarks, geographical and systematical, and a few plates with Latin descriptions, appended to Flinders's account of his voyage [the *Voyage*]; and 3d, 15 folio plates of plants [the *Illustrationes Florae Novae Hollandiae*], also a fragment. These were expressions of intentions, and no more. That full account of the Botanical discoveries made during Flinders's expedition, which the public had a right to expect, has never appeared. The dried plants, or what remains of them, are to be found in the cases of the British Museum; and the beautiful plates made by Mr. Ferdinand Bauer with such care and admirable skill, after slumbering for years at the Admiralty, are now (we understand) also at the [British]Museum. No publication of them has taken place in this country beyond the few cases mentioned; what their merits are we chiefly know in consequence of their having been partially made public in VIENNA [Lindley's capital letters]!'[18] No doubt this is a vitriolic critique, but the fact remains that Ferdinand's finished watercolours (see Chapter 12) were in England, just a fraction of his work, but even this had not been published. A more complete publication of Ferdinand's graphical record, however, would have been possible only in collaboration with the United Imperial Royal Natural History Cabinets in Vienna, an idea probably unpalatable to Lindley.

Lindley implicitly referred to the publications of Endlicher, in particular to the latter's *Iconographia Generum Plantarum*, where more of Ferdinand's redrawn tracings had appeared than in any other work. However, Ferdinand's graphical record of the Norfolk pine was also published in Vienna in two plates illustrating a less-well-known work. Entitled *Die Coniferen nach Lambert, Loudon und anderen* it refers, among others, to Lambert's *The genus Pinus* (see Chapter 9), but is a largely independent work by Franz Antoine the son,[19] which appeared in 1840–1846 in Vienna. He was the son of the homonymous imperial royal court gardener in Vienna, who had previously laid out the strictly private garden of Francis I, next to the Hofburg, now the Burggarten (see Chapter 12). Remarkably Antoine the son produced all the lithographs for his publication by himself. Two of them are based on Ferdinand's drawings of the Norfolk pine. These had been made available to Antoine the son by Eduard Fenzl, who had become Endlicher's successor as keeper of the herbarium of the United Imperial Royal Natural History Cabinet, then kept in the newly erected Imperial Royal Botanical Museum (see Chapter 19). The first lithograph shows catkins and cones of this tree as well as dissections of them; the other offers an impression of a juvenile and a mature tree with their markedly diverse architecture. Ferdinand had added to the image of the stem base of the mature specimen a representation of a man to indicate size, but his identity, possibly again his servant, remains unknown. The height of the specimen can be estimated to be approximately thirty metres, significantly beyond the maximum recorded size of the Norfolk pine. The value of these two plates is increased by the fact that Ferdinand's pencil drawings and specimens of this spectacular gymnosperm were lost in the Oberhöflein fire (see Chapter 19).

Along similar lines – that is, focused on a single plant group such as the conifers – is Schott's *Rutaceae Fragmenta Botanica*, another slim volume in large format, published in Vienna in 1834. It is a rare work to be purchased by the happy few, never reprinted, not yet digitized and known to the present author only from the copies in the Staatsbibliothek in Berlin and the Österreichische Nationalbibliothek in Vienna. Three lithographs included in this work are based on pencil drawings or pen-and-ink tracings by Ferdinand; two of them illustrate plants from Norfolk Island – yellowwood, *Sarcomelicope simplicifolia*, and the endemic *Melicope littoralis* – its basionyms validated by Endlicher in his *Prodromus*. The third lithograph shows flintwood, *Geijera salicifolia*, from the mainland of Australia; in this case no specimen was available to Schott and he based his description exclusively on Ferdinand's pencil drawing or pen-and-ink tracing marked 'Broad Sound'.

Although Vienna had never been a centre for the study of the Australian flora, thanks to Endlicher it had been for a very brief spell the centre for the study of the flora of Norfolk Island.

LEIPZIG

Maximilian I Joseph, King of Bavaria, had suggested to his son-in-law Francis I, Emperor of Austria, to agree to an addition to the list of persons accompanying the latter's daughter Archduchess Leopoldine on her bridal voyage to Rio de Janeiro.[20] This is how it came to be that Carl Friedrich Philipp Martius, an assistant to the Royal Bavarian Academy of Sciences in Munich, aged 22, and zoologist Johann Baptist Spix got the permission to travel to the Portuguese colony of Brazil. On their extensive travels they penetrated deep into the Interior eagerly collecting, sketching and recording their many observations before ascending and descending the Amazon River and returning to Europe.

In contrast to the reception at the Admiralty Brown and Ferdinand had experienced in 1805 (see Chapter 11), Martius and Spix were greeted on the very day of their return to the Bavarian capital in 1821 by the news that Maximilian I Joseph had made them Knights of the Order of Merit of the Bavarian Crown and thus had elevated them to nobility. A few days later Martius was elected full member of the Royal Bavarian Academy of Sciences and appointed assistant curator of the botanic garden in Munich, with a career rising smoothly and steadily upwards. Before this general background Martius decided to produce the first world monograph of the palm family, of which he had seen many spectacular representatives in Brazil. The end product was his monumental *Historia Naturalis Palmarum*, published over a period of thirty years in Leipzig, the first fascicle appearing in 1823 and the last in 1853. With its weight of around twenty-eight kilograms and its book-block dimensions of sixty by forty-three centimetres, the three-volume set is a superlative work, which could be financed only by subscription. Luxury copies contain coloured lithographs, the standard edition black-and-white lithographs; hybrids are also known, like the copy in the library of the Botanical Museum Berlin-Dahlem. Needless to say the *Historia Naturalis Palmarum* was praised by reviewers, among them Goethe – minister, poet, connoisseur and bibliophile in Weimar.

The taxonomic part of this work is unconventional in comprising lithographs of two kinds: one group illustrates individual palm species with their fronds, inflorescences, infrutescences, flowers and fruits, including many dissections. The second group focuses on the architecture of the individual palm species, showing solitary specimens embedded in the landscape. Although the latter approach is open to critique since the landscapes were often added by graphical artists in Munich, who had never been to the tropics, precisely these views made the *Historia Naturalis Palmarum* a highly popular and sought-after work. However, Martius had a problem: although he had seen many neotropical palms with his own eyes and made pencil drawings of some of them, which are now conserved in the Bayerische Staatsbibliothek in Munich, he had never travelled in the palaeotropical regions. As a consequence he looked out for illustrations documenting Old World palms, both published and unpublished.

Endlicher must have made available to Martius in Munich several of Ferdinand's pencil drawings or the respective pen-and-ink tracings based on them. These would almost certainly have been given under a loan agreement. This is how it came to pass that three palms documented by Ferdinand during the voyage in the *Investigator* found their way into the *Historia Naturalis Palmarum:* the sand palm, *Livistona humilis*; the elegant palm, *Ptychosperma elegans*; and the Norfolk Island palm, *Rhopalostylis baueri*. All three are represented in six conventional botanical illustrations but also in three views: two of them show scenes on the Sir Edward Pellew Islands in the Gulf of Carpentaria, one with representations of the sand palm and the elegant palm, the other with the elegant palm only; the third is a landscape on Norfolk Island. In the scene on that island an image of the Norfolk Island palm is combined with (1) the representations of three specimens of the Norfolk pine, one of them mature, (2) two Norfolk tree ferns, *Cyathea brownii*, with (3) *Freycinetia baueri* climbing as a liana on their stems (see Chapter 11), and (4) the New Zealand flax, *Phormium tenax*. Whereas the basis of the two views from the Gulf of Carpentaria must have been lost in the Oberhöflein fire, that of the scene on Norfolk Island[21] survived in the Naturhistorisches Museum for a simple reason: it was filed under the ferns, which in 1943 were not transported to Oberhöflein Castle but to another safe storage site (see Chapter 19). As a matter of fact Ferdinand's surviving pen-and-ink drawing shows three specimens of Norfolk tree fern of different sizes with *Freycinetia baueri* climbing on them, together with two human figures, which are difficult to interpret. One of them may have been Ferdinand's servant. Clearly the anonymous graphical artist in Munich had (1) produced a mirror image of Ferdinand's pen-and-ink drawing, and then (2) deleted one representation of the Norfolk tree fern, (3) added the representation of a solitary Norfolk Island palm instead, and (4) added furthermore those of the three Norfolk pines, both almost certainly based on other drawings by Ferdinand that were destroyed in the Oberhöflein fire. By contrast (5) the New Zealand flax may have been copied from another source. Needless to say, the relative proportions of the trees in this landscape are totally obliterated and the one remaining human figure is of a doubtful size.

Ferdinand would not have been amused to see what had been done with his pencil drawings in Munich or Leipzig. They were totally changed as in the scene on Norfolk Island (see above), simplified or rendered as mirror images and incorrectly coloured, since the colourists in Leipzig were unable to crack Ferdinand's Australian colour code. An example may illustrate this point: while Ferdinand's watercolour of the sand palm, then still at the Admiralty, shows dark brownish fruits, the colourists in Leipzig represent some in bright red and some, irritatingly, in dark blue colours. Furthermore, Ferdinand would have been envious of the print run of Martius's *Historia Naturalis Palmarum*, which was much higher than of his *Icones Florae Novae Hollandiae* and the *Flora Graeca*. However, he would at the same time have been flattered by the long list of subscribers to Martius's magnum opus featuring the representatives of feudal Europe in the early to mid-nineteenth century – Franz I and his fourth wife, Karoline Augusta, Princess of Bavaria; Alexander I, Emperor and Autocrat of all Russias; Maximilian I Joseph and his second wife, Karoline Friederike Wilhelmine, Princess of Baden; Willem I, King of the Netherlands; Friedrich Wilhelm III, King of Prussia; Friedrich I August, King of Saxony; Wilhelm I, King of Württemberg. The aristocracy was also represented, among them Alexander Nikolaievich Prince Golitzin, minister of education in St Petersburg, Josef Count Esterhazy (Jószef Esterházy gróf Galántai) in Vienna, who had Metternich as his father-in-law, and Count Sternberg in Prague (see Chapter 17).

This is how it came to be that at least nine lithographs based on Ferdinand's graphic record from the voyage in the *Investigator* were published in 1837 and 1838 in Leipzig and made their way into the private libraries of the reigning families in central Europe as well as into the Royal Libraries in Berlin, Bonn and Dresden.

Ferdinand's pencil drawings and /or the pen-and-ink tracings for the six conventional illustrations of the three palm species could not be found among the collection of Martius's palm drawings in the Bayerische Staatsbibliothek. They were probably sent back from Munich to Vienna and later inserted among the palm specimens in the herbarium of the Naturhistorisches Museum; all this material was also destroyed in the Oberhöflein catastrophe (see Chapter 19).

UTRECHT

This disaster was also the ultimate fate of Ferdinand's pencil drawings of cycads prepared during the voyage on the *Investigator* and the pen-and-ink tracings produced later. Endlicher must have sent them to Friedrich Anton Wilhelm Miquel, then working as a physician in a clinic in Rotterdam, later Professor of Botany and Director of the Botanical Garden at Utrecht University. He made use of the graphic record when arranging for the illustrations to accompany his *Monographia Cycadearum*. This work was published in Utrecht in 1842, sixteen years after the death of Ferdinand. In this case composite lithographs with figures prepared by Ferdinand and other illustrators were the end result.

In summary, only a small fraction of Ferdinand's drawings made during the voyage on the *Investigator* – maybe some seven per cent – was published posthumously. He would have been appalled at the outcome: sometimes those published were truncated, and they were mostly printed in black and white only, with all his attention to colour documentation neglected, worlds apart from the sophistication of his watercolours done for the Admiralty or the coloured engravings published in the *Flora Graeca* and his own *Illustrationes Florae Novae Hollandiae*. Ferdinand, no doubt a perfectionist, would have been so frustrated that he would have liked this chapter to be illustrated by one of his watercolours prepared for Sibthorp (Fig. 18.1.), not by any of the posthumously published images.

ENDNOTES

1 Unless otherwise stated, this chapter is based on Mabberley & Moore (1999) and Lack (2003).
2 A. Mendez (pers. comm., 2015)
3 Smith (1841)
4 Mabberley (1985)
5 Lack (1999c)
6 Ackers (2007)
7 Urban (1906)
8 Scholler (1953)
9 Schmutzer (2011)
10 Schott (1827)
11 Ubrizsy Savoia (2002)
12 Riedl-Dorn (1998)
13 Mills (2012)
14 Green (1994)
15 Hogland (pers. com., 1983)
16 Stearn (1947)
17 Uppsala, Universitetsbibliotek, Handskrifts- och Musikenheten, Dörfler Brevsamlingar, Franz Bauer to J. F. Jacquin, 9 June 1838 (unpublished letter).
18 Lindley (1844)
19 Martz (2001)
20 Lack (2010)
21 Riedl-Dorn (2014)

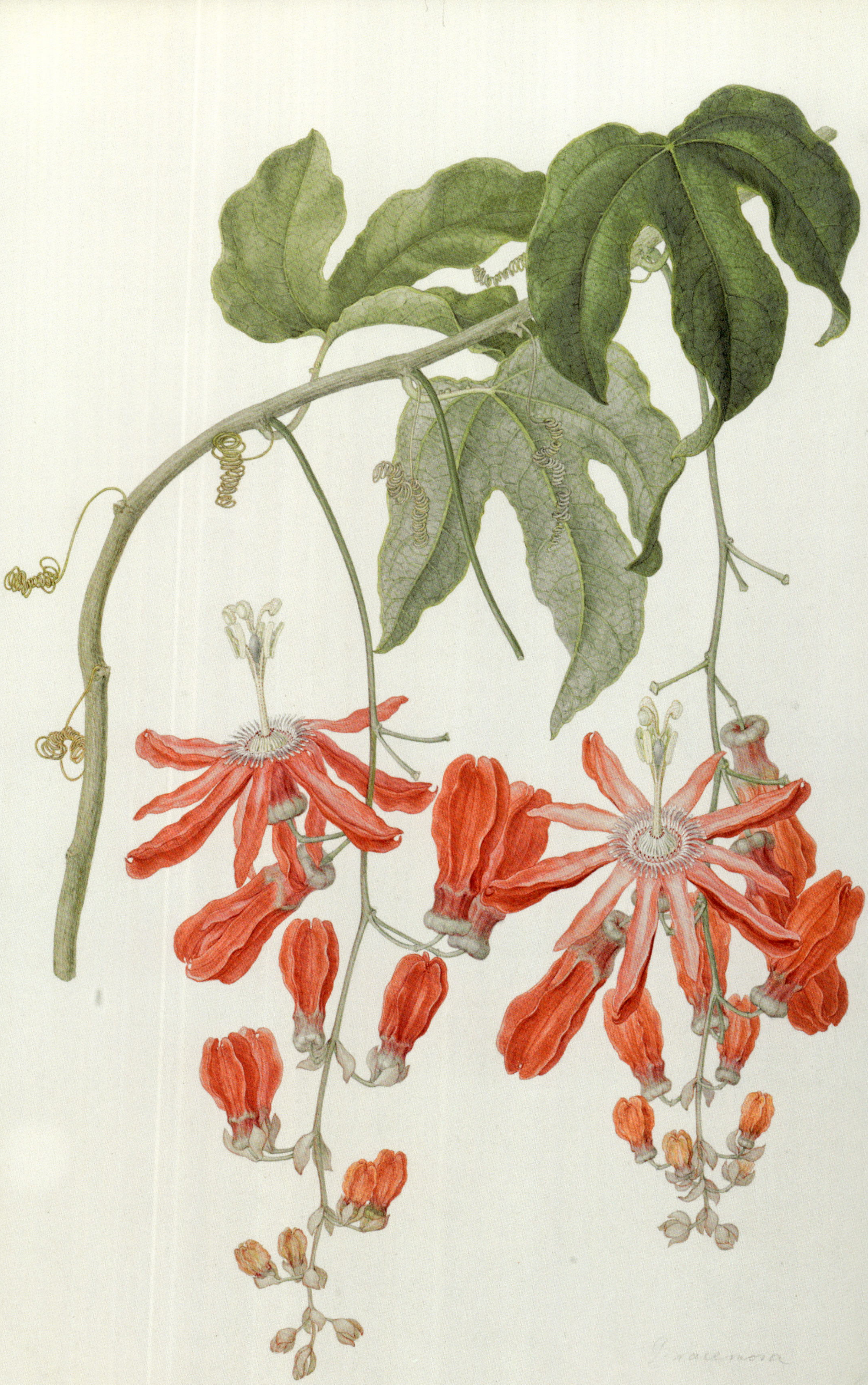
P. racemosa

CHAPTER 19

Fortunate circumstances

THE INEVITABLE EROSION

Time is a strange thing. Everyone responsible for a library, an archive or a museum is well aware of the fact that inevitably some erosion of a collection takes place over a long period of time – single objects become untraceable, are transferred or get lost. Needless to say, this must be avoided by the keeper in charge under any conditions, but as a rule the reality is different from the ideal world. Taking into consideration firstly that Ferdinand, Joseph and Franz died one hundred and eighty-nine, one hundred and eighty-four and one hundred and seventy-five years ago, respectively, and that secondly their drawings, watercolours, paintings, manuscripts, letters and collections have been widely dispersed, only a small fraction of their materials is untraceable, has been transferred, evacuated or got lost.

This is best interpreted as a mixture of fortunate circumstances and the special attention given to the work of the three brothers by keepers and conservators. No doubt the high prestige attributed to the acquisitions paid for by Francis I, Emperor of Austria, and Ernest Augustus, King of Hanover, played an important role. This equally applies to the materials given by Queen Victoria to the British Museum, overruling the two codicils to Banks's testament (see Chapter 17), and those donated to Oxford University by Sibthorp, who had created a foundation with his alma mater as beneficiary (see Chapters 6 and 9). The owners of strictly private collections, like the Reigning Prince of Liechtenstein, the Royal Society, the Linnean Society and the Royal Horticultural Society, were always aware of the value of the Bauer material they conserve. Even single books from libraries known to have been dispersed, like that of Empress Consort Maria Feodorovna, could be spotted by the author on the international art market (see Chapter 3).

FIG. 19.1. Red passion flower, *Passiflora racemosa*. Watercolour by Ferdinand Bauer, signed, c.1825. Berlin, Stiftung Preußischer Kulturbesitz, Staatsbibliothek zu Berlin, Handschriftenabteilung, Libri Picturati A 102, volume A: f. 37 (deposited in Cracow, Biblioteka Jagiellońska, Berlinka).

UNTRACEABLE

Owing to many factors we are very well informed about the works of the Bauer brothers that were finished and delivered to their various commissioners.[1] Thanks to the bureaucracy in Vienna we know many details of Ferdinand's estate, Elizabeth's list of contents for the lettered portfolios and the auction catalogue provides information of similar depth for Franz (see Chapter 17). However, the auction dispersed many items and only all those purchased by Watson for Ernest Augustus and most of those acquired by Brown are traceable (see Chapter 17). By contrast nothing is known of the whereabouts of Franz's tulips and camellias in colour, Ferdinand's pencil drawings of apples and pears, hyacinths, and birds in England and Australia, and Joseph's 'Theseus recognized by his father' and his' Pythagoras addressing his disciples'. The same applies to Franz's Etruscan vases and ornaments, and a few of Ferdinand's views from the Apennine Peninsula, as well as three of Ferdinand's flower pieces, two reported from the Naturhistorisches Museum (see Chapter 11) and one purchased at the auction by Brown. The exemplars for a fraction of the copper engravings and lithographs are equally untraceable, among them several of Franz's watercolours for the *Philosophical Transactions* and Ferdinand's watercolours for Mikan's *Delectus Florae et Faunae Brasiliae*. It is also very likely that initially many more colour-coded pencil drawings of plants and animals recorded by Ferdinand for Sibthorp existed than those acquired by Oxford University, which are now in the Sherardian Library, but they also remain so far untraceable.

TRANSFERRED

During the almost two centuries covered by this chapter, very many materials by the Bauer brothers have been transferred – moved over some distance from one building to another without change of ownership. This is best exemplified by the three largest collections: those at the Natural History Museum, the Naturhistorisches Museum, and the Sherardian Library, Oxford. Franz's eight portfolios marked 'Kew Plants' were accessioned in 1841 by the British Museum, then in Montagu House in Bloomsbury, London, Ferdinand's two hundred and fifty-two watercolours of plants and of animals from Australia were transferred in 1843 from the Admiralty to Montagu House, to which were added later a further forty-nine watercolours of plants from Australia by Ferdinand bequested by Brown (see Chapter 17). When the new museum building by Alfred Waterhouse in South Kensington was ready to receive the natural history collections and libraries, the combined materials by Franz and Ferdinand were moved in.

In Vienna the situation was more complex. When in 1844 the botanical part of the United Imperial Royal Natural History Cabinets moved out of the Augustinian wing of the Hofburg due to lack of space, the herbarium and library were deposited in the newly erected K. K. Botanisches Museum (the Imperial Royal Botanical Museum) situated within the premises of the botanical garden of Vienna University. There the herbarium and the botanical library were not affected by the fire in the Hofburg, which broke out on 31 October 1848 as a consequence of the revolution, destroying parts of the zoological collections and library,[2] though Ferdinand's zoological specimens survived. Only when the spectacular building of the K. K. Naturhistorisches Hof-Museum (the Imperial Royal Natural History Court Museum) erected by Gottfried Semper on the Ring, forming part of Vienna's Imperial Forum modelled on Rome, was ready to receive the natural history collections and libraries, did the herbarium in the strict sense move in. Curiously, the decision was taken for the three-dimensional botanical objects, such as seeds, fruits, wood specimens and material in jars, as well as the botanical library, to remain in the Imperial Royal Botanical Museum and became property of Vienna University. This resulted in the anomaly that the herbarium specimen of the Phillip Island Glory collected by Ferdinand and his pencil drawing of this species are now in the Naturhistorisches Museum, while the seeds are kept in the herbarium of Vienna University (see Chapter 10).

Ferdinand's graphical record of the flora and fauna of the Levant commissioned by Sibthorp had been bequeathed by the latter to his alma mater, but in order to be studied and published the botanical material also had to be moved to Norfolk[3] (see Chapter 9). As is well known, James Sowerby and his family in London were commissioned to engrave the copper plates, to take the prints and to colour them. As a consequence, Ferdinand's watercolours for the respective instalment had to be sent from Norwich to London, where a master copy was made, subsequently transferred on the polished copper plates, which were engraved, and then proof prints were taken. As the final step this proof print was coloured and sent together with Ferdinand's watercolour back to Norwich for checking. Afterwards, the proof print with Smith's annotations was sent back to London, where the scientific name and plate number were added and Smith's corrections made. In this way, high quality was achieved and maintained for more than twenty years. When Smith died in Norwich on 17 March 1828, just two years after Ferdinand, twelve fascicles and a total of six hundred coloured copper engravings had been published. For the completion of the great work by Brown and later by Lindley Ferdinand's watercolours had to be moved from Norwich to London, where the procedure established between Smith and the Sowerbys was maintained. Before Platt closed the accounts on 29 December 1840, two weeks after Franz's death, he paid for the last transportation: Ferdinand's graphical record and the nine hundred and sixty-six engraved copper plates were sent from London to Oxford. While the latter were melted down after Bohn had organized a reprint of the complete work in ten volumes, the former remained at Oxford. Years before Hawkins had already written, 'Were they [Ferdinand's watercolours] even not so superior as they really are to the coloured engravings, they ought to be carefully preserved as public record belonging to the University of Oxford. But I will venture to assist that as works of Art they are superior to any thing of the kind existing and will continue one of the most valuable treasures of the University'.

Ferdinand's graphical record of his tour of the Levant with Sibthorp has been perfectly preserved and continues to be one of the glories of Oxford. Having been first deposited in the Bodleian Library, Ferdinand's material ended up in the Sherardian Library, located in the building of the Department of Plant Sciences in Park Road and belonging to the Bodleian Libraries of Oxford University. In marked contrast Sibthorp's estate, bequeathed to Oxford University in 1795, was sold because of its poor income for £4500 in 1914.[4] The proceeds were subsequently invested in approved securities with the income directed towards financing the Professorship of Rural Botany, now the Sibthorpian Chair at Oxford University.

EVACUATED FROM BERLIN TO CRACOW

When a company runs into financial difficulties, it is standard practice to ask an accountant for a report, to redefine priorities and to mobilize the reserves, which often means selling parts of the property. This was precisely the situation in which the Council of the Horticultural Society found itself in 1859: in order to pay the debts it was decided to sell the house in Regent Street and parts of the fine library, which among others comprised some 1500 plant drawings.[5] These included Ferdinand's 'Passion Flowers' (Fig. 19.1.), which were auctioned in the same year by Messrs Sotheby & Wilkinson in London. Friedrich Wilhelm IV, King of Prussia, acquired the collection and donated it to the Königliche Bibliothek zu Berlin, situated on the Forum Fridericianum (see Chapter 7). This led to calling this acquisition 'The Berlin Passion Flowers'. Later the Königliche Bibliothek was moved to the huge building complex erected by Ernst von Ihne on the boulevard Unter den Linden and subsequently renamed Preussische Staatsbibliothek.

When actions of war are likely to affect a city, it is standard procedure to evacuate the more valuable objects to safe storage sites. This was done by the Preussische Staatsbibliothek during the Second World War, and the greater part of its holdings left Berlin in various directions. Together with other manuscript material – among them the music autograph of *Die Zauberflöte* by Mozart, Ferdinand's colleague in the Jacquin household – the 'Berlin Passion Flowers' were brought in late October to early November 1941 to Fürstenstein Castle (now Zamek Książ, Poland),[6] the largest fortress in the province of Silesia, then belonging to the Deutsches Reich. Because this building was to be used by the Nazi Party, the contents of the repository was moved on 2 July 1944 to the nearby Benedectine monastery in Grüssau (now Krzeszów, Poland). There, five hundred and five boxes were stored on the right-hand side gallery of the Church of the Assumption of Lady Mary, where they could not be seen from below. This evacuation was successful – the two volumes with the shelf mark 'Libri Picturati A 102' survived the end of the Second World War intact.

After the German-speaking monks of the Benedictine Monastery had been expelled on 12 May 1946 in accordance with the Potsdam agreement, the extensive building was repopulated with Polish-speaking Benedictine nuns from Lviv (now in the Ukraine), which had become Soviet territory. In early August 1946, after the departure of the Soviet troops, the Polish authorities transferred the evacuated material belonging to the Preussische Staatsbibliothek to repositories in Cracow and concentrated them in December 1948 in the Biblioteka Jagiellońska, where this admittedly very mixed collection was stored under the nicknames 'Berlinka' (the girl from Berlin) or 'Pruski skarb' (the Prussian treasure). Ferdinand's Passion Flowers' (see Chapeter 12) are still there, albeit pending negotiations on a possible restitution. However, after the fall of the Berlin Wall, Chancellor Helmut Kohl declared the whole matter *Kanzlersache*,[7] meaning a topic on the agenda to be dealt with exclusively by the Chancellor with no other authority entitled to start negotiations. For good reasons the whole matter has been shelved since – smooth German–Polish cooperation within the framework of the European Union was rightly considered to be of far superior relevance than the restitution of library items from Cracow to Berlin.

EVACUATED FROM VIENNA TO VADUZ

From 1806 until c.1910 the fourteen massive volumes of the *Codex Liechtenstein* (see Chapter 2) had been kept in the library of the Reigning Prince of Liechtenstein in his fine city residence in Vienna's Herrengasse. Before this palace was demolished in 1913, the shelves and the holdings were moved to the ground floor of the summer palace in the Rossau Quarter of Vienna, where Joseph had installed the princely gallery in the *piano nobile* and the second floor on the orders of Johann I Joseph, beginning in c.1806 (see Chapter 12).

The occupation of Austria by the Deutsche Wehrmacht on 12 March 1938 and the infamous Anschluss, the de facto annexation of the country to the Deutsches Reich, applauded by many and deplored by few, had wide-ranging consequences for Franz I, 13th Reigning Prince of Liechtenstein, his art collection and his library. By that time Franz I, formerly the Imperial and Royal Ambassador representing the Austro-Hungarian Monarchy in St Petersburg, a Knight of the Order of the Golden Fleece and now aged eighty-four, lived in Feldsberg. Due to the Treaty of Saint-Germain-en-Laye becoming operational on 16 July 1920 this town had become part of Czechoslovakia and was now officially called Valtice. Since that day Franz I's very extensive possessions lay in two separate countries, Czechoslovakia and Austria, whereas he was the sovereign of a third, the Principality of Liechtenstein. Starting from 17 April 1930 his grand-nephew Prince Franz Joseph, then a young man of twenty-three years who had grown up on the family's estates in Styria and Moravia, represented him in the Principality. A few days after the Anschluss, on 30 March 1938, Franz I declared Franz Joseph Prince Regent, and after the former's death on 25 July 1938 Franz Joseph became as Franz Joseph II the 14th Reigning Prince of Liechtenstein with the responsibility for huge properties falling on his shoulders. Realizing that very major political storms were ahead, Franz Joseph II wisely moved to the Principality and took, as the first Reigning Prince ever, permanent residence in Vaduz Castle. The Munich Agreement concluded on 1 October 1938 led to the occupation of parts of Czechoslovakia by the Deutsche Wehrmacht, among them the Feldsberg – Eisgrub area, and soon later to their de facto annexation by the Deutsches Reich. This was followed by the occupation of the western and central parts of Czechoslovakia by the Deutsche Wehrmacht on 15 March 1939 and their transformation into a 'Protectorate of Bohemia and Moravia'.

For the German authorities Franz Joseph II, an extremely wealthy man with extensive properties almost exclusively in the enlarged Deutsches Reich and its sphere of influence, was an anomaly. On the one hand he and his employees, among them Gustav Wilhelm, since 1940 the director of the Princely Collections[8] and thus one of the successors of Joseph, moved often and freely in and out of the Deutsches Reich.[9] On the other hand the Principality had declared its neutrality and politically followed the example of Switzerland, with which it shared the Swiss franc as currency, while many in Vienna and elsewhere thought that the tiny principality effectively belonged to the Deutsches Reich. This created a considerable amount of suspicion among the new authorities in Vienna and Berlin. Consequently there was limited respect for the Reigning Prince as a sovereign. This was quickly felt after the Anschluss when the summer palace in the Rossau district of Vienna with its contents was put on the 'Reichsliste der unverzichtbaren Kulturgüter' (List of indispensable art objects of the Deutsches Reich) – items that were strictly banned from export. The arguments put forward by the solicitors of Franz Joseph II, that the palace and its contents were the private property of a sovereign, were wiped off the table.

Evacuation of the contents of the Rossau Palace was formally ordered by the Reichsstatthalter (Governor of the Deutsches Reich) in Vienna on 30 November 1942. Needless to say he only permitted transfers within the borders of the Deutsches Reich, though the process had already started before then. This resulted in repositories in the country house of the Liechtenstein family near Mödling (Schloss Liechtenstein), where major parts of the library and the archives were stored, in Feldsberg, Eisgrub and elsewhere. However, when the tide turned against the Deutsche Wehrmacht after the battle of Stalingrad (now Wolgograd in the Russian Federation) and the Soviet front approached from the east, it became evident that it would be far better to transfer the contents of these safe storage sites in a westerly direction. This second stage of evacuation was partly successful, though not entirely, e.g. due to lack of transport vehicles the repository near Mödling could not be evacuated. Franz Joseph II was faced with additional worries: if the Soviet army reached his estates in former Czechoslovakia, would he be regarded as a sovereign or rather as a Czechoslovak or, worse still, as a German citizen? Would his moveable possessions, including his huge and extremely valuable art collection and library, be confiscated? Would his real estate be confiscated or put under state administration in the Soviet sphere of interest? As a rich landowner and member of the ruling class could he even face imprisonment? All this made the transfer of Franz Joseph II's moveable possessions further west, ideally to the Principality, a matter of increasing urgency, in particularly since these were assets which – in case of financial problems due to confiscation of the estates in Czechoslovakia – could eventually be converted into cash.

Over the years several requests on various levels in the hierarchy were made to get an exemption from the strict export ban, but in the end all proposals were rejected: for example, an export permit by Baldur von Schirach, Gauleiter and Reichsstatthalter in Vienna, given in late June 1944, was deliberately torpedoed in August 1944 by Hans Delbrügge, the Regierungspräsident (president of the administration) of the Reichstatthalterei in Vienna, then residing in Prince Metternich's former office in Ballhausplatz 2. Talks held by Prince Karl Alfred, one of the younger brothers of the Reigning Prince, in the Ministry of the Interior in Berlin turned out to be equally futile. For the German authorities it was absolutely unthinkable to allow the Princely Collections to leave the Deutsches Reich, even under the assurance in writing that their return would be guaranteed. So in autumn 1943 Franz Joseph II in person started to smuggle in his car several of the most precious paintings in small format out of the Deutsches Reich – among them Leonardo's 'Ginevra de' Benci' (see Chapter 12) and Hans Memling's 'The Virgin and Child', the latter now in the National Gallery of Canada in Ottawa.[10] However, there was no way to transfer heavy or large items, such as Rubens's 'Ascension of Mary' (see Chapter 12), to the Principality without permission from the German authorities. Meanwhile Joseph Steegmann, a solicitor based in Berlin with excellent connections to the potentates in the system and in the Nazi party and at the same time married to a Swiss national with contacts in Vaduz, had found a remedy to circumvent the export ban, at least partially. He successfully organized in Berlin an export permit valid from 7 November 1944 of items of the Princely Collections regarded to be of inferior quality and relevance. Needless to say, a speedy transfer of the so-called second set to the Principality was organized. By contrast, material of top quality, the first set, had to remain behind, with lists for both categories drawn up by the director and submitted.

According to personal information that the present author received from Wilhelm,[11] the *Codex Liechtenstein* had been put on the list of items of top quality, almost certainly by Hans Bohatta, the librarian of the Princely Collection and an expert on incunables. This was a considerate and well-funded decision. If the *Codex Liechtenstein* together with the bulk of the library had been evacuated to the repository near Mödling, the fourteen volumes would almost certainly have been destroyed or transported to the Soviet Union, as were a few volumes of Jacquin's *Florae Austriacae Icones* belonging to this library (see Chapter 3). It is not quite clear when precisely the *Codex Liechtenstein* left the Rossau Palace and if it had been first deposited at the safe storage site in the defunct Carthusian monastery at Gaming. According to information from Wilhelm, the fourteen volumes were definitely among the items stored later, deep in the salt mines in Lauffen near Ischl in the Salzkammergut, known for their stable temperature and humidity. This must have happened at the latest on 27 February 1945. For good reason this depository was anathema to Wilhelm – a mottled mixture of works of art had been deposited there, material from the Kunsthistorisches Museum, the Naturhistorisches Museum and the Princely Collections, as well as looted art and objects of most dubious provenance, such as material commandeered by the Deutsche Wehrmacht in occupied regions. As rightly felt by Wilhelm, this amalgam would raise endless questions and numerous problems were it to have been detected by Allied troops.

Again Steegmann was the man to help out: towards the very end of the war, on 21 March 1945, he organized in Vienna permission to transport the material of the Princely Collections of top quality from the evacuation site in Lauffen to another evacuation site, Königsegg Castle on Reichenau Island in Lake Constance. This transfer on 30 and 31 March 1945 was done by buses hired in the Principality that had had their seats removed and Liechtenstein flags affixed to their roofs to mark them as neutral vehicles in case of air attack. Under very difficult conditions, the small convoy crossed ruined cities like Munich and luckily reached the destination without losses. The idea behind the move to Reichenau Island was clear: to try to smuggle the material by boat across Lake Constance to Switzerland, the distance to the Swiss shore being only one kilometre. However, even during the final weeks of the Second World War the border between the Deutsches Reich and Switzerland was under strict control by the German authorities; boats with searchlights were patrolling day and night, and therefore this plan had to be immediately abandoned.

As a consequence Wilhelm decided on 4 April 1945 to transfer the materials, all of the first set, from Königsegg Castle to Meiningen, a village on the right bank of the River Rhine on the border between de facto Deutsches Reich and Switzerland and much nearer to the Principality. In a dramatic attempt he made use of his contacts with the removal company Weiss in Bregenz, which was prepared to attempt a transport by rail to the Principality. The price agreed on was unconventional – three Liechtenstein passports for the owners of the company. In April 1945 this bargain had its logic. Nobody could know where the victorious Soviet army, then engaged in the Vienna Offensive under Fyodor Ivanovich Tolbukhin, Marshal of the Soviet Union, would stop on its march westward – on the River Enns, which it did, on the River Salzach or on the River Rhine. By contrast, it was certain that the Swiss authorities would permit holders of Liechtenstein passports entry into their country as well as to Switzerland. This last trip was highly successful – three railway wagons sealed by German customs arrived at Schaan station in the Principality. It is hard to believe, but true, that the German customs office had agreed to clear this part of the Princely Collections beyond the borders of the Deutsches Reich, which was what happened. Wilhelm had found out that this was to be done on the basis of the list for objects of the second set, which erroneously had not been ticked, not on the basis of the list of objects of the first set, which was not available to the German customs office. After successfully gaining access to the sealed-off depository from the ceiling, Wilhelm spent several nights in renumbering the paintings and other objects of art and in removing labels giving the name of the artist - in order to minimize the number of questions asked during the customs inspection, which took place on 21 April 1945 and went well.

Six days later Austria declared its independence and on 7 and 8 May 1945 the Deutsche Wehrmacht surrendered unconditionally in Reims and in Berlin. As an element pertaining to the top quality items and together with what Wilhelm in a conversation with the present author called the 'succus', i.e. the meat, of the Princely Collections, the *Codex Liechtenstein* had reached the Principality and found safe storage in a place that it had never been to before – Vaduz Castle. A flight full of risks and critical moments had come to its end.

Decades later the well-known antiquarian Hans Peter Kraus of New York, a specialist in the trade of incunables, manuscripts and rare books, heard about the *Codex Liechtenstein* and tried to buy it. Since this treasure had been in the Liechtenstein family for generations, the offer was rejected. According to Wilhelm, who had become private secretary to the Reigning Prince in 1945, one of the other reasons was the fact that Franz Joseph II, since 1948 a Knight of the Order of the Golden Fleece, enjoyed leafing through the volumes. They reminded him of the gardens of his youth in Styria, Moravia and in Austria below the River Enns, where the *Codex Liechtenstein* had originated. There may even have been a family connection – after all one of the great-grandfathers of his mother, Archduchess Elisabeth Amalie, was Francis I, Emperor of Austria. The latter was the man who had bought major parts of Ferdinand's estate in 1826 (see Chapter 17), and had probably also enjoyed looking at fine botanical illustrations kept in his private library in the Hofburg, famous for its botanical books.[12]

LOST FOR EVER – THE OBERHÖFLEIN FIRE

In 1881 the new building of the Imperial Royal Natural History Court Museum, had been opened by Franz Joseph I, Emperor of Austria, who was also later to act as godfather to Franz Joseph, subsequently Franz Jospeph II, the 14th Reigning Prince of Liechtenstein. Six years later Günther Beck Ritter von Managetta, then head of the Department of Botany, decided to integrate Ferdinand's botanical drawings and tracings from the voyage in the *Investigator* into the herbarium. This meant that they were treated like herbarium specimens, sent on loan, annotated and even exchanged, while Ferdinand's other botanical drawings, like those referring to the *Codex Liechtenstein* (see Chapter 2) and other pencil drawings, remained separate. From the conservator's point of view, filing the drawings and tracings between the herbarium specimens was a somewhat problematical decision, but for the day-to-day routine in the herbarium it had many positive effects, one of them being the much more frequent consultation of Ferdinand's drawings.

After the Anschluss, but before the outbreak of the Second World War, the authorities of the Naturhistorisches Museum took the prudent decision to remove their most spectacular items – the absolutely stunning collection of jewels and meteorites – to safe storage.[13] Among them was the so-called 'Bouquet of Flowers' in a vase of berg crystal, consisting of 2102 cut diamonds and a further 761 cut precious stones, approximately 50 centimetres high and weighing 2.8 kilograms, described by Goethe and a gift from the great Maria Theresa to her husband. Other top items included a gold nugget weighing 548 grams from Miass in the Ural Mountains, a gift from Nikolas I, Emperor and Autocrat of all Russias, to Ferdinand I, Emperor of Austria, and the famous Hraschina meteorite, of which a fragment is at the Natural History Museum. The removal of these objects to safe storage was done to protect the material from being transferred out of Vienna and to inhibit easy access by the new potentates.[4]

The Department of Botany began to evacuate its rooms on the top floor of the building in 1942 by removing the orchid and grass herbarium to safe storage sites, because these two collections were particularly rich in type material. When reports on the extensive losses of the herbarium and library of the Botanisches Museum Berlin-Dahlem due to an allied air attack on 1–2 March 1943 reached Vienna, the decision was taken to vacate the rooms completely. The remainder of the herbarium, including Ferdinand's specimens as well as his pencil drawings and pen-and-ink tracings integrated into it, comprised at least three million specimens; this collection was divided up and stored in six repositories in what was then called Gau Niederdonau, the long-established name Niederösterreich (Lower Austria, the former Archduchy of Austria below the River Enns) being forbidden by the German authorities. To the great delight of all involved, this rescue operation was successful. When the hostilities came to an end and the armistice was signed, nothing had been lost.

A disaster happened a few weeks later, in late May 1945, at one of the evacuation sites, Oberhōflein Castle. This property of the Suttner family had been selected for good reason – it was situated in the Waldviertel, in densely wooded hilly country with no industrial activity in the vicinity and to the west of Weinviertel (see Chapter 1), not far from a railway station on the line from Retz to Drosendorf, with massive medieval walls offering shelter. However, fire broke out in one of the halls of Oberhöflein Castle before midnight,[15] was detected late, could not be extinguished and destroyed the contents of this one room. Within hours about one-sixth of the herbarium of the Naturhistorisches Museum was reduced to ash, including Ferdinand's specimens, pencil drawings and pen-and-ink tracings, which had been integrated into this part of the collection. When the list of plant families destroyed in the Oberhöflein fire was published in 1948, it became clear that material of several groups of particular importance for the flora of Australia had been lost – among them major parts of the petaloid monocots, the aroids, the palms and the Protea family.

No eyewitness report of the fire is known to exist, and no member of the museum staff was present when the disaster happened; not even the date of the calamity is known. Information on the fire reached Vienna only in July 1945 and no official investigation seems ever to have been made. Allegations exist that the fire had been started on purpose,[16] but this has never been substantiated. Later Franz Petrak, a famous mycologist and the man in charge of the evacuation site in Oberhöflein, published a brief report, but according to Karl Heinz Rechinger, then the director of the Department of Botany, the former was shocked about what he had seen and did not provide further details.[17] All this is understandable: in the weeks following the armistice the tiny village of Oberhöflein, only six kilometres from the Czechoslovak border, was flooded by refugees and displaced persons, largely German native speakers from Czechoslovakia expelled or fleeing from their homes, as well as Soviet troops. Extensive looting, arson and the usual atrocities after the end of a war took place in the Waldviertel '(Forest Quarter)', as in the other eastern parts of the by then re-established Republic of Austria. During these difficult months sheer survival was clearly the first priority, not an investigation into the loss of herbarium specimens.

Although the Oberhöflein fire was a disaster for the Department of Botany of the Naturhistorisches Museum, it destroyed only a part of the entirety of Ferdinand's pencil drawings, pen-and-ink-tracings and collections. The quantity is difficult to estimate, but maybe around three hundred drawings plus the pertinent tracings, together with an unknown number of herbarium specimens. Undoubtedly, it is a lucky fact that significant losses occurred nowhere else – not in the other evacuation sites of the Naturhistorisches Museum nor in Göttingen; not in London, where evacuation was finished before bombs had hit the roof of the Botany Department of the Natural History Museum[18] nor in Oxford.

ENDNOTES

1 Lack (2003)
2 Riedl-Dorn (1998)
3 Lack (1999c)
4 A. J. Dorcy (pers. comm. 4 ix 1995)
5 Lack (1999b)
6 Schochow (2003)
7 T. Brandis (pers. comm, c.2006)
8 A. Wilhelm (1998)
9 G. Wilhelm (1998)
10 Call No. 6191.
11 Lack (2000a)
12 Lack (2008a)
13 Michel (1948 b)
14 Michel (1948 a)
15 Petrak (1948)
16 www.burgen-austria.com/archive.php?id=1485 (accessed December 2014)
17 K. H. Rechinger (pers. comm., c. 2000)
18 Stearn (1981)

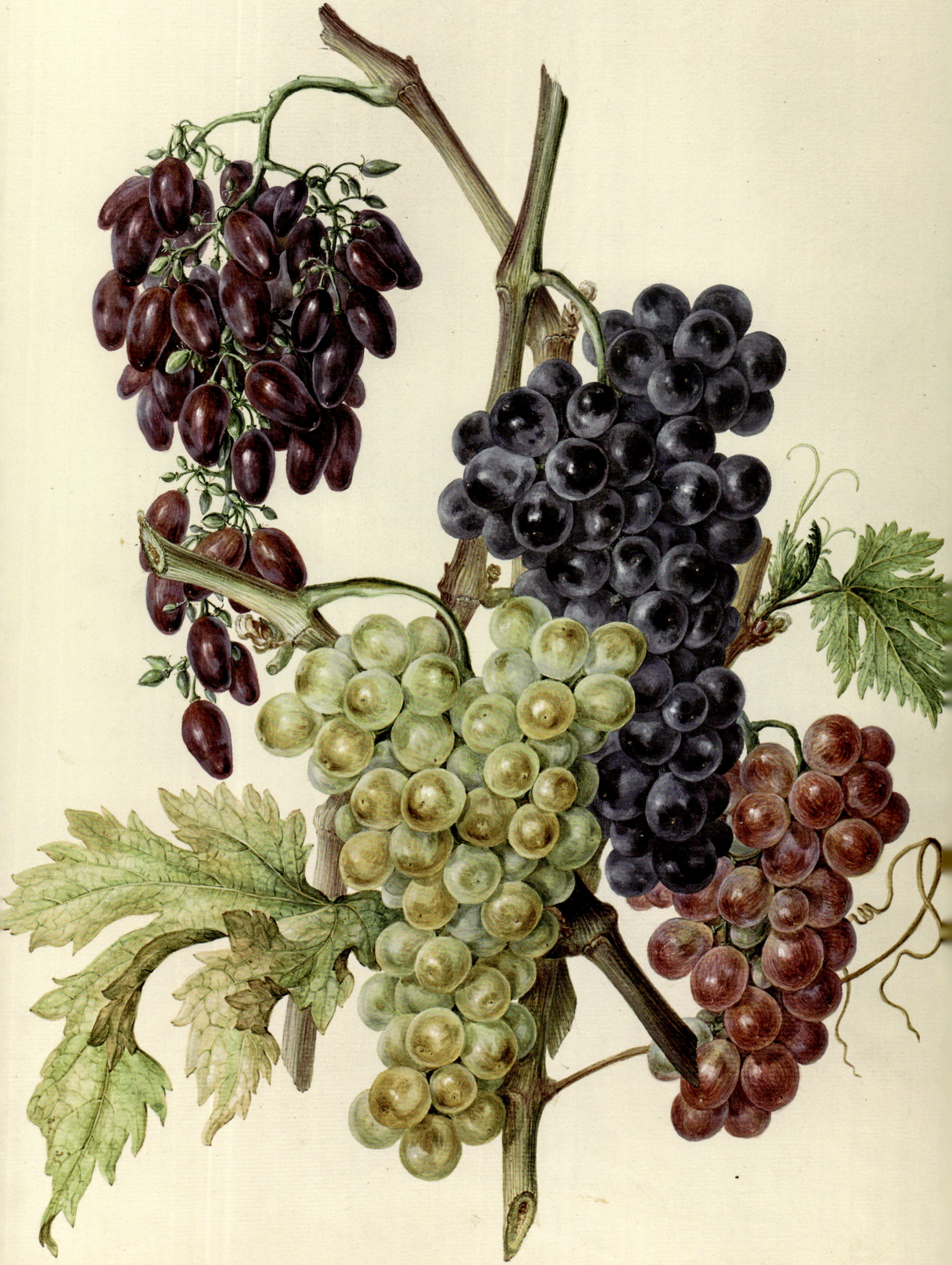

CHAPTER 20

Afterlife

OBITUARIES

Obituaries can be extremely valuable sources of information, about the person deceased but even more so about his or her family and the circle of his or her colleagues and friends. This general observation also applies to the Bauer brothers. No obituaries of Johann and Joseph are known to exist, reflecting in a sense the very different – and definitely minor – achievements of the two elder brothers, whereas necrologies have been published commemorating Franz and Ferdinand. However, these differ substantially.

On 26 December 1840, ten days after Franz's funeral, an anonymous obituary appeared in the literary journal *Athenaeum*, published in London.[1] Its wording and phrases are strongly reminiscent of *Kew and its gardens* and therefore suggest Scheer as author, which is taken as granted in this context. Clearly his text is full of sympathy, describing Franz as 'a most amiable and intelligent gentleman of the true old school with whom we had had the good fortune to become personally acquainted' and as a 'fine-hearted old man'. Furthermore, Scheer stresses Franz's unselfishness in writing, 'Of Mr. Bauer . . . it has been truly said, "that nothing prevented his acquiring an extraordinary degree of fame, except his remarkably unobtrusive modesty – he worked rather for the credit of others, than for his own"'.

FIG. 20.1. Grape, *Vitis vinifera*. Watercolour by the Bauer brothers, c.1783. Watercolour, ink. *Codex Liechtenstein*, volume 6: f. 161. – Vienna, Fürstliche Sammlungen.

There is ample evidence that Franz must have conversed with Scheer about his youth in Feldsberg, since the latter stressed the very early death of the father and the important role played by the mother Therese in developing his artistic talent and that of his brothers. Scheer underlines the key role played by Banks, who induced Franz to remain in England by offering a generous annuity for life, adding that Franz's only other income 'was fifteen guineas, which the editor of one of the cheap publications of the present day [*The Penny Magazine*] sent to him for some short papers on the smut in wheat'. Franz's long association with Everard Home is described by Scheer as an 'intimacy of the most lasting and most useful kind'. According to this obituary it was due to Home's suggestion to George IV that a botanical museum be established 'at Kew, which was to be intrusted [sic] to Mr. Bauer. The house now belonging to the King of Hanover was purchased for this purpose – the shelves were prepared – all the botanical books in the King's library were to be removed there, and some had, in fact, been sent down [to Kew], when, unfortunately, a dispute arose respecting the land, to which the Woods and Forests [Department] laid claim; and some artillery wagons driving off with the bookcases gave Mr. Bauer the first intimation that the plan had been abandoned'. This is a very interesting statement showing how far plans for the future museum – not the botanic garden – at Kew had advanced and what a prominent role Franz was to play as the first director of that botanical museum. Scheer's obituary ends on a poetic tone describing Franz's end as 'final departure from this world under circumstances the most consolatory, full of resignation, faith and hope, and free from sufferings, save the increasing debilities of old age'. As evident from Franz's letter to Jacquin the son in June 1838, the truth may have been quite different (see Chapter 16). Scheer's obituary was reprinted verbatim in *Annals and Magazine of Natural History*[2] and the ephemeral *The English Journal. A Miscellany of Literature, Sciences and Fine Arts,*[3] two journals published in London.

Since Franz had been a fellow of both the Linnean Society and the Royal Society, obituaries appeared in the *Proceedings of the Linnean Society*[4] and the *Proceedings of the Royal Society,*[5] the latter an abbreviated version of the former. Not unexpectedly, the text is full of admiration for Franz and very informed, making reference for example to Stoerck's *De usu Pulsatillae nigricantis* (see Chapter 2) and ending with the statement: 'It is this rare and previously almost unexampled union of the observer and the artist that has placed Mr. Bauer in the first rank of scientific draughtsmen. His paintings . . . are no less perfect as models of artistic skill and effect, then as representatives of natural objects. Of all his predecessors, Ehret alone approaches him in these particulars: among his contemporaries, none but his brother Ferdinand can be recognized as his equal'. An amended version of this text with notes on the auction of Franz's effects appeared in January 1842 in the *Gentleman's Magazine,*[6] a monthly digest of news and commentary published in London.

As evident from the title page of the *Illustrations of orchidaceous plants* (see Chapter 13), Franz had also been a Fellow of the Horticultural Society of London. Therefore it is no surprise that an obituary also appeared in the gardening journal *Gardener's Magazine,*[7] a repeat of Scheer's text first published in the *Athenaeum*. The last note from one of Franz's contemporaries came from Kew. In a short contribution entitled 'Notes on Kew and Kew gardens' and published in the family journal *Leisure Hour* the Reverend Thomas T. Haverfield recorded Franz as a man of 'the most extraordinary scientific talent, and perhaps the finest painter of flowers in water colours that ever existed'.[8] His personality is described as 'modest and retiring – one who preferred public utility to personal reputation. . .'. On a critical note Haverfield added that Franz's language was adhering to his German intonation, 'though he lived for a full half century in England'. Clearly the Reverend had no feeling for language being an essential component of the identity of every individual, in particular when they are permanently obliged to speak a second tongue. No doubt it was natural and logical for Franz to adapt, but he saw no necessity to hide his Austrian background. In any case, the author of these recollections must have received a lasting impression from Franz – when the text appeared the latter had already been dead for almost 22 years.

Not a single obituary of Ferdinand appeared in Vienna, possibly because he had committed the unpardonable mistake of leaving the 'kaiserliche Haupt- und Residenzstadt', even today regarded by some as a form of high treason. Furthermore, Ferdinand seems to have worked in his last twelve years in Vienna almost exclusively for commissioners in England, and his social circle may well have been smaller than that of Franz. For persons such as Jacquin the son or Trattinick it would have been natural to publish a necrology for Ferdinand, but this they apparently did not do. It was in Regensburg that an extremely brief anonymous note on Ferdinand's death appeared in the journal *Flora.*[9]

However, a belated obituary – or rather a first brief biography – of Ferdinand appeared in the *Journal of Botany* edited by William Jackson Hooker in London,[10] albeit no less than seventeen years after the former's death. Its author Jan (Johann, John) Lhotsky (Lgocki) had been born in Lemberg, received his doctorate from Jena University, travelled in Australia and was to die in London in 1866.[11] He claims to have based his text on 'original family documents in my possessions', of which only Ferdinand's letters now in the archives of the Linnean Society of London survive (see Chapter 10). The whereabouts of the other material, if it existed, could unfortunately not be traced, nor could it be clarified how Lhotsky acquired it. The most plausible source is Franz, the addressee of Ferdinand's letters from Australia (see Chapter 10). In any case Lhotsky's text is extremely detailed, in particular on Ferdinand's youth (see Chapters 2–4) and contains very accurate statements, such as 'Rev. N. Boccius, Superior of the convent and hospital of *Fratrum Misericordiae* [Lhotsky's italics] in Feldsperg [sic]', while most of the more recent authors incorrectly called Boccius an abbot and rather vaguely referred to the monastery of the Brothers of Mercy. Another very precise statement is the reason given for Sibthorp to come to Vienna – to study the 'unique manuscripts of Dioscorides in the Imperial Library' (i.e. the *Codex Vindobonensis* and the *Codex Napolitanus*, see Chapter 5). By contrast, the inaccuracies in Lhotsky's account are very minor, for example reporting Sibthorp as having arrived in Vienna in 1784, while he was at that time still at Göttingen University.

Like Scheer's obituary for Franz published in the *Athenaeum*, Lhotsky's text is full of praise for Ferdinand having 'portrayed botanical subjects more admirably' than anyone else, with the author's intention given as 'to plant . . . a cypress on . . . [his] grave'. In addition, Lhotsky stresses Ferdinand's activity as a botanical collector, for example of *Salvia veneris* on the island of Cyprus, information he could only have received from Franz or from the consultation of the *Flora Graeca*. However, it cannot be excluded that Ferdinand and Lhotsky met in Vienna: the latter had attended Vienna University in 1816 and was again in this city from 1822 to 1828, albeit probably in prison – for having published in 1820 in Leipzig the two confused political pamphlets *Uiber Aufklärung, Bildung, Entwicklung als Höchstes im Leben der Menschheit* and *Beiträge zu einer Politik*.

It seems that no biographer of Ferdinand has ever given any detailed thought as to how Lhotsky could have collected all the very accurate information on, for example, Ferdinand's youth embedded in his obituary. An unpublished letter by Franz sent to Lhotsky[12] and kept in the archive of the Linnean Society together with the manuscript is indicative. Franz tells Lhotsky that he had read the manuscript of the latter's biography of his younger brother and approves of its contents, apart from the minor critique that the town of Feldsberg belonged to (the Archduchy of) 'Austria' (below the River Enns) and not to (Margraviate of) 'Moravia'. It seems plausible to assume that it was Franz who had suggested to Lhotsky to write the text and the latter had reasons to do so. Schreibers in Vienna had suggested to Lhotsky a voyage to Australia on which he should collect plants as Ferdinand had done before him. Lhotsky has also been reported to have been the first to record the toponym 'kembery' from the natives, which became Canberra, the name of the capital of Australia. Back in London on 4 April 1839 at the latest, Lhotsky treasured friendly feelings for his compatriot – after all they were both subjects of the Emperor of Austria and the distance between the places of births of Lhotsky and Ferdinand was significantly smaller than from Lemberg to Munich or Berlin, let alone London. Further biographical information, such as Ferdinand's botanical excursions to the 'Alps of Austria and Styria' in his late years, may well originate from Franz, just like the note that 'even in an advanced period of life, [Ferdinand] Bauer made numerous sketches after the celebrated flower-pieces of Van Huysum' for which no proof anywhere else could be traced. In a city like Vienna Ferdinand could choose – there was a flower piece by Jan van Huysum in the Imperial Gallery[13] and another in the Liechtenstein Gallery,[14] which was in the charge of his brother Joseph (see Chapter 12). Furthermore, Lhotsky claims that Ferdinand had been employed until 1813 by the Admiralty and that in Hietzing 'he worked hard to in executing and completing his drawings of New Holland plants and animals', confirming the views expressed in Chapter 11 and 12. There is even an interesting hypothesis: Lhotsky is 'inclined to think' that Ferdinand had assisted Westall (see Chapter 10) in producing his views of the Australian coasts in oil, now kept in the collection of the Ministry of Defence Art Collection in London. However, for this assumption no proof has otherwise been found.[15]

Lhotsky's epitaph for Ferdinand is perfectly worded and reads, 'he sought not to idealize nature, but to seize the ideal features of nature', which could be translated with less elegance into 'he recorded the statistical mean values in his documentation of plants and animals'. On the same line is Lhotsky's statement that Ferdinand had 'only (so to speak) transcribe nature, and his transcripts were ever alike faithful and elegant'.

POSTHUMOUS PRAISE

With only a small fraction of their pictorial record published and that in extremely small print runs (except those printed in the *Philosophical Transactions*), comments on the work of Franz and Ferdinand were by necessity small in numbers. They either came from the few individuals who had access to the drawings and watercolours in Göttingen, London, Oxford and Vienna or, alternatively, those who were able to consult the extremely rare prints.

Endlicher in Vienna belonged to the first group. In the preface of his *Prodromus Florae Insulae Norfolkicae*[16] he called Ferdinand 'celeberrimus plantarum pictor' (the most famous painter of plants), stressing his industry in collecting plants, his skill in drying them and his 'divine plane . . . ingenio' (almost divine genius) of drawing them. Michele Tenore, professor of botany and director of the Royal Botanic Garden of Naples University, belonged to the second group. In order to consult a copy of the *Flora Graeca* he had to travel to Paris, where Benjamin Delessert, a wealthy banker and Maecenas of botanical pursuits, had subscribed to the exceedingly expensive work. Here is a summary of his review spanning no less than twenty-one printed pages: 'La classica terra delle scienze e delle arti, la patria di Teofrasto, Omero, Aristotile, Fidia, Prassitele, ben si meritava il dono della più magnifica Flora del mondo. Tal' è di fatti la Flora Graeca del Sibthorp; sia che considerar se ne vogliano gl'illustri scienziati che vi hanno dato opera, sia per la superiorità della sua artistica esecuzione'[17] (The classical territory of the sciences and the arts, the country of Theophrastus, Homer, Aristotle, Phidias, Praxiteles well deserved the gift of the most splendid Flora of the world. This in fact is the Flora Graeca of Sibthorp; considering both what the illustrious scientists collaborating in this work achieved and what concerns the superiority of its artistic execution). Tenore's accolade, published in the *Rendiconto delle Adunanze e de Lavori dell' Accademia delle Scienze Sezione della Società Reale Borbonica di Napoli,* is echoed by Thomas Martyn, his former homologue at Cambridge University, who had written earlier, 'This classical work may truly be denominated fortunate except in the premature death of the excellent author [Sibthorp; in fact only the collector of the materials]; for no man could have been selected more able to ascertain the plants of Greece than Sibthorp; or so fit to be the editor [i.e. author] of them as Smith; no draughtsman could have been chosen like [Ferdinand] Bauer; no engraver like Sowerby. In short, the whole work is deserving the highest praise'.[18] Joseph Dalton Hooker (see Chapter 16), soon to become his father's successor as director of the Royal Botanic Garden at Kew, was more brief in his comment on the *Flora Graeca,* calling it simply 'the greatest botanical work that has ever appeared'.[19]

It was William Griffith, assistant surgeon and for a short time acting superintendent of the Botanical Garden in Calcutta, who was the first to praise Franz for his palynological studies. On 31 October 1843 he wrote from Calcutta to an uncle in England, with the letter posthumously published, 'Mr. Bauer is not half appreciated yet: he is considered a very great artist, what is that to what he was? But he did not fight for his own hand, though he worked hard enough in all conscience. Mr. Bauer in fact preceded [sic] all in the train of discovery: he saw in 1797, that [sic] others did not see till 30 years after. For instance, the elongation of the pollens' inner membrane [cell wall] into a tube, the first stage towards the complete knowledge we now have of vegetable embryogeny. Unfortunately, Mr. Bauer drew, but did not write, and when I recall to mind a remark of Mr. Brown, that it is was a disadvantage to be able to draw, I always fancy he had [Franz] Bauer in his mind's eye; for had he been a writer and not a drawer, before 1800, in great probability we should have known nearly as much of embryology as we do now'.[20] Indeed this is a very wise and most appropriate comment on the lack of impact of Franz's palynological studies (see Chapter 14).

In one of his characteristic footnotes to a paper published in the following year Brown wrote '. . . I may be permitted to add how fortunate I consider myself in having so long enjoyed the friendship and so often been indebted for the important assistance of these two distinguished men [Franz and Ferdinand], whose merits in the branch of art which they cultivated have never been equalled . . .'.[21] This is indeed high praise from a brilliant, but very critical, mind, appropriately called Jupiter Botanicus by Martius.[22]

The last and similar accolade was very brief. It came from the famous bibliographer Georg August Pritzel, curator at the Königliche Bibliothek in Berlin,[23] who had noted in late 1871 in the second edition of his standard work *Thesaurus literaturae botanicae* about Franz simply 'der grösste Pflanzenmaler' (the greatest plant painter). This was the most qualified praise possible coming from a man who had seen more botanical illustrations than anyone before and after him – he had analysed some 40,000 works in Berlin, Geneva, London, Paris, Vienna and elsewhere for his great bibliography, including Franz's material purchased for Ernest Augustus, then already in Göttingen. Pritzel's praise was followed by a long gap in time, which effectively ended only after the Second World War with the publication of Wilfrid Blunt's *The art of botanical illustration,*[24] which devoted a whole chapter to the Bauer brothers. This work stands at the beginning of what may be called the Bauerian Renaissance. No attempt is made here to analyse the subsequent studies on the Bauer brothers, incomplete and biased towards Ferdinand as they are; they are listed in the bibliography.

It is interesting to note that three exhibitions were recently focused on Ferdinand: 'An exquisite eye' in the Museum of Sydney, 'The Flora Graeca Story' in the Bodleian Library at Oxford and 'Ferdinand Lucas Bauer' in the Naturhistorisches Museum in Vienna, all accompanied by homonymous catalogues.[25] By contrast a single exhibition was dedicated to Joseph, Franz and Ferdinand, albeit only on their early work. It took place in the Liechtensteinisches Landesmuseum in Vaduz and had the second edition of *A Garden for Eternity* (see Preface) as catalogue. So far not a single exhibition was aimed at a cross section of the works accomplished by all three brothers. No attempt is made here to list the plethora of exhibitions which have included works by Ferdinand, but two stand out – that which took place in the Art Gallery of South Australia in Adelaide entitled 'The Encounter 1802'[26] and the one in the Künstlerhaus in Vienna entitled 'Die Entdeckung der Welt. Die Welt der Entdeckungen';[27] both took place in 2002.

ENIGMAS

Enigmas remain. The most important gap in our understanding is the use of drawing aids in the graphical work of Franz and Ferdinand. Thanks to the detailed auction catalogue of Franz's belongings we possess a list of his microscopes, micrometres etc., and from the inventory of Ferdinand's estate we know for certain that the latter owned both a camera obscura and a camera lucida (see Chapter 17), but none of these instruments survived. Considering the life-long collaboration between the two brothers it is extremely likely that both knew how to make use of most of these optical aids, which of course does not prove that they used them. The hypothesis may be permitted that Franz and Ferdinand possessed the competence to adapt these instruments to their specific requirements, maybe by making use of an additional lens or mirror, though this again is only a hypothesis. Staggering precision has been repeatedly mentioned in this text, e.g. in Ferdinand's representation of the blue swimmer crab, *Portunus pelagicus*[28] (see Chapter 10). However, the fact remains that we have no record of an eye-witness on the actual use of these instruments by Franz and Ferdinand which is surprising – neither Sibthorp, nor Banks, Brown, Flinders, Good, Home or Lindley referred to them. By contrast, the camera obscura used by Antonio Canal survives in the Museo Correr in Venice[29] and it is extremely likely that his nephew Canaletto (see Chapter 3) made use of a similar instrument when in Vienna in 1759–1761. Significantly, both artists were born in Venice. Since Canaletto was commissioned by Joseph Wenzel to record the summer palace in the Rossau district of Vienna in two oil paintings (see Chapter 3), the former could even have met the prince's court painter at the time, Lucas Bauer. In short, it cannot be excluded that the knowledge of the camera obscura, then no more a secret but described e.g. in Francesco Algarotti's, *Saggio sopra la pittura* published in 1756 allegedly in Venice, could have been passed to the specialists in Vienna, e.g. Lucas Bauer. Needless to say this again is highly conjectural, in particular since he died soon afterwards. The occasional or persistent use of drawings aids, like the camera obscura and, later, the camera clara (see Chapter 10) by Franz and Ferdinand, as yet not proven, would be in agreement with the admittedly somewhat controversial, if not provocative Hockney-Falco thesis. This claims that advances in realism and accuracy in the history of Western art are primarily the result of optical aids, such as the camera obscura, the camera lucida and mirrors,[30] rather than solely due to what is usually called 'artistic technique and skill'. Further research by experts of optical aids would be necessary to solve this enigma.

The meagre written record left behind by the Bauer brothers is a minor enigma. Was writing a type of self-expression they felt disinclined to? This remains unclear, since losses must have been significant – there is for example a letter 'No. 14' sent by Franz to Brown which has survived (see Chapter 13), while those before and after this number are untraceable. Fewer letters by Ferdinand are known, but he was definitely not a person who totally refrained from writing letters (see Chapter 10). By contrast we have only the extensive texts written by Joseph when describing and interpreting the paintings of the gallery he was in charge of, but otherwise no correspondence whatsoever could be traced so far.

Another enigma is the fact that all three brothers remained unmarried, which is the more surprising since at least Joseph and Franz had a continuous income from permanent positions. Apart from the fact that Franz adopted very late in his life Elizabeth Baker, possibly some forty years his junior, absolutely nothing is known about women in the lives of the Bauer brothers. Interesting information is provided by Moritz Alois von Becker, a historian who had been in his earlier years a tutor to the family of the Reigning Prince of Liechtenstein, and was the first to praise the *Codex Liechtenstein* after Sibthorp. In his much later chronicle of Feldsberg[31] Becker reported that all three Bauer brothers remained bachelors and were 'Sonderlinge' (eccentrics), 'everyone in a different way', one-sided characters 'devoid of a sense of family'. There may be some truth in this, but the fact remains that Becker's text was published no less than 61 years after the death of Ferdinand and may at best reflect hearsay collected in Feldsberg and Vienna about Joseph, Ferdinand and, possibly, Johann. Anyway, the reason Becker gives for the poorly specified eccentricities of the Bauer brothers, the early death of their mother Therese, is totally wrong – as a matter of fact she died when her youngest son had reached the age of thirty (see Chapter 8). In addition it seems highly unlikely that Becker had any first-hand information on Franz, who had left Feldsberg almost a century before the chronicle was printed and who is known to have been the centre of an informal club that met regularly in his home in Kew (see Chapter 16).

Lhotsky described Ferdinand as a 'plain straightforward man, full of application and energy', which may have been true, as also he may have had a 'deep sense of gratitude towards those friends and patrons, who had done him service', a character trait approaching, possibly and occasionally, that of an inferiority complex. Another surprise is the fact that apart from William Hooker, who declared himself on the title page of *Paradisus londinensis* 'pupil of Francis Bauer'[32] and is not to be mistaken for William Jackson Hooker, no other disciples of Franz or Ferdinand are known. This is indeed astonishing when considering that the two brothers had stayed for many years in England and had definitely belonged to the botanical establishment of their time. Since working in watercolours requires a high degree of concentration, the lack of assistants working for Franz and Ferdinand is unsurprising.

The financial situation of the Bauer brothers remains a further enigma. How did it come that Ferdinand who never had a permanent position like his three elder brothers managed to buy himself a house and garden in Hietzing near Vienna and died a man in comfortable circumstances while the final balance of Franz's estate may have been at best nil? Since Franz had inherited money twice, from his brothers Ferdinand and Joseph, this is even more astonishing. Again we have no clue.

No portrait of Ferdinand has ever been mentioned in the literature and the present author has been unable to trace one. This continues to be an enigma. Why did Ferdinand refrain from having his portrait painted as Franz had twice at Kew? We simply do not know. Needless to say, Ferdinand could have produced a very accurate portrait of himself in front of a mirror, but this he – like his elder brothers – apparently did not consider. From modern botanical illustrators it was learned that as a rule they feel totally disinclined to produce a portrait or a self-portrait, being so focused on the accurate delineation of plants, which necessitates a totally different mental approach. The lack of a portrait by Ferdinand has led in a curious way to an imbalance in the presence of the two brothers in England: while Franz's early portrait in oil is permanently exhibited in a prominent place, the lobby of the building housing the Herbarium, Library and Archives of the Royal Botanic Gardens Kew, Ferdinand is absent. This situation is mirrored in the Library and Archives of the Natural History Museum, where a reproduction of Franz's portrait at Kew has been placed in the Reading Room near the reference works while Ferdinand is again missing.

The main employers of Franz and Ferdinand have fared better: an anonymous portrait of Boccius in oil can be seen in the Convent of the Brothers of Mercy in Vienna. A bust of Jacquin in bronze has been placed in the pantheon of Vienna University, the Arkadenhof of its main building on the Ring, and in stone on the balustrade of the Naturhistorisches Museum. A life-size statue in marble of Banks seated is exhibited in the main hall of the Natural History Museum with several portraits in oil of him also in London, for example in the Royal and Linnean Societies and in the Council Chamber of the Royal Horticultural Society.[33] Of Joseph's superiors Johann I Joseph is commemorated in a slightly over life-sized statue, and that in a place of outstanding splendour, the Feldherrnhalle (Hall of the Commanders) in the Heeresgeschichtliches Museum in Vienna.

THE MYSTERY OF TIME

With the exception of Denmark House and 23 Kew Green in Kew, the houses where Joseph, Franz and Ferdinand lived no longer exist. There is no plaque to commemorate them on those existing houses, nor on the houses erected after demolition of the earlier buildings – in Kew, London, Hietzing or elsewhere, and the same applies to Feldsberg, where the three brothers had spent their formative years.

Of the very many conservatories Franz and Ferdinand must have known, only three stand largely unaltered. One of them is at least partly still used for overwintering tender plants – the orangery in Schönbrunn – although much of the remaining area is used for putting on events. The large Palm House erected in the 1820s on the Hietzing end of the Schönbrunn complex has become part of the zoological garden. By contrast, the orangery in the Royal Botanic Gardens at Kew is now a café and a restaurant. Other buildings on the grounds, such as the Pagoda, Hanover House and what is now called Kew Palace, as well as the numerous buildings of the palace of Schönbrunn, still remind us of the times when the Bauer brothers lived. The same is true of Feldsberg Castle, where Franz claimed to have been born. Stripped of parts of its gallery and original furniture, which largely ended up in the Fürstliche Sammlungen in Vienna and in Vaduz, with several empty frames in Feldsberg, this stately home is only a shadow of the great days it had experienced under Joseph Wenzel and his successors. However, the connoisseur will enjoy the exquisite parquet floors, the spectacular over-doors and the fine chapel which remained untouched.[34] The orangery in Eisgrub has long since been replaced by a glass and cast iron conservatory and the medicinal garden of the defunct convent of the Brothers of Mercy in Feldsberg was in part built over long ago. The Royal Botanic Gardens Kew, the Botanic Garden of Vienna University and the Botanic Garden of Oxford University have considerably extended their grounds, but at least parts of their old walls still exist. Only the area of the gardens and parks of Schönbrunn have remained basically unaltered since the time of the Bauer brothers.

Otherwise mainly buildings with essentially transcendental purposes such as churches have withstood the changes of time, and these may best act as places of commemoration for the Bauer brothers. The Parish Church of the Assumption of Lady Mary in Feldsberg, where all Bauer children were baptised, is one of them; St Anne Church in Kew, where Franz's epitaph stands is another; and the Parish Church in Hietzing with all its baroque splendour, where the requiem for Ferdinand was sung, is a third. Special mention is made here of a single spacious room which has also remained unaltered – the Great Gallery in the summer palace of the Liechtenstein family in the Rossau district of Vienna (see Chapter 12).

Considering the increase in the world population since the birth of the Bauer brothers, Ferdinand would not be able to record or collect the plants exactly in the places he had found them previously. At least one of them, the Phillip Island Glory, is lost forever, while numerous localities have totally changed in Australia, similarly along the coasts of the Mediterranean Sea, where interference by humans has been particularly strong. However, there are still unspoiled spots where Ferdinand would be able to record the same plants that he had seen more than two centuries ago – on Mount Athos or along the partly virgin coasts of the Gulf of Carpentaria in northern Australia, on the high mountains of Crete or on the Islands of the La Recherche Archipelago off the south coast of Australia. In very many other places, such as in the vast tracks of land eaten up by huge metropolitan agglomerations such as Istanbul, Sydney, Cape Town or Rome, he would simply be petrified. In a similar way Franz may be horrified by what has become of London, a megacity of more than fifteen million inhabitants, with the village of Kew, where he lived for half of a century, having become part of it. Maybe Ferdinand, the wanderer, would find less change nearer to his birth place: his last collecting sites, Leopoldsberg north of the city centre of Vienna and Eichkogel near Mödling are now sites of nature conservation and protection, just like Raisten near Feldsberg with its vineyards. Together with his elder brothers he had enjoyed the grapes from this region and depicted them (Fig. 20.1.).

ENDNOTES

1 Anon. (1840b)
2 Anon. (1841d)
3 Anon. (1841e)
4 Anon. (1841a)
5 Anon. (1841b)
6 Anon. (1842)
7 Anon. (1841c)
8 Haverfield (1862)
9 Anon. (1826)
10 Lhotsky (1843)
11 Kruta (1977)
12 London, Linnean Society, archive, Franz Bauer to Jan Lhotsky, 17 March1839 (unpublished letter)
13 GG 559
14 GE 540
15 Findley (1998), Radford (2002)
16 Endlicher (1833)
17 Tenore (1842)
18 Martyn in Coxe (1811)
19 Hooker (1855 – 1859)
20 Griffith (1847)
21 Brown (1844)
22 Mabberely (1985)
23 Stafleu (1973)
24 Blunt (1950)
25 Watts, Pomfrett & Mabberley (1997), Lack (1999d), Riedl-Dorn, Köberl & Stuart (2014)
26 Thomas (2002), Mabberley (2002)
27 Seipel (2002), Riedl-Dorn (2002)
28 Lack (1998b)
29 Inv. No. Cl. XXIX s.n. 30
30 e.g. Hockney (2009)
31 Becker (1887)
32 Salisbury (1806)
33 J. Compton (pers. com., 16 xii 2014)
34 Jeřábek (2003)

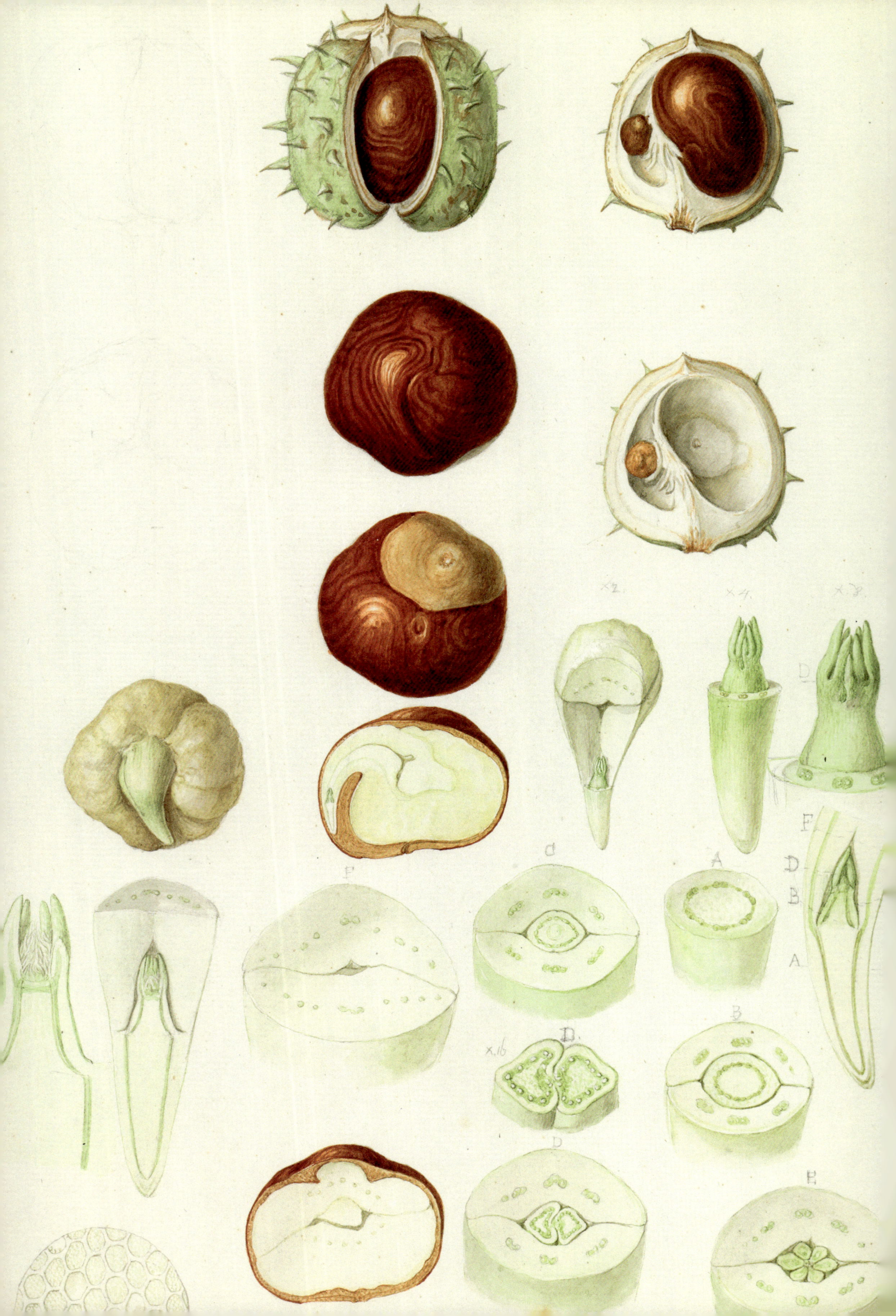

F
D
B
A
C
F
A
B
x.16
D.
D
E

EPILOGUE

Our story effectively began with the birth of Dorothea Bauer in 1750, when Francis I Stephen was Emperor of the Holy Roman Empire and George II King of Great Britain and Ireland. When her brother Franz died in 1840, the Holy Roman Empire had vanished, Ferdinand I had become Emperor of Austria, and Victoria, Queen of Great Britain and Ireland, had just married. The drastic political changes, which had happened during an interval of ninety years, are echoed in similar changes in the fields of arts and sciences. Johann Sebastian Bach had died a few months before Dorothea was born, and Verdi started to work on his opera *Nabucco* in the very year in which Franz died, just to mention an example from music. Less than three years after the birth of Dorothea, Linnaeus had published his *Species Plantarum* applying for the first time binomial nomenclature to plant, fungal and algal diversity and following his own artificial order in plant diversity. Endlicher had almost finished the publication of his *Genera plantarum*, aiming at a natural order in plant diversity, in the year when Franz died. Some of these deep changes are also visible in the work of the Bauer brothers. The first title pages painted for Boccius in Feldsberg in the years 1776–1778, with their pendant flower garlands and *rocaille* ornaments, belong to the late rococo period, Ferdinand's passion flowers, his swan-song, and Franz's late human pathology with its attention to anatomical detail are simply timeless, extra dry documentations of facts, not attributable or associable to any point in this period .

It is notable that over this interval from 1771 to 1840 the Bauer brothers remained very much on course – Joseph as a member of the court of the Reigning Prince of Liechtenstein, Franz and Ferdinand effectively as kind of 'court painters' working largely for their employers or commissioners, albeit in the case of Franz with an outstanding amount of freedom. In short, they were neither obliged nor prepared to make compromises despite all the changes around them but continued in doing it their way and reaching an outstanding degree of sophistication. Very rightly the Bauer brothers have been called by Peter Green in a book review published in the *Journal of the Adelaide Botanical Garden* in 1977 'the greatest botanical artists of all time; not only because of the exceptionally high quality of the lifelike and aesthetically pleasing plant portraits they painted, but because they included in their compositions careful, painstaking, botanically significant and accurate dissections and enlargements of the flower parts. It may be argued that it is the very addition of these drawings of selected and magnified parts which marks off botanical art from plant portraiture, yet, with the Bauers, the whole forms a composition of such quality and beauty as to represent one of the highest forms of art'. On a similar line is Mabberley's comment in his short biography of Ferdinand 'There is something magical in the way the animals, but particularly the plants, have a three-dimensional quality that the observer can almost embrace'.

However, in the social hierarchy of their time it was Joseph Bauer who ranked highest. This becomes evident in a document kept in the Österreichisches Staatsarchiv, first shown in the exhibition 'Europe in Vienna', which took place in the Belvedere Palace in Vienna in early 2015. When the organisers of the Congress of Vienna prepared in September 1814 a list of sites recommended for a visit to the souvereigns, ministers and diplomats who had come together in the Kaiserliche Haupt- und Residenzstadt, they indicated among others the following institutions: the Imperial Gallery, the Imperial Library, the Imperial Royal United Natural History Cabinets, the Botanical Garden of Vienna University and the Gallery of the Reigning Prince of Liechtenstein. These had Füger, Ossolinski, Schreibers, Jacquin the son and Joseph as their respective directors, and there is no doubt that the latter had to receive the prominent guests. As a Fellow of the Royal Society and as the first resident artist at the Royal Garden at Kew, Franz probably

Horse chestnut, *Aesculus hippocastanum*. Watercolour by Franz Bauer, signed, 1819. Paper, watercolour. – London, Natural History Museum, Library and Archives, Franz Bauer Miscellany (loose).

ranked second in social status, followed by Ferdinand, who effectively was just a freelance natural history illustrator, who had worked for different commissioners. Today the achievements of the Bauer brothers are ranked differently. Joseph is best regarded as one of the many directors of an art collection, albeit one of the finest and most diverse then in existence. Franz had produced an extensive pictorial documentation of plant and animal diversity on a global scale with an emphasis on orchids and had become a pioneer in his microscopical observations, e.g. of the cell nucleus. However, Ferdinand is today regarded as the most prominent of the three.

This becomes evident for example in Joachim Rees's *Die verzeichnete Fremde*, a weighty *Habilitationsschrift* which appeared in Paderborn in March 2015. Using the approach of an art historian, Rees analyses the aims and achievements of the illustrators engaged to produce a pictorial documentation of humans, animals, plants, landscapes, artefacts etc. observed during the major expeditions undertaken between c.1730 and c.1830, starting with the Second Russian Expedition to Kamchatka headed by Vitus Jonassen Bering and ending with the circumnavigation of the globe in the *L'Astrolabe* headed by Jules-Sébastien César Dumont d'Urville. In the long list of illustrators, from Johann Christian Berckhan and Johann Wilhelm Lürsenius on the expedition first mentioned to those engaged on the circumnavigations flying British, French and Russian colours, e.g. Sydney Parkinson, Georg Adam Forster, Guillaume Prévost and Ludwig York Choris, Ferdinand ranks prominent. According to Rees, Ferdinand occupies a special position among his colleagues – because of his colour code and his dissociation of *disegno* and *colore*. Comparable to the very accurate topographical measurements expressed as numbers by Flinders, Ferdinand followed his earlier ingenious idea to break down the hues of the rainbow into numbers, thereby refining his earlier approach first used in Feldsberg and subsequently during his travels with Sibthorp in the Levant.

Expressing information as numbers was and is considered both trendsetting and eminently modern. Very rightly the famous physicist and mathematician William Thomson, later the 1st Lord Kelvin and one of Banks' successors as President of the Royal Society, is cited by Rees in having stated 'I often say that when you can measure what you are speaking about, and express it in numbers, you know something about it; but when you cannot measure it, when you cannot express it in numbers, your knowledge is of a meagre and unsatisfactory kind: it may be the beginning of knowledge, but you have scarcely advanced to the state of science, whatever the matter may be'. When Ferdinand added no less than 330 different numbers, probably his record, to his large-size pencil drawing of the Blue Swimmer Crab, he probably neither considered these generalities nor the novelty of his practice. However, he must have been aware of the fact that no one else could interpret his numbers, apart from probably his brother Franz. In fact his approach provided a form of self-protection – only he, Ferdinand, could be commissioned to produce coloured repeats based on the pencil drawings prepared in the field. It seems that Ferdinand was strictly focused on numbers, maybe even obsessed by them, having written many more numbers in his life than words; furthermore his capacity of attributing numbers to the different hues was so excellent that he only very occasionally could be shown to have corrected a number. Although even more sublime colour charts than the system used on the voyage in the *Investigator* were published before 1800, e.g. Christian Friedrich Prange's *Farbenlexikon*, Halle, 1782 and the exceedingly rare, anonymous *Wiener Farben-Kabinett*, Vienna, 1794, both works illustrating more than 5500 shades of colour, Ferdinand was the only travelling illustrator of his time who consistently followed this approach during his whole life.

It is grace to die in time. The Bauer brothers were born at a time when an image could only be produced by an artist, and when dissemination of an image was only possible either by copying by hand or by making a copper engraving to be printed in an effectively limited print-run. During their lifetime they experienced one major revolution in their field - copper engraving was superseded by the cheaper lithograph which made significantly larger printruns possible. Only Franz lived long enough to experience a second revolution – the invention of heliography, the first in a series of techniques inducing the light to produce a permanent image, soon followed by the daguerreotype. Is it not significant that according to the auction catalogue of Franz's belongings his ultimate acquisition to his collection was a daguerreotype showing the Pont Neuf in Paris, dated 1840, the year of his death? As a matter of fact a new era had just started, in which everyone, with the help of a camera, could produce an image. This soon would begin to make the conventional illustrator, in particular the portraitist of human figures, redundant. It is a curious historical fact that the one and the same person who reached, together with his younger brother Ferdinand, the utmost refinement in the pictorial representation of natural history objects was the very person, who became the most important witness for heliography and saw in his very last year the rise of the daguerreotype.

The Bauer brothers had seen a lot: Joseph the staggering collection of art of all sorts assembled by the aristocracy in Rome, the treasures acquired by the Liechtenstein family and also the even larger and more spectacular collection brought together by the Habsburg family in Vienna. Franz had documented botanical, mycological and, admittedly to a smaller extent, zoological diversity – *ex situ*, as observed by him in parks (see p. 468), gardens, conservatories, fields, herbaria, the anatomical cabinet and the dissection room. In addition he had penetrated with his microscope deeply into the unknown, recorded the cell nucleus, pollen diversity and pollen germination. By contrast Ferdinand had largely refrained from using the microscope and had instead travelled the world to observe botanical, zoological and, to a limited extent, mycological diversity *in situ,* producing after his arrival home a painted record of a refinement similar to that of his elder brother. In addition Ferdinand was a genuine collector of a considerable number of botanical and zoological specimens which he took with him back to Vienna. Both Franz and Ferdinand were witnesses to the explosion of knowledge concerning the diversity in plants, fungi and animals and had contributed to it. For the first time they had recorded a considerable number of genera and species new to science.

In a sense Franz and Ferdinand were single-minded specialists with no time or talent for self-promotion. Enveloped in nuance and, as mentioned by several contemporaries, indefatigable at their work, which can serve as a motto for their lives, they produced a huge oeuvre of breathtaking quality and staggering quantity. It seems that documenting the, at the time, not even remotely comprehensible diversity of natural history objects was for them a very intensive process. Possibly this was in part something like a continuous worship of god and the wonders of his creation. Maybe this attitude is best seen in context with their unselfishness, their preparedness to serve others, their dedication to projects commissioned by their superiors – probably as a result of early imprinting on their youthful minds. It would be ridiculous to claim that Franz and Ferdinand had no ego, but they definitely had egos different to those of people like Sibthorp, Brown or Home. Being prepared to submit their talents to the egos of others made Franz and Ferdinand peripheral figures in academia and the competitive world of scientists where ruthless behaviour is not infrequent and often helps in making a career. It is also precisely their somewhat submissive behaviour which makes the Bauer brothers unattractive to most biographers. Very appropriately in this context, the anonymous obituary published in the *Athenaeum* noted about Franz, admittedly with a slightly hagiographical touch, 'Few men, perhaps, devoted their existence so entirely to an uninterrupted observation of natural history objects, and enriched science with so many discoveries, and still fewer have been so indifferent to all ordinary considerations; for Bauer [and in fact we may say Franz and Ferdinand] regarded his [her] talents merely as a stepping-stone to that tranquillity and happiness which ought to be the ultimate object of all science and all acquirements'.

Apart from scientific accuracy the plant and animal drawings and watercolours of Franz and Ferdinand transmit also an additional message – there is beauty in the living world or, as is so nicely worded by John Keats, one of their contemporaries, when he wrote in the first stance of his *Endymion*, 'A thing of beauty is a joy for ever'. It is this sublime harmony of art and science which had made Sibthorp write from Naples to his father in Oxford about Ferdinand's work 'I never saw Beauty & Accuracy so fully combined' and similar words apply to the works of Franz.

This union of art and science is encapsulated in the vast oeuvre of the Bauer brothers, the subject of this book. At the same time their work became for a highly specialized profession, the natural history illustrators, the great model to follow. The penultimate sentence of Marlene Norst's book (see Preface) reads 'Australia can consider itself fortunate that it had in Ferdinand Bauer an artist dedicated to the portrayal of its native flora and fauna'. This may be paraphrased into the world can consider itself fortunate that it had the Bauer brothers as artists dedicated to the portrayal of its flora and fauna. Sensitive to beauty, they possessed the special amalgam of knowledge, aesthetic sense and technical skill which made them unique recorders of the diversity of living nature setting the golden standard – forever.

REFERENCES

UNPUBLISHED SOURCES

This list comprises written sources only. For pencil drawings, watercolours, paintings by the Bauer brothers see Appendix 1.

BRNO
Moravský zemský archiv, Matriky Valtice, Sign. 16/3530, 65/3531, 65/3533

LEIDEN
Universiteitsbibliotheek, Codex BPL 303

LINCOLN
Lincolnshire Archives, Sibthorp Papers, 3 SIB/1/1. – John Sibthorp to Humphrey Sibthorp, 10 xi 1785

LONDON
Linnean Society, Archive. – Franz Bauer to Jan Lhotsky, 17 iii 1839

Natural History Museum, Library & Archives – Brown Correspondence 1: f. 39 (Sir James Everard Home to Robert Brown, 3 xii 1840), f. 42 (W. Meyer to Robert Brown, 11 xii 1840), f. 214 (Sir James Everard Home to Robert Brown, 17 xii 1840); MS B 97; Draft for Franz Bauer's epitaph in W. Meyer's hand

University College London, Library, MS Add 254, Lord Odo Russell Collection. – Franz Bauer to Joseph Franz Edler von Jacquin, 21 v 1792, 19 viii 1814; Nikolaus Joseph Baron von Jacquin to Sir Joseph Banks, 28 ii 1807

MADRID
Real Jardin Botánico, Archivio, Joseph Franz Baron von Jacquin to Mariano La Gasca y Segura, 7 iii 1816

PITTSBURGH
Hunt Institute for Botanical Documentation, Archive. – Franz Bauer to Joseph Franz Edler von Jacquin, 12 ix 1791

POTSDAM
Landeshauptarchiv, Rep 37, (Alt) Friedland, No. 382. – H. C. Itzenplitz to H. C. von Friedland, 20 xii 1792, 4 iv 1793, 10 v 1793, 21 viii 1793

SACRAMENTO
California State Library, Sutro Branch. – Franz Bauer to Sir Joseph Banks, 28 ii 1807

UPPSALA
Universitetsbiblioteket, Handskrifts- och musikenheten, Dörflers Brefsammlingar, Franz Bauer to Joseph Franz Baron Jacquin, 9 vi 1838

WIEN
Fürstliche Sammlungen, Archiv, Majorats-Kassa Rechnung 1788, No. 524

Stadt- und Landesarchiv -Bestand Patrimoniale Verwaltung und Justiz A 13 Klosterneuburg über Hietzing, Verlassenschaftsakt Ferdinand Bauer, Erbsteuer Ausweis; Franziszäischer Kataster 1817 – ca 1831, B1 – Parzellenprotokoll, Hietzing, P1 – Pläne, Hietzing; Grundbücher, Grundbuch Klosterneuburg 20 / 99 b: f. 421, 21, Art. 1392, 1393; Mag. Zivilgericht, Abh., Fasz. 2, 7425/41, Verlassenschaft Joseph Bauer

BIBLIOGRAPHY

Ackers, G. (2007). The fine ferns of *Flora Graeca*. *Pteridologist*, 4: 203–208.

Allen, M. (1967). *The Hookers of Kew 1785–1911*. London.

Allford, J. M. (1999). List of published works of John Lindley. Pp. 197–218 In: Stearn, W. T. (ed.), *John Lindley 1799–1865. Gardener – botanist and pioneer orchidologist*. Woodbridge, Suffolk.

Andersson, L. (1998). A revision of the genus *Cinchona* (Rubiaceae – Cinchoneae). *Mem. New York Bot. Gard.*, 80.

Andrews, H. C. (1801). *Bauera rubioides*. *Bot. Repos.*, 3: t. 198.

Anon. (1794). *Wiener Farben-Kabinett*. – Wien.

Anon. (1799). *Rede bey Gelegenheit da der wohlehrwürdige Vater Norbert Boccius, des Ordens des Heiligen Johann von Gott Exprovincial und wirklicher Prior seine fünfzigjährige Ordensprofeßion den 28sten Heumonat 1799 zu Feldsperg feyerlich erneuerte*. Pressburg.

Anon. (1806) We are happy to inform... *Ann. Bot. (König & Sims)*, 2: 593–594.

Anon. (1826). Am 17. März verstarb ... *Flora*, 9: 240.

Anon. (1830). Achte Versammlung der Naturforscher und Ärzte zu Heidelberg im September 1829. *Isis*, 1830: 449–736.

Anon. (1831). Gestorben. *Wiener Zeitung*, 1831: 1508.

Anon. (1835). Verzeichnis der zusammengesetzten Mikroskope, welche von S. Plössl, Opticus in Wien seit 1826, auf Bestellung verfertigt worden sind. *Z. Phys. Verwandt. Wiss.*, 3: 91–93.

Anon. (1836). *Österreichische National-Encyclopädie* 4. Wien.

Anon. (1840 a). *Verzeichnis der von Herrn Jos. Freih. v. Jacquin hinterlassenen Bücher, botanischen, chemischen und physikalischen Inhalts*. Wien.

Anon. (1840 b). Francis Bauer, Esq. *Athenaeum*, 687: 1025–1026.

Anon. (1841a). Francis Bauer. *Proc. Linn. Soc. London*, 1: 101–104.

Anon. (1841b). Francis Bauer. *Abstr. Pap. Philos. Trans. Roy. Soc. London*, 4: 342–344.

Anon. (1841c). Francis Bauer. *Gard. Mag. & Reg. Rural Domest. Improv.*, 7: 186–188.

Anon. (1841d). Francis Bauer, Esq. *Ann. Mag. Nat. Hist.*, 7: 77–78.

Anon. (1841e). Francis Bauer. *Engl. J.*, 1: 222–223.

Anon. (1841f). *A catalogue of the botanical drawings, books, microscopes, microscopical preparations, & c. of the late Francis Bauer, ... including several works by his brother, Ferdinand Bauer ... which will be sold by auction by Messr. Christie & Manson, ... 8 King Street, St. James Square ... November 24th, 1841*. [London].

Anon. (1842). Francis Bauer, Esq. F. R. S. *Gentleman's Mag.*, 172: 107–109.

Anon. (1873). *Catalog der von Dr. Andreas Andresen hinterlassenen Kunstsammlung*. Leipzig.

Anon. (1893). *Verzeichnis der Handschriften im preussischen Staate* 1 (2). Berlin.

Anon. (1904). *The history of the collections contained in the Natural History Departments of the British Museum* 1. London.

Anon. (1978). *English drawings and watercolours ... which will be sold at auction by Christie, Manson & Woods Ltd. ... at their Great Rooms 8 King Street, St. James's, London ... on July 19, 1978*. London.

Anon. (1985). *Important 19th century pictures and continental drawings of the 19th and 20th centuries ... which will be sold at Christie's Great Rooms on ... 20 March 1985*. London.

Anon. (1999). 169. Ferdinand Bauer (Austrian 1760–1826). Pp. 98–100 in Anon., *19th century British and European paintings and watercolours. 11 am Tuesday 22nd June 1999* [auction catalogue Phillips, New Bond Street, London]. London.

Baker, H. G. (1951). A palynological treasure house. *Nature*, 167: 457–460.

Bancher, E., Hölzl, J. & Kotlan, F. (1968). Georg Simon Plössl. *Technikgeschichte*, 35: 127–147.

Banks, J. (1817). Notes relative the first appearance of the *Aphis lanata*, or the apple tree insect, in this country. *Trans. Hort. Soc. London*, 2: 161–169.

Barrow, J. H. (1840). *The Mirror of Parliament for the third session of the fourteenth Parliament of Great Britain and Ireland Session of 1840*, 3. London.

Bastin, J. (1973). Dr Joseph Arnold and the discovery of *Rafflesia arnoldi*. *J. Soc. Bibliogr. Nat. Hist.*, 6: 305–372.

Bauer, Franz (1819). Microscopic observations on the red snow ... in a letter to W. T. Brade. *Quart. J. Sci. Lit. Arts*, 7: 222–229.

Bauer, Franz (1820). Some experiments on the fungi which constitute the colouring matter of the red snow discovered in Baffin's Bay. *Philos. Trans. Roy. Soc. London*, 110: 165–172.

Bauer, Franz (1823). The Croonian Lecture. Microscopical observations on the suspension of the muscular motions of the *Vibrio tritici*. *Philos. Trans. Roy. Soc. London*, 113: 1–16.

Bauer, Franz (1833a). The smut balls or pepper brand. *Penny Mag.*, 2: 126-128, 180-182.

Bauer, Franz (1833b). The grain worms. *Penny Mag.*, 2: 300–302, 334–335.

Bauer, Franz (1837). On making drawings of microscopic objects. Pp. 221–226 In: Goring, C. R. & Pritchard, A., *Micrographia: containing practical essays on reflecting, solar, oxy-hydrogen gas microscopes; micrometers; eyepieces, & c. & c.* London.

Bauer, Franz (1839). [Letter dated 26 February 1839]. *Lit. Gaz. J. Belles Lettres Arts Sci.*, 1154: 137–138.

Bauer, Franz (1840). On the Ergot of Rye. *Trans. Linn. Soc. London*, 18: 475–482.

Bauer, Franz (1849). Observations on Ergot. *Proc. Linn. Soc. London*, 1: 52.

Baumstark, R. (1990). Katalog. – Pp. 83–166 In: Baumstark, R. (ed.), *Joseph Wenzel von Liechtenstein. Fürst und Diplomat im Europa des 18. Jahrhunderts*. Einsiedeln.

Bayly, M. J. & Kellow, A. V. (2004). (1650) Proposal to reject the name *Veronica decussata* (Plantaginaceae). *Taxon*, 53: 571–572.

Beale, N. & E. (2011). *Echoes of Ingen Housz. The long lost story of the genius who rescued the Habsburgs from smallpox*

and became the father of photosynthesis. Salisbury.
Beasley, A. W. (2000). *Home away from Home*. Wellington.
Becker, M. A. (1887). Feldsberg in Niederösterreich. *Blätt. Ver. Landeskunde Niederösterreich, N. F.*, 20: 336–411.
Berger, E. (2004). *Historische Gärten Österreichs. Garten- und Parkanlagen von der Renaissance bis um 1930*. 3. Wien.
Birke, V. (1981). *Guido Reni: Zeichnungen. Ausstellung Graphische Sammlung Albertina* 275.
Birke, V. (2005). Johann Jebmeyer (1770–1858) "Hofpflanzenmaler" in Wien. *Römische Mitt.*, 47: 395–409.
Bleichmar, D. (2011). The geography of observation: distance and visibility in eighteenth-century botanical travel. Pp. 373–395 In: Daston, L. & Lunbeck, E. (ed.), *Histories of Scientific Observation*. Chicago.
Blunt, W. (1950). *The art of botanical illustration*. London (= *The New Naturalist* [14]).
Blunt, W. (1976). Introduction. Pp. [9]–[22] In: Stearn, W. T., *The Australian Flower Paintings of Ferdinand Bauer*. London.
Brain, P. & R. (1986). *Disappearing Devon*. Instow.
Brichtová, D. (2002). *Schloss Mikulov*. Mikulov.
Britten, J. (1899). Francis Bauer's 'Delineations of Exotick Plants'. *J. Bot.*, 37: 181–183.
Britten, J. (1901). Francis Bauer's 'Delineations of Exotick Plants'. *J. Bot.*, 39: 107–108.
Brown, R. (1813). On *Woodsia*, a new Genus of Ferns. – *Trans. Linn. Soc. London*, 11: 170–174.
Brown, R. (1819). List of plants collected by the officers & c., in Capt. Ross's Voyage, on the Coasts of Baffin's Bay. Pp. cxli–cxliv In: Ross, J., *A Voyage of Discovery for the Purpose of Exploring Baffin's Bay*. London.
Brown, R. (1821). An account of a new genus of plants, named *Rafflesia*. *Trans. Linn. Soc. London*, 13: 201–324.
Brown, R. (1831). *Observations on the organs and mode of fecundation in Asclepiadeae and Orchideae*. London.
Brown, R. (1833). On the organs and mode of fecundation in Asclepiadeae and Orchideae. *Trans. Linn. Soc. London*, 16: 685–746.
Brown, R. (1844). On the female flower and fruit of *Rafflesia arnoldii* and the *Hydnora africana*. – *Trans. Linn. Soc. London*, 19: 221–247.
Bruce, M. R. (1970). John Sibthorp. *Taxon*, 19: 353–361.
Bruwier, M. – C. (1994). *Présence de l'Égypte dans les collections de la Bibliothèque Universitaire Moretus Plantin*. Namur.
Camper, P. (1782). *Abhandlung über den besten Schuh*. Wien.
Caneva, G. (1992). *Il mondo di cerere nella Loggia di Psiche. Villa Farnesina. Sede dell' Accademia Nazionale dei Lincei*. Roma.
Carter, H. B. (1988). *Sir Joseph Banks*. London.
Cavanagh, T. (1990). Australian plants cultivated in England, 1771–1800. Pp. 273–283 In: Short, P. S. (ed.), *History of Systematic Botany in Australasia*. South Yarra, Vic.
Chambers, N. (ed.) (2000). *The Letters of Sir Joseph Banks. A selection, 1768–1820*. London.
Clusius, C. (1983). *Fungorum in Pannoniis observatorum brevis historia et Codex Clusii*. Budapest.
Collins, J. (2004). *Papacy and Politics in Eighteenth-century Rome. Pius VI and the arts*. Cambridge.
Correa de Serra, J. (1802). On the *Doryanthes*, a new Genus of Plants from New Holland, next akin to the *Agave*. *Trans. Linn. Soc. London*, 6: 211–213.
Coxe, W. (1811). *Literary Life and Selected Works of Benjamin Stillingfleet*. London.
Cribb, P. (1999). Lindley's life-long love affair with orchids. Pp. 106–141 In: Stearn, W. T. (ed.), *John Lindley 1799–1865. Gardener – botanist and pioneer orchidologist*. Woodbridge, Suffolk.
Criste, C. (1905). *Feldmarschall Johannes Fürst von Liechtenstein*. Wien.
Cunningham, A. (1837). Florae insularum Novae Zelandiae precursor: or a specimen of the botany of the islands of New Zealand. *Compan. Bot. Mag.*, 2: 358–378.
Cuveland, M. de (2006). *Natur im Aquarell. Meisterwerke der Wiener Hofmaler Johann und Joseph Knapp* München.
Daston, L. & Galison, P. (2007). *Objectivity*. New York.
De Melo, N. F., Cervi, A. C. & Guerra, M. (2001). Karyology and cytotaxonomy of the genus *Passiflora* L. (Passifloraceae). *Plant Syst. Evol.*, 226: 69–84.
Denney, M. (2010). The name behind *Cyclamen repandum*. *Cyclamen*, 34: 54–58.
Desmond, R. (2007). *The History of the Royal Botanic Gardens Kew*, edn. 2. Kew.
Dierkens, A. (ed.) (1993). *Henri-Joseph Redouté et l'expédition de Bonaparte en Égypte*. Bruxelles.
Diment, J. A., Humphries, C. J., Newington, L. & Shaughnessy, E. (1984). Catalogue of the natural history drawings commissioned by Joseph Banks on the *Endeavour* voyage 1768–1771, part 1. *Bull. Brit. Mus. (Nat. Hist.) Hist. Ser.*, 11.
Diment, J. A., Humphries, C. J., Press, J. R., Shaughnessy, E. & Newington, L. (1987), Catalogue of the natural history drawings commissioned by Joseph Banks on the *Endeavour* voyage 1768–1771, part 2. *Bull. Brit. Mus. (Nat. Hist.) Hist. Ser.*, 12.
Dolezal, H. (1973). Zur Editionsgeschichte der Icones Plantarum Rariorum von Nikolaus Joseph von Jacquin. Pp. 136–172 In: *Festschrift für Claus Nissen*. Wiesbaden.
Duncan, G. (2012). *The genus Lachenalia*. Kew.
Ecott, T. (2004). *Vanilla. Travels in search of the luscious substance*. London.
Edmonson, J. (2002a). The 13th Earl of Derby's botanical and horticultural library. Pp. 136–143 In: Fisher, C. (ed.), *A Passion for Natural History. The life and legacy of the 13th Earl of Derby*. Liverpool.
Edmondson, J. (2002b). 75. Pitch pine & Cembran pine. P. 154 In: Fisher, C. (ed.), *A passion for natural history. The life and legacy of the 13th Earl of Derby*. Liverpool.
Edwards, P. I. (1976). Robert Brown (1773–1858) and the natural history of Matthew Flinders' voyage in H. M. S. Investigator, 1801–1805. *J. Soc. Hist. Nat. Hist.*, 7: 385–407.
Edwards, P. I. (ed.) (1981). The Journal of Peter Good. Gardener on Matthew Flinders Voyage to Terra Australis 1801–03. *Bull. Brit. Mus. (Nat. Hist.) Hist.* Ser., 9.
Elliott, B. (2004). *The Royal Horticultural Society. A History 1804–2004*. Chichester.
Endersby, J. (2008). *Imperial nature: Joseph Hooker and the practices of Victorian science*. Chicago.
Endlicher, S. (1833), *Prodromus florae insulae Norfolkicae*. Vindobona.
Endlicher, S. (1836). Bemerkungen über die Flora der Südseeinseln. *Ann. Wiener Mus. Naturgesch.*, 1: 129–190.
Estensen, N. (2002). *The Life of Matthew Flinders*. Sydney.
Everett, D. (2013). *The genus Tulipa. Tulips of the world*. Kew.
Farjon, A. & Filer, D. (2013). *An Atlas of the World's Conifers. An analysis of their distribution, biogeography, diversity and conservation status*. Leiden.
Fettweis, G. B. L. (2001). Über Beiträge aus den Ländern der Habsburger Monarchie zur Entwicklung der Montanwissenschaften und damit auch der Geowissenschaften im 16. und 18. Jahrhundert. *Mitt. Österr. Ges. Wissenschaftsgesch.*, 21: 1–16.
Findlay, E. (1998). *Arcadian Quest: William Westall's Australian sketches*. Canberra.
Firla, M. (2004). *Angelo Soliman – ein Wiener Afrikaner im 18. Jahrhundert*. Baden (= *Katalogblätter des Rollettmuseums Baden* 48).
Fitzinger, L. (1840). Nekrolog. Joseph Franz Freyherr von Jacquin. *Mitth. K. K. Landwirthschaftsges.*, 9: 133–143.
Flowers, E. (1967). Menzies, Sir Charles (1783–1866). *Australian Dictionary of Biography* 2. Carlton, Vict.
Foley, M. J. Y. (2003). *Salvia veneris* Hedge (Lamiaceae / Labiatae) in Cyprus and its relationship to its taxonomically closest allies. *Karca Arboretum Mag.*, 7: 1–6.
Ford, B. J. (2001). The Royal Society and the microscope. *Notes & Rec. Roy. Soc. London* 55: 29–49.
Frodl, G. (2002). Malerei. Pp. 257–264 In: Frodl, G. (ed.), *19. Jahrhundert*. München (=Fillitz, H. (ed.), *Geschichte der bildenden Kunst in Österreich* 5).
Frodl, G. & Frodl - Schneemann, M. (2010). *Die Blumenmalerei in Wien*. Wien.
Fusina, S. (2002). *Bauer, Bauer & Co*. Lodi.
Gage, A. T. & Stearn, W. T. (1988). *A Bicentenary History of the Linnean Society of London*. London.
Galland, R. B. (2007). Popliteal aneurysms: from John Hunter to the 21st century. *Ann. Roy. Coll. Surgeons Engl.*, 89: 466–471.
Gams, J. (1972a). Einführung. Pp. 1–27 In: Anon., *Österreichische Künstler und Rom vom Barock zur Sezession*. Wien.
Gams, J. (1972b). Katalog. Pp. 99–244 In: Anon., *Österreichische Künstler und Rom vom Barock zur Sezession*. Wien.
George. A. (2006). An Australian convict-made travelling desk from 1805. *Australiana*, 28: 16–19.
Giardina, G., Raimondo, F. M. & Spadaro, V. (2007). A catalogue of plants growing in Sicily. *Bocconea*, 20: 5–582.
Grabner, S. (2002). Zeichnung und Aquarell im beginnenden 19. Jahrhundert. Pp. 377–395 In: Frodl, G. (ed.), *19. Jahrhundert*. München (= Fillitz, H. (ed.), *Geschichte der bildenden Kunst in Österreich* 5).
Grant, A. (1999). *Instow. A history*. Instow.
Grant, J. R. (1994). Typification of six Robert Brown Brassicaceae names from arctic North America. *Novon*, 4: 250–253.
Granville, A. B. (1820). A case of the human foetus found in the ovarium, of the size it usually acquires at the end of the fourth month. *Philos. Trans. Roy. Soc. London*, 110: 101–107.
Graves, A. (1905). *The Royal*

Academy of Arts. A Complete Dictionary of Contributors and their Work from its Foundation in 1769 to 1904, 1. London.

Green, P. S. (1977). A selection of Australian flower paintings by Ferdinand Bauer. *J. Adelaide Bot. Gard.*, 1: 145–149.

Green, P. S. (1994). *Flora of Australia* 49. *Oceanic Islands* 1. Canberra.

Griffith, W. (1847). *Posthumous papers bequeathed to the honourable, the East India Company; and printed by order of the government of Bengal. Journal of travels in Assam, Burma, Bootan, Affghanistan and the neighbouring countries.* Calcutta.

Gruber, J. W. (1991). Does the Platypus lay eggs? The history of an event in science. *Arch. Nat. Hist.*, 18: 51–123.

Grzechca-Mohr, U. (2011). Eliza, Arbeiten auf Papier: das graphische Werk der Landgräfin Elizabeth (1770– 1840). *Jahrb. Hochtaunuskreis*, 19: 177–185.

Hadjikyriakou, G. N. (2007). *Aromatic and Spicy Plants in Cyprus.* Nikosia.

Hall, M. B. (1984). *All Scientists Now. The Royal Society in the nineteenth century.* Cambridge.

Harris, H. (1999). *The Birth of the Cell.* New Haven.

Harris, S. (2007). *The Magnificent Flora Graeca. How the Mediterranean came in the English garden.* Oxford.

Harris, S. (2011). John Sibthorp: teacher of botany. *Oxford Plant Systematics*, 17: 16–17.

Haupt, H. (1990). Diplomatie und Repräsentation im Dienste des Kaiserhauses. Die Öffentlichen Einzüge des Fürsten Joseph Wenzel von Liechtenstein. Pp. 24–53 In: Baumstark, R. (ed.), *Joseph Wenzel von Liechtenstein. Fürst und Diplomat im Europa des 18. Jahrhunderts.* Einsiedeln.

Haverfield, T. T. (1862). Notes on Kew and Kew gardens. *Leisure Hour,* 1862 (570): 766–767.

Heine, H. (1963). On the correct spelling of the generic name *Schwenckia* D. van Royen ex L. (Solanaceae), with a note about Martin Wilhelm Schwencke. *Kew Bull.*, 16: 465–469.

Helm, J. (1965). Die Pillnitzer Centurienbände „Plantae selectae vivis coloribus depictae". *Act. Hist. Leopoldina*, 3.

Heltman, H. (2002). Johann Hedwig (1730–1799), ein bedeutender Arzt und Botaniker aus Siebenbürgen. *Orvostörténeti Közlem.*, 2002: 121–130.

Henry, B. (1975). *British Botanical and Horticultural Literature before 1800.* Oxford.

Hoare, M. E. (1975). *The Tactless Philosopher. Johann Reinhold Forster (1729–98).* Melbourne.

Hockney, D. (2009). *Secret Knowledge. Rediscovering the Lost Techniques of the Old Masters.* London.

Hoeppe, G. (1999). *Blau. Die Farbe des Himmels.* Heidelberg.

Home, E. (1786). An account of Mr. Hunter's method of performing the operation of the popliteal aneurism. *London Med. J.*, 7: 391–406.

Home, E. (1787). Supplement to the account of Mr. Hunter's method of performing the operation of the popliteal aneurysm. *London Med. J.*, 8: 126–136.

Home, E. (1800). Some observations on the head of the *Ornithorhynchus paradoxus. Philos. Trans. Roy. Soc. London,* 90: 432–436.

Home, E. (1802a). A description of the anatomy of the *Ornithorhynchus paradoxus. Philos. Trans. Roy. Soc. London,* 92: 67–84.

Home, E. (1802b). Description of the anatomy of the *Ornithorhynchus paradoxus. Philos. Trans. Roy.* Soc. *London,* 92: 348–364.

Home, E. (1808). An account of some peculiarities in the anatomical structure of the Wombat, with observations on the female organs of reproduction. *Philos. Trans. Roy. Soc. London,* 98: 304–312.

Home, E. (1814). *Lectures on comparative anatomy in which are explained the preparations in the Hunterian Collection illustrated by engravings* 1. London.

Home, E. (1816). Farther observation on the feet of animals whose progressive motion can be carried out against gravity. *Philos. Trans. Roy. Soc. London,* 106: 322–330.

Home, E. (1817a). On the passage of the ovum from the ovarium to the uterus in women. *Philos. Trans. Roy. Soc. London,* 107: 252–261.

Home, E. (1817b). The distinguishing characters between the ova of the Sepia, and those of the vermes testacea, that live in water, explained. *Philos. Trans. Roy. Soc. London,* 107: 207–301.

Home, E. (1817c). Some account of the nests of the Java swallow, and of the glands that secrete the mucus of which they are composed. *Phil. Trans. Roy. Soc. London,* 107: 332–338.

Home, E. (1818a). The Croonian Lecture. On the changes the blood undergoes in the act of coagulation. *Philos. Trans. Roy. Soc. London,* 108: 172–184.

Home, E. (1818b). Some additions to the Croonian Lecture, on the changes the blood undergoes in the act of coagulation. *Philos. Trans. Roy. Soc. London,* 108: 185–198.

Home, E. (1819a). The Croonian lecture. On the conversion of pus into granulations or new flesh. *Philos. Trans. Roy. Soc. London,* 109: 1–10.

Home, E. (1819b). On the ova of the different tribes of *Opossum* and *Ornithorhynchus. Philos. Trans. Roy. Soc. London,* 109: 234-240.

Home, E. (1820a). A farther investigation of the component parts of the blood. *Philos. Trans. Roy. Soc. London,* 110: 1–10.

Home, E. (1820b). Observations on the human uretha, showing its internal structure, as it appeared in the microscope of F. Bauer, Esq. *Philos. Trans. Roy. Soc. London,* 110: 183–195.

Home, E. (1821). On the black rete mucosum of the negro, being a defence against the scorching effect of the sun's rays. *Philos. Trans. Roy. Soc. London,* 111: 1–6.

Home, E. (1822a). On the anatomical structure of the eye; illustrated by microscopical drawings by F. Bauer. *Philos. Trans. Roy. Soc. London,* 112: 76–85.

Home, E. (1822b). Observations on the changes the egg undergoes during incubation in the common fowl, illustrated by microscopical drawings. *Philos. Trans. Roy. Soc. London,* 112: 339-356.

Home, E. (1823a). On the double organs of generation of the lamprey, the conger eel, the common eel, the barnacle and earth worms, which impregnate themselves; though the last from copulating, appear mutually to impregnate one another. *Philos. Trans. Roy. Soc. London,* 113: 140–151.

Home, E. (1823b). *Lectures on comparative anatomy in which are explained the preparations in the Hunterian Collection illustrated by engravings* 3. London.

Home, E. (1824a). The Croonian Lecture. On the internal structure of the human brain, when examined in the microscope, as compared with that of fishes, insects and worms. *Philos. Trans. Roy. Soc. London,* 114: 1–10.

Home, E. (1824b). Some curious facts respecting the walrus and seal, discovered by the examination of specimens brought to England by different ships lately returned from the polar circle. *Philos. Trans. Roy. Soc. London,* 114: 233–241.

Home, E. (1824c). An account of the organs of generation of the Mexican Proteus, called by the natives axolotl. *Philos. Trans. Roy. Soc. London,* 114: 419–423.

Home, E. (1825). Observations on the changes the ovum of the frog undergoes during formation of the tadpole. *Philos. Trans. Roy. Soc. London,* 115: 81–86.

Home, E. (1826a). On the coagulation by heat of the fluid blood in an aneurysmal tumour. *Philos. Trans. Roy. Soc. London,* 116: 189–201.

Home, E. (1826b). On the production and formation of pearls. *Philos. Trans. Roy. Soc. London,* 116: 338–341.

Home, E. (1827). Croonian Lecture for 1826 [On the propagation of the common oyster, and the large fresh-water mussel]. *Philos. Trans. Roy. Soc. London,* 117: 39–48.

Home, E. (1828). *Supplement to the foregoing lectures on comparative anatomy in which are explained the preparations in the Hunterian Collection illustrated by engravings.* London.

Home, E. (1830a) *A short tract on the formation of tumours and the peculiarities that are met with in the structure of those that have become cancerous.* London.

Home, E. (1830b). A report on the stomach of the zariffa. *Philos. Trans. Roy. Soc. London,* 120: 85–86.

Hooker, J. D. (1855–1859). *Flora Tasmaniae.* London (= Hooker, J. D., *The botany of the Antarctic voyage of H. M. discovery ships Erebus and Terror in the years 1839–1843* vol. 3).

Hooker, W. J. (1832). *Pterostylis banksii. Bot. Mag.*, 59: t. 3172.

Hooker, W. J. (1835). *Microtis parviflora, Microtis media. Bot. Mag.*, 72: t. 3377 - 3378.

Hunter, J. (1794). *A treatise on the blood, inflammation and gun-shot wounds.* London.

Ibáñez, V. (1992). *Trabajos cientificos y correspondencia de Tadeo Haenke.* Madrid (= *La Expedición Malaspina 1789–1794*, 4).

Jacquin, J. F. v. (1787). *Lacerta vivipera. Nov. Act. Helv. Phys.-Math.*, 1787: 33–34.

Jacquin, J. F. v. (1828). Über eine einfache praktische Methode, das Vergrößerungsverhältnis zu bestimmen. *Z. Phys. Math.*, 4: 1–18.

Jacquin, N. J. v. (1792). *Anleitung zur Pflanzenkenntnis nach Linné's Methode.* Wien.

Jacquin, N. J. v. (1800). *Fragmenta botanica* 1. Viennae.

Jäggi, J. (1893). Der *Ranunculus bellidiflorus* des Joh. Gessner. *Ber. Schweiz. Bot. Ges.*, 3: 77–96.

Jeřábek, T. (2003). *Valtice. Schloss.* Libice nad Cidlinou.

Jirásek, V. (1962). The botanic garden of the Charles University in Prague. *Act. Horti Pragensis,* 1962: 5–10.

Jones, D. L. (2006). *A complete guide to native orchids of Australia including the island territories.* Sydney.

Jones, D. L., Clements, M. A. & Molloy, B. P. J. (2002). A synopsis of the subtribe Pterostylidinae. Pp. 125–144 In: Jones, D. L. & Clements, M. A. (ed.) *A review of Pterostylis (Orchidaceae)*. Victoria (= *Australian Orchid Research* 4).

Kaiser, W. & Piechocki, W. (1968). Medizinisch-botanischer Unterricht im 18. Jahrhundert an der Universität Halle. *Hercynia*, 5: 201–229.

Kamin-Kajfež, V. (2010). Franc Kavčič (Caucig) in egyptomanija. *Ann. Hist. Sociolog.*, 20: 363–372.

Kaplan, A. & Stulik, D. C. (2013). The first scientific investigation of Niépce's images from UK and US collections: image substrate. *Imaging Sci. J.*, 61: 629–646.

Karig, K.P. (2010). Neufund einer *Ophrys tenthredinifera* auf Zypern.– *J. Eur. Orch.* 42: 420–422.

Ker, J. B. (1817). A revision of the genus *Amaryllis*. *J. Sci. Arts (London)*, 2: 342–371.

Ker, J. B. (1818). Of the three species of the natural order Orchideae, represented in Plate VI. *J. Sci. Arts (London)*, 4: 199–206.

Kerr, R. (1989). Chinese Porcelain at Spring Grove Diary: Sir Joseph Banks's Manuscript. *Apollo*, 1989: 30–34.

Kippes, E. (2000). *Die Entwicklung der Region Feldsberg unter der Herrschaft des Hauses Liechtenstein vom 18. bis zum 20. Jahrhundert: Feudalismus–Demokratisierung–Nationalismus.* Wien.

Knall-Brskovsky, U. (1994). Fanti, Gaetano (Ercole Antonio). *Dizionario Biografico degli Italiani* 44: 633–635.

Knapp. S. (2003). *Potted Histories. An artistic voyage through plant illustration*. London.

Knight, T. A. (1818). On the mode of propagation of the *Lycoperdon cancellatum*, a species of fungus, which destroys the leaves and branches of the pear tree. *Trans. Hort. Soc. London*, 2: 178–183.

Koehler, S., Singer, R. B. & Amaral, M. C. E. (2012). Taxonomic revision of the neotropical genus *Christensonella* (Maxillariinae, Orchidaceae). *Bot. J. Linn. Soc.*, 168: 449–472.

Körner, S. (2005). Die Gärten des Fürsten Aloys I. von Liechtenstein. Gartenkunst in gesellschaftlichen Umbruchzeiten. *Jahrb. Ver. Fürstentum Liechtenstein*, 104: 85–136.

Kofler, M. (2004). Einleitung. P. [7] In: Rozmann, K. & Müller-Kaspar, U. (ed.), *Franz Caucig. Ein Wiener Künstler der Goethe-Zeit in Italien*. Ruhpolding.

Konig, C. (1804). On vegetable monstrosities, with some account of a pretended *Ranunculus bellidiflorus*. *Ann. Bot. (König & Sims)*, 1 (2): 368–376.

Koning, J. de & Kiehn, M. (1998). *In meo hortulo. In meinem Garten / In meo hortulo. In mijn tuin.* Leiden.

Koreny, F. (1985). *Albrecht Dürer und die Tier- und Pflanzenstudien der Renaissance*. München.

Koukal, V. (1995). *Valtice a řád milosrdných bratří. Historie a osobnosti*. Praha.

Kräftner, J. (ed.) (2004). *Liechtenstein Museum Wien. Die Sammlungen*. München.

Kräftner, J. (ed.) (2008). *Liechtenstein Museum Wien. Oasen der Stille. Die grossen Landschaftsgärten in Mitteleuropa*. Wien.

Krahnke, H. (ed.) (2001). Die Mitglieder der Akademie der Wissenschaften zu Göttingen 1751–2001. *Abh. Akad. Wiss. Göttingen, Phil.-Hist. Kl.*, ser. 3, 246 (= *Abh. Akad. Wiss. Göttingen, Naturwiss. Kl.*, ser. 3, 50).

Kravec, T. P. (1949). Dokumenty po istorii izobretenija fotografii. *Trudy Archiva Akad. Nauk SSSR*, 7.

Kreutz, C. A. J. (2004): *Die Orchideen von Zypern. The orchids of Cyprus. Beschreibung, Lebensweise, Verbreitung, Gefährdung, Schutz und Ikonographie. Description, pattern of life, distribution, threat, conservation and iconography.* Landgraaf.

Kronfeld, E. M. (1909). Mineralogisch – chemische Bemerkungen von der europäischen Studienreise Jacquin d. J. 1788–1790. *Arch. Ges. Naturwiss. Technik*, 1: 158–165.

Kronfeld, E. M. (1919). Die Kapuzinerlinde. *Mitt. Deutsch. Dendrolog. Ges.*, 1919: 270–273.

Kronfeld, E. M. (1921). Jacquin des Jüngeren botanische Studienreise 1788–1790. Aus den unveröffentlichten Briefen herausgegeben. *Beih. Bot. Centralbl.*, 38 (2): 132–176.

Kruta, V. (1977). *Dr. John Lhotsky: the turbulent Australian writer, naturalist and explorer.* Melbourne.

Kühnel, J. (1960). *Thaddaeus Haenke. Leben und Wirken eines Forschers. Veröff. Coll. Carolinum*, 9.

Kurdiovsky, R. (2003). Gaetano Fanti. Pp. 541–542 In: Anon., *Saur. Allgemeines Künstlerlexikon* 36. München.

Lack, H. W. (1996). Die frühe botanische Erforschung der Insel Kreta. *Ann. Naturhist. Mus. Wien*, 98 B: 183–236.

Lack, H. W. (1997a). Die Briefe von Ferdinand Bauer aus Australien und ihre Toponyme. *Ann. Naturhist. Mus. Wien*, 99 B: 591–614.

Lack, H. W. (1997b). Die Frontispize von John Sibthorps "Flora Graeca". *Ann. Naturhist. Mus. Wien*, 99 B: 615–654.

Lack, H. W. (1998a). Jacquin's ‚Selectarum stirpium americanarum historia'. The extravagant second edition and its title pages. *Bot. Mag.*, ser. 6, 15: 194–214.

Lack, H. W. (1998b). Recording form in early nineteenth century botanical drawings. Ferdinand Bauer's 'cameras'. *Bot. Mag.*, ser., 6, 15: 254–274.

Lack, H. W. (1999a). Eine unbekannte Wiener Bilderhandschrift: Die 'Phytantologia Eikonike' des Johann Jakob Well. *Ann. Naturhist. Mus. Wien*, 101 B: 531–564.

Lack, H. W. (1999b). The Berlin Passion Flowers. Ferdinand Bauer's sawn song. *Bot. Mag.*, ser. 6, 16: 286–296.

Lack, H. W. [with Mabberley, D. J.] (1999c). *The Flora Graeca Story. Sibthorp, Bauer, and Hawkins in the Levant*. Oxford.

Lack. H. W. (1999d). *The Flora Graeca story: Oxford's finest botanical treasure*. Oxford.

Lack, H. W. (2000a). *Ein Garten für die Ewigkeit. Der Codex Liechtenstein*. Bern. (English version: *A Garden for Eternity. The Codex Liechtenstein*. Berne).

Lack, H. W. (2000b). Lilac and horse-chestnut: discovery and rediscovery. *Bot. Mag.*, ser. 6, 17: 109–141.

Lack, H. W. (2000c). Die Berufung von Nikolaus Joseph Jacquin an die Universität Wien. *Ann. Naturhist. Mus. Wien*, 102 B: 375–388.

Lack, H. W. (2003). Ferdinand, Joseph und Franz Bauer: Testamente, Verlassenschaften und deren Schicksale. *Ann. Naturhist. Mus. Wien*, 104 B: 479–551.

Lack, H. W. (2004). An annotated catalogue of the printed illustrations by Franz Bauer (1758–1840). *Arch. Nat. Hist.*, 31: 80–111.

Lack, H. W. (2006a). I disegni botanici a matita di Goethe e Bauer in Sicilia. *Quad. Seminario Storia Sci.*, 9: 133–141.

Lack, H. W. (2006b). *Florilegium Imperiale. Botanische Schätze für Kaiser Franz I. von Österreich*. München. (English version: *Florilegium Imperiale.Botanical Illustrations for Francis I of Austria*. Munich).

Lack, H. W. (2008a). *Ein Garten Eden. Meisterwerke der botanischen Illustration / Garden of Eden. Masterpieces of botanical illustration / Un Jardin d'Eden. Chefs-d'oevre de l'illustration botanique*. Köln.

Lack, H. W. (2008b). *Franz Bauer. Das gemalte Zeugnis der Natur*. Wien (English version: *Franz Bauer. The painted record of nature*. Wien).

Lack, H. W. (2009). *Alexander von Humboldt und die botanische Erforschung Amerikas*. München. (English version: *Alexander von Humboldt and the botanical exploration of the Americas*. Munich).

Lack, H. W. (2010). *Carl Friedrich Philipp von Martius. The book of palms / Das Buch der Palmen / Les livre des palmiers*. Köln.

Lack, H. W. (2011a). Bilder aus Asphalt. Zur Rolle von Kontaktverfahren bei der Erfindung der Fotografie. *Bildwelten des Wissens*, 8: 64–71.

Lack, H. W. (2011b). Sir Joseph Banks, Franz Bauer und der Wirtswechsel des Getreiderostes (*Puccinia graminis* Pers.). *Ann. Naturhist. Mus. Wien*, 112 B: 253–264.

Lack, H. W. (2014). New World Explorations: Brazil. Pp. 132–153 In: Fraser, S. M. & Sellers, V. B. (eds.), *Flora Illustrata. Great Works from the LuEsther T. Mertz Library of the New York Botanical Garden*. Bronx, New York.

Lack, H. W. (2015). Pierre-Joseph Redouté, the 'Raphael of flowers' / Pierre-Joseph Redouté, der 'Raffael der Blumen' / Pierre-Joseph Redouté, le 'Raphaël des fleurs'. Pp. 6–41 In: Dressendörfer, W. & Lack. H. W., *Pierre-Joseph Redouté, Selection of the most beautiful flowers / Auslese der schönsten Blumen / Choix des plus belles fleurs*. Köln.

Lack, H. W. & Ibáñez. V. (1997). Recording colour in late eighteenth century botanical drawings: Sydney Parkinson, Ferdinand Bauer and Thaddäus Haenke. *Bot. Mag.*, ser. 6, 14: 87–100.

Lambert, A. B. (1804). Observations on the *Zizania aquatica*. *Trans. Linn Soc. London*, 7: 264–265.

Landon, Robbins H. (1981). *Joseph Haydn. Sein Leben in Bildern und Dokumenten*. Wien.

Landon, Robbins H. (1991). *Mozart and Vienna*. London.

Lange, P. J. de, Gardner, R. O., Sykes, W. R., Crowcroft, G. M., Cameron, E. K., Stalker, F., Christian, M. L. & Braggins, J. E. (2005). Vascular flora of Norfolk Island: some additions and taxonomic notes. *New Zealand J. Bot.*, 43: 563–596.

Leins, P. (1994). Der Botanische Garten der Ruprecht-Karls-Universität Heidelberg (1593–1993). *Heidelberger Jahrb.*, 38: 47–75.

Lhotsky, J. (1843). Biographical sketch of Ferdinand Bauer, natural history painter to the

expedition of Captain Flinders, R. N., to Terra Australis. *London J. Bot. (Hooker)*, 2: 106–113.
Lindley, J. (1821). *Digitalium monographia*. London.
Lindley, J. (1821–1826). *Collectanea botanica*. London.
Lindley, J. (1844). The Botany of the Antarctic Voyage of H. M.'s Discovery Ships Erebus and Terror in the years 1839–1843, By Joseph D. Hooker [book review]. *Gard. Chron.*, 1844: 446–447.
Link, J. H. F. (1822). *Enumeratio plantarum hortii regii botanici berolinensis altera* 2. Berolini.
Livesley, B. & Pentelow, G. M. (1978). The burning of John Hunter's papers: a new explanation. *Ann. Roy. Coll. Surgeons Engl.*, 60(2): 79–84.
Lord, J. M. (1991). Pollination and seed dispersal in *Freycinetia baueriana*, a dioecious liane that has lost its pollinator. *New Zealand J. Bot.*, 29: 83–86.
Lorenz, H. (1990). Fischer von Erlachs "Schloss der Rosse". Pp. 71–81 In: Baumstark, R. (ed.), *Joseph Wenzel von Liechtenstein. Fürst und Diplomat im Europa des 18. Jahrhunderts*. Einsiedeln.
Lorenz, H. (1999a). 19. Wien I., Stadtpalais Kaunitz-Liechtenstein. Pp. 251–252 In: Lorenz, H. (ed.), *Barock*. München (= Fillitz, H. (ed.), *Geschichte der bildenden Kunst in Österreich* 4).
Lorenz, H. (1999b). 23. Wien IX., Gartenpalais Liechtenstein. Pp. 255–256 In: Lorenz, H. (ed.), *Barock*. München (= Fillitz, H. (ed.), *Geschichte der bildenden Kunst in Österreich* 4).
Mabberley, D. J. (1985). *Jupiter Botanicus. Robert Brown of the British Museum*. Braunschweig.
Mabberley, D. (1999a). *Ferdinand Bauer. The nature of discovery*. London.
Mabberley, D. (1999b). Robert Brown on *Rafflesia*. *Blumea*, 44: 343–350.
Mabberley, D. J. (2002). Ferdinand Bauer. Pp. 50–101 In: Thomas, S. (ed.), *The encounter, 1802. Art of the Flinders and Baudin voyages*. Adelaide.
Mabberley, D. J. (2011). A note on some adulatory botanical plates distributed by Sir Joseph Banks. *Kew Bull.*, 66: 475–477.
Mabberley, D. J. & Moore, D. T. (1999). Catalogue of the holdings of the Natural History Museum (London) of the Australian botanical drawings of Ferdinand Bauer (1760 – 1826) and cognate materials relating to the Investigator voyage of 1801–1805. *Bull. Nat. Hist. Mus. London (Bot.)*, 29: 81–226.
Mabberley, D., Pignatti-Wikus, E. & Riedl-Dorn, C. (2007). An extinct tree 'revived'. *Bot. Mag.* ser. 7, 24: 190–195.
Mabberley, D. & San Pío Aldarén, M. Pilar (2011). *La carta de colores de Haenke de la Espedición de Malaspina: un enigma. Haenke's Malaspina colour-chart: an enigma*. Madrid.
Madrenas, J., Potter, P. & Cairns, E. (2005). Giving credit where credit is due: John Hunter and the discovery of erythrocyte sedimentation rate. *Lancet*, 366: 2140–2141.
Madriñán, S. (2013). *Nikolaus Joseph Jacquin's American plants. Botanical expedition to the Caribbean (1754–1759) and the publication of the 'Selectarum Stirpium Americanarum Historia'*. Leiden.
Maiwald, V. (1904). *Geschichte der Botanik in Böhmen*. Wien.
Martinez – Azorin, M. et al. (2011). The identity of *Albuca caudata* Jacq. (Hyacinthaceae) and a description of a new related species: *A. bakeri*. *PhytoKeys*, 5: 5–19.
Martz, J. (2001). Über die Geschichte und Bedeutung der Gärtnerfamilie Antoine. *Hist. Gärten. Mitt. Österr. Ges. Hist. Gärten*, 2001 (1): 8–12.
Mattingley, C. (2010). *A brilliant touch. Adam Forster's wildflower paintings*. Canberra.
May, T. W. & Pascoe, I. (1996). History of the taxonomic study of Australian fungi. Pp. 171–206 In: Anon., *Fungi of Australia* 1 A. Canberra.
Meikle, R. D. (1977, 1985). *Flora of Cyprus* 1-2. Kew.
Meyer, C. A. (1831). *Verzeichniss der Pflanzen, welche während der, auf allerhöchsten Befehl, in den Jahren 1829 und 1830 unternommenen Reise im Caucasus und in den Provinzen am westlichen Ufer des Caspischen Meeres gefunden und eingesammelt worden sind*. St. Petersburg.
Meynell, G. (1981). Banks papers in the Kent Archives Office, including notebooks by Joseph Banks and Francis Bauer. *Arch. Nat. Hist.*, 10: 77–88.
Meynell, G. (1982). Kew and the Royal Gardens Committee of 1838. *Arch. Nat. Hist.*, 10: 469–477.
Meynell, G. (1983). Francis Bauer, Joseph Banks, Everard Home and others. *Arch. Nat. Hist.*, 11: 209–221.
Michel, H. (1948a). Bergungsmaßnahmen und Widerstandsbewegung. *Ann. Naturhist. Mus. Wien*, 56: 3–6.
Michel, H. (1948b). Mineralogische Abteilung. *Ann. Naturhist. Mus. Wien*, 56: 14–15.
Miller, H. S. (1970). The herbarium of Aylmer Bourke Lambert. Notes on its acquisition, dispersal, and present whereabouts. *Taxon*, 19: 489–553.
Mills, K. (ed.) (2012). *Allan Cunningham: journal of a botanist on Norfolk Island in 1830: a transcription of the botanist's journal while on the island accompanied by contemporary notes by the author*. Jamberoo, N. S. W.
Moheit, U. (1993). *Alexander von Humboldt. Briefe aus Amerika 1799–1804*. Berlin (= *Beitr. Alexander-von-Humboldt-Forschung* 16).
Money, N. P. (2007). *The Triumph of the Fungi. A rotten history*. Oxford.
Moore, D. T. (1998). The pencil landscape drawings made by Ferdinand Bauer in Norfolk Island, from August 1804 to February 1805 in The Natural History Museum, London. *Arch. Nat. Hist.*, 25: 213–220.
Moore, D. T. (2005). Original papers concerning Robert Brown's collections: Manuscripts, diaries and note books in Britain. Pp. 143–165 In: Wege, J. et al. (ed.), *Matthew Flinders and his scientific gentlemen*. Welshpool, DC, Western Australia.
Moyal, A. (2001). *Platypus*. Crows Nest, NSW.
Moyal, A. (2008). *Koala. A historical biography*. Collingwood.
Müller-Kaspar, U. (2004). Antikensammlungen. Pp. 65–92 In: Rozmann, K. & Müller-Kaspar, U. (ed.), *Franz Caucig, Ein Wiener Künstler der Goethe-Zeit in Italien*. – Ruhpolding.
Nais, J. (2001). *Rafflesias of the World*. Kota Kinabalu.
Nelson, E. C. (1994). Nicolaas Witsen's letter of 1698 to Martin Lister about a Dutch expedition to the South Land (Western Australia): the original text and a review of its significance for the history of Australian natural history. *Arch. Nat. Hist.*, 21: 147–167.
Nelson, E. C. (2007). The original material of two Turkish species of *Erica* (Ericaceae) described and named by Richard Anthony Salisbury (1761 – 1829). *Turk. J. Bot.*, 31: 463–466.
Nickelsen, K. (2006). *Draughtsmen, botanists and nature: The construction of eighteenth-century botanical illustrations*. Dordrecht (= *Archimedes. New Studies in the history and philosophy of science and technology* 15).
Nicolson, D. H. & Fosberg, F. R. (2004). *The Forsters and the botany of the second Cook expedition (1772–1775)*, edn. 2. (= *Regnum Veg.* 139).
Nissen, C. (1951). *Die botanische Buchillustration. Ihre Geschichte und Bibliographie* 2. Stuttgart.
Noack, F. (1927). *Das Deutschtum in Rom* 1. Stuttgart.
Norst, M. J. (1988). *Austrians and Australia*. Potts Point, NSW.
Norst, M. J. (1989). *Ferdinand Bauer. The Australian Natural History Drawings*. London (= *Art in Natural History* 1).
Norst, M. J. (1990). Recognition and renaissance: Ferdinand Bauer 1760– 1826. *Australian Nat. Hist.*, 23: 296–305.
O'Brien, P. (1987). *Joseph Banks: a life*. London.
Olsen, P. (2010). *Upside Down World. Early European impressions of Australia's curious animals*. Canberra.
Oppenheimer, J. (1946). *New Aspects of John and William Hunter*. New York (= *Historical Library Yale Medical Library* 12).
Page, A. T. & Stearn, W. T. (1988). *A bicentenary History of the Linnean Society of London*. London.
Palla, E. (1891). Über die Entwicklung und Bedeutung der Zellfäden im Pollen von *Strelitzia reginae*. *Ber. Deutsch. Bot. Ges.*, 9: 85–90.
Palla, R. (2013). *Der Kapitän und der Künstler*. Köln.
Panista, M. & Tzanoudakis, D. (2010). Floristic diversity on small islets: Leros islet's group (East Aegean Islands). *Phytologia Balcanica*, 16: 271–284.
Péron, F. (1998). Mémoire sur les établissements anglais à Nouvelle Hollande, à la Terre de Diemen et dans les archipels du grand océan pacifique'. Présentation, edition et notes de Rober Martin. *Rev. Inst. Napoléon*, 176: 1–187.
Pertlik, F. (2013). Johann Fibig (* ~1720 † 1792). Edition einer mineralogisch orientierten Denkschrift 'Studium der Naturgeschichte' und Kommentar zum 'Handbuch der Mineralogie'. *Mitt. Österr. Miner. Ges.*, 159: 171–186.
Petrak, F. (1948). Botanische Abteilung. *Ann. Naturhist. Mus. Wien*, 56: 10–13.
Petrie, C. (1959). The bicentenary of the Younger Pitt. *Quarterly Rev.*, 297: 254–265.
Petz – Grabenbauer, M. (1996). Der botanische Garten der Universität Wien als wissenschaftliche Forschungsstätte unter Joseph Franz von Jacquin. *Mitt. Österr. Ges. Wissenschaftsgesch.*, 16: 1–20.
Petz – Grabenbauer, M. (2006). Impressionen zu Zauber und Zauberflöte. Mozart und der Salon der Jacquins im Botanischen Garten der Universität Wien. *Mensch, Wissenschaft, Magie. Mitt. Österr. Ges. Wissenschaftsgeschichte*, 24: 57–94.
Pfundheller, J. (1881). *Der Blumenkaiser. Österreichisches Zeit- und Kulturbild*. Wien.
Pierson, S. (2007). *Collectors, Collections and Museums: The field of Chinese Ceramics in Britain, 1560–1960*. Bern.
Pignatti-Wikus, E., Riedl-Dorn, C. & Mabberley, D. J. (2000a).

Ferdinand Bauer's field drawings of endemic Western Australian plants made at King George Sound and Lucky Bay, December 1801 – January 1802. I: Families Brassicaceae, Goodeniaceae p. p., Lentibulariaceae, Campanulaceae p. p., Orchidaceae, Pittosporaceae p. p. , Rutaceae p.p., Stylidaceae, Xyridaceae. *Rendiconti Lincei Sci. Fis. Mat. Nat.*, ser. 9, 11: 69–109.

Pignatti-Wikus, E., Riedl-Dorn, C. & Mabberley, D. (2000b). Ferdinand Bauer's field drawings of endemic Western Australian plants made at King George Sound and Lucky Bay, December 1801–January 1802. II: Families Apiaceae, Asteraceae, Campanulaceae p. p., Dilleniaceae, Fabaceae, Goodeniaceae p. p., Iridaceae, Laxmanniaceae, Loganiaceae, Menyanthaceae, Pittosporaceae p. p., Primulaceae, Proteaceae, Rutaceae p. p., Scrophulariaceae, Solanaceae, Sterculiaceae, and Tremandraceae. *Rendiconti Lincei Sci. Fis. Mat. Nat.,* ser. 9, 11: 201–244.

Prange, C. F. (1782). *Farbenlexikon.* – Halle.

Press, V. (1987). Goethe und das Haus Liechtenstein. *Jahrb. Ver. Fürstentum Liechtenstein*, 87: 39–76.

Press, V. (1990). Fürst Joseph Wenzel von Liechtenstein (1696–1772). Ein Aristokrat zwischen Armee, Kaiserhof und Fürstenhaus. Pp. 10–53 In: Baumstark, R. (ed.), *Joseph Wenzel von Liechtenstein. Fürst und Diplomat im Europa des 18. Jahrhunderts.* Einsiedeln.

Prevelakis, E. & Kalliataki Merticopoulou, K. (ed.) (1996). *Epirus, Ali Pasha and the Greek Revolution: Consular reports of William Meyer from Preveza*, 1–2. Athens (= *Monuments of Greek History* 12).

Queckett, E. J. (1840). Observations on the ergot of Rye, and some other grasses. *Trans. Linn. Soc. London*, 18: 453–473.

Queckett, E. J. (1849). Observations on the anatomical and physiological nature of the ergot. *Proc. Linn. Soc. London*, 1: 4–7.

Qvist, G. (1981). *John Hunter (1728–1793).* London.

R., S. E. v. (1782). Lustreise durch Oesterreich und Mähren nach Brün im September 1782. Pp. 237–274 In: Bernoulli, J. (ed.), *Sammlung kurzer Reisebeschreibungen und anderer zur Erweiterung der Land- und Menschenkenntniß dienender Nachrichten* 12. Berlin.

Radford, R. (2002). William Westall and the landscape tradition. Pp. 102–123 In: Thomas, S. (ed.), *The Encounter, 1802. Art of the Flinders and Baudin voyages.* Adelaide.

Raphael, S. (1970). The publication dates of the Transactions of the Linnean Society, Series I, 1791–1875. *Biol. J. Linn.* Soc., 2: 61–76.

Rauch, A. (2006). Klassizismus und Romantik: Europas Malerei zwischen zwei Revolutionen. Pp. 318–479 In: Toman, R. (ed.), *Klassizismus und Romantik. Architektur. Skulptur. Malerei. Zeichnung.* [Köln].

Rees, J. (2015). *Die verzeichnete Fremde. Formen und Funktionen des Zeichnens im Kontext europäischer Forschungsreisen 1770–1830.* – Paderborn.

Reinikka, M. A. (1995). *A history of the orchid.* Portland, Or.

Renkema, H. W. & Rendle, A. B. (1930). Aylmer Bourke Lambert and his 'Description of the Genus *Pinus*'. *J. Linn. Soc., Bot* , 48: 439–466.

Resemann, A. (2006). Zeichnungen des Klassizismus und der Romantik. Pp. 480–500 In: Toman, R. (ed.), *Klassizismus und Romantik. Architektur. Skulptur. Malerei. Zeichnung.* [Köln].

Riedl-Dorn, C. (1989). *Die Grüne Welt der Habsburger.* (= *Veröff. Naturhist. Mus. Wien N. F.* 23).

Riedl-Dorn, C. (1998). *Das Haus der Wunder. Zur Geschichte des Naturhistorischen Museums in Wien.* Wien.

Riedl-Dorn, C. (1999). Ferdinand Lucas Bauer (1760–1826) – erster österreichisches Maler in Australien. *Novara. Mitt. Österr. – Südpazifisch. Ges.* 2: 9–48.

Riedl-Dorn, C. (2000). Historic notes. Pp. 201–206 In: Pignatti, E., Riedl-Dorn, C. & Mabberley, D. J., Ferdinand Bauer's field drawings of endemic Western Australian plants made at King George Sound and Lucky Bay, December 1801–January 1802. II. Families Apiaceae, Asteraceae, Campanulaceae p. p., Dilleniaceae, Fabaceae, Goodeniaceae p. p., Iridaceae, Laxmanniaceae, Loganiaceae, Menyanthaceae, Pittosporaceae p. p., Primulaceae, Proteaceae, Rutaceae p. p., Scrophulariaceae, Solanaceae, Sterculiaceae, and Tremandraceae. *Rendiconti Lincei Sci. Fis. Mat. Nat.*, ser. 9, 11: 201–244.

Riedl-Dorn, C. (2001). Österreichische naturforschende Reisende im südlichen Afrika. Pp. 436–445 In: Eisenhofer, S. (ed.), *Spuren des Regenbogens. Kunst und Leben im südlichen Afrika / Tracing the Rainbow. Art and Life in Southern Africa.* Stuttgart (= *Kataloge des Oberösterreichischen Landesmuseums Linz N. F.* 162).

Riedl-Dorn, C. (2002). Ferdinand Lucas Bauer. Pp. 67–83 In: Seipel, W. (ed.), *Die Entdeckung der Welt. Die Welt der Entdeckungen. Österreichische Forscher, Sammler, Abenteurer.* Milano.

Riedl-Dorn, C. (2014). Ferdinand Lucas Bauer. Pp. 8–39 In: Riedl-Dorn, C., Köberl, C. & Stuart, D. (ed.), *Ferdinand Lucas Bauer.* Wien.

Riedl-Dorn, C., Köberl, C. & Stuart, D. (ed.) (2014). *Ferdinand Lucas Bauer.* Wien.

Rigby, N. (2005). ‚Not at all a particular ship': adopting vessels for British voyages of exploration 1768–1801. Pp. 13–23 In: Wege, J. et al. (ed.), *Mathew Flinders and his scientific gentlemen. The expedition of HMS Investigator to Australia, 1801–05.* Welshpool DC, Western Australia.

Römer, U. (1993). Bauer, Lucas. Pp. 553–554 In: Anon., *Saur. Allgemeines Künstlerlexikon* 7. München.

Rogasch, W. (2001). *Schlösser & Gärten in Böhmen & Mähren.* Köln.

Rozmann, K. (1978). *Franc Kavčič / Caucig.* Ljubljana.

Rusconi, M. (1821). *Amours des salamandres aquatiques et development de tetard de ces salamandres depuis l'oeuf jusqu'à l'animal.* Milan.

Russell, A. (1954). John Hawkins, F. G. S., F. R. H. S., F. R. S., 1761–1841. A distinguished Cornishman and early mining geologist. *J. Roy. Inst. Cornwall, N. S.*, 2: 98–106.

Sabine, J. (1822). Account of the newly produced Hybrid *Passiflora. Trans. Hort. Soc. London*, 4: 258–268.

Salisbury, R. A. (1805a). Description of *Bauera rubioides. Ann. Bot. (König & Sims),* 1: 512–516.

Salisbury, R. A. (1805b). *Paradisus londinensis* 1(1). London.

Satterthwait, D. R. (1982). Passifloraceae. Pp. 147–158 In: *Flora of Australia 8.* Canberra.

Savage, S. (1930). Appendix. Extracts from the Smithian correspondence. Pp. 465–466 In: Renkema, H. W. & Rendle, A. B., Aylmer Bourke Lambert and his 'Description of the Genus *Pinus*'. *J. Linn. Soc., Bot.*, 48: 439–466.

Sawyer, F. C. (1971). A short list of the libraries and lists of MSS. and original drawings in the British Museum (Natural History). *Bull. Brit. Mus. (Nat. Hist.) Hist.*, Ser. 4: 77–204.

Schechter, D. C. & Bergan, J. J. (1986). Popliteal aneurysm: a celebration of the bicentennial of John Hunter's operation. *Ann. Vasc. Surgery*, 1: 118–126.

Scheer, F. (1840). *Kew and its Gardens.* London.

Schifter, H. (1992). Die Vögel aus dem Nachlaß Ferdinand Lucas Bauer im Naturhistorischen Museum Wien und die darunter befindlichen Typen. *Ann. Naturhist. Mus. Wien*, 93 B: 41–57.

Schmuck, T. (2009). *Baltische Genesis. Die Grundlegung der Embryologie im 19. Jahrhundert.* Aachen (= Riha, O. (ed.) *Relationes* 2).

Schmutzer, K. (2011). *Der Liebe zur Naturgeschichte halber. Johann Natterers Reisen in Brasilien 1817–1836.* Wien (= *Österr. Akad. Wiss., Math. Naturwiss. Kl., Veröff. Komm.. Gesch. Naturwiss. Math. Med.* 64).

Schochow, W. (2003). *Bücherschicksale. Die Verlagerung der Preußischen Staatsbibliothek. Veröff. Hist. Komm. Berlin* 102.

Scholler, H. (1953). Carl Franz Anton Ritter von Schreibers. *Ann. Naturhist. Mus. Wien,* 59: 23 – 48.

Schott, H. (1827) Fasciculus plantarum brasiliensium. Pp. [403]–410 In: Sprengel, C. , *Systema Vegetabilium*, edn.16, 4(2). Gottingae.

Schrire, B. (2008). Going . . . Going . . . is it gone ? *Streblorrhiza speciosa* – the Philip Island Glory Pea. *Bot. Mag.*, ser. 6, 24: 196–197.

Schütz, K. (2005). Bernardo Bellotto in Wien und München (1759–1761). Pp. 101–107 In: Seipel, W. (ed.), *Bernardo Bellotto genannt Canaletto. Europäische Veduten.* Milano.

Schultes, J. A. (1825). Fortsetzung des Auszuges aus einem Schreiben an Se. Excellez, den Hrn. Grafen de Bray, k. baier. Gesandten zu Paris. *Flora*, 8(1) Beilage 1: 1–48.

Scrase, D. (1983). *Flowers of three centuries. One hundred drawings & watercolors from the Broughton collection.* Washington, D.C.

Scrase, D. (1997). *Flower Drawings.* Cambridge.

Seemann, B. (1869). Obituary of Frederick Scheer. *J. Bot.*, 7: 268–270.

Seipel, W. (ed.) (2002). *Die Entdeckung der Welt. Die Welt der Entdeckungen. Österreichische Forscher, Sammler, Abenteurer.* Milano.

Sherwood, S. & Rix, M. (2008). *Treasures of botanical art. Icons from the Shirley Sherwood and Kew collections.* Kew.

Simpkins, D. M. (1974). Biographical sketch of James Sowerby, written by his son, James de Carle Sowerby, 1825. *J. Soc. Bibliogr. Nat. Hist.*, 6: 402–415.

Simpson, E. (1849). *The History of Kew, in the county of Surrey.* London.

Sims, J. (1804). *Bauera rubioides. Bot. Mag.*, 19: t. 715.

Sitwell, S. (1956) The Romance of the Flower Book. Pp. 1–5 In: Sitwell, S. & Blunt, W., *Great Flower Books 1700–1900.* London.

Smith, B. (1988). The intellectual and artist framework of Pacific exploration in the eighteenth century. Pp. 123–128 In: Eisler, W. & Smith, B. (eds.). *Terra Australis: The furthest shore.* Sydney.

Smith, J. (1840). Observations on the cause of ergot. *J. Trans. Linn. Soc. London* 18: 449–452.

Smith, J. (1841). Notice of a plant which produces perfect seeds without any apparent action of pollen. *Trans. Linn. Soc. London*, 18: 509–512.

Smith, J. (1849). Observations on the cause of the Ergot. *Proc. Linn. Soc. London*, 1: 1–2.

Smith, J. (1866). *Ferns: British & Foreign*. London.

Smith, J. (1875). *Historia Filicum; an exposition of the nature, number, and organography of ferns.* London.

Smith, R. C. (1983). Nicephore Niepce in England. Pp. 43–50 In: Anon., *History of Photography* 7.

Spix, J. P. von & Martius, C. F. P. v. (1823). *Reise in Brasilien, auf Befehl Sr. Majestät Maximilian Joseph I., Königs von Bayern, in den Jahren 1817–1820 mit Karten und Abbildungen*. 1. München.

Stafleu, F. A. (1963). L'Héritier de Brutelle: the man and his work. Pp. xiii–xliii In: L'Héritier de Brutelle, Ch. – L., *Sertum anglicum* 1788. Pittsburgh, Penns. (= Lawrence, H. M. (ed.), *The Hunt Facsimile* Series 1).

Stafleu, F. A. (1973). Pritzel and his Thesaurus. *Taxon*, 22: 119–130.

Stafleu, F. A. (1981). Nikolaus Freiherr von Jacquin und die systematische Botanik seiner Zeit. *Anz. Österr. Akad. Wiss., Phil. Hist. Kl.*, 117: 287-310 (= *Veröff. Komm. Gesch. Math. Naturwiss. Med.* 32).

Stearn, W. T. (1947). Endlicher's "Genera plantarum", "Iconographia generum plantarum" and „Atakta Botanika". *J. Arnold Arbor.*, 28: 424–429.

Stearn, W. T. (1960a). An introduction to Robert Brown's 'Prodromus Florae Novae Hollandiae'. Pp. v–lii In: Brown, R., Prodromus florae Novae Hollandiae. Facsimile edition. Weinheim (= Cramer, J. & Swann, H. K. (ed.), *Historiae naturalis classica* 6).

Stearn, W. T. (1960b). Franz & Ferdinand Bauer, masters of botanical illustration. *Endeavour*, 19: 27–35.

Stearn, W. T. (1976). Preface. P. [7] In: Stearn, W. T., *The Australian Flower Paintings of Ferdinand Bauer*. London.

Stearn, W. T. (1981). *The Natural History Museum at South Kensington. A history of the British Museum (Natural History) 1753–1980.* – London.

Stearn, W. T. (1993). Francis Bauer (1758–1840). Kew's first botanical artist. Pp. 7–18 In: Stewart, J. & Stearn, W. T. (ed.), *The Orchid Paintings of Franz Bauer*. London.

Stearn, W. T. (1999). The life, times and achievements of John Lindley 1799–1865. Pp. 15–72 In: Stearn, W. T. (ed.), *John Lindley 1799–1865. Gardener – botanist and pioneer orchidologist.* Woodbridge, Suffolk.

Steiner, E. & Schulz, P. (2005). Plössl-Mikroskope – ein Vergleich mit modernen Geräten. *Ann. Naturhist. Mus. Wien.*, 107 B: 39–55.

Stewart, J. & Stearn, W. T. (1993). A selection of Franz Bauer's orchid paintings. Pp. 39–153 In: Stewart, J. & Stearn, W. T. (ed.). *The Orchid Paintings of Franz Bauer.* London.

Strid, A. & B. (ed.) (2009, 2010, 2012, 2013). *Sibthorp & Smith Flora Graeca. An annotated re-issue* 1–5. Königstein.

Stulik, D., Kaplan, A. & Khanjian, H. (2013). The first scientific investigation of Nièpce's images from UK and US collections: image layer and image formation. *Imaging Sci. J.*, 61: 602–628.

Tausch, H. & Johannsen, R. H. (2004). Aloys Hirt. Verzeichniß der bekanntesten jeztlebenden Künstler in Rom. Pp. 299–366 In: Sedlarz, C. (ed.), *Aloys Hirt. Archäologe, Historiker, Kunstkenner*. Hannover-Laatzen (= Wiedemann, C. (ed.), *Berliner Klassik. Eine Großstadtkultur um 1800.* 1).

Tenberge, K. B. (1999). Biology and life strategy of the ergot fungi. Pp. 25–56 In: Křen, V. & Cvak, L. (ed.), *Ergot. The genus Claviceps.* Amsterdam (= Hardman, R. (ed.), *Medicinal and aromatic plants – Industrial profiles* 6).

Tenore, M. (1842). Annotazione alla Flora Greca. *Rendiconto Accad. Sci. Soc. Borbon. Napoli*, 1: 82–103.

Thomas, S. (ed.) (2002). *The encounter, 1802. Art of the Flinders and Baudin voyages.* Adelaide.

Thomson, J. (1834). On the mummy cloth of Egypt; with observations on some manufactures of the ancients. *London Edinburgh Phil. Mag. & J. Sci.*, 5: 355–365.

Tietze, H. (1919). *Die Kunstsammlungen der Stadt Salzburg*. Wien (= Dvořák, M. (ed.), *Österreichische Kunsttopographie* 16).

Tsintides, T., Christodoulou, C. S., Delipetrou, P. & Georghiou, K. (eds.) (2007). *To kokkino biblio tēs chlōridas tēs Kyprou. The red data book of the Flora of Cyprus.* Lefkosia.

Ubrizsy Savoia, A. (2002). *Rapporti italo – ungheresi nella nascita della botanica in Ungheria*. Roma.

Urban, I. (1906). Vitae itineraque collectorum botanicorum, notae collaboratorum biographicae, Florae Brasiliensis ratio edendi chronologia, systema, index familiarum. Martius, C. F. P. de, Eichler, A. G. & Urban, I. (ed.), *Flora Brasiliensis* 1 (1). Monachii.

Vallance, T. G., Moore, D. T. & Groves, E. W. (2001). *Nature's Investigator. The diary of Robert Brown in Australia, 1801–1805.* Canberra.

Verdcourt, B. (1986). The discovery of some original drawings of Ferdinand Bauer at Kew. *Kew Mag.*, 3: 176–178.

Vogelsang-Eastwood, G. (2000). Textiles. Pp. 268–298 In: Nicholson, P. T. & Shaw, I. (ed.), *Ancient Egyptian materials and technology*. Cambridge.

Wacha, G. (1999). Dallinger (D. von Dalling, Tallinger, Thallinger). Pp. 527–529 In: *Saur. Allgemeines Künstler-Lexikon* 23. München.

Wagner, W. (1972, 1973). Die Rompensionäre der Wiener Akademie der Bildenden Künste 1772–1848 nach den Quellen im Archiv der Akademie. *Römische Hist. Mitt.*, 14: 65–109, 15: 13–66.

Walker, C. C. (1994). Frederick Scheer (1792–1868) of Kew. *Kew Mag.*, 11: 74–81.

Wardroper, J. (2002). *Wicked Ernest: the truth about the man who was almost Britain's king.* – London.

Ware, G. (2010). Franz Bauer and John Smith – The eyes and hands of William Hooker: The significance of their work in *Genera Filicum. Hardy Fern Found. Quarter.*, 20: 59–72.

Watts, P., Pomfrett, J. A. & Mabberley, D. (1997). *An exquisite eye. The Australian flora and fauna drawings 1801–1820 of Ferdinand Bauer.* Glebe, NSW.

Watzka, C. (2008). Der Hospitalsorden des Heiligen Johann von Gott in der Habsburgermonarchie 1605–1781. Pp. 106-32 In: Specht, H. & Andraschek-Holzer, R. (ed.), *Bettelorden in Mitteleuropa. Geschichte, Kunst, Spiritualität.* St. Pölten (= *Beiträge zur Kirchengeschichte Niederösterreichs 15; = Geschichtliche Beilage zum St. Pöltner Diözesanblatt* 32).

Wege, J., George, A., Gathe, J., Lemson, K. & Napier, K. (ed.) (2005). *Matthew Flinders and his scientific gentlemen.* Welshpool, DC.

Weissenbacher, G. (1998). *In Hietzing gebaut. Architektur und Geschichte eines Wiener Bezirks* 2. Wien.

Werner, P. (2007). *Roter Schnee oder Die Suche nach dem färbenden Prinzip.* Berlin (= *Beitr. Alexander-von-Humboldt-Forschung* 28).

Wheeler, A. (1986). Catalogue of the natural history drawings commissioned by Joseph Banks on the *Endeavour* Voyage 1768–1771. Part 3. *Bull. Brit. Mus. (Nat. Hist.) Hist. Ser.* 13.

Wheeler, A. & Moore, D. T. (1994). The animal drawings of Ferdinand Bauer in the Natural History Museum, London. *Arch. Nat. Hist.*, 21: 309–344.

Whitehead, P. (1988). Natural history drawings on British eighteenth century expeditions to the Pacific. Pp. 135–140 In: Eisler, W. & Smith, B. (ed.), *Terra Australis. The farthest shore.* Sydney.

Wiench, P. & Römer, U. (1993). Bauer, Joseph Anton. P. 549 In: *Saur. Allgemeines Künstler Lexikon* 7. München.

Wilhelm, A. (1998). Vorwort. *Jahrb. Hist. Ver. Fürstentum Liechtenstein*, 1998: 3.

Wilhelm, G. (1998). Der Weg der Liechtenstein-Galerie von Wien nach Vaduz. *Jahrb. Hist. Ver. Fürstentum Liechtenstein*, 1998: 5–48.

Willander, A. (1980). *Musikgeschichte der Stadt Baden.* Baden bei Wien.

Wimmer, C. A. (2014). Österreichische Barockgärten im europäischen Vergleich. *Gartenkunst*, 26: 235–274.

Wittmack, L. (1924). Zwei österreichische Pflanzenmaler in England. Pp. 37–42 In: *Beiträge zum landwirtschaftlichen Pflanzenbau, insbesondere Getreidebau. Festschrift zum 70. Geburtstag Professors Dr. h. c. Franz Schindler*. Berlin.

Wodehouse, R. P. (1935). *Pollen grains. Their structure, identification and significance in science and medicine*. New York.

Wolf, A. (1875). *Fürstin Eleonore Liechtenstein 1745–1812. Nach Briefen und Memoiren ihrer Zeit.* Wien.

Wolný, G. (1858). *Kirchliche Topographie von Mähren.* 2. Abteilung. Brünner Diözese 2. Brünn.

Wright, R. (1988). Appendix II. Shipping movements, 1788–1814. Pp. 210–224 In: Nobbs, R. (ed.). *Norfolk Island and its first settlement, 1788–1814.* North Sydney.

Załęski, K. (2005). 65. Bernard Belotto. Edelmann mit einem Mohren. P. 227 In: Seipel, W. (ed.), *Bernardo Bellotto genannt Canaletto. Europäische Veduten.* Milano.

Zatloukal, P., Krejčiřík, P. & Zatloukal, O. (ed.) (2012), *The Lednice-Valtice estate.* Praha.

APPENDIX 1 CATALOGUE RAISONNÉ

CATALOGUE RAISONNÉ OF THE WORKS OF THE BAUER BROTHERS

This appendix lists all works of the Bauer brothers, either signed or attributed to them, which are currently traceable. It excludes prints based on these works. For drawings and watercolours by the Bauer brothers currently not traceable see the text, in particular Chapter 19. The abbreviation 'p. p.' stands for a heterogeneous volume or collection, either comprising material by more than one of the three brothers or material by one of the three brothers together with works by other illustrators.

FERDINAND BAUER

ADELAIDE
Art Gallery of South Australia, Acc. No. 20009P38

BERLIN
Staatsbibliothek Preussischer Kulturbesitz, Handschriftenabteilung, Libri Picturati A 102 (presently preserved in the Biblioteka Jagiellońska, Cracow)

GÖTTINGEN
Niedersächsische Staats- und Universitätsbibliothek, Handschriftenabteilung, gr 2° MS Histor. Nat. 94 (p.p.)

INSTOW
Tapeley Estate, A miscellany of estate views

LONDON
British Museum, Print Room 1960-7-16-2 (fide Moore 1998)

Linnean Society, Archives, Illustration of *Lambertia formosa* and *Brunonia australis*

Natural History Museum, Library & Archives, Pencil drawings referring to *The genus Pinus*, Pencil landscape drawings from Norfolk Island, Finished drawings from the voyage in the *Investigator*, Unfinished drawings of plants cultivated in the Royal Garden at Kew (p. p.)

Lindley Library, Cingalese Drawings (p.p.), watercolours published in *Digitalium monographia*.

OXFORD
Ashmolean Museum, A miscellany of Oxford views (e.g. PD. As No. 366, fide Mabberley 1999a)

Sherardian Library, one of the Bodleian Libraries, MS 238 – 245, 247, 408 (cf. Lack 1999c)

PRESCOTT
Knowsley Hall, Pencil drawings and watercolours referring to *The genus Pinus*

TRURO
Cornwall County Record Office, Hawkins Papers (p.p.) (cf. Lack 1999c)

WIEN
Fürstliche Sammlungen, Bibliothek, C I I Codex Liechtenstein (p. p.)

Naturhistorisches Museum, Archiv, Early pencil drawings of the Bauer brothers, Pencil drawings from the voyage in the *Investigator*, Miscellaneous pencil drawings.

FRANZ BAUER

BRNO
Konvent milosrných bratří, Knihovna, Views of convents (p.p.)

CAMBRIDGE
Fitzwilliam Museum, PD. 173–1973, 174–1973, 58–1975

GÖTTINGEN
Niedersächsische Staats- und Universitätsbibliothek, Alte Drucke, GR 2 Bot V, 9231 Rara (title page)

Niedersächsische Staats- und Universitätsbibliothek, Handschriftenabteilung, gr 2° MS. Histor. Nat. 94 (p.p.)

LONDON
Natural History Museum, Library & Archives, Unfinished drawings of plants cultivated in the Royal Garden at Kew (p.p.), Finished drawings of plants cultivated in the Royal Garden at Kew, Germination of wheat, Diseases of corn, Bauer Miscellany, Series of original watercolour drawings by W. Clift, Franz Bauer, Mrs. Marsh, and others (p. p.)

Royal Society, Picture Library, numerous drawings and watercolours (e.g. RS 8685)

University of London, Senate House Library, GB 96 MS 1015 (p. p.)

MAIDSTONE
Kent Archives Office, U 951. Z 33

MILANO
Biblioteca Nazionale Braidense, I SC VI 59 (title page; fide Madriñán 2013)

STUTTGART
Württembergische Landesbibliothek, Alte Drucke, Nat. G. fol. 763 (title page)

WASHINGTON, DC.
Library of Congress , QK98.J3 (Office) (title page)

WIEN
Fürstliche Sammlungen, Bibliothek, C I I Codex Liechtenstein (p.p.)

Österreichische Nationalbibliothek, Sammlung von Handschriften und alten Drucken, 177687 – E (title page)

Naturhistorisches Museum, Archiv, Early pencil drawings of the Bauer brothers (p.p.); Miscellaneous pencil drawings (p.p.)

JOSEPH BAUER

BRNO
Konvent milosrných bratří, Knihovna, Views of convents (p. p.)

CLERMONT-FERRAND
Musée d'art Roger-Quilliot, Inv. No. 73.25.1

LONDON
British Museum, Reg. No. 1978, 0121.276, Reg. No. 1978, 0121. 277, Reg. No. 1978, 0121. 280.

WIEN
Akademie der Künste, Kupferstichkabinett, Handzeichnungen, Miscellaneous drawings from Rome (e.g. HZ 16502)

Fürstliche Sammlungen, Bibliothek, C I I Codex Liechtenstein (p.p.)

Fürstliche Sammlungen, Gemälde, Inv. No. GE 1360

Fürstliche Sammlungen, Graphik, Inv. No. GR 119, 120, 121.

Naturhistorisches Museum, Archiv, Early pencil drawings of the Bauer brothers (p.p.)

APPENDIX 2 TIMELINE

GENERAL HISTORY

1756 Seven Years' War begins

1760 George II, King of Great Britain and Ireland, dies; his grandson becomes successor as George III

1763 Seven Years' War ends

1765 Francis Stephen of Lorraine, Emperor of the Holy Roman Empire, dies; his son Joseph II becomes successor

1776 United States Declaration of Independence

1780 Maria Theresa, Queen of Bohemia and Hungary, dies

1783 Peace of Paris concluded

1789 French Revolution begins

1790 Joseph II, Emperor of the Holy Roman Empire, dies; his brother becomes his successor as Leopold II

1792 Leopold II, Emperor of the Holy Roman Empire dies; his son becomes his successor as Francis II

1799 Napoleon First Consul

1804 Francis II, Emperor of the Holy Roman Empire, declares himself Franz I, Emperor of Austria; Napoleon crowns himself Emperor of the French

1805 Battle of Trafalgar, Battle of Austerlitz

1806 Holy Roman Empire dissolved

1809 Battle of Aspern – Essling

1813 Battle of Leipzig

1814 Congress of Vienna begins

1815 Battle of Waterloo, Congress of Vienna ends

1820 George III, King of the United Kingdom, dies; his son becomes his successor as George IV

1830 George IV, King of the United Kingdom, dies; his brother becomes his successor as William IV

1835 Franz I, Emperor of Austria, dies; his son becomes his successor as Ferdinand I

1837 William IV, King of the United Kingdom, dies; his niece becomes his successor as Victoria I

1848 Ferdinand I, Emperor of Austria, resigns; his nephew becomes his successor as Franz Joseph I.

JOSEPH BAUER

1756 born in Feldsberg

c.1770 starts to work for Norbert Boccius, OH

c.1780 attends Imperial Royal United Academy of Arts in Vienna

1781 leaves for Rome on a stipend from Franz Joseph I, Reigning Prince of Liechtenstein

1788 returns to Vienna, becomes valet of Aloys I Joseph, Reigning Prince of Liechtenstein

c.1801 appointed court painter by Aloys I Joseph

c.1806 appointed director of the Liechtenstein Gallery by Johann I Joseph, Reigning Prince of Liechtenstein

c.1810 opening of the Liechtenstein Gallery

1831 dies in Rossau near Vienna

FRANZ BAUER

1758 born in Feldsberg

c.1772 starts to work for Norbert Boccius, OH

c.1782 attends Imperial Royal United Academy of Arts in Vienna

works for Nikolaus Joseph Edler von Jacquin in Vienna

1788 departs from Vienna accompanying Joseph Franz Edler von Jacquin for an official tour across Europe; arrives in London

1789 is engaged by Sir Joseph Banks as natural history illustrator on the basis of an annuity for life

1790 settles down permanently at Kew and becomes first resident plant illustrator at the Royal Garden

c.1794 starts to publish *Delineations of exotick plants*

1802 observes for the first time nuclei in plant cells; starts to study pollen diversity

1803 discontinues publication of *Delineations of exotick plants*

c.1807 starts to study the life cycle of wheat and fungal infections of cereal grasses

c.1815 begins to collaborate with Sir Everard Home

1817 publishes *Strelitzia depicta*

1819 publishes first scientific paper

1820 Sir Joseph Banks dies

1821 elected Fellow of the Royal Society

1823 Fourth volume of Home's *Lectures on comparative anatomy* appears

1828 Home's *Supplement to the foregoing lectures on comparative anatomy* appears

1830 begins to publish *Illustrations of orchidaceous plants*

c.1835 starts to prepare fern illustrations for William Jackson Hooker

1838 Hooker's *Genera filicum* start to appear

1840 adopts Elizabeth Baker; dies at Kew

FERDINAND BAUER

1760 born in Feldsberg

c.1774 starts to work for Norbert Boccius, OH

c.1784 attends Imperial Royal United Academy of Arts in Vienna, works for Nikolaus Joseph Edler von Jacquin in Vienna

1786 leaves with John Sibthorp for a tour of the Levant in March

1787 arrives with John Sibthorp in Oxford in December, starts to prepare watercolours for the projected *Flora Graeca*

1792 begins to prepare watercolours of the animals of the Levant

1794 discontinues work for John Sibthorp, moves to London

c.1796 starts to produce illustrations for Lambert's *The genus Pinus*

1801 embarks with Matthew Flinders and Robert Brown on the voyage of the *Investigator*

1802 circumnavigation of Australia completed

1804 departs from Sydney for a stay of six months on Norfolk Island

1805 returns on the *Investigator* to London

1806 starts to prepare watercolours documenting plants from Australia; *Flora Graeca* begins to appear

1811 starts to work on his *Illustrationes Florae Novae Hollandiae*

1814 returns to Vienna

1816 discontinues work on the *Illustrationes Florae Novae Hollandiae*

c.1817 begins to prepare watercolours documenting animals from Australia

1818 purchases a house in Hietzing near Vienna

c.1820 starts work on passionflowers

1821 John Lindley's *Digitalium monographia* appears

1826 dies in Hietzing

INDEX

It is recommended that the introductory notes to the five indices – (1) General, (2) Persons, (3) Place names, (4) Books, journals and manuscripts, and (5) Plants, fungi, algae, animals – be read first.

Page spans indicate that the subject is mentioned within the span rather than indicating extensive subject coverage.

Fig. numbers are followed by the page the illustration and caption occur on. Where the illustration and caption are on different pages, illustration page numbers are in *italic* and caption page numbers in normal font.

GENERAL

Information on the key persons of 'The Bauers' is presented only in the general index and in summary form; for all others see the Index of Persons. For institutions still extant only the modern names are given.

PERSONS

This index lists most of the persons mentioned in the text, with the exception of the three principal figures of the book: Joseph Bauer (1756 – 1831), Franz Bauer (1758 – 1840) and Ferdinand Bauer (1760 – 1826). For biographical data of the more important persons see Dramatis personae, p. 8.

BOOKS, JOURNALS AND MANUSCRIPTS

This index lists all titles of books, journals and manuscripts mentioned in the text.

PLACE NAMES

This index lists many localities mentioned in the text; the emphasis is on those places visited by the Bauer brothers or where the scenes of major events of their lives. Since the Bauer brothers lived in Feldsberg, Vienna, Rome, London, Kew and Oxford there are very many passing references to these towns, so none of them are mentioned.

PLANTS, FUNGI, ALGAE, ANIMALS

This index lists all scientific names of plants and animals mentioned in the text.